Methods in Enzymology

Volume 108
IMMUNOCHEMICAL TECHNIQUES
Part G
Separation and Characterization
of Lymphoid Cells

METHODS IN ENZYMOLOGY

EDITORS-IN-CHIEF

Sidney P. Colowick   Nathan O. Kaplan

Methods in Enzymology

Volume 108

Immunochemical Techniques

Part G
Separation and Characterization
of Lymphoid Cells

EDITED BY

Giovanni Di Sabato
DEPARTMENT OF MOLECULAR BIOLOGY
VANDERBILT UNIVERSITY
NASHVILLE, TENNESSEE

John J. Langone
DEPARTMENT OF MEDICINE
BAYLOR COLLEGE OF MEDICINE
HOUSTON, TEXAS

Helen Van Vunakis
DEPARTMENT OF BIOCHEMISTRY
BRANDEIS UNIVERSITY
WALTHAM, MASSACHUSETTS

1984

ACADEMIC PRESS, INC.

(Harcourt Brace Jovanovich, Publishers)

Orlando   San Diego   New York   London
Toronto   Montreal   Sydney   Tokyo

ACADEMIC PRESS, INC.
Orlando, Florida 32887

United Kingdom Edition published by
ACADEMIC PRESS, INC. (LONDON) LTD.
24/28 Oval Road, London NW1 7DX

LIBRARY OF CONGRESS CATALOG CARD NUMBER: 54-9110
ISBN 0-12-182008-4

Printed and bound by CPI Group (UK) Ltd, Croydon, CR0 4YY
Transferred to Digital Print 2011

Table of Contents

Section I. Surgical Techniques in Immunology

Section II. Methods for the Separation and Purification of Populations and Subpopulations of Lymphoreticular Cells

Section III. Methods for the Study of Surface Immunoglobulin

Section IV. Methods for the Detection, Purification, and Biochemical Characterization of Lymphoid Cell Surface Antigens

Contributors to Volume 108

Article numbers are in parentheses following the names of contributors.
Affiliations listed are current.

ORESTE ACUTO (56), *Division of Tumor Immunology, Dana-Farber Cancer Institute and the Department of Pathology, Harvard Medical School, Boston, Massachusetts 02115*

J. P. ALLISON (55), *Science Park Research Division, University of Texas System Cancer Center, Smithville, Texas 78957*

DAVID W. ANDREWS (53), *Department of Biochemistry and Molecular Biology, Harvard University, Cambridge, Massachusetts 02138*

ETTORE APPELLA (47), *National Cancer Institute, National Institutes of Health, Bethesda, Maryland 20205*

A. BASTEN (22), *Clinical Immunology Research Centre, University of Sydney, Sydney 2006, Australia*

JACK R. BENNINK (24), *The Wistar Institute, Philadelphia, Pennsylvania 19104*

PAUL R. BERGSTRESSER (60), *Department of Dermatology, University of Texas Health Science Center at Dallas, Dallas, Texas 75235*

RICHARD D. BERLIN (33), *Department of Physiology, University of Connecticut Health Center, Farmington, Connecticut 06032*

M. ROSA BONO (53), *Institut Pasteur, Laboratoire Immunogenetique Humaine, Paris, France*

ARNE BØYUM (9), *Norwegian Defence Research Establishment, Division for Environmental Toxicology, N-2007 Kjeller, Norway*

STEVEN J. BURAKOFF (54), *Division of Pediatric Oncology, Dana-Farber Cancer Institute, Harvard Medical School, Boston, Massachusetts 02115*

JOHN E. COLIGAN (43), *Laboratory of Immunogenetics, National Institute of Allergy and Infectious Diseases, National Institutes of Health, Bethesda, Maryland 20205*

P. CRESWICK (22), *Clinical Immunology Research Centre, University of Sydney, Sydney 2006, Australia*

SUSAN E. CULLEN (45, 49), *Department of Microbiology and Immunology, Washington University School of Medicine, St. Louis, Missouri 63110*

BRUCE H. DAVIS (33), *Department of Pathology, State University of New York, Upstate Medical Center, Syracuse, New York 13210*

JOSÉ A. LÓPEZ DE CASTRO (52), *Departamento de Immunología, Fundación Jiménez Díaz, 28040 Madrid, Spain*

STEFANELLO DE PETRIS (41), *Department of Zoology, University College, London WC1E 6BT, England*

GIOVANNI DI SABATO (6), *Department of Molecular Biology, Vanderbilt University, Nashville, Tennessee 37235*

PAUL J. DURDA (59), *New England Nuclear Corporation, Boston, Massachusetts 02118*

BRUCE E. ELLIOTT (7), *Cancer Research Laboratories, Department of Pathology, Queen's University, Kingston, Ontario K7L 3N6, Canada*

LORRAINE FLAHERTY (46), *Laboratory of Immunology, Center for Laboratories and Research, New York State Department of Health, Albany, New York 12201*

JUDITH FORAN (54), *Division of Pediatric Oncology, Dana-Farber Cancer Institute, Harvard Medical School, Boston, Massachusetts 02115*

LUCIANA FORNI (41), *Basel Institute for Immunology, 4005 Basel, Switzerland*

DANIEL L. FRIEDMAN (6), *Department of Molecular Biology, Vanderbilt University, Nashville, Tennessee 37235*

JUSTINE S. GARVEY (26), *Department of Biology, Syracuse University, Syracuse, New York 13210*

VICTOR GHETIE (13, 40), *Department of Immunology, Victor Babes Institute, R-76201 Bucharest, Romania*

BEPPINO C. GIOVANELLA (34), *The Stehlin Foundation for Cancer Research, Houston, Texas 77002*

BRUCE GLICK (1), *Poultry Science Department, Mississippi State University, Mississippi State, Mississippi 39762*

S. H. GOLUB (36), *Department of Surgery, Division of Oncology, UCLA School of Medicine, Los Angeles, California 90024*

JOAN C. GORGA (54), *Department of Biochemistry and Molecular Biology, Harvard University, Cambridge, Massachusetts 02138*

DIANA K. GOROFF (25), *Department of Microbiology, Georgetown University Schools of Medicine and Dentistry, Washington, D.C. 20007*

PAUL D. GOTTLIEB (59), *Department of Microbiology, University of Texas at Austin, Austin, Texas 78712*

J. DIXON GRAY (36), *Department of Surgery, Division of Oncology, UCLA School of Medicine, Los Angeles, California 90024*

DAVID G. HAEGERT (38), *Department of Pathology, McGill University and Montreal General Hospital, Montreal, Quebec H3G 1A4, Canada*

M.-L. HAMMARSTRÖM (16), *Department of Immunology, University of Stockholm, S-10691 Stockholm, Sweden*

S. HAMMARSTRÖM (16), *Department of Immunology, University of Stockholm, S-10691 Stockholm, Sweden*

KURT HANNIG (18), *Max-Planck-Institut für Biochemie, 8033 Martinsried, Federal Republic of Germany*

ERNIL HANSEN (18), *Institut für Anästhesiologie, Universität München, Klinikum Grosshadern, 8000 München 70, Federal Republic of Germany*

DAN-PAUL HARTMANN (32), *Department of Pathology, Georgetown University Schools of Medicine and Dentistry, Washington, D.C. 20007*

MATTHEW F. HEIL (26), *Department of Medicine, New York Medical College, Valhalla, New York 10595*

U. HELLSTRÖM (16), *Department of Immunology, University of Stockholm, S-10691 Stockholm, Sweden*

HERBERT B. HERSCOWITZ (25), *Department of Microbiology, Georgetown University Schools of Medicine and Dentistry, Washington, D.C. 20007*

L. A. HERZENBERG (19), *Department of Genetics, Stanford University, Stanford, California 94305*

MAY-KIN HO (31), *Immunology Research Program, E.I. duPont de Nemours & Co., No. Billerica, Massachusetts 01862*

ROGER A. HUBBARD (14), *Department of Biochemistry, Medical University of South Carolina, Charleston, South Carolina 29425*

JAMES T. HUNTER (34), *The Stehlin Foundation for Cancer Research, Houston, Texas 77002*

PATRICIA P. JONES (44), *Department of Biological Sciences, Stanford University, Stanford, California 94305*

DOLORES V. JUAREZ (60), *Department of Dermatology, University of Texas Health Science Center at Dallas, Dallas, Texas 75235*

ELLIOTT KAGAN (32), *Department of Pathology, Georgetown University Schools of Medicine and Dentistry, Washington, D.C. 20007*

JAMES F. KAUFMAN (53), *Basel Institute for Immunology, CH-4005 Basel 5, Switzerland*

NORMAN J. KLEIMAN (6), *Department of Molecular Biology, Vanderbilt University, Nashville, Tennessee 37235*

PETER KNUDSEN (53), *Beth Israel Hospital, Harvard Medical School, Cambridge, Massachusetts 02138*

ROBERT KORNGOLD (24), *The Wistar Institute, Philadelphia, Pennsylvania 19104*

MICHELLE LETARTE (57, 58), *Research Institute, The Hospital for Sick Children, Toronto, Ontario M5G 1X8, Canada*

DAVID A. LITVIN (28), *Department of Microbiology and Immunology, Albert Einstein College of Medicine, Bronx, New York 10461*

F. LOOR (37), *Laboratoire d'Immunologie, Universite Louis Pasteur, F 67048 Strasbourg, France*

JORDAN E. LUHR (25), *Department of Microbiology, Georgetown University*

Schools of Medicine and Dentistry, Washington, D.C. 20007

MICHAEL A. LYNES (46), Department of Biology, Williams College, Williamstown, Massachusetts 01267

MICHAEL G. MAGE (11, 12), Laboratory of Biochemistry, National Cancer Institute, National Institutes of Health, Bethesda, Maryland 20205

W. LEE MALOY (43), Laboratory of Immunogenetics, National Institute of Allergy and Infectious Diseases, National Institutes of Health, Bethesda, Maryland 20205

JOHN J. MARCHALONIS (14), Department of Biochemistry, Medical University of South Carolina, Charleston, South Carolina 29425

RICHARD M. MCCARRON (25), Neuroimmunology Branch, National Institute of Neurological, and Communicative Disorders and Stroke, National Institutes of Health, Bethesda, Maryland 20205

W. ROBERT MCMASTER (51), Department of Medical Genetics, The University of British Columbia, Vancouver, British Columbia V6T 1W5, Canada

RICHARD G. MILLER (8), Department of Immunology, University of Toronto and the Ontario Cancer Institute, Toronto, Ontario M4X 1K9, Canada

BARBARA B. MISHELL (29), Department of Microbiology and Immunology, University of California, Berkeley, Berkeley, California 94720

ROBERT I. MISHELL (29), Department of Microbiology and Immunology, University of California, Berkeley, Berkeley, California 94720

VERA B. MORHENN (35), Department of Dermatology, Stanford University Medical Center, Stanford, California 94305

D. E. MOSIER (27), Institute for Cancer Research, Fox Chase Cancer Center, Philadelphia, Pennsylvania 19111

MARY A. MURPHY (25), Department of Microbiology, Georgetown University Schools of Medicine and Dentistry, Washington, D.C. 20007

RODERICK NAIRN (48), Department of Microbiology and Immunology, The University of Michigan Medical School, Ann Arbor, Michigan 48109

D. G. NEWELL (39), Public Health Laboratory, Southampton General Hospital, Southampton S09 4XY, England

MALGORZATA NIEZGODKA (59), Department of Immunochemistry, Polish Academy of Sciences, Ludwik Hirszfeld Institute of Immunology and Experimental Therapy, Czerska 12, 53-114 Wroclaw, Poland

IMRE OLÁH (1), Second Department of Anatomy, Semmelweis University Medical School, 1093 Budapest, Hungary

JANET M. OLIVER (33), Department of Pathology, University of New Mexico Medical School, Albuquerque, New Mexico 87131

D. R. PARKS (19), Department of Genetics, Stanford University, Stanford, California 94305

M. A. PELLEGRINO (55), Department of Therapeutics, Hybritech Incorporated, San Diego, California 92121

P. PERLMANN (16), Department of Immunology, University of Stockholm, S-10691 Stockholm, Sweden

EVA A. PFENDT (35), Department of Dermatology, Stanford University Medical Center, Stanford, California 94305

HUGH F. PROSS (7), Departments of Radiation Oncology and Microbiology and Immunology, Queen's University, Kingston, Ontario K7L 3N6, Canada

EDWARD B. REILLY (59), Massachusetts Institute of Technology Center for Cancer Research, Cambridge, Massachusetts 02139

YAIR REISNER (17), Department of Biophysics, The Weizmann Institute of Science, Rehovoth 76100, Israel

PAUL L. ROMAIN (15, 56), Division of Tumor Immunology, Dana-Farber Cancer Institute and the Department of Medicine, Harvard Medical School, Boston, Massachusetts 02115

DAVID L. ROSENSTREICH (28), Department of Medicine and Microbiology and Immunology, Albert Einstein College of Medicine, Bronx, New York 10461

C. RUSSO (55), Department of Medicine,

Cornell Medical College, New York, New York 10021

JANET A. SAWICKI (47), Wistar Institute of Anatomy and Biology, Philadelphia, Pennsylvania 19104

STUART F. SCHLOSSMAN (15, 56), Division of Tumor Immunology, Dana-Farber Cancer Institute and the Department of Medicine, Harvard Medical School, Boston, Massachusetts 02115

SAMUEL F. SCHLUTER (14), Department of Biochemistry, Medical University of South Carolina, Charleston, South Carolina 29425

NATHAN SHARON (17), Department of Biophysics, The Weizmann Institute of Science, Rehovoth 76100, Israel

FUNG-WIN SHEN (21), Memorial Sloan-Kettering Cancer Center and Sloan-Kettering Division, Cornell Graduate School of Medical Sciences, New York, New York 10021

ETHAN M. SHEVACH (23), Laboratory of Immunology, National Institute of Allergy and Infectious Diseases, National Institutes of Health, Bethesda, Maryland 20205

KEN SHORTMAN (10), The Walter and Eliza Hall Institute of Medical Research, Royal Melbourne Hospital, Victoria 3050, Australia

JOHN SJÖQUIST (13, 40), Department of Medical and Physiological Chemistry, The Biomedical Center, Uppsala University, S-751 23 Uppsala, Sweden

MARK J. SOLOSKI (50), Department of Molecular Biology and Genetics, Subdepartment of Immunology, Johns Hopkins School of Medicine, Baltimore, Maryland 21205

TIMOTHY A. SPRINGER (31), Department of Membrane Immunochemistry, Dana-Farber Cancer Institute, Boston, Massachusetts 02115

DAVID STEINMULLER (3, 4, 5), Histocompatibility Laboratory, Transplantation Society of Michigan, Ann Arbor, Michigan 48104

HENRY C. STEVENSON (20), Biological Therapeutics Branch, Biological Response Modifiers Program, National Cancer Institute, Frederick, Maryland 21701

JACK L. STROMINGER (53, 54), Department of Biochemistry and Molecular Biology, Harvard University, Cambridge, Massachusetts 02138, and Dana-Farber Cancer Institute, Harvard Medical School, Boston Massachusetts 02115

PIERSON J. VAN ALTEN (2), Department of Anatomy, College of Medicine, University of Illinois at Chicago, Chicago, Illinois 60680

LUIGI VARESIO (30), Biological Response Modifiers Program, National Cancer Institute-FCRF, Frederick, Maryland 21701

ELLEN S. VITETTA (42, 50), Department of Microbiology, The University of Texas Health Science Center at Dallas, Dallas, Texas 75235

DENNIS M. WONG (30), Division of Blood and Blood Products, National Center for Drugs and Biologics, Bethesda, Maryland 20205

DOROTHY YUAN (42), Department of Microbiology, The University of Texas Health Science Center at Dallas, Dallas, Texas 75235

Preface

Previous volumes of "Immunochemical Techniques" have dealt primarily with various aspects and practical applications of the antigen–antibody reaction. This, the seventh of the series, is the first of a set of volumes describing methods for the study of the biology of lymphoid cells. Future volumes will include methods for the study of serum immunoglobulins, thymic hormones, lymphokines, as well as methods for the study of B and T cell functions.

The first section of this volume includes some of the surgical techniques used in immunological research. We feel that the presentation of these procedures serves a useful purpose, particularly in view of the fact that they seldom have been described in detail elsewhere.

The number of methods for the separation of populations and subpopulations of lymphoid cells has increased at an exponential rate over the past few years. It would have been difficult even to approach completeness in the section describing these methods. We have, therefore, limited our choices to those methods that illustrate the application of one or more basic principles of cell separation. Our intent in doing so is to provide the reader with methods that can either be used directly or be adapted to their particular experimental needs.

Similarly, in the second half of the volume dealing with methods for the study of surface immunoglobulin and lymphoid cell surface antigens, we have chosen topics that offer the reader a variety of approaches to the study of these molecules. Methods for the study of surface antigens of transformed lymphoid cells will be included in a future volume of this Immunochemical Techniques series.

We acknowledge the many helpful suggestions offered by numerous authors and colleagues, especially regarding the general planning of the volume and the choice of topics to be included. In particular, we want to thank Drs. Sidney Colowick and Nathan Kaplan for their continuous encouragement and never-failing advice. The patient and diligent secretarial assistance of Cindy Young is also gratefully acknowledged.

GIOVANNI DI SABATO
JOHN J. LANGONE
HELEN VAN VUNAKIS

METHODS IN ENZYMOLOGY

EDITED BY

Sidney P. Colowick and Nathan O. Kaplan

VANDERBILT UNIVERSITY
SCHOOL OF MEDICINE
NASHVILLE, TENNESSEE

DEPARTMENT OF CHEMISTRY
UNIVERSITY OF CALIFORNIA
AT SAN DIEGO
LA JOLLA, CALIFORNIA

METHODS IN ENZYMOLOGY

EDITORS-IN-CHIEF

Sidney P. Colowick   Nathan O. Kaplan

VOLUME VIII. Complex Carbohydrates
Edited by ELIZABETH F. NEUFELD AND VICTOR GINSBURG

VOLUME IX. Carbohydrate Metabolism
Edited by WILLIS A. WOOD

VOLUME X. Oxidation and Phosphorylation
Edited by RONALD W. ESTABROOK AND MAYNARD E. PULLMAN

VOLUME XI. Enzyme Structure
Edited by C. H. W. HIRS

VOLUME XII. Nucleic Acids (Parts A and B)
Edited by LAWRENCE GROSSMAN AND KIVIE MOLDAVE

VOLUME XIII. Citric Acid Cycle
Edited by J. M. LOWENSTEIN

VOLUME XIV. Lipids
Edited by J. M. LOWENSTEIN

VOLUME XV. Steroids and Terpenoids
Edited by RAYMOND B. CLAYTON

VOLUME XVI. Fast Reactions
Edited by KENNETH KUSTIN

VOLUME XVII. Metabolism of Amino Acids and Amines (Parts A and B)
Edited by HERBERT TABOR AND CELIA WHITE TABOR

VOLUME XVIII. Vitamins and Coenzymes (Parts A, B, and C)
Edited by DONALD B. MCCORMICK AND LEMUEL D. WRIGHT

VOLUME XIX. Proteolytic Enzymes
Edited by GERTRUDE E. PERLMANN AND LASZLO LORAND

VOLUME 87. Enzyme Kinetics and Mechanism (Part C: Intermediates, Stereochemistry, and Rate Studies)
Edited by DANIEL L. PURICH

VOLUME 88. Biomembranes (Part I: Visual Pigments and Purple Membranes, II)
Edited by LESTER PACKER

VOLUME 89. Carbohydrate Metabolism (Part D)
Edited by WILLIS A. WOOD

VOLUME 90. Carbohydrate Metabolism (Part E)
Edited by Willis A. Wood

VOLUME 91. Enzyme Structure (Part I)
Edited by C. H. W. HIRS AND SERGE N. TIMASHEFF

VOLUME 92. Immunochemical Techniques (Part E: Monoclonal Antibodies and General Immunoassay Methods)
Edited by JOHN J. LANGONE AND HELEN VAN VUNAKIS

VOLUME 93. Immunochemical Techniques (Part F: Conventional Antibodies, Fc Receptors, and Cytotoxicity)
Edited by JOHN J. LANGONE AND HELEN VAN VUNAKIS

VOLUME 94. Polyamines
Edited by HERBERT TABOR AND CELIA WHITE TABOR

VOLUME 95. Cumulative Subject Index Volumes 61–74 and 76–80
Edited by EDWARD A. DENNIS AND MARTHA G. DENNIS

VOLUME 96. Biomembranes [Part J: Membrane Biogenesis: Assembly and Targeting (General Methods; Eukaryotes)]
Edited by SIDNEY FLEISCHER AND BECCA FLEISCHER

VOLUME 97. Biomembranes [Part K: Membrane Biogenesis: Assembly and Targeting (Prokaryotes, Mitochondria, and Chloroplasts)]
Edited by SIDNEY FLEISCHER AND BECCA FLEISCHER

VOLUME 98. Biomembranes [Part L: Membrane Biogenesis (Processing and Recycling)]
Edited by SIDNEY FLEISCHER AND BECCA FLEISCHER

Section I

Surgical Techniques in Immunology

[1] Methods of Bursectomy

By BRUCE GLICK and IMRE OLÁH

The bursa of Fabricius is a dorsal diverticulum of the proctadael region of the cloaca.[1,2] A fortunate observation elevated the bursa to a prominent position.[3-5] Bursectomy was shown to markedly reduce or eliminate humoral immunity.[5,6] The historical and recent experiments with the bursa have been reviewed.[2,7] This chapter will critique the various methods of bursectomy.

Surgical Bursectomy: Posthatch

A variety of anesthetics have been tested in the neonatal chicken.[8] We routinely administer sodium thiopental (Dipentol, Diamond Laboratories) intraperitoneally (ip) or intravenously (iv) to newly hatched chicks or iv at later ages. Newly hatched chicks will tolerate between 0.025 and 0.1 ml of sodium thiopental (30 mg/ml) when administered ip. Administering the lower volume allows one to determine the degree of sensitivity of each group of chicks. The chicks are anesthetized within 3 min of the ip injecttion. Intravenous injection of sodium thiopental should be at the level of 1 mg/30 g of body weight with a 20-sec interval separating each 0.1 ml injected. Anesthesia immediately follows the iv injection.

With the thumb and forefinger of the left hand the tail is raised to a vertical position, the few feathers beneath the tail are removed, and the area is swabbed with 70% ethanol (EA) (Fig. 1). A superficial-lateral incision, 3–6 mm, at the base of the tail and just above the upper lip of the vent is made.[2-5] A minimum of bleeding is experienced. Too large an incision may allow the yolk sac to enter the orifice. Curved forceps must be used to break through the remaining connective tissue and enter the body cavity. Gentle pressure with the left hand, applied to the dorsal

[1] J. Jolly, *Arch. Anat. Microsc. Morphol. Exp.* **16**, 363 (1915).
[2] B. Glick, in "Avian Biology" (D. S. Farner, J. F. King, and K. C. Parkes, eds.), Vol. 7, pp. 443–500. Academic Press, New York, 1983.
[3] B. Glick, Ph.D. Dissertation, Ohio State University, Columbus (1955).
[4] T. S. Chang, B. Glick, and A. R. Winter, *Poult. Sci.* **34**, 1187 (1955).
[5] B. Glick, T. S. Chang, and R. G. Jaap, *Poult. Sci.* **35**, 224 (1956).
[6] M. D. Cooper, R. D. A. Peterson, M. A. South, and R. A. Good, *J. Exp. Med.* **123**, 75 (1966).
[7] B. Glick, *BioScience* **33**, 187 (1983).
[8] M. R. Fedde, *Poult. Sci.* **57**, 1376 (1977).

METHODS IN ENZYMOLOGY, VOL. 108

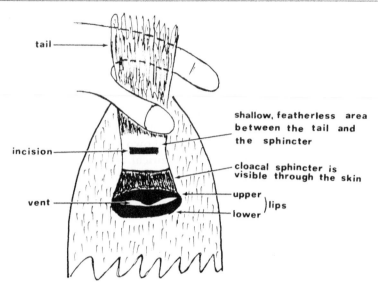

FIG. 1. Posterior aspect. Location of incision with reference to tail and upper lip of vent.

surface of the chick, will force the apex of the bursa into the newly made orifice (Fig. 1). The exposed bursa is grasped with curved forceps and is extended beyond the orifice until the bursal duct is viewed. The duct is severed at its point of attachment with the cloaca and large intestine (Fig. 1). The incision may be sutured; however, without sutures the incision will completely heal within 1 week. The bursa must be removed intact since damage to its serosa could lead to the release of bursal lymphocytes into the body cavity. Lymphoid cells may be identified along the bursal duct (Fig. 2). Therefore, it is important to sever the duct flush with its attachment to the alimentary canal. Additional lymphoid cells, apparently T cells,[9] are observed dorsal to the duct. Surgical bursectomy performed at hatch is more effective than bursectomy performed at later ages.[10-12] However, numerous investigators reported that chicks bursectomized at hatch responded to a secondary and tertiary injection of antigen.[13] Enhanced immunosuppression occurs at hatch when bursectomy is followed within 24 hr by irradiation (650 R).[6]

[9] S. Odend'hal and J. E. Breazile, *J. Reticuloendothel. Soc.* **25,** 315 (1979).
[10] T. S. Chang, M. S. Rheins, and A. R. Winter, *Poult. Sci.* **36,** 735 (1957).
[11] A. P. Mueller, H. R. Wolfe, and R. K. Meyer, *J. Immunol.* **85,** 172 (1960).
[12] M. A. Graetzer, W. P. Cote, and H. R. Wolfe, *J. Immunol.* **91,** 576 (1963).
[13] B. Glick, *Int. Rev. Cytol.* **48,** 345 (1977).

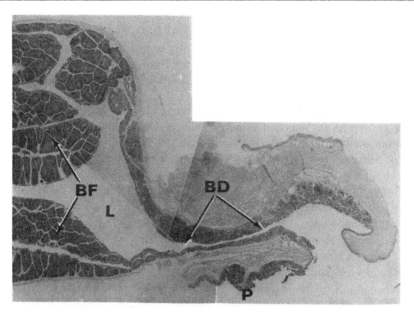

FIG. 2. Lymphocytes collect along the bursal duct (BD) which leads into the bursa's lumen (L). Bursal follicles (BF). Lymphoid tissue in the proctacdael area (P). Magnification: ×7.

Hormonal Bursectomy

With the observation that androgen injections of neonates produced bursa regression, Meyer *et al.*[14] administered 19-nortestosterone *in ovo* and arrested the lymphoid development of the bursa. In his laboratory a variety of androgens were evaluated. It was found that 0.5 mg of androsterone or 1.5 mg of androstane-3,17-dione or 1.5 mg of dehydroepiandrosterone administered on the fifth day of embryonic development produces bursaless embryos.[15] On the fifth day of embryonic developoment, a hole is made at the pointed end of the egg with a 23-gauge needle. With a 20-gauge needle, 19-nortestosterone (0.63 mg in 0.1 ml corn oil) is injected into the albumin. The hole is sealed with paraffin or transparent tape. Precipitin response to bovine serum albumin is eliminated in chicks hatched from these eggs while delaying the injection of 19-nortestosterone

[14] R. K. Meyer, Rao M. Appaswamy, and R. L. Aspinall. *Endocrinology (Baltimore)* **64**, 890 (1959).
[15] R. L. Aspinall, R. K. Meyer, and M. A. Rao, *Endocrinology (Baltimore)* **68**, 944 (1961).

until the twelfth day of incubation fails to eliminate the precipitin response.[11,16] The injection of 2.5–4.0 mg of testosterone propionate (TP) into the allantoic cavity on the eleventh or twelfth day of embryonic development prevents the appearance of lymphoid tissue in the bursa of Fabricius.[17] The egg is candled (passage of an incandescent wavelength of light through the egg) to reveal the air cell, which is in the large end of the egg, and its semipermeable membrane below which develops the embryo. The air cell is outlined with a pencil. Then, an area about 3 mm below the semipermeable membrane is wiped with 70% EA and pierced with a 23-gauge needle. Testosterone propionate (50 mg/ml, 0.05–0.1 ml) dissolved in autoclaved sesame oil is injected into the allantoic cavity. The hole is covered with paraffin or transparent tape. Hatchability of the injected eggs will be at least 50%.

Hormonal bursectomy may be effected by dipping eggs in steroid solutions.[18,19] The dipping technique will either eliminate the bursa in the hatched chick or markedly reduce its size with general hypertrophy of epithelial cells and inhibition of bursal lymphopoiesis.[20] Chicks hatched from eggs dipped in TP, unlike their TP-injected counterparts, produce excessive quantities of immunoglobulin M (IgM), the pentameric Ig, and markedly reduced levels of IgG.[21,22] This dysgammaglobulinemic model allowed us to conclude that, at least, IgM synthesis has a bursal independent pathway.[21–23] The pointed end (2–4 cm) of a 3-day-old embryonated egg is dipped for 5 sec into an EA-TP solution and then returned to the setting tray. The EA-TP solution is maintained at 18°. Concentrations of TP below 1.2 g% (1.2 g TP in 100 ml EA) will have varying effects on bursal development while EA containing 2 g% TP will eliminate the bursa in 80% of the hatched chicks. Bioassay and isotope procedures reveal that the albumin of eggs which are dipped in EA solutions (0.5–0.8 g% in TP) contains between 74 and 111 μg of TP within 60 min of shell exposure.[24,25]

Surgical Bursectomy in Ovo

In ovo bursectomy between 17 and 19 days of embryonic development lead to the proposal that the bursal microenvironment is necessary for the

[16] A. P. Mueller, H. R. Wolfe, R. K. Meyer, and R. L. Aspinall, *J. Immunol.* **88**, 354 (1962).
[17] N. L. Warner and F. M. Burnet, *Aust. J. Biol. Sci.* **14**, 580 (1961).
[18] B. Glick and C. R. Sadler, *Poult. Sci.* **40**, 185 (1960).
[19] B. Glick, *Endocrinology* (*Baltimore*) **69**, 984 (1961).
[20] B. Glick and F. C. McDuffie, *J. Reticuloendothel. Soc.* **17**, 119 (1975).
[21] K. G. Lerner, B. Glick, and F. C. McDuffie, *J. Immunol.* **107**, 493 (1971).
[22] D. S. V. Subba Rao, F. C. McDuffie, and B. Glick, *J. Immunol.* **120**, 783 (1978).
[23] B. Glick, *Proc. Soc. Exp. Biol. Med.* **127**, 1054 (1968).
[24] J. A. Wilson and B. Glick, *J. Miss. Acad. Sci.* **12**, 308 (1966).
[25] J. A. Wilson, N. Mitlin, and B. Glick, *Poult. Sci.* **50**, 655 (1971).

transformation of IgM to IgG.[26,27] Eggs are prepared for *in ovo* bursectomy by candling, wiping the lower half of the shell with 70% EA, and then cutting a 1.5 × 2-cm window in the region covering the caudal portion of the embryo. The caudal portion of the embryo is pulled through a small orifice in the vascular chorioallantoic membrane. The bursa is removed as described in the section on surgical bursectomy. The embryo is returned to the interior and the opening sealed with transparent tape. All the procedures are performed aseptically. Survival should be approximately 70%.

Experiments with monospecific IgM serum (anti-μ-serum) and IgG serum (anti-γ-serum) revealed that IgM appeared in the bursal cells by day 14 of embryonic development, IgG by day 21, and occasional bursal cells possessed IgM and IgG on the twenty-first day of embryonic development.[28] The injection of anti-μ (0.1–0.2 ml of 1.9 mg antibody) into a chorionic vessel (see cyclophosphamide procedure for *in ovo* iv injections) on day 13 of embryonic development followed by bursectomy at hatch will suppress synthesis of IgM and IgG.[29]

A bursal independent pathway for the synthesis of IgM was identified in our laboratory employing hormonal bursectomy.[2,13] Verification and extension of our data have come from *in ovo* bursectomies performed prior to 72 hr of incubation.[30–33] The *in ovo* bursectomies performed at or before the 18 somites stage eliminated the bursa and one-third of the large intestine in hatched chicks, but did not eliminate B cells or IgM synthesis. A paraphrase of Fitzimmons' *in ovo* bursectomy[30] procedure follows.

Eggs are incubated for approximately 3 days and then placed small end up for 1 hr. They are wiped with 70% EA. Albumin and shell membranes are collected from a single egg for later use. An orifice about 1 cm in diameter is cut at the small end of the shell. This will reveal the embryo (Fig. 3a). The embryo is brought to the top of the hole by the addition of albumin. Using a microscope, a small slit is made in the chorion posterior

[26] P. J. Van Alten, W. A. Cain, R. A. Good, and M. D. Cooper, *Nature (London)* **217,** 358 (1968).

[27] M. D. Cooper, W. A. Cain, P. J. Van Alten, and R. A. Good, *Int. Arch. Allergy* **35,** 242 (1969).

[28] P. W. Kincade and M. D. Cooper, *J. Immunol.* **106,** 371 (1971).

[29] P. W. Kincade, A. R. Lawton, D. E. Bockman, and M. D. Cooper, *Proc. Natl. Acad. Sci. U.S.A.* **67,** 1918 (1970).

[30] R. C. Fitzimmons, M. F. Garrod, and I. Garnett, *Cell. Immunol.* **9,** 377 (1973).

[31] D. X. Dixon and R. C. Fitzimmons, *Dev. Comp. Immunol.* **4,** 713 (1980).

[32] B. D. Janković, Z. Knezević, K. Isaković, K. Mitrović, B. M. Markovič, and M. Rajčevič, *Eur. J. Immunol.* **5,** 656 (1975).

[33] B. D. Janković, K. Isaković, B. M. Markovič, M. Rajčevič, and Z. Knezević, *Exp. Hematol.* **4,** 246 (1976).

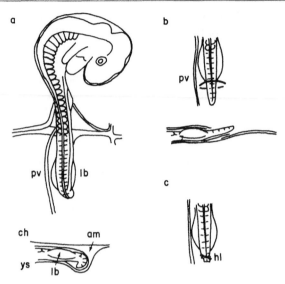

FIG. 3. (a) Embryo (72-hr) with side view of tail area to show extraembryonic membranes. am, Amnion; ch, chorion; lb, leg bud; pv, posterior vitelline vein; ys, yolk sac membrane. (b) Tail region with membranes cut and tail protruding. (c) Tail tied with hair loop and tip amputated. hl, Hair loop. Reproduced with permission of Fitzimmons et al.[30]

to the leg bud. The tip of a knife is inserted to one side of the midline and the opening is extended to the opposite side. The vitelline circulation and yolk sac should not be damaged. Cut the amnion and draw out the tip of the tail to the top of the chorion (Fig. 3b). Desiccation is prevented by the addition of albumin or Ringer's solution. A loop of fine hair, previously rinsed in 70% EA, is positioned over the free end of the tail, posterior to the leg buds, and tightened with forceps. The loop will prevent blood flow through the paired dorsal aorta. The ends of the hair are trimmed. The tail posterior to the loop is severed and removed (Fig. 3c). The tail stump is returned beneath the chorion. Following the addition of albumin to fill the air space, an appropriate size of double shell membrane is placed over the opening. In a few minutes, the albumin on the underside of the membrane will dry and assist sealing. The sealing is completed by applying a thin layer of paraffin over the edges of the double shell membrane. The eggs are now ready for normal incubation.

Cyclophosphamide

The ontogeny of humoral immunity is suppressed when chicks are injected on the first 3 days after hatch with 4 or 6 mg of cyclophosphamide

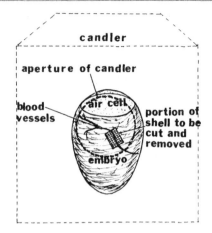

FIG. 4. A fertile egg candled to view location of blood vessels.

(Cy).[34] Cy eliminates lymphocytes from the bursa but not the thymus and magnifies the presence of bursal secretory cells.[35-38] An aqueous solution of 0.1 or 0.2 ml Cy (20 mg/ml) is injected intramuscularly the first 3 days after hatching. Higher doses will lead to excessive mortality during the first 3 weeks of age.[34] The ability of Cy to suppress bursal lymphopoiesis and humoral immunity may depend on the breed or strain of chickens.[39]

Chemical bursectomy with Cy has also been effected with intravenous injections at 12 to 14, 14 to 16, and 16 to 18 days of embryonic development.[40,41] The greatest mortality occurs in the youngest embryos with the most effective humoral immune suppression observed in the 16- to 18-day group. Eggs are candled at 16 days of embryonic development and a major vein is outlined in pencil (Fig. 4). The area is disinfected with 70% EA. A variable drill is used to cut the square piece of shell which overlays the vein. The shell window is removed with curved forceps. The exposed membrane is cleared with mineral oil. Cy (0.1 ml, 2 mg or 1 mg/10 g body weight) is injected with a 30-gauge needle (bevel up) attached to a tuberculin syringe. Allow several seconds before removing the needle. Cover

[34] S. P. Lerman and W. P. Weidanz, *J. Immunol.* **105,** 614 (1970).
[35] B. Glick, *Transplantation* **11,** 433 (1971).
[36] T. J. Linna, D. Frommel, and R. A. Good, *Int. Arch. Allergy* **42,** 20 (1972).
[37] P. Toivanen, A. Toivanen, and R. A. Good, *J. Immunol.* **109,** 1058 (1972).
[38] I. Oláh and B. Glick, *Experientia* **34,** 1642 (1975).
[39] B. T. Rouse and A. Szenberg, *Aust. J. Exp. Biol. Med.* **52,** 873 (1974).
[40] J. Eskola and P. Toivanen, *Cell. Immunol.* **13,** 459 (1974).
[41] O. Lassila, J. Eskola, and P. Toivanen, *J. Immunol.* **123,** 2091 (1979).

the site with transparent tape. Since multiple iv injections are tedious and produce high mortality, we have experimented with air cell application of Cy.[42] The surface at the round end (immediately above the air cell) of the egg is disinfected with 70% EA. A 20-gauge needle is used to penetrate the round end. An aqueous solution of Cy (0.1 ml, 20 mg/ml) is applied to the air cell membrane by inserting into the hole a 20-gauge needle attached to a tuberculin syringe. The hole is covered with transparent tape permitting the application of future doses of Cy. Application of Cy (2 mg) at 16, 17, and 18 days of embryonic development leads to an approximate 20% reduction in hatchability and suppresses bursal lymphopoiesis. A saline solution of 30 ml of colchicine (1 mg/ml) when applied to the anal lip of chicks on the first 4 days after hatching may be more efficient in inhibiting bursal lymphopoiesis than the ip injection of Cy to newly hatched chicks.[43,44]

[42] B. Glick, *Fed. Proc., Fed. Am. Soc. Exp. Biol.* **40,** 1102 (1981).
[43] T. Romppanen and T. E. Sorvari, *Am. J. Pathol.* **100,** 193 (1980).
[44] T. Romppanen, M. K. Viljanen, and T. E. Sorvari, *Immunology* **42,** 391 (1981).

[2] Thymectomy and Engraftment of Thymic Tissue

By Pierson J. Van Alten

The essential role of the thymus in the development of immunocompetence was clearly established by a number of investigators in the early 1960s. These observations showed that ablation of the thymus in newborn rodents later resulted in a number of severe immunologic deficiencies.[1,2] The consequences of neonatal thymectomy primarily consist in a depletion of lymphocytes in peripheral blood and lymphoid tissues and diminished competence of cellular immune responses. To better understand the role of the thymus in developmental immunobiology it became necessary to use thymic grafting in an attempt to repair the immunologic defect resulting from neonatal thymectomy.[3] In this chapter we will describe in detail the methods for performing surgical thymectomy and immunological reconstitution by engraftment of thymic tissue.

[1] R. A. Good, A. P. Dalmasso, C. Martinez, O. K. Archer, J. C. Pierce, and B. W. Papermaster, *J. Exp. Med.* **116,** 773 (1962).
[2] J. F. A. P. Miller, *Lancet* **2,** 748 (1961).
[3] J. F. A. P. Miller, A. H. E. Marshall, and R. G. White, *Adv. Immunol.* **2,** 111 (1962).

Surgical Procedure for Removal of the Thymus

The preferred surgical method for removal of the thymus from new-born mice and rats is by aspiration. The surgical procedure has been well detailed for newborn rodents by Sjödin *et al.*[4] It is similar to that described for adults, and is currently still used.[5]

The procedure, for both newborn and adult rodents, must be performed under magnification. For this purpose, a binocular dissecting microscope having a working distance of at least 8 cm and a magnification of 20 to 30 times is most suitable. The binocular microscope will allow one to have a comfortable working distance and enough visualization so that all organs within the neck and mediastinum can be clearly seen and identified. The operating board on which the animal is placed should be sloped with the head end being about 2 cm higher. The animal should be securely held to the board by rubber bands. The rubber bands must not interfere with respiratory and circulatory functions but, at the same time, must immunobilize the animal so that there is no undesirable movement during the operation. The whole procedure should be carried out so that the operator is comfortably seated with the hands supported, thus avoiding any unnecessary movements which may result in the rupture of large blood vessels or injury to nerves within the mediastinum.

The removal of the thymus by aspiration depends on having a good vacuum system which will assure an even negative pressure at the tip of the pipet. This is best achieved with a water vacuum pump connected to a standard 500-ml vacuum flask with a side arm. Attached to the aspiration flask is tubing which will not collapse from the vacuum. Connected to the end of the tubing is a tapered glass pipet. The diameter of the glass tubing used to make the pipet should not be greater than 7 mm. The aspiration pipets are prepared by heating the glass tubing and drawing it out to the desired diameter, curving it slightly, and cutting it off. The drawn out end is carefully ground on a fine carborundum stone and then fine-polished to remove all sharp edges. The diameter of the tapered end of the pipet should be approximately 1.0 mm for newborn mice and 1.5 mm for newborn rats.

Surgical instruments should be of high quality with fine, precisely crafted tips for the critical, delicate dissections. They should include microdissection scissors with straight 10 mm length blades, number 3 and 5 type jeweler's forceps, and a 95 mm forceps with curved tips to use for retraction. The tips of the forceps should be ground down to blunt them so

[4] K. Sjödin, A. P. Dalmasso, J. M. Smith, and C. Martinez, *Transplantation* **1**, 521 (1963).
[5] H. R. Smith, T. M. Chused, P. A. Smathers, and A. D. Steinberg, *J. Immunol.* **130**, 1200 (1983).

that delicate tissues will not be injured. All tissues in the newborn animal are especially friable and thus easily damaged with sharp points. The incision should be closed with a 5-0 silk suture using circle taper needles and then treated with spray-on wound dressing.

Hypothermia may be used to anesthetize the animal. It can be induced by placing the newborn animal in the refrigerator or on ice, but protecting it from becoming wet. The animals must be cooled until they stop moving and then operated on quickly before they warm up. The drawback of hypothermia is that the lights used during the operation often produce heat which warms the animal and may make it move. If the animal moves at a critical moment during the procedure, injury may occur. Hypothermia is advantageous because animals quickly recover by gently warming them under a lamp. Ether can also be used as an anesthetic for newborn animals. The animal is placed in a closed desiccator jar containing surgical sponges moistened with ether. The animal must be anesthetized deeply enough so that all spontaneous movements cease. The major difficulty with ether is excessive mucus secretion which, however, can be avoided by a subcutaneous injection of 0.01 mg of atropine sulfate prior to anesthesia. The advantage of the ether anesthesia is that the animal is unlikely to move while being operated on and the recovery period is short. Pentobarbital (Nembutal) or other injectable anesthetics are difficult to use since the difference between an anesthetic and lethal dose is rather small. Also, with injectable anesthetics the recovery time is longer.

All operations should be carried out under sterile conditions. The instruments, including the glass cannulas, should be immersed in tincture of zephirin chloride or 70% alcohol. Seventy percent alcohol also is used to swab the neck and thorax just prior to making the incision. If the operator keeps his hands clean it is not necessary to wear gloves.

Neonatal thymectomy of mice or rats must be performed within 24 hrs after birth to assure the maximum depression of the T cell system. The following is a detailed description of the technique for aspiration of the thymus. The anesthetized animal is securely strapped to the animal board with head elevated and close to the operator. The skin is swabbed with 70% alcohol and a midline longitudinal incision through the skin and superficial fascia from the angle of the mandible to the fourth rib is made with microdissecting scissors. By careful blunt dissection the submaxillary glands are freed and reflected anteriorly; the sternohyoid muscles are separated, exposing trachea and sternum. With the point of the microdissecting scissors kept along the dorsal aspect of the sternum, a cut is made along the middle of the sternum to the level of the fourth rib. This incision requires great care in order to avoid puncturing the heart or the arch of the aorta. The sternum is now lifted ventrally and laterally with the curved forceps and the fascia overlying the mediastinum is split, exposing the thymus. The suction pipet is placed over the anterior end of one thymic lobe with gentle manipulation, and by maintaining a constant negative pressure, one and then the other lobe is aspirated. Excessive negative pressure and too deep penetration of the pipet into the mediastinum may cause injuries to the heart or aortic arch. Following aspiration of both thymic lobes, the mediastinum should be carefully inspected for thymic remnants which may be clinging to the connective tissue. As one develops skill with manipulating the pipet and having the correct suction pressure, complete thymectomy will usually be accomplished with little difficulty.

Before closing the incision, the animal is removed from the operating board, the skin and underlying fascia are brought together with a pair of blunt forceps, and the incision is closed with one or two interrupted 5-0 silk sutures. The animal should be carefully cleansed with warm saline to remove blood from the skin. Spray-on wound dressing is then applied to the incision with a cotton-tipped applicator. The animal should be gently warmed under a lamp for 1–2 hr and returned to the mother after full recovery. Cannibalism by the mothers can be greatly reduced if the mothers are adequately handled prior to delivery so that they become docile.

The technique for thymectomy in adult rodents is similar to that for the newborn. The pipets used should have a distal diameter of approximately 2.2 mm for mice and 2.7 mm for rats. As with the newborn, the diameter of the pipet should be slightly smaller than the thymic lobe. With adult animals, care should be taken to avoid pneumothorax by carrying out the operation as quickly as possible and by squeezing the thorax slightly while closing the incision. Closure of the skin incision is best accomplished by use of 2 or 3 wound clips (9 or 11 mm) followed by swabbing with spray-on wound dressing with a cotton-tipped applicator.

Effect of Thymic Ablation on Immune Functions

From numerous observations made in the 1960s it became apparent that neonatal removal of the thymus is associated with depletion of the lymphocyte population and impaired immune function. In examining its embryological development it became evident that the thymus is the first organ in which lymphoid cells develop. The hypothesis was then formulated that lymphocytes first develop in the thymus and then migrate to peripheral lymphoid tissues.[6] The data presented in Table I summarize a number of cellular immune functions which are affected following neonatal thymectomy.

Although initially it did not appear that adult thymectomy had any

[6] R. Pahwa, S. Ikehara, S. G. Pahwa, and R. A. Good, *Thymus* 1, 27 (1979).

TABLE I

SOME CHANGES IN IMMUNE FUNCTIONS FOLLOWING NEONATAL REMOVAL OF THYMUS

Function	Activity	References[a]
Allograft rejection	Decreased	Miller (1)
Lymphoid tissue	Atrophy	Good et al. (2)
Induction of graft-versus-host disease	Decreased	Dalmasso et al. (3)
Susceptibility to graft-versus-host disease	Increased	Good et al. (2)
Tolerance: susceptibility and duration	Increased	Hilgard et al. (4)
Proliferative response to T cell mitogens and allogenic cells	Decreased	Meuwissen et al. (5)
Autoimmune disease	Retarded	Steinberg et al. (6)
Autoimmune disease	Accelerated	Theofilopoulos and Dixon (7)
B cell proliferation and autoimmune disease	Increased	Smith et al. (8)
Suppressor cell activity	Decreased	Cantor and Gershon (9)
Helper cell activity	Decreased	Haines and Siskind (10)
Leukemogenesis	Decreased	Zielinski et al. (11)
Major histocompatibility restricted self recognition	Decreased	Kruisbeek et al. (12)
Secretory and serum IgA production	Decreased	Ebersole et al. (13)
Spontaneous occurrence of diabetes mellitus	Decreased	Like et al. (14)

[a] References: (1) J. F. A. P. Miller, *Lancet* **2**, 748 (1961). (2) R. A. Good, A. P. Dalmasso, C. Martinez, O. K. Archer, J. C. Pierce, and B. W. Papermaster, *J. Exp. Med.* **116**, 773 (1962). (3) A. P. Dalmasso, C. Martinez, and R. A. Good, *Proc. Soc. Exp. Biol. Med.* **110**, 205 (1962). (4) H. R. Hilgard, K. Sjödin, C. Martinez, and R. A. Good, *Transplantation* **2**, 638 (1964). (5) H. J. Meuwissen, P. J. Van Alten, and R. A. Good, *Transplantation* **7**, 1 (1969). (6) A. D. Steinberg, J. B. Roths, E. D. Murphy, R. T. Steinberg, and E. S. Raveche, *J. Immunol.* **125**, 871 (1980). (7) A. N. Theofilopoulos and F. J. Dixon, *Immunol. Rev.* **55**, 179 (1981). (8) H. R. Smith, T. M. Chused, P. A. Smathers, and A. D. Steinberg, *J. Immunol.* **130**, 1200 (1983). (9) H. Cantor and H. Gershon, *Fed. Proc., Fed. Am. Soc. Exp. Biol.* **38**, 2058 (1979). (10) K. A. Haines and G. W. Siskind, *J. Immunol.* **124**, 1878 (1980). (11) C. C. Zielinski, S. D. Waksal, and S. K. Datta, *J. Immunol.* **129**, 882 (1982). (12) A. M. Kruisbeek, S. O. Sharrow, B. J. Mathieson, and A. Singer, *J. Immunol.* **127**, 2168 (1981). (13) J. L. Ebersole, M. A. Taubman, and D. J. Smith, *J. Immunol.* **123**, 19 (1979). (14) A. A. Like, E. Kislauskis, R. M. Williams, and A. A. Rossini, *Science* **216**, 647 (1982).

major effect on the immune competence of the animal, as methods for detecting immunologic functions became more refined and precise, it became evident that some of the regulatory functions of the immune system were affected. In Table II data are summarized showing that the suppressor cell activities of thymic dependent cells are most severely affected following adult thymectomy. An overview of the effect of adult thymec-

TABLE II
SOME CHANGES IN IMMUNE FUNCTIONS FOLLOWING REMOVAL OF ADULT THYMUS

Function measured	Activity	References[a]
Antibody response to T cell-independent antigens	Increased	Kerbel and Eidinger (1)
Carrier-specific helper cell activity	Increased	Kappler et al. (2)
Generation of cytotoxic T cells	Increased	Simpson and Cantor (3)
Induction of tolerance	Decreased	Nachitigal et al. (4)
Suppressor cell activity	Decreased	Reinisch et al. (5)
Precursor of splenic T cells (L 1,2,3+)	Decreased	Cantor and Boyse (6)
Idiotype-specific transplantation resistance	Decreased	Daley et al. (7)
Macrophage-mediated antibody-dependent cellular cytotoxicity	Decreased	Parthenais and Haskill (8)
Tumor metastasis	Increased	Parthenais and Haskill (9)

[a] References: (1) R. S. Kerbel and D. Eidinger, *Eur. J. Immunol.* **2**, 114 (1972). (2) J. W. Kappler, P. C. Hunter, D. Jacob, and E. Lord, *J. Immunol.* **113**, 27 (1974). (3) E. Simpson and H. Cantor, *Eur. J. Immunol.* **5**, 337 (1975). (4) D. Nachtigal, I. Zan-Bar, and M. Feldman, *Transplant. Rev.* **26**, 87 (1975). (5) C. L. Reinisch, S. L. Andrew, and S. F. Schlossman, *Proc. Natl. Acad. Sci. U.S.A.* **7**, 2989 (1977). (6) H. Cantor and E. Boyse, *Cold Spring Harbor Symp. Quant. Biol.* **39**, 23 (1977). (7) M. J. Daley, H. M. Gebel, and R. G. Lynch, *J. Immunol.* **120**, 1620 (1978). (8) E. Parthenais and S. Haskill, *J. Immunol.* **123**, 1329 (1979). (9) E. Parthenais and S. Haskill, *J. Immunol.* **123**, 1334 (1979).

tomy and the possible effects on the regulatory control of the immune system in relation to idiotypic-specific transplantation resistance is presented by Daley el al.[7]

A major point which throughout the 1960s was difficult to resolve was the relationship of the thymus to antibody formation. For example, following thymectomy, the serum immunoglobulin level, measured by immunoelectrophoresis, was unaffected. Also when thymectomized animals were injected with *Brucella abortus, Salmonella typhosa,* or *Salmonella paratyphi,* antibody production was comparable to that of normal mice.[8] In contrast, when *Salmonella typhimurium,* influenza viruses, sheep erythrocytes, or serum proteins were injected, there was a significant reduction in the amount of antibody produced by thymectomized animals.[9]

[7] M. J. Daley, H. M. Gebel, and R. C. Lynch, *J. Immunol.* **120**, 1620 (1978).
[8] G. Mosser, R. A. Good, and M. D. Cooper, *Int. Arch. Allergy Appl. Immunol.* **39**, 62 (1970).
[9] J. F. A. P. Miller, *Thymus* **1**, 3 (1979).

Studies in birds led to the hypothesis that there is a division of labor among lymphocytes, and that those involved in antibody production are not dependent on the thymus for their maturation.[10] Such a division at the cellular level was shown by Roitt et al.[11] when they identified cell surface characteristics distinguishing T from B cells. The interrelationship of the T and B cell systems was clarified by the observation of Claman et al.[12] who showed that when irradiated mice were ingrafted with both thymic and bone marrow cells they produced more antibody than if they received either thymic or bone marrow cells. From these studies it became clear that collaboration of T and B cells was necessary for full immune function.

Ingraftment of Thymic Tissue

Much knowledge of thymic function has been derived from research using thymic grafts for the reconstitution of the immunobiological functions of neonatally thymectomized or athymic (e.g., nude) animals. From such studies it became apparent that, in addition to grafting whole thymic tissue, reconstitution of immune functions could be obtained by injection of thymic cell suspensions or by implanting cell impermeable diffusion chambers containing thymic cells, into the peritoneal cavity (cf. Pahwa et al.[6]).

Thymic implants into athymic recipients should be performed within 2 to 3 weeks after birth to avoid postthymectomy wasting disease.[13] The thymic tissue to be grafted is usually obtained from donors 1–5 days of age. The entire thymus is removed aseptically and placed in a small petri dish containing Hanks' balanced salt solution. The thymus is cut into three or four pieces using a sterile razor blade. The recipient animal is anesthetized, immobilized on an animal board, and the skin on the back disinfected. A midline incision of approximately 2 cm is made on the back. Using a small spatula, the thymic tissue is lifted from the petri dish and carefully tucked beneath the skin on both sides of the incision. The skin may be closed by using 9 mm wound clips or two or three interrupted 5-0 silk sutures and then applying spray-on wound dressing to the incision with a cotton-tipped applicator. For the ingraftment of slices of thymic tissue into the peritoneal cavity, a small midline incision through the

[10] M. D. Cooper, R. D. A. Peterson, M. A. South, and R. A. Good, J. Exp. Med. 123, 5 (1966).
[11] I. M. Roitt, M. F. Greaves, G. Torrigianti, J. Brostoff, and J. H. L. Playfair, Lancet 2, 367 (1969).
[12] H. N. Claman, E. A. Chaperon, and R. F. Triplett, Proc. Soc. Exp. Biol. Med. 122, 1167 (1966).
[13] O. Stutman, E. J. Yunis, C. Martinez, and R. A. Good, J. Immunol. 98, 79 (1967).

abdominal skin and underlying fascia is made. The peritoneal cavity is opened, the donor tissue is placed bilaterally, and then 0.1 to 0.2 ml of Hanks' media is injected into the abdominal cavity. The peritoneum is closed with continuous 6-0 silk suture and the skin incision with interrupted 5-0 silk or 9 mm wound clips.

Reconstitution of athymic animals can be accomplished using thymic cell suspensions (in Ringer's lactate saline or Hanks' media) prepared from donors 1–5 days of age. The thymus is removed aseptically and cut into small pieces in a petri dish containing a small amount of media. The content of the petri dish is transferred into a glass homogenizer and the tissue fragments are gently squeezed against the wall of the homegenizer with the aid of a loosely fitting pestle. The pestle is slowly lowered. The single-cell suspension rises to the top of the homogenizer and is transferred into a test tube. The cell suspension is centrifuged and washed three times, the mononuclear cells are counted in a hemocytometer, and the viability assessed by trypan blue exclusion. The cells are diluted to 25 $\times$ 10^6 per 0.1 ml. Between 1 and 4 intraperitoneal injections, totaling 50–300 $\times$ 10^6 viable thymocytes, are administered with a 27-gauge needle between 7 and 15 days of age.[14]

Thymic tissue enclosed in diffusion chambers has been found to restore immunocompetence in some cases in neonatally thymectomized mice. Restoration, however, is not consistent nor does it occur in athymic nude mice.[15] Diffusion chambers are prepared using moistened discs of Millipore or Amicon filters, 13 mm in diameter cemented (Millipore cement) to both sides of Lucite rings having external diameters of 14 mm and internal diameters of 10 mm. The filter discs have average pore size of 0.45 and the Lucite rings have a thickness of 2 mm and contain an access hole through which cell suspensions may be injected. The chamber has a capacity of 0.15 ml. After the cement dries the chambers are sterilized with ethylene oxide for at least 2 hr. Following sterilization, while still wrapped, they are set aside for 72 hr to lose residual vapors. Thymic cell suspensions, prepared as described above, are injected with a 24-gauge needle into two diffusion chambers. The access hole is sealed with cement and when dried the chambers are placed bilaterally along with 0.2 ml of media within the peritoneal cavity of anesthetized animals. The introduction of the cell suspension into the chambers and their implantation into the animals is performed under sterile conditions.

Since thymic function of athymic animals could be restored, in some cases, with diffusion chamber implants containing thymic tissue, it be-

[14] E. J. Yunis, H. R. Hilgard, C. Martinez, and R. A. Good, *J. Exp. Med.* **121,** 607 (1965).
[15] O. Stutman, *Ann. N.Y. Acad. Sci.* **249,** 89 (1975).

came apparent that whole thymic tissue was not needed for T cell development. Further, it had also been shown that thymic tissue, irradiated or devoid of thymocytes, could partially restore the immune function of thymectomized animals.[16] Thus, it was postulated that precursors of lymphoid cells circulate to the thymus which provides an environment for differentiation into competent T cells.[17] Subsequently, when Bigger *et al.*[18] implanted embryonic thymic anlage into neonatally thymectomized mice, they observed complete development of thymic structure and restoration of immune function. Recently, Hong *et al.*[19] showed that 9 day organ-cultured fragments of normal thymic tissue of newborn mice, when transplanted into athymic nude mice, restore their T cell competence. The method for organ culturing of thymic tissue is as follows. The thymus is obtained from normal newborn mice, removed aseptically, and placed into sterile petri dishes containing Ham's F12 media supplemented with 100 units penicillin, 100 μg streptomycin per ml, and with 10% horse serum or fetal bovine serum. The thymus is cut into fragments of approximately 2 mm in diameter and placed onto sterile stainless-steel organ culture grids in 35 mm culture dishes. Sufficient Ham's F12 supplemented media is added so that the surface of the grid is moistened. The cultures are incubated at 37° in 5% CO_2–95% air. The cultures are fed at 3 to 4 day intervals by replacement of approximately 80% of the media. Hong *et al.*[19] cultured the thymic fragments for 9 days and then grafted them under the kidney capsule of athymic nude mice. For engraftment of tissue fragments or cellular pellets under the kidney capsule, a small incision is made on the left dorso-lateral body wall of the recipient animal. The kidney is located and carefully manipulated to the incision, thus exposing its surface. With a finely drawn tungsten needle or ultramicroknife, a small incision is made in the capsule. With a blunt microneedle the subcapsular tissue is carefully pushed aside and the tissue to be grafted is placed beneath the capsule. The kidney is then returned to its normal position. The body wall and skin are closed with 5-0 silk suture or 9 mm wound clips. An important advantage to grafting under the kidney capsule is that the grafts have a protected environment and are easily recoverable for histological examination. Generally, it is believed that it is the epithelial reticular cells of the thymus which provide the microenvironment for

[16] E. F. Hays, *Blood,* **29,** 29 (1967).
[17] D. Metcalf and M. A. S. Moore, "Haemopoietic Cells." North-Holland Publ., Amsterdam, 1971.
[18] W. D. Bigger, B. H. Park, and R. A. Good, *Annu. Rev. Med.* **24,** 135 (1972).
[19] R. Hong, S. D. Horowitz, C. Fortman, and D. W. Martin, Jr., *Clin. Immunol. Immunopathol.* **23,** 448 (1982).

differentiation of functional T cells.[20] Further support for such a conclusion has been provided by Jordan *et al.*[21] These authors implanted under the kidney capsule of mice thymic monolayer cultures derived from explants of newborn mouse thymus. The nonlymphoid cells developing had the characteristics of epithelial-like cells. In view of the fact, however, that the original monolayer, when cultured to confluence, was found to be about 80% macrophages,[22] it can not be ruled out that also macrophages might have played a role in the differentiation of the lymphoid cells (see also Beller *et al.*[23]).

Conclusion

In reviewing the results obtained with the thymectomy of newborn mice and rats and the engraftment of thymic tissue into athymic animals, it is evident that these methods have contributed the definitive data demonstrating the lymphopoietic function of the thymus. They have also provided fundamental insights into the nature and function of the small lymphocyte and greatly contributed to the present understanding of the interlinkage and regulatory relationships of the various thymic dependent cells with other cellular elements comprising the host defense system. In recent years experimental approaches using *in vivo* systems have been used less extensively due to the evolution of *in vitro* techniques developed by Mishell and Dutton[24] and Marbrook.[25] Nevertheless, thymectomy and implantation of thymic tissues are currently used in elucidating the role of the immune system in autoimmune diseases, tumor immunity, and the waning of immunocompetence with aging.

[20] H. Cantor and I. L. Weissman, *Prog. Allergy* **20,** 1 (1976).
[21] R. K. Jordan, D. A. Crouse, and J. J. T. Owen, *J. Reticuloendothel. Soc.* **26,** 373 (1979).
[22] R. K. Jordan and D. A. Crouse, *J. Reticuloendothel. Soc.* **26,** 385 (1979).
[23] D. I. Beller, A. G. Farr, and E. R. Unanue, *Fed. Proc., Fed. Am. Soc. Exp. Biol.* **37,** 91 (1978).
[24] R. I. Mishell and R. W. Dutton, *J. Exp. Med.* **126,** 423 (1967).
[25] J. Marbrook, *Lancet* **2,** 1279 (1967).

[3] Skin Grafting

By David Steinmuller

Introduction

Before histocompatibility testing by *in vitro* methods was commonplace, Medawar[1] described the rejection of skin allografts (grafts between genetically dissimilar donors and recipients of the same species) as "perhaps the completest single measurement of the inborn diversity of mice or men." Indeed skin grafts still are used commonly to measure genetic differences between individuals[2,3] and, of most importance to cellular immunologists, competence of the thymus-dependent immune system.[4] The immunological[5] and medical[6] aspects of skin grafting are reviewed elsewhere, as well as techniques of clinical skin grafting.[7] The purpose here is to describe techniques of transplanting and evaluating skin grafts in mice and rats, the most widely used experimental animals in cancer research, immunology, and immunogenetics laboratories. Reference should be made to the classic paper by Billingham and Medawar[8] or the later version by Billingham[9] for the application of virtually the same methods to skin grafting hamsters, guinea pigs, and rabbits.

The simplest way of skin grafting small rodents, whose skin is so mobile and loosely adherent to the underlying body wall, apparently would be to excise a full-thickness piece of skin and suture or clip into place a graft of comparable size and thickness. In fact, this is the least elegant and desirable method because it subjects the graft to ischemic necrosis; revascularization can occur only from the cut edges of the surrounding host skin, and, especially with large grafts, there is massive

[1] P. B. Medawar, *Harvey Lect.* **52,** 144 (1958).
[2] F. E. Ward, N. R. Mendell, H. F. Seigler, J. M. MacQueen, and D. B. Amos, *Transplantation* **26,** 194 (1978).
[3] M. Jonker, J. Hoogeboom, A. van Leeuwen, K. T. Koch, D. B. van Oud Alblas, and J. J. van Rood, *Transplantation* **27,** 91 (1979).
[4] J. F. A. P. Miller, *Ann. N.Y. Acad. Sci.* **99,** 340 (1962).
[5] G. D. Snell, J. Dausset, and S. Nathenson, "Histocompatibility." Academic Press, New York, 1976.
[6] A. B. Cosimi, J. F. Burke, and P. S. Russell, *Surg. Clin. North Am.* **58,** 435 (1978).
[7] R. Rudolph, J. C. Fisher, and J. L. Ninnemann, "Skin Grafting." Little, Brown, Boston, Massachusetts, 1979.
[8] R. E. Billingham and P. B. Medawar, *J. Exp. Biol.* **28,** 385 (1951).
[9] R. E. Billingham, *in* "Transplantation of Tissues and Cells" (R. E. Billingham and W. K. Silvers, eds.), p. 1. Wistar Inst. Press, Philadelphia, Pennsylvania, 1961.

nonspecific degeneration and sloughing of necrotic tissue, which makes it difficult to assess initial graft "take" and subsequent immunologically mediated rejection. The recommended procedure is to prepare a graft "bed" by excising host skin down to but not through the richly vascularized superficial fascia immediately over the *panniculus carnosus,* the subcutaneous muscle responsible for skin twitching is most mammals that is absent from most regions of the skin of primates.[8] This layer of fascia contains the main arteries, veins, and lymphatics of the skin, and opposing a trimmed graft (see below) to it with an appropriate pressure dressing facilitates healing and revascularization and minimizes nonspecific necrosis.

Source and Preparation of Grafts

The best sources of murine skin grafts are the ears, the tail, and the belly. Since thin grafts heal-on better than thick ones, the thin skin of the ventrum should be used in preference to the significantly thicker skin of the dorsum. With major histocompatibility differences, grafts of similar size and thickness from these three sources have similar survival times. However, the source of the graft can influence the survival of skin allografts involving minor histocompatibility differences,[10,11] a phenomenon that may result from regional differences in the number of Langerhans cells, the antigen-presenting cells of the epidermis.[12] Moreover, because of anteroposterior differences in the take and growth of skin grafts in the mouse, at least,[13] the positioning of the graft bed should also be standardized.

An economical advantage of ear and tail skin grafts is that they can be obtained from the living animal, sparing the donor for other uses. In order to obtain ear skin, the donor is sacrificed or anesthetized and the hair around the base of the ear is clipped (Oster small animal electric clippers with size 40 cutting blades are recommended). Sharp surgical scissors are used to cut off the pinna cleanly at its base while holding it taut at its tip. This projecting portion of the external ear is largely cartilaginous, and the cut site will not bleed significantly or require medication or dressing as long as the cut is not too low, i.e., not on the scalp per se. Forceps are used to peal apart the two sides of the pinna. The skin is placed, raw-side down, on saline (0.9% sodium chloride)-drenched filter paper (a 9-cm

[10] B. J. Mathieson, L. Flaherty, D. Bennett, and E. A. Boyse, *Transplantation* **19**, 525 (1975).
[11] J. Sena, S. S. Wachtel, and G. Murphy, *Transplantation* **21**, 412 (1976).
[12] P. R. Bergstresser, C. R. Fletcher, and J. W. Streilein, *J. Invest. Dermatol.* **74**, 77 (1980).
[13] L. Kubai and R. Auerbach, *Transplantation* **30**, 128 (1980).

diameter circle of Whatman No. 50 placed in the inverted lid of an ordinary plastic petri dish works nicely). The cartilage layer almost invariably sticks to one side, and it must be removed before grafting because it can delay the rejection of allografts with minor histocompatibility differences.[11,14] This is done by holding the skin down with forceps and scraping it heavily with a dulled scalpel blade (a No. 20 blade is deliberately dulled with a sharpening stone; if the blade is at all sharp, it will cut through the skin). For body skin, the donor is sacrificed, stretched belly-up, and spread-eagle on an operating board. The ventral fur is removed with electric clippers (it is not necessary to shave the skin with a razor). Unless only a small piece is required, the skin is excised with scissors or scalpel from the groin to the chin as close to the legs as possible. The skin is then placed, raw-side up, on the board. Again using the deliberately dulled scalpel blade, all the subcutaneous fat and muscle is carefully scraped away. The skin can then be cut into smaller pieces for grafting. A simple way to do this is to place it, raw-side down, on a piece of dry filter paper, which in turn is placed on a cork board. Then, using a sharp cork borer, circles of skin of appropriate diameter (e.g., 1–2 cm for mice and 2–3 cm for rats) are cut out. The grafts are immediately placed, raw-side down, on saline-drenched filter paper. They can be stored in this fashion in an ordinary refrigerator if not immediately used. With body skin grafts, it is important to avoid using areas of skin in which the hairs are in the "anagen" or active growth phase because grafts of such skin are particularly susceptible to ischemic necrosis.[9,15] It is easy to recognize "active" skin because it is considerably thicker and darker than surrounding "resting" skin. Grafts of tail skin are recommended for the occasional grafter because they are the simplest to prepare. The donor is sacrificed or anesthetized and the base of the tail near the anus is scrubbed with 70% alcohol. With rats, the entire tail should be cleaned with soap and water. (If one wants to spare the donor, one can amputate part of the tail and control bleeding with a tourniquet or wound clip.) With a No. 10 surgical blade a radial incision is made at the base of the tail, connected by a longitudinal incision to the tip, and the skin is pulled off the entire tail. The skin is placed, dermal-side down, on saline-drenched filter paper in a petri dish, taking care to flatten it and unroll the edges. The skin is then cut into segments of desired length, e.g., 1–2 cm for mice and 2–3 cm for rats. No further trimming is required. Tail skin has the advantage of being very easy to distinguish from the surrounding skin after grafting.

[14] M. Polackova and V. Viklicky, *Folia Biol.* (*Prague*) **22**, 362 (1976).
[15] C. G. Ramselaar and E. J. Ruytenberg, *Br. J. Exp. Pathol.* **62**, 252 (1981).

Preparation of the Recipient and Grafting

Anesthesia. For mice, a dilute solution of sodium pentobarbital is prepared by mixing one part of a commercial solution of approximately 60 mg/ml with 9 parts of 10% ethanol in distilled water. This is administered intraperitoneally at a dose of 0.1 ml/10 g body weight, or less if the recipients show respiratory distress at the full dose. For rats, an aqueous solution of 36 mg/ml of chloral hydrate administered intraperitoneally at 0.75–1.0 ml/100 g body weight is preferable. Mice can be weighed to the nearest gram on a top-loading balance. A triple-beam balance with an animal cage (e.g., Ohaus model 730) is usually required for rats. When rodents are under general anesthesia, it is extremely important to help them maintain their normal body temperature by keeping them warm (a 60-W light bulb is suitable for this purpose) particularly during the recovery period. Care should be exerted, however, not to overheat them.

Preparation of the Recipient. The anesthetized animal is tied to an operating board with rubber bands passed through syringe needle adapters (A. H. Thomas, Phila., PA, No. 8960-D52) or some other small tubular device that permits one to slip the loops around the limbs and keep them taut (Fig. 1). For the occasional grafter, the rubber bands can be kept tight by fixing them to a wooden board with metal push-pins. A more elegant operating board can be constructed from 0.5 in. Plexiglas with cutouts for holding the rubber bands (Fig. 1). The lateral wall of the thorax is the best site for skin grafts because it provides a firm foundation where bandages and dressings can be secured without interfering with the animal's normal movements. Electric clippers are used to remove the fur from this area (a small, portable vacuum cleaner is handy for collecting fur and keeping hair from blowing around). The skin is then cleaned with a gauze sponge drenched with 70% alcohol or some other antiseptic solution, taking care to wet-down the surrounding fur to keep hair out of the operating field.

Grafting. A cut is made at a point midway between the lower end of the rib cage and the scapula. The skin is cut down to the vascular fascia lying just above the subcutaneous muscle (Fig. 2). For mice, small curved iris scissors are used; for adult rats, larger curved scissors are required. Some blood in the bed is permissible, but not clots. Thus, if a large vessel is cut, the bleeding must be stopped with pressure from a cotton swab or surgical sponge. It is not necessary to fit the bed exactly to the intended graft. In fact, an "open fit" is preferable (e.g., Fig. 1, middle) because grafts are less subject to technical failure from sliding over the surrounding intact skin. The latter soon contracts around the graft anyway, and the remaining raw areas of the bed are reepithelialized by migration and pro-

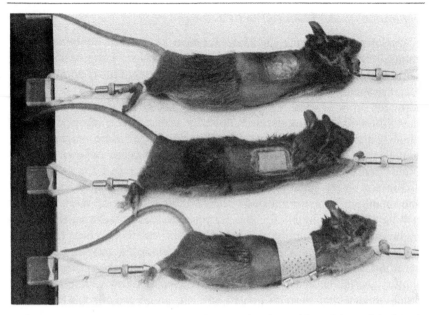

FIG. 1. Three stages in skin grafting mice, showing the position of the graft bed (top), placement of the graft (middle; in this case a rectangle of tail skin), and a Band-Aid dressing secured with 2 wound clips (bottom). Note the needle adaptors and rubber bands tying the animals to the Plexiglas operating board, which was covered with white paper to provide contrast for the photograph.

liferation of cells from the cut edges of the host skin.[8] Placement of the graft is facilitated by keeping the bed moist with sterile saline. Unlike rabbits, where prophylaxis with antibiotics is necessary, mice and rats are remarkably resistant to topical infection, and sterile instruments, operating masks, and gloves, etc. are not required—just common-sense cleanliness. Grafts are best handled with smooth rather than serrated forceps; stainless steel jeweler's forceps (e.g., Accurate Surgical and Scientific Instruments, Westbury, NY, No. JF-4) are highly recommended for very thin ear and belly skin grafts, which tend to curl or wrinkle when transferred from the petri dish to the graft bed, and must be flattened out without any rolled-under edges. With grafts of body skin, particularly if chronic rejection is anticipated, later identification will be easier if the graft is placed on the bed so that its hairs point in a different direction from those of the host.

Dressings. The classic method calls for a nonadhering dressing over the graft held in place with a plaster of Paris bandage wrapped around the thorax.[8,9] With mice, a simple plastic adhesive bandage works just as well

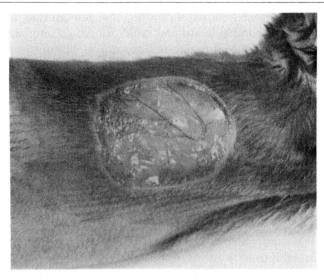

FIG. 2. A closer view of a graft bed showing the highly vascularized superficial fascia overlying the subcutaneous muscle. If the site of the excision were rolled between the fingers, it would be obvious that the defect does not extend through the skin to the interior body wall. Note the pigmented bases of hair follicles at 9–10 o'clock that should have been removed when the bed was prepared because they could give rise to epithelium that might undermine the graft or confuse the appraisal of graft survival.

and is simpler to apply as long as a very flexible brand is used (3/4-in. Johnson & Johnson Band-Aid No. 5634 plastic strips are highly recommended). A special tool is made by filing down the serrations of an ordinary 5-in. dressing forceps. With a wooden applicator stick, white Vaseline (Chesebrough-Ponds) is smeared over the pad and the adhesive surfaces of the Band-Aid. The latter is then grasped by its pad with the smooth-tipped dressing forceps and carefully positioned with its pad centered over the graft. The Band-Aid is held in place with the operator's thumb and index finger at the belly and backbone, respectively. The mouse is then turned over, being careful not to slide the graft off the bed, and the two free ends of the Band-Aid are opposed. The ends of the Band-Aid are then rolled up with forceps toward the belly, where they are secured in place *to the skin* with two 9-mm Clay Adams "Auto Clips" (Fig. 1, bottom; Clay Adams wound clips, appliers and removers are available from Fisher Scientific, Curtis Matheson, A. H. Thomas, etc.). The mouse is removed from the operating board, dried with a towel or tissue, and placed in a warm location (e.g., under a gooseneck lamp) until it recovers from the anesthesia.

An analogous procedure is followed for the skin grafting of rats, except that a larger operating board and firmer dressings and bandages are required. The immediate dressing is a square of Vaseline-impregnated gauze (e.g., Chesebrough-Ponds, Greenwich, CN, No. 5-4146) large enough to overlap the graft bed all around. In order to keep the dressings from slipping posteriorly, the skin around the Vaseline gauze is liberally painted with a surgical adhesive, e.g., *mastisol,* prepared by dissolving 40 g gum mastic (Sigma Chemical, St. Louis, MO, No. G 0878) and 1.25 ml castor oil in 100 ml benzene. A 1- to 2-in.-wide gauze bandage is rapidly wound three or four times around the thorax, followed by a 1-ft (or more, depending on the animal's size) length of 1.5-in.-wide, extra-fast setting plaster bandage (e.g., Johnson & Johnson "Specialist" No. HRI 8137-007363). To prepare the latter, 1.5 × 12-in. strips are cut from the larger commercial roles, rolled up around a 1.5-in.-long piece of plastic tubing, and secured with a rubber band. To apply the plaster dressing, a dressing forceps is used to immerse the small roll in *hot* water for about 5 sec; then the wet plaster dressing is wrapped around the thorax over the gauze dressing, smoothing the plaster with the fingers. The animal is released from the operating board and placed under a 60-W light bulb, where the plaster bandage will set in 2–4 min. Gnawing of the dressing by the animal may be prevented by painting the dressing with a saturated solution of picric acid.

Appraisal of Graft Rejection

Primary Inspection. Although thin, well-trimmed skin grafts transplanted orthotopically as described here are well vascularized in 5–6 days,[8] at this time the union of the graft with its bed is still submaximal, and one should leave the dressings and bandages on for 8 days. If one must inspect the grafts earlier (e.g., at 6 days postgrafting to detect accelerated rejection on preimmunized hosts), the dressings may have to be replaced. Even at 8 days, one should remove the dressing very carefully, with forceps if necessary, because part of the graft may be stuck to it by dried tissue exudate and the graft may be pulled off with the dressing. To remove the dressings, the animal is anesthetized by placing it in a closed glass jar or other container with a wad of cotton or some gauze pads wet with ether or, preferably, a nonflammable, nonexplosive inhalant anesthetic such as methoxyflurane, enflurane, or halothane. The Band-Aid or plaster cast is cut with bandage scissors. A sturdy pair is required for the latter; a "baby" pair for the former (e.g., Storz Instrument Co., St. Louis, MO, No. N-5258). For mice, the wound clips securing the Band-Aids are first removed with the Clay Adams Auto Clip remover. Then the bandages are peeled off, taking care not to dislodge the graft. Because of the intense hyperkeratosis accompanying the healing-on process, pieces of cuticle, or even an entire dead cast of the graft epithelium with the original hairs, may come off with the dressing, revealing firm, white regenerating epidermis.[8,9] If the edge of the graft has pulled-off and there is any bleeding, the animal almost certainly will traumatize the graft. In this case, the area should be rebandaged immediately. This may occur even with well healed-on grafts. However, if an animal is going to traumatize the graft, it does so very soon. Hence, it is wise to inspect grafts a few hours after the dressings are removed, when it is easy to detect trauma-mediated technical failures, which later might be mistaken for immunologically mediated rejection.

Appraisal of Rejection. It is relatively easy to assign end-points for acutely rejecting skin grafts in terms of the complete, grossly visual breakdown of the surface epidermis. The grafts should be inspected daily. After lightly anesthetizing the hosts, if necessary, the graft surfaces should be probed with the tips of a fine forceps. Totally degenerate epidermis is pasty and easily picked-off just before the graft surface becomes a dry scab or eschar. It is more difficult to assign precise end points for chronically rejecting grafts, for example, when dealing with immunosuppressed hosts or certain minor histocompatibility differences. The loss of color in pigmented skin allografts is not a reliable sign of rejection because melanocytes may be lost from viable graft epidermis.[8] Chronically rejecting allografts usually contract considerably, so it is wise to use relatively large grafts if this type of rejection is anticipated. One can consider such grafts "rejected" when their entire surface is hairless and looks like a smooth, shiny scar, with no sign of underlying dermal ridges, which are usually seen at earlier stages with reflected light and a 5× magnifying visor or lens. More details and a description of the histologic changes in rejecting skin allografts may be found elsewhere.[16,17] Because of the subjective aspect of assessing end-points, it is wise to assign random numbers to groups of test and control recipients and appraise graft survival "blindly."

Other Methods

The method described is reliable and has withstood the test of time. However, its possible disadvantage for the needs of some investigations is that the grafts are not visible until the dressings are removed. It is possible

[16] P. B. Medawar, *J. Anat.* (*London*) **78,** 176 (1944).
[17] P. B. Medawar, *J. Anat.* (*London*) **79,** 157 (1945).

to suture a thin skin graft to a bed on the superficial fascia of the lateral thoracic wall and leave it uncovered and thus visible from the day of grafting, until rejection.[18] However, this procedure requires trimming the teeth and toenails of the recipients to prevent the animals from destroying the graft, and the method is practicable probably only with mice or very young rats. The method of Bailey and Usama[19] as modified by Bailey[20] is a simple, ingenious technique for grafting thin slices of tail skin orthotopically to the tail itself, where they are protected by a length of glass tubing and also visible at all times. This method is especially useful for histocompatibility testing or for comparing the survival of antigenically different test grafts on the same host because as many as 8 small (e.g., 5×2 mm) grafts can be placed on a single adult mouse tail. Obviously, the method is not appropriate for studies requiring larger grafts or other host sites. For skin grafting chickens or other birds, the article by Billingham[21] should be consulted.

[18] P. H. Sugarbaker and A. E. Chang, *J. Immunol. Methods* **31,** 167 (1979).
[19] D. W. Bailey and B. Usama, *Transplant. Bull.* **7,** 424 (1960).
[20] D. W. Bailey, *Tranplantation* **1,** 70 (1963).
[21] R. E. Billingham, *in* "Transplantation of Tissues and Cells" (R. E. Billingham and W. K. Silvers, eds.), p. 31. Wistar Inst. Press, Philadelphia, Pennsylvania, 1961.

[4] Graft-versus-Host Reactions

By DAVID STEINMULLER

Introduction

Graft-versus-host (GVH) reactions have been aptly defined by Elkins[1] as the immunological responses of donor lymphoid cells to foreign histocompatibility antigens expressed by the host. Thus, just as the ability to reject foreign skin grafts is a measure of immune competence at the organismic level, the ability to mediate GVH reactions is an *in vivo* measure of immune competence at the cellular level. In fact, GVH reactions may be considered a special form of adoptive immunity (see this volume [5]) in which cell-surface antigens of the host itself are the targets. Indeed, systemic GVH reactions can result in GVH disease,[2,3] the complex syndrome

[1] W. L. Elkins, *Prog. Allergy* **15,** 78 (1971).
[2] P. L. Weiden, *in* "Biology of Bone Marrow Transplantation" (R. P. Gale and C. F. Fox, eds.), p. 37. Academic Press, New York, 1980.
[3] J. A. Hansen, J. M. Woodruff, and R. A. Good, *in* "Immunodermatology" (B. Safai and R. A. Good, eds.), p. 229. Plenum, New York, 1981.

METHODS IN ENZYMOLOGY, VOL. 108

resulting from the combined effects of GVH reactions, secondary infectious complications,[1] and aberrations of the immune system, that is the major obstacle to the more widespread clinical application of bone marrow transplantation.

The three prerequisites[4] for the induction of GVH reactions are (1) the graft must include a significant number of immunologically competent lymphocytes; (2) the host must express histocompatibility antigens foreign to the lymphocyte donor; and (3) the host must be, at least temporarily, incapable of rejecting the grafted cells. In the latter regard, permissive hosts are (1) embryonic and neonatal animals[4]; (2) F_1 hybrids produced by mating two highly inbred strains, the lymphocyte donor being one of the parent strains[4]; (3) hosts with inherited or congenital immunologic deficiencies (e.g., athymic mice or children with severe combined immunodeficiency[3]); and (4) hosts deliberately immunosuppressed by total-body irradiation or cytotoxic drugs such as cyclophosphamide.[2] The immunopathology of GVH disease (e.g., "runt" disease in neonates, "allogeneic" disease in F_1 hybrids, "secondary" disease in radiation chimeras, and acute and chronic GVH disease after clinical bone marrow transplantation) is reviewed extensively elsewhere.[1-6] This chapter is concerned with the application of acute GVH reactions as "tools of research,"[5] specifically as *in vivo* measures of the immunologic competence of lymphoid cells from various sources.

Sources of Effector Cells

Because acute GVH reactions are mediated by thymus-dependent lymphocytes,[7] the GVH potency of a lymphoid-cell inoculum depends on its content of mature T cells.[8] Thus, mouse bone marrow, which, in contrast to human bone marrow, contains very few T cells,[8] is a poor source of GVH-inducing cells compared to splenic or lymph node cells, whereas fatal GVH disease can be produced with cells obtained from the thoracic duct, a rich source of T cells, even with only minor histocompatibility differences.[9] Methods of preparing single-cell suspensions from various lymphoid organs and the thoracic duct are described elsewhere in this volume [6]. The appropriate dose is an empirical matter depending on the type of assay (systemic versus regional, see below), the cell source

[4] R. E. Billingham, *Science* **130**, 947 (1959).

[5] M. Simonsen, *Prog. Allergy* **6**, 349 (1962).

[6] S. C. Grebe and J. W. Streilein, *Adv. Immunol.* **22**, 119 (1976).

[7] A. P. Dalmasso, C. Martinez, and R. A. Good, *Proc. Soc. Exp. Biol. Med.* **110**, 205 (1962).

[8] D. A. Vallera, A. Filipovich, C. C. B. Soderling, and J. H. Kersey, *Clin. Immunol. Immunopathol.* **23**, 437 (1982).

[9] R. Korngold and J. Sprent, *J. Exp. Med.* **148**, 1687 (1978).

(the proportion of mature T cells, as just indicated), the immune status of the donor (whether the donor is "normal" or has been specifically immunized against host histocompatibility antigens[1]), and the nature and extent of donor–host genetic disparity (much higher doses of cells usually are required to induce GVH reactions with minor histocompatibility differences[1]), and an experimental protocol may require testing different dosages. However, in general, GVH reactions usually can be elicited with 10^7 or fewer unseparated spleen or lymph node cells, at least when major histocompatibility differences are involved.[1,5,6]

Systemic GVH Assays

Mortality and Gross Pathology. When appropriate numbers of allogeneic T cells are injected intravenously or intraperitoneally into permissive perinatal or adult hosts (see above), the simplest way of assaying the consequences is to record the mortality and the well-known pathological changes associated with GVH disease, such as runting or weight loss and the development of skin lesions, hunched posture, and diarrhea.[10-12] However, this relatively crude way of measuring GVH reactions is subject to numerous variables, such as the overall health of the hosts to start with, the fastidiousness of animal care and management, and, especially, complications of infectious diseases.[13]

The Discriminant Spleen Assay. This more precise and objective assay was designed by Simonsen[5] to take advantage of the marked splenomegaly that characteristically precedes the atrophy of lymphoid tissue when adult T cells are injected into allogeneic embryos or neonates. To conduct the assay with newborn mice or rats, litters 1–10 days old that are being well cared for by their mothers are selected. Half the litter is used as test animals and half as uninjected controls (one group can be marked by cauterizing the tips of their tails with an electrocauterizer or with the end of a scalpel heated with a Bunsen burner). Cells (10^7) (or some other desired number) are injected intraperitoneally with a 25-gauge $\frac{7}{8}$-in. needle. Even this small volume cannot be injected directly into the tight belly of a newborn rodent without some of the fluid escaping, so the tip of the needle has to be inserted through an arm or leg muscle and be pushed subcutaneously until it passes into the peritoneal space. To do this, the newborns are carefully anesthetized with ether or some other inhalant

[10] R. E. Billingham, V. Defendi, W. K. Silvers, and D. Steinmuller, *J. Natl. Cancer Inst. (U.S.)* **28,** 365 (1962).

[11] P. Stastney, V. A. Stembridge, and M. Ziff, *J. Exp. Med.* **118,** 635 (1963).

[12] D. W. van Bekkum and M. J. de Vries, "Radiation Chimaeras." Academic Press, New York, 1967.

[13] D. Keast and M. N. I. Walters, *Immunology* **15,** 247 (1968).

(see below) or very young litters, which are poikilothermic, are placed in a refrigerator for a few minutes. Splenomegaly peaks 8–10 days after donor-cell injection.[14] The entire litter is sacrificed at this time by cervical dislocation or by overanesthetizing it. First each entire animal is weighed to the nearest milligram; then the spleen is excised and weighed, being careful first to remove all adherent fat. The spleen weight is expressed as a percentage of the body weight for each test and control host. An *index of spleen enlargement* for each test host is then calculated by dividing its spleen/body weight ratio by the mean ratio for the littermate controls. This index permits comparing results between litters because the variation in spleen weight between litters is greater than that within litters.[14]

Inhibition of Colony-Forming Units (CFU). A disadvantage of the Simonsen assay is that it requires newborn hosts. However, systemic GVH reactivity can also be measured in terms of the ability of allogeneic lymphoid cells to inhibit hematopoietic CFUs in lethally irradiated adult hosts.[15] The assay was designed for inbred animals where one can use syngeneic bone marrow (i.e., marrow from other, histocompatible donors of the same strain) as a source of hematopoietic stem cells, but one can perform it also with outbred animals if one obtains bone marrow cells from the host before irradiation for later injection. The assay is conducted by exposing hosts to a lethal dose of total-body, gamma irradiation (750–900 rads for mice or rats) and then giving them 2×10^6 syngeneic (or autologous) bone marrow cells intravenously alone (the control) or with the desired number of test lymphoid cells. Six days later 0.2 μCi of ^{59}Fe citrate is injected intraperitoneally into the hosts. After 17 hr, the hosts are sacrificed, their spleens are removed, and the radioactivity present is determined in a liquid scintillation counter, using a scintillation cocktail appropriate for counting aqueous samples.

Regional GVH Assays

Intracutaneous. One can obtain a semiquantitative measure of GVH reactivity in most laboratory animals by injecting a few million allogeneic lymphocytes directly into the skin and scoring the resultant reactions like delayed-type hypersensitivity tests.[6] The best sites for injection are the flank for guinea pigs, the belly for rats, and the back for mice. First the hair is removed with electric clippers (Oster small animal clippers with size 40 cutting blades are recommended); then the area is shaved with a razor or depilated with a cream such as Nair (Carter Products, NY). In

---

[14] M. Simonsen, J. Engelbreth-Holm, E. Jensen, and H. Poulsen, *Ann. N. Y. Acad. Sci.* **73**, 834 (1958).

[15] H. Blomgren and B. Andersson, *Cell. Immunol.* **3**, 318 (1972).

rats and guinea pigs, the desired dose (e.g., 10^7 cells) is injected intradermally in 0.1 ml using a 27-gauge needle with an intradermal bevel; with mice 30 μl is injected using a 30-gauge needle and a Hamilton No. 705 syringe (Hamilton Co., Reno, NV). The cells must be injected intradermally, not subcutaneously, by inserting the needle very superficially bevel-up. If a bleb does not form by pushing the syringe plunger the needle is in the subcutaneous tissue. Replicate injections are given each host at 3–4 separate sites. The reactions peak at 2–3 days, and the injection sites should be assessed at 24-hr intervals for erythema, induration, and necrosis (the latter may occur only with presensitized cells), assigning scores according to size and intensity, such as +/− for a small, erythematous spot without swelling and 5+ for a large indurated and necrotic reaction. Streilein and Billingham[16] recommend the following scale for mice: 0, no response; 1+, barely perceptible swelling; 2+, swelling 3–4 mm in diameter, site soft to palpation; 3+, swelling 5 mm or more in diameter, site firm; 4+, large reaction with indurated core; 5+, site necrotic or ulcerated. At least 3 sites on each animal should be read daily and the results expressed as a mean score. Moreover, because of the subjective nature of the assay, it is wise to code-number the subjects and score them "blindly" in experiments involving different dosages and treatment groups. A disadvantage of the assay is that if the hosts are able to respond to donor histocompatibility antigens, the skin reactions may include a host-versus-graft component. One can preclude this with inbred animals by using F_1 hybrid hosts, as discussed later for the PLN enlargement assay.

Intrarenal. The injection of lymphoid cells from inbred, parent-strain donors beneath the renal capsule of adult F_1 hybrid hosts results in a marked local, invasive-destructive lesion.[17] There is an initial burst of donor-cell proliferation in the injected kidney that peaks at 6 days, which can be measured by the incorporation of ^{125}IUdR,[18] followed by an increase in renal weight that peaks at 14 days. For the assay, the injected and the uninjected contralateral (control) kidneys either are weighed or the incorporation of ^{125}IUdR is determined with a gamma counter, or both, and the results expressed as ratios of injected to control values.[18] The intrarenal GVH reaction is of great interest to transplantation immunologists because its histopathology mimics that of acute renal allograft rejection.[6] However, the assay is difficult because the kidney must be surgically exteriorized for inoculation,[17] and it is not as sensitive as the PLN enlargement assay described below.

[16] J. W. Streilein and R. E. Billingham, *J. Exp. Med.* **131**, 409 (1970).
[17] W. L. Elkins, *J. Exp. Med.* **120**, 329 (1964).
[18] W. L. Elkins, *Transplantation* **9**, 273 (1970).

Popliteal Lymph Node (PLN) Enlargement Assay

Introduction. The injection of parent-strain lymphoid cells below the hind footpads of F_1 hybrid rats[19] or mice[20] results in an enlargement of the single lymph node draining the site, the PLN, which can be weighed to provide a simple, reliable measurement of regional GVH reactivity. The assay is easier and more sensitive than the intrarenal assay,[19] and more objective than the intracutaneous assay. Its only disadvantage is that it is applicable only to inbred animals where F_1 hybrids can be produced by mating genetically homozygous parents. Such hybrids are genetically tolerant of parent-strain cells; if they are not used, host-versus-graft reactions contribute to the reaction.[6] One cannot suppress the host component with radiation, for example, because radiosensitive host lymphocytes are required for the full expression of PLN enlargement.[6]

Injecting the Host. With rats, the desired number of cells (e.g., 10^7 donor lymph node or spleen cells) is injected in 0.1 ml using a 25-gauge, $\frac{7}{8}$-in. needle attached to a 1 ml or smaller syringe. With mice, the cells are injected in 50 μl using the same size needle but a Hamilton No. 705 syringe. The animal is anesthetized by placing it in a closed glass jar or other container with a wad of cotton or gauze pads wet with ether or, preferably, a nonflammable, nonexplosive inhalant anesthetic such as methoxyflurane, enflurane, or halothane. The hind foot of the anesthetized animal is grasped by the toes and the tip of the needle is inserted just in front of the footpads. The tip of the needle is pushed 6–10 mm toward the heel before delivering the dose. The needle should be in the subcutaneous, not the intradermal, tissue because the former injection site gives the best results.[19] When the needle is withdrawn, a brief pressure is applied at the puncture site with a cotton applicator or piece of gauze to prevent leakage. Although one can inject both hind feet (e.g., with different cell populations or dosages[19]), it is useful to leave one foot uninjected as a control (see below).

Harvesting the Node. PLN enlargement usually peaks 7 days after inoculation of the footpad in both rats and mice and this is the optimal time for weighing the nodes, though one can weigh them earlier if one wants time-course data. The PLN is a spherical node embedded in adipose tissue in the middle of the popliteal fossa[21] (Fig. 1). To remove it, the animal is sacrificed with an overdose of anesthesia (or with mice, by cervical dislocation). The animal is pinned belly-down on an operating board and the legs are rotated so that the soles face upward. The hair is

[19] W. L. Ford, W. Burr, and M. Simonsen, *Transplantation* **10**, 258 (1970).
[20] V. W. Twist and R. D. Barnes, *Transplantation* **15**, 182 (1973).
[21] Y. Kawashima, M. Sugimura, Y. C. Hwang, and N. Kudo, *Jpn. J. Vet. Res.* **12**, 69 (1964).

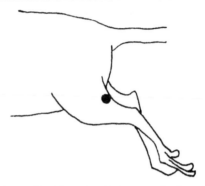

FIG. 1. Location of the popliteal lymph node in rats and mice (after Kawashima *et al.*[21]; the node actually is more medial than suggested by this sketch).

removed from the site with electric clippers and the area is soaked with 70% alcohol. A *superficial* incision is made with blunt-tipped scissors (the tip of the scissors should run just under the skin) from mid-thigh nearly to the heel, passing over the middle of the popliteal region, and the skin from the incision is retracted to reveal the node, which is quite obvious on the injected side (Fig. 2). The incision is kept open by pinning-down or clamping the retracted skin, and the node is removed with smooth-tipped, fine forceps (curved jeweler's forceps are highly recommended; e.g., Storz Instrument Co., St. Louis, MO, No. E-1947-7). The node must be freed of

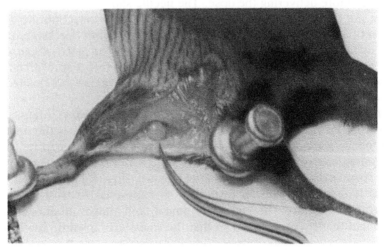

FIG. 2. The enlarged left popliteal lymph node of an F_1 hybrid mouse 7 days after the injection of 10^7 allogenic spleen cells below the left footpad (the node was freed from the surrounding fat for the photograph).

fat before placing it in saline (0.9% sodium chloride) in a small petri dish. If the node does not sink, fat is still attached to it. If necessary, the node can be excised from an anesthetized, living animal by making a smaller incision and suturing the retracted skin together after removing the node.

Weighing the Node. It is very important to remove excess moisture before weighing. This can be done by gently rolling the node on a stack of 2 or more pieces of hard filter paper (e.g., Whatman No. 50). Each node is weighed to 0.01 mg with an analytical balance. A convenient way of doing this is to first weigh an empty piece of weighing paper, then place the nodes one by one on the paper and record the change in weight each time. The results for each injected host are expressed as the *PLN enlargement ratio* calculated by dividing the weight of the node from the injected side by that of the node from the uninjected, control side. One should test at least 3 hosts with each dose and type of cells and calculate the mean enlargement ratio ± the standard error or standard deviation. With organ enlargement assays in general, the significance of differences between test and control groups is best determined by analysis of variance.

GVH Reactions in Birds

GVH reactions in chickens and other birds can be assayed by measuring spleen enlargment[5] or by counting and measuring the white "pocks" that appear on the chorioallantoic membrane (CAM) following inoculation of the embryo with allogeneic blood or lymphoid cells. See Longenecker *et al.*[22,23] for details of the CAM assay.

[22] B. M. Longenecker, S. Sheridan, G. R. J. Law, and R. F. Ruth, *Transplantation* **9,** 544 (1970).

[23] B. M. Longenecker, F. Pazderka, G. R. J. Law, and R. F. Ruth, *Cell. Immunol.* **8,** 1 (1973).

[5] Adoptive Transfer

By DAVID STEINMULLER

The term, adoptive immunity, was coined by Billingham *et al.*[1] in 1954 to describe the transfer of immunity from one animal to another by live, immunocompetent cells as opposed to immune serum or preformed antibody (the term passive transfer should be reserved for the transfer of

[1] R. E. Billingham, L. Brent, and P. B. Medawar, *Proc. R. Soc. London, Ser. B* **143,** 58 (1954).

METHODS IN ENZYMOLOGY, VOL. 108

immunity by the latter). In fact, although they did not refer to it as such, Landsteiner and Chase[2] clearly demonstrated the principle of adoptive transfer 12 years earlier when they showed that delayed-type hypersensitivity to picryl chloride could be transferred from sensitized to normal guinea pigs with peritoneal exudate cells but not with serum. Adoptive transfer was originally applied only to delayed-type hypersensitivity and transplantation immunity, but with the subsequent expansion of cellular immunology, it has been used to transfer virtually all types of immunity, including tumor immunity, auto-immunity, antimicrobial immunity and anti-idiotypic immunity (see the table).

The first requirement for a successful adoptive transfer is that the primary host be adequately immunized and lymphoid cells be transferred at an appropriate time. Second, from the standpoint of histocompatibility, the secondary host must accept the transferred cells, at least temporarily. For the adoptive transfer of humoral immunity, appropriately sensitized cells can rapidly release preformed antibody into the circulation of histoincompatible hosts before the transferred cells are destroyed by allograft immunity.[3] However, for the sustained cell function required for cell-mediated immunity, the primary and secondary hosts either must be totally histocompatible (syngeneic, i.e., members of the same highly inbred strain), or the secondary hosts must be unresponsive to foreign histocompatibility antigens expressed by the transferred cells. The latter category includes F_1 hybrids (given cells from one of the inbred parent strains) or deliberately immunosuppressed hosts. However, in either case, the results of an adoptive transfer could be confounded by graft-versus-host reactions (e.g., by the induction of nonspecific suppressor cells[4]) if the inoculum contains T cells capable of reacting with foreign histocompatibility antigens of the host. Hence, at least for cell-mediated reactions, it is definitely preferable to conduct adoptive transfer experiments with inbred animals.

Other factors that must be considered for a successful adoptive transfer are (1) the *source* (blood, peritoneal exudate, spleen, lymph nodes, thoracic duct) and *type* (bulk T cells or the appropriate subset for delayed-type hypersensitivity, allograft immunity, tumor immunity; B cells for humoral immunity; "activated" macrophages for certain types of tumor immunity; etc.) of the transferred cells, (2) the *dose* of cells transferred, (3) the *route* of administration, and (4) the relative *timing* of cell transfer and antigen challenge of the secondary host. In most cases, the optimal

[2] K. Landsteiner and M. W. Chase, *Proc. Soc. Exp. Biol. Med.* **49**, 688 (1942).
[3] S. Harris, T. N. Harris, and M. B. Farber, *J. Immunol.* **72**, 148 (1954).
[4] E. Parthenais and W. S. Lapp, *Scand. J. Immunol.* **7**, 215 (1978).

conditions for adoptively transferring immunity to any particular antigen challenge must be determined empirically, and it is difficult to generalize here because of the great variety of immune responses that can be transferred. Representative examples of the successful adoptive transfer of various types of immunity are summarized in the table. The spleen and lymph nodes are the most common sources of transferred lymphocytes, particularly regional nodes draining the site of a graft or local reaction in the primary host. The number of cells is quite variable though usually 10–50 million unseparated cells suffice. Marchal et al.[5] have evidence that delayed-type hypersensitivity to xenogeneic erythrocytes can be adoptively transferred by one single cell. Most investigators prefer the intravenous route (see below), but satisfactory results usually can be obtained by the technically easier intraperitoneal route as well. Cell transfer and antigen challenge usually are concurrent, though sometimes the cell transfer precedes or follows the administration of antigen by one or more days.

Although the intravenous route of administration is preferred, it presents difficulties in terms of the paucity of superficial veins in rodents, surgical access, and possible asphyxiation of the host due to lung embolization. The latter is particularly a problem with tissue culture expanded lymphocytes, which tend to be sticky and agglutinate, especially if lectin is present in the culture media. Hence, at least for intravenous injection, it is especially important to wash cells thoroughly and ensure that clumps, aggregates, and general debris are removed before administration. The tail vein is the preferred intravenous route in the mouse and, though more controversial, in the rat as well.[6] The secret to success is making sure the vein is dilated before before injection, especially with pigmented strains. To this end, investigators place the mice under a heat lamp. This procedure, however, can greatly agitate the animals without achieving dilation of the veins. A more effective method is to grasp the mouse by the scruff of the neck with one hand and hold the entire tail under running hot water while firmly stroking the sides of the tail from the base to the tip with the thumb and forefinger of the other hand. If the water is hot enough, the lateral tail veins dilate in less than 10 sec, whereupon the mouse is rapidly placed in a restraining cage (available from most laboratory supply firms) with the tail pulled out for injection. A 27-gauge needle with the bevel up is used for this purpose. One should start at least halfway down the tail so that if one misses the vein one can try again more proximally. The needle should be inserted absolutely parallel to the vein so that it does not pierce

[5] G. Marchal, M. Seman, G. Milon, P. Truffa-Bachi, and V. Zilberfarb, J. Immunol. 129, 954 (1982).
[6] M. V. Barrow, Lab. Anim. Care 18, 570 (1968).

REPRESENTATIVE EXAMPLES OF ADOPTIVELY TRANSFERRED IMMUNITY[a]

Type of immunity	Antigen challenge	Host species	Transferred cells				Reference
			Source[b]	No. $\times 10^{-6}$	Route[c]	Time[d]	
Delayed-type hypersensitivity	Picryl chloride	Guinea pig	PE	?	ip	-2	2
	Picryl chloride, oxazolone	Mouse	S, LN	50–60	iv	0	e
	Sheep or chicken erythrocytes	Mouse	CTC	1–2[f]	sc	0	5
Transplantation	Skin allograft	Mouse	S, LN	40–200[g]	ip	0 or -3	1
	Skin allograft	Mouse	S, LN	5–200	ip, iv	+1, -3	h
	Skin allograft	Mouse	S	100	iv	+1	i
	Heart allograft	Rat	LN, TD	5–20	iv	0	j
Tumor (syngeneic)	Mammary carcinoma	Rat	PE	5–10	iv	+1	k
	Hepatoma	Guinea pig	S	200	iv	0	l
	Plasmacytoma	Mouse	S	50	iv	+5	m
	Lymphoma	Mouse	TCE	25–50	iv	+5	n
Autoimmunity	Central nerve tissue	Rat	S	20	iv	-[o]	p
	Central nerve tissue	Mouse	LN	60	iv	-[o]	q
	Ovary	Mouse	S	40	ip	-[o]	r
	Erythrocytes	Mouse	Hy[s]	25–50	ip	-[o]	t
Antiviral	Herpes simplex virus	Mouse	PB	5	ip	-1	u
	Herpes simplex virus	Mouse	S	50	iv	+1	v
	Lymphocytic choriomeningitis virus	Mouse	S	30–70	iv	+4 or +7	w
	Rabies virus	Mouse	S	60	iv	+4	x
	Moloney sarcoma virus	Mouse	S, LN	2–100	ip	-7	y
	Influenza virus	Mouse	CTC	3	iv	+1	z
	Influenza virus	Hamster	S	1–2	ip	+1	aa
Antibacterial	Shigella paradysenteriae	Rabbit	LN	200–300	iv	-[o]	3
	Listeria monocytogenes	Mouse	S	10	iv	-4	bb
	Rickettsia conorii	Mouse	S	100	iv	-1	cc
Antifungal	Nocardia asteroides	Mouse	S	50	iv	-2	dd
Antiparasitic	Trichinella spiralis	Mouse	LN	20	iv	0	ee
	Schistosoma mansoni	Mouse	S	40	iv	+42	ff
	Nippostrongylus brasiliensis	Rat	TD	100	iv	0	gg
	Plasmodium bershei	Mouse	S	20	iv	-1	hh
Antidiotypic	Myeloma cells	Mouse	S	35–50	ip	0	ii
	Hybridoma cells	Mouse	S	10–30	iv	-5	jj

[b] CTC, Cloned T cells; LN, lymph nodes; PB, peripheral blood; PE, peritoneal exudate; S, spleen; TCE, tissue culture expanded T cells; TD, thoracic duct.

[c] ip, Intraperitoneal; iv, intravenous; sc, subcutaneous.

[d] Days relative to antigen challenge; e.g., −1 = 1 day before challenge, +1 = 1 day after challenge, 0 = same time as challenge.

[e] W. Ptak, J. W. Rosenstein, and R. K. Gershon, *Proc. Natl. Acad. Sci. U.S.A.* **79**, 2375 (1982).

[f] One to two *cells*, not millions, injected beneath the footpad with antigen.

[g] Estimation based on the number of spleen and lymph node fragments transferred.

[h] J. Warren and G. Gowland, *J. Pathol.* **100**, 38 (1970).

[i] A. E. Chang and P. H. Sugarbaker, *Transplantation* **29**, 381 (1980).

[j] B. M. Hall, S. Dorsch, and B. Roser, *J. Exp. Med.* **148**, 890 (1978).

[k] C. M. Boyer, J. W. Kreider, and G. L. Bartlett, *Cancer Res.* **42**, 2211 (1982).

[l] S. Shu, P. A. Steerenberg, J. T. Hunter, D. H. Evans, and H. J. Rapp, *Cancer Res.* **41**, 3499 (1981).

[m] M. B. Mokyr, J. C. Hengst, and S. Dray, *Cancer Res.* **42**, 974 (1982).

[n] T. J. Eberlein, M. Rosenstein, and S. A. Rosenberg, *J. Exp. Med.* **156**, 385 (1982).

[o] No antigen was injected into the secondary hosts.

[p] C. Bolton, G. Allsopp, and M. L. Cuzner, *Clin. Exp. Immunol.* **47**, 127 (1982).

[q] C. B. Pettinelli, R. B. Fritz, C. H. Chou, and D. E. McFarlin, *J. Immunol.* **129**, 1209 (1982).

[r] S. Sakasuchi, T. Takahashi, and Y. Nishizuka, *J. Exp. Med.* **156**, 1565 (1982).

[s] Monoclonal antibody-secreting hybridoma cells.

[t] L. A. Cooke, N. A. Staines, A. Morgan, C. Moorhouse, and G. Harris, *Immunology* **47**, 569 (1982).

[u] S. Kohl and S. L. Loo, *J. Immunol.* **129**, 370 (1982).

[v] S. C. Mosensen and H. K. Anderson, *Infect. Immun.* **33**, 743 (1981).

[w] C. J. Pfau, J. K. Valenti, D. C. Pevear, and K. D. Hunt, *J. Exp. Med.* **156**, 79 (1982).

[x] B. S. Prabhakar, H. R. Fischman, and N. Nathanson, *J. Gen. Virol.* **56**, 25 (1981).

[y] H. Kimura, Y. Yamasuchi, and T. Fujisawa, *Gann* **73**, 446 (1982).

[z] Y. L. Lin and B. A. Askonas, *J. Exp. Med.* **154**, 225 (1981).

[aa] C. R. Crawford, R. Jennings, N. Bradford, and C. W. Potter, *Clin. Exp. Immunol.* **48**, 739 (1982).

[bb] R. A. Barry and D. J. Hinrichs, *Infect. Immun.* **35**, 560 (1982).

[cc] I. N. Kokorin, E. A. Kabanova, E. M. Shirokova, and G. E. Abrosimova, *Acta Virol. (Prague)* **26**, 91 (1982).

[dd] R. L. Deem, B. L. Beaman, and M. E. Gershwin, *Infect. Immun.* **38**, 914 (1982).

[ee] D. Wakelin and M. M. Wilson, *Parasitology* **74**, 215 (1977).

[ff] J. V. Weinstock, S. W. Chensue, and D. L. Boros, *J. Immunol.* **130**, 423 (1983).

[gg] Y. Nawa and H. R. P. Miller, *Cell. Immunol.* **42**, 225 (1979).

[hh] J. J. Ferraroni and C. A. Speer, *Infect. Immun.* **36**, 1109 (1982).

[ii] S. Bridges, *J. Immunol.* **121**, 479 (1978).

[jj] T. F. Kresina, Y. Baine, and A. Nisonoff, *J. Immunol.* **130**, 1478 (1983).

[kk] L. Ortiz-Ortiz, W. D. Weigle, and D. E. Parks, *J. Exp. Med.* **156**, 898 (1982).

it again laterally. Practice is required to learn the correct level for inserting the needle. Barrow[6] has described a modification of this technique for the intravenous injection of lightly anesthetized rats. Anderson[7] recommends the lingual vein for the rat, hamster, and other small laboratory animals. It is also possible to inject cells into the venous sinus on the ventral side of the base of the penis of male rodents. The ear veins of rabbits are easy to dilate and inject. A veterinarian should be consulted for the cutdowns required for injecting cells into the deeper veins of larger animals, such as the jugular vein in the guinea pig.

A specialized form of adoptive transfer which is not mentioned in the table and which is frequently used to study tumor immunity is the *Winn test*.[8] Single cells of the tumor (maintained as such *in vitro*, in the ascites form *in vivo*, or enzymatically or mechanically dispersed) must be used in this method. An appropriate number of tumor cells (usually 5×10^3 to 5×10^6 depending on viruluency) are mixed in a small volume (e.g., 0.2 ml) with an equal or greater number (e.g., 5-fold) of immune lymphoid cells, and the mixture injected subcutaneously into the back or axilla. The test is controlled by injecting tumor cells alone or with nonimmune lymphoid cells. Tumor growth is measured periodically by calculating tumor area as the product of two perpendicular diameters measured with vernier calipers, a method that correlates well with tumor weight.[9]

[7] J. M. Anderson, *Science* **140**, 195 (1963).
[8] H. J. Winn, *J. Immunol.* **86**, 228 (1961).
[9] L. L. Perry, B. Benacerraf, R. T. McCluskey, and M. I. Greene, *Am. J. Pathol.* **92**, 491 (1978).

Section II

Methods for the Separation and Purification of Populations
and Subpopulations of Lymphoreticular Cells

[6] Preparation of Single-Cell Suspensions from Lymphoid Organs

By Norman J. Kleiman, Daniel L. Friedman, and Giovanni Di Sabato

Lymphoid cells may be obtained from various sources: peripheral blood, lymph, and lymphoid organs. In the case of peripheral blood and lymph, the cells are already in single-cell suspension and the techniques for the separation of the various populations and subpopulations of cells are discussed in other articles of this volume. Organs from which lymphoid cells are obtained are usually spleen, thymus, and lymph nodes. This chapter will be limited to the description of techniques as applied to the mouse and the rat.

Preparation of Lymphocytes from Spleen

Removal of Spleen. Mice of 18–20 g or rats of 180–300 g are sacrificed by cervical dislocation or by decapitation. The abdominal skin is swabbed with 70% ethanol. A small dorsoventral incision through the skin of the left flank is made. The skin is pulled apart, exposing the spleen which is easily removed by making a small incision through the peritoneum (Fig. 1, B). Some fatty or connective tissue attached to the organ is easily excised. The organ is put in a Petri dish containing 5–20 ml of phosphate buffered saline (PBS) with 5–10% fetal calf serum.

Removal of the Thymus. The skin of the abdomen is swabbed with 70% ethanol. An incision through the skin and superficial fascia is made with scissors along the midline. The thymus is then exposed by cutting the rib cage with a longitudinal cut through the middle of the sternum. (Care must be taken at this point not to cut or puncture the heart or the great vessels for this will cause hemorrhage that will make difficult the subsequent stages of the operation.) The thymus appears as a bilobate, whitish organ situated on the upper mediastinum, overlaying the base of the heart and the great vessels (Fig. 1, A). The organ is easily removed with forceps.

Removal of Lymph Nodes.[1–3] The location of lymph nodes in the mouse is shown in Fig. 1.[1] The largest and easiest to find lymph nodes are

[1] T. B. Dunn, *J. Natl. Cancer Inst.* **14**, 1281 (1954).
[2] Y. Kawashima, M. Sugimura, Y.-C. Hwang, and N. Kupo, *Jpn. J. Vet. Res.* **12**, 69 (1974).
[3] G. H. Higgins, *Anat. Rec.* **30**, 243 (1925).

METHODS IN ENZYMOLOGY, VOL. 108

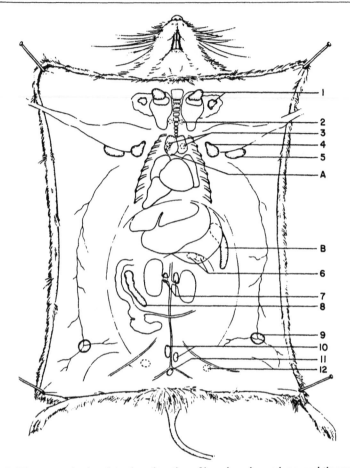

Fig. 1. Diagrammatic sketch to show location of lymph nodes, spleen, and thymus in the mouse. (1) Superficial cervical nodes. Lateral node is partly obscured by parotid salivary gland tissue. (2) Deep cervical nodes. (3) Thoracic nodes lying behind the thymus. (4) Axillary node. (5) Brachial node. (A) Thymus. (B) Spleen. (6) Pancreatic nodes lying at margin of pancreas and behind stomach. (7) Renal nodes. (8) Mesenteric node. (9) Inguinal node. (10) Lumbar nodes. (11) Candal node. (12) Sciatic node lying in the back below sciatic nerve. Reproduced from *Journal of the National Cancer Institute.*

the axillary, bronchial, inguinal, and mesenteric. The superficial cervical lymph nodes (2–3 on each side; not shown in figure) are located above the submaxillary salivary glands. They are found by making an incision from the neck to the chin and reflecting the skin on each side. The smallest are the deep cervical, renal, and caudal. The average weight of the combined inguinal and axillary nodes of the mouse is 16–18 mg at 6 weeks and 23–28 mg at 14 weeks.

Preparation of Cell Suspension

A single cell suspension is obtained by cutting the organ into small fragments with scissors and then gently pressing it through a fine, stainless-steel wire mesh (40 × 40 mesh, United Surgical, Largo, Florida 33540) with the aid of a plastic, disposable syringe piston.

Another method (suggested by Dr. David Steinmuller) that may be used for the preparation of single cell suspensions involves nylon mesh (110 μm pore size Nitex HC-3-110, Tetco, Inc., Elmsford, N.Y.). The nylon mesh sheet can be cut into small squares which are sterilized 3–4 to a package and discarded after use. The lymph nodes and spleen are thoroughly teased apart and the crude suspension is pipetted through the nylon mesh. Usually, it is necessary to wet the center of the mesh to make the suspension drip through by gravity. Apparently, this method results in less damage to the cells than passing tissue fragments through a stainless steel mesh with a pestle.

The single-cell suspension is transferred to a test tube with the aid of a Pasteur pipet and centrifuged in a clinical centrifuge at about 300 g for 5 min. The cells are then resuspended in 0.1 M Tris–HCl pH 7.2–0.8% ammonium chloride in order to lyse the red cells. The cells are centrifuged again, washed once in PBS, and counted. In order to remove clumps and small cell aggregates, passage of the cell suspension through a sufficiently fine nylon or stainless-steel mesh is often adequate. Another simple and effective method consists in centrifuging the cell suspension at low speed for a few seconds or allowing the clumps to sediment by gravity for a few minutes.

Cell Number Determination Using the Hemocytometer

Counting cells in a hemocytometer remains a useful and inexpensive method for enumerating cells and is a standard method for calibration of the Coulter particle counter. The hemocytometer consists of a special slide, which, together with its cover slip, forms two chambers of known depth (0.1 mm), one on each side of the slide. The slide has calibrated squares ruled off on its surface. The cells within a known area are counted under the microscope, allowing the number of cells in a suspension to be calculated.

Cells are diluted with Hanks' solution to between 5 × 10^5 and 2 × 10^6 cells/ml. The suspension is mixed immediately prior to filling the precleaned hemocytometer chamber. A small drop of diluted cell suspension is placed at the edge of the coverslip so that it is spontaneously taken into the chamber by capillary action. It is important that the drop size be large enough to fill the chamber, but not so large as to overflow. The cells are allowed to settle for a few minutes before counting.

There are nine 1-mm² square sections in the ruled area of the Neubauer type hemocytometer. For counting lymphocytes, the 1-mm² sections at the four corners are used. Each of these sections is subdivided further into 16 squares. At 100× magnification, usually an entire corner section can be brought into the field of vision. The cells are most easily visualized and counted under phase contrast. By convention, cells on the top or left-hand lines are not counted, those on the lower and right-hand lines are included. Routinely, eight 1-mm² sections, the four corner sections on each side of the hemocytometer, are counted. The average cell number for a single 1-mm² section is then calculated. Since this represents the number of cells in 0.1 mm³, to obtain the cells/ml of diluted sample, this value is multiplied by 10^4. Statistically, it is necessary to count a total of approximately 400 cells in order to obtain 90% accuracy.[4]

The major disadvantage of hemocytometer counting is that it is time consuming and becomes impractical when large numbers of samples must be counted. Also, significant problems of reproducibility arise if the chamber is not filled properly or if the chamber is not scrupulously clean. Additional counting errors of approximately 10% are inherent in the method.[4,5] In addition to measurements of total cell number, the hemocytometer is useful in the determination of the viable cell number (see below).

Cell Number Determination Using an Electronic Particle Counter

Use of an electronic particle counter circumvents the obvious disadvantages of hemocytometer counting. This method is simple, rapid, and highly reproducible.[6,7] The most widely used instrument, the Coulter counter, will be discussed, though several other instruments are commercially available. The principle on which the Coulter counter operates is the difference in conductivity between cells and the medium in which they are suspended. A vacuum system pulls a precise volume of cell suspension (usually 0.5 ml) through an aperture (usually 100 μm in diameter) in a glass tube. The passage of a cell through the aperture causes a drop in the voltage that is continuously applied across the aperture. The resulting pulse is amplified, recorded, and can be observed on an oscilliscope. The pulse height on the oscilliscope is approximately proportional to the volume of the cell.

The lymphocyte sample that is to be counted must be (1) largely free of

[4] K. X. Sanford, W. R. Earle, V. J. Evans, H. K. Waltz, and J. E. Shannon, *J. Natl. Cancer Inst.* **11,** 773 (1951).

[5] J. Berkson, T. B. Magath, and M. Hurn, *Am. J. Physiol.* **128,** 309 (1939).

[6] M. Harris, *Cancer Res.* **19,** 1020 (1959).

[7] G. Brecher, M. Schneiderman, and G. Z. Williams, *Am. J. Clin. Pathol.* **26,** 1439 (1956).

connective tissue and other debris which might clog the aperture, (2) suspended in a solution in which the cells are relatively stable, and (3) diluted to a concentration range which is optimal for accurate counts on the Coulter counter. Lymphocytes are relatively stable when diluted in Hanks' solution and kept at 0° though they do slowly undergo lysis. They are less stable when diluted with saline or kept at room temperature. For best results, suspensions should be diluted to between 10,000 and 80,000 cells/ml. If necessary, lower cell numbers can be counted with appropriate corrections for background. Cell number determinations greater than 80,000/ml become less precise because of the large corrections for coincidence.

The optimal settings for a particular Coulter counter and a particular cell type must be determined at the outset. These settings can then be used indefinitely. The optimal settings are those which will encompass all of the cells within a sample without counting debris. The aperture current and sensitivity settings are first adjusted so that pulses on the oscilloscope pattern are about one-half to three-fourths of the height of the screen. The cell number of a constant sample is then determined as a function of the threshold setting which is varied over its entire range. As the threshold setting is lowered, the cell count will increase, as more and more cells are included in the count. Theoretically, a plateau is reached at a threshold setting that is low enough to include all of the cells, but too high to include debris. In practice, depending upon the amount and size of debris in the sample, the plateau may be only a slight, transient, decrease in the slope of the curve. From this plateau and from comparisons with careful hemocytometer counts, a threshold value is chosen. The settings found to be optimal on a model Fn Coulter counter with bovine thymocytes are aperture, 4; sensitivity, 0.707; and threshold, 35.

When counting cells, the following precautions should be observed:

1. Samples should be mixed gently immediately before counting to avoid errors due to the settling of cells. Vigorous shaking should be avoided since this may result in high counts due to the presence of bubbles.

2. The aperture should be free of debris, and the glass aperture tube should be clean. The aperture is visualized using a side microscope focused on the aperture and displayed on a screen.

3. The background count should be less than 200 cells per 0.5 ml. If necessary, the diluting solutions should be filtered through bacterial filters.

4. The aperture tube should be flushed out from previous samples and should be free of air bubbles.

5. Triplicate counts of the sample should agree to within less than 5%.

The operation of the Coulter counter is simple. A stopcock is opened to allow vacuum to pull a column of mercury through a capillary system. As the mercury passes a contact, the display is zeroed. The stopcock is then closed and the mercury moves back through the capillary to equilibrate with atmospheric pressure. A section of mercury column is calibrated to contain 0.5 ml and contacts at either end of this section are activated and deactivated as the mercury passes by. As the mercury moves it creates a negative pressure in the aperture tube. Exactly 0.5 ml of solution is counted as it is sucked through the aperture.

Triplicate counts are averaged, corrected for coincidence (by consulting the chart provided by the manufacturer), and multiplied by the dilution factor to give cells/0.5 ml.

Determination of Viability by Dye Exclusion

This method is based upon the observation that viable cells are impermeable to certain dyes, whereas nonviable cells allow the dye to enter the cell and to stain its contents. The most commonly used dye is trypan blue, though a variety of other dyes have been successfully used.[8,9] Unless care is taken, high values are obtained with trypan blue, since it slowly damages cells. Nigrosin is being used with increasing frequency due to its lower toxicity.[10] We will describe the use of trypan blue and nigrosin.

Stock Solutions

Trypan Blue. Dissolve 0.5 mg of trypan blue in 100 ml of Hanks' solution. Heat to dissolve, cool, and filter.

Nigrosin. Dissolve 0.5 mg of nigrosin in 100 ml of Hanks' solution. Stir well and filter.

To 80 μl of cells (approximately 1–2 × 10⁶/ml) add 20 μl of trypan blue, or to 70 μl of cells add 30 μl of nigrosin. Place an aliquot in the hemocytometer, wait 4 or 5 min, and count the fraction of cells which is stained darkly. The count is most easily made under phase contrast at 100× magnification, under which the lymphocytes will be highly refractile and will be clearly distinguished from stained cells. Lightly stained cells should not be scored as dead cells as they may be damaged by the dye. It is crucial, especially with trypan blue, to complete the count within a few minutes. At least 200 cells should be counted. A slide and coverslip may

[8] H. J. Phillips and J. E. Terryberry, *Exp. Cell Res.* **13,** 341 (1957).
[9] J. H. Hanks and J. H. Wallace, *Proc. Soc. Exp. Biol. Med.* **98,** 188 (1958).
[10] J. P. Kaltenbach, M. H. Kaltenbach, and W. B. Lyons, *Exp. Cell Res.* **15,** 112 (1958).

be used in place of the hemocytometer, but care must be exercised so that the cells are not damaged by squashing.

Acknowledgments

The advice of Dr. David Steinmuller in the preparation of this article is gratefully acknowledged. During the preparation of this article DLF was supported by NSF Grant 82-09671 and GDS by NSF Grant 80-11018.

[7] Rosetting Techniques to Detect Cell Surface Markers on Mouse and Human Lymphoreticular Cells

By BRUCE E. ELLIOTT and HUGH F. PROSS

Introduction

A wide variety of techniques have been used to detect cell surface antigens and receptors on subpopulations of lymphoreticular cells. Among these are rosetting techniques which involve the specific binding of indicator cells or immunoabsorbent beads to distinct cell populations to form so called "rosettes." The main advantages of rosetting assays are (1) they are simple, rapid, well established, and provide a read-out which is visible under the light microscope; (2) they are easily amenable to double labeling techniques (e.g., erythrocytes and fluorescent stain[1]); (3) they can provide an amplification of the signal in that an intact indicator cell or bead binds via a ligand to a specific "acceptor" molecule (antigen or receptor) on the cell surface; and (4) they offer a convenient means of separating rosetting cells by density[2] or sedimentation velocity.[3]

Despite the wide use of rosetting assays there remain certain disadvantages of rosetting procedures. Unlike assays involving soluble ligands,[4] quantitation of the relative amount of ligand by rosetting cells is difficult, although a crude estimate can be obtained by counting the numbers of indicator cells bound.[5] Dissociation of rosettes is another problem, particularly if the acceptor molecule tends to cap on the cell surface. Treatments which reduce capping (e.g., addition of metabolic inhibitors[3]), can stabilize rosettes. Fixation[5] is also helpful in stabilizing rosettes,

[1] R. F. Ashman and M. C. Raff, *J. Exp. Med.* **137,** 69 (1973).

[2] C. R. Parish and J. A. Hayward, *Proc. R. Soc. London, Ser. B* **187,** 65 (1974).

[3] B. E. Elliott, J. S. Haskill, and M. A. Axelrad, *J. Exp. Med.* **141,** 599 (1975).

[4] M. R. Locken and L. A. Herzenberg, *Ann. N.Y. Acad. Sci.* **254,** 163 (1975).

[5] B. E. Elliott and J. S. Haskill, *Eur. J. Immunol.* **3,** 68 (1973).

although care must be taken in interpreting results from fixed preparations. Furthermore, variations among laboratories in the technique can lead to contradictory results.

The basic principle of the rosetting assay, namely the binding of an indicator cell or bead via a ligand to an acceptor molecule on a cell surface, can be adapted in several ways. Table I presents examples of some of the more commonly used forms of the assay. These include spontaneous rosettes in which the ligand is endogenously present on the indicator cell surface (e.g., erythrocyte determinants), specific antibody rosettes in which the ligand is an antibody (specific for a cell surface

TABLE I
ROSETTE TYPES[a]

Type	Indicator and coating	Cell type detected
Spontaneous		
E or nE	Sheep RBC, untreated or neuraminidase treated	Most human T cells
H	Human RBC, autologous or heterologous	Human T subpopulation
M	Mouse RBC	Human IgM-bearing B cells
Receptor-mediated		
EA ($Fc_{\gamma,\mu}$)	Ox RBC + rabbit anti-ox RBC, of appropriate class, or sheep RBC plus monoclonal anti-SRBC	Cells with receptors for Fc of class used
EAC	Ox RBC, rabbit IgM anti-Ox RBC, and fresh mouse serum as a source of complement	Cells with receptors for C3b
Specific ag	Ox RBC, latex beads, polyacrylamide beads coupled with protein antigen by agents such as tannic acid, carbodiimide (EDCI), or chromic chloride	Lymphoid cells with receptors for antigen
Antibody-mediated		
Immunoglobulins, surface markers	Specific antibody against surface structures (such as immunoglobulins or monoclonal antibody (MAB)-defined markers) is bound to indicator cells or beads as above	smIg B cells, cells with adsorbed Ig, cells with the ag defined by the MAB
Lymphocyte–tumor Conjugates	Tumor cells, untreated or neuraminidase treated	Nonspecific in unpurified murine spleen or human PBL (see text)

[a] For specific references see Table II and text.

antigen) coupled to the surface of an erythrocyte or bead, and specific antigen rosettes, in which the ligand is an antigen (e.g., human serum albumin[6]) coupled to the indicator, specific for a cell surface receptor. A detailed description of all these techniques is beyond the scope of this review. We will concentrate on the following main aspects, based on experience in our laboratories: (1) a generalized technique for detecting rosette-forming cells, (2) a detailed description of immunobead indicator methods and antigen-specific rosettes, (3) a brief discussion of lymphocyte–tumor cell conjugates, and (4) an evaluation of separation techniques.

A Generalized Rosette Method

Reagents and Methodology

Lymphocyte Preparation. Human peripheral blood is depleted of monocytes by treatment with carbonyl iron[7] as follows: 100 mg of carbonyl iron (Type E, A.D. Mackay, Darien, CT) is added to 10 ml heparinized whole blood and incubated at 37° for 30 min with shaking every 7–8 min. The blood is then passed slowly over a large (2.5 kg) magnet (Sargeant-Welch, Toronto) into an empty tube. This is repeated 2 more times. The monocyte-depleted mononuclear cells (less than 2% monocytes) are then separated from red cells and granulocytes by Ficoll–Isopaque density gradient centrifugation[8] as follows: 5 ml of blood are layered on 3 ml of Ficoll–Isopaque (F.Ip.) solution (room temperature) in 12-ml centrifuge tubes [108 ml of 9% (w/v) Ficoll (Pharmacia, Uppsala)] plus 45 ml of 32.8% (w/v) sodium metrizoate (Isopaque, Nygaard Chemicals, Oslo or Accurate Chemicals, New York) or Lymphoprep (Nygaard Chemicals), specific gravity 1.077. The gradients are centrifuged at room temperature for 12 min at 800 g. The interface cells and fluid are mixed with at least an equal volume of medium–5% fetal calf serum (FCS) (to dilute residual F.Ip) and centrifuged once at 200 g for 5 min, followed by 2 washes at 150 g for 2 min. The resulting lymphocytes are counted using trypan blue exclusion and resuspended to a concentration of 2×10^6/ml in medium.

Murine lymphocytes (e.g., from the spleen) may be prepared using F.Ip.[2] also, but the specific gravity must be adjusted to 1.09 (see below). Most investigators obtain splenic lymphocytes by simply teasing apart the spleen using blunt forceps, followed by resuspension of the cells and debris in 10 ml of medium. This is allowed to stand for 10 min to let the

[6] D. Sulitzeanu and J. S. Haskill, *J. Immunol. Methods* **2,** 11 (1972).
[7] G. Lundgren, C. F. Zukoski, and G. Moller, *Clin Exp. Immunol.* **3,** 817 (1968).
[8] A. Bøyum, *Scand. J. Clin. Lab. Invest.* **21,** Suppl. 97, 77 (1968).

debris fall to the bottom of the tube, or else passed through gauze or glass wool. The cells are then washed three times in medium with centrifugation at 150 g for 5 min each time. This preparation may be used as is if the presence of red cells is unimportant, or if there is little erythrocyte contamination. If red cells must be removed, the cell pellet can be treated with 0.5 ml distilled water for 15 sec followed by rapid addition of 10 ml of medium and two washes with medium. Red cells can also be lysed using Tris-buffered ammonium chloride or hemolytic Gey's solution as described in detail elsewhere.[9]

Rosette Formation for Enumeration. The nature of the cell receptor detected by rosette formation depends on the indicator cell used, while the efficiency of the assay is influenced by the medium, the serum, the temperature of incubation, and the temperature of the counting chamber.[10] These individual differences are documented in Table II. In the case of EA and EAC rosettes (to detect Fc receptor- and complement receptor-bearing cells, respectively) the methods are applicable to both murine and human lymphocytes. Spontaneous rosettes such as autologous rosettes using human RBC, M rosettes to detect human IgM-bearing B cells, or E rosettes to detect human T cells are only formed by human lymphocytes. For enumeration of rosettes, the procedure is as follows: 100 μl of lymphocytes (2 × 10^6/ml) is added to 100 μl of 1% indicator cells in the appropriate medium in a small test tube (e.g., 5 × 35-mm plastic LP/2 precipitin tubes, Luckham Labs, Burgess Hill, Sussex, England). The tubes are preincubated if necessary and then centrifuged at 200 g for 5 min. After incubation, all but about 100 μl of the supernatant is removed and replaced with 100 μl of a suitable stain such as Wright's stain or crystal violet. A drop of the preparation is then placed in a hemocytometer or rosette chamber[10] (Fig. 1) or on a slide and (gently) covered with a coverslip (see below). Lymphocytes having 3 or more RBC firmly apposed to the surface are counted as rosettes. RBC which are not firmly attached will move away from the lymphocyte if the microscope stage is tapped gently. An example of E rosetting human peripheral blood lymphocytes is shown in Fig. 2.

Rosette Formation for Depletion. One of the advantages of the rosette technique is the ease with which rosette-forming cells can be separated from nonrosetting cells. To do this, rosettes are formed using the same relative proportions of lymphocytes and indicator cells in equal volumes of up to 5 ml each in 12-ml centrifuge tubes (or multiples thereof, depend-

[9] B. B. Mishell and S. M. Shiigi, *in* "Selected Methods in Cellular Immunology" (B. B. Mishell and S. M. Shiigi, eds.), p. 22. Freeman, San Francisco, California, 1980.
[10] H. F. Pross, L. A. Gallinger, P. Rubin, and M. G. Baines, *Int. Arch. Allergy Appl. Immunol.* **66**, 365 (1981).

ing on the total volume to be depleted). In our experience, the use of larger (i.e., 50 ml) tubes to handle the entire separation in one step is not as efficient in terms of rosette separation or cell yields. The tubes are centrifuged and incubated as described in Table II. After rosette formation is complete, 5 ml of the supernatant is removed and the rosettes are gently resuspended in the remaining 5 ml and applied to a F.Ip. gradient as described below.

Special Considerations

Delicate Rosettes. Some rosette preparations are extremely easily disrupted, especially E and H rosettes. For this reason, these rosettes should be counted in a hemocytometer or rosette chamber[10] (Fig. 1) so that the weight of the coverslip is not on the rosettes. In contrast to hemocytometers, rosette chambers offer the additional advantages of being disposable, inexpensive, sealable (with paraffin wax or nail polish), and compatible with light microscopy under oil immersion. Delicate rosettes should be resuspended extremely gently, such as by twirling the tube between the fingers and thumb, and by using wide-bore, short, Pasteur pipets to transfer the preparation to the counting chamber. The pipet is placed in the tube and the tube and pipet gently tilted to allow the suspension to come up by capillary action.

Double Rosettes. It is possible to use "double rosetting" to detect cells having more than one receptor, e.g., Tγ cells.[11] To do this, the indicator cells must be distinguishable from each other, such as by size (human vs Sheep RBC), fluorescent dye, or type of indicator (beads vs red cells). The assay conditions used are usually those favoring the more delicate of the rosettes being detected, but care must be taken that the lymphocyte/indicator cell proportions and the assay conditions used result in the correct counts when used in single rosette assay controls. To avoid these technical problems, double marker-bearing cells are usually assessed by enriching for one rosette forming cell (RFC) type by density gradient centrifugation, followed by rerosetting with the other indicator cell under study. The technique depends on obtaining virtually 100% RFC in the cell pellet followed by efficient dissociation of the first indicator cell so that all of the relevant receptors are exposed for the second indicator cell. This technique is usually used to detect T cells (ERFC) having receptors for the Fc portion of the different antibody classes.[12]

[11] A. A. Bom-van Norloos, H. G. Pegels, R. H. J. van Oers, J. Silberbusch, T. M. Feltkamp-Vroom, R. Goudsmit, W. P. Ziehlmaker, A. E. G. von dem Borne, and C. J. M. Melief, *N. Engl. J. Med.* **302**, 933 (1980).

[12] L. Moretta, M. Ferrarini, M. C. Mingari, A. Moretta, and S. R. Webb, *J. Immunol.* **117**, 2171 (1976).

TABLE II
ROSETTES METHODOLOGY

Type	H[a]	E	E_{29}	EA_γ	EA_μ	EAC	M	MAB
Indicator[b] and coating	HRBC	nSRBC[c]	SRBC	ORBC IgG[d]	ORBC IgM	ORBC IgM/C	MRBC	ORBC mIgG[e]
Medium	15% HSA	FCS	FCS	PBS	PBS	PBS	FCS	HBSS
Preincubation (37°)[f] (min)	—	15	5	—	—	—	—	—
Centrifugation (200 g)(RT)(min)	5	5	5	2	2	2	5	—
Incubation[g]	2 hr (ice)	2 hr (ice)	1 hr (29°)	—	1	—	1 hr (23°)	1 hr (37°)
Percentage[h]	48–72	60–80	45–65	15–35	1	12–25	4–14	—

[a] For definitions see Table I and text.

[b] Red cells at 1% in medium specified, lymphocytes at 2×10^6/ml in RPMI-5% FCS. The FCS is absorbed with the appropriate RBC prior to use. Human serum albumin (Connaught Laboratories, Toronto, Canada, or Hyland Laboratories, Costa Mesa, CA) is prepared from a stock solution of 25% as supplied from the company. The use of HSA in "autologous" rosette preparations markedly enhances rosette formation. HRBC, SRBC, ORBC, MRBC, human, sheep, ox, mouse RBC, respectively.

[c] nSRBC, Neuraminidase (NASE)-treated SRBC [Z. Bentwich, S. D. Douglas, E. Skutelsk. and H. G. Kunkel, *J. Exp. Med.* **137**, 1532 (1973)]–500 U NASE (Boehringer-Mannheim, Montreal, Quebec, Canada) plus 1 ml 10% SRBC, 60 min, 20°. Sheep RBC can be used without NASE treatment [M. Jondal, G. Holm, and H. Wigzell, *J. Exp. Med.* **136**, 207 (1972)], but the percentages obtained are less than with nRBC. If the rosettes are incubated overnight on ice the percentages are approximately the same.

[d] Rabbit anti-ox RBC or IgG or IgM class (Cordis Laboratories, Miami, FL) is bound to ox RBC with a subagglutinating concentration of antibody at room temperature [A. F. LoBuglio, R. S. Cotran, and J. H. Jandl, *Science* **158**, 1583 (1967)]. EA indicator cells may also be prepared using SRBC and monoclonal anti-SRBC (R. S. Kerbel and B. E. Elliott, this series, Vol. 93). For EAC indicator cells [C. Bianco, R. Patrick, and V. Nussenzweig, *J. Exp. Med.* **132**, 702 (1970)], mouse C3b is bound to EA(IgM) by incubation of 2% EA with an equal volume of fresh mouse serum (as a source of complement) for 20 min at 37°.

[e] Purified rabbit anti mouse IgG is coupled to OxRBC by chromic chloride [T. Egeland and T. Lea, *J. Immunol. Methods* **55**, 213 (1982)].

[f] Optimum T_μ counts (30–50%) require preincubation of lymphocytes at 37° overnight [L. Moretta, M. Tarrini, M. L. Durante, and M. C. Mingari, *Eur. J. Immunol.* **5**, 565 (1975)].

[g] Percents are approximate means for fresh peripheral blood lymphocytes. Optimum counts of HRFC are obtained using cold counting chambers. Suboptimal incubation temperature of 29° is necessary for ERFC to be of the high-affinity type [W. H. West, C. W. Sienknecht, A. S. Townes, and R. B. Herberman, *Clin. Immunopathol.* **5**, 60 (1976)]. Conversely, optimum MRFC counts are only obtained at 23° [S. Gupta and M. H. Grieco, *Int. Arch. Allergy Appl. Immunol.* **49**, 734 (1975)]. E and H RFC require careful resuspension and should be counted in hemocytometers or rosette chambers [L. A. Gallinger, H. F. Pross, and M. G. Baines, *Int. J. Cancer* **26**, 139 (1980); H. F. Pross, L. A. Gallinger, P. Rubin, and M. G. Baines, *Int. Arch. Allergy Appl. Immunol.* **66**, 365 (1981)]. Percentage MAB-RFC depends on the monoclonal antibody used [T. Egeland and T. Lea, *J. Immunol. Methods* **55**, 213 (1982)].

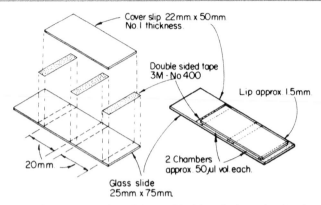

FIG. 1. Details of the construction of the modified Cunningham chamber for rosette and conjugate enumeration. (Reproduced, with permission of the publisher, from Pross *et al.*[10])

Detection of Cell Surface Markers Using Antibody-Coated Latex Spheres

The microrosette assay employing anti-mouse globulin coupled erythrocytes has been used by several laboratories to detect cell surface deter-

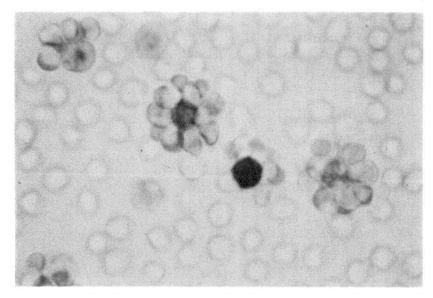

FIG. 2. E rosettes were made from human peripheral blood lymphocytes and scored as described in the text. Both rosetting and nonrosetting cells are shown. Magnification: ×400.

minants.[13,14] As summarized in Table I, these assays involve an indirect detection method: the test cells are incubated with a first antibody which is specific for a cell surface determinant, washed, and then incubated with erythrocytes coupled with an antibody directed against the first antibody (e.g., an anti-mouse immunoglobulin reagent). This form of rosette assay is not in widespread use, however, presumably because of the need for freshly coupled erythrocytes each day that an assay is performed.

Several laboratories have used fluorescent latex spheres as indicator cells.[14-16] The advantage of the fluorescent bead approach is increased sensitivity, increased signal to noise ratio, and ease of separation by fluorescent activated cell sorting.

In this section the method of Kierans and Longenecker will be described.[16] These investigators have found that the best spheres for antibody coupling are Fluoresbrite, fluorescent monodispersed carboxylated latex microspheres (Polysciences Inc., Paul Valley Industries Park, Warrington, PA). They have used successfully both 0.26- and 0.57-μ-diameter fluorescent latex spheres for coupling. The activation process is as follows: a latex suspension (50 μl of 2.5% solid stock material) is added to a 1.5-ml plastic Eppendorf tube, to which 1 ml of 1 N HCl is added. The suspension is then spun in an Eppendorf 5412 centrifuge at high speed for 10 min so that the spheres pellet. Smaller spheres may require a higher centrifugation force (13,000 g) in a Sorvall centrifuge. After removing the supernatant, the spheres are washed twice in phosphate-buffered saline (PBS, pH 7.2). This should bring the pH of the suspension back to approximately 7.2. The pellet is then resuspended in 0.75 ml of 0.35 M mannitol–0.01 M NaCl at 4°. Following this, 100 μl of a fresh 10 mg/ml solution of (1-ethyl-3-(3-dimethylaminopropyl)carbodiimide–HCl (Sigma Chemicals, St. Louis, MO) in mannitol is added and the suspension is mixed for 30 sec. The tubes are then quickly centrifuged for 10 min at high speed and the supernatant is withdrawn (even if some spheres are still in suspension). To this pellet is added 1 ml of a 1 mg/ml solution of the protein to be coupled to the spheres (the antibody solution) in PBS. The antibodies used in our studies have been monoclonal antibodies derived either from an ascites fluid or culture supernatant and purified on Protein A columns (Pharmacia, Uppsala, Sweden) to separate them from nonimmunoglobulin proteins[17] (which would decrease the efficiency of the cou-

[13] C. R. Parish and I. F. C. McKenzie, J. Immunol. Methods 20, 173 (1978).

[14] F. Indiveri, B. S. Wilson, M. A. Pellegrino, and S. Ferrone, J. Immunol. Methods 29, 101 (1979).

[15] T. J. Higgins, C. R. Parish, P. M. Hogarth, I. F. C. McKenzie, and C. J. Hammerling, Immunogenetics 11, 467 (1980).

[16] M. W. Kieran and B. M. Longenecker, J. Immunol. Methods 66, 349 (1984).

[17] P. L. Ey, S. J. Prowse, and C. R. Jenkin, Immunochemistry 15, 429 (1978).

pling procedure). The protein–sphere mixture is sonicated for 30 sec in order to break up clumps and to let the protein coat the entire sphere. It is important that the bead mixture be kept on ice when sonicating to prevent denaturation of the protein. The mixture is then incubated overnight while shaking at 4°. Once the incubation is complete, the spheres are washed twice in PBS/5% FCS and 10 mM sodium azide. The spheres are then stable at 4° in the dark for several months.

The cell–sphere reaction is carried out as follows. The antibody-coupled spheres (200 μl of the stock solution) plus 50 μl of medium containing approximately 10^6 test cells are pipetted into the wells of a 96-well flat-bottomed plate and then spun for 2 min at room temperature at 500 g. The larger spheres require a shorter centrifugation time. The cells are incubated on ice for 1 hr and then resuspended in the well by pipetting. The contents are then layered onto 1 ml of FCS and centrifuged at 200 g for approximately 3 min in order to separate the beads from the cells. The supernatant is removed, the pellet resuspended, and the procedure is repeated. The cells are then transferred to a glass slide, covered with a coverslip, and viewed by fluorescent microscopy. It is also possible to analyze the cells using the fluorescence activated cell sorter.

The results in Fig. 3 show an example of staining with two different types of antibody-coupled beads, analyzed by a fluorescence activated cell sorter (Becton-Dickinson, Sunnyvale, CA). The beads coupled with monoclonal anti-Ly-6.2 antibody were found to bind specifically to B10.BR (Ly-6.2$^+$) lymphoblasts (Fig. 3B), but not to A.TH (Ly-6.2$^-$) lymphoblasts (Fig. 3A). Likewise, beads coupled with monoclonal anti-Ia-19 antibody were found to bind specifically to a subpopulation of Ia-bearing CBA spleen cells corresponding to the proportion of B cells present (Fig. 3D), but not to the Ia-negative tumor cell line, MDAY-D2 (obtained from Dr. R. S. Kerbel) (Fig. 3C). The difference in fluorescence intensity of labeled cells in Fig. 3C and D represents the difference in size of the beads used for coupling. The magnitude of the signal indicated by the mean fluorescence intensity and the low background in this system is ideal for analyzing and sorting cell populations by the fluorescence-activated cell sorter. It is particularly useful in detecting low frequency events. We have also adapted this process for detecting antigen-binding T lymphocytes, activated against specific antigens, the methodology for which will be reviewed elsewhere.[17a]

Antigen-Specific Rosettes

In addition to the spontaneous rosettes described above, for which the ligand and receptor are poorly defined, many laboratories have adopted

[17a] B. E. Elliott, R. Palfree, and S. Mundinger, submitted for publication.

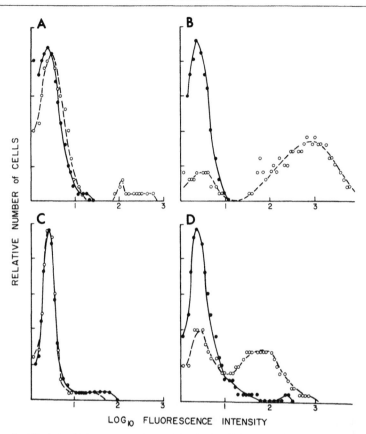

Fig. 3. (A) A.TH (Ly-6.2⁻ lymphoblasts, untreated (●), or stained with anti-Ly-6.2 antibody-coupled latex spheres (○); (B) B10.BR (Ly-6.2⁺) lymphoblasts, untreated (●), or stained with anti-Ly-6.2-coupled spheres (○); (C) MDAY-D2 tumor cells (Ia⁻), untreated (●) or stained with anti-Ia-19 antibody-coupled spheres (○); (D) CBA spleen cells (Ia-19⁺), untreated (●) or stained with anti-Ia-19 coupled spheres (○). Anti-Ly-6.2 antibody (clone 8/106-389) and anti-Ia-19 (clone 116/32-R5) were provided by Dr. Tada and Dr. U. Hammerling, respectively. The latex spheres (Polysciences Inc., Warrington, PA) were 0.57 μm (A and B) and 0.25 μm (C and D) in diameter. Cells (10,000 per group) were analyzed by a fluorescence-activated cell sorter.

the rosette assay to detect antigen-specific rosette forming immunocompetent cells. These methods include direct rosette assays, in which the lymphocytes recognize and bind antigenic determinants on the erythrocyte membrane.[1,5] This approach has been used primarily to study erythrocyte antigen-binding immunocompetent cells.[1,3,18]

In addition to direct rosetting methods, it is also possible to couple

[18] J. F. Bach, J. Y. Muller, and M. Dardenne, *Nature (London)* **227,** 1251 (1970).

erythrocytes directly with various haptens (e.g., tri- or di-nitrophenyl) or proteins (e.g., human serum albumin, ovalbumin, and fowl γ-globulin) to detect antigen-specific lymphocytes.[6,19,20] The methods of coupling vary depending on the reagents to be conjugated[21,22] and a detailed description of rosette preparation, fixation, and enumeration is described elsewhere.[23]

The following general points will be made here: First, the frequency of antigen-specific rosettes in normal lymph nodes and spleens is low (10^{-4}–10^{-5}). For functional and analytic studies it is often necessary to immunize the animal in order to increase the proportion of immunocompetent cells (by about 10-to-100-fold)[3] and/or to enrich for rosette-forming cells. Second, in experimental protocols involving hapten or protein-coupled erythrocytes to detect RFC, it is necessary to include controls with uncoupled erythrocytes. The background frequency of spontaneous rosettes must be subtracted in order to obtain the true frequency of antigen specific rosettes. Third, in experiments with antigen-coupled erythrocytes additional controls for specificity must include specific inhibition of rosette formation with soluble free forms of the coupled antigen. In experiments involving rosetting with uncoupled erythrocytes test cells can be incubated with antigenically unrelated types of red cells which can be distinguished by size (e.g., donkey and sheep RBC[3]). The incidence of rosette-forming cells binding both types of indicator cell would represent the level of nonspecific binding, providing the RBC can be shown to be serologically non-cross-reactive. Fourth, passively acquired antibody can artificially increase the proportions of RFC, particularly in analyses carried out using immune cells. This nonspecific binding can be reduced by removing phagocytic and adherent cells from the lymphocyte preparation by adherence to plastic or by other methods,[7] since adherent cells are the most likely to contribute to nonspecific binding. In addition, fixation and staining procedures allow the visualization of RFC under high power magnification so that the lymphocyte morphology of the rosette-forming cell can be confirmed.[3]

Lymphocyte–Tumor Cell Conjugates

There has been considerable interest recently in the phenomenon of conjugate formation between tumor cells and lymphocytes[24-29] as an assay

[19] S. Kontiainen and S. Andersson, *J. Exp. Med.* **142**, 1035 (1975).

[20] D. Sulitzeanu and M. A. Axelrad, *Immunology* **24**, 803 (1973).

[21] V. J. Pasenen and O. Mäkelä, *Immunology* **16**, 399 (1969).

[22] M. D. Rittenberg and K. L. Pratt, *Proc. Soc. Exp. Biol. Med.* **132**, 575 (1969).

[23] B. E. Elliott, *in* "Immunological Methods" (I. Lefkovitz and B. Pernis, eds.), p. 241. Academic Press, New York, 1979.

[24] E. Grimm and B. Bonavida, *J. Immunol.* **123**, 2861 (1979).

to detect the frequency of lymphocytes recognizing tumor cells, and as an intermediate step in the determination of lymphocytes capable of tumor cell lysis (lytic conjugates) such as NK cells[25-28] or cytolytic T cells.[24,29] Unlike rosettes, in which three or more red cells surround a lymphocyte, a single lymphocyte bound to a tumor cell is sufficient to declare it as a conjugate. Conjugate formation can be shown using either human or murine spleen cells and is carried out as follows[10]: 100 μl of 2×10^6 tumor cells, obtained from a cell line such as K562 in man or YAC in mouse (or any particular cell line which has been used for *in vitro* immunization), are added to 100 μl of 2×10^6/ml lymphocytes in RPMI–5% FCS. The mixture is centrifuged at 200 g for 5 min and incubated for 30 min on ice. About half of the supernatant is then removed and a drop of stain such as crystal violet is added. The conjugates are then resuspended gently (as described above for E rosettes) and counted in a rosette chamber. The number of conjugates obtained varies markedly, depending on the tumor target cell line used. It is our experience that the majority of conjugates formed by normal, unpurified human peripheral blood or murine spleen cells are nonspecific in nature, and consist of T cells, monocytes, natural killer (NK) cells, and null cells[10,25,26] With purification techniques such as nylon wool passage or gradient separation, the number of conjugates more closely correlates with functional activity such as NK[26,27] or cytotoxic T cell[29] lysis. The correlation between binding and cytotoxicity is best determined using the lytic conjugate assay,[24,25,28] in which the proportion of bound tumor cells which has been killed by the adherent lymphocytes is scored.

Separation Techniques

Ficoll–Isopaque. The separation of rosette-forming from nonrosette-forming cells can easily be accomplished by density gradient centrifugation through Ficoll–Isopaque. For depletion of human RFC, F.Ip. of specific gravity 1.077 is prepared as described above for the separation of human mononuclear cells from granulocytes and red cells.[8] For murine RFC separation, the method of Parish *et al.*[30] is used. Ficoll–Isopaque of specific gravity 1.09 is prepared by adding 12 parts of 14% (w/v) Ficoll in

[25] P. Rubin, H. Pross, and J. Roder, *J. Immunol.* **128**, 2553 (1982).
[26] J. Roder and R. Kiessling, *Scand. J. Immunol.* **8**, 135 (1978).
[27] T. Timonen, E. Säkselä, A. Ranki, and P. Haayry, *Cell. Immunol.* **48**, 133 (1979).
[28] S. Targan, E. Grimm, and B. Bonavida, *J. Clin. Lab. Immunol.* **4**, 168 (1980).
[29] J. E. Ryser, B. Sordat, J. C. Cerottini, and K. T. Brunner, *Eur. J. Immunol.* **7**, 110 (1977).
[30] C. R. Parish, S. M. Kirov, N. Bowern, and R. V. Blanden, *Eur. J. Immunol.* **4**, 808 (1974).

distilled water to 5 parts of 32.8% (w/v) Isopaque. The F.Ip. is stored at 4° in the dark, but must be brought to 20° prior to use.

Three to five milliliters of the rosetted preparation (at about 2×10^6 cells/ml for human RFC or up to 2×10^7 cells/ml for murine RFC) is layered onto 3 ml of the F.Ip. solution in 12-ml tubes. For murine RFC separations the tubes should be polycarbonate to avoid breakage on centrifugation. The gradients are then centrifuged at 20° for 20 min at 800 g (human RFC depletions) or 2000 g (murine RFC depletions). The pellet and interface cells are removed, resuspended in at least an equal volume of medium, and washed once at 300 g for 10 min to remove residual F.Ip. Two more washes are done at 150 g for 5 min before the cells are used. The red cells which are adherent to the rosetting lymphocytes may be removed by hypotonic lysis as described above. It should be pointed out, however, that hypotonic lysis may affect both functional and surface marker assays, and that the effects of red cell lysis should be checked in each experimental system before the method is adopted for routine use.[31] Aliquots of both the interface (depleted) and pellet (enriched) cells should be removed for rerosetting to determine the efficiency of the rosette depletion.

Rosette Separation by Velocity Sedimentation. Another method of rosette separation is velocity sedimentation at unit gravity[3,23] (see this volume [8]). The advantage of rosette separation at unit gravity is that it is very gentle, and that relatively small rosettes (e.g., with 4–5 RBC bound) can readily be separated from single cells. Rosettes can also be separated according to differences in size (i.e., the number of RBC bound). This method is also recommended for the separation of fragile rosettes. The fragility of the rosettes can be lessened by incorporating 10 mM sodium azide in the medium to minimize capping of cell surface determinants. The sedimentation time (1–1.5 hr) is significantly shorter than that required to separate single cells by size, so the single cell band remains very narrow at the top of the gradient (2–4 mm/hr) while the rosettes fall at a faster rate (6–20 mm/hr). The yields (>80%) of rosettes can be optimized by adding erythrocytes to the fraction tubes before centrifugation, in order to form a good pellet. This approach has been used to separate antigen-specific immunocompetent cells from normal and immune spleen cell populations.[3,5,23]

Rosette Separation Techniques—Absolute and Functional Yields. The yield of cells obtained after rosette depletions is important for both practical and theoretical reasons. From a practical point of view, it is desirable to know how many cells to put on a gradient in order to have enough cells

[31] S. Sklar, M. Richter, G. Ettin, and M. Richter, *Immunopharmacology* 2, 349 (1980).

to work with after depletion. As a general rule, one obtains at the interface of F.Ip. separations about 50–75% of the number expected from calculating the input number minus the proportion of rosetting cells. The *functional* "yield" after depletion depends, of course, on what proportion of the functioning cells under study have the receptor being depleted. Since most functional and receptor assays are performed on a per cell basis, it is essential to keep a complete account of input and total pellet and interface cell numbers if accurate conclusions are to be drawn as to the functional role of a particular receptor-bearing cell. The proportion of cells of one type having a particular receptor of another type, can be calculated from rosette depletion experiments as follows[32,33]:

$$K_2 = \frac{K_1 - aK_1}{1 - b}$$

In this equation a is the proportion of the cell type under study which forms rosettes of the type being depleted; b is the proportion of the total lymphocyte population forming such rosettes; K_1 is the predepletion proportion of the lymphocyte type under study and K_2 is the postdepletion proportion. K_1 and K_2 could also represent pre and post depletion *function* as long as the expression used is proportional to effector cell number (e.g., lytic units[34]). A self-evident but important fact which is apparent from this equation is that rosette depletion will have no effect on the proportions of another cell type unless the proportion a of rosettes in that population is greater than the proportion b in the total cell population. The extremes of enrichment and depletion occur at $a = 0$ and $a = 1.0$, respectively. If $a = b$, rosette depletion will have absolutely no effect on markers or functions measured in terms of proportions. The usual conclusion from such an experiment is to state that the cells being studied do not bear the receptor which was depleted. In actual fact, failure to show a change in the proportions of a particular cell type after rosette depletion indicates that the proportion of that cell type having the rosetting receptor equals the proportion of rosettes in the total cell population. In the case of human ERFC depletion, for example, equality of pre- and postdepletion cell function or cell proportion indicates that approximately 60% of the cells in question are E rosette-forming.

[32] H. F. Pross, M. Jondal, and M. G. Baines, *Int. J. Cancer* **20,** 353 (1977).
[33] H. F. Pross and M. G. Baines, *in* "Natural Cell-Mediated Immunity Against Tumors" (R. B. Herberman, ed.), p. 151. Academic Press, New York, 1980.
[34] H. F. Pross, M. G. Baines, P. Rubin, P. Shragge, and M. S. Patterson, *J. Clin. Immunol.* **1,** 51 (1981).

General Conclusions

In this review we have broadened the conventional definition of rosette assays to include tumor-lymphocyte conjugates and the detection of surface markers using antibody-coated immunobeads. The latter approach marks a relatively new technical advance and, combined with the increasing repertoire of monoclonal antibodies has proved to be very useful. Nevertheless conventional rosette assays are still routinely used for distinct purposes (as summarized in Tables I and II). The generalized rosette methods described in this article are meant only as guidelines, which should be sufficient to allow the individual investigator to develop the assays according to his own needs. These conventional rosette methods allow the identification of a considerable number of different lymphocyte types, especially in man. When used in conjunction with antigen-specific rosette or immunobead assays, the result is a powerful tool of fundamental importance to both research and diagnostic immunology.

Acknowledgments

We would like to thank Ms. Susan Mundinger and Mrs. Mabel Chau for technical assistance, and graduate students Lucy Gallinger Curtis and Peter Rubin for many hours spent perfecting rosette and lymphocyte–tumor conjugate assay techniques. We also thank Drs. M. Kieran and M. Longenecker for providing us with the description of their fluorescent sphere technique and preliminary data, and Drs. U. Hammerling and T. Tada for gifts of monoclonal antibodies. This work was supported by grants from the National Cancer Institute of Canada, the Medical Research Council of Canada, the Ontario Cancer Treatment and Research Foundation, and Physicians' Services Incorporated. Dr. Elliott is Research Scholar of the National Cancer Institute of Canada and Dr. Pross is a Career Scientist of the Ontario Cancer Treatment and Research Foundation.

[8] Separation of Cells by Velocity Sedimentation

By RICHARD G. MILLER

Introduction

In many problems in biochemistry and cell biology it would be useful to fractionate a complex population of viable cells into component subpopulations differing in function. Since cell function is often correlated with physical parameters, such as cell size, useful separations can often be obtained by fractionating on the basis of such parameters. The procedure could be used preparatively to obtain a purified subpopulation for further experiments or, analytically, to study directly the properties of a particular subpopulation under varying conditions. The analogy with molecular separations, in which various physically based techniques such as centrifugation and electrophoresis are used both preparatively and analytically, is obvious.

This chapter will be restricted to separation of cells by velocity sedimentation in the earth's gravitational field. (For previous reviews see Miller.[1]) However, this procedure is only one of several available procedures for obtaining cell separation on the basis of physical differences in cells. If one limits oneself to physical methods for which the basis of the separation is reasonably well understood and which, at least potentially, can be applied analytically, four methods stand out: sedimentation, equilibrium density-gradient centrifugation, countercurrent distribution, and electrophoresis. All four methods have been concisely reviewed by Shortman.[2,3] They are all bulk processes in that they separate a population of cells into subpopulations essentially at one time.

These methods underwent rapid development in the 1960s and early 1970s and, except for the introduction of Percoll as a material for performing density separations,[4] have undergone little significant development since. More recent developments have focused on cell separation on the basis of biological markers. Electronic cell sorting is the most popular of these. It analyzes and separates cells one at a time, usually on the basis of fluorescent cell surface markers.[5-7] The major advantage of this approach is the potentially much greater biological specificity. Apart from the sophisticated apparatus required, the major disadvantage is the relatively slow cell separation rate. Thus, to process 10^9 cells (approximately 1 g)

[1] This article is a revised and updated version of an article that first appeared in R. G. Miller, *New Tech. Biophys. Cell Biol.* **1** (1973); *in* "In Vitro Methods in Cell Mediated and Tumor Immunity" (B. R. Bloom and J. R. David, eds.). Academic Press, New York, 1976.

[2] K. Shortman, *Annu. Rev. Biophys. Bioeng.* **1**, 93 (1972).

[3] K. Shortman, *Contemp. Top. Mol. Immunol.* **3**, 161 (1974).

[4] H. Pertoft, K. Rubin, L. Kjellen, T. C. Laurent, and B. Klingeborn. *Exp. Cell Res.* **110**, 449 (1977).

[5] W. A. Bonner, H. R. Hulett, R. G. Sweet, and L. A. Herzenberg, *Rev. Sci. Instrum.* **43**, 404 (1972).

[6] L. A. Herzenberg and L. A. Herzenberg, *In* "Handbook of Experimental Immunology" (D. M. Weir, ed.), 2nd ed., Vol. 2, Chapter 22. Blackwell, Oxford, 1973.

[7] R. G. Miller, M. E. Lalande, M. J. McCutcheon, S. S. Stewart, and G. B. Price, *J. Immunol. Methods* **47**, 13 (1981).

would typically take more than 5 days of continuous sorting (not feasible), whereas the same cell number can be handled with one of the bulk procedures in a few hours.

Theory

Consider a sphere (volume V, radius r, density ρ) falling through a viscous medium under the action of the earth's gravitational field. The net gravitational force on the particle is given by $(\rho - \rho')gV$, where ρ' is the density of the viscous medium and g is the acceleration due to gravity (980 cm/sec^2). The net gravitational force is opposed by a viscous drag force set up in opposition to the motion of the particle. For a sphere, the drag force is given by $6\pi\eta r v$ (Stokes' law) where η is the coefficient of viscosity and v is the velocity of the sedimenting particle. An equilibrium state is soon reached in which the net gravitational force is exactly balanced by the viscous drag force and the particle falls at a constant velocity, its terminal velocity, s. Equating the two forces, one finds that s is given by Eq. (1):

$$s = \frac{2(\rho - \rho')gr^2}{9\eta} \tag{1}$$

Under physiological conditions, the vast majority of mammalian cells have radii varying from 2.5 to 10 μm, and densities varying from 1.05 to 1.10 g/cm^3. If they are falling through an aqueous medium with a density of 1.0 g/cm^3, one can see that size variations can give rise to about a 16-fold variation in terminal velocity whereas density variations give rise to only about a 2-fold variation in terminal velocity. Thus, variations in s will primarily be due to variations in size.

Cells are seldom perfectly spherical and some, such as erythrocytes, are not even approximately so. However, the above expression still holds true[8] with an error of less than 10% for both prolate and oblate spheroids with axial ratios $a/b < 3$ providing that one takes r to be that of an equivalent sphere, i.e., $(3V/4\pi)^{1/3}$. Equation (1) assumes the cell to be moving at terminal velocity and, for typical cells under typical conditions, the time required to reach this speed is, at most, a few microseconds.

If one knew the density and volume of a cell, one should, in principle, be able to calculate its s value directly from Eq. (1). This has been verified experimentally for sheep erythrocytes,[8] antibody-producing cells,[9] and mouse L cells.[10] Under conditions representative of actual experimental

[8] R. G. Miller and R. A. Phillips, *J. Cell. Physiol.* **73**, 191 (1969).
[9] R. A. Phillips and R. G. Miller, *Cell Tissue Kinet.* **3**, 263 (1970).
[10] H. R. MacDonald and R. G. Miller, *Biophys. J.* **10**, 834 (1970).

conditions used for separating cells by sedimentation, $\rho = 1.01$ g/cm³ and $\eta = 1.567$ cP (water at 4°). Thus, we have

$$s = 5.0(\rho - 1.01)r^2 \tag{2}$$

in which s is in mm/hr, ρ in g/cm³, and r in μm. A typical nucleated cell has a density of 1.06 g/cm³, for which this expression becomes $s = r^2/4$.

Sedimentation and density separation are not completely independent procedures. Sedimentation separates cells primarily on the basis of size, rather than density. However, cell density is roughly inversely proportional to cell size, since the nucleus of a cell is much more dense than the cytoplasm and larger cells tend to have a smaller nucleus-to-cytoplasm ratio. Nonetheless, there are significant exceptions to this relationship; e.g., cytoplasmic granules, such as are found in granulocytes, can markedly increase cytoplasmic density. Thus, sedimentation and density separation can often be complementary.

Methods

The objective of the procedure to be described is to form a thin layer of cell suspension on top of a fluid column, let the cells sediment under the influence of gravity for an appropriate length of time, and collect fractions containing cells that have moved different distances. The procedure has been designed such that the cells are affected as little as possible during the separation. In particular, their viability is not affected. Any changes resulting from metabolic activity should be greatly reduced by lowering the temperature. For this reason, we have usually performed separations in a cold room at 4°, although we have also obtained satisfactory results at room temperature. The fluid column in which the cells sediment should be a buffered isotonic saline such as PBS (Dulbecco's phosphate-buffered saline). Any further additives should not affect the cells.

The following sections describe two closely related practical systems for meeting the above objectives and discuss their limitations.

Apparatus. The fluid column as described will be unstable to convection and mechanical jarring. Stability can be maintained by introducing a shallow density gradient into the fluid column. The choice of gradient material is important. It should have sufficiently high molecular weight so as not to create an appreciable tonicity gradient, and it should also be nontoxic to cells. Materials such as fetal calf serum, bovine serum albumin (BSA) (Cohn fraction 5), or Ficoll have proved to be satisfactory. Materials such as sucrose are not satisfactory on the basis of both tonicity and toxicity problems.

The gradient is made as shallow as possible, with ρ as low as possible to ensure that separation is on the basis of differences in sedimentation

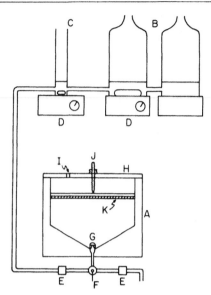

FIG. 1. The "staput" sedimentation chamber: A, sedimentation chamber; B, gradient maker; C, intermediate vessel; D, magnetic stirrers; E, flow-rate regulators; F, three-way valve; G, flow baffle; H, chamber lid; I, vent hole in chamber lid; J, screw; K, cell band shortly after loading is completed.

ιαιe rather than differences in density. A typical gradient might vary from 0 to 2% BSA in PBS which covers the density range 1.0074 to 1.0113 g/cm^3. The equivalent serum gradient would be 0 to 30% serum in PBS.

The next problem is the formation of a cell layer on top of the gradient. This can be achieved with the apparatus of Fig. 1 ("staput").[8] It has three main parts: the sedimentation chamber (A), the gradient maker (B), and a small intermediate vessel (C). The sedimentation chamber should be transparent, as it is a big advantage to be able to see cell bands during the separation process. We have used sedimentation chambers made of Lucite, polycarbonate, or glass. The last two may be steam sterilized. The angle in the cone of the sedimentation chamber is nominally 30°, but this is not critical and cone angles as low as 15° to the horizontal appear to be satisfactory. All tubing interconnections are made using silicone tubing (Silastic, Dow Corning) because of the very low tendency for cells to stick to silicone rubber.

Cells are loaded into the chamber through (C) and are lifted into their starting position (indicated by K in Fig. 1) by the gradient (loaded via gravity) introduced through the gradient maker (B). The small stainless-steel baffle (G) at the bottom of the chamber is used to deflect incoming fluid during the loading procedure. Without this baffle it is very difficult to

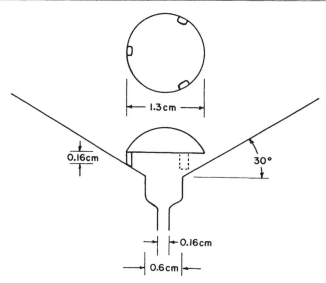

Fig. 2. Detail of chamber inlet and flow baffle for apparatus of Fig. 1 using a sedimentation chamber 12 cm in diameter.

load a gradient without mixing. Figure 2 gives details of the baffle and inlet construction for a chamber 12 cm in internal diameter. Returning to Fig. 1, the chamber has a lid (H), which serves to keep out dust. The screw (J) mounted in the lid has a sharpened point. This enables one reproducibly to fill the chamber to the same volume by stopping the filling process when the screw tip just touches the rising gradient. E represents a two-way needle valve, and F a three-way tap. After the cells have sedimented for an appropriate length of time, fractions are collected through the bottom of the chamber.

If the concentration of cells in the starting band exceeds a certain critical limit, a phenomenon called "streaming" takes place, and useful separations cannot be achieved. The highest concentration at which steaming does not take place (the "streaming limit") is characteristic of the cells being sedimented. For cells having a volume of about 100 μm^3, the streaming limit is about 5×10^6 cells/ml. Thus, for a given chamber, the maximum cell load is determined by the streaming limit and the maximum allowable thickness of the starting band for the resolution desired (see the next section for a detailed discussion of both streaming and resolution). To process more cells, one must use a chamber of large diameter. We have successfully used chambers ranging in diameter from 11 to 39 cm. Chamber size is more conveniently designated by volume in cube centimeters per millimeter of length in the cylindrical part of the chamber, since this is the parameter required to determine resolution and calculate sedimentation velocities. Call this the chamber constant a. Then the chamber diameter range of 11 to 39 cm corresponds to a values of 9.5 to 119 cm^3/mm.

One problem was encountered with the 39-cm-diameter chamber. The total volume of gradient in this chamber is about 20 liters so that to collect fractions in a reasonable length of time we have used rather high drain rates (up to 400 ml/min). At these drain rates, swirling occurred similar to that seen in a draining bathtub. This problem was solved by attaching to the chamber lid a vane that splits the chamber into two sections almost to the bottom. This problem did not occur for a chamber 24 cm in diameter. A complete cell-separating apparatus, made of glass, with chamber diameters of approximately 11, 17, or 25 cm is available commercially from O. H. Johns Scientific, Toronto, Canada.

The detailed protocol we use for performing a separation in a chamber 12 cm in diameter ($a = 11.4$ cm^3/mm) is as follows.

1. Set up apparatus in coldroom as in Fig. 1. We use gradient bottles and an intermediate vessel with internal diameter of 7 and 2.2 cm, respectively.

2. Prepare at least 300 ml each of 1 and 2% BSA in PBS. Bring to a temperature of 4°.

3. Fill all the connecting tubing (made of silicone rubber) of the apparatus on the chamber side of (C) with PBS, making sure to get rid of all air bubbles.

4. Center the flow baffle (G) inside the cone of the sedimentation chamber. It is important that this be centered carefully.

5. Clamp the lines between all three gradient chambers. Load 300 ml of 1% BSA into the left-hand bottle of (B) and 300 ml of 2% BSA into the right-hand bottle.

6. Load top layer (30 ml of saline) into the sedimentation chamber through the intermediate vessel (C). Note that the total volume of the top layer will be whatever fluid is put in (C) plus the volume of the tubing between (C) and the sedimentation chamber (typically less than 5 ml). Check that the flow baffle is still correctly centered. The function of this top layer is to prevent disturbance of the cell band by erratic movements of the rising fluid meniscus as the chamber is filling.

7. Load cells in 20 ml of 0.2% BSA in PBS through (C). The cell concentration must be below the streaming limit for the cells being loaded. Clamp the line the instant the buffer chamber empties.

8. Rinse the buffer chamber twice with 0.2% BSA in PBS. Use a 50-ml syringe and a piece of tubing to do this. Check that no air bubbles entered the lines during the rinse.

9. Fill (C) with the buffer gradient (0.35% BSA in saline) to the level of the fluid in the gradient bottles. Adjust the needle valve for a flow rate of 2–3 ml/min. Turn on magnetic stirrers.

10. Remove all clamps and record time (t_1); the gradient will load itself. It will rise rapidly from 0.35 to 1.0% BSA and slowly thereafter to 2.0% BSA. The reasons for using a gradient of this shape are described in the next section. Once the cells have been lifted off the bottom, the flow rate can be increased. Continuously adjust by eye to a rate just below that at which the cell band is disturbed. Small disturbances will settle out. Loading should be as rapid as possible (10–15 min). The time elapsed between loading cells and starting the gradient should also be as short as possible (4–8 min).

11. Record time cell band reaches top of cone (t_2).

12. After an appropriate sedimentation time (4 hr is usually adequate), start unloading the chamber through the bottom at a rate of about 30 ml/min. Record time (t_3). Discard the cone volume.

13. Collect remainder of gradient in equal-sized fractions (e.g., 15 ml). Record time first fraction started (t_4) and last fraction finished (t_5). Also record number and volume of last fraction.

Occasionally, one wishes to separate very large or very sticky cells. There is a reasonable probability that some of these will sediment onto or, in passing, stick to the cone during the loading procedure. These cells may either be permanently lost, or even worse, come off randomly during the draining procedure and contaminate all fractions. To avoid these possibilities, it is desirable to have a procedure in which cells can be layered on top of a preformed gradient. A system for doing this is shown in Fig. 3, which also illustrates the operating procedure.

The sedimentation chamber ("muffin") is a completely closed cylinder with entrance ports on opposite sides of the top and bottom, as depicted. The chamber is mounted between two stands and is free to rotate about axis A which is perpendicular to the plane determined by the entrance ports. To load, the chamber is first rotated to an angle of approximately 30° with respect to the horizontal and filled with saline. Next, a linear gradient (e.g., 1–2% BSA) is introduced through the bottom port. The gradient displaces the saline already in the chamber through the top port. Next, cells (in 0.5% BSA, for example) are loaded through the top port followed by, if desired, an overlay (e.g., 0.3% BSA). Then the chamber is slowly and smoothly rotated to a horizontal position and left in this position for the duration of the sedimentation time (e.g., 4 hr). It may seem surprising that one can rotate the chamber in this way without complete disruption of the gradient. However, even mild density gradients are remarkably stable and all fluid appears to preserve its relative

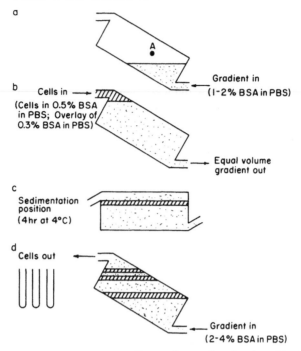

FIG. 3. Operating procedure for "muffin" sedimentation chamber. (a) The chamber is tilted to approximately 30° from the horizontal about axis A and a gradient is introduced through the lower port. (b) The cells and overlay are introduced through the upper port. (c) The chamber is carefully rotated to horizontal for cell sedimentation. (d) The chamber is again rotated to 30° from the horizontal, and fractions are collected by displacing the chamber contents through either the upper or lower port with an appropriate gradient (collection through the upper port is illustrated).

position during the rotation providing this is performed without jerking. To drain the chamber, one first rotates it back to its original position. It can then be unloaded through either the top or bottom. This is best done by displacing the contents with an appropriate gradient, e.g., 0.3–0% BSA for displacement through the bottom.

We have successfully used Lucite chambers of this design with a diameter of 14 cm and depth of either 3.8 or 8 cm and a glass chamber of this design with diameter 13 cm and depth 8 cm. Problems were experienced with a much larger chamber, 39 cm in diameter and 13 cm deep, in that it was difficult to rotate it by hand sufficiently smoothly to prevent disturbance of the gradient during rotation although some kind of mechanical device could almost certainly be used to avoid this problem.

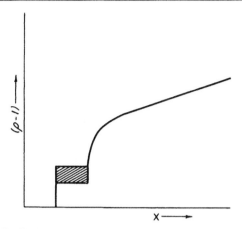

FIG. 4. Density distribution for attenuated step gradient immediately after loading. The part contributed by the cells is shown hatched.

Streaming, Resolution, and Cell-Load Limitations. The major limitation to separation of cells by velocity sedimentation is the phenomenon of streaming. This occurs if the cell concentration (cells/ml) in the cell band loaded into the chamber exceeds a certain number, called the streaming limit. Shortly after loading, large numbers of filaments ("streamers") a few millimeters long can be seen hanging down from the cell band. When viewed from above, the cell band has a mottled appearance. Streaming does not appear to be a cell-aggregation phenomenon because when streaming is set up in a small cuvette and viewed with a microscope, all cells in streamers appear to be single and to move relatively independently of one another.

Streaming is probably due to some kind of density inversion phenomenon taking place at the interface between the cell band and the gradient. Tulp and Bont[11] reasoned that adding a small quantity of viscous material to the cell band to make its viscosity slightly greater than that of the underlying gradient would prevent accumulation of cells at the interface and thus would prevent any such inversion phenomenon. Using this procedure, we have not been able to increase the streaming limit. Streaming can, however, be considerably reduced by introducing a "rounding off" of the sharp edge of the gradient immediately below the band of cells. This is called a buffered-step gradient and is illustrated in Fig. 4. For sheep erythrocytes, the streaming limit using the sheer-step gradient is 5 ×

[11] A. Tulp and W. S. Bont, *Anal. Biochem.* **67,** 11 (1975).

10^6/ml. The streaming limit increases to about 1.5×10^7/ml when a buffered step gradient is used.[8]

Buffered-step gradients of a wide shape range can be generated by appropriate choice of the areas of the two bottles in a two-bottle gradient generating system. Let the gradient be taken from bottle 1, which has area A_1 and contains material of density ρ_1; bottle 2 has area A_2 and contains materials of density ρ_2. Then

$$\rho(x) = \rho_2 - (\rho_2 - \rho_1)(1 - x/x_0)A_2/A_1 \qquad (3)$$

where x is distance measured from the top of the gradient and x_0 is the depth of the gradient when the bottles are empty.

Both sheer-step and buffered-step gradients can be generated with the apparatus of Fig. 1. The gradient bottles of (B) are filled to the desired level with 1 and 2% BSA in PBS. For a sheer-step gradient, cells (in 0.2% BSA in PBS) are loaded into the sedimentation chamber via (C) which is then rinsed out with saline and used as a subsidiary mixing vessel for the 1–2% BSA gradient from (B). For a buffered-step gradient, the cells are loaded as above and chamber (C) is again rinsed, but this time filled to the same level as the bottles of (B) with 0.35% BSA in PBS. When flow resumes, a gradient similar to that in Fig. 4 will be generated. The precise details of the shape of the buffered-step gradient do not seem to be important; the increase in the streaming limit appears to be about the same whether the gradient is sharply or gradually rounded off.

The streaming limit varies with the cells being sedimented. All tests made to date are consistent with the streaming limit being inversely proportional to V, the average cell volume. Thus, large cells have a lower streaming limit than small cells.

The maximum cell load that can be separated on a given chamber is determined by the resolution desired and the streaming limit. What is the intrinsic resolution of the system, i.e., at what point do further reductions in the width of the starting band yield no significant improvement in resolution? This is a more difficult question.

Consider a population of cells, all of which have the same density and volume. Load these as a very thin band and measure the sedimentation profile after different sedimentation times. Then one can define the intrinsic resolution of the system as δs where δs is the full width at half height of the sedimentation profile. In various experiments (see, e.g., Fig. 5) approximating this situation using both cells and plastic beads, it has been found that $\delta s/s$ is a constant, independent of both s and the sedimentation time (for time ranges of 3–6 hr), having the value of 0.18. (This corresponds to an intrinsic coefficient of variation for the sedimentation distribution of 0.076.) This intrinsic dispersion is disturbingly large, much larger than one would expect on the basis of diffusion alone, and we have been unable to find a satisfactory explanation for it. However, all attempts by ourselves and others to improve on it have been unsuccessful.

To get the optimum resolution while processing the maximum number of cells possible for a cell population with an s value of 5 mm/hr and a sedimentation time of 4 hr proceed as follows. The width of the band in the sedimentation chamber resulting from the intrinsic dispersion after 4 hr of sedimentation will be $4 \times 5 \times 0.18 = 3.6$ mm. Assuming that the intrinsic dispersion in the separation and the loss of resolution resulting from a starting band of finite width are independent (which, experimentally, seems approximately true), the sedimentation profile of the 5 mm/hr cell would be only $\sqrt{2} \times 3.6$ mm wide if the starting band were also 3.6 mm wide. Clearly, using the above assumption about how to add dispersive factors, there is little point in making the starting band much narrower than about 1.8 mm.

Suppose, in the above example, one wished to process 3×10^8 cells and that the streaming limit for this population is 10^7 cells/ml. The cells can be loaded in no less than 30 ml. To get a starting band 1.8 mm wide implies a chamber constant, a, of $30/1.8 = 16.7$ cm^3/mm corresponding to a chamber 14.6 cm in diameter.

Preparation of Cell Suspensions. The method used for preparation of cell suspensions is critical to the success of any cell-separation experiment because of the possibility of generating artifacts from dead cells, cellular debris, or cellular aggregates. Our experience has been largely limited to cell suspensions of lymphoid or myeloid origin from mice. For a solid organ, such as thymus or spleen, the procedure we have found to be the best (modified from that of Shortman, Williams, and Adams[12]) is as follows. Place the organ in the middle of a saucer-shaped wire screen (60 mesh) and cut it into several hundred pieces with sharp scissors. Rub the pieces through the screen with a glass pestle, keeping the screen center in buffered saline containing 0.2% BSA. Put this initial suspension into a tube and allow the large aggregates to settle out for 3–5 min and discard. Next, the cell suspension is gently layered over a few milliliters of 6% BSA in PBS in a plastic or siliconized glass tube. Within 10–15 min, many of the remaining aggregates will settle into the 6% BSA. Next, the overlay is removed with a Pasteur pipet and gently layered into a second tube containing a few milliliters (2–3 cm deep) of 6% BSA in PBS. The cells are spun through the BSA at a force not exceeding 250 g for the minimum time required to pellet all of them. The pellet is then resuspended in the solution to be used for cell loading in the sedimentation chamber, as described

[12] K. Shortman, N. Williams, and P. Adams, *J. Immunol. Methods* **1,** 273 (1972).

above. The resulting suspension is more than 90% viable by either fluorescein diacetate[13] or trypan blue[14] viability tests and contains less than 1% cell aggregates (often much less).

Cell suspensions with similar viability can be made from spleen or thymus by enzymatic digestion using, for example, trypsin. We have not used such methods for fear of altering the chemical and biological properties of the cells. Meistrich,[15] in studies of mouse sperm cell differentiations, has shown that if a cell suspension prepared mechanically from mouse seminiferous tubules is treated with trypsin (0.25%, w/v) and deoxyribonuclease (20 μg/ml) at room temperature for a short period of time (less than 3 min) all nonviable cells are destroyed with no apparent loss of viable cells (as measured by trypan blue uptake). The same procedure appears to work for suspensions of lymphoid and myeloid cells but, again, this procedure undoubtedly alters the biological properties of the cells.

Counting and Size Analysis. In a typical cell-separation experiment it is necessary to determine the cell concentration in many fractions and therefore some kind of automated cell counter is called for. The best currently available is an electronic cell detector of the Coulter type.[16] The heart of this instrument is a small aperture, typically about 100 μm in diameter, which is mounted in a glass tube and immersed in saline containing the cells to be counted. Electrodes are mounted on either side of the aperture, and electrical current flows from one electrode to the other through the aperture. To count cells, a known volume of fluid is sucked through the aperture. Each time a cell goes through the aperture, it increases the electrical resistance of the aperture and a pulse is generated, which can be amplified and counted in appropriate electronics. This instrument allows very rapid determination of cell concentration per fraction. Unless very carefully cleaned beforehand (e.g., by using the BSA wash procedure described in the preceding section), accurate counts of cell suspensions prepared from solid tissues are difficult to obtain in the cell counter because of the debris included in the suspension. However, in a sedimentation separation, the debris is largely left behind by the sedimenting cells, and this problem does not occur. Cell suspensions initially at 4° appear to produce clogging of the aperture if they are warmed to room temperature before counting.

Using the Coulter counter, it is possible to get much more information than just the cell concentration. The magnitude of the signal produced by a cell traversing the aperture is proportional, over a restricted but useful

[13] B. Rotman and B. W. Papermaster, *Proc. Natl. Acad. Sci. U.S.A.* **55,** 134 (1966).
[14] A. M. Pappenheimer, *J. Exp. Med.* **25,** 633 (1917).
[15] M. L. Meistrich, *J. Cell. Physiol.* **80,** 299 (1972).
[16] W. H. Coulter, U.S. Patent 2,656,508 (1953).

range, to the volume of the cell detected. One can get a direct measure of the volume distribution of the cells in the sample by connecting the Coulter counter to a pulse-height analyzer. This instrument analyzes each incident pulse and adds the count to one of a set of serially arranged memory locations ("channels") according to the size of the pulse. At the completion of a run, the memory can be displayed visually or printed out to produce a graph of pulse number versus pulse size, i.e., cell number versus cell volume.

The rise time of the signal produced by a cell traversing an aperture will be determined, among other things, by aperture length. Commercial apertures tend to be so short (<100 μm) that the rise time is too short to be handled with fidelity by standard electronics. This problem can be overcome by making one's own apertures using ruby watch jewels 200 μm long. These can be obtained very cheaply from Erismann-Schinz, S.A., La Neuveville, Switzerland. The jewels should be directly fused to a soft glass tube having a coefficient of thermal expansion close to but slightly greater than that of the watch jewel.

The electronics of most commercial cell counters of the Coulter type is such as to produce some distortion of the recorded volume spectrum, but these problems can be overcome by appropriate redesign of the circuitry.[17,18] Some further distortion and dispersion of the volume spectrum is inherent in the nature of the events taking place in the detection aperture but, although very complex, these processes are largely understood[17,18] and do not detract from the usefulness of the method. The coefficient of variation of the volume distribution measured for a uniform population of particles of volume 100 μm^3 detected in an aperture 100 μm in diameter and 200 μm long is about 0.1. Volume dispersions greater than this are due to true dispersion in the sample itself.

If one wishes to measure absolute volumes with the volume spectrometer outlined above, it is necessary to have some good volume standards in the cell-volume region being investigated. We have found the most satisfactory volume standards to be plastic beads (styrene divinylbenzene copolymer beads, 6–14 μm, Dow Chemical). Uniform beads in the volume range of cells cannot be obtained commercially at a reasonable price. However, the Dow beads can be fractionated by velocity sedimentation using the techniques described in this article to give a large number of samples, each of quite uniform volume. The problem remains of how to measure the volume of the beads in each fraction. Ideally, one would like an accuracy of about 5%. This would require a diameter measurement accuracy of 5/3%, not obtainable by optical measurement because of dif-

[17] N. B. Grover, J. Naaman, S. Ben-Sasson, and F. Dolijanski, *Biophys. J.* **9**, 1398 (1969).
[18] R. G. Miller and L. J. Wuest, *Ser. Haematol.* **2**, 128 (1972).

fraction problems. Electron microscopy is difficult because spherical aberration makes it almost impossible to make measurements at this level of precision even if the beads are put on a grid of known spacing at the time of the photograph. A solution to the problem is to calculate the absolute volume of the bead standard from the bead density as measured on a sucrose density gradient and the bead sedimentation velocity as measured by the methods described in this chapter. Alternatively the number of spheres in a volume containing a known weight of particles of known density can be counted. Absolute accuracies of 5% can be obtained.

Calculation of s Values. Homogeneous populations of cells can be characterized by their s value (sedimentation velocity in mm/hr) providing that all separations are related to a standard set of conditions. The standard conditions we have adopted are a temperature of 4° and a gradient in which the average density in the region through which the cells sediment is very close to 1.010 g/cm³. Under these conditions, absolute s values can be measured with a precision of 5%.

In calculating s values from experimental data, the distance a cell sediments is inferred from the volume of fluid through which it falls. To do this accurately, some conventions and corrections that are not altogether obvious are required. Consider a separation done in the apparatus of Fig. 1, i.e., the cells have been both loaded and drained through the bottom of a conical chamber. On draining, one cone volume of fluid is discarded and the rest of the chamber contents are collected as fractions of equal volume, v. Number the fractions in the order of collection, the last fraction having number N_f and volume V_f.

Define the $s = 0$ point as the middle of the input cell band. Treating fraction number as a continuous variable, the corresponding position, in terms of fraction number, is given by Eq. (4).

$$N(s = 0) = N_f - \frac{(V_0 + \frac{1}{2}V_{cb} - V_f)}{v} - \frac{1}{2} \qquad (4)$$

where V_0 is the total volume of fluid above the cell band and V_{cb} is the volume of the cell band. The final subtraction of 1/2 is necessary when one treats N as a continuous variable: A particular fraction, when collected, extends from $N - 1$ to N on a continuous scale, but we wish to consider integer values of N, and the s value associated with them, as representing the average properties of the cells in the fraction having that N value.

Both loading and unloading of the chamber take finite time, during which the cells will sediment. When a cell sediments while in the cone, the volume of fluid fallen through per unit distance fallen is smaller than while in the cylindrical region. Thus, if one calculates the s value solely on the basis of the volume of the fluid fallen through, one will underestimate the s value. This is most easily allowed for by correcting the time the cells spend in the cone during loading and unloading by an appropriate factor. If the volume flow rate is constant and the cells start precisely at the apex of the cone, one can show that the actual time the cells spend in the cone during loading and unloading should be reduced by 40%. Using this correction factor and also allowing for the fact that cells in the last fraction collected sediment for a longer time than the cells in the first fraction collected, the s value (in mm/hr) for fraction N is given by Eq. (5).

$$ s(N) = \frac{(N_f - N - \frac{1}{2})v - (V_0 + \frac{1}{2}V_{cb} - V_f)}{a[(t_4 - t_1) - 0.4(t_4 + t_2 - t_3 - t_1) + N(t_5 - t_4)/N_f]} \tag{5} $$

where the times t_1 to t_5 (all in hours) are, respectively, the times at which loading started (t_1), the cells reach the top of the cone (t_2), draining started (t_3), first fraction started (t_4), and last fraction finished (t_5), a is the chamber constant, i.e., the milliliters of fluid per millimeter of length in the cylindrical region of the chamber. Equation (5) is still not exact because of variations in the loading rate and the finite thickness of the cell band, but should be correct to within 1–2% if the cells spend less than 10% of the total sedimentation time in the conical portion of the chamber.

Calculation of accurate s values for the sedimentation chamber of Fig. 3 is much more difficult because a simple analytic expression for the cone correction factor cannot be obtained. Assuming the chamber to be unloaded from the top, and its entire contents to be collected

$$ s(N) = \frac{(N - \frac{1}{2})v - (V_0 + \frac{1}{2}V_{cb})}{a[(t_2 - t_1) + fN(t_3 - t_2)/N_f]} \tag{6} $$

where t_1 is the time the cells are loaded and the chamber is rotated to the sedimentation position, t_2 the time the chamber is rotated back and the first fraction started, t_3 the time the last fraction, N_f, is finished, and f an undetermined constant, which should be approximately equal to l/D, l, and D being chamber depth and diameter, respectively.

In establishing standard conditions for sedimentation, temperature control is of critical importance. For the sedimentation equation [Eq. (1)] it is seen that s is inversely proportional to the coefficient of viscosity. The viscosity of the BSA in PBS solutions used does not differ appreciably from water. The most viscous material used, 2% BSA in PBS, has a viscosity only about 2% greater than that of distilled water. However, the viscosity of water decreases 36% between 4 and 20°, the rate of change being greatest near 4°. It has been verified experimentally[8] that the s value for sheep erythrocytes increases accordingly.

Applications

This section describes some representative sedimentation experiments chosen to illustrate artifacts and limitations of the method, relative merits of measuring activity or enrichment profiles for cell types of interest, and some applications of biological interest.

Figure 5 shows the results of an experiment in which mouse bone marrow cells were sedimented and one fraction from the separation run a second time under identical conditions in an attempt to determine the

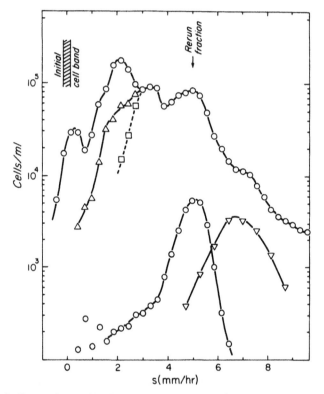

FIG. 5. Sedimentation profile of mouse bone marrow cells. One fraction from the separation (indicated by the arrow) was run a second time under identical conditions. (O) Total cells/ml as measured in an electronic cell detector for the initial separation (upper curve) and the rerun fraction (lower curve); (△) nucleated cells/ml as measured visually with a microscope (where different from total cell count) in the initial separation; (□) viable cells/ml as measured visually using FDA viability test (where different from total cell count) in the initial separation; (▽) doublets/ml as measured visually (mainly granulocyte doublets) in the initial separation.

ultimate resolution of the system. The starting bands were 1 mm thick, the sedimentation times were 4 hr, and the sedimentation chamber was of the type shown in Fig. 3 ("muffin").

Reproducible sedimentation profiles can be obtained over more than a 100-fold variation in cell concentration. Therefore, the results have been plotted on a logarithmic scale. The peaks in the total cell count profile in the regions of $s = 2$, 3, and 5 mm/hr are primarily mature red cells, lymphoid cells, and myeloid cells, respectively. However, some of the remaining profile is artifact.

Cell counts performed with the electronic cell detector indicate a peak at $s = 0.2$ mm/hr. Counts performed visually with a microscope indicate almost no cells in this region. This peak corresponds to cell debris. It is almost completely absent when the FCS wash procedure is used. The shoulder in the profile in the region of $s = 7$ is also mostly artifact. It arises from cell doublets (primarily granulocyte doublets) present in the initial cell suspension, as shown by the doublet profile in the figure, which was determined by microscopic examination of the fractions. Theoretically [Eq. (2)], a doublet formed between two spheres of the same volume should sediment $2^{2/3}$ ($=1.59$) times as fast as the single spheres themselves. This has been verified experimentally using plastic spheres of known volume. The shoulder in the nucleated cell profile in the $s = 2$ region appears to consist almost entirely of nonviable cells (mostly lymphocytes) as determined by the fluorescein diacetate viability test. The magnitude of this peak increases with increasing harshness of the procedures used to prepare the cell suspension.

The s values of the various peaks in the profile are reproducible to an accuracy of at least 5% from one run to another, provided the separation conditions, particularly temperature, are maintained constant. Thus, for the conditions used (defined earlier), mouse red cells have a modal s value of 2.0 ± 0.1 mm/hr.

The mean s value of the rerun fraction is within 1% of its initial value. This indicates that the sedimentation properties of the cells do not change throughout the separation procedure. However, the rerun peak is very broad ($\delta s/s = 0.20$), broader than can be explained on the basis of diffusion or the finite widths of the starting bands. Similar results ($\delta s/s \geq 0.18$) are obtained if one repeats the experiment with Dow beads, which clearly cannot change in any way during the two separations. Thus, the broad width of the rerun peak must be taken as a measure of the intrinsic resolution of the system. Any measurement (e.g., a functional assay for cells of a particular type) that leads to a sedimentation profile with a width less than this limit must be viewed with suspicion.

Consider next an experiment in which the sedimentation profile of a particular subclass of cells is determined using an assay of cell function. The cells investigated were mouse spleen antibody-forming cells (AFC) making IgG antibody against sheep erythrocytes. These were obtained 4 days after a second immunization of the mouse with sheep erythrocytes. A single-cell suspension was made from mouse spleen and fractionated by sedimentation. Each fraction was assayed for its content of IgG AFC using the Jerne plaque assay (for detail, see Phillips and Miller[9]).

Figure 6 shows the results of one such experiment. The starting band was 1.7 mm thick, sedimentation time was 3.5 hr, and the sedimentation chamber was of the type shown in Fig. 1. Nucleated and total cell count profiles are similar to those in Fig. 5 except that spleen contains far fewer myeloid cells than bone marrow so that there is not a peak in the $s = 5$ mm/hr region. However, there is a pronounced red cell peak at 2.0 mm/hr, and a lymphocyte peak at 3.1 mm/hr. Similarly, there is a debris peak at

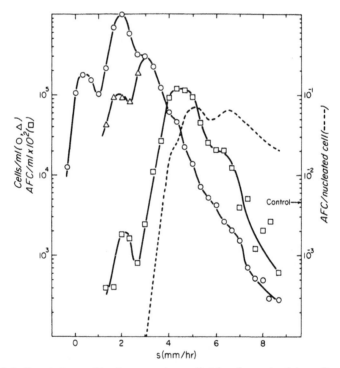

FIG. 6. Sedimentation profile of mouse spleen cells taken from mice 4 days after a second immunization with sheep erythrocytes. (○) Total cells/ml; (△) nucleated cells/ml (when different from the total cell count; (□) antibody-forming cells (AFC)/ml. The dashed line is antibody-forming cells per nucleated cell referred to right-hand axis, on which the control value is indicated.

$s = 0.3$ mm/hr and a peak of nonviable nucleated cells (mostly lymphocytes) in the $s = 2$ mm/hr region.

The AFC content of each fraction was measured and is expressed on the figure both as an activity profile (AFC/ml) and as an enrichment profile (AFC/nucleated cell). Maximum activity is found around $s = 4.4$ mm/hr; maximum enrichment is found in two peaks at higher s values.

Calculation of the recovery of the cell of interest is very simple with an activity profile. Thus, in Fig. 6, one merely adds up the number of AFC/ml in each fraction (i.e., the values plotted), multiplies by the fraction volume, and compares the result with the total number of AFC loaded. For the experiment in Fig. 6, the recovery was 90%.

If one is using an assay that depends upon cell function for a response to occur, it is essential to have complete knowledge of the dose–response curve (the ideal being linearity) and to know that it is independent of relative enrichment or depletion of other cell types present in the unfractionated tissue. It was verified that the Jerne plaque assay, under the conditions used for the experiment of Fig. 6, had a linear dose–response curve independent of the number of AFC plated, total number of cells plated, or relative enrichment and depletion of particular cell types. Given all this, it is simple to construct either activity or enrichment profiles. However, very few functional assays have all these properties, and to get meaningful profiles it may be necessary to measure dose–response curves on individual fractions. By doing this, it is sometimes possible to obtain meaningful activity and enrichment profiles even using a nonlinear functional assay.

The experiment in Fig. 6 has been repeated more than 10 times. Although the modal sedimentation velocity of the IgG AFC was the same (4.4 ± 0.2 mm/hr) in all experiments, the peak was always much broader than that routinely obtained with other types of cells such as red cells or lymphocytes, and the shape tended to vary from one experiment to the next. An explanation for the broadness of the peak can be found from the fact that AFC are a population of dividing cells. If a cell has a volume $2V_0$ just before mitosis, then its daughters just after mitosis will have volumes V_0. In a randomly dividing population, if N_0 cells are entering mitosis, then $2N_0$ cells should be leaving mitosis (see Fig. 7a). If one assumes that the density of a cell is independent of position in the cell cycle, then [from (Eq. (2)] when the volume doubles, the sedimentation rate should increase by $2^{2/3}$ ($=1.59$). Such a population of cells should have sedimentation profile resembling that shown in Fig. 7b.

Thus, the variability in shape of the IgG AFC profile of Fig. 6 can be explained by noting that 4 days after immunization, the time at which the profile was measured, corresponds to the time of maximum number of

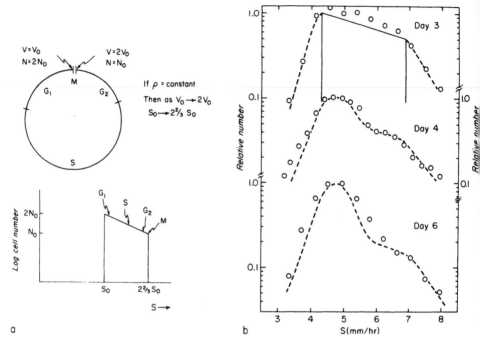

FIG. 7. Relationship of cell growth and division cycle to sedimentation profile. (a) Above: Cell cycle model. M represents the period during which the cell is in mitosis, S the period when the cell is replicating its DNA, and G_1 and G_2 the two intervening periods. Below: Theoretical sedimentation profile for a homogeneous population of cells all in completely asynchronous exponential growth and having cell density independent of position in the cell cycle. (b) IgG antibody-forming cell activity profile measured 3, 4, and 6 days after a second immunization with sheep erythrocytes. The solid line on the day 3 profile is the theoretical curve shown in (a). The dashed line in all three curves represents the theoretical curve broadened by the known finite resolution of the sedimentation equation and includes a progressively increasing noncycling component on days 4 and 6.

IgG AFC. One might expect cell division to stop at about this time. However, if one compares the IgG AFC profiles 3 days and 6 days after immunization, then the AFC should be mostly in or out of cell cycle, respectively (see Fig. 7b). The day 3 profile fits the theoretical distribution rather well. The day 6 profile suggests that, when a cell stops dividing, it does so with a volume close to that of a cell just leaving mitosis. The variability seen on day 4 can be explained as a variable composite population of dividing and nondividing cells. Note that this model for the sedimentation distribution of AFC was based entirely on an analysis of the activity profile. Such an analysis is almost impossible starting from the enrichment profile, since it is determined by the distribution of other (irrelevant) cells as well as that of the AFC.

The definitive test of the cell-cycle model just outlined would be to demonstrate directly that different parts of the profile are in different parts of the cell cycle by scoring for, e.g., mitotic AFC and AFC undergoing DNA synthesis. This direct experiment is difficult because AFC are a minority population in the fractions collected. Thus, in Fig. 6, the most purified fraction is only about 8% AFC (corresponding to an 18-fold enrichment over unfractionated cells), and the most purified fractions obtained in any experiments were still only 28% AFC. (The highest enrichment over control in the series was 110-fold and was not correlated with purification.[9])

A direct test of the model is most simply made by investigating a homogeneous population of cells all of which are in cycle. To this end, the sedimentation profile of mouse L cells, in asynchronous exponential growth in suspension culture, was investigated.[10]

The sedimentation profile from one experiment is shown in Fig. 8. The peak at $s = 4.5$ corresponds to cell fragments and the profile beyond $s = 18$ is largely clumps and polyploid cells. The limits of the theoretical sedimentation distribution, assuming no density variation through the cell cycle and perfect resolution for the sedimentation separation, are indi-

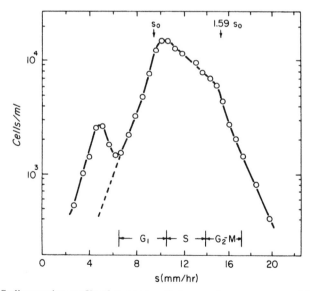

Fig. 8. Sedimentation profile of mouse L cells in asynchronous exponential growth. The dashed line extrapolates the true cell distribution into the region of cell debris.

cated on the figure by arrows. The cells were pulse labeled with [³H]thymidine before separation, and the fractions were scored for percentage labeled cells and percentage mitosis after fractionation. By defining S phase to be the region of the distribution in which the percentage of [³H]thymidine-labeled cells exceeds the asynchronous control and by excluding the outermost 4% of the cell concentration distribution as being clumps and polyploid cells at large s, and debris at small s, one can establish the boundaries of G_1, S, and G_2–M, as shown in Fig. 8. The percentage of cells found in each phase in this manner is in excellent agreement with the values found by other methods.[10]

From Eq. (1), one can see that if all cells have the same density, then $s = kV^{2/3}$, or $\log s = k' + 2/3 \log V$ so that a plot of $\log s$ versus $\log V$ would have a slope of 2/3. Volume distributions were obtained for the L cell fractions using the volume spectrometer described earlier. A plot of $\log s$ versus $\log V_0$, where V_0 is the modal volume of the volume distribution obtained for each fraction, does indeed[10] fit a straight line with slope of about 2/3. However, this plot fails to utilize much of the information contained in the volume distributions. The latter vary in shape as well as modal volume, and this shape variation appears to reflect real properties of the cells being analyzed. This information can all be represented on a contour map in which cell sedimentation velocity and volume are the independent variables and contours of equal cell number are plotted as illustrated in Fig. 9.

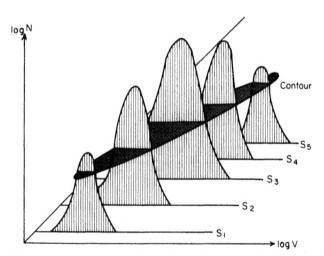

Fig. 9. Schematic representation of the volume distributions for several adjacent fractions from a velocity sedimentation separation indicating how the data are to be combined to form a contour map ("fingerprint") as in Fig. 10.

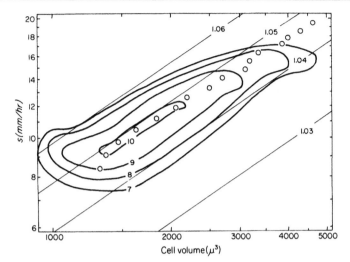

Fig. 10. Fingerprint of mouse L cells in asynchronous exponential growth.

Figure 10 shows such a contour map or "fingerprint" for mouse L cells in exponential growth. The contours of equal cell number represent 2-fold changes in cell concentration. Thus, contour 10 represents 2^{10} cells per milliliter per s interval per V interval. The data points give modal volume as a function of s and can be fitted reasonably well with a straight line parallel to the slanted lines shown, which are contours of equal density calculated from Eq. (1). From Fig. 10, the mean density of the cells is about 1.048 g/cm^3, not significantly different from the density of 1.051 g/cm^3 (MacDonald and Miller[10]) measured on a continuous Ficoll density gradient,[19] and density is approximately independent of position in the cell cycle. However, some very small cells appear to have an anomalously high density, and some very large cells appear to have an anomalously low density. Fingerprints of more complex cell suspensions show much more complicated structure and may give information difficult to obtain in any other way. In particular, such fingerprints may be useful in distinguishing between various myeloproliferative disorders of human bone marrow.[20]

The basic procedures outlined in this chapter have been successfully applied in studies of the cerebellum, pituitary gland, lung, liver, hemopoietic system, sperm development, and various tumor systems as well as cellular immunology.

[19] R. M. Gorczynski, R. G. Miller, and R. A. Phillips, *Immunology* **19,** 817 (1970).
[20] R. Moon, R. A. Phillips, and R. G. Miller, *Ser. Haematol.* **2,** 163 (1972).

[9] Separation of Lymphocytes, Granulocytes, and
Monocytes from Human Blood Using
Iodinated Density Gradient Media

By ARNE BØYUM

Introduction

Cell separation techniques are based on differences in physical and/or biological properties of cells. The physical characteristics such as particle (cell) density (d) and size (or radius = r) determine the sedimentation rate (u) in a liquid medium, as given by the general sedimentation law [Eq. (1)].

$$u = \frac{2r^2(d - d_0)G}{9\eta} \tag{1}$$

in which G denotes the centrifugal acceleration or the acceleration of gravity, d_0 the density, and η the viscosity of the medium through which the particle falls.

In order to use this equation to calculate sedimentation rates of blood cells or other cells, it may be necessary to introduce a factor correcting for deviation from spherical shape. In any case, it demonstrates that the sedimentation rate depends strongly on cell size. Nevertheless, techniques for separation of the major subgroups of white blood cells, are mostly based on distinct density differences. Mononuclear cells (monocytes and lymphocytes) have a lower density than granulocytes and erythrocytes. By layering the blood over a separation fluid of intermediate density, it is possible to separate the blood cells into two fractions by centrifugation. For these separations, compounds originally developed as iodinated X-ray contrast media are often used as the gradient material,[1-3] but other gradient media are also available.[4] After centrifugation, the mononuclear cells float on top of the separation fluid, whereas the other cells have sedimented to the bottom of the tube. In a second step, granulocytes can be separated from the erythrocytes by washing and dextran

[1] A. Bøyum, Scand. J. Clin. Lab. Invest. 21, Suppl. 97, 77 (1968).
[2] A. Bøyum, Tissue Antigens 4, 269 (1974).
[3] A. Bøyum, Scand. J. Immunol. 5, Suppl. 5, 9 (1976).
[4] J. T. Kurnick, L. Østberg, M. Stegagno, A. K. Kimura, A. Ørn, and O. Sjøberg, Scand. J. Immunol. 10, 563 (1979).

sedimentation at unit gravity. It is even possible to do this separation using a one-step procedure.[5,6]

Monocytes have a lower density than other cells in human blood and may be separated by density gradient centrifugation. The average monocyte purity is reported to vary from 60 to 90%, with lymphocytes as the main contaminant. The variability in purity is due to the overlapping densities of these two cell populations.[7] However, the monocyte purity can be improved by using a hyperosmotic density gradient medium.[8] When the osmolality increases, the cells expel water and shrink. The density of the cells then increases, they move faster, and may even pass the density barrier that was present initially. These events can take place only if the cell membrane is semipermeable, allowing free passage of water, whereas there is no or only negligible (net) transport of solutes across the cell membrane. The basis for using these principles for cell separation is that the various cell types have different osmotic sensitivities as a result of, for instance, a different water content. In hypertonic solutions a cell with a high water content will increase its density more than a cell with a low water content. However, an apparently different osmotic sensitivity could also be due to a delayed passage of water through the cell membrane. In any event, the lymphocytes that overlap in density with monocytes appear to be more sensitive to an increase of osmolality than the monocytes. Thus with the correct combination of density and osmolality all of the cells loaded onto the gradient, except for the low-density fraction of the monocytes, move to the bottom during centrifugation. A monocyte fraction of 95–98% purity can then be collected from the top of the gradient layer.

Materials and General Methods

1. Ficoll 400 and dextran 500 (Pharmacia, Uppsala, Sweden). Dextran is used as a 6% solution in 0.9% NaCl.

2. Methylcellulose (Methocel, standard, 25 cp, Dow Chemicals, Midland, MI) is used as a 2% solution in 0.9% NaCl.

3. Iron particles in suspension (Lymphocyte Separating Reagent, Technicon Instruments Co, Tarrytown, N.Y.).

4. Lymphoprep (9.6% Isopaque = sodium metrizoate, 5.6% Ficoll, density = 1.077 g/ml), Nyegaard & Co, Oslo, Norway. Modifications of

[5] D. English and B. R. Andersen, *J. Immunol. Methods* **54,** 43 (1974).
[6] J. Guidalli, P. J. M. Philip, P. Delque, and P. Sudaka, *J. Immunol. Methods* **54,** 43 (1982).
[7] H. Loos, B. Blok-Shut, R. van Dorn, R. Hoksbergen, A. B. de la Riviere, and L. A. Meerhof, *Blood* **48,** 731 (1976).
[8] A. Bøyum, *Scand. J. Immunol.* **17,** 429 (1983).

this separation fluid, in which the iodinated X-ray contrast medium (Isopaque) has been replaced by Hypaque are available as Ficoll–Paque (Pharmacia), LSM solution (Bionetics, Kensington, Md), or Histopaque (Sigma).

5. Nycodenz, a nonionic, iodinated gradient medium (Nyegaard & Co). A 27.6% stock solution has a density of 1.148 g/ml and osmolality of 290 mOsm. By mixing Nycodenz with NaCl solutions of varying concentration, the final density and osmolality can be adjusted independently of each other over a wide range of densities. To obtain a separation fluid with a chosen density, V parts of Nycodenz are mixed with V_1 parts of a NaCl solution, V_1 being calculated from Eq. (2):

$$Vd + V_1 d_1 = (V + V_1) d_2 \qquad (2)$$

where d is the density of Nycodenz (1.148 g/ml), d_1 the density of the NaCl solution, and d_2 the density of the mixture. The greatest accuracy is obtained by using weighed fractions of the two components rather than by measuring volumes. Still it is often necessary to adjust the density before use. The desired osmolality is obtained by varying the concentration of NaCl. An alternative procedure for preparing solutions of defined osmolality is to prepare two solutions of the same density, and high and low osmolalities respectively. By mixing these solutions in appropriate proportions, the required osmolality can be obtained. A ready made solution (1.068 g/ml, 335 mOsm) for monocyte separation is available.

6. Percoll (Pharmacia). A gradient medium for separation of mononuclear cells and granulocytes is prepared by mixing nine volumes of the stock solution with one volume of 9% NaCl. The density is measured and the suspension is further diluted with 0.9% NaCl to the desired density (1.077 g/ml) as described for Nycodenz [Eq. (2)].

Human blood is collected in vacutainers containing EDTA as anticoagulant.

Removal of Erythrocytes by NH₄Cl Lysis

Pelleted cells (granulocytes and erythrocytes) are resuspended in 0.83% NH_4Cl, 10 mM HEPES buffer (pH 7), and incubated for 7 min at 37°. Thereafter the cells are centrifuged for 7 min at 650 g and the supernatant removed.

Measurement of Density

The density of solutions can be measured using a 10 or 5 ml pycnometer. It is important to check the accuracy of the pycnometer by weighing it with and without water, and to make allowance for the temperature. Alternatively, one can determine the density electronically with a densimeter (DMA 40, Anton Paar, Graz, Austria), based on the variation of the natural frequency of a hollow oscillator, when filled with liquids of varying densities.

Measurement of Osmolality

The osmolality can be determined from freezing point depression measurements using an osmometer (e.g., Advanced Instruments Inc., Needham Heights, MA).

Defibrination to Remove Platelets

Defibrination is carried out by shaking 12 ml of blood in a tube containing 25 glass beads 2 to 3 mm in diameter. Larger volumes can be defibrinated by manual swirling of 100-ml flasks (10 min) containing 50 ml of blood and 15 glass beads (5 mm in diameter).

Cell Separation Procedures

Separation of Blood Leukocytes

Unfractionated leukocytes may be obtained simply by leaving blood treated with an anticoagulant to sediment in a test tube or beaker at room temperature. The plasma layer containing approximately 70% of the leukocytes may be collected after 1–2 hr when the red cells have settled. The sedimentation is more rapid if an erythrocyte-aggregating agent is added, as follows: (1) 10 parts blood + 1 (or 2) parts dextran (6%); (2) 10 parts blood + 0.5 (or 1) part methylcellulose (2%). The sedimentation of erythrocytes takes 15–40 min depending on the height (50–80 mm) of the blood cell column. The sedimentation tends to slow down with a tube diameter of less than 10 mm. The red cell contamination amounts to 1–5 per leukocyte. The separated leukocytes may be used for functional studies. More often, however, this procedure is used as a first step in techniques designed to separate subgroups of white cells.

Separation of Mononuclear Cells and Granulocytes

Principle. Mononuclear cells (monocytes and lymphocytes) have a lower density than erythrocytes and granulocytes. Thus, after centrifugation of blood cells loaded on a separation fluid of appropriate density (1.077 g/ml), the mononuclear cells remaining at the top are easily collected. The granulocytes in the bottom fraction may next be separated from the erythrocytes by dextran sedimentation.

Separation of Cells from Whole Blood. The Isopaque–Ficoll (IF) technique is shown in Fig 1. Defibrinated or treated with anticoagulant human

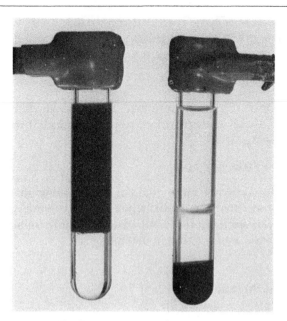

Fig. 1. The Isopaque–Ficoll separation technique. Equal parts of anticoagulated or defibrinated blood and 0.9% NaCl are mixed, and 6–8 ml of this mixture is layered over 3 ml of Isopaque–Ficoll in a centrifuge tube with inner diameter of 12–14 mm. After centrifugation at room temperature for 20 min at 1900 rpm (600 g at the bottom of the tube) mononuclear cells (and platelets) band at the top of the separation fluid, while erythrocytes and granulocytes form a pellet at the bottom of the tube. The mononuclear cells are collected and cells at the bottom are suspended in 6 ml of 0.9% NaCl and centrifuged (7 min at 650 g). The supernatant is removed, and the original blood volume (3 ml) is restored by the addition of 0.9% NaCl or medium; 0.3 ml of 6% dextran is then added. The supernatant containing the granulocytes is harvested when the red cells have settled.

blood is mixed with an equal volume of 0.9% NaCl. Using a Pasteur pipet or a graduated 10-ml pipet, 6 to 8 ml of this mixture is layered over 3 ml of IF in a centrifuge tube with an inner diameter of 12 to 14 mm, by keeping the pipet against the tube wall 2 to 4 cm above the fluid meniscus. To minimize mixing of the two layers the blood should flow out of the pipet continuously. After centrifugation for 20 min at 600 g the blood cells will have separated into two fractions: a white layer consisting of mononuclear cells and platelets at the interface region, and a pelleted fraction containing the erythrocytes and granulocytes. The supernatant is first removed down to 3–4 mm above the white band. The mononuclear cells are then collected with a Pasteur pipet, together with the top half of the separation fluid. The cells seem to be located mostly around the periphery of the tube, but to ensure complete removal of cells the pipet should be moved over the whole cross-sectional area. The granulocytes can be iso-

TABLE I
ISOLATION OF MONONUCLEAR CELLS FROM WHOLE BLOOD[a]

Sample	Differential counts (%)				Yield of mononuclear cells (%)	Erythro- cytes/100 cells
	Granulo- cytes	Lympho- cytes	Mono- cytes	Baso- phils		
EDTA-blood	0.8	81.2	17.4	0.6	98 ± 5	2.6
Defibrinated blood	0.9	91.2	7.7	0.2	55 ± 3	33

[a] Four milliliters of EDTA-blood or defibrinated blood was mixed with 4 ml 0.9% NaCl. Each mixture (8 ml) was layered over 3 ml Isopaque–Ficoll and centrifuged at 600 g for 20 min. The table gives differential counts of white cells in the top fraction as mean of 20 separations. The erythrocyte contamination is given in percentage of total number of leukocytes and erythrocytes. The yield (±SE) was calculated as the number of cells recovered, in percentage of cell numbers in undefibrinated blood.

lated from the bottom fraction by washing once with 0.9% NaCl, reconstitution of the original volume (4 ml), and dextran sedimentation. However, an alternative procedure is recommended for granulocyte isolation (see below).

As shown in Table I, an almost pure suspension of mononuclear cells is obtained with EDTA-blood, composed of approximately 81% lymphocytes and 18% monocytes. The yield of mononuclear cells is close to 100% as compared to 50 to 70% with undiluted blood. To minimize cell loss during washing it is probably an advantage to use conical tubes. A lower cell yield (55%) and higher erythrocyte contamination are obtained with defibrinated blood. Defibrination causes a selective loss of monocytes, thus yielding a purer lymphocyte suspension. It has been reported that the distribution of lymphocyte subpopulations is not affected by defibrination.[9] Platelet contamination is avoided by using defibrinated blood. This procedure has been further standardized for tissue typing.[10] Any type of anticoagulant can be used.

Larger volumes (30 to 50 ml) may be separated in one tube of appropriate diameter. However, the yield of mononuclear cells may decrease, because it becomes more difficult to remove the cells efficiently from the interface region.

[9] H. E. Johnsen and M. Madsen, Scand. J. Immunol. 8, 239 (1978).
[10] E. Thorsby and A. Bratli, in "Histocompatibility Testing-1970" (P. I. Terasaki, ed.), p. 655. Munksgaard, Copenhagen, 1970.

This technique is generally applicable to blood lymphocyte isolation, but in some species (cow, goat, rabbit), erythrocyte removal by the standard procedure is not satisfactory. This difficulty is overcome by increasing the speed of centrifugation to 1200 g. Another possibility is to enhance the aggregation and thereby the sedimentation of red cells by adding methylcellulose to the Ip-Ficoll solution to a final concentration of 0.1%. This is a convenient procedure for isolating lymphocytes from mouse blood.

Separation of Cells from Leukocyte-Rich Plasma. For separation of cells from large blood volumes (>50 ml), it is preferable to remove erythrocytes by dextran sedimentation in an initial step, although this entails about a 30% loss of leukocytes. Up to 8 ml of the leukocyte-rich plasma layered over 3 ml of the separation fluid may be separated in one tube. After centrifugation, the mononuclear cells are found at the interface region. The granulocytes at the bottom are contaminated by 2–5 erythrocytes per granulocyte. If desired, the red cells can be removed by NH_4Cl lysis. With citrated blood the three different separation fluids gave approximately similar purity and yields of mononuclear cells (Table II). Similar results were obtained with EDTA-treated blood. In other experiments it was found that Isopaque–Ficoll and Percoll solutions tended (borderline significance) to yield slightly more pure granulocyte suspensions than Nycodenz–NaCl; this was even more promounced when using

TABLE II

SEPARATION OF MONONUCLEAR CELLS AND GRANULOCYTES FROM LEUKOCYTE-RICH PLASMA[a]

Sample	Osmolality (mOsm)	Differential counts (%)			Cell yield	
		Granulo-cytes	Lympho-cytes	Mono-cytes	Granulo-cytes cells	Mono-nuclear cells
Nycodenz–NaCl	314	A 0.1 ± 0.1	77.7 ± 3.8	21.8 ± 3.8		88 ± 9
		B 96.7 ± 1.1	2.2 ± 0.6	1.0 ± 0.7	64 ± 9	
Isopaque–Ficoll	312	A 1.3 ± 0.9	75.3 ± 3.7	22.7 ± 4.2		97 ± 16
		B 99.7 ± 0.2	0.3 ± 0.2		57 ± 7	
Percoll	325	A 1.3 ± 0.7	70.4 ± 3.9	28.1 ± 4.4		105 ± 14
		B 99.0 ± 0.4	1.0 ± 0.4		57 ± 9	

[a] Ten parts of citrated blood were mixed with 1 part dextran 6% (in 0.9% NaCl). After sedimentation of red cells, 2 ml of the leukocyte-rich plasma was layered over 3 ml of the separation fluid and centrifuged for 20 min at 600 g. Mononuclear cells were harvested from the interface region, and granulocytes from the bottom of the tube. A, Top fraction. B, Bottom fraction. The cell yield is the number of cells recovered as percentage of the total number of cells layered on the separation fluid. Mean values (±SE) from five separations.

TABLE III
ISOLATION OF MONONUCLEAR CELLS AND GRANULOCYTES FROM
LARGE BLOOD VOLUMES[a]

| Fraction | Differential counts (%) | | | | Erythrocytes/ 100 cells |
	Granulo- cytes	Lympho- cytes	Mono- cytes	Baso- phils	
Top	0.6	77.3	20.4	1.7	3.7
Bottom	99.2	0.7	0.05	0.05	70.0

[a] Five hundred milliliters of citrated blood was mixed with 50 ml Dextran 6%. After sedimentation of red cells, the plasma layer was removed and centrifuged for 8 min at 500 g. The pellet was resuspended in 16 ml of the supernatant, and 8 ml was layered over 3 ml Isopaque–Ficoll and centrifuged for 20 min at 600 g. Average of five separations with blood from three individuals.

citrated blood from the blood bank that was stored 3–5 hr after withdrawal.

For large volume of leukocyte-rich plasma, the most practical procedure may be to concentrate the cell suspension before separation with IF. This is demonstrated in Table III which shows the results when the cells were harvested from 500 ml blood. The purity of the cells recovered from the top and bottom fractions was similar to that obtained with smaller blood volumes, and approximately 60% of the cells in blood was recovered.

It is inevitable that some cell loss occurs during repeated washings, particularly for granulocytes. This loss can partly be prevented if the washing fluid contains 2–5% EDTA-plasma.

Isolation of Lymphocytes

Principle. Lymphocytes and monocytes copurify when blood is fractionated using the Isopaque–Ficoll method (Tables I, II, and III). By capitalizing on the ability of monocytes to engulf collodial iron particles, their density can be increased sufficiently to enable them to sediment through the gradient medium, whereas the lymphocytes remain at the top.

The procedure is as follows: 10 ml of heparinized blood is centrifuged for 10 min at 600 g. The leukocyte layer resting on top of the erythrocyte pellet is removed (1 ml) and mixed with 1 ml of the collodial iron suspension (this volume [30]). To ensure a high concentration of heparinized plasma the buffer in which the iron particles are suspended may be re-

TABLE IV
PREINCUBATION WITH IRON PARTICLES TO REMOVE MONOCYTES[a]

| | Differential counts (%) | | | | |
Sample	Granulo-cytes	Lympho-cytes	Monocytes	Basophils	Lymphocyte yield
Control	0.1 ± 0.04	78.3 ± 3.8	20.9 ± 3.8	0.7 ± 0.2	106 ± 8
Incubation with particles	2.6 ± 1.2	96.1 ± 1.2	1.0 ± 0.2	0.3 ± 0.1	55 ± 6

[a] Leukocytes from whole blood obtained by centrifugation were incubated with iron particles and thereafter separated with Isopaque–Ficoll in parallel with a control group in which 4 ml of a 1 : 1 mixture of heparinized blood and 0.9% NaCl was loaded on the gradient medium. The lymphocyte yield is calculated as the number of cells recovered in percentage of lymphocytes in unseparated (uncentrifuged) blood. Mean ± SE from six separations.

placed by plasma. The mixture is incubated in a shaking bath at 37° for 30 min. If needed, cells from several 10-ml tubes can be mixed, but the height of the mixture should not exceed 8–10 mm. After 30 min the cell suspension is mixed with an equal volume of 0.9% NaCl and 4–6 ml is layered on top of 3 ml IF solution and centrifuged. As shown in Table IV an almost pure suspension of lymphocytes is obtained from the interface.

Purification of Monocytes

Principle. Monocytes have a lower average density than lymphocytes but, because the densities of the two types of cells overlap, it is difficult to establish a satisfactory separation with good yield based on density differences alone. However, the separation of monocytes and lymphocytes can be improved by increasing the osmolality of the gradient medium. The cells then expel water, shrink, and their density increases. In this respect lymphocytes are more sensitive than monocytes, and thus they sediment further during centrifugation, whereas the monocytes remain at the top of the gradient.

Isolation of Monocytes from Leukocyte-Rich Plasma Using Nycodenz-NaCl Solutions of Varying Density and Osmolality. The separation technique is illustrated in Fig. 2. Leukocyte-rich plasma (2–6 ml) is layered over 3 ml of Nycodenz–NaCl solution and centrifuged at room temperature for 15 min at 600 g. The tubes are then carefully removed from the centrifuge. There is no distinct band at the interface after centrifugation, but the separation fluid itself has a grayish color (Fig. 2), caused by nonpelleted platelets. First, the clear plasma down to 3–4 mm above the

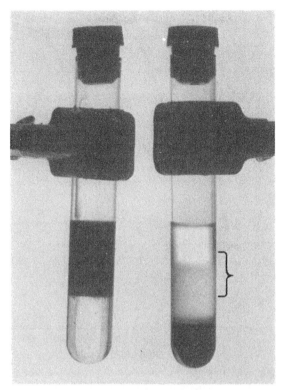

FIG. 2. Illustration of the monocyte separation technique. A sample of 3 ml of EDTA-treated blood (as shown) or 2–6 ml of leukocyte-rich plasma is layered over 3 ml of the separation fluid, and centrifuged for 15 min at 600 g. After centrifugation the clear plasma is removed down to 3–4 mm above the interface. Thereafter, as indicated by the brace, the remaining plasma together with slightly more than half the volume of the separation fluid are collected. The cells are counted and smears made for differential counting.

interface between plasma and the separation fluid is removed. Next, the remaining plasma together with slightly more than half the volume of the separation fluid is collected and the cells are counted before making smears for microscopic examination. Erythrocytes and granulocytes have sedimented to the bottom during centrifugation, and the cells in the upper half of the Nycodenz solution are almost exclusively lymphocytes and monocytes. The results with Nycodenz–NaCl solutions of 3 different densities are shown in Fig. 3. The osmolality is increased stepwise by 2.5–4%, which theoretically would increase the density of the cells by 0.001–0.002 g/ml. This is sufficient to cause a striking improvement of monocyte purity to 95–98%, at each density level tested. As the density increases, however, it is necessary to adjust the osmolality to a higher level.

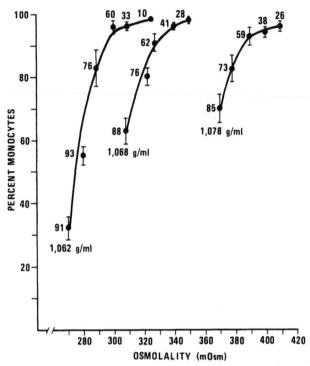

FIG. 3. Monocyte separation with Nycodenz–NaCl solutions at three different densities. Monocyte separation from leukocyte-rich plasma (1.5 ml) layered over Nycodenz–NaCl solutions at 3 different densities, with varying osmolality. Mean values (± SE) from five separations. The monocyte yield, indicated at each experimental point, is defined as the percentage of the total number of monocytes applied on top of the separation fluid which is recovered at the interface region after centrifugation.

Calculations by means of the sedimentation law [Eq.(1)] has confirmed that the small density increase resulting from a 3% increase of osmolality would enable the lymphocytes to reach the bottom of the tube with the present centrifugation procedure, provided their initial density is equal to that of the gradient medium. Altogether, it appears that the low-density fractions of the lymphocytes are somewhat more sensitive to an increase of osmolality than the monocytes. However, the monocytes are not completely unaffected, and there is a concomitant decrease of monocyte yield as the osmolality and purity increase. The variability in several separations, using Nycodenz solutions with a density of 1.062 g/ml, is further demonstrated in Table V. The wide range of yield of monocytes reflects the variability of monocyte concentration in blood. There is no difference

TABLE V
VARIABILITY OF THE MONOCYTE SEPARATION PROCEDURE[a]

Osmolality (mOsm)	Number of observations	Monocyte purity (%)	Monocyte yield (%)	Number of monocytes recovered from 1 ml plasma (10^{-3})
300	45	93 (83–100)	50 (19–98)	311 (84–788)
310	31	95 (87.5–98.5)	24 (9–60)	141 (38–589)

[a] Leukocyte-rich plasma was obtained by dextran sedimentation, and 1.5–2 ml was layered over 3 ml of Nycodenz separation fluid (density of 1.062 g/ml), and centrifuged for 15 min at 600 g. Mean values and ranges are shown.

in terms of either purity or yield of monocytes whether 1.5 or 6 ml plasma is separated in one tube.

Occasionally the monocyte fraction may be contaminated by a few (1–5%) granulocytes, possibly due to swirling and upward streaming of the low-viscous Nycodenz solution during deceleration. To avoid or reduce this contamination a steady, smoothly running centrifuge should be used.

Isolation of Monocytes from Defibrinated Blood. The purity of monocytes using defibrinated blood is essentially similar to that obtained with EDTA-treated blood (Table VI). However, the yield of monocytes is reduced by more than 50%, probably due to a selective loss of monocytes during defibrination. The advantage of defibrination is that the platelets

TABLE VI
MONOCYTE SEPARATION FROM DEFIBRINATED BLOOD[a]

Osmolality (mOsm)	Differential counts (%)		Monocyte yield (%)	"Erythrocytes" per 100 cells
	Lymphocytes	Monocytes		
288	15 ± 5	85 ± 5	78 ± 6	57 ± 10
300	4 ± 1	96 ± 1	39 ± 6	73 ± 4
309	3 ± 2	97 ± 2	17 ± 5	73 ± 1

[a] Following dextran sedimentation of defibrinated blood, 1.5 ml of the plasma was layered over Nycodenz with a density of 1.061 g/ml and centrifuged. Mean values (± SE) from five experiments.

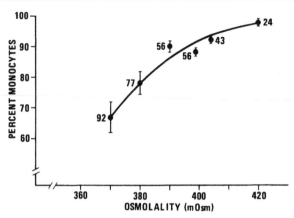

Fig. 4. Monocyte separation from whole blood. Three milliliters of EDTA-blood was layered over 3 ml of Nycodenz–NaCl solution (density 1.078 g/ml), and centrifuged for 15 min at 600 g. Mean values from 4–5 separations.

are removed. On the other hand, there is a considerable increase of erythrocyte contamination. It appears that the defibrination procedure causes deformation of some erythrocytes and these tend to cosediment with the mononuclear cells.

Isolation of Monocytes from Whole Blood by a One-Step Procedure. Three milliliters of EDTA-blood is layered over 3 ml of a Nycodenz–NaCl solution with density of 1.078 g/ml (or 1.068 g/ml), and centrifuged. As shown in Fig. 4, the overall pattern is similar to that obtained with leukocyte-rich plasma, the purity being improved by increasing the osmolality.

Anticoagulant. A high purity has only been obtained with EDTA as anticoagulant, and with leukocyte-rich plasma from defibrinated blood. With heparinized blood the purity is markedly reduced, and with citrated blod there is no monocyte enrichment at all.

Removal of Platelets from Monocyte Preparations. The platelet contamination in the monocyte suspension can be reduced by differential centrifugation. To this end 10 ml of anticoagulated blood is centrifuged at 200 g for 10 min. The supernatant which contains the majority (70–80%) of the platelets is removed and centrifuged at 2000 g for 20 min to obtain platelet-depleted plasma in which to resuspend the cells. If required, the procedure may be repeated before the separation of monocytes. Platelets can be removed from leukocyte-rich plasma, obtained by dextran sedimentation by centrifugation for 10 min at 100 g before separation using the Nycodenz separation fluid. Still another possibility is to remove the platelets as a final step, after Nycodenz separation. A washing fluid consisting of 0.9% NaCl containing 5% of EDTA-plasma is suitable for this purpose, and centrifugation is carried out at 80 g for 10 min. It is inevitable that some loss of cells occurs during differential centrifugation procedures.

Viability. Viability has proved satisfactory, as evaluated by phagocytosis of latex particles and *E. coli.* The phagocytic capacity is not influenced by Nycodenz solutions up to 600 mOsm.

Standard Separation Procedures. Monocyte separation from EDTA-blood is carried out directly by using a Nycodenz–NaCl solution with density of 1.068 or 1.078 g/ml. With leukocyte-rich plasma a density of 1.062 or 1.068 g/ml is chosen. Recipes for making separation fluids of appropriate densities and osmolalities are given in Table VII. It may be preferable to have complete control of density and osmolality from measurements. This is best done by preparing two stock solutions for each density level, one with a high and one with a low osmolality: (1) 11.48 g of Nycodenz 27.60% + 14.27 g of 0.7% NaCl; (2) 11.48 g of Nycodenz 27.60% + 14.80 g of 1% NaCl.

The densities are measured, adjusted to 1.062 g/ml, and the osmolality is determined. By mixing these two stock solutions, separation fluids of the desired osmolality (300, 310 mOsm) may be obtained. Stock solutions at the 1.068 and 1.078 g/ml density may be prepared in a similar way by

TABLE VII
RECIPES FOR NYCODENZ–NaCl SEPARATION FLUIDS[a]

NaCl solution		Final density (g/ml)	Final osmolality (mOsm)
Concentration (%)	Weight (g)		
0.85	14.51	1.062	300
0.91	14.65	1.062	310
1.02	12.65	1.068	330
1.08	12.75	1.068	340
0.90[b]	9.57	1.077	310
1.4	10.1	1.078	390
1.47	10.14	1.078	400
1.53	10.2	1.078	410

[a] The grams of a NaCl solution to be mixed with 10 ml (11.48 g) of Nycodenz (27.6%) for obtaining a separation fluid of desired density and osmolality.

[b] Gradient medium for separation of mononuclear cells (lymphocytes and monocytes).

changing the concentrations of the NaCl solution used for diluting Nycodenz (Table VII).

For separations from small blood volumes, 3 ml of EDTA-blood is layered over 3 ml of the separation fluid. For larger volumes, dextran sedimentation is preferable, and 6 ml of the cell-rich plasma may be separated in one 12-ml tube. Even larger cell numbers may be separated in one tube if the cells are concentrated by centrifugation before separation with Nycodenz. This separation is carried out by centrifugation for 15 min at 600 g, and the monocytes are recovered from the interface region. In case of unsatisfactory separation it may be necessary to adjust centrifugation conditions with respect to speed and check the accuracy of density and osmolality measurements.

[10] Analytical and Preparative Equilibrium Density Separation of Lymphoid Cells on Albumin and Metrizamide

By KEN SHORTMAN

Introduction

The buoyant density of a cell reflects its average chemical composition, with the major determinant being the relative water content or relative dry weight. Cell buoyant density is fundamentally independent of cell size, although large cells often tend to have a relatively higher water content and for this reason are often of low buoyant density. Certain cell types may readily be separated on this basis (e.g., mature polymorphs from mononuclear cells in blood). However, B and T lymphocytes and their specialized subsets generally show overlapping distribution patterns and, except in special cases, are difficult to separate from each other by this parameter. Buoyant density separation finds its greatest use in the separation of distinct developmental states or activation states within any one cell lineage. The method is capable of very high precision and resolution and the buoyant density of a given cell type can be reproduced to $\pm$ 0.0003 g/cm^3 over a period of years. High resolution gradients generally produce more physically distinct peaks and subsets than can be defined by our present immunological criteria, and many of these may not be physiologically significant. Separation into around 15 fractions usually gives the maximum usable biological information but more fractions are needed to resolve all the minor peaks. Once the cell distribution pattern is estab-

METHODS IN ENZYMOLOGY, VOL. 108

lished a great deal can be accomplished with simple nongradient neutral density cuts involving a rapid centrifugation in defined density media. These same simple density cuts may be used for eliminating dead cells and erythrocytes from lymphoid cell suspensions. These aspects have been reviewed elsewhere.[1-5]

General Considerations

Osmolarity and pH Control

The physical properties of a cell vary with external conditions and cell buoyant density is very sensitive to such changes. For this reason the osmolarity, pH, temperature, and composition of the gradient media need to be defined very precisely for reproducible results. The direct osmotic contribution of the gradient substance itself is reduced by choosing a material of high molecular weight. In the case of albumin (or say Ficoll) the direct osmotic pressure contribution is minimal, but in the case of metrizamide it is very important and in high density solutions most of the normal salts must be omitted to maintain osmotic balance. Osmolarity can also be markedly affected by salt or water contaminants in the substance used to make the gradient (especially with albumin), by the anions or cations added to adjust the pH of the medium (in the case of albumin at neutral pH), by the breakdown of the gradient material (by light in the case of metrizamide), or by interaction of the gradient material with water in the gradient salt solution (as in the case of Ficoll). It should also be noted that serum osmolarity varies widely with species, and while it is possible to separate, say, mouse cells at human tonicity, optimum results are obtained at the osmolarity appropriate to the species. pH has considerable indirect osmotic effects, causing swelling or shrinkage of cells. Due to the selective permeability of erythrocytes to anions and the Donnan equilibrium across the membrane, a lowering of pH (as in unneutralized albumin media) produces a change in the charge of intracellular hemoglobin which in turn causes the erythrocyte to accumulate anions, and thus to swell and become less dense. Other cells with a damaged but continuous membrane will show a somewhat similar behavior, and will swell at low pH. For this reason selective removal of red cells and certain dam-

[1] K. Shortman, *Annu. Rev. Biophys. Bioeng.* **1**, 93 (1972).

[2] K. Shortman, *Contemp. Top. Mol. Immunol* **3**, 161 (1974).

[3] K. Shortman, H. von Boehmer, J. Lipp, and K. Hopper, *Transplant. Rev.* **25**, 163 (1975).

[4] K. Shortman, J. M. Fidler, R. A. Schlegel, G. J. V. Nossal, M. Howard, J. Lipp, and H. von Boehmer, *Contemp. Top. Immunobiol.* **5**, 1 (1976).

[5] K. Shortman, N. Williams, and P. Adams, *J. Immunol. Methods* **1**, 273 (1972).

aged cells from lymphoid cell suspensions is best carried out at neutral, rather than acid pH. Under the stresses of density gradient centrifugation normal lymphocytes also show some changes in density with pH, although this effect is less than with damaged cells or erythrocytes. These points have been the subject of more detailed papers and reviews. [1,2,5-7]

Cell–Cell Interaction and Resolution

Overt cell aggregation or less visually obvious cell to cell interaction is a major factor limiting density separation of cells, and determines the maximum effective cell load on a gradient. Unfortunately most macromolecules used in gradient media markedly increase cell interaction at neutral pH. This can be reduced by working at a somewhat acid pH (e.g., using unneutralized albumin at pH 5.1, near its isoelectric point), by adding dispersing agents, or by using a medium with an inherent dispersing effect (such as metrizamide). The mechanical aspects of the system can help overcome cell interaction problems. It is helpful to keep the cells as dispersed as possible from the beginning, by directly introducing them into the gradient rather than adding them as a concentrated layer. Relatively high centrifugal fields, (say 4000 g), higher than needed just to move cells through the viscous gradient, will help pull apart cells differing in buoyant density. Interaction of sedimenting cells with the centrifuge tube walls (since the cells move in a radial field and the tube walls are parallel) is also a complicating effect; this can be overcome by starting centrifugation with cells near the bottom of the tube, dispersed in the dense medium, so most float upward away from the tube walls. These problems have been considered in detail by us elsewhere,[1-3,8] and some alternative solutions given by Leif.[9]

Choice of Gradient Medium

The above considerations of osmolarity control, pH control, and reduction in cell association, together with convenience in handling and preparation, determine the choice of gradient medium. Albumin is less convenient than many media since salt and water may first need to be eliminated from some commercial preparations, because preparation of concentrated solutions takes time, and because the final media are susceptible to bacterial contamination unless kept frozen. In addition, the

---

[6] N. Williams, N. Kraft, and K. Shortman, *Immunology* **22**, 885 (1972).

[7] N. Williams and K. Shortman, *Aust. J. Exp. Biol. Med. Sci.* **50**, 133 (1972).

[8] K. Shortman, *Aust. J. Exp. Biol. Med. Sci.* **46**, 375 (1968).

[9] R. C. Leif, *Methods Cell Sep.* **1**, 181 (1977).

use of albumin at neutral pH requires neutralization and then care to maintain osmotic balance, and introduces the problem of aggregation at pH 7. Nevertheless, unneutralized at pH 5.1, albumin gives excellent resolution even at high cell loads, good viable cell recovery, only moderate viscosity, good control of osmolarity, and linear density gradients. Where it is necessary to work at pH 7 (for example to maximize the density of erythrocytes or damaged cells) metrizamide is an excellent choice, being inherently a good dispersant, and providing media of low viscosity which are easy to prepare. Disadvantages are the need to balance the significant osmotic contribution of metrizamide itself, and the need to protect it from light. Percoll is another choice, allowing relatively easy preparation of isoosmotic low viscosity media. However the nonlinearity of density gradients can be a problem with Percoll, due to the way the material itself redistributes even in a moderate centrifugal field and for this reason it becomes impossible to define the effective density limits of simple density cuts. Ficoll is another alternative offering relatively easy preparation of medium and presenting no problems in defining densities. However it is of high viscosity, it enhances aggregation to the point where additional dispersing agents and low pH are required for good resolution, and its interaction with water markedly increases the effective osmotic pressure of the balance salt solutions in the high density regions of gradients.[6] In this survey we will use unneutralized albumin at acid pH for continuous gradients, with metrizamide as an alternative choice for neutral pH gradients or density cuts.

Continuous Gradients, Discontinuous Gradients,
  and Simple Density Cuts

A continuous density gradient, with a large number of fractions cut after centrifugation to equilibrium, will always provide the maximum information, the maximum resolution, and will have the maximum cell load. It is also the most work. For this reason discontinuous gradients, consisting of four or five layers of differing density media, are often used. The concentration of cells at the interfaces after centrifugation is considered an added attraction, being visually pleasing and aiding the collection of 4 or 5 fractions. It should be remembered that these bands of cells are quite arbitrary cuts, essentially artifacts of the technique, rather than natural subpopulations. In addition, since it is these sharp narrow zones of density change which carry almost all the cells and constitute the separation zones, the cell capacity of discontinuous gradients for effective separation is much less than for continuous gradients.[1,2,9] Cell capacity is actually increased by mixing or diffusing the sharp interfaces, thereby increasing the actual volume which acts as a true gradient.[5] In many applications one or two rapid density cuts, using only a single density solution but dispersing cells into the medium and avoiding concentration of cells at sharp interfaces, can be even simpler than discontinuous gradients and at the same time can allow high cell loads with very clear separation between cells above or below a given buoyant density. This approach is most effective if the actual density distribution of the cells required is known from a prior continuous gradient analysis, since a rational decision on the exact density of the cuts can then be made.

Preparation and Calibration of Gradient Media

Osmolarity Control

The media given below are designed to be isoosmotic with mouse serum[6] (308 mos$M$ equivalent to 0.168 M NaCl). Media isoosmotic with human serum (269 mos$M$ equivalent to 0.147 M NaCl) may be obtained by further dilution; alternatively the appropriate concentrations for human osmolarity are given in parenthesis. In these protocols isoosmotic conditions are maintained by the principle of mixing isoosmotic solutions of subcomponents. The final osmolarity of the stock gradient media should be checked using a vapor pressure osmometer, with 0.308 M mannitol as a standard (0.269 M mannitol for human osmolarity). Depending on the albumin preparation, it may be possible to eliminate the initial dialysis and drying steps if the final osmolarity checks indicate salt contaminants or excessive moisture are not a problem.

Density Determinations

The density of the light and dense media for continuous gradients, of the media used for density cuts, and of the standards used for continuous gradient analysis, are determined at 4° either by standard gravimetric procedures, or by using a digital density meter (K. Paar, K.G., Graz, Austria). For gravimetric determination 25 ml weighing bottles are filled with 4° equilibrated solutions, with care to avoid bubbles in viscous media. The actual weighing may be performed rapidly at room temperature, the weight changes due to condensation of water on the bottle surface being negligible if large volume bottles are used. The bottles are volume calibrated by weighing dry and empty, then weighing filled with distilled water, using a similar approach of preequilibration at 4°, then rapid weighing at room temperature. The absolute density of media is then calculated.

Buffered Balanced Salt Solution

This isoosmotic balanced salt solution, buffered at pH 7.2 with HEPES and some phosphate, is used both for initial preparation of cell suspensions and for preparing density media at neutral pH. It is most conveniently prepared by mixing a series of 5 times isoosmotic concentrated stock solutions of the individual components, as follows. The values for mouse tonicity are given first, for human tonicity in parentheses. (1) NaCl, 0.840 M (0.735 M), 242 ml; (2) KCl, 0.840 M (0.735 M), 6 ml; (3) CaCl$_2$, 0.560 M (0.490 M), 6 ml; (4) MgSO$_4$, 0.840 M (0.735 M), 2 ml; (5) potassium phosphate buffer 5 times isoosmotic, see below, 4 ml; (6) HEPES buffer 5 times isoosmotic, see below, 12 ml; (7) distilled water, 1088 ml. Final volume, 1360 ml.

The 5 times isoosmotic potassium phosphate buffer is prepared by mixing solutions of KH$_2$PO$_4$, 0.840 M (0.735 M), with K$_2$HPO$_4$, 0.560 M (0.490 M), until a 50-fold diluted sample gives a pH of 7.2. The 5 times isoosmotic HEPES buffer is prepared by neutralizing N-2-hydroxyethyl-piperazine-N-2-ethanesulfonic acid (Calbiochem), 1.68 M (1.47 M), with NaOH, 1.68 M (1.47 M), until a 50-fold diluted sample gives a pH of 7.2.

Unbuffered Balanced Salt Solution

This unbuffered isoosmotic balanced salt solution is used for preparing acid pH albumin density media, where the albumin tiself near its isoelectric point serves as a strong buffer at pH 5.1. It is most conveniently prepared by mixing a series of 5 times isoosmotic concentrated stock solutions of the individual components, as follows. Values for mouse tonicity are given first, for human tonicity in parentheses. (1) NaCl, 0.840 M (0.735 M), 242 ml; (2) KCl, 0.840 M (0.735 M), 8 ml; (3) CaCl$_2$, 0.560 M (0.490 M), 6 ml; (4) MgSO$_4$, 0.840 M (0.735 M), 2 ml; (5) KH$_2$PO$_4$, 0.840 M (0.735 M), 2 ml; (6) distilled water, 1040 ml. Final volume, 1300 ml.

Albumin Density Media, pH 5.1

Albumin Source. Bovine plasma albumin, Fraction V powder, unneutralized (e.g., Armour Pharmaceutical Company) is used. Most preparations contain sufficient salt contaminants to markedly affect the osmolarity of concentrated albumin media, and salts as well as absorbed water usually must be removed before use. This may not be necessary if these contaminants are low or in fortunate balance; vapor presssure osmometer readings would be needed to verify this if the salt removal step is omitted.

Dissolving and Handling Albumin Solutions. Unneutralized albumin powder takes a day or two to dissolve and the more concentrated viscous solutions are awkward to handle. Solutions should be prepared in a cold room to reduce any microbial growth. The required amount of albumin powder is layered above all but 50 ml of the required amount of liquid in a conical flask, sealed with Parafilm at the top to prevent evaporation. Low speed wrist action shaking is used initially, followed by mixing with a magnetic stirring bar driven by a relatively powerful motor. The last 50 ml of liquid is added in two stages to the top of the medium, to effect solution of residual floating albumin. All albumin solutions are stored frozen in sealed containers to prevent evaporation. Thawing should be carried out at warm, not hot temperatures. After thawing, extensive mixing is required to restore a homogeneous solution. Small quantities of media may be sterilized before use by passage under syringe pressure through a Millipore filter (0.45 μm with a prefilter).

Removal of Salt and Water. A 15–20% solution of albumin in distilled water is dialyzed for 2 days against 4 changes of 10–20 volumes of distilled water. Frequent mixing, especially of the viscous albumin solution within the dialysis bag, is essential; an air bubble within the bag facilitates mixing by inversion. To minimize bacterial growth dialysis is at 4° using precooled water, and a drop of chloroform is placed in the water. After dialysis, the solution is sterilized by filtration (Millipore, 0.45 μM using a prefilter). The solution is first freeze-dried, then the powder further dried over P_2O_5 in a vacuum desiccator for several days to remove the significant level of residual water left by freeze drying.

Final Density Media. Salt- and water-free albumin powder is dissolved, using the technique above, to produce a convenient stock solution, say 35% w/w. For each 100 g of powder is added 4 ml water (5.0 ml for human osmolarity) to compensate for the slight osmotic pressure of albumin itself, together with a final volume of 182 ml (181 ml for human osmolarity) of unbuffered balanced salt solution of appropriate osmolarity. The density of the stock solution is checked. Light and dense solutions for density gradient generation, or solutions for density cuts, may then be prepared by dilution with unbuffered balanced salt solution.

Metrizamide Density Media, pH 7.2

Metrizamide [2-(3-acetamido-5-N-methylacetamido-2,4,6-triiodobenzamido)-2-deoxy-D-glucose], molecular weight 789.1, is obtained from Nyegaard and Co. A/S, Oslo. The centrifugation grade crystalline material suffices for most applications, but an analytical grade is also available. Solutions must be protected from light and must not be heated above 55°.

A theoretically isoosmotic stock solution is first prepared by dissolving in distilled water to 0.308 M (or 0.269 M for human osmolarity). To avoid forming sticky lumps, metrizamide (24.3 g) is gradually added to ~50 ml water and stirred with a magnetic rod at room temperature. When dissolved the final solution is made to 100 ml. This 0.308 M stock has a density of about 1.13 g/cm^3 at 4°. Dilutions to the appropriate density are then made in buffered balanced salt solution. The media are sterilized if necessary by Millipore filtration, and are stored, normally frozen, in light-proof containers, sealed to prevent evaporation.

Continuous Gradient Procedures

Cell Input

The lymphoid cell preparation to be separated should be a single cell suspension, with most coarse clumps and fine debris eliminated[6]; extensive prewashing of the cells is not necessary. Since damaged cells are very dense and form a pellet (as do erythrocytes at neutral pH) these components usually need not be eliminated before separation; however they do contribute to the total cell load and an excessive level of damaged cells causes aggregation and poor separation. For acid pH albumin gradients of 15 ml volume, up to 4×10^8 cells can be separated with good resolution, and loads of 10^9 cells can be accommodated with reduced but adequate separation. For neutral pH metrizamide gradients the cell load should be 2- to 3-fold lower. The cells, in the form of a pellet, are dispersed directly into the 7.5 ml dense albumin (or metrizamide) used for generating the gradient.

Gradient Generation

The apparatus used for generating the gradient is shown in Fig. 1. It is designed for effective mixing of viscous materials without frothing. Alternatively, more complex and more precise systems have been described by Leif.[9] The mixing chamber is made from acrylic or other clear material, with a central exit tube and two entry tubes. The acrylic mixing paddle with flexible coupling to a shaft rotates so as to generate a downward screw action thrust. The stirrer is driven at a constant speed of around 110 rpm, using a synchronous motor. A peristaltic pump with three channels each pumping at around 0.25 ml/min/channel is used, one channel to pump in the light density medium through the two chamber inlets, two channels in parallel to pump out the mixture from the single outlet at 0.5 ml/min. A pump using relatively hard tubing and producing minimun puls-

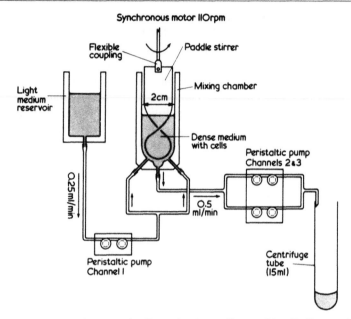

FIG. 1. An apparatus for preparing linear density gradients, with cells dispensed into the dense medium. From Shortman.[8]

ing is required for the more viscous media (e.g., a Perspex 3-channel stack unit, from LKB or Werner Meyer, Lucerne). The exit line is run to the top of the centrifuge tube. The entire apparatus is kept in a cold-room, in a simple hood with a UV lamp for sterile applications. For sterile work the tubes and chambers are soaked in chlorhexidine, rinsed with sterile water, then dried with alcohol before use.

A typical wide range gradient suitable for lymphocyte separation would run from 1.065 to 1.100 g/cm^3 for mouse osmolarity media, or 1.055–1.090 g/cm^3 for human osmolarity media. Narrower range gradients are used to give better separation in particular regions of interest.[8] The inlet line is first filled with light medium to near the entry point, and a small amount of dense medium is run into the start of the exit line. The pelleted cells are dispersed into about 5 ml of the dense medium, the suspension transferred to the mixing chamber, and any residual cell suspension rinsed across with the remaining medium. The final volume of cells in dense medium is 7.5 ml for a 15 ml gradient. Stirring is commenced, then light medium is pumped into the chamber as the developing gradient is pumped out and into the centrifuge tube, flowing down the tube wall.

Centrifugation

The centrifuge tube [e.g., 16 mm diameter × 100 mm (TW) Du Pont Cat. No. 03125] containing the gradient is capped to prevent evaporation, and centrifuged at 4000 g for 30 min at 4°, using a swing-out head (e.g., the HB-4 head of a Sorvall RC2-B centrifuge). The initial acceleration and final deceleration should be slow to prevent swirling; a modification to the centrifuge may be needed to allow slow initial acceleration. This speed should be adequate both to bring the cells to equilibrium and to separate any doublets or clumps of cells of differing density.[8]

Fraction Collection

The apparatus for collection of fractions by upward displacement is given in Fig. 2. Alternative systems have been devised by Leif.[9] Fractions are normally collected at 4°. The apparatus may be sterilized if needed, as for the gradient generation system, and may be contained in a simple sterile hood. In Fig. 2 the dense nonaqueous displacing liquid is bromo-benzene, and since this dissolves many plastics, the slightly complex plumbing is to avoid contact between peristaltic pump tubing and this solvent; a dense, low viscosity silicone oil would be a simpler alternative. The centrifuge tube is carefully removed to the clear tube holder (e.g., of acrylic) and a nylon or teflon exit cap, with a conical exit chamber and an O-ring seal is screwed on. The bottom of the tube is pierced by a needle on a screw thread, through a Neoprene seal. Once the dense gradient me-

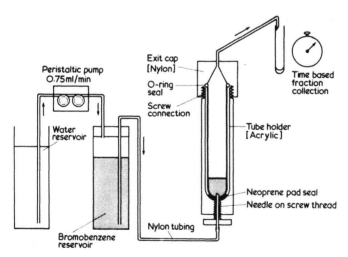

FIG. 2. Collection of fractions after centrifugation of cells in continuous density gradients. From Shortman.[8]

dium has filled the needle and displaced air, the tube (nylon) carrying bromobenzene is connected. Bromobenzene is then pumped in to displace the aqueous gradient upward out the conical exit and into a collection tube. The bromobenzene is itself displaced from a reservoir by water pumped in at the top by a peristaltic pump. The same pump as used for gradient generation may be employed, with several channels in parallel to obtain the appropriate rate. The gradient is pumped out at a fixed rate around 0.75 ml/min, and fractions (10–25) are collected on a time basis, and the volume of each calculated. Fractions are usually of equal volume, but it may be useful to cut smaller fractions in a region of interest, or a region where most cells band. Unless the gradient extends well beyond the cell banding regions (a wasteful approach) the first and last fractions will concentrate all viable cells outside the gradient limits, as well as dead and damaged cells. These fractions may therefore give false peaks in density distribution profiles, but they are nevertheless needed for book-keeping purposes and to establish the gradient limits.

Density Measurements of Fractions

For many purposes the density of individual or selected fractions may be determined by using a digital density meter (K. Paar, K. G., Graz, Austria), by inclusion of standard density marker beads (Pharmacia) in the gradient, or (if the cell concentration is not too high) by refractometry after establishing a calibration curve relating refractive index to medium density or concentration. However, for the most accurate density distribution analyses, the precise density of each fraction must be measured to five decimal places, so that the density increment covered by each fraction can be determined with precision.

Density estimates of this accuracy can readily be made by determining the buoyant position of small samples of each fraction on a non aqueous density gradient,[8] as illustrated in Fig. 3. All operations are in a cold room at 4°. A continuous, linear nonaqueous gradient spanning a somewhat wider density range than the albumin or metrizamide gradients is generated into a 50-ml graduated burette tube sealed at the bottom, and enclosed in a noncirculating water jacket to serve as a temperature buffer. A conventional, glass, linear gradient mixing system is used, consisting of two solvent reservoirs cylindrical in shape and of equal diameter, connected at the base, the dense reservoir serving as mixing chamber and source of the developing gradient. The light (1.045 g/cm³) and the dense (1.10 g/cm³) nonaqueous solutions are made by mixing bromobenzene A.R. (density 1.52 g/cm³) and petroleum spirit A.R., bp 80–100° (density 0.79 g/cm³); the bulk mixtures are stored in sealed bottles in the cold

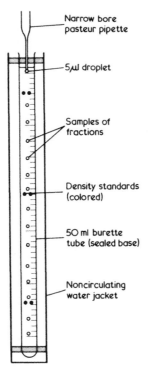

FIG. 3. Measurement of the mean density of aqueous density gradient fractions by determining the buoyant density of droplet samples in a nonaqueous gradient. From Shortman.[8]

room. The nonaqueous gradients are conveniently made the day before use, and stored in the cold room with the burette top sealed to prevent evaporation. Fractions from the aqueous gradients containing the cells are briefly mixed and stood for 10 min to allow air bubbles to rise. A small sample is withdrawn from the center of each fraction with a fine bore Pasteur pipet, using the absolute minimum depression of the bulb, and with care to exclude minute air bubbles and to maintain a liquid column to the pipet tip. The outside of the pipet tip is wiped clean and a small droplet (~5 μl) is released into the nonaqueous gradient by placing the tip under the miniscus of the nonaqueous gradient, expressing a droplet of the aqueous medium, then lifting the tip above the miniscus to release the droplet. The droplet then sinks till it reaches the region corresponding to its buoyant density. Droplets of close to the same size are released from each fraction in order, beginning with the most dense. In addition, duplicate droplets from each of 4 spaced density standards are released in appropriate order. A fixed time (15–30 min) after all droplets are applied the buoyant position of fraction samples and standards is read off the burette scale. Note that since, due to dehydration, droplets slowly shift position after the initial rapid sedimentation, the simultaneous reading of all fractions and standards is important. The density versus the burette reading of the standards is then plotted, and the mean density of individual fractions read from this plot.

The standards used for calibrating the nonaqueous gradient must be prepared and conserved very carefully if absolute accuracy is to be maintained. A series of solutions of albumin (or metrizamide), differing by intervals of 0.01 g/cm³ or less, is prepared as described previously, and colored (prior to density estimation) with a few drops of concentrated methyl violet solution. Samples (3–5 ml) of each standard are sealed in glass ampules and stored frozen as primary standards. Periodically an ampule of the primary standard is thawed and thoroughly mixed, the ampule opened, and the contents dispersed into a series of small plastic tubes tightly stoppered with plastic plugs. These are also stored frozen, but they are used within several months because a slow freeze-drying effect gradually increases their density. Each secondary standard is used only a few times before being discarded. Sealed primary standards keep for years in the frozen state.

Recovery and Counting of Cells

Cells are recovered from each fraction after density measurements by dilution at least 5-fold in buffered balanced salt solution, centrifugation to a pellet, and reconstitution to a fixed volume with an appropriate medium. Thorough mixing is needed after dilution, and the centrifugal force may need to be higher than usual to recover all cells. If Coulter cell counts are performed, care must be taken that fractions which concentrate erythrocytes do not, because of coincidence, give a false high lymphocyte count. If hemocytometer counts are used replicate counts may be needed to ensure that statistical error does not distort the distribution profile. In both cases it is important to remember that most damaged cells will concentrate in the bottom or pellet fractions, and some vacuolated, damaged cells may also be found in the top fraction.

Calculation of Density Distribution Profiles and Expression of Results

The simplest mode of expressing results, assuming all fractions are of equal volume, is a plot of *total* cells per fraction (or *total* number of cells of a given type or *total* functional potential per fraction) versus fraction number. If assays made on each fraction are on a "per cell" basis (such as counts of the proportion of cells of a given morphological type on smears, or functional capacity of the culture of a fixed number of cells), this must be multiplied by the total viable cells per fraction, to give the actual distribution by buoyant density of that particular cell type. Enrichment plots (activity per cell in different fractions) are also useful, but may tend to focus interest on the highest enrichment regions where sometimes only a few, atypical cells might band.

However, density distributions expressed this way on a per fraction or per volume basis assume the density gradient is itself perfectly linear, and this is seldom the case. Nonlinear gradients can produce artifacts, flat regions generating troughs, steep regions of density change generating peaks, in what is in reality a smooth, continuous profile. Nonlinear gradients can arise from the difficulties of mixing and handling viscous gradient materials, although this problem can be eliminated with careful use of the equipment described above. However, even if a perfectly linear gradient is generated, redistribution of cells during centrifugation will itself, if cell loads are high, distort the gradient shape. For this reason any precise analysis must be expressed as a true density distribution profile, plotting cells per density increment (rather than cells per volume or per fraction) against fraction density. The density increment of each fraction must be determined accurately, based on density measurements on each fraction to 5 decimal places as described above. To do this the mean density per fraction is first plotted against cumulative fraction volume, and a line (or curve) drawn through the points in the middle of each fraction range. The limits of each fraction are drawn in, and from the intercept with the density curve the density at the beginning and end of each fraction determined. From this the density increment covered by each fraction can be obtained. A computer program may readily be devised to handle the raw data, to make the density increment calculations and then to produce the final plot of cells (or function) per density increment against density.

It is also important, regardless of the mode of plotting results, to compare the total recoveries of cells or activity in all fractions with the input onto the gradient, determined from an unseparated sample of the original suspension. Recoveries of viable cells should be 75–100%. Low recoveries may suggest a selection for more durable cells against more fragile (for example small lymphocytes against activated blasts) and so indicate bias in the results. Alternatively, low recovery of biological activity may indicate a need for cell interaction to generate a given function; if so, remixing fractions should reconstitute the "lost" activity.

Simple Density Cut Procedures

General Applications

Once the density distribution of a cell population is established using continuous gradient procedures, any given density band of cells can be obtained by two simple neutral density "cuts" in two successive solutions of defined density, selecting first cells denser than the light limit, then cells lighter than the dense limit. For many purposes a single spin dividing cells into those lighter and denser than a given medium will suffice. Even without prior continuous gradient analysis, a series of trial test density cuts can often establish the appropriate conditions for future work. The important variables for consistent separation are (1) the same close control over medium density and osmolarity as used in continuous gradients, (2) adequate centrifugation speed in a swing-out head, and (3) avoidance of high cell concentrations during separation, by dispersing cells directly into the medium and by the elimination of all sharp interfaces between media.

Cell Input

The maximum cell input for clean separation is a little less in the density cut procedure than with a continuous gradient. A maximum of 5×10^8 cells can be applied with pH 5.1 albumin, and 2×10^8 cells for pH 7.2 metrazamide; these maximum loads are reduced if the level of damaged cells is high, or if cross contamination must be kept to a minimum. The maximum load drops about 2-fold if the medium interfaces are not diffused before centrifugation.

Damaged Cell and Red Cell Removal

To eliminate as a pellet cells mechanically damaged by the procedure of teasing out lymphoid tissues, a density cut in pH 5.1 albumin at 1.093–1.094 g/cm^3 (1.088 g/cm^3 human osmolarity) may be used; this will not remove erythrocytes, nor will it remove cells which have died in culture or have been killed with antibody and complement.[5] To eliminate in the pellet all dead cells regardless of origin, as well as erythrocytes, a density cut in pH 7.2 metrizamide at 1.091–1.092 g/cm^3 (for mouse osmolarity) is used.

Procedure

The procedure is illustrated in Fig. 4. The cell suspension is centrifuged (400 *g*/7 min), all but a trace of the supernatant removed and the cell

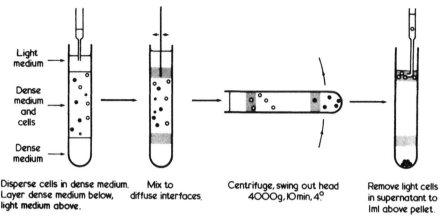

Disperse cells in dense medium.   Mix to   Centrifuge, swing out head   Remove light cells
Layer dense medium below,   diffuse interfaces.   4000g, l0min, 4°   in supernatant to
light medium above.   lml above pellet.

FIG. 4. Simple density cuts for separating cells in albumin or metrizamide density media. The technique is also used for removing damaged or red cells as a pellet from lighter density viable cells. From Shortman et al.[5]

pellet dispersed in 7 ml medium of the appropriate density, using a vortex mixer. This suspension is transferred to the centrifuge tube. The trace of supernatant associated with the cell pellet makes the average density of this suspension a little less than that of the density separation medium. Of the density separation medium, without cells, 5 ml is then layered below the cell containing layer, and 4 ml of a lighter density medium (usually balanced salt solution) layered above. The upper and lower fluid interfaces are then stirred, using a rod or Pasteur pipet, to produce a more diffuse zone, a gradual gradient. The tube is then centrifuged (4000 g, 10 min, 4°) in a swing-out head. Most of the light density cells are then in the upper interface, but some may remain in the albumin medium. The denser cells pellet. Using a Pasteur pipet with a bent tip, all the upper liquid zones, including the upper and lower interface, down to the last 1 ml above the cell pellet is removed, diluted around 5-fold, thoroughly mixed, and centrifuged to recover the light density cells. The cells in the pellet, including the most dense 1 ml of medium, are likewise washed and recovered if required, or rejected when the technique is used for damaged cell removal.

As with continuous gradients, it is important to check both light and dense fractions for total and viable cell recoveries, adjusting the density of the medium if required, before adopting the procedure for routine use.

[11] Separation of Lymphocytes on Antibody-Coated Plates

By MICHAEL G. MAGE

Introduction

Use of antibody-coated polystyrene tissue culture dishes (plating) for cell separation is a simple, rapid, preparative, inexpensive, and sterile procedure.[1,2] It has become a routine method in many laboratories for the separation of immunoglobulin-positive (Ig$^+$) and immunoglobulin-negative (Ig$^-$) lymphocytes. Plating can be used as a general method of cell separation for situations where a single cell suspension can be obtained and an antibody is available that binds to a subpopulation of cells in the mixture. This method has been used for the separation of human T lymphocyte subsets,[3] mouse T cells,[4] guinea pig bone marrow neutrophils,[5] lectin-binding T cells,[6] and preparation of monolayers of tumor target cells for microcinematography.[7]

Separation of Ig$^+$ and Ig$^-$ Lymphocytes

The following procedure describes the separation of Ig$^+$ and Ig$^-$ mouse spleen lymphocytes. It can be readily adapted to other species by using the appropriate anti-Ig. All reagents are sterilized by filtration and sterile procedures are used throughout. Briefly, the procedure consists of placing a cell suspension in a tissue culture dish that has been precoated with anti-Ig. On settling, the Ig$^+$ cells bind to the antibody-coated surface and the nonadherent Ig$^-$ cells are removed. If removal of splenic adherent cells (macrophages) is required, it is accomplished by incubating the suspension of spleen cells in an uncoated polystyrene tissue culture dish at 37° prior to separating the Ig$^+$ and Ig$^-$ lymphocytes.[8]

---

[1] M. G. Mage, L. L. McHugh, and T. L. Rothstein *J. Immunol. Methods* **15**, 47 (1977).

[2] M. G. Mage, L. L. McHugh, and T. L. Rothstein, *Transplant. Proc.* **8**, 399 (1976).

[3] S. M. Payne, S. O. Sharrow, G. M. Shearer, and W. E. Biddison, *Int. J. Immunopharmacol.* **3**, 227 (1981).

[4] L. J. Wysocki and V. L. Sato, *Proc. Natl. Acad. Sci. U.S.A.* **75**, 2844 (1978).

[5] W. H. Evans and M. Mage, *Exp. Hematol.* **6**, 37 (1978).

[6] B. J. Fowlkes (N.I.H.), personal communication.

[7] T. L. Rothstein, M. G. Mage, G. Jones, and L. L. McHugh, *J. Immunol.* **121**, 1652 (1978).

[8] This volume [27].

METHODS IN ENZYMOLOGY, VOL. 108

Reagents

Specifically Purified Anti-Mouse Ig

This reagent is prepared from hyperimmune anti-Ig by adsorption and elution from an affinity column of mouse Ig.[9,10] The eluate from the affinity column is neutralized and dialyzed against phosphate buffered saline (PBS). After centrifugation, the supernatant is sterilized by filtration and stored at 4°. Before use, this stock anti-Ig solution is diluted in sterile PBS to an A_{280} of 0.2. Such dilute sterile solutions have remained active after storage at 4° for over a year.

RPMI 1640 Medium

RPMI 1640 medium without bicarbonate, with 0.02 M HEPES and 10% fetal calf serum (FCS) or newborn calf serum.

Low Ionic Strength, Isoosmotic Medium[11]

This medium is used to prepare stable single cell suspensions. It is made by mixing in the indicated proportions the following solutions:

Solution A, 25 ml, isoosmotic (0.308 mos$M$ for mouse) buffered balanced salt solution, such as Hanks.

Solution B, 50 ml, isoosmotic HEPES buffer, pH 7.2 (mix 0.336 M HEPES and 0.336 M NaOH to pH 7.2).

Solution C, 800 ml, sorbitol, 0.308 M.

Solution D, 125 ml, glucose, 0.308 M.

Normal Goat Ig

Normal goat Ig is prepared from serum.[9] Following dialysis against PBS, it is sterilized by filtration and stored at 4°.

Procedure

Preparation of Antibody-Coated Dishes

Polystyrene tissue culture dishes, 100 mm diameter, (e.g., Falcon 3003), are incubated for 1 hr at room temperature, or overnight at 4° with 4 ml of anti-Ig solution. The solution containing unbound antibody is recovered and can be reused up to 4 additional times. The dishes are washed 5 times with PBS. They can be stored with PBS for at least 1 month at 4°.

[9] B. A. L. Hurn and S. M. Chandler, this series, Vol. 70 [5].
[10] D. M. Livingston, this series, Vol. 34 [91].
[11] H. von Boehmer and K. Shortman, *J. Immunol. Methods* **2**, 293 (1973).

Precoated dishes are commercially available (Seragen Inc., Boston, MA), but have not been tested by the author.

Preparation of a Stable Single Cell Suspension

Single cell suspensions from lymphoid organs are prepared as described.[12] We have found that for spleen cells, having a suspension of single cells stable for several hours at room temperature without forming clumps or aggregates is important for successful separation because aggregates contain mixed populations of cells and tend to dislodge cells from the bottom of the dish. Such suspension is prepared by centrifuging the cells (1300 rpm for 5 min at 4°) and resuspending the pellet in 30 ml of low ionic strength, isoosmotic medium.[11] In this medium, dead cells and debris form a clump or clot which is removed by filtration through sterile gauze. The filtered suspension is centrifuged, and the pellet is resuspended in RPMI 1640 medium, counted, and diluted to 2×10^7 cells/ml.

Preparation of Ig^- Cells

The spleen cell suspension, 1×10^8 cells in 5 ml of medium, is pipetted into an antibody-coated dish. The cells are allowed to settle. After 0.5 hr at room temperature, the nonadherent cells are resuspended by manually swirling the dish for 30 sec. The suspension of nonadherent cells is poured into a second antibody-coated dish for another 0.5 hr incubation. The nonadherent cells are again resuspended and poured into a third antibody coated dish for another 0.5 hr incubation. At the end of these three cycles, the nonadherent cells are recovered, counted, and tested for residual Ig^+ cells by immunofluorescence.[13]

Preparation of Ig^+ Cells

When the Ig^+ cells are to be used *in situ* as a B cell monolayer,[14,15] the first antibody coated dish is gently washed 5 times with PBS immediately after resuspending and pouring off the nonadherent cells. When Ig^+ cells are to be released from the dish, the first of the three dishes is precoated with antibody that has been diluted with normal goat Ig. (The dilute antibody solution is prepared by adding normal goat Ig solution to the anti-Ig solution to produce a 1/7 dilution of anti-Ig in normal goat Ig, based on the A_{280} of the two solutions.) The suspension of spleen cells is added

---

[12] This volume [6].

[13] L. A. Herzenberg and L. A. Herzenberg, *in* "Handbook of Experimental Immunology" (D. M. Weir, ed.), 3rd/ed., Chapter 22.1. Blackwell, Oxford, 1978.

[14] M. G. Mage and L. L. McHugh, *J. Immunol.* **115**, 911 (1975).

[15] M. G. Mage, this volume [12].

to the dish precoated with dilute antibody. After incubation for 0.5 hr at room temperature, the nonadherent cells are resuspended and poured off as previously described. The dish, with its adherent monolayer of Ig^+ cells, is gently washed five times with PBS, and then receives 10 ml of medium. To release the Ig^+ cells, medium is pipetted forcefully against the bottom of the dish, producing a clear area of about a 1 to 2 cm diameter. This operation is repeated until the entire monolayer has been dislodged.[1] Alternatively, normal mouse serum may be added to the medium in the dish, followed by an incubation for 2 hr at 37°. At the end of this time, approximately half of the cells can be released by swirling the dish and the remainder by gently pipetting.

The recovered population of adherent cells is over 90% Ig^+, and less than 1% Thy-1[+]. It is over 90% viable, and can be used for generation of antibody plaque-forming cells in Mishell–Dutton culture.[16] The nonadherent population is over 95% Ig^- (Fig. 1a) and contains about 20% of cells that are both Ig^- and Thy-1[-]. The depletion of Ig^+ cells is much greater than the depletion of plaque forming precursors, implying that this "Ig^-" population contains pre-B cells and/or residual Ig^+ cells that have more cell divisions in Mishell–Dutton culture than do the B cells that have been separated.[16] The nonadherent Ig^- population can participate in a mixed lymphocyte culture, can generate cytotoxic T lymphocytes (CTL) *in vitro*,[1] and produce a graft-versus-host reaction *in vivo*.[1] It can be further fractionated into Lyt-2[+] and Lyt-2[-] subpopulations.[17] The latter separation is an example of the use of anti-Ig coated dishes as a general method of cell separation. Its description follows.

Separation of Lyt-2[+] and Lyt-2[-] Lymphocytes from Mouse Spleen

The Ig^- cell population prepared by the previous procedure is centrifuged and the pellet is resuspended in monoclonal anti-Lyt-2 antibody (1 ml per 10[8] cells). The optimum concentration of antibody is determined empirically. Monoclonal antibody populations bind to one determinant only. In antibody excess, each IgG antibody molecule may have only one of its two combining sites bound to the cell surface antigen. Dissociation of this bond would then lead to dissociation of the antibody. For this reason, use of antibody excess does not improve the separation and may actually diminish it.

Two monoclonal anti-Lyt-2 antibodies have been successfully used in this laboratory. The first, a mouse anti-Lyt-2.2 (with specificity for a

[16] J. J. Mond, M. G. Mage, T. L. Rothstein, D. E. Mosier, R. Asofsky, and W. E. Paul, *J. Immunol.* **125,** 1526 (1980).
[17] M. Mage, B. Mathieson, S. Sharrow, L. McHugh, U. Hammerling, C. Kannelopoulis-Langevin, D. Brideau, Jr., and C. Thomas, III, *Eur. J. Immunol.* **11,** 228 (1981).

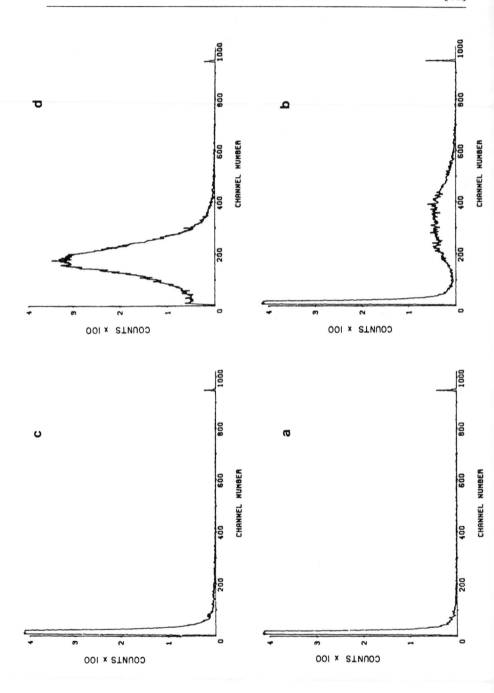

Lyt-2 allotypic determinant found in BALB/c and certain other inbred mouse strains) is produced by clone 19/138.[17,18] This hybridoma is available from the American Type Culture Collection, Rockville, Md. The optimum concentration appears to be about 10-fold higher than the 50% cytotoxic titer on BALB/c thymocytes. We have used it as a 1/1000 dilution of serum from a mouse carrying the cell line. The second antibody, a rat anti-Lyt-2, clone 53-6.72 reacts with Lyt-2$^+$ cells from all strains of mice.[19] It also is available from ATCC. We use it as a 1/10 dilution of tissue culture supernatant.

The Ig$^-$ cells are incubated with the monoclonal anti-Lyt-2 antibody for 45 min at 0° with occasional shaking. Approximately one-third of the Ig$^-$ cells can bind the anti-Lyt-2 (Fig. 1b). Unbound antibody is removed by washing twice with cold medium. The pellet is resuspended to 1.0 to 1.5 × 10^7/ml, and 5 ml is placed in an anti-Ig coated dish for 40 min at 4°. The Lyt-2$^+$ cells, now coated with anti-Lyt-2 antibody, bind to the anti-Ig antibody on the surface of the dish. (In our experience, hyperimmune goat anti-mouse Ig cross reacts with rat Ig sufficiently well to be used for binding either rat or mouse antibodies.) After 40 min the dish is swirled, the nonadherent cells poured onto a second anti-Ig dish for another 40 min and the first dish is washed gently 5 times with PBS. Ten milliliters of medium is added, the dish is allowed to warm to room temperature for 30 min and the adherent Lyt-2$^+$ cells are removed by pipetting medium onto the surface of the dish. The nonadherent and released adherent cells can be analyzed by flow microfluorometry,[13] using either a fluorescein-labeled anti-Ig or a fluorescein-labeled anti-Lyt-2 (this volume [19]). The nonadherent Lyt-2$^-$ cells are less than 5% Lyt-2$^+$ (Fig. 1c) and have been used successfully as helper cells *in vitro*.[17] The released adherent cells, over 96% Lyt-2$^+$ (Fig. 1d), have been used for the generation of CTL *in vitro* and for the demonstration of synergy between Lyt-2$^+$ cells and Lyt-2$^-$ cells in producing a graft-versus-host reaction *in vivo*.[17]

In order for cells to be separated on the basis of their own or their investigator-added surface Ig, a sufficient number of bonds must be formed with the dish-bound anti-Ig for the cell to remain bound despite

[18] G. J. Hammerling, U. Hammerling, and L. Flaherty, *J. Exp. Med.* **150**, 108 (1979).
[19] J. A. Ledbetter and L. A. Herzenberg, *Immunol. Rev.* **47**, 63 (1979).

FIG. 1. Analysis by flow microfluorometry of the separation by plating of mouse Ig$^-$ spleen lymphocytes into Lyt-2$^+$ and Lyt-2$^-$ subpopulations. The fluorescent reagent is fluorescein-labeled goat antibody to mouse Ig. (a) Ig$^-$ spleen cells. (b) Ig$^-$ spleen cells after incubation with anti-Lyt-2. (c) Lyt-2$^-$ cells, nonadherent to the anti-Ig plate. (d) Lyt-2$^+$ cells, adherent to the anti-Ig plate and released by pipetting.

the shear forces generated on swirling the dish to resuspend the non-adherent cells. These conditions exist in the case of a surface antigen present at high levels, such as immunoglobulin. In this case, in fact, the cells cannot be released without damage.[1] In order to release them in a viable condition, the dish is precoated with anti-Ig that has been diluted with normal Ig as previously described. By so doing, the antibodies bound to the dish are spaced further apart and the cells will bind to the dish-bound anti-Ig with fewer bonds. When the cell surface antigen used for the separation is present at low concentration, the number of bonds formed with the dish can be maximized by coating the dish with undiluted specifically purified anti-Ig. Since, on occasion, as with the Lyt-2 antigen, patching and capping[20] occur at room temperature, the separation is greatly improved at 4°.[17] This treatment was used for the separation of subsets of human T cells with *O*KT4 or *O*KT8 antibodies.[3]

Double Direct Separation

As an alternative to incubating cells in suspension with antibody to a cell surface antigen, the antibody can be placed on the anti-Ig coated dish, giving it a double coating of antibody. If the first (anti-Ig) antibody is an anti-Fc,[21] it will bind to the Fc portion of the second (anti-cell) antibody, leaving both combining sites of the latter unhindered to react with the cell surface antigen when the cell suspension is placed in the dish.

---

[20] F. Loor, L. Forni, and B. Pernis, *Eur. J. Immunol.* **2**, 203 (1972).
[21] M. G. Mage, this series, Vol. 70 [6].

[12] Cell Separation on Cellular Monolayers

By Michael G. Mage

Introduction

Cytotoxic T lymphocytes (CTL),[*,1] their precursors (CTLp),[2,3] some graft-versus-host (GVH) antigen reactive cells,[4,4a] and some suppressor T cells[5,6] and helper T cells[7] have been shown to be capable of binding specifically to target cells bearing the relevant surface antigens by means of antigen-specific receptors.[8] Binding of these antigen-reactive cells to target cell monolayers has been used to demonstrate the specificity of the interaction.[1] The antigen-specific binding has been used to prepare cell populations that lack reactivity to a given antigen. Populations enriched for antigen reactive cells have been prepared by elution of the cells bound to the monolayer.

Alloreactive CTL have been bound to monolayers prepared from cells that adhere naturally to plastic surfaces, such as macrophages,[1] fibroblasts,[9] and some virus-transformed cell lines bearing tumor-associated specific antigens.[10] Virus infected and haptenylated macrophage monolayers[11] have been used to bind CTL specific for antigen-modified MHC antigens. Techniques employing antibodies,[12] poly(L-lysine) (PPL),[13] lec-

---

* Abbreviations: CTL, cytotoxic T lymphocytes; CTLp, cytotoxic T lymphocytes precursors; GVH, graft-vs-host; PLL, polylysine; NK, natural killer; Ts, suppressor T cells; Th, helper T cells; PBS, phosphate-buffered saline; RBC, red blood cells; Con A, concanavalin A; CRBC, chicken red blood cells; PHA, phytohemagglutinin; ADCC, antibody-dependent cytotoxic cells; SRBC, sheep red blood cells; FCS, fetal calf serum; MHC, major histocompatibility complex.

[1] B. D. Brondz, *Transplant. Rev.* **10**, 112 (1972).
[2] H. Y. Schnagl and W. Boyle, *Nature (London)* **279**, 331 (1979).
[3] A. Kelso and W. Boyle, *Cell. Immunol.* **67**, 355 (1982).
[4] M. G. Mage and L. L. McHugh, *J. Immunol.* **115**, 911 (1975).
[4a] B. Bonavida and E. Kedar, *Nature (London)* **249**, 658 (1974).
[5] B. Argyris and A. Contellessa, *Transplantation* **28**, 372 (1979).
[6] B. D. Brondz, A. V. Karaulov, I. F. Abronina, and Z. K. Blandova, *Scand. J. Immunol.* **13**, 517 (1981).
[7] O. Werdelin, O. Braendstrup, and E. M. Shevach, *J. Immunol.* **123**, 1755 (1979).
[8] B. D. Brondz, A. V. Andreev, S. G. Egorova, and G. I. Drizlikh, *Scand. J. Immunol.* **10**, 195 (1979).
[9] G. Berke and R. H. Levey, *J. Exp. Med.* **135**, 972 (1972).
[10] S. Pan, P. J. Wettstein, and B. B. Knowles, *J. Immunol.* **128**, 243 (1982).
[11] U. Kees, A. Mullbacher, and R. V. Blanden, *J. Exp. Med.* **148**, 1711 (1978).
[12] M. G. Mage, L. L. McHugh, and T. L. Rothstein, *J. Immunol. Methods* **15**, 47 (1977).
[13] D. Stulting and G. Berke, *J. Exp. Med.* **137**, 932 (1973).

METHODS IN ENZYMOLOGY, VOL. 108   ISBN 0-12-182008-4

tins,[14] and lectins together with fixed cells[15] have been used to form mono-layers of allogeneic target cells for specific binding of alloantigen reactive T cells. The last technique has also been used for the specific binding of natural killer (NK) cells.[16] Antibody-coated monolayers have also been used for the separation of cells with receptors for the Fc portion of the IgG molecule.[17,18]

Fibroblast Monolayers

These have been used to bind alloreactive CTL,[9] but have not been found to bind CTLp or GVH reactive cells.[10,19,20] While these cells (in common with most somatic cells) have class I MHC antigens(H-2K, L, D in mouse), they lack class II (I region) antigens.

Procedure

Mouse embryo fibroblasts, prepared by standard tissue culture meth-ods[21] to establish primary cultures, are trypsinized[22] and used to establish further cultures in plastic dishes (e.g., Falcon 3002) or flasks (e.g., Falcon 3024). The cells are allowed to multiply sufficiently to form a confluent monolayer.

Macrophage Monolayers

Peritoneal macrophage monolayers have been used to bind CTL,[1] suppressor T cells (Ts),[6] and helper T cells (Th).[7] They have class I surface antigens and variable amounts of class II antigens, depending on the state of their activation.[23]

Procedure

Peritoneal macrophages are prepared by injecting 3 ml of thioglycol-late medium (Difco, Detroit MI) intraperitoneally into mice, 3 days prior

[14] M. G. Mage, L. L. McHugh, and T. L. Rothstein, *Transplant. Proc.* **8**, 399 (1976).
[15] A. Silva, M. O. de Landazuri, J. Alvarez, and J. M. Kreisler, *J. Immunol. Methods* **23**, 303 (1978).
[16] W. H. Phillips, J. R. Ortaldo, and R. B. Herberman, *J. Immunol.* **125**, 2322 (1980).
[17] R. L. H. Bolhuis, H. R. E. Schuit, A. M. Nooyen, and C. P. M. Ronteltap, *Eur. J. Immunol.* **8**, 731 (1978).
[18] R. L. H. Bolhuis and H. Schellekens, *Scand. J. Immunol.* **13**, 401 (1981).
[19] M. Mage and L. L. McHugh, *J. Immunol.* **111**, 652 (1973).
[20] W. R. Clark and A. K. Kimura, *Transplantation* **16**, 110 (1973).
[21] M. M. Basher, this series, Vol. 58 [9].
[22] J. A. McAteer and W. H. J. Douglas, this series, Vol. 58 [10].
[23] B. R. Smith and K. A. Ault, *J. Immunol.* **127**, 2020 (1981).

to washing out the peritoneal cavity with phosphate-buffered saline (PBS).[11] The recovered cells are washed and resuspended in appropriate tissue culture medium (e.g., RPMI 1640 with 10% fetal calf serum (FCS)) at the density of 3×10^7/ml and 5 ml of the suspension is placed in a 10-cm plastic tissue culture dish that is then incubated at 37° overnight in a 5% CO_2 humidified atmosphere. At the end of the incubation the dish is washed 5 times with sterile PBS or medium to remove nonadherent cells. It has been reported that nonspecific adsorption of lymphocytes to the monolayer can be reduced by treating the monolayer with Pronase (Calbiochem) (25 μg/ml in serum-free medium) for 30 min at 37°, followed by removal of the Pronase and neutralization of the remaining enzyme on the monolayer with 50% bovine serum.[8]

PLL-Bound Monolayers

The use of PLL-coated surfaces was introduced by Kennedy and Axelrad[24] to form monolayers of red blood cells (RBC) for assaying antibody plaque-forming cells. PLL-coated surfaces have been used to form monolayers of spleen cells,[4,25,26] thoracic duct lymphocytes,[27] and tumor cells[13] that were used in turn to bind to CTL. The advantage of this technique is its simplicity. The disadvantage is the relatively high amount of spontaneous release of monolayer cells during the subsequent separation procedures.[13,14,28]

Procedure

PLL (Sigma Chem. Co., St. Louis MO, molecular weight 30,000 or higher) is dissolved in PBS to a concentration of 50 μg/ml. Five milliliters of this solution is placed in a 10-cm tissue culture dish. After a 1-hr incubation at room temperature, the dish is washed five times with PBS and left at 4° with PBS in it until ready for use. Alternatively, the dish is incubated with PLL for 10 min, followed by a 10 min centrifugation at 700 rpm.[29] A single cell suspension[30] from spleen[31] or tumor cells[13] is centrifuged and resuspended in protein-free medium twice. Soluble proteins present in the medium compete for the PLL and thus interfere with the binding of the cells. The cells are finally resuspended to a concentration of

[24] J. C. Kennedy and M. A. Axelrad, *Immunology* **20**, 253 (1971).
[25] E. Kedar and B. Bonavida, *J. Immunol.* **115**, 1301 (1975).
[26] J. R. Neefe and D. H. Sachs, *J. Exp. Med.* **144**, 996 (1977).
[27] P. M. Chisholm and W. L. Ford, *Eur. J. Immunol.* **8**, 438 (1978).
[28] J. R. Neefe and D. H. Sachs, *Cell. Immunol.* **29**, 129 (1977).
[29] E. Kedar, W. R. Clark, and B. Bonavida, *Transplantation* **25**, 146 (1978).
[30] This volume.
[31] M. G. Mage, this volume [11].

3×10^7 per ml and 5 ml of this suspension is placed in the dish. After a 1-hr incubation at room temperature, the dish is gently washed with PBS 5 times to remove nonadherent cells[31] and stored at 4° with protein-free medium for use the same day.

Lectin-Bound Monolayers

These are prepared in the same fashion as the PLL-bound monolayers, except that concanavalin A (Con A) (Pharmacia) (100 μg/ml) is used instead of PLL. Lymphocyte monolayers formed with Con A have been found to be bound to the dish more tightly than monolayers bound with PLL.[14]

Antibody-Bound Monolayers

Specifically purified antiimmunoglobulin binds B cells to polystyrene surfaces more tightly than Con A or PLL. B cells have both class I and class II MHC antigens on their surface and B cell monolayers have been shown to bind some GVH antigen-reactive cells.[4] The preparation of the antibody-coated dishes and the separation of Ig$^+$ from Ig$^-$ cells is described in detail elsewhere [11].[31] When the separation is carried out to prepare B cell monolayers, the load of unfractionated spleen cells placed in the antibody-coated dish should be about 2×10^8 in 5 ml. Following formation of the monolayer, the dish is gently washed 5 times with PBS and stored at 4° in HEPES medium with 10% FCS[31] until used the same day.

PLL-Fixed RBC-Lectin-Bound Monolayers

This method of preparing target cell monolayers was introduced by Silva et al.[15] to deal with the problems of release of target cells from the PLL-coated dishes and nonspecific binding to the monolayer. It also allows the formation of monolayers of cells that bind more tightly to lectins than to PLL. Dishes are coated with PLL, followed by addition of chicken red blood cells (CRBC). The CRBC are fixed with glutaraldehyde, coated with phytohemagglutinin (PHA), and then with target cells that have receptors for PHA. The method has been used for the binding of murine CTL to allogeneic tumor target cells,[15] of immune human CTL to allogeneic leukocytes,[32] and of human NK and ADCC cells to a variety of

---

[32] S. Kato, P. Ivanyi, E. Lacko, B. Breur, R. DuBois, and V. P. Eijsvoogel, *J. Immunol.* **128**, 949 (1982).

human tumor cell monolayers,[16,33,34] to study the effect of differentiation-inducing stimuli on the ability of tumor cells to bind to NK cells[35] and to demonstrate the existence of kidney-specific CTL by differential binding of canine alloimmune CTL to kidney and leukocyte monolayers.[36] The following procedure, based on the work by Silva *et al.*,[15] illustrates the use of the method to form monolayers of P815 murine mastocytoma cells.

Materials

1. Polystyrene tissue culture dishes, 35 mm diameter (Falcon). (Other sizes of polystyrene tissue culture plates or flasks may be used, in which case the volumes of the reagents used are changed in proportion to the surface area of the flask or plate.)

2. PLL. 30,000 MW (Sigma Chemical Co., St. Louis, MO), 50 μg/ml in PBS.

3. CRBC less than 1 week old, washed, 1% suspension in PBS.

4. Glutaraldehyde, 0.2% in PBS.

5. Glycine, 0.1 M in PBS.

6. PHA, (PHA-HA16, Burroughs-Wellcome, Greenville, NC); 5 ml stock solution, diluted 1/50 with PBS just before use.

7. P815 cells,[15] 3 × 10[7]/ml in PBS (American Type Culture Collection, Rockville, MD).

8. Erythrocytes syngeneic to the target cells to be used, 1% suspension in PBS.

Procedure

The polystyrene plates are incubated with a series of reagents in the following order. After each incubation the dishes are washed three times with PBS in order to remove unbound reagent.

1. PLL, 1 ml/dish, 1 hr at room temperature.

2. CRBC, 1 ml/dish, 1 hr at room temperature.

3. Glutaraldehyde, 1 ml/dish, 10 min at room temperature.

4. Glycine, 1 ml/dish, 10 min at room temperature (to inactivate any unreacted glutaraldehyde). At this point the dishes can be stored (with sterile PBS in them) for up to 1 month at 4°, and rewashed with PBS just before use.

[33] R. L. H. Bolhuis and H. Schellekens, *Scand. J. Immunol.* 13, 401 (1981).
[34] M. O. deLandezuri, A. Silva, J. Alvarez, and R. Herbeman, *J. Immunol.* 123, 252 (1979).
[35] M.-C. Dokkelar, U. Testa, W. Vainchenker, Y. Finale, C. Tetand, P. Salem, and T. Tursz, *J. Immunol.* 128, 211 (1982).
[36] P. A. Vegt, W. A. Buurman, C. J. Van der Linden, A. J. J. Daeman, J. M. Greep, and J. Jeekel, *Transplantation* 33, 465 (1982).

5. PHA, 1 ml/dish, 30 min at 37° in 5% CO_2.

6. P815 target cells, 1 ml/dish, 30 min at 37° in 5% CO_2.

7. Syngeneic RBCs, 1 ml/dish, 30 min at 37° in 5% CO_2 (to fill any interstices between the target cells, and thus reduce nonspecific binding of cells in the responding cell population.[15] The dishes are now ready for immediate use for binding alloreactive CTL in the responding cell population.

SRBC Ghost-IgG Anti-SRBC Monolayer

This type of monolayer[17,18] has been used for the specific adsorption of cells with Fc receptors.

Materials

1. PLL (previous procedure, this article).
2. SRBC, 1×10^8/ml in PBS.
3. IgG fraction[37] of rabbit anti-SRBC.[17,18]

Procedure

Polystyrene tissue culture flasks (Falcon 3013) are coated with PLL followed by treatment with SRBC in the manner described in the previous procedure. The RBCs are lysed by incubation for 30 sec in distilled water, immediately followed by three washings with PBS. The flask is incubated with 5 ml of an appropriate dilution of IgG anti-SRBC for 45 min at 37° and washed three times with PBS.

Adsorption of Responding Cells to Monolayers

T Cells (CTL, T Suppressor, and T Helper Cells)

Successful binding of T cells to cellular monolayers usually requires incubation at 30°[8,11] or at 37°.[1] The length of incubation ranges from 30 min[27] to 4 hr.[7,9] Binding of CTL to target cells is an active process. Microcinematography studies[38,39] have shown the CTL crawling around on the target cells, moving from one to another. Incubation at 30° rather than at 37° has been used for binding CTL in the absence of target cell lysis.[8,11] Centrifugation of the CTL on the monolayers has also been used to promote binding.[4,40] The following procedure illustrates the adsorption

---

[37] B. A. L. Hurn, and S. M. Chandler, this series, Vol. 70 [5].

[38] T. L. Rothstein, M. G. Mage, G. Jones, and L. L. McHugh, J. Immunol. **121,** 1652 (1978).

[39] H. Ginsburg, W. Ax, and G. Berke, Transplant. Proc. **1,** 551 (1969).

[40] E. Kedar, M. O. de Landazuri, B. Bonavida, and J. Fahey, J. Immunol. Methods **5,** 97 (1974).

of CTL to monolayers and the removal of a nonadherent population depleted of CTL activity.

Procedure

Five milliliters of a single cell suspension[30] of spleen cells[31] from an appropriately immunized animal[4,29] at 1×10^7/ml in RPMI 1640 medium with 10% FCS is placed in a 100-mm-diameter tissue culture dish precoated with a monolayer of fibroblasts, macrophages, or B cells prepared as described in this article. If the subsequent incubation is not carried out in a CO_2-enriched atmosphere, it is recommended to use medium without bicarbonate[31] to prevent excess alkalinity.

The dish is incubated for 45 min at 37° to allow the cells in suspension to settle onto the monolayer. After the incubation, the dish is centrifuged at 700 rpm for 5 min at room temperature, and swirled to resuspend the nonadherent cells. This operation can be repeated several times on additional monolayer to increase the degree of depletion.[7]

FcR-Positive Lymphocytes

These cells can be bound to the SRBC ghost-IgG anti-SRBC monolayer in the same fashion as the CTL described above, except that the incubation time is reduced to 10 min at 37°, followed by centrifugation at 400 g for 5 min. Following collection of the unadsorbed cells and washing of the dish, the adsorbed cells are collected after they have detached by incubating the dishes overnight at 37°.[17] FcR$^+$ lymphocytes have also been separated on monolayers of anti-SRBC coated *unlysed* SRBC.[41] In this case, the FcR$^+$ cells are recovered after treatment of the monolayers with buffered isotonic NH$_4$Cl to lyse the SRBCs (this volume [6]).

Elution of Adsorbed Cells from the Monolayer

Elution with EDTA

Elution with EDTA has been used to remove peritoneal CTL from monolayers of PLL-bound target cells[13] and for the removal of rat lymph node CTL from monolayers of thoracic duct lymphocytes.[27] After removal of the nonadherent cell population, the dish is gently washed 3–4 times with medium (e.g., RPMI-1640 with 10% FCS) and 5 ml of 5 mM EDTA in medium is added to the dish.[13] The dish is incubated at 37° for 10 min with rocking[13] at 5 cycles/min. The eluted cells are then removed for assay.

[41] E. Kedar, M. O. de Landazuri, and B. Bonavida, *J. Immunol.* **112**, 1231 (1974).

Elution with Pronase

The washed monolayer (with the absorbed cells) is incubated for 30 min at 37° with Pronase (Calbiochem) (25 μg/ml) containing 1% Viocase (GIBCO, Grand Island, NY) in medium without FCS. The plate is rocked for 5 min and the detached cells are removed. The pronase is neutralized by adding 50% bovine serum to the removed cells. The monolayer is then retreated in the same manner, except that the Pronase concentration is 100 μg/ml and the incubation is for 10 min.[6,8]

Measuring the Extent of Contamination of Fractionated Cell Population by Spontaneously Released Monolayer Cells

When the monolayer cells are larger than or morphologically distinct from the cells to be fractionated, they can be distinguished by microscopic examination. Alternatively, the monolayer cells can be radioactively labeled and the radioactivity determined. In one study[16] these two methods were in good agreement. Indirect immunofluorescence[42] using appropriate alloantisera[26] may be used to identify monolayer cells that are allogeneic to the separated cells.

[42] L. A. Herzenberg and L. A. Herzenberg, *in* "Handbook of Experimental Immunology" (D. M. Weir, ed.), 3rd ed., Chapter 22.1. Blackwell, Oxford, 1978.

[13] Separation of Cells by Affinity Chromatography on Protein A Gels

By Victor Ghetie and John Sjöquist

General Remarks

Protein A from *Staphylococcus aureus* (SpA)[1] reacts with the Fc region of cell-surface IgG.[2,3] The interaction between SpA and cell-surface IgG has been used to develop four main techniques to isolate lymphocytes and other cells.

[1] Abbreviations: CRBC, chicken red blood cells; ES, erythrocytes coated with protein A; Sa, *Staphylococcus aureus;* SpA, protein A of *Staphylococcus aureus;* SRBC, sheep red blood cells.
[2] V. Ghetie, H. A. Fabricius, K. Nilsson, and J. Sjöquist, *Immunology* **26,** 1081 (1974).
[3] G. Dorval, K. I. Welsh, and H. Wigzell, *Scand. J. Immunol.* **3,** 405 (1974).

1. Binding of *Staphylococcus aureus* (Sa) to IgG-bearing cells or production of rosettes with ES and IgG-bearing cells[4,5] followed by separation of the unbound Sa or the ES from the rosetted cells by density gradient centrifugation.[6,7]

2. The use of magnetic albumin microspheres containing SpA coated with IgG antibody directed against specific cell surface antigens. These microspheres adhere to cells bearing the antigen which can be removed by applying a magnet.[8,9]

3. The use of monolayers of Sa or ES on plastic dishes which specifically bind IgG-bearing cells.[10,11] Unbound cells are removed by washing. Monolayer-adherent cells are detached by treatment with lysostaphin (for Sa) or ammonium chloride (for ES).[10] Alternatively, the monolayers may be treated with specific antibody directed against a membrane antigen[10] or IgG present on the cell surface.[12]

4. Affinity chromatography of cells on SpA-Sepharose gels.[13–15] This material may be used to bind cells having IgG on their surface or cells coated with an IgG antibody specific for a particular membrane antigen. The cells are removed from the column mechanically or by elution with soluble IgG.[13] The degree of purification achieved by affinity chromatography on SpA-Sepharose is comparable with that obtained with Sa or ES rosette or by specific adherence to Sa to ES monolayers.[13]

Until now, methods based on the interaction of immobilized SpA with IgG-bearing cells have not been used extensively for cell separation. Many authors still prefer immobilized antibody as a tool for cell isolation. However, since in recent years SpA-Sepharose 6MB has become commercially available, affinity chromatography may become the method of choice for separation of IgG-bearing cells.

[4] V. Ghetie, K. Nilsson, and J. Sjöquist, *Scand. J. Immunol.* **3**, 397 (1974).

[5] V. Ghetie, K. Nilsson, and J. Sjöquist, *Eur. J. Immunol.* **4**, 500 (1974).

[6] V. Ghetie, K. Nilsson, and J. Sjöquist, *Proc. Natl. Acad. Sci. U.S.A.* **71**, 4831 (1974).

[7] V. Ghetie, G. Stålenheim, and J. Sjöquist, *Scand. J. Immunol.* **4**, 471 (1975).

[8] K. I. Widder, A. E. Senyei, H. Ovadia, and P. Y. Paterson, *Clin. Immunol. Immunopathol.* **14**, 395 (1979).

[9] K. I. Widder, A. E. Senyei, H. Ovadia, and P. Y. Paterson, *J. Pharm. Sci.* **70**, 387 (1981).

[10] V. Ghetie and J. Sjöquist, *J. Immunol.* **115**, 659 (1975).

[11] A. A. Nash, *J. Immunol. Methods* **12**, 149 (1976).

[12] P. G. Bundesen and J. Gordon, *J. Immunol. Methods* **30**, 179 (1979).

[13] V. Ghetie, G. Mota, and J. Sjöquist, *J. Immunol. Methods* **21**, 133 (1978).

[14] A. Millon, M. Houdayer, and J. J. Metzger, *Dev. Comp. Immunol.* **2**, 699 (1978).

[15] S. Bartlett, E. Hulten, and P. Vretblad, *Immunol. 80 [Eighty], Int. Congr. Immunol., 4th, 1980* 19.5.01 (1980).

Preparation of SpA-Sepharose 6BM Gels

SpA and SpA-Sepharose 6MB (1 mg SpA/ml packed gel) may be obtained commercially (Pharmacia, Uppsala, Sweden). SpA can also be prepared from Sa strain Cowan-1[16] or the methicillin-resistant strain A676.[17] SpA-Sepharose 6MB may be prepared as follows[13]: 5 g of CNBr-activated Sepharose 6MB is swelled and incubated overnight at 4° with 10 mg SpA in 10 ml (final volume) of 0.1 M carbonate-bicarbonate buffer, pH 8.8 containing 0.5 M NaCl (coupling buffer). The gel is washed and resuspended for 2 hr in 1 M glycine buffer pH 9.0. The excess blocking reagent is removed by washing sequentially with coupling buffer, 0.1 M acetate buffer pH 4.0 containing 0.5 M NaCl and again coupling buffer. The gel is then resuspended in an appropriate volume of culture medium (e.g., TC-199 or RPMI-1640) containing sodium azide (1 mg/ml) and stored at 4° until used. By using radiolabeled SpA[18] it was found that about 2 mg of SpA are bound to 1 ml of packed gel. SpA-Sepharose 6MB with lower or higher SpA content can be prepared by modifying the SpA/Sepharose ratio. However, a high degree of substitution is not recommended because it may cause nonspecific absorption of cells. Moreover, the elution of bound cells may become difficult due to multipoint interaction between cells and adsorbent.

Preparation of SpA-Sepharose 6MB Columns

SpA-Sepharose 6MB is resuspended in tissue culture medium and packed into a column of appropriate size. We have used small-size plastic columns (0.5 × 3 cm) with an enlarged upper part (Bio-Rad, Richmond, Virginia) or medium-size Pharmacia columns (1 × 7 cm). Both types of columns are provided with a nylon screen at the bottom allowing the free passage of cells. Columns may be sterilized by washing for 30 min at room temperature with 1% formaldehyde in phosphate-buffered saline followed by extensive washing with sterile medium. SpA-Sepharose 6MB columns equilibrated with sterile tissue culture medium may be kept indefinitely at 4°. The cell fractionation, however, must be performed at room temperature since at 4° the viability of cells may be affected.

Before using an SpA-Sepharose 6MB column for cell separation, its ability to interact with cell-bound IgG should be checked by the following procedure. Suspensions of 2% chicken (CRBC) and sheep (SRBC) red blood cells are mixed in 2 : 1 ratio. The suspension is treated with sub-

[16] H. Hjelm, K. Hjelm, and J. Sjöquist, *FEBS Lett.* **28,** 73 (1972).
[17] R. Lindmark, J. Movitz, and J. Sjöquist, *Eur. J. Biochem.* **74,** 623 (1977).
[18] P. Biberfeld, V. Ghetie, and J. Sjöquist, *J. Immunol. Methods* **6,** 249 (1975).

agglutinating doses of anti-SRBC rabbit IgG (5 μg/ml suspension). After a 30-min incubation at room temperature, the cells are washed several times with TC-199 medium. The suspension (0.75 ml) is applied to a column of SpA-Sepharose 6MB (1 × 7 cm). The column is closed and kept for 15 min at room temperature. The nonadherent cells are washed off with medium. The percentage of CRBC and SRBC in the eluate is determined (CRBC are nucleated). The column is then washed with 20 ml of medium and the gel is gently suspended with the aid of a Pasteur pipet. The eluted cells are collected and the percentage of CRBC and SRBC is again determined. If the SpA-Sepharose 6MB gel is correctly prepared the results presented below should be obtained.

Sample	CRBC (%)	SRBC (%)
Initial mixture	67	33
Nonadherent cells	95	5
Adherent cells	5	95

The column is finally washed with 30 ml of medium, 10 ml of 0.2 M acetic acid (to remove the antibody bound to the column) and again with 50 ml of medium.

SpA-Sepharose 6MB columns are easy to prepare and may be used many times over. The results are reproducible. Physical entrapping and nonspecific adsorption to SpA-Sepharose 6MB are negligible.

Affinity Chromatography of Cells on SpA-Sepharose 6MB Column

Cells (no more than 2 × 10^7/ml packed gel) suspended in a volume of medium smaller than the void volume of the column (0.2 ml for the small-size column) are applied to the column. The column is closed and kept for 15 min at room temperature. Nonadherent cells are washed off with medium at a flow rate of approximately 150 ml/hr (20 ml/hr for the small-size column). The completeness of the removal can be checked under the microscope. Specifically bound cells are eluted by one of the following procedures:

1. Two milliliters of dog IgG (Miles Laboratories, Elkhart, IN) (for the small-size column) in TC-199 medium (20 mg/ml) is passed through the column. The column is closed, kept for 1 hr at 37°, and then washed with TC-199 medium containing dog IgG (5 mg/ml) until no cells are eluted (about 15 ml for the small-size column).

2. The closed column is stirred on a Vortex mixer for 3 min at the lowest speed by inserting the bottom end of the column into the rubber shaker. Immediately after shaking, the column is eluted with medium at room temperature (about 10 ml for the small-size column).

3. Five milliliters of medium (for the small-size column) is added. The gel is resuspended in the large upper part of the column with a coiled stainless-steel wire and simultaneously washed with about 10 ml of medium at room temperature.

After each run the column is washed with 25 ml of 0.1 M glycine buffer, pH 3.0 containing 0.15 M NaCl (sterilized through a Millipore filter, if necessary) to remove any bound protein. The column is then equilibrated by extensive washing with medium and kept at 4° (either under sterile conditions or with sodium azide) until further used. Chromatography on SpA-Sepharose 6MB is probably the simplest and cheapest of all the techniques using SpA for separation of cells.

Cells eluted off the column with dog IgG and cultured for 24 hr have the same viability (65%) as unfractionated cells. On the other hand, cells recovered by resuspending the gel have a lower viability (45%). The number of cells eluted with the latter procedure, however, appears to be higher (Table I). Dog IgG does not elute all types of IgG-bearing cells bound to the column. Only 30% of mouse thymocytes coated with rabbit anti-Thy 1 antibody are eluted and CRBC coated with rabbit IgG antibodies are not eluted at all from SpA-Sepharose 6MB.[13] Although SpA is a B

TABLE I

ELUTION OF MOUSE SPLEEN LYMPHOCYTES COATED
WITH RABBIT ANTI-MOUSE IgG FROM A
SpA-SEPHAROSE 6MB COLUMN

Method of elution	Cells eluted (%)	Viability (%)[a]		
		Before elution	After elution	After culture
Dog IgG[b]	61	95	89	65
Stirring	74	95	86	ND[c]
Resuspending	100	95	83	45

[a] Determined by trypan blue method.
[b] Only 10% of the mouse lymphocytes were eluted with fetal calf serum or rat IgG.
[c] Not determined.

TABLE II
Isolation of Ig-Bearing Mouse Lymphocytes
by Affinity Chromatography on
SpA-Sepharose 6MB

Cell fraction	Ig-bearing cells[a] (%)
Before chromatography[b]	48.8 ± 1.7
After chromatography	
Nonadherent to column	10.9 ± 1.2[c]
Adherent to column	90.3 ± 1.8

[a] As determined by ES rosette assay.
[b] The cells were treated with rabbit anti-mouse IgG.
[c] By rechromatography the percentage of ES rosette forming cells became less than 2%.

and T cell mitogen,[19] mouse spleen lymphocytes fractionated on SpA-Sepharose 6MB do not show increased ability to incorporate [^{3}H]thymidine.

Applications

Affinity chromatography on SpA-Sepharose 6MB has been used for the isolation of mouse spleen B lymphocytes,[13] porcine lymphocytes,[14] guinea pig cells bearing labile bound IgG (L cells),[20] and human peripheral blood lymphocytes.[15] Data on the isolation of mouse B lymphocytes after treating the cells with rabbit anti-mouse IgG are presented in Table II. Clearly, the percentage of Ig-bearing cells (B cells) is greatly decreased in the nonadherent cell population, while the adherent cell population is considerably enriched in IgG-bearing cells. The viability of the fractionated cells was not significantly affected.

In order to remove the IgG anti-mouse Ig antibodies from the cell surface, the cells are cultivated in an appropriate medium (with fetal calf serum) for 24 hr. The antibodies are almost completely shed from the surface within the first 6 hr of incubation. Culturing is a necessary step in order to obtain "clean" cells, since removal of anti-Ig antibody from the mouse lymphocyte surface causes modulation of the corresponding receptors.[20] It is also required for the cells to resynthesize their lost receptors.

[19] J. J. Langone, *Adv. Immunol.* **32**, 158 (1981).
[20] M. A. Dobre, I. Moraru, and E. Mandache, *Rev. Roum. Biochim.* **18**, 258 (1981).
[21] M. C. Raff and S. De Petris, *Fed. Proc., Fed. Am. Soc. Exp. Biol.* **32**, 48 (1973).

TABLE III
SEPARATION OF Ig-BEARING MOUSE LYMPHOCYTES (B CELLS)
BY DIFFERENT TECHNIQUES

	Ig-bearing cells (%)		Degree of enrichment
Technique	Before fractionation	After fractionation	
Affinity chromatography on SpA Sepharose 6MB	47	91	1.9
ES-rosette separation	41	82	2.0
Adherence to ES monolayers	39	83	2.1

Limitation of the Procedure

The degree of purification of Ig-bearing cells achieved with SpA-Sepharose 6MB is comparable to that obtained with ES-rosette (this volume [7]) or on ES-monolayer (this volume [12]) (Table III).

Separation of well-defined cell populations on SpA-Sepharose 6MB can generally be achieved only if antibodies (from species with SpA-reacting IgG) against specific surface antigens are available. An IgG antibody directed against a specific antigen is added to a population containing various types of cells. The cells are then applied to a SpA-Sepharose 6MB column. The nonadherent population is eluted with medium, as described above. The antibody-bearing subpopulation is eluted from the column with IgG or by mechanical treatment. The subpopulation enriched in cells bearing the specific antigen has interacted with antibody and SpA. This is an unavoidable consequence of methods using antibody as a tool for cell separation. The antibody present on the surface of the eluted cells may affect their normal function. For this reason, affinity chromatography of cells on SpA-Sepharose 6MB is the method of choice for *depletion* of a cell population rather than for the enrichment of a population in cells bearing a specific antigen.

Acknowledgments

Our investigation on the use of SpA in cell separation was supported by grants from the Swedish Medical Research Council and the Romanian Academy of Medical Sciences, and was done in cooperation with Drs. K. Nilsson, G. Stalenheim (Uppsala), G. Mota, I. Moraru, M. A. Dobre, and E. Mandache (Bucharest).

[14] Separation of Lymphoid Cells Using Immunoadsorbent Affinity Chromatography

By ROGER A. HUBBARD, SAMUEL F. SCHLUTER, and JOHN J. MARCHALONIS

Introduction

A key to the understanding of the immune system is the ability to separate and purify lymphoid cells so as to study their characteristics, functions, and interactions. A major advance in achieving this aim has been the development of techniques which take advantage of the discriminatory power of antibodies to select for cells bearing a particular type of surface antigen. In many cases this results in the separation of functional lymphocyte subsets since these are usually characterized by particular surface antigen markers. For example, anti-IgM antibodies can be used to separate B cells, most of which carry surface immunoglobulin molecules, from T cells, which do not carry surface immunoglobulin molecules. In this chapter we will describe the use of immunoadsorbent techniques to separate lymphoid cells. In these methods antibodies specific for a lymphocyte surface component are bonded to solid phase matrices to provide a selective anchor for isolating given subpopulations. Other methodologies using specific antibodies are described elsewhere in this volume.

General Considerations

The ideal technique should allow the quantitative recovery of a highly purified, viable, unaltered cell population and have the capacity to handle a minimum of 10^8 cells (i.e., one mouse spleen or thymus).[1] No present technique can satisfy all these criteria. Immunoadsorbent techniques are generally rapid and simple, capable of efficient and specific retention of cells even when these comprise a small fraction of the cell suspension, and allow the recovery of the unbound cells in a viable, unaltered state. However, the problem of the recovery of the bound cells has not been adequately solved[1] and reduced yields and viabilities have to be tolerated.

The type of matrix used in these techniques should satisfy the criteria of being inert, nontoxic, and exhibiting low nonspecific adsorption of cells. Additionally, coupling of protein to the matrix should be easily

---

[1] R. S. Basch, J. W. Berman, and E. Lakow, *J. Immunol. Methods* **56,** 269 (1983).

METHODS IN ENZYMOLOGY, VOL. 108

accomplished. Many materials ranging from dextran to magnetic beads have been used as immunoadsorbent matrices, but we will confine our discussion to Sepharose and glass or plastic beads.

All techniques for the enrichment of lymphocyte subsets should begin with the separation of the total lymphoid population from other cell types that might sterically interfere with the immunoadsorbent separation reaction. This can be easily done using the various starting materials commonly employed as good sources of lymphocytes, namely, peripheral blood, thoracic duct lymph, lymph nodes, thymus, or spleen. The composition of the media used to suspend the cells should be chosen to best maintain their viability but it should not contain any material that might interfere with the binding of the cells to the immunoadsorbent (e.g., materials cross reactive with the target antigen).

An important consideration regarding separation methods such as these is the possible impact of the technique itself on the functional characteristics of the cells. Specifically, interaction of immobilized selecting antibodies with lymphocytes can lead to the disappearance of epitopes through capping phenomena, loss, sometimes irreversibly, of receptor functions,[2] physical "stripping" off of epitopes,[1] or specific or nonspecific lymphocyte activation.

Philosophies of Lymphocyte Separation

As we have mentioned, the selectivity of these methods is provided by antibody molecules that are directed against subset specific antigenic determinants. These determinants delineate the various lymphocyte subpopulations which are classically referred to as T, B, and null cells. Thus B cells can be conveniently retrieved from a mixed population using antiimmunoglobulin, T cells by using antibodies to Thy 1 antigens, null cells are taken as those that remain after removal of both T and B cells. Additionally, many monoclonal reagents are commercially available which can further categorize T cells on the basis of Lyt antigens, for example, into killer, helper, or suppressor cells.[3] It is certain that in the future monoclonal antibody technology will allow us to define and separate lymphocyte subclasses to an even finer degree.

These antibody reagents can be used to specifically deplete a population of a particular cell type (negative selection) or, alternatively, to specifically select and purify a given cell type from the population (positive

---

[2] L. Moretta, M. C. Mingari, and C. A. Romanzi, *Nature (London)* **272**, 618 (1978).

[3] A. Matossian-Rogers, P. Rogers, and L. A. Herzenberg, *Cell. Immunol.* **69**, 91 (1982).

selection). Each selection procedure has inherent advantages and drawbacks. Negative selection procedures are the easier to perform, but are incapable of producing a defined, purified population of cells. This can only be achieved using positive selection procedures. However, with the positive selection procedures it is difficult to recover the immunoadsorbent bound cells with good yields and viabilities. Attempts to solve this problem have ranged from the use of mechanical agitation to release the cells, enzymes to digest the matrix[4] or the use of a gelatin matrix which is melted to release the cells,[5] but these mechanical techniques are not fully successful since nonspecifically trapped cells are also released and significant cell death occurs. The best method to use, as with all affinity techniques, is to specifically desorb the bound cells by competitive elution. This can be accomplished by using free antigen to block the immobilized antibody or, alternatively, free antibody to cover the antigenic sites on the cells. However, this is not always possible either because these reagents are unavailable or because the cells are bound too tightly through extensive multivalent attachments. Another possible way to solve this problem is to leave the cells attached and perform functional assays *in situ*.

Probably the most important parameter determining the success of immunoadsorbent techniques is the level of antibody substitution on the matrix. This will be influenced by whether positive or negative selection is desired. For negative selection the level of antibodies must be high enough to allow adequate crosslinking and thus retention of the bound cells. For positive selection the level of antibodies must be low enough so that desorption of the cells is not made too difficult. The antibody valency can be varied by changing the antibody concentration in the coupling solution and/or by mixing antibody and irrelevant immunoglobulin in different proportions. For best results the correct level of antibody substitution should be determined by the experimenter for each antibody/cell selection procedure.

Immunoadsorbent Procedures

In this chapter we will describe the method of immunoadsorbent affinity chromatography with reference to the procedures we use to separate Ig$^+$ and Ig$^-$ cells. We will identify the important parameters so that experimenters can adapt the protocols to their own requirements. We should

[4] E. F. Gold, R. Kleinman, and S. Bon-Efraim, *J. Immunol. Methods* **6**, 31 (1974).
[5] S. F. Schlossman and L. Hudson, *J. Immunol.* **11**, 313 (1973).

emphasize that these methods are very easy to perform and tolerant of a wide range of conditions.

Initial Cell Preparation

As it is important in these methods that single cell suspensions be used in order to avoid nonspecific trapping or clogging of the affinity columns, we shall briefly describe the procedures we use to prepare single cell suspensions.

To isolate the total mononuclear cell population from peripheral blood the method of Bøyum[6] is used. Briefly, venous blood is drawn aseptically into a syringe containing sufficient heparin (bovine lung, Upjohn) to yield a final concentration of 10 U/ml of blood. Heparinized blood diluted 1 : 2 with phosphate-buffered saline (PBS), pH 7.4, is then layered over Ficoll–Hypaque (Lymphoprep, Accurate Chemical and Scientific Corp, Amityville, NY) and centrifuged at 400 g for 45 min at 25°. The mononuclear cells at the plasma–Ficoll interface are withdrawn with a Pasteur pipet and washed 3 times with PBS. The cells are then resuspended to a concentration of 5×10^6/ml in RPM1 1640 (Gibco, Grand Island, NY) supplemented with 5% Nu-serum (Collaborative Research, Bethesda, MD), antibiotics, 10 mM HEPES, and 25 mM NaHCO$_3$ (complete medium).

To remove naturally adherent cells, i.e., monocyte–macrophages, the mononuclear cell suspension is incubated in plastic tissue culture flasks for 2 hr at 37° in a humidified CO$_2$ incubator (for this and other techniques for the removal of macrophages please refer to appropriate articles within this volume [27]). Nonadhering cells are typically 95% lymphocytes as assessed through microscopic morphology and membrane markers.

For the isolation of splenic lymphocytes, spleens are removed aseptically and transferred to a 100-mm petri dish containing 5–10 ml of complete medium. The spleens are carefully deencapsulated using a sterile scalpel and forceps and then placed in a small nylon mesh bag specifically designed for this purpose. With the aid of a sterile spatula and forceps, the spleen cells are carefully teased through the mesh. The cells are then washed 3 times in complete medium and the final pellet is resuspended in 0.16 M NH$_4$Cl, 0.1 mM Na EDTA, 10 mM KHCO$_3$ (ACK) to lyse erythrocytes. Hemolysis is achieved by gentle agitation of the tube for 5 min at room temperature, followed by a centrifugation at low speed (100 g, 10 min) in order that red cell ghosts and other cellular debris will remain in

---

[6] A. Bøyum, *Scand. J. Clin. Lab Invest.* **21**, Suppl. 97, 1 (1968).

the supernate. The cells are then washed 3 times in complete medium and resuspended to $5 \times 10^6/\text{ml}$; monocyte–macrophages are removed as discussed previously.

Typical viabilities after lysing red cells are around 85–90%. If this level of viability is unacceptably low, one may elect to use the Ficoll–Hypaque method to isolate splenic mononuclear cells. When using this technique, the worker is cautioned that after centrifugation the mononuclear cells will form a diffuse zone in the upper part of the Ficoll–Hypaque rather than the distinct interface seen when human peripheral blood is treated in this way. Some high density small lymphocytes may collect in the lower reaches of this zone and, in fact, may pellet with the erythrocytes. One may wish to experiment with Ficoll–Hypaque formulations of higher density if the yield of a particular lymphocyte subset of interest appears low. Ultimately, the decision between the ACK lysis method or the Ficoll–Hypaque method should be based upon the individual's criteria of yield and viability.

Immunadsorbent Affinity Chromatography

Beads are generally used for affinity chromatography since they pack evenly and allow uniform flow. For use with lymphoid cells the beads should have a mean diameter of approximately 200 μm so that they pack leaving sufficient space between beads to allow cells to pass freely without being physically trapped.[7] We have used three types of materials that have proven satisfactory with regard to these requirements. (1) Sepharose 6MB (Macrobeads; 250–300 μm) obtained from Pharmacia Fine Chemicals. (2) Glass beads (200–225 μm) obtained from Sigma Chemical Company (glass beads Type IV) or from the 3M company (Super-Brite Glass Beads). (3) Plastic beads (nominal diameter of 160–300 μm) obtained from Nunc Plastics (Biosilon Beads).

The major advantage of affinity chromatography over other methods of immunoseparation (panning) is the enormous increase in surface area provided above that which is feasible using a flat plastic surface. This would provide an obvious advantage when fractionating weakly binding cells; as these cells are eluted, they will be somewhat retarded by their interaction, albeit weak, with protein immobilized on the beads. Nonbinding cells would, of course, be eluted much faster thereby affecting separation. The panning technique would probably not isolate a weakly binding cell.

[7] H. Wigzell, *Scand. J. Immunol.* **5,** Suppl. 5, 23 (1976).

For application to the affinity columns the concentration of cells should be no more than 5×10^7 cells/ml since higher concentrations can result in mechanical disturbances and clogging of the column.[7] Even so, a significant amount of nonspecific trapping and retention of cells on the columns can occur when target cells become attached, this despite the fact that there is very little nonspecific adsorption of lymphoid cells to control immunoglobulin coated beads.[7,8] Thus, contamination of the recovered bound cell fraction may result. These trapped cells can be released by very gently mixing the beads in the column with a Pasteur pipet and then washing with media. This procedure must be gentle or significant loss of the specifically bound cells will result.

Equipment. For columns we use 5 or 10 ml glass pipets plugged with an amount of glass wool just sufficient to support the bed material. In order to prevent nonspecific adherence of the lymphocytes, the columns, including the glass wool, must be siliconized. Ready made siliconizing solutions are available commercially (Sigmacote, Sigma; Silicone solution, Serva Fine Chemicals Inc; Surfasil, Pierce Chemical Co.) or can be made by dissolving 1–2 ml of dimethyldichlorosilane (Sigma) in 100 ml of benzene. The siliconizing solutions are toxic so one should wear gloves and work in a fume hood when treating glassware. A few milliliters of solution is added to the column (with the glass wool plug in place) and swirled around so that all surfaces are wetted. After discarding the remaining solution, the column is dried and then washed extensively with water. If desired the siliconized columns can easily be sterilized by autoclaving. Suitable columns are also available commercially (e.g., Pharmacia chromatographic column K9/15) but these usually cannot be autoclaved and need to be sterilized by other means (e.g., with ethylene oxide). With the commercial columns one should ensure that the disc supporting the bed material will allow free passage of the cells (approximately 80 μm mesh). Disposable plastic syringes can also be used as columns. We use Tygon tubing for connection to the column.

Immunoadsorbent Affinity Chromatography Using Sepharose 6MB

Preparation of the Gel. Proteins can be coupled to Sepharose by introducing reactive groups into the gel through a reaction with cyanogen bromide.[9] Under mild conditions these groups will react with the primary amino groups of proteins to form covalent crosslinks. Sepharose 6MB is supplied as a freeze-dried powder already activated with cyanogen bromide.

[8] G. L. Manderino, G. T. Gooch, and A. B. Stavitsky, *Cell. Immunol.* **41**, 264 (1978).

[9] R. Axen, J. Porath, and S. Ernback, *Nature (London)* **214**, 1302 (1967).

One gram of CNBr activated Sepharose 6MB (approximately 3 ml swollen gel) is allowed to swell in 1 mM HCl for 15 min. Low pH conditions are used to prevent base catalyzed hydrolysis of the reactive groups on the gel. The gel is washed with a total of 200 ml of 1 mM HCl on a sintered glass funnel. Several aliquots are used and the supernatant medium allowed to flow through between successive additions. The gel is then washed with 10 ml of coupling buffer and immediately transferred to 3 ml of the same buffer (giving a gel : buffer ratio of 1 : 1) containing the immunoglobulin (see below).

The amount of immunoglobulin added to the gel for coupling will depend on whether purified whole immunoglobulin fraction, affinity purified antibodies, or monoclonal antibodies are used and also on the degree of binding strength desired between the cells and the beads. Too high a level of immunoglobulin can also increase the chance of nonspecific adsorption. Using the globulin fraction of rabbit anti-mouse IgM serum for the separation of Ig$^+$ and Ig$^-$ cells, we obtain good results using 10 mg of protein per gram of dry Sepharose. As immunoglobulins are coupled with an efficiency of 80–95%, this results in about 3 mg of globulin per ml of swollen gel. If pure antibodies are used the protein should be reduced to approximately 1–5 mg/g dry Sepharose.[8,10]

The immunoglobulin adjusted to the appropriate concentration (i.e., 10 mg/3 ml or 1–5 mg/3 ml) is dialyzed against coupling buffer, usually 0.1 M NaHCO$_3$, 0.5 M NaCl, pH 8.0. The high salt concentration in this buffer minimizes protein–protein interactions and nonspecific adsorption. Borate, HEPES or phosphate can also be used for the coupling buffer but Tris or other materials containing amino groups should not be used since they will couple to the gel. Before adding to the gel, the A_{280} of the solution is measured. After the washed reactive gel and the immunoglobulin solution are mixed (see above), the suspension is rotated end over end (using a magnetic stirrer will damage the beads) for 2 hr at room temperature or, if more convenient, overnight at 4°. To remove the unbound ligand the gel is washed in a sintered glass funnel with 20 ml of 0.1 M acetate or 0.1 M citrate, 0.5 M NaCl, pH 4.0, followed by 20 ml of coupling buffer. This cycle is repeated 3 times. The A_{280} of the washings is measured and the coupling efficiency is determined by comparing the total A_{280} units ($A_{280} \times$ volume) recovered to the A_{280} units initially added to the gel. The coupling efficiency should be 80–95%. A number of reactive groups remain on the gel after coupling and these are blocked by incubating the gel for 2 hr at room temperature with 0.2 M glycine (pH 8)

---

[10] A. Marshak-Rothstein, P. Fink, T. Gridley, D. H. Raulet, M. J. Bevan, and M. L. Gefter, *J. Immunol.* **122**, 2491 (1979).

or 1 M ethanolamine (pH 8). The blocking agent is removed by washing with 50 ml of coupling buffer. The gel is now ready for use and can be stored up to several years at 4° in buffer containing 0.05% sodium azide.

Affinity Chromatography of Cells

The following is the procedure we use to separate Ig^+ and Ig^- cells. Except where noted, all procedures are carried out at room temperature.

To pour the column the slurry should be fairly thick but still allow air bubbles to rise; an approximately 75% settled gel suspension is about right. A little buffer is poured through the column to remove any trapped air bubbles and with a small amount of buffer remaining in the column the gel suspension is added and allowed to pack under gravity flow. The column is then washed extensively with at least 10 bed volumes of media. Before applying the cells the media is allowed to flow through the column until the level just touches the top of the gel. We control the flow by clamping the outlet tubing. The cell suspension ($1-5 \times 10^7$ cells/ml; 10^8 cells maximum) is added and allowed to pass through under gravity flow, followed immediately by 2 bed volumes of media. This fraction comprises the nonbound cells. After washing the column with 10–20 bed volumes of media, the bound cells are specifically eluted with 1 mg/ml of purified mouse IgM in media. Release of the bound cells arises through competition between the cell surface immunoglobulin and the free immunoglobulin for the gel immobilized antibody. Alternatively it is possible to recover the cells by mechanical agitation: The gel is poured into a petri dish containing a few milliliters of cold (4°) media and agitated by swirling. After allowing the beads to settle the released cells are recovered in the supernatant medium. The gel is regenerated by washing with 2–3 bed volumes of 0.1 M glycine, 0.5 M NaCl, pH 2.5, followed by at least 10 bed volumes of media and stored at 4° in the presence of 0.05% sodium azide.

As judged by immunofluorescent staining[11] with rabbit anti-mouse IgM, we recover about 80–90% of the Ig^- cells in the unbound fraction.[12] No Ig^+ cells can be detected in this fraction. Conversely, 80–90% of the bound cells eluted with mouse IgM possess surface immunoglobulin. These comprise about 35–40% of the Ig^+ cells applied to the column. Desorption of the cells by mechanical agitation results in a higher yield (approximately 60–70%) but the extent of contamination with Ig^- cells is significantly increased.

Remarks. We obtain complete adsorption of target cells by allowing the suspension to flow uninterrupted through the column. Other workers

---

[11] D. DeLuca, in "Antibody as a Tool: The Applications of Immunochemistry" (J. J. Marchalonis and G. W. Warr, eds.), p. 189. Wiley, New York, 1982.

[12] G. W. Warr, J. C. Lee, and J. J. Marchalonis, *J. Immunol.* **121,** 1767 (1978).

have found that to obtain complete adsorption it is necessary to incubate the cells for about 10–20 min after they have entered the column.[13] There are two problems that may arise if this is done. There is an increased chance that cellular metabolic processes such as capping or shedding of surface antigens will lead to reduced efficiency of separation of cells. Thus it may be necessary to inhibit these processes using sodium azide (0.02%) or by lowering the temperature to 4°. Short exposure to sodium azide appears not to damage cells.[14] In addition, increasing the time of contact between the cells and the beads increases the degree of crosslinking making elution of bound cells more difficult. This may be desirable if the aim is merely to specifically deplete a cell population.

As explained earlier, cells bound to the immunoadsorbent are best recovered by competitive elution. However, if this is not possible the cells must be removed by mechanical agitation. Gentle swirling as described above may be adequate, or more vigorous action required such as vortexing the beads for 10 sec.[10] These actions usually result in a loss of cell viability (e.g., 90% compared to 98%[10]).

Immunadsorbent Affinity Chromatography Using Glass and Plastic Beads

Procedure. This technique is similar to panning in that immunoglobulin is noncovalently attached to the glass or plastic beads by simply incubating the two together. The method we use is basically that of Wigzell[7] with the incorporation of only a few changes. Columns and glass wool packing are siliconized as described previously. Prior to use, glass beads should be acid washed with any common laboratory acid cleaning solution followed by extensive washing with distilled water. After finally resuspending the beads in PBS they can be autoclaved, if desired. The flask containing the beads is then swirled to form an easily pourable slurry and the column is then poured. Polystyrene plastic beads are supplied sterile and can be used directly as supplied. They are simply resuspended in PBS to form a slurry and then poured into a column.

Once either type of column is poured, they are ready to be coated with antibody. The clamp or valve at the bottom of the column is opened to drain off excess PBS followed by the addition of protein solution in PBS. For most proteins, a concentration of 1 mg/ml will saturate the bead surfaces.[7]

One void volume of added antibody solution will ensure that each

[13] "Cell Affinity Chromatography. Principles and Methods." Pharmacia Fine Chemicals Handbook, Uppsala, Sweden, 1981.
[14] G. E. Schrempf-Decker, D. Baron, and P. Wernet, *J. Immunol. Methods* **32**, 285 (1980).

bead in the column is optimally exposed to protein. Coating is then achieved by incubating the column overnight at 4° or by a 30-min incubation at 45° followed by a 30-min incubation at 4° (alternatively, coating of the beads can be done when the beads are in a slurry and the column can be poured thereafter). After coating, the column is washed free of unbound protein with at least three void volumes of PBS; the A_{280} of the effluent should be monitored to determine if the washing is complete. To preclude nonspecific attachment of cells, the final wash of the column should be with PBS + 5% FCS. If added serum presents a problem, PBS + 1% gelatin may be used. After incubation of the column for 1 hr at 4° in the final washing PBS + serum or gelatin, the cell separation can be carried out.

At either 4° or at room temperature cell suspensions of 1 to 5×10^7/ml are added to a column having a bed volume of 1 ml per 5×10^6 cells. Elution is begun and the elution speed is adjusted such that a nonadsorbed cell would travel down the column no faster than 1 cm/min. This rate can be approximated by collecting one fraction each minute of a given volume. This volume can be arrived upon for a particular column by dividing the known void volume by the column bed height. Cells in each fraction can then be assayed for desired markers or functions.

For recovery of the positively selected cells the same considerations and methods apply as were discussed for Sepharose beads in the previous section. Thus the cells can be recovered by competitive elution with a pertinent ligand or by mechanical agitation.

[15] Use of Anti-Fab Columns for the Isolation of Human Lymphocyte Populations

By PAUL L. ROMAIN and STUART F. SCHLOSSMAN

A number of methods based on the cell surface and physical characteristics of peripheral blood mononuclear cells (PBMC) have been used to separate these cells into T, B, monocyte, and null cell populations. Rosetting of PBMC with sheep erythrocytes will positively select a population consisting mostly of T lymphocytes. A small number of contaminating E rosette-positive non-T non-B cells may also be present. These include cells previously termed null cells, that are highly effective at natural killing and ADCC. The reciprocal E rosette-negative population is quite heterogeneous, usually including only 20% B lymphocytes, as well

METHODS IN ENZYMOLOGY, VOL. 108

as monocytes, E rosette-negative (E⁻), surface immunoglobulin (sIg) negative, "null" cells, and variable numbers of T cells.[1] Despite its heterogeneity, this cell mixture is often used as the source of B cells for *in vitro* assays of immune cell function. Depletion of phagocytic or adherent mononuclear cells from this preparation often leads to enrichment for B cells to approximately 30–50% of the total population.[2] A highly enriched preparation of B cells can be isolated, however, by taking advantage of the unique presence of surface membrane Ig on B lymphocytes. Early reports using anti-Ig immunoabsorbent columns demonstrated that B cells with sIg can be specifically retained from lymphoid cell suspensions, although nonspecific retention of cells may occur.[3,4] Such nonspecific retention was shown to be minimized by use of a Sephadex G-200 anti-Fab column from which B cells could be eluted by enzymatic digestion of the soluble immunoabsorbent.[5] This method was further adapted to obtain relatively purified populations of human B cells by elution from such columns with human γ-globulin.[6] T cells, monocytes, and null cells can then be more easily enriched for by other techniques.

This chapter will focus on a detailed description of the methods for preparing and using such specific column chromatography for the isolation of B cells. By this method, B cells are selected by binding of their sIg to rabbit anti-human F(ab')₂ attached to these columns.

Preparation and Purification of Rabbit Anti-F(ab')₂ Sera

Reagents

Human γ-globulin, F. II (Miles Laboratories, Elkhart, IN)
Sodium acetate, 0.1 M, pH 4.5
Crystallized pepsin (Miles)
Sodium sulfate
Phosphate buffered saline (PBS), pH 7.4
Complete Freund's adjuvant (CFA, Difco Laboratories, Detroit, MI)
Sepharose 4B (Pharmacia Chemicals, Piscataway, NJ)
Cyanogen bromide (Eastman Kodak, Rochester, NY)
Sodium hydroxide, 1 N

---

[1] K. C. Anderson, J. D. Griffin, M. P. Bates, B. Slaughenhoupt, S. F. Schlossman, and L. M. Nadler, *J. Immunol. Methods* **61,** 283 (1983).

[2] P. L. Romain, G. R. Burmester, R. W. Enlow, and R. J. Winchester, *Rheumatol. Int.* **2,** 121 (1982).

[3] H. Wigzell, K. G. Sundquist, and T. O. Yoshida, *Scand. J. Immunol.* **1,** 75 (1972).

[4] M. Crone, C. Koch, and M. Simonsen, *Transplant. Rev.* **10,** 36 (1972).

[5] S. F. Schlossman and L. Hudson, *J. Immunol.* **110,** 313 (1973).

[6] L. Chess, R. P. MacDermott, and S. F. Schlossman, *J. Immunol.* **113,** 1113 (1974).

Borate-buffered saline (BBS), pH 8.3
Glycine–HCl buffer 0.1 M, pH 2.5
Phosphate buffer 2 M, pH 8.0

Procedure

1. Dissolve 5 mg of pepsin in 0.1 M sodium acetate buffer (pH 4.5).
2. In the above solution dissolve 500 mg of human γ-globulin and incubate at 37° for 20 hr.
3. After incubation, adjust the digested solution to pH 8.0, then drop-wise add 50 ml of Na_2SO_4 (25 g/100 ml). (The white precipitate formed contains the $F(ab')_2$.)
4. Dissolve the precipitate in water, then dialyze it against 0.1 M sodium acetate, followed by PBS and bring it to a concentration of 5 mg/ml in PBS.
5. Emulsify 1 mg (0.2 ml) of $F(ab')_2$ in CFA and inject it intramuscularly weekly for 3 weeks into rabbits.
6. One week after the final immunization, the rabbits may be bled and sera collected.
7. To purify the above antisera, first prepare activated Sepharose by mixing 20 ml of 50 mg/ml CnBr with 30 ml of Sepharose 4B for 12 min, maintaining the pH at 11.0 with 1 N NaOH. (Note: These steps should be performed with appropriate care in a chemical fume hood.)
8. Wash the activated Sepharose in the borate buffered saline, then add to it 200 mg of human gammaglobulin for 18 hr at 4°.
9. After washing the conjugated Sepharose exhaustively with PBS, pack it into a 1 × 40-cm glass column to a height of 10 ml.
10. Pass 100 ml of the rabbit anti-$F(ab')_2$ sera through the packed column. Anti-$F(ab')_2$ antibody will be retained by the γ-globulin conjugated Sepharose column.
11. Elute the retained anti-$F(ab')_2$ from column with the glycine–HCl buffer, collecting it in 2 M phosphate buffer.
12. Dialyze the purified antibody against PBS, concentrate it to approximately 10 mg/ml, and store at −70°.

Preparation of Sephadex G-200 Anti-$F(ab')_2$
Immunoabsorbent Columns

Reagents

Sephadex G-200 (Pharmacia)
CnBr
NaOH, 1 N
BBS, pH 8.3

Purified rabbit anti-human F(ab')$_2$ (see above)
PBS
Medium 199 (Gibco, Grand Island, NY)
Fetal calf serum (FCS) (Microbiological Associates, Walkersville, MD)
Ethylenediaminetetraacetic acid (EDTA), 2.5 mM
Penicillin–streptomycin solution (Gibco)

Procedure

1. Activate 60 ml of Sephadex G-200, sieved to achieve uniform bead size (88–120 μm), with 100 mg CnBr for 10 min while maintaining the pH at 10.2 with 1 N NaOH (using appropriate safety precautions).

2. Wash the activated Sephadex G-200 with 500 ml–1 liter of BBS, then add 20 mg of the purified anti-F(ab')$_2$ and incubate at room temperature for 4 hr without mechanical stirring.

3. Wash the Sephadex G-200 anti-F(ab')$_2$ conjugate with PBS over a sintered glass funnel without suction, mixing every 15 min with a glass rod.

4. Fit disposable 12-ml syringes with polyethylene sintered disks (Bell Art Products, Benawalk, NJ), and pack them with 8–10 ml of the anti-F(ab')$_2$ conjugated Sephadex.

5. Wash the columns with medium 199 containing 5% FCS, 2.5 mM EDTA, and 1% penicillin–streptomycin.

The columns are now ready for use in fractionation. It should be noted that the anti-F(ab')$_2$ should be purified and tested for antibody activity prior to conjugation to the Sephadex G-200. In addition, with proper CnBr activation a 20–30% loss in volume of Sephadex will result. Once preparation is completed the anti-F(ab')$_2$-conjugated Sephadex can be stored in sodium azide at 4° for up to 3 months without loss of activity. If the conjugate is stored in azide it should be rewashed extensively in medium (as in step 5 above) prior to its use in separating cells for functional studies.

Cell Preparation and Fractionation

Reagents

Ficoll–Hypaque (Pharmacia)
Hanks' balanced salt solution (HBSS, Microbiological Assoc.)
Medium 199 (complete, as described in step 5 above)
Human AB serum, pooled from multiple donors, or
Human γ-globulin (Miles)
Minimum essential medium (MEM) (Gibco)

Procedures

1. PBMC are isolated in a standard fashion by Ficoll–Hypaque density gradient centrifugation, washed, depleted of mononuclear phagocytes, and resuspended in complete medium at $10–20 \times 10^6$ cells/ml.

2. A 5–10 ml amount of the lymphocyte suspension is then added to the column at room temperature (a total of $50–100 \times 10^6$ cells per 8–10 ml column is usually optimal).

3. Eluates are then collected (containing sIg-negative cells) by stepwise elution with 15 ml amounts of medium at a flow rate of 0.3–0.5 ml/min until the effluent is virtually cell free.

4. To elute the retained (sIg$^+$) cells the column is then washed for 30 min with two 15 ml volumes containing either human γ-globulin at 10 mg/ml or 20% human AB serum in MEM. During the incubation and elution by competitive binding with human Ig containing wash, the column can be gently mixed by drawing the column material up and down in a Pasteur pipet.

5. The eluted cells are then collected in medium, washed, and analyzed before use.

It should be noted that total cell recovery should be relatively high since cells are not nonspecifically retained on Sephadex G-200. The monocyte depletion step is extremely important since residual monocytes may be found in both the Ig$^-$ and Ig$^+$ preparations. Contamination with monocytes (or T cells) of the Ig$^+$ cells may be diminished to a degree by preincubation of the cells to be passed through the column at 37° for 1 hr to remove cytophilic Ig.[7] Phagocytosis-dependent monocyte depletion methods (such as carbonyl iron) (this volume [30]) are preferable since some B cells may be isolated with the monocytes during adherence procedures and significant numbers of B cells bind to nylon wool columns. Given the above qualifications, the sIg$^+$ population usually consists of 70–80% B cells, with the remaining cells divided about evenly between null cells and monocytes, with a small fraction (3–4%) of T cells.

Use of Anti-Fab Columns for Isolation of T and Null Cells

By combining the use of the anti-Fab column with separation techniques described elsewhere in this volume (including E rosetting), it is possible to rapidly and efficiently separate cells into enriched functional T, B, and null cell fractions.[8–10] Recent advances in the production of a

[7] P. Stashenko, L. M. Nadler, R. Hardy, and S. F. Schlossman, *J. Immunol.* **125**, 1678 (1980).

[8] R. E. Rocklin, R. P. MacDermott, L. Chess, S. F. Schlossman, and J. R. David, *J. Exp. Med.* **140**, 1303 (1974).

variety of monoclonal antibodies which identify distinct subsets of PBMC have made the isolation of even more highly purified subsets of cells possible, including B cell preparations which are consistently >90% pure. Nonetheless, these techniques require the availability of a large quantity and variety of monoclonal reagents.[1,11] The anti-Fab column technique is particularly useful for the isolation of a highly enriched (or depleted) B cell preparation. It has the inherent advantage that it positively selects a distinct cell population on the basis of a well-characterized surface structure of known functional importance.

Acknowledgments

This work was supported in part by National Institutes of Health Grant AI 12069. Dr. Romain is the recipient of National Institutes of Health Clinical Investigator Award K08-AM01181.

---

[9] P. M. Sondel, L. Chess, R. P. MacDermott, and S. F. Schlossman, *J. Immunol.* **114**, 982 (1975).
[10] R. P. MacDermott, L. Chess, and S. F. Schlossman, *Clin. Immunol. Immunopathol.* **4**, 415 (1975).
[11] J. D. Griffin, R. Beveridge, and S. F. Schlossman, *Blood* **60**, 30 (1982).

[16] Fractionation of Human Lymphocytes on *Helix pomatia* A Hemagglutinin-Sepharose and Wheat Germ Agglutinin-Sepharose

By U. Hellström, M.-L. Hammarström, S. Hammarström, and P. Perlmann

Introduction

Lectins are frequently used to characterize different cell types and to separate cells from each other. The rationale for this use are differences in expression of cell surface bound glycopeptides and glycolipids on cells of different origin or different stage of differentiation. Different techniques applied in this context include cell agglutination,[1] rosetting,[2] and affinity

---

[1] Y. Reisner, M. Linker-Israeli, and N. Sharon, *Cell. Immunol.* **25**, 129 (1976).
[2] Y. Reisner, S. Ikehara, M. Z. Hodes, and R. A. Good, *Proc. Natl. Acad. Sci. U.S.A.* **77**, 1164 (1980).

chromatography with lectins, either coated on nylon fibers,[3] plastic tubes, or Petri dishes[4] or covalently bound to Sepharose particles.[5] More recently fluorochrome-labeled lectins have also been used to analyze cell populations with the fluorescence-activating cell sorter (FACS).[6] For review see Sharon.[7]

In our studies, we have used two lectins, *Helix pomatia* A hemagglutinin (HP) and wheat germ agglutinin (WGA) to characterize and fractionate lymphocytes from humans as well as other species. In humans thymus-derived lymphocytes (T cells), comprising both circulating T cells[8] and thymocytes,[9] have surface carbohydrates which become accessible for interaction with the lectin HP after neuraminidase treatment of the cells.[8] In healthy adults, only a smaller fraction of the peripheral bone marrow-derived lymphocytes (B cells) interacts with HP.[10] In contrast, a large proportion of the B cells in cord blood[10] and malignant B cells from patients with chronic lymphocytic leukemia (CLL) have receptors for HP after enzyme treatment.[11] Similarly, in all other species tested, HP also binds preferentially to T cells. Thus, HP covalently coupled to Sepharose 6MB could be used to separate lymphocytes into T and B cell-enriched fractions in man,[5] mouse,[12] rat,[13] cow,[14] and horse.[15] Both in man and mouse, a single major surface glycoprotein (gp 150 kd in man and

---

[3] G. M. Edelman, V. Rutishauser, and C. F. Malette, *Proc. Natl. Acad. Sci. U.S.A.* **68**, 2153 (1971).

[4] D. H. Boldt and R. D. Lyons, *J. Immunol.* **123**, 808 (1979).

[5] U. Hellström, S. Hammarström, M-L. Dillner, H. Perlmann, and P. Perlmann, *Scand. J. Immunol.* **5**, Suppl. 5, 45 (1976).

[6] B. J. Fowles, M. J. Waxdal, S. O. Sharrow, C. A. Thomas, R. Asofsky, and B. J. Mathieson, *J. Immunol.* **125**, 623 (1980).

[7] N. Sharon, *Adv. Immunol.* **34**, 213 (1983).

[8] S. Hammarström, U. Hellström, P. Perlmann, and M-L. Dillner, *J. Exp. Med.* **138**, 1270 (1973).

[9] J. Zeromsky, S. Hammarström, U. Hellström, P. Biberfeld, and P. Perlmann, *Scand. J. Immunol.* **16**, 83 (1982).

[10] U. Hellström, P. Perlmann, E.-S. Robertsson, and S. Hammarström, *Scand. J. Immunol.* **7**, 191 (1978).

[11] U. Hellström, H. Mellstedt, P. Perlmann, G. Holm, and D. Pettersson, *Clin. Exp. Immunol.* **26**, 196 (1976).

[12] O. Haller, M. Gidlund, U. Hellström, S. Hammarström, and H. Wigzell, *Eur. J. Immunol.* **8**, 765 (1978).

[13] R. H. Swanborg, U. Hellström, H. Perlmann, S. Hammarström, and P. Perlmann, *Scand. J. Immunol.* **6**, 235 (1977).

[14] B. Morein, U. Hellström, L.-G. Axelsson, C. Johansson, and S. Hammarström, *Vet. Immunol. Immunopathol.* **1**, 27 (1979).

[15] H. Broström, U. Hellström, I. Ziwerts, N. Obel, and P. Perlmann, *Vet. Immunol. Immunopathol.*, in press.

gp 130 kd in mouse) is responsible for almost all HP-binding to the cells.[16] Minor HP-binding components, present on some lymphocytes but not on others were gp 100 kd and gp 200 kd and some glycolipids.[17,18] Thymocytes appear to contain considerable lower amounts of a less sialylated form of gp 150 kd than PBL. In the B cell compartment gp 150 kd appears to mark a separate lineage of B cells.[18]

In contrast to HP, WGA reacts with all circulating lymphocytes.[19,20] This binding cannot, as in the case of HP, be attributed to the presence of a single major cell surface component. On the contrary, WGA binds to a large number of lymphocyte surface glycopeptides.[21] However, there are clear-cut differences in the avidity with which WGA binds to either T, B, or null cells, i.e., lymphocytic cells lacking common T or B cell markers.[19,20,22] Nevertheless the differences in avidity for WGA between cells within the human T cell compartment are sufficient to permit fractionation into two subpopulations different with respect to mitogenic or mixed lymphocyte responsiveness and certain suppressor cell activities.[19,23–25]

Reagents

Lectins

Helix pomatia A hemagglutinin (HP) was prepared on insolubilized hog A + H blood-group substance as described earlier.[26] The HP preparations were analyzed for their content of hemagglutinin N/ml by a colorimetric method.[26] HP is a hexameric glycoprotein containing six homoge-

[16] B. Axelsson, A. Kimura, S. Hammarström, H. Wigzell, K. Nilsson, and H. Mellstedt, *Eur. J. Immunol.* **8,** 757 (1978).
[17] B. Axelsson, U. Hellström, S. Hammarström, H. Mellstedt, P. Perlmann, and D. Pettersson, *Hum. Lymphocyte Differ.* **1,** 285 (1981).
[18] B. Axelsson *et al., Eur. J. Immunol.* (submitted).
[19] U. Hellström, M.-L. Dillner, S. Hammarström, and P. Perlmann, *J. Exp. Med.* **144,** 1381 (1976).
[20] D. H. Boldt, R. P. MacDermott, and E. P. Jorolan, *J. Immunol.* **114,** 1532 (1975).
[21] M-L. Dillner-Centerlind, B. Axelsson, S. Hammarström, U. Hellström, and P. Perlmann, *Eur. J. Immunol.* **10,** 434 (1980).
[22] D. H. Boldt, *Mol. Immunol.* **17,** 47 (1980).
[23] S. Hammarström, U. Hellström, M.-L. Dillner, P. Perlmann, H. Perlmann, B. Axelsson, and E.-S. Robertsson, *in* "Affinity Chromatography" (O. Hoffman-Ostenhof, M. Breitenbach, F. Koller, D. Kraft, and O. Scheiner, eds.), p. 273. Pergamon, Oxford, 1978.
[24] T. Lehtinen, P. Perlmann, and U. Hellström, *Scand. J. Immunol.* **12,** 309 (1980).
[25] M-L. Dillner-Centerlind, U. Hellström, E-S. Robertsson, S. Hammarström, and P. Perlmann, *Scand. J. Immunol.* **12,** 13 (1980).
[26] S. Hammarström and E. A. Kabat, *Biochemistry* **8,** 2696 (1969).

neous carbohydrate binding sites (molecular weight 79,000 ± 4000).[27,28] The specificity range of HP is α-D-GalNAcp $>$ α-D-GlcNAcp $\approx$ β-D-GalNAcp $>$ β-D-GlcNAcp $>$ α-D-Galp.[29,29a] HP is not mitogenic for neuraminidase-treated or untreated human peripheral blood lymphocytes.[30]

Highly purified wheat germ agglutinin (WGA) was purchased from Pharmacia Fine Chemicals AB, Uppsala, Sweden. It was purified by affinity chromatography, using D-GlcNAc coupled to epoxy-activated Sepharose. WGA is a dimeric protein containing two or four carbohydrate binding sites (molecular weight $\sim$ 36,000).[31,32] The binding site is complementary to a sequence of three β-(1 $\rightarrow$ 4)-linked D-GlcNAc units (N,N',N''-triacetylchitotriose).[31,33] Under certain conditions WGA is mitogenic for human lymphocytes.[21]

HP and WGA were trace labeled with carrier-free Na^{125}I (Radiochemical Centre, Amersham, U.K.) using the chloramine-T procedure of Hunter and Greenwood[34] as described earlier.[35]

Lectin-Sepharose Gels

The lectins were covalently coupled to cyanogen bromide-activated Sepharose 6MB particles (Pharmacia Fine Chemicals AB). This matrix was developed for cell fractionation and consists of particles of large and relatively uniform size (250–315 μm in diameter). Columns packed with these particles give high flow rates, resulting in minimal nonspecific trapping of cells.[5]

Coupling was performed as follows: 1 g of lyophilized Sepharose 6MB particles was swollen in 1 mM HCl and washed on a glass filter with 1 mM HCl, followed by 0.1 M NaHCO$_3$–0.5 M NaCl, pH 8.6. The gel ($\sim$3 ml) was then conjugated with HP or WGA, containing trace amounts of ^{125}I-labeled lectin (6 ml solution in 0.1 M NaHCO$_3$–0.5 M NaCl, pH 8.6

[27] S. Hammarström and E. A. Kabat, *Biochemistry* **10**, 1684 (1971).

[28] S. Hammarström, A. Westöö, and I. Björk, *Scand. J. Immunol.* **1**, 295 (1972).

[29] S. Hammarström, L. A. Murphy, I. J. Goldstein, and M. E. Etzler, *Biochemistry* **16**, 2750 (1977).

[29a] Abbreviations of sugars: α,β,-D-GalNAcp, α- or β-linked 2-acetoamido-2-deoxy-D-galactopyranoside; α,β,-D-GlcNAcp, α- or β-linked 2-acetoamido-2-deoxy-D-glucose; α-D-Galp, α-linked D-galactopyranoside.

[30] M.-L. Dillner, S. Hammarström, and P. Perlmann, *Exp. Cell Res.* **96**, 374 (1975).

[31] Y. Nagata and M. M. Burger, *J. Biol. Chem.* **249**, 3116 (1974).

[32] R. H. Rice and M. E. Etzler, *Biochem. Biophys. Res. Commun.* **59**, 414 (1974).

[33] I. J. Goldstein, S. Hammarström, and G. Sumdblad, *Biochim. Biophys. Acta* **405**, 53 (1975).

[34] W. M. Hunter and F. C. Greenwood, *Nature (London)* **194**, 495 (1962).

[35] S. Hammarström, *Scand. J. Immunol.* **2**, 53 (1973).

containing 20 mg WGA or either 3.6 or 20 mg HP depending on whether the gel was to be used for fractionation of PBL or of T cells, respectively). The reaction mixture was stirred by slow vertical rotation overnight at 4°. The gel was then washed extensively with the coupling buffer. Remaining active groups were blocked by incubation for 2 hr at 20° with 0.5 M glycine in 0.1 M NaHCO$_3$, pH 8.3. To remove excess blocking reagent and physically adsorbed HP or WGA, the gel was washed three times with alternatively low and high pH buffer solutions (0.1 M acetate buffer, pH 4.0, and 0.1 M borate buffer, pH 8.0, both containing 0.5 M NaCl). Before use the gels were washed with Tris-buffered Hanks' salt solution, pH 7.4 (TH) and equilibrated overnight at 4° in TH containing 0.2% human serum albumin (HSA) and 0.02% NaN$_3$ (TH-HSA-NaN$_3$). Seventy to eighty-five percent of the lectins were bound to the gel under these conditions.

A degree of substitution of ~1 mg HP/ml swollen gel was found to be optimal to retain the HP-binding cells present in 25 × 10^6 neuraminidase-treated human PBL. For fractionation of the corresponding number of T cells on HP or WGA gels a higher degree of substitution (~5 mg/ml gel) was needed. A lower degree of substitution gave incomplete retention of binding cells, while a higher degree of substitution made it difficult to elute the cells from the columns. Cells could, however, still be eluted from moderately overcharged columns by mechanically stirring the gel or by increasing the concentrations of the competitive sugar haptens used for elution (see below).

When prepared as above the gels could be used several times. After fractionation, the gels were washed with phosphate-buffered saline, pH 7.2 (PBS) and stored at 4° in PBS + 0.02% NaN$_3$. On the day before use, the gels were washed three times by using alternatively low and high pH buffer solutions as described above. Thereafter, the gels were equilibrated by slow vertical rotation overnight at 4° with TH-HSA-NaN$_3$.

In our hands commercially available HP-Sepharose and WGA-Sepharose gels could not be used more than once. These gels also needed washing with high and low pH buffers before use since they often contain small amounts of free lectin. The presence of free lectin in the gel results in unsatisfactory fractionation of the cells.

We have used a rapid batch technique[25] to check that the lymphocyte preparations bind to and can be released from the Sepharose beads by addition of competitive sugar haptens. One tenth of equilibrated Sepharose 6MB particles was mixed with 2.5 × 10^6 lymphocytes (neuraminidase treated for HP-Sepharose or untreated for WGA-Sepharose) suspended in 0.1 ml of TH-HSA-NaN$_3$. The Sepharose suspensions were incubated for 15 min (HP) or 3 min (WGA) and thereafter directly trans-

ferred with some buffer to a microscopic slide. The number of lymphocytes bound per bead circumference was then scored in a phase contrast
microscope. Gels binding 20–25 lymphocytes per bead circumference
gave good separation of cells. In this batch technique the appropriate
sugar solutions can also be added to ascertain that bound cells can be
dissociated from the Sepharose beads.

Cells

Human PBL were purified from defibrinated blood by gelatin sedimentation, treatment with colloidal iron and magnet, followed by Ficoll–
Paque gradient centrifugation.[36] Purified human T cells comprising cells
with both high- and low-avidity receptors for sheep erythrocytes[25] were
prepared either by rosetting with neuraminidase-treated sheep erythrocytes in neat calf serum at 4° overnight[25] or by passage of PBL over a
column charged with human IgG complexed with rabbit anti-human IgG.[37]
Such cells are devoid of both B cells and T cells with receptors for the Fc
portion of IgG ($Fc\gamma R^+$ cells). The ability of T cells prepared by the rosetting procedure to bind to WGA[25] or HP[38] was markedly reduced. The lectin
binding capacity of the lymphocytes could be restored if the cells were
kept in 50% human normal serum (HNS) at 37° overnight in air + 5%
CO_2.[25] For successful fractionation of T cells obtained by rosetting with
sheep erythrocytes such an incubation is necessary.

To expose cellular receptors for HP, neuraminidase treatment of the
lymphocytes is needed.[8] Optimal conditions for treatment with neuraminidase are 10 μg Clostridium perfringens neuraminidase (type VI, 1–3
U/mg NAN-lactose substrate, Sigma Chemical Co., St. Louis, Mo.) to 25
× 10^6 lymphocytes in 2 ml TH + 0.1% HSA for 45 min at 37°. As shown
by binding experiments with ^{125}I-labeled HP, the same number of HP-
binding receptors on PBL (e.g., 0.6–1.1 × 10^6 molecules per cell[39]) was
obtained after treatment with Vibrio comma neuraminidase (Behringwerke AG., Marburg), 0.05 U per 25 × 10^6 lymphocytes in 2 ml TH
+ 0.1% HSA for 45 min at 37°.

---

[36] H. Perlmann, P. Perlmann, G. Pape, and G. Halldén, Scand. J. Immunol. 5, Suppl. 5, 57
(1976).
[37] H. Wigzell, Scand. J. Immunol. 5, Suppl. 5, 23 (1976).
[38] U. Hellström et al., unpublished (1982).
[39] U. Hellström, M-L. Dillner, S. Hammarström, and P. Perlmann, Scand. J. Immunol. 5, 65
(1976).

Fractionation

Medium

TH supplemented with 0.2% HSA and 0.02% NaN$_3$ was used throughout. Since both HP and WGA interact with serum glycoproteins which compete with the binding of the lymphocytes to the lectins, it is absolutely necessary to exclude fetal calf serum (FCS) or HNS from the fractionation medium. Moreover, HP is hexavalent in native form, and causes rapid capping of the membrane receptors for HP. To prevent capping NaN$_3$ was included in the medium. In this medium, fractionation can be performed at room temperature. WGA also induces capping of its cellular receptors although not as rapidly as HP. TH-HSA-NaN$_3$ was therefore also used for fractionation of lymphocytes on WGA Sepharose.

For specific elution of cells bound to HP-Sepharose, different concentrations of D-GalNAc were used.[5] Similarly D-GlcNAc was used to elute cells bound to WGA Sepharose.[19] Up to a concentration of 0.4 M these sugars do not affect the viability and functional capacities of the cells.

Procedures

Figures 1 and 2 show two experiments designed to determine the sugar concentrations needed for elution of lymphocytes bound to either HP- or WGA-Sepharose columns. Neuraminidase-treated (80 × 10^6) human PBL were applied to 3 ml HP-Sepharose column (1 mg HP/ml) and the nonretained cells were washed off with medium (Fig. 1). The column was eluted with 0.001, 0.01, 0.1, and 1.0 mg/ml of D-GalNAc. Six percent of the cells passed through the column. Almost no cells were eluted at 0.001 and 0.01 mg/ml while the majority of the cells were eluted at 0.1 mg/ml (15%) and at 1.0 mg/ml of D-GalNAc (55%). After these elutions no cells were left on the column as shown by subsequent treatment of the gel matrix with higher sugar concentrations (5 mg/ml) or by microscopy. The total cell recovery was 79%.

A 1-ml column of WGA-Sepharose (5 mg WGA/ml) was used to fractionate 25 × 10^6 human T cells (Fig. 2). Retained cells were eluted with 0.1, 1.0, and 10.0 mg D-GlcNAc/ml. Forty percent of the cells passed the column, 6% of the cells were eluted with 0.1 mg/ml, and 19 and 20% with 1.0 and 10.0 mg/ml, respectively. The cell recovery was 85% and no cells were left on the gel after the last elution step.

Based on these results the following fractionation procedures were adopted. HP-Sepharose (~1 mg/ml) was transferred to a small column (K9/15, Pharmacia Fine Chemicals AB) supplied with a 80-μm nylon net.

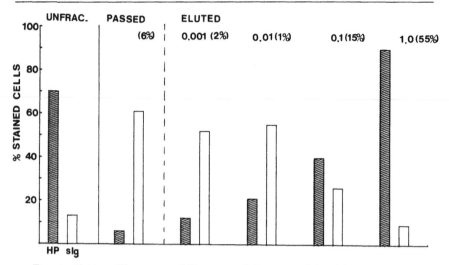

FIG. 1. Eighty million neuraminidase-treated human peripheral blood lymphocytes were fractionated on HP-Sepharose (1 mg HP/ml gel). After a 15-min incubation, nonretained cells (passed) were collected. Retained cells (eluted) were eluted with 0.001, 0.01, 0.1, and 1.0 mg D-GalNAc/ml. The yield of the cells in each fraction are given in parentheses at the top of the figure. The frequency of cells with cellular receptors for HP (HP+ cells, hatched bars) and cells with surface bound immunoglobulin (sIg+ cells, open bars) were quantitated by immunofluorescence.[5] The percentage of HP+ or sIg+ cells in the unfractionated population is shown to the left in the figure (in these experiments no precautions to avoid staining of FcγR-bound immunoglobulin were taken).

The gel was washed with 10 ml TH-HSA-NaN$_3$. Neuraminidase-treated PBL (25 × 10^6) in 0.2 ml TH-HSA-NaN$_3$ were allowed to penetrate into 1 ml HP-Sepharose gel. The cells were allowed to bind to the column for 10–15 min. Nonretained cells were washed off with 50 ml TH-HSA-NaN$_3$. Subsequently the column was eluted with 50 ml of 0.1 mg/ml D-GalNAc in TH-HSA-NaN$_3$ (fraction EI). Immediately thereafter, ~10 ml of the same medium containing 1.0 mg/ml of D-GalNAc was allowed to penetrate the column. The flow was stopped for 5–10 min to increase the efficiency of dissociation. Approximately 40 ml of the same sugar solution was then added and the gel beads were suspended by stirring of the gel. The solution was finally drained off (fraction EII).

For fractionation of T-lymphocytes on WGA-Sepharose, 25 × 10^6 lymphocytes, in 0.2 ml TH-HSA-NaN$_3$, were applied per 1 ml WGA-Sepharose gel (~5 mg WGA/ml, packed in a K9/15 column) and incubated on the column for 3 min. Loosely bound cells were then washed off with 50 ml TH-HSA-NaN$_3$. Immediately thereafter, a solution of 25 mg/ml D-GlcNAc in TH-HSA-NaN$_3$ (~10 ml) was allowed to penetrate the gel and the flow stopped for 5–10 min. Bound cells were then eluted with the remaining sugar solution (~40 ml) as described above.

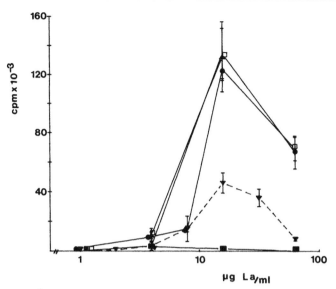

FIG. 2. Twenty-five million human peripheral T cells, depleted of FcγR⁺ cells, were fractionated on WGA-Sepharose (5 mg WGA/ml gel). Forty percent of the cells were recovered in the passed, nonretained fraction. Stepwise elution with 0.1, 1.0, and 10.0 mg D-GlcNAc/ml released 6, 19, and 20% of the cells added to the column, respectively. Unfractionated and fractionated T cells were analyzed for their capacity to proliferate in response to the lectin leucoagglutinin (La) from *Phaseolus vulgaris*. Two hundred thousand T cells were cultured in conical 96-well tissue culture plates (total volume 0.2 ml) for 3 days with different concentrations of La in RPMI-HEPES, supplemented with antibiotics, glutamine, and 0.4% human serum albumin and thereafter pulsed for 16 hr with 1 μCi of [³H]thymidine/well.[25] Results are expressed as mean cpm ± SD from triplicates. (▼) Unfractionated T cells; (■) passed cell fraction, cells eluted with 0.1 mg D-GlcNAc/ml (▲), 1.0 mg D-GlcNAc/ml (●), 10.0 mg D-GlcNAc/ml (□). Abscissa: μg La/ml and 10⁶ lymphocytes. Ordinate: (cpm/0.2 × 10⁶ lymphocytes) × 10⁻³.

To remove trace amounts of lectin which may have been released from the column during fractionation, the passed cell fraction was washed with 1 mg D-GalNAc/ml (HP-column) or with 25 mg D-GlcNAc/ml (WGA-column). All cell fractions were finally washed with TH-HSA, counted, and tested for viability (this volume [6]).

Some Representative Results

Fractionation of Neuraminidase-Treated Human Peripheral Blood Lymphocytes on HP-Sepharose

B-lymphocytes with surface bound immunoglobulin (sIg⁺) were strongly enriched in the fraction which was not retained on the column (passed fraction,P)[5] (see also Fig. 1). However some sIg$^+$ cells were also recovered in fraction EI, i.e., the fraction containing cells with low avidity receptors for HP. Some of these sIg$^+$ cells also displayed HP receptors as shown by double staining experiments.[10] Since staining due to Fc receptor bound immunoglobulin was excluded in this analysis we concluded that a small fraction of the B-lymphocytes in adult blood are indeed HP$^+$. Recent studies with a specific chicken antiserum against gp150 demonstrated that the HP$^+$sIg$^+$ cells also carried this major HP-binding glycoprotein.[18] Some sIg$^+$ cells were recovered in the cell fraction binding strongly to the HP-Sepharose column (fraction EII). However, Fc receptor bound IgG was responsible for this staining[5] and mature sIg$^+$ B cells seem to be absent from this fraction. Cells lacking ordinary B and T cell markers (null cells)[40] were recovered in the passed and EI fractions.

Of the T cells, as characterized by receptors for sheep erythrocytes (E$^+$ cells), 45–60% were recovered in fraction EII[5] and to a minor extent (10–15%) in EI.[5] Studies of T cell functions, such as the ability to respond to mitogenic lectins, indicated that the distribution of this activity in the fractions parallelled that of E$^+$ cells.[5,23] Cells with Fc receptors for IgG (FcγR$^+$) and cells with receptors for activated complement component C3 (C3R$^+$) were found in all fractions but were enriched in the fractions P and EI.[41,42] The antibody-dependent cellular cytotoxicity (ADCC) and natural killer activity (NK) parallelled the distribution of FcγR$^+$ and C3R$^+$ cells.[23,41,42]

Fractionation of Neuraminidase-Treated Human T-Lymphocytes on HP-Sepharose

As mentioned above, fractionation of T cells prepared by E-rosetting required HP-Sepharose gels charged with higher concentrations of HP ($\sim$5 mg/ml). The reason for this is probably that the E-rosetting procedure leads to a decrease in the density of cellular receptors for HP, most likely as a consequence of the slight activation of T cells which this procedure induces.[25] Due to the higher degree of substitution of the gel, higher concentrations of D-GalNAc are needed for elution, i.e., 1 mg/ml (EI) and 5 mg/ml (EII). Table I summarizes the results of these experiments. As can be seen, approximately 5% of the cells passed the column, approximately 10% were weakly bound (EI), and approximately 60% were

[40] R. Kiessling, G. Petranyi, K. Kärre, M. Jondal, D. Tracey, and H. Wigzell, *J. Exp. Med.* **143**, 772 (1976).
[41] P. Perlmann, H. Perlmann, B. Wåhlin, and S. Hammarström, *Immunopathol., Int. Symp., 7th, 1976* p. 321 (1976).
[42] G. R. Pape, M. Troye, and P. Perlmann, *Scand. J. Immunol.* **10**, 109 (1979).

TABLE I

PHENOTYPIC AND MORPHOLOGICAL CHARACTERIZATION OF HUMAN PERIPHERAL T-LYMPHOCYTES
FRACTIONATED ON HP-SEPHAROSE COLUMNS

Fraction[a]	Yields[b] ($n = 22$)	Cells (%) with different characteristics[c]						
		$Fc\mu R^+$ ($n = 13$)	$Fc\gamma R^+$ ($n = 13$)	OKT3$^+$/Leu-1$^+$ ($n = 6$)	Mϕ ($n = 6$)	HLA-DR$^+$ ($n = 6$)	LGL ($n = 6$)
U		50 ± 17	17 ± 11	95 ± 5	0.4 ± 0.3	4 ± 1	7 ± 5
P	5 ± 5	37 ± 21	18 ± 14	85 ± 6	0.5 ± 0.5	8 ± 4	13 ± 9
EI	10 ± 8	10 ± 8	35 ± 16	93 ± 5	0.4 ± 0.3	3 ± 2	11 ± 13
EII	58 ± 13	49 ± 13	14 ± 8	96 ± 4	0.1 ± 0.1	3 ± 2	4 ± 5

[a] U, Unfractionated; P, passed (nonretained); EI, elution with D-GalNAc (1 mg/ml); EII, elution with D-GalNAc 5 mg/ml.
[b] Percentage (mean ± SD) of cells recovered in fractions.
[c] Mean ± SD from 6 to 22 different experiments. T cells with receptors for the Fc portion of IgM ($Fc\mu R^+$) or IgG ($Fc\gamma R^+$) were assessed by rosetting with IgG or IgM-sensitized bovine erythrocytes, respectively.[24] OKT3 or Leu-1 monoclonal antibodies were used in indirect immunofluorescence to determine the frequencies of T cells.[43] Monocytic cells (Mϕ) and large granular lymphocytes (LGL[45]) were quantitated from Giemsa-stained smears. HLA-DR$^+$ cells were assessed by indirect immunofluorescence using mouse monoclonal antibodies against human HLA-DR framework determinants.[46]

strongly bound to the column (EII). Cells with Fcγ receptors (FcγR$^+$) were enriched in the fraction EI, while the highest proportion of FcμR$^+$ cells was found in fraction EII. When these cell fractions were analyzed with the monoclonal anti-T cell antibodies OKT4 and Leu3a or OKT8 and Leu2a[43] no clear cut differences in the relative proportions of positive cells were seen.[38] However they differed considerably in their capacity to regulate B cell differentiation and antibody production. Helper and suppressor cell functions in either antigen induced (specific) or pokeweed mitogen induced (polyclonal) B cell differentiation systems are under investigation and will be described elsewhere.[44]

Table I also shows that approximately 5% of the cells applied to the column lacked the T cell-specific antigens defined by the monoclonal antibodies OKT3 or Leu1.[43] These cells were preferentially recovered in fractions P and EI which also were relatively enriched in large granular lymphocytes (LGL[45]), HLA-DR$^+$ cells,[46] and monocytic cells. Cells carry-

[43] E. L. Reinherz and S. F. Schlossman, *Immunol. Today* **2**, 69 (1981).
[44] U. Hellström, E-S. Robertsson, M. Wikén, and P. Perlmann, *Scand. J. Immunol.* (submitted for publication).
[45] T. Timonen, E. Saksela, A. Ranki, and P. Häyry, *Cell. Immunol.* **48**, 133 (1979).
[46] F. Indiveri, A. K. Ng, C. Russo, V. Quaranta, M. A. Pellegrino, and S. Ferrone, *J. Immunol. Methods* **39**, 343 (1980).

ing the monocyte/null cell associated antigen M1 defined by the monoclonal antibody OKM1[47] had a distribution similar to that of LGL (data not shown[38]).

Fractionation of Antigen-Activated Human T-Lymphocytes on HP-Sepharose

T-lymphocytes obtained by sheep erythrocytes rosetting from a donor immunized with tetanus toxoid were cultured for 5, 6, and 7 days with or without antigen in the presence of autologous monocytic cells. All cells in the antigen activated cultures were HP+, as determined by indirect immunofluorescence (Fig. 3). Using the less sensitive direct immunofluores-

[47] J. Breard, E. L. Reinherz, P. C. Kung, G. Goldstein, and S. F. Schlossman, *J. Immunol.* **124**, 1943 (1980).

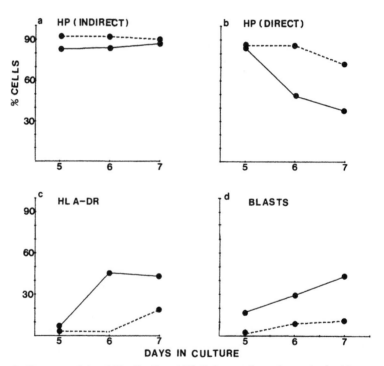

FIG. 3. Human peripheral T-cells (1 × 10⁶/ml) from a donor immunized with tetanus toxoid were cultured with (●——●) or without (●---●) antigen in the presence of 5% autologous monocytes for 5, 6, and 7 days, respectively. The cultures were analyzed for frequency of HP+ cells by indirect (a) or direct (b) immunofluorescence,[23] for frequency of HLA-DR+ cells (c) (indirect immunofluorescence), and for frequency of blast cells in Giemsa-stained smears (d).

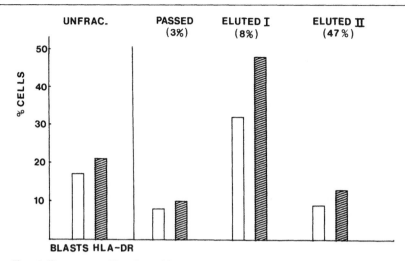

FIG. 4. Tetanus toxoid-activated human T-lymphocytes were harvested after 6 days in culture, treated with neuraminidase, and fractionated on HP-Sepharose (5 mg HP/ml gel). The yields of cells in the different fractions are given at the top of the figure. Open bars show the percentage of blast cells and hatched bars the percentage of HLA-DR⁺ cells in the different fractions.

cence technique we found that the percentage of HP⁺ cells in the antigen activated cultures decreased from 85% at day 5 to 40% at day 7 (Fig. 3b), indicating that the density of the cellular HP-receptors decreased with time of culture. In contrast, in these cultures the percentage of HLA-DR⁺ cells and blast cells increased with time (Fig. 3c and d). Figure 4 shows one experiment in which tetanus toxoid-activated T cells were fractionated on HP-Sepharose (5 mg/ml) after treatment with neuraminidase. Approximately 10% of the cells were recovered in fraction EI and 50% in fraction EII. Blast cells and HLA-DR⁺ cells were enriched in fraction EI, i.e., the fraction containing cells with low density HP receptors. In contrast, cells driven into DNA synthesis and proliferation were highly enriched in fraction EII as assessed by [³H]thymidine uptake (data not shown[38]).

Fractionation of Human T-Lymphocytes on WGA-Sepharose

Although all T cells bind WGA, immunofluorescence studies showed that 15–25% of the T cells are more readily stained with WGA than the majority of the cells. Thus, at low WGA concentrations a plateau in the titration curve was obtained.[19] On the basis of this difference in reactivity it was possible to divide T cells into two subpopulations, one which passed through the WGA-Sepharose column and another which was retained and could be eluted with the competitive sugar hapten D-GlcNAc.[19] That the two cell populations differed in avidity for WGA was demonstrated by immunofluorescence. Ten times higher concentrations of FITC-WGA were needed to stain all cells in the passed fraction as compared to the cells in the eluted fraction.[38]

The two T cell fractions were functionally different. The eluted fraction contained cells giving a strong proliferative response to La[19] (Fig. 2) and concanavalin A.[23] This response could not be suppressed by the addition of FcγR[+] cells.[25] The response to mitogenic lectins of the passed cell fraction was more complex. No proliferative response was seen if the passed fraction was derived from FcγR[+]-depleted T cells, i.e., cells obtained by passage of PBL over an immunoglobulin/antiimmunoglobulin complex column[37] (Hellström *et al.*[19] and Fig. 2). If the passed fraction was derived from T cells obtained by sheep erythrocyte rosetting, we found that it gave a significant proliferative response to mitogens.[25] This response was susceptible to suppression by FcγR[+] cells.[25] A possible interpretation of these latter findings is that certain T cells which become conditioned to respond to La by E-rosetting have low avidity WGA receptors. However, further analysis of the cellular composition of these fractions is necessary before establishing the reasons for these differences.

Cells proliferating in mixed leukocyte culture (MLC) were found in both WGA-Sepharose column fractions.[24] In contrast, if T cells were activated by alloantigen in MLC for 5 days and then subjected to WGA-Sepharose column fractionation we found that the fraction enriched in cells with high avidity WGA receptors contained most of the progenitor cells which proliferate in MLC and which act as effector cells in cell mediated lympholysis (CML).[24]

Comments

As exemplified above, lymphocytes can be fractionated into different subsets by affinity chromatography on insolubilized lectins. The method is rapid and allows separation of large number of cells. Functionally active cells in good yields (total recovery 70–90%) are obtained. However, to achieve satisfactory fractionations requires strict adherence to the experimental protocols such as the degree of substitution of the gels with lectin, the number of cells/ml gel, incubation times, and sugar concentrations applied for elution. This is particularly important when cell populations differing only in relative binding affinity to the lectin are fractionated.

The principle of affinity chromatography is readily applicable when some cell populations contain lectin binding surface structures while others lack such structures or express them in concentrations which are too low to assure retention on the matrix. This appears to be the case when *Helix pomatia* A hemagglutinin is used to separate mature T cells from mature B cells in human blood. Here, the T cells contain large amounts of the major HP-binding surface glycoprotein (gp 150) which is absent from the majority of the circulating B cells.[16,17] However, HP affinity can also be used to separate cells which bind to the lectin with different avidities. Analysis on the fluorescence activating cell sorter of human peripheral blood T cells has indicated that they are heterogeneous in their HP-binding capacity.[44] As demonstrated above, fractionation of such cells on HP-Sepharose columns also showed that 10–15% of them had lower avidity for HP than the majority of the T cells. These differences in binding may have different causes. Thus, the cell surface concentration of gp 150 may differ with the stage of maturation or functional differentiation of the cells. This is probably the explanation for the shift in HP binding of antigen activated T cells where the blast transformed (large) cells were primarily found in the low avidity HP fraction while the proliferating (smaller) cells were enriched in the high avidity fraction. However, qualitative differences in HP-binding structures may also account for differences in binding to the columns of cells within the T cell compartment. It has been shown elsewhere[17] that different malignant B cells, reflecting different stages of maturation, vary in their expression of primarily two HP-binding glycoproteins, gp 150 and gp 200. It is possible that similar differences in expression of HP-binding surface components also occur in T cell differentiation. However, further experiments are needed to establish how qualitative and/or quantitative differences in HP-binding surface components affect the binding of different T cell subsets to HP-Sepharose.

The molecular basis for fractionation of T cell subsets on WGA-columns is even more complicated. WGA binds to at least 10 glycopeptides present on the surface of human T cells.[21] A heterogeneous expression of WGA-binding membrane proteins on human T cell populations has recently been reported.[48] It remains to be established if such a heterogeneity also is responsible for the separation of functionally different T cells on WGA-columns found by us.

In addition to HP and WGA, several other lectins have been used for characterization of lymphocytes. Thus, peanut agglutinin (PNA) and *Limulus polyphemus* agglutinin (LPA) have been utilized to identify T cell

[48] M. R. Torrisi and P. P. Da Silva, *Proc. Natl. Acad. Sci. U.S.A.* **79,** 5671 (1982).

subsets.[49] The ontogeny of human thymocytes has been studied by comparing their reactivity with PNA with that of monoclonal antibodies defining T cell-associated differentiation antigens.[50] For a comprehensive review of these and similar experiments see Sharon.[7]

Acknowledgments

This study was supported by Grants B-UR 3485-100 and B-UR 2032 from the Swedish Natural Science Research Council. We thank Eva-Stina Robertsson, Bernt Axelsson, and Margareta Wikén for helpful assistance and discussions.

[49] T. Nakano, Y. Imai, and T. Osawa, *J. Immunol.* **125,** 1928 (1980).
[50] S. Berrih, C. Boussuges, J.-P. Binet, and J.-F. Bach, *Cell. Immunol.* **74,** 260 (1982).

[17] Fractionation of Subpopulations of Mouse and Human Lymphocytes by Peanut Agglutinin or Soybean Agglutinin

By YAIR REISNER and NATHAN SHARON

Introduction

The application of lectins, a class of carbohydrate-binding and cell-agglutinating proteins of nonimmune origin, to the isolation and characterization of soluble glycoproteins and for probing cell surface sugars is well established.[1,2] They are now being used to an increasing extent also for cell separation.[3] In principle, any population of single cells may be sorted into subpopulations with the aid of lectins, provided there are differences in their cell surface sugars. Most work on this subject, however, has been done with mammalian cells, particularly lymphocytes.[3]

Among the several lectins shown to date to be useful for cell separation and identification, peanut agglutinin (PNA)[3a] is by far the most popular one. Following our demonstration in 1976 that PNA can be used for

[1] H. Lis and N. Sharon, *in* "The Antigens" (M. Sela, ed.), Vol. 4, p. 429. Academic Press, New York, 1977.

[2] H. Lis and N. Sharon, *in* "Biology of Carbohydrates" (V. Ginsburg and P. W. Robbins, eds.) Vol. 2, pp. 1–85. Wiley, New York, 1984.

[3] N. Sharon, *Adv. Immunol.* **34,** 213 (1983).

[3a] Abbreviations: BSA, bovine serum albumin; FACS, fluorescent-activated cell sorter; FCS, fetal calf serum; FITC, fluorescein isothiocyanate; PBS, phosphate-buffered saline; PHA, phytohemagglutinin from *Phaseolus vulgaris;* PNA, peanut agglutinin; SBA, soybean agglutinin.

fractionation of mouse thymocytes into immature (cortical) and mature (medullary) cell subpopulations,[4] this lectin has been used for a variety of purposes in numerous laboratories, with murine and human lymphocytes as well as with lymphocytes of other species. The application of soybean agglutinin (SBA), although rather limited, is of great importance since it has been shown that it can be used for the fractionation of cells for bone marrow transplantation in humans across histocompatability barriers.[5,6]

In this chapter we describe different procedures for the fractionation of murine and human lymphocytes by PNA and SBA.

Methodology

Several techniques are available for cell separation by lectins, all of which afford high yields of fully viable and immunologically functional cells. If necessary, each cell fraction can be repurified, by the same or by a different technique. Care should be taken, however, to avoid prolonged contact (over 30 min) between the lectin and the cells, since this may result in nonspecific binding or uptake of the lectins, thus making it impossible to remove the lectin from the cells with specific sugars. Also, when working with mitogenic lectins, prolonged contact may lead to lymphocyte stimulation. However, the lectins commonly used for cell separation are nonmitogenic, or may be mitogenic only under special conditions (e.g., when polymerized or when the cells have been treated by sialidase[1]).

Selective Agglutination. Selective agglutination is the technique of choice when working with mixtures of cells that differ markedly in the extent of lectin they can bind and when a high proportion of the cells may be agglutinated. The cell aggregates are separated from the unagglutinated cells by sedimentation at unit gravity in an appropriate medium: 20–50% fetal calf serum (FCS) or 5% bovine serum albumin (BSA) and are then dissociated into single cells by suspension in a solution of a sugar for which the lectin is specific. In addition to its simplicity, this method can easily be scaled up for work with large numbers of cells (10^{10} or more). Only poor separation by this technique can be achieved, if at all, when the number of cells is relatively small (less than 10^8), the percentage of lectin-positive cells is low (less than 10–20%), or the density of lectin receptors on the cells is low ($<10^5$ per cell). In such cases, separation by other techniques should be used.

[4] Y. Reisner, M. Linker-Israeli, and N. Sharon, *Cell. Immunol.* **25,** 129 (1976).
[5] Y. Reisner, N. Kapoor, D. Kirkpatrick, M. S. Pollack, B. Dupont, R. A. Good, and R. J. O'Reilly, *Lancet* **2,** 327 (1981).
[6] Y. Reisner, N. Kapoor, D. Kirkpatrick, M. S. Pollack, S. Cunningham-Rundles, B. Dupont, M. Z. Hodes, R. A. Good, and R. J. O'Reilly, *Blood* **61,** 341 (1983).

Mixed Rosetting. The rosetting method makes use of lectin-mediated formation of rosettes between cells having a receptor for a particular lectin and erythrocytes that also bind that lectin. Subagglutinating concentrations of the lectin are used. The rosettes are separated from the nonrosetting cells by centrifugation on a Ficoll–Hypaque density gradient or by sedimentation at unit gravity in an appropriate medium such as a solution of BSA. They are then dissociated by a specific sugar, and the erythrocytes removed by differential centrifugation or osmotic shock. (See also this volume [7–10].)

Affinity Chromatography. For cell separation by affinity chromatography, the lectin is immobilized by covalently binding it to a solid support. The same affinity adsorbent can be used repeatedly and reproducibly. Separation is usually performed on a column of the immobilized lectin to which a suspension of cells is applied. The unbound cells are washed out with buffer, and the lectin-bound cells are eluted with a specific sugar. Cells which specifically bind a particular lectin can also be isolated by affinity chromatography on columns to which antibodies to the lectin are covalently bound.[7] (See also this volume [16].)

Flow Microfluorimetry. This technique is used primarily for analytical purposes, but it is also applicable for cell separation. The cells are sorted one by one in a fluorescent activated cell sorter (FACS), usually according to the amount of fluorescent lectin bound. The major disadvantages of flow microfluorimetry are the length of time (hours) required to separate even small numbers (10^7–10^8 cells) of cells and the high cost of the equipment. (See also this volume [19].)

Reagents

Lectins. Lectins are obtained from commercial sources or purified by affinity chromatography according to published procedures (reviewed in Lis and Sharon[8]). In our laboratory, PNA is isolated according to Lotan *et al.*,[9] except that affinity chromatography is carried out with Sepharose-bound D-galactose as adsorbent, instead of Sepharose-*N*-ε-aminocaproyl-β-D-galactopyranosylamine. The D-galactose is coupled to Sepharose by the divinylsulfone method as described by Iglesias *et al.*[10] SBA is isolated by a modification[11] of the method described by Gordon *et al.*[12,13] The SBA

[7] C. Irlé, P.-F. Piguet, and P. Vassalli, *J. Exp. Med.* **148**, 32 (1978).
[8] H. Lis and N. Sharon, *J. Chromatogr.* **215**, 361 (1981).
[9] R. Lotan, E. Skutelsky, D. Danon, and N. Sharon, *J. Biol. Chem.* **250**, 8518 (1975).
[10] J. L. Iglesias, H. Lis, and N. Sharon, *Eur. J. Biochem.* **123**, 247 (1982).
[11] C. L. Jaffe, S. Ehrlich-Rogozinski, H. Lis, and N. Sharon, *FEBS Lett.* **82**, 191 (1977).
[12] J. A. Gordon, S. Blumberg, H. Lis, and N. Sharon, *FEBS Lett.* **24**, 193 (1972).
[13] H. Lis and N. Sharon, Vol. 28, p. 360.

should preferably be a fresh preparation; in particular, it should be free of polymeric lectin, which can be revealed by polyacrylamide gel electrophoresis in basic gels.[14] The polymers are poorly soluble, and when present, poor cell separations may result. If necessary, the SBA should be fractionated by gel filtration on Sephadex G-100, and only the species of MW 120,000 should be used. Fluorescent derivatives of PNA and SBA are commercially available, or can be prepared by reacting the lectins with fluorescein isothiocyanate (FITC)[15] and repurifying the FITC-lectin by affinity chromatography.

Immobilized PNA.[16] Sepharose 6MB activated with CNBr (1 g; Pharmacia, Sweden) is swollen in 1 mM HCl and then exhaustively washed with the same solution. PNA (10 mg) is dissolved in 5 ml sodium bicarbonate (pH 8.3, 0.1 M) containing 0.5 M NaCl (coupling buffer) and added to the swollen gel. The mixture is slowly rotated head to tail for 2 hr at room temperature. The beads are allowed to settle and the supernatant is removed. Analysis of the supernatant for the presence of PNA (by measuring the OD at 280 nm with PNA as standard) shows that less than 2% of the starting amount of the lectin is present. The derivatized beads are treated with 10 ml of 0.2 M glycine, pH 8.0 to block the unreacted CNBr, after which they are washed in the coupling buffer followed by washing in 0.1 M acetate buffer containing 0.5 M NaCl, pH 4.0. This sequence of washing in alkaline and acid buffers is repeated twice. The beads are then resuspended in sterile PBS and stored at 4°.

Antibodies. These are obtained from commercial sources. For the characterization of mouse cells: anti-Thy-1, H-2, Lyt-1, Lyt-2, Lyt-3, IgG, and IgM. For human cells: anti-OKT1 and OKT3 (pan-T antibodies); anti-OKT8 (anti-suppressor T cell antibody); anti-OKT4 (anti-helper T cell antibody); for early (immature) T cells, anti-OKT6; and for B cells, anti-IgG and IgM.

Buffers. Phosphate buffered saline (PBS): 0.01 M sodium phosphate pH 7.2–7.4 and 0.15 M NaCl. PBS–BSA, PBS containing 1% w/v BSA. Lysing buffer, prepared by adding 10 ml of 0.17 M Tris–HCl to 90 ml of 0.16 M ammonium chloride, and adjusting the pH to 7.2 by dropwise addition of 1 M NaOH.

Analysis of the Separated Cells

The cells are counted in a hemocytometer, and cell viability is assessed by the trypan blue exclusion test. The immunological properties of

---

[14] R. Lotan, H. Lis, and N. Sharon, *Biochem. Biophys. Res. Commun.* **62**, 144 (1975).
[15] H. F. Clark and C. C. Shephard, *Virology* **20**, 642 (1965).
[16] M. Rosenberg, E. Gazit, and N. Sharon, *Hum. Immunol.* **7**, 67 (1983).

the cell fractions obtained are established by standard immunological methods.[17,18] These include *in vitro* assays such as proliferative responses to polyclonal mitogens (PHA, concanavalin A, pokeweed mitogen, and lipopolysaccharide), or alloantigens (in the mixed lymphocytes culture), generation of cytotoxic lymphocytes, and antibody production against specific antigens. Assays *in vivo* include graft-versus-host activity (Simonsen spleen assay[19]) and delayed type hypersensitivity. Surface markers are evaluated by serological cytotoxicity test,[20] by fluorescent staining with monoclonal antibodies using the sandwich technique, and, less frequently, with fluorescent lectins. Fluorescent cells are enumerated by fluorescent microscopy or by cytofluorimetry.

Only selected data on the properties of the fractionated cells are given. These should suffice for evaluation of the quality of the separation.

Fractionation of Mouse Thymocytes by Selective Agglutination with PNA[4]

Preparation of Thymocytes. Mice, age 6–10 weeks, are killed by cervical dislocation; their thymuses are removed, minced in cold PBS, and passed through a fine stainless-steel mesh (150–200 μm) to obtain a single cell suspension. Erythrocytes are removed by treatment with lysing buffer at 0° for 10 min. The cells are washed twice in PBS–BSA, and resuspended in the same buffer. The average yield per thymus is about 10^8 cells. The thymocyte suspension (8×10^8 cells/ml, 0.25 ml) is incubated in a 5-ml plastic tube with PNA (0.25 ml, 1 mg/ml PBS) for 20 min at room temperature. At the end of the incubation, the cells are layered gently with a Pasteur pipet on top of a solution of heat-inactivated FCS in PBS (20%, 8 ml) in a conical 12-ml glass tube. After 15–30 min at room temperature most of the thymocytes sediment, whereas the unagglutinated single cells remain on top. The bottom and top fractions (each of about 0.5–1 ml) are removed separately with a Pasteur pipet and transferred into 15-ml plastic tubes containing D-galactose (0.2 M in PBS, 5 ml) to dissociate the agglutinated cells. After 10–15 min at room temperature, 5 ml of PBS is added and the cells are collected by centrifugation, resuspended in 5 ml of

[17] J. S. Garvey, N. E. Cremer, and D. H. Sussdorf, eds., "Methods in Immunology—A Laboratory Text for Instruction and Research," 3rd ed. Benjamin, Reading, Massachusetts, 1977.

[18] B. B. Mishell and S. M. Shiigi, eds., "Selected Methods in Cellular Immunology." Freeman, San Francisco, California, 1980.

[19] M. Simonsen, J. Engelbreth-Holm, E. Jensen, and H. Poulsen, *Ann. N.Y. Acad. Sci.* **73**, 834 (1958).

[20] "N.I.H. Lymphocyte Microcytotoxicity Technique, NIAID Manual for Tissue Typing Techniques," N.I.H. Publ. 80,545, pp. 39–41 (1979).

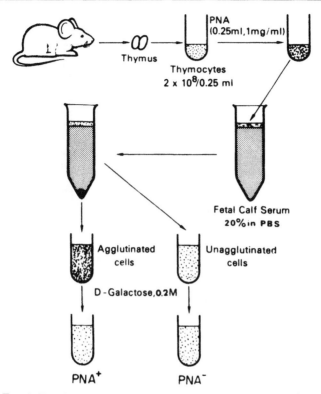

FIG. 1. Fractionation of mouse thymocytes by peanut agglutinin.

a 0.2 M solution of D-galactose in PBS, and collected again by centrifugation (200 g, 5 min). Finally the cells are washed twice with PBS. A scheme of the fractionation procedure is given in Fig. 1.

The agglutinated fraction (PNA⁺ cells) contains 1.3–1.4 × 10^8 cells (85–90% of cells recovered) and the unagglutinated fraction (PNA⁻ cells) about 0.15 × 10^8 cells (total yield 70–80%). More than 95% of the cells in the PNA⁺ fraction bear the immature cortical phenotype TL⁺Lyt-1⁺2⁺3⁺, whereas the PNA⁻ fraction contains about 10–20% of such cells and it is mainly comprised of cells bearing the more differentiated phenotypes TL⁻Lyt-1⁻2⁺3⁺ (~12%) and TL⁻Lyt-1⁺2⁻3⁻ (~60%) commonly found among mature circulating T cells. The efficiency of the fractionation, as reflected by the mixed lymphocyte reaction,[21] is shown in Fig. 2. Staining of frozen sections of mouse thymus has confirmed that the PNA⁺ cells are located in the cortex and absent from the medulla.[22]

[21] T. Umiel, M. Linker-Israeli, M. Itzchaki, N. Trainin, Y. Reisner, and N. Sharon, *Cell. Immunol.* **37,** 134 (1978).

[22] M. L. Rose and F. Malchiodi, *Immunology* **42,** 583 (1981).

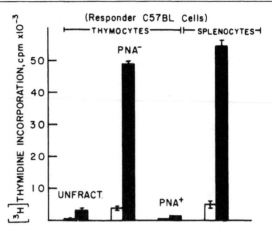

FIG. 2. Activity in the mixed lymphocyte reaction of C57BL mouse thmyocytes fraction-ated by PNA and of splenocytes. Equal amounts of unfractionated and fractionated cells are used. Irradiated stimulator cells: (□) C57BL; (■) CBA. (Data from Umiel et al.[21])

In the past few years numerous studies using the agglutination method have described in detail the differences between the cortical immunoin-competent (PNA$^+$) and the medullary immunocompetent (PNA$^-$) thymo-cytes, with respect to their surface markers, immunological and biochem-ical properties, and functional development (reviewed in Sharon[3]).

Fractionation of Human Thymocytes by Selective Agglutination with PNA[23]

Fresh normal thymuses are obtained from patients who had part of their thymus removed during corrective cardiac surgery. A portion of the thymus (5–10 g) is immediately teased in a solution of FCS (2%) in PBS. If necessary, the thymus tissue may be kept for several hours in Hanks' balanced salt solution at 4°. The crude preparation of thymocytes is pressed through a fine stainless-steel mesh (150–200 μm) to give a single cell suspension. At least 2×10^8 cells should be obtained, with a viability of over 95%. The fractionation procedure is the same as that described for mouse thymocytes, with the following changes. The concentration of the starting thymocyte suspension is 4×10^8 cells/ml, PNA is 2 mg/ml, and FCS, 50%. The total cell recovery after fractionation is 60–70%; of these, 50–70% are in the agglutinated (PNA$^+$) fraction, and the rest in the unag-glutinated (PNA$^-$) fraction. Cell viability in each of the fractions is >95%.

Examination of separated cell fractions by cytofluorimetry with FITC-

[23] Y. Reisner, M. Biniaminov, E. Rosenthal, N. Sharon, and B. Ramot, *Proc. Natl. Acad. Sci. U.S.A.* **76**, 447 (1979).

PNA reveals higher cross contamination in the PNA⁻ fraction (10–25% of PNA⁺ cells) compared to the PNA⁺ fraction (5–10% of PNA⁻ cells).[24]

Characterization by T cell-specific monoclonal antibodies[24,25] shows that the majority (70–80%) of the agglutinated thymocytes bear the immature phenotype OKT6⁺OKT4⁺OKT8⁺OKT3⁻. This phenotype is exclusively found in the thymic cortex by staining of frozen sections of thymuses with the corresponding antibodies, as well as with PNA.[24] The unagglutinated thymocytes are mainly comprised (60–80%) of cells bearing the more mature phenotype OKT3⁺OKT6⁻, which can further be divided into OKT4⁺OKT8⁻ cells, and OKT4⁻OKT8⁺ cells. These cell phenotypes are confined to the thymic medulla. The two cell fractions also differ markedly in their proliferative response to alloantigens and PHA (in the presence of monocytes or interleukin-2). In the latter assay, the PNA⁻ cells give a response which is usually 5–10 times higher than that of the PNA⁺ cells.[23,24] The residual response of the PNA⁺ cells appears to be due to the presence of 10 to 20% of PNA⁺OKT1⁺ cells which respond strongly to PHA.

Separation of Mouse Splenocytes by Selective Agglutination with SBA[26]

Spleens from mice (aged 6–10 weeks) killed by cervical dislocation are removed and dispersed by pressing through a stainless-steel mesh (150–200 μm) into PBS to obtain a single cell suspension. Erythrocytes are removed with lysing buffer at 0° for 10 min. The cells are washed twice in PBS–BSA and resuspended in the same buffer. The average yield is 10^8 cells/spleen. The splenocyte suspension (2×10^8 cells in 0.5 ml PBS) is incubated in polystyrene tubes (17 × 100 mm) with SBA (0.5 ml, 2 mg/ml) for 5–10 min at room temperature. The cells are then gently layered with a Pasteur pipet on top of heat-inactivated FCS (50%, 40 ml) or BSA (5% in PBS, 40 ml) in a conical glass tube. After 15 min at room temperature, most of the agglutinated lymphocytes sediment, whereas the unagglutinated single cells remain on the surface. The bottom and top fractions (about 1 ml each) are removed separately with a Pasteur pipet, transferred to 15-ml conical plastic tubes, and the cells are suspended in D-galactose (0.2 M in PBS, 10 ml). After 10 min at room temperature the cells are collected by centrifugation (200 g, 5 min), and washed twice with D-galactose (0.2 M in PBS, 10 ml) to give a single cell suspension. Finally, the cells are washed twice with PBS. About 8.3×10^7 and 3.6×10^7 cells are obtained from the agglutinated and unagglutinated fractions, respec-

[24] T. Umiel, J. F. Daley, A. K. Bhan, R. H. Levey, S. F. Schlossman, and E. L. Reinherz, *J. Immunol.* **129,** 1054 (1982).

[25] S. Berrih, C. Boussuges, J. P. Binet, and J. F. Bach, *Cell. Immunol.* **74,** 260 (1982).

[26] Y. Reisner, A. Ravid, and N. Sharon, *Biochem. Biophys. Res. Commun.* **72,** 1585 (1976).

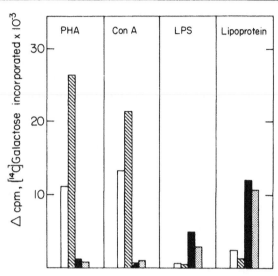

FIG. 3. Efficiency of separation of mouse splenocytes by SBA as evaluated by mitogenic response of the cells to T cell mitogens [PHA and concanavalin A (Con A)] and B cell mitogens [lipopolysaccharide (LPS) and lipoprotein]. Cell fractions: ▨, SBA⁻; ■, SBA⁺; □, nylon nonadherent; ▨, nonlysed by anti-Thy-1 and complement. (Data from Rosenfelder et al.[27])

tively (total yield 60%). The trypan blue exclusion test (this volume [6]) shows that each of the fractions contains more than 95% viable cells.

The unagglutinated (SBA⁻) fraction is comprised of 78% Thy-1⁺ cells, as shown by the anti-Thy-1 cytotoxicity assay; a direct fluorescence assay with anti-IgG and anti-IgM antisera shows about 5% B cell contamination. Assays of various T and B cell functions *in vitro* and *in vivo* also reveal that, compared to the unseparated splenocytes, the unagglutinated fraction is enriched for T cells and depleted of B cells, whereas the agglutinated fraction is largely comprised of B cells and myeloid cells. This is illustrated in Fig. 3, in which the mitogenic responses of splenocytes fractionated by SBA are compared to those separated by conventional techniques[27] (see also Van Eijk et al.[28]).

Separation of Antibody Helper and Antibody Suppressor Human T Cells by Mixed Rosetting with SBA[29] (This Volume [19])

Heparinized human blood is diluted with an equal volume of Hanks' balanced salt solution and 30-ml portions are layered over 15 ml of Ficoll–

[27] G. Rosenfelder, R. V. W. Van Eijk, and P. F. Muhlradt, *Eur. J. Biochem.* **97**, 229 (1979).
[28] R. V. W. Van Eijk, G. Rosenfelder, and P. F. Muhlradt, *Eur. J. Biochem.* **101**, 185 (1979).
[29] Y. Reisner, S. Pahwa, J. W. Chiao, N. Sharon, R. L. Evans, and R. A. Good, *Proc. Natl. Acad. Sci. U.S.A.* **77**, 6778 (1980).

Hypaque (Lymphoprep, Nyegaard, Oslo, Norway) in 50-ml plastic centrifuge tubes. The tubes are centrifuged for 30 min at 400 g at room temperature. Peripheral blood mononuclear cells are collected from the interface, washed twice with PBS–BSA, and resuspended in the same buffer. The yield is 1.5–3 × 10⁷ cells/15 ml of whole blood. The peripheral blood mononuclear cell suspension (4 × 10⁸ cells/ml, 0.5 ml) is incubated in a 17 × 100-mm polystyrene tube for 5 min at room temperature with autologous erythrocytes (4 × 10⁹ cells/ml, 50 μl) and SBA (2 mg/ml, 0.5 ml). The cells are then layered, with a Pasteur pipet, on top of 8 ml of 5% BSA in PBS in a 15-ml conical plastic tube. After 20 min at room temperature, most of the rosetted and agglutinated cells sediment, whereas the unagglutinated single cells remain on the top of the BSA solution. The bottom and top fractions (0.5–1 ml each) are removed separately with a Pasteur pipet and transferred to 15-ml conical plastic tubes. The cells are then suspended in 10 ml of 0.2 M D-galactose in PBS. After 5 min at room temperature the cells are collected by centrifugation (200 g, 5 min) and washed twice more with D-galactose to yield a single-cell suspension. Finally, the erythrocytes are lysed by treatment with lysing buffer, and the lymphocytes are recovered by washing twice with RPMI-1640 medium.

The total yield of cells is 60–80%, 75–85% of which are in the agglutinated (SBA⁺) fraction. Each fraction contains more than 95% viable cells, as judged by trypan blue dye exclusion. The agglutinated fraction contains T cells (55%), B cells (14%), and monocytes (22%), while the unagglutinated fraction contains T cells (88%) and "null" cells (7%).[29] Mixing of peripheral blood lymphocytes with T cells from the fractions separated by SBA, and subsequent testing for specific antibody production against sheep red blood cells, reveals a marked enrichment for helper and suppressor T cells in the agglutinated and unagglutinated fraction, respectively (Table I).

Fractionation of Human Cord Blood Lymphocytes by Affinity
  Chromatography on Immobilized PNA[16]

Cord blood samples from normal full-term delivery newborns are collected under sterile conditions in heparin (100 U/ml) and processed within 2–3 hr. The heparinized whole blood is diluted with an equal volume of PBS. Eight-milliliter aliquots are layered on 3 ml Ficoll–Ipaque (Pharmacia, Sweden) and centrifuged for 40 min at 1200 rpm. The cells of the interface are collected and washed twice with PBS. The yield is about 1.5–2 × 10⁶ mononuclear cells per ml of whole blood. For affinity chromatography, Pasteur pipets plugged with glass wool and filled with 1 ml of PNA-coated beads are used. Cells (50 × 10⁶ in 1 ml PBS) are applied on

TABLE I
ANTIBODY PRODUCTION BY PERIPHERAL BLOOD
MONONUCLEAR (PBM) CELLS: EFFECT OF ADDITION OF T
CELLS SEPARATED BY SBA[a]

	Anti-SRBC[b] plaque-forming cells/ 10^6 cultured cells			
Cells	Expt. 1	Expt. 2	Expt. 3	Expt. 4
PBM	399	99	400	434
SBA⁻T	0	0	0	0
SBA⁺T	0	0	0	0
PBM + SBA⁻T	201	46	182	101
PBM + SBA⁺T	2466	489	1207	968

[a] Data from Reisner et al.[29]
[b] SRBC, Sheep red blood cells.

the column which is then washed with 30–40 ml PBS to remove the unbound (PNA⁻) cells. Subsequently, 1 ml of D-galactose (0.2 M in PBS) is passed through the column and the flow stopped for 5 min to allow more effective interaction of the sugar with PNA. The lectin bound (PNA⁺) cells are collected by elution of the column with 40 ml of D-galactose (0.2 M in PBS). The cells are collected by centrifugation (200 g, 5 min), washed once in 30 ml PBS, and then once in 30 ml RPMI 1640 medium. Total recovery is 75–90% of which 22% of the cells (on the average) are in the PNA⁺ fraction and 78% in the PNA⁻ fraction; the cells are fully viable (>97%). PNA⁺ cord blood mononuclear cells respond poorly to PHA (1 μg/ml), concanavalin A (3 μg/ml), and pokeweed mitogens (1/40, w/v), the level of thymidine incorporation into these cells being 25, 15, and 15%, respectively, of that of the PNA⁻ cells. The ability of the PNA⁺ cord blood lymphocytes to respond to an allogeneic stimulus in the mixed lymphocyte culture system is also considerably lower than that of the PNA⁻ cells (about 15%). Serological and cytofluorimetric studies suggest that over 50% of the PNA⁺ cells do not bear surface antigens of mature T or B cells. It has been suggested that these "null" cells may represent early forms of either the B or the T lineage.[30]

Separation of Thymocytes with PNA in the FACS[31]

Thymocytes are prepared as described above for fractionation by selective agglutination. The cells are centrifuged, resuspended at the density

[30] R. Maccario, L. Nespoli, G. Mingrat, A. Vitiello, A. G. Ugazio, and G. R. Burgio, *J. Immunol.* **130**, 1129 (1983).
[31] C. Wei-Feng, R. Scollay, and K. Shortman, *J. Immunol.* **129**, 18 (1982).

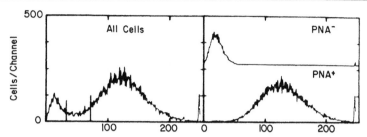

FIG. 4. Fluorescence distribution of CBA mouse thymocytes labeled with FITC-PNA. Left panel, unfractionated cells; the vertical lines denote position of cuts used for sorting purposes. Right panel, fractionated cells. (From Wei-Feng et al.[31])

of 10^7 cells/0.1 ml of a solution of FCS (10% v/v) in PBS, containing FITC-PNA (5 μg/ml), and incubated at 4° for 20 min. Under these conditions, no detectable agglutination occurs. After staining, the cells are washed and resuspended in the FCS-PBS solution for sorting on a FACS. Low-angle light scatter is used to eliminate dead cells from the sorted fractions. The fluorescence profile obtained from unfractionated labeled thymocytes is shown in Fig. 4; a cut between the peaks in the middle of the trough gives 85% high-fluorescence cells (PNA$^+$) and 15% low-fluorescence cells (PNA$^-$). To reduce cross-contamination in sorting experiments, two cuts can be made (see Fig. 4) to yield 10% of PNA$^-$ cells and 80% of PNA$^+$ cells. The cells in the middle fraction (10%) are discarded. The average purity of both PNA$^+$ and PNA$^-$ fractions as estimated by reanalysis in the FACS under the same conditions is over 98.5%.

Concluding Remarks

Most of the methods described in this chapter have been used for a variety of purposes in different laboratories, especially for studies of the properties and maturation of mouse and human thymocytes, as well as for characterization of mouse splenocytes. It is quite likely that PNA and SBA, and perhaps other lectins, may be used for the isolation of other subpopulations of murine or human lymphocytes, once the suitable conditions are established. In particular, there is a need for a simple and rapid method for the separation of human peripheral blood lymphocytes into T and B cells.

Studies of the immune system of animals other than mouse or man have been hampered by the lack of suitable lymphocyte surface markers. From the limited amount of work on the interaction of lectins with lymphocytes of rat, guinea pig, cattle, sheep, monkey, and chicken, it appears that lectins may prove to be useful tools for the investigation of the immune system of these animals.

[18] Electrophoretic Separation of Lymphoid Cells

By Ernil Hansen and Kurt Hannig

Introduction

The dissection of a complex system into its components, their separate analysis, and the controlled reconstitution is a scientific principle that has been used quite successfully in biochemistry. Its application in immunology calls for techniques for large-scale separation of highly pure and functionally intact lymphocyte subpopulations. Since most lymphocyte functions are mediated through their cell surface, a separation based on differences in cell membrane properties, such as surface charge, seems to be especially promising.

Preparative electrophoresis of cells became possible with the development of free-flow electrophoresis by K. Hanning.[1-3] The application of this method to the separation of lymphocytes is closely connected with the name of our collegue, the late K. Zeiller. He and a number of workers in other laboratories have demonstrated the potential of free-flow electrophoresis to efficiently separate murine T and B cells, as well as lymphocyte subpopulations at different stages of activation or differentiation (reviewed in refs. 4–9). Although most investigations have dealt with lymphocytes from mice and rats, lymphoid cells from humans,[5,10-12] non-

[1] K. Hannig, Z. Anal. Chem 181, 244 (1961).
[2] K. Hannig, in "Methods in Microbiology" (J. R. Norris and D. W. Ribbons, eds.), Vol. 5, Chapter 8, p. 513. Academic Press, New York, 1971.
[3] K. Hannig, Tech. Biochem. Biophys. Morphol. 1, 191 (1972).
[4] K. Zeiller, Behring Inst. Mitt. 52, 11 (1972).
[5] P. Häyry, L. C. Anderson, C. Gahmberg, P. Roberts, A. Ranki, and S. Nordling, Isr. J. Med. Sci. 11, 1299 (1975).
[6] G. V. Sherbet, ed., "The Biophysical Characterization of the Cell Surface," Chapter 4, p. 36. Academic Press, New York, 1978.
[7] T. G. Pretlow and T. P. Pretlow, Int. Rev. Cyto. 61, 85 (1979).
[8] K. Shortman, H. v. Boehmer, J. Lipp, and K. Hopper, Transplant Rev. 25, 163 (1975).
[9] K. Shortman, J. M. Fiedler, R. A. Schlegel, D. J. V. Nossal, M. Howard, J. Lipp, and H. v. Boehmer, Contemp. Top. Immunobiol. 5, 1 (1976).
[10] G. Stein, H. D. Flad, R. Pabst, and F. Trepel, Biomedicine 19, 388 (1973).
[11] G. Stein, Biomedicine 23, 5 (1975).
[12] P. H. Chollet, P. Hervé, J. Chassagne, M. Masse, R. Plagne, and A. Peters, Biomedicine 28, 119 (1978).

human primates,[13] chicken,[14] guinea pigs,[5] rabbits,[15] and cats[16] have also been separated. Other separation methods based on differences in cell surface charge, such as density gradient electrophoresis, have been developed, and have been reviewed recently.[17] Only free-flow electrophoresis however has found wide application, and will be discussed here.

We have developed an analytical version of a free-flow electrophoresis apparatus.[18] It is equipped with an optical detection system, and permits rapid, semiautomatic evaluation of large numbers of cell samples. This chapter however, will deal exclusively with preparative cell electrophoresis.

Recently we have described the electrophoretic separation of antibody-labeled cells, a method that greatly expands the separation potential of cell electrophoresis.[19] Its application to the separation of human lymphocyte subpopulations is also presented here.

Free-Flow Electrophoresis and the Separation of Murine Lymphocytes

Principle

Charged particles move in an electric field. Because of net negative charge, mainly due to sialic acid residues,[20] lymphocytes are deflected toward the anode.

In free-flow electrophoresis a buffer film is streaming in laminar flow between two glass plates. The cell suspension is injected continuously at a point at the upper part of the separation chamber and carried downward. On their way the cells are deflected by an electric field perpendicular to the buffer flow, and can thus be fractionated (Fig. 1). Under the conditions used, the separation is strictly according to cell surface charge density, and independent of density or size.[21] A detailed account of the theory of cell electrophoresis is given elsewhere.[2,3,6]

---

[13] F. R. Seiller, R. Johannsen, H. H. Sedlacek, and K. Zeiller, *Transplant. Proc.* **6**, 173 (1974).

[14] W. Dröge, R. Zucker, and K. Hannig, *Cell. Immunol.* **12**, 186 (1974).

[15] N. Sabolovic, D. Sabolovic, and A. M. Guilmin, *Immunology* **32**, 581 (1977).

[16] F. Dumont and G. Reichart, *Comp. Immunol. Microbiol. Infect. Dis.* **2**, 23 (1979).

[17] N. Catsimpoolas, ed., "Methods of Cell Separation," Vol. 1 and 2. Plenum, New York, 1977 and 1979.

[18] K. Hannig, *Electrophoresis* **3**, 235 (1982).

[19] E. Hansen and K. Hannig, *J. Immunol. Methods* **51**, 197 (1982).

[20] J. N. Mehrishi and K. Zeiller, *Br. Med. J.* **1**, 360 (1974).

[21] K. Hannig, H. Wirth, B. H. Meyer, and K. Zeiller, *Hoppe-Seyler's Z. Physiol. Chem.* **356**, 1209 (1975).

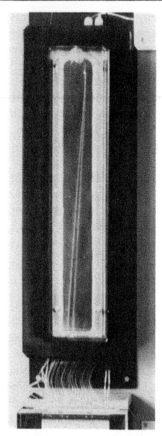

FIG. 1. Path of cells in the separation chamber during free-flow electrophoresis of human and rabbit erythrocytes.

Apparatus

Apparatuses for free-flow electrophoresis are commercially available from Bender & Hobein, München, Desaga, Heidelberg, and Hirschmann, München. In the Vap 5 model, developed in our laboratory, the two parallel glass plates of the separation chamber are 0.5 mm apart, and the chamber is 50 cm high and 10 cm wide. On both sides there are ion exchange membranes that separate the electrophoresis buffer from the electrode chambers. At the bottom of the separation chamber the buffer stream is split into 90 channels and over a multichannel peristaltic pump led to the cooled collection tubes. The buffer flow is thus regulated by the speed of the peristaltic pump. A roller pump for sample application is located at the upper part of the apparatus. This pump can be used to blow air bubbles into an ice-cooled, sealed sample tube, thereby mixing the cell suspension and forcing it through the inlet tubing.

Glass Wall Coating. Band broadening effects are compensated best if the zeta-potential of the glass walls in the separation chamber is approximated to that of the cells.[22] This is achieved by coating the glass plates with an appropriate protein solution, such as 3% bovine serum albumin (BSA)[22a] or lysed erythrocytes. When the standard (see later) reveals a skewness of the electrophoretic distribution toward the cathode, especially after the glass plates have been cleaned, the protein coat should be renewed.

Sterilization. The separation chamber and the electrode chambers can be sterilized by incubation with 3.5% formaldehyde for 4 hr, followed by a rinse with 5 liters of sterile, distilled water.

Cell Preparation (This Volume [6,8,9,24])

A prerequisite of any cell separation technique is a single-cell suspension. Lymphocytes are physiologically present in a single-cell status, and so lymphoid tissues can be disrupted mechanically, without the need for enzymatic treatment that might possibly damage cell membrane structures and alter surface charge.

Lymph nodes, spleen, or thymus are teased with forceps and a needle. Larger amounts of tissue can be processed with the aid of a steel sieve or with a mixer producing shear forces. Cell preparation should take place on ice and in a cold balanced salt solution supplemented with 1% BSA or 10% fetal calf serum.

Erythrocytes do not usually interfere with the electrophoretic separation of spleen cells, and prior lysis is unnecessary. In electrophoresis they are separated from nucleated cells in the region of high electrophoretic mobility (EPM). Nonlymphoid spleen cells may be eliminated by density gradient centrifugation or 1 g velocity sedimentation prior to electrophoresis.

Thoracic duct lymphocytes can be obtained by lymph drainage according to Bollmann.[23] Bone marrow is flushed out of femur and tibia after resection of the epiphyses. Cell clusters are disrupted by pipetting. Small

---

[22] K. Zeiller, R. Löser, G. Pascher, and K. Hannig, *Hoppe-Seyler's Z. Physiol. Chem.* **356,** 1225 (1975).

[22a] Abbreviations: BSA, bovine serum albumin; EPM, electrophoretic mobility; TRITC, tetramethylrhodamine B isothiocyanate; FITC, fluorescein isothiocyanate; SRBC, sheep red blood cells; RBLA, rat B lymphocyte-specific antigen; RTLA, rat T lymphocyte-specific antigen; PHA, phytohemagglutinin; Con A, concanavalin A; ASECS, antigen-specific electrophoretic cell separation.

[23] J. L. Bollmann, J. C. Cain, and J. H. Grindlay, *J. Lab. Clin. Med.* **33,** 1349 (1948).

COMPOSITION OF SEPARATION AND ELECTRODE BUFFERS

Buffer	Electrophoresis buffer[a]		Electrode buffer[b]	
	g/liter	mmol/liter	g/liter	mmol/liter
Triethanolamine	2.2	15	11.2	75
Potassium acetate	0.4	4	2.0	20
Glycine	18.0	240	—	—
Glucose	2.0	11	—	—

[a] Adjusted to pH 7.2 with acetic acid; adjusted to isoosmolarity with sucrose.
[b] Adjusted to pH 7.2 with acetic acid.

lymphocytes can be prepared for electrophoresis by 1 g velocity sedimentation.[5,24]

Tissue remains and cell debris are removed by filtration through layers of gauze or a piece of paper towel. Cells are washed by centrifugations for 10 min at 100 g. Cell aggregates and dead cells are removed by filtration through a wet plug of cotton wool in a Pasteur pipet. Various other methods for dead cell removal have been published.[25,26] Cell yield for the various lymphoid tissues shows a high degree of variation, but approximate values may be given to help in planning an electrophoresis experiment. For the preparation of about 3×10^8 nucleated cells—a number of cells that can be separated by electrophoresis in less than 1 hr—the following amounts of lymphoid tissues are needed: the thymus of 1 rat, the spleens of 1–2 rats, the cervical and mesenteric lymph nodes of 3–4 rats, the thoracic duct lymph of 2 rats drained for 12 hr, or the bone marrow from femura and tibiae of 6 rats with 1 g separation of small cells. For mice the numbers are approximately 4-fold higher.

Separation Medium

The composition of the electrophoresis buffer is given in the table. Despite an intensive search for other combinations,[22] this buffer has been used by us and others for many years, and has remained basically unchanged. Other buffers for the electrophoretic separation of lymphocytes have been described.[27,28]

Requirements set on the electrophoresis buffer are low ionic strength

[24] K. Zeiller and E. Hansen, *J. Histochem. Cytochem.* **26**, 369 (1978).
[25] K. Shortman, N. Williams, and P. Adams, *J. Immunol. Methods*, **1**, 273 (1972).
[26] C. R. Parish, S. M. Kirov, N. Bowern, N. Blanden, and R. V. Blanden, *Eur. J. Immunol.* **4**, 808 (1974).
[27] H. v. Boehmer, K. Shortman, and G. J. V. Nossal, *J. Cell. Physiol.* **83**, 231 (1974).
[28] F. Dumont, and R. C. Habbersett, *J. Immunol. Methods* **53**, 233 (1982).

and conductivity, at physiological pH and osmolarity.[3,6] The reasons for a low ionic strength are reduction in Joule's heat and subsequent thermoconvection, increase in the absolute and relative EPM of the particles to be separated, and increased thickness of the ion cloud around the particles effective for electrophoresis, thus "smoothening" the cell surface. A conductivity of 800–1000 μmho/cm is desirable.

Physiological pH is maintained by buffers such as triethanolamine or HEPES. Osmolarity is adjusted by addition of sugars or glycine. The latter is preferred by us because of its negligible effect on the viscosity of the medium, and the induced increase in the dielectric constant it causes.[22] The buffer should be isoosmotic to the serum of the species, i.e., about 310 mOsm/liter for mouse and 290 mOsm/liter for rat and for human lymphocytes.[29] The addition of calcium or magnesium ions (0.5 mmol/ liter) can help preserve viability of the cells, but reduces their EPM by 15% and enhances cell aggregation.[22]

The low ionic strength milieu has a destabilizing effect on the cell membrane. Therefore (1) the cells must be transferred to the electrophoresis buffer gradually. The cells are first suspended into a 1 : 1 mixture of electrophoresis buffer and protein-free balanced salt solution before being transferred to pure electrophoresis buffer. Cell clumps sometimes formed by released chromatin can be dissolved by addition of traces of deoxyribonuclease.[22] The major cell loss occurs upon the first contact with the medium of low ionic strength. Cells already damaged during cell preparation seem to be especially affected. Further reduction in cell viability is observed only if exposure to the electrophoresis buffer exceeds 1 hr. (2) The exposure time of the cells to the electrophoresis medium must be as short as possible. Lymphocytes are transferred to the separation medium in aliquots. Only as many cells as can be separated in 1 hr are prepared at any one time. The separated cells are collected in tubes containing 1 ml of 1% BSA or 5% fetal calf serum in balanced salt solution. After about 30 min of electrophoretic separation the collection tubes are replaced, and the separated cells are transferred immediately to medium of physiological ionic strength.

A 5-fold concentrated buffer without glycine or other osmoexpanders is used as the electrode buffer (see the table). Both separation and electrode buffers are autoclaved, filtered through membrane filters, and can be stored for weeks at 4°.

Procedure

Electrophoresis is carried out at 4–5° with a buffer flow rate of 300–600 ml/hr. This results in an exposure time of 3–6 min in the electric field

[29] N. Williams, N. Kraft, and K. Shortman, *Immunology* **22**, 885 (1972).

for a cell. The applied voltage is raised to about 1000 V, which corresponds to an effective electric field strength of 70–100 V/cm. With the low ionic strength medium this leads to a current of about 200 mA.

Standard. Before and after an electrophoresis of lymphocytes, freshly prepared or glutardialdehyde-fixed erythrocytes are separated. The electrophoretic distribution of these particles is analyzed for position, band resolution, and skewness. This standard serves as a control for the stability of the separation conditions. Consistent electrophoretic cell separations are obtained, when peaks are located in constant positions relative to the peak of the standard.

The cell sample is adjusted to a concentration of $2–6 \times 10^7$ nucleated cells per milliliter, and applied at a rate of 5–6 ml/hr. With lymphocytes only about 20 (of the 90 available) fractions have to be collected, starting with the peak of the erythrocyte standard. Cell numbers in the fractions are determined in a Coulter Counter with thresholds set to distinguish nucleated cells, and are expressed as percentage of total recovered cells (this volume [6]).

From the numbers of fractions between the peak of separated cells and the peak of cells subjected to the apparatus without an electric field, the electrophoretic migration path can be calculated, considering a slit width of 1.12 mm/fraction at the bottom of the separation chamber. The exposure time for the cells in the electric field can be derived from the buffer flow rate, the dimensions of the separation chamber, and the position of the sample inlet. Absolute values of EPM can be calculated from the migration path (D), the effective electric field strength (H), and the exposure time (τ). EPM $= D/H\tau$ (cm^2 V^{-1} sec^{-1}). However, since the EPM is also dependent on the electrophoretic separation medium and other factors, it is not a useful parameter for comparison of electrophoretic data.

As an example of free-flow electrophoresis of murine lymphocytes the procedure for the electrophoretic separation of lymph node cells from the rat is described in some detail.

Free-Flow Electrophoresis of Rat Lymph Node Cells

Materials

Balanced salt solution (Puck G tissue culture medium, Difco)
NaCl 138 mmol/liter, KCl 5.4 mmol/liter, MgSO$_4$ 0.6 mmol/liter, CaCl$_2$ 0.1 mmol/liter
Na$_2$HPO$_4$ 0.2 mmol/liter, KH$_2$PO$_4$ 1.1 mmol/liter, glucose 6.1 mmol/liter, pH 7.3, 290 mOsm/liter
1% bovine serum albumin (BSA, Armour, deionized) in Puck G
Electrophoresis buffer (see the table) pH 7.2, 290 mOsm/liter
Electrode buffer (see the table) pH 7.2

Procedure

1. A rat, 8 weeks of age, is killed by cervical dislocation. Cervical and mesenteric lymph nodes are removed. They are transferred into ice-cold Puck G medium supplemented with 1% BSA and teased with forceps and a needle in a Petri dish on ice.

2. The supernatant is removed with a pipet and replaced by fresh buffer. Teasing is continued until the supernatant stays clear.

3. The cell suspension is filtered through layers of gauze. Cells are washed twice in cold Puck G buffer at 4° for 10 min at 100 *g*.

4. The cells are resuspended in 20 ml of cold Puck G solution; 20 ml of cold electrophoresis buffer is added dropwise with gently shaking.

5. After centrifugation the cell pellet is resuspended in electrophoresis buffer and washed once more for 10 min at 100 *g*.

6. The cells are suspended in separation medium with a plastic pipet and filtered through a wet plug of cotton wool in a Pasteur pipet.

7. Cell concentration is determined in a Coulter counter, and cell viability tested by Trypan Blue exclusion (this volume [6]).

8. During cell preparation the electrophoresis apparatus is run to attain stability of separation conditions. Prior to the electrophoresis of lymphocytes, for a standard a preparation of erythrocytes is separated and its electrophoretic distribution analyzed.

9. The lymphoid cells are subjected to free-flow electrophoresis at a rate of 5 ml/hr. Electrophoresis is carried out at 5° with a buffer flow of 550 ml/hr at a voltage of 1000 V.

10. Fractions are collected in 10-ml plastic tubes containing 1 ml of 1% BSA in Puck G medium. Aliquots are taken for Coulter counter analysis.

Results. Figure 2A shows the electrophoretic separation of rat lymph node cells, and Fig. 2B that of spleen cells under the same conditions. A bimodal distribution of nucleated cells is observed in both cases. Immunofluorescence with antibodies against rat B lymphocyte-specific antigens (RBLA) and rat T lymphocyte-specific antigens (RTLA)[30] identifies the cells of high EPM as T lymphocytes, and the cells of low EPM as B lymphocytes. They can be isolated with over 95% purity.

Bimodal electrophoretic distributions are also obtained with lymphocytes from thoracic duct lymph,[4,27] Peyer's patches,[31] and peripheral

---

[30] K. Zeiller, and G. Pascher, *Eur. J. Immunol* **8**, 469 (1978).

[31] F. Dumont, A. Ashmed, and R. Habbersett, *in* "Cell Electrophoresis: Clinical Application and Methodology" (A. W. Preece and D. Sabolovič, eds.), p. 157. Elsevier/North-Holland Biomedical Press, Amsterdam, 1979.

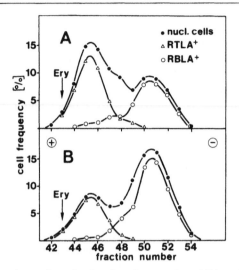

FIG. 2. Free-flow electrophoresis of rat lymphocytes from (A) lymph nodes; (B) spleen. Both RBLA and RTLA were detected by direct immunofluorescence using FITC- or TRITC-conjugated antisera raised in rabbits.

blood.[5,32] This efficient separation of murine T and B cells by free-flow electrophoresis has been verified in a number of investigations, with characterization of the separated cells by various tests including cell surface antigens, SDS–PAGE analysis of cell membrane proteins, Fc receptors, complement receptors, rosette formation with SRBC, graft-versus-host reaction, mixed lymphocyte reaction, response to T or B cell-specific mitogens, cytotoxicity, cell cooperation, and adoptive immune response.[4–9] Free-flow electrophoresis of murine lymphocytes, however, is not limited to the separation of T and B cells. Subsets of T and B lymphocytes have been isolated by electrophoresis,[28,31,33] and successful separation of lymphocytes, both of T[8,32,33] and B[4,9,34] lineages at various stages of differentiation has been achieved. In the electrophoretic separation of spleen cells, B cells are found in the region of low EPM. After immunization, however, the precursors of antibody forming cells are found in the region of high EPM. The observed change in surface charge density of lymphocytes during differentiation and activation may be related to cell adhesion phenomena.[6,35]

[32] K. Zeiller, G. Pascher, G. Wagner, H. G. Liebich, E. Holzberg, and K. Hannig, *Immunology* **26,** 995 (1974).
[33] K. Zeiller, R. K. Schindler, and H. G. Liebich, *Isr. J. Med. Sci.* **11,** 1242 (1975).
[34] K. Zeiller, G. Pascher, and K. Hannig, *Immunology* **31,** 863 (1976).
[35] U. Galili, P. Häyry, and E. Klein, *Cell. Immunol.* **48,** 91 (1979).

The electrophoretic distribution of murine thymocytes or bone marrow small lymphocytes is unimodal with a distinct shoulder in the region of higher EPM. Free-flow electrophoresis of thymus cells results in the separation of cortical, cortisone-sensible thymocyte of high Thy-1 antigen content with low EPM from medullary, PHA-responsive thymocytes of low Thy-1 antigen content with high EPM.[4,8,32,36] Free-flow electrophoresis of small bone marrow lymphocytes, deprived of most nonlymphoid cells through 1 g velocity sedimentation, separates the dominant lymphoid cell population of immature precursors from mature T and B cells.[5,24,37,38] Cell loss in free-flow electrophoresis of lymphocytes is mainly due to the low ionic strength of the separation medium, and amounts to 15–20%. This cell loss can be shown to be nonselective by assaying the pooled fractions of recovered cells.

Cell viability, as judged by Trypan Blue exclusion test, is over 90% in all relevant fractions. Dead cells tend to acquire higher negative surface charge. Functional integrity of the separated cells has been documented in a number of functional assays, *in vitro* and *in vivo*.

Antigen-Specific Electrophoretic Cell Separation (ASECS) and the Isolation of Human Lymphocyte Subpopulations

Introduction

The limitations of preparative electrophoretic lymphocyte separation stem mainly from the modest differences in surface charge density of lymphocytes. This has been a serious drawback in the application of free-flow electrophoresis to the separation of human lymphocytes. Here, differences in EPM even between T and B cells are too low to allow effective separation.

A preparative electrophoresis of human lymphocytes isolated from peripheral blood by Ficoll–Hypaque density gradient centrifugation is presented in Fig. 4A. As shown by others,[5,10–12] a rather homogeneous distribution is observed. The shoulder in the region of lower EPM is caused by monocytes present in the preparation.[11,19,39] The position in the electrophoretic separation of B cells bearing surface Ig is not significantly different from the position of the peak of the other cells (i.e., mainly T lymphocytes). Only a slight enrichment of Ig-posiitive B cells and PHA-

[36] D. Sabolovič and F. Dumont, *Immunology* **24**, 601 (1973).
[37] K. Zeiller and E. Hansen, *Cell. Immunol.* **44**, 381 (1979).
[38] D. G. Osmond, R. G. Miller, and H. von Boehmer, *J. Immunol.* **114**, 1230 (1975).
[39] E. M. Levy, S. Silverman, K. Schmid, and S. R. Cooperband, *Cell. Immunol.* **40**, 222 (1978).

reactive T cells is observed at the extremes of the distribution curve, representing a minimal proportion of the respective cell population.[10] It should be mentioned however that separation of lymphocytes from monocytes in the region of high EPM may be so efficient, that the mitogenic response to Con A or PHA after electrophoretic separation is diminished, because of the removal of accessory cells.[40] This deficiency can be compensated by the addition of monocyte preparations to these fractions.

The obvious failure of cell electrophoresis to separate human T and B cells and the need for further dissection of other lymphocyte subpopulations as well has led us to develop a new preparative method we call ASECS.[19] It combines the high specificity of antibody reactions with the high separation capacity of continuous free-flow electrophoresis. The method is based on the fact that immunoglobulins bear a lower net negative charge than cell surfaces. The EPM of a cell population can thus be decreased by reaction with appropriate antibodies. This effect of immunoglobulins on cells including lymphocytes in a number of analytical investigations.[6]

Principle

Lymphocytes are prepared for ASECS by incubating them with a monospecific antiserum that reacts with a certain subpopulation of cells (Fig. 3). The antibodies bound to the cell surface will cover some of the net negative surface charge of the respective cell population. A second, and eventually a third, fluorescence-labeled antibody is added for augmentation of the effect and for convenient detection of the antigen-positive cells by immunofluorescence. The masking of negatively charged groups on the cell surface leads to a decreased deflection of the cells in the electric field.

Method

Essential points for the performance of ASECS are noncapping conditions, sandwich techniques, and optimal concentrations of antibodies. Most of the analytical investigations about antibody effects on the EPM of cells have been carried out at room temperature or at 37°. Some of the discrepancies in these studies may originate from this fact, since it remains unclear to what extent capping of the antibodies on the cell surface may have occurred. In order to obtain a reproducible decrease in EPM, it is important to avoid antibody-induced redistribution of surface antigens. Therefore, cells are kept at 4° during incubation with the antibody and separation. Addition of blocking agents like sodium azide can thus be avoided.

[40] H. A. Abramson, *J. Gen. Physiol.* **12**, 711 (1929).

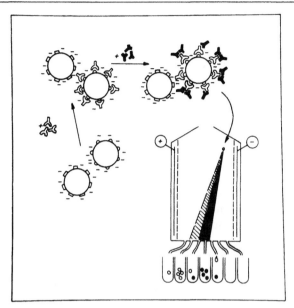

FIG. 3. Principle of ASECS. The cell population to be separated is specifically labeled with an appropriate antibody followed by attaching a second antibody raised against the first (sandwich method). As a consequence their negative charge is reduced. Using this decrease in negative surface charge, the labeled cells can then be separated in free-flow electrophoresis from unlabeled cells.

A second and possibly a third antibody may be used to enhance the slowing effect on EPM, and by using fluorescence-labeled antibodies the detection of the antibody-reactive cells is facilitated. Sandwich techniques are especially helpful when working with monoclonal antibodies, because of the low intensity of staining. A third antibody may then be applied, as long as specificity is unaffected. The number of antigen-positive cells is determined in the electrophoretic fractions by immunofluorescence microscopy. The isoelectric point of immunoglobulins is less changed by conjugation with rhodamine rather than fluoresceine. For instance, in an experiment in which we subjected to free-flow electrophoresis human erythrocytes, FITC-labeled, TRITC-labeled and unlabeled goat anti-rabbit-IgG immunoglobulins, we observed mean electrophoretic migration paths of 40, 20, 7, and 2 mm, respectively. Thus, the difference in charge density between lymphocytes and bound immunoglobulins can be better maintained by using a rhodamine label.

Especially with sandwich techniques, the formation of soluble immunocomplexes, and therefore nonspecific uptake of antibody via Fc receptors, has to be considered. This reaction, however, does not contribute to a change in the EPM of cells.[19] Nevertheless, precautions must be taken by removing aggregates from antibody preparations by high-speed centrifugation before use, and by washing the cells 3 times between incubations with antiserum. Antibody fragments may be used, if available. Second and third antibodies directed against Ig may cross react with surface Ig on B cells. These antisera therefore must be absorbed before use with immobilized human Ig or human buffy coat preparations.

In order to attain maximal slowing effect on EPM, as much antibody as possible must be bound on the cell surface, without undesirable nonspecific staining and cell aggregation however. For every antibody combination or new batch of antiserum, optimal concentrations must be determined before use. Prior to a set of ASECS experiments, therefore, human lymphocytes are incubated with various dilutions of the selected first and second, and, if appropriate, third antibody and then the percentage of antigen-positive cells and cell aggregation is determined by immunofluorescence microscopy. The optimal concentration of antiserum is thus determined. The antiserum concentration necessary to achieve optimal uptake of antibody by the lymphocytes may be about 10-fold higher than that necessary for detection of surface antigens by immunofluorescence. The same antiserum can be used several times over, since only a minimal proportion of it is absorbed on the cells.

A convenient source of human lymphocytes is peripheral defibrinated blood. Defibrination is preferable to addition of anticoagulants.[41] Platelets that may form aggregates with nucleated cells can be removed by centrifugation. Mononuclear cells are isolated by centrifugation in a one-step density gradient of Ficoll–Hypaque according to Bøyum.[42] About 1×10^6 nucleated cells are recovered per 1 ml of blood.

Separation of Human T and B Lymphocytes

Materials

Glass beads
Ficoll–Hypaque (LSM lymphocyte separation medium, Mediapharm)
Balanced salt solution (Puck G tissue culture medium, Difco)
Electrophoresis buffer (see the table)
Electrode buffer (see the table)
Rabbit anti-human IgM antiserum (Behring Institut)
TRITC-conjugated goat anti-rabbit IgG immunoglobulin (Nordic Immunology)

[41] H. E. Johnsen and M. Madsen, *Scand. J. Immunol.* **8**, 239 (1978).
[42] A. Bøyum, *Scand. J. Clin. Lab. Invest.* **21**, 31 (1968). See also this volume [9].

Procedure

1. Defibrination of blood: 100 ml of blood is drawn by venipuncture and defibrinated by gentle shaking in three 50-ml plastic tubes containing 10–20 glass beads each.

2. Removal of platelets: The blood is filtered through layers of gauze to remove the clot, and centrifuged for 10 min at 200 g at room temperature. The supernatant containing most of the platelets is discarded.

3. Preparation of mononuclear cells: The sediment is diluted with 3 volumes of Puck G solution. Portions of 6 ml of resuspended cells are carefully layered on top of 3 ml of LSM (d = 1.077 kg/liter) in 10-ml plastic tubes. After centrifugation at room temperature for 30 min at 400 g, the mononuclear cells forming a band at the interface are removed with a Pasteur pipet. The cell sample is diluted by addition of at least the same volume of Puck G solution, and centrifuged for 30 min at 400 g. The cells are washed twice in cold Puck G medium. These and all following centrifugations are carried out at 4° for 10 min at 100 g. The cells are resuspended in 2 ml of Puck G, and the cell concentration is determined with a Coulter counter. Cell viability is tested by Trypan Blue exclusion.

4. Incubation with first antibody: Antisera are centrifuged for 30 min at 18,000 g prior to use; 5 × 10^7 nucleated cells are centrifuged in Eppendorf tubes at 100 g for 10 min, and resuspended in 1 ml of the proper dilution of rabbit anti-human IgM antiserum, previously found to give optimal staining (1 : 4 dilution used by us). The sample is incubated for 30 min at 4°, with occasional mixing. After centrifugation for 10 min at 100 g the antiserum is removed, and the cells washed twice in Puck G.

5. Incubation with second antibody: The cell pellet is resuspended in 1 ml of the proper dilution of TRITC-conjugated goat anti-rabbit IgG immunoglobulin (1 : 2 dilution used by us), and incubated for 30 min at 4° in the dark. After centrifugation the antiserum is removed. The cells are washed once in a 1 : 1 mixture of Puck G and electrophoresis buffer, and once in electrophoresis buffer.

6. Electrophoretic separation: Cells are resuspended in 2 ml of electrophoresis buffer, and filtered through a wet plug of cotton wool in a Pasteur pipet. Cells are counted and viability is determined. The percentage of cells bearing surface IgM is evaluated by immunofluorescence microscopy. Conditions for the electrophoretic separation are 4°, an electric field of 900–1000 V, a buffer flow rate of 500 ml/hr, and a sample application of 6 ml/hr. Fractions are collected in 10-ml tubes containing 1 ml of 1% BSA in Puck G buffer. Aliquots are taken for Coulter counter analysis and for immunofluorescence microscopy to determine the electrophoretic distribution of nucleated cells and of fluorescence-labeled B cells.

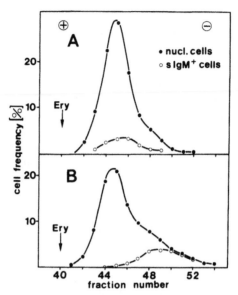

Fig. 4. Free-flow electrophoresis of human peripheral blood lymphocytes. (A) separation of untreated cells. Identification of B cells by indirect immunofluorescence with antiserum against IgM. (B) separation of antibody-labeled cells (ASECS) using rabbit anti-human IgM antiserum and TRITC-conjugated goat anti-rabbit IgG immunoglobulin. Cell donor and separation conditions are identical to A.

Figure 4B shows the result of such an experiment. B cells are pooled in the region of low EPM, T cells in the region of high EPM.

Separation of Human T Cell Subpopulations

We describe here the application of ASECS in the separation of human T cell subpopulations using the monoclonal antibody OKT8. This antibody labels the human suppressor/cytotoxic T lymphocyte population making up about 30% of human blood lymphocytes.[43]

Antibodies. Monoclonal antibody OKT8 (Ortho Pharmaceutical Corp.); rabbit anti-mouse IgG immunoglobulin (Miles-Yeda Ltd.); TRITC-conjugated goat anti-rabbit IgG immunoglobulin (Nordic Immunology).

[43] Y. Thomas, J. Sosman, O. Irigoyen, S. M. Friedman, P. C. Kung, G. Goldstein, and L. Chess, *J. Immunol.* **125,** 2402 (1980).

Procedure. The procedure follows that described for the separation of T and B lymphocytes, except for the following changes.

Incubation with first antibody: Antisera are centrifuged for 30 min at 18,000 g prior to use. Nucleated cells (5×10^7) are centrifuged in an Eppendorf tube at 100 g for 10 min, and resuspended in 1 ml of the proper dilution of OKT8 monoclonal antibody (1:20 dilution of OKT8 used by us). The mixture is incubated for 30 min at 4°, with occasional mixing. After centrifugation for 10 min at 100 g, the antiserum is removed and the cells washed twice in Puck G.

Incubation with second and third antibody: The cell pellet is resuspended in 1 ml of the proper dilution (1:2) of rabbit anti-mouse IgG immunoglobulin that had previously been absorbed with 1/10 volume of human buffy coat. The sample is incubated for 30 min at 4°. After centrifugation, the antiserum is removed and the cells are washed twice in Puck G buffer.

The pelleted cells are resuspended in 1 ml of the proper dilution (1:2) of TRITC-conjugated goat anti-rabbit IgG immunoglobulin previously absorbed with 1/10 volume of human buffy coat. The preparation is incubated for 30 min at 4° in the dark. After centrifugation the antiserum is removed, and the cells are washed once in a 1:1 mixture of Puck G and electrophoresis buffer, and once in electrophoresis buffer. Figure 5 shows the preparative free-flow electrophoresis of these cells. OKT8$^+$ cells are pooled in the region of low EPM, whereas OKT4$^+$ lymphocytes of the inducer/helper type are enriched in the reion of high EPM.

Results

We have applied ASECS to the separation of human T and B lymphocytes and to the isolation of human T cell subpopulations.[19]

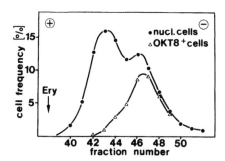

FIG. 5. ACECS of human peripheral blood lymphocytes using monoclonal antibody OKT8, rabbit anti-mouse IgG immunoglobulin, and TRITC-conjugated goat anti-rabbit IgG immunoglobulin.

After the reaction of human mononuclear cells with anti-IgM antibody and a fluorescence-labeled second antibody, B cells show a shift to the cathode by 4–5 fractions (Fig. 4B), compared to unlabeled cells (Fig. 4A). This corresponds to a reduction in mean EPM of 15–20%. In some fractions in the region of low EPM over 95% of the cells had surface IgM (sIgM). A pool of fractions containing more than half of the sIgM+ cells recovered from electrophoretic separation was about 90% pure. Cell recovery was about 65% for all cells and about 80% for the antibody-labeled B cells, possibly reflecting a higher resistance of antibody-coated cells to the separation medium of low ionic strength. Cell viability was greater than 90% in all fractions.

Very similar results were obtained with the electrophoretic separation of cells labeled with monoclonal antibody OKT8 (Fig. 5), or the comparable monoclonal antibody T811.[19,44] The corresponding T cell subpopulation could be isolated in high yield and purity. The viability was also high.

Flow cytometric analysis has shown that in ASECS the cells with the strongest fluorescence intensity have the lowest EPM.[19] Thus, electrophoresis separates antibody-labeled cells according to their antigen density.

Comments

Although most properties of cells are not impaired by antibodies, it should be kept in mind that in some cases bound antibody may influence certain lymphocyte functions.[45,46] After ASECS, antibodies can be removed from the cells by incubation at 37°, which results in capping and shedding of antigen–antibody complexes from the lymphocyte surface.

The problem of bound antibodies can also be overcome by the fact that ASECS allows both positive and negative selection. To isolate for instance the suppressor/cytotoxic T cell subpopulation after antibody labeling, OKT8+ cells can be collected in the region of low EPM. The same cell population without bound antibodies can be recovered in the region of high EPM, if monoclonal antibody OKT4 and antiserum directed against B cells are used on the cells before electrophoresis.

Compared to other antibody-dependent cell separation methods, like immunoadsorbent chromatography or immunorosetting, the absence of a carrier is a major advantage of ASECS, thus avoiding nonspecific interactions. Some attempts have been made recently to use antibodies coupled

[44] E. P. Rieber, J. Lohmeyer, D. J. Schendel, and G. Riethmüller, *Hybridoma* **1,** 59 (1981).
[45] J. Braun and E. R. Unanue, *Immunol. Rev.* **52,** 3 (1980).
[46] J. P. van Wauwe and J. G. Goossens, *Cell. Immunol.* **77,** 23 (1983).

to plastic microspheres for the electrophoretic separation of human lymphocytes.[47] Aggregation and damage of the microsphere-labeled lymphocytes, however, was observed. In addition, multivalent binding may render dissociation of the particles from the cells difficult. We find it neither necessary nor desirable to add cell-particle interactions to the cell-antibody reaction for electrophoretic cell separation.

A promising approach to the amplification of the differences in cell surface charge induced by bound antibody may be the prior modification of the antibodies used. Charged groups may be introduced or neutralized, and charged peptides or proteins may be coupled to immunoglobulin,[48] without interfering with antibody specificity.

Another important advantage of ASECS in comparison to other immunospecific cell separation methods is its high separation capacity. While, for instance, in fluorescence-activated cell sorting the cell flow rate is limited to about 2000 cells/sec, in free-flow electrophoresis the antibody/labeled cells can be separated at a rate of 100,000 cells/sec.

The introduction of ASECS has greatly enhanced the potential of free-flow electrophoresis. This technique opens possibility of electrophoretic separation of human lymphocyte subpopulations on a preparative scale and should also be helpful for the separation of murine lymphocyte subclasses.

[47] A. Smolka, D. Kempner, and A. Rembaum, *Electrophoresis* **3**, 300 (1982).
[48] G. B. Olson, M. McFadden, and B. H. Bartels, *in* "Electrophoresis 81" (R. C. Allen and P. Arnaud, eds.), p. 933. de Gruyter, Berlin, 1981.

[19] Fluorescence-Activated Cell Sorting: Theory, Experimental Optimization, and Applications in Lymphoid Cell Biology

By D. R. PARKS and L. A. HERZENBERG

Introduction

Most of the questions being asked in contemporary cell biological research can be framed in terms of the properties of the individual cells in the system and the interactions among those cells. Flow cytometry and sorting now provide the most informative and powerful methods for the analysis and separation of cell populations in such research. The analytical strength of flow cytometry lies in its ability to make quantitative,

METHODS IN ENZYMOLOGY, VOL. 108

multiparameter measurements on statistically adequate numbers of cells to define the properties of a cell population or its component subpopulations. The techniques can be applied to any sample that can be obtained as a single cell suspension. Although the same methodology and instrumentation can be used for analysis and sorting of all kinds of particles, including subcellular components like chromosomes[1] or whole nuclei, we limit this chapter to consideration of cells.

Multiparameter analysis, combining measurements of intrinsic cell properties like light scatter with quantitative assessment of investigator-controlled features such as cell surface immunofluorescence, makes it possible to evaluate particular cell populations even in complex mixtures. Monoclonal antibodies have vastly increased the range and effectiveness of immunofluorescence measurements, while flow cytometry and sorting have been important aids in the development and characterization of new monoclonal antibodies.

The FACS[1a] makes quantitative, correlated multiparameter measurements on each cell. These measurements may be the ultimate data in an experiment or may be used to define sorting criteria yielding sorted cells for further work. Other cell separation techniques rely on nonspecific physical properties, like buoyant density or nylon adherence, or on application of a specific reagent, such as in antibody affinity binding, in ways that are neither quantitative nor multiparameter.

At present some experiments require more sorted cells than FACS techniques can supply in a reasonable time. Typical running rates of 5000 cells per second mean that it takes a day to run 10^8 cells. Some increase in cell analysis and sorting rate may be expected in the future, but the main openings will probably come from improvements in experimental techniques allowing smaller cell samples to be used.

Flow cytometric techniques are now used in a wide variety of basic and clinical research programs and are coming into use for clinical diagnostic purposes. The array of uses ranges from chromosome sorting to identification of freshwater algae, and books,[2] conference proceedings,[3]

---

[1] A. V. Carrano, M. A. Van Dilla, and J. W. Gray, *in* "Flow Cytometry and Sorting" (M. Melamed, P. Mullaney, and M. Mendelsohn, eds.), p. 421. Wiley, New York, 1979.

[1a] Abbreviations: FACS, fluorescence-activated cell sorter; PI, propidium iodide; PE, phycoerythrin; APC, allophycocyanin; PMT, photomultiplier tube; dhfr, dihydrofolate reductase; AHH, aryl hydrocarbon hydroxylase; BP, benzo[*a*]pyrene; TCDD, 2,3,7,8-tetrachlorodibenzo-*p*-dioxin.

[2] M. Melamed, P. Mullaney, and M. Mendelsohn, eds, "Flow Cytometry and Sorting." Wiley, New York, 1979.

[3] Automated Cytology Conference Proceedings, *J. Histochem. Cytochem.* **22**(7) (1974); **24**(1) (1976); **25**(7) (1977); **27**(1) (1979).

and review articles[4-6] as well as the journal *Cytometry* are available as sources of information.

The discussion in this chapter focuses on sorting-capable instruments and their use in lymphoid cell analysis, immunochemical and enzymological research, studies of gene amplification and hybridoma selection, and other rare cell sorting problems. The emphasis is on information relevant to the analysis and sorting of viable, immunofluorescent stained cells. We have tried to discuss theory and practical considerations in ways that will be clear to those not already using flow cytometry but with enough rigor and detail to be informative to current users and to facilitate applications beyond those described specifically in the applications sections.

History of the Instrumentation

The history of flow cytometry and cell sorting has been reviewed by Melamed and Mullaney[7] and by Herzenberg et al.[8] The following sketch of the development of the instrumentation is based largely on those reviews.

Ideas for cell counting and measurement in flow have been around for a long time, but reliability and reproducibility of measurements was not easily achieved until laminar-flow guidance of a small cell-containing stream in moving cell-free fluid[9] was applied to give stable flow in relatively large channels. Several microscope-based systems were developed to measure light scatter, light absorption and/or fluorescence using arc-lamp illumination and the epifluorescence principle.[10,11] Van Dilla et al.[12] at Los Alamos made the first use of an argon-ion laser in a system where the axes of cell flow, laser beam path, and fluorescence detection were mutually orthogonal.

Electrostatic sorting utilizing stabilized drop formation and individual drop charging, as used in current cell sorting machines, developed from the ink-jet deflection work of Sweet.[13] Fulwyler first applied this tech-

[4] K. A. Ault, *Diagn. Immunol.* **1**, 2 (1983).

[5] M. R. Loken and A. M. Stall, *J. Immunol. Methods* **50**, R85 (1982).

[6] L. A. Herzenberg and L. A. Herzenberg, in "Handbook of Experimental Immunology" (D. M. Weir, ed.), 3rd ed., p. 22.1. Blackwell, Oxford, 1978.

[7] M. R. Melamed and P. F. Mullaney, in "Flow Cytometry and Sorting" (M. Melamed, P. Mullaney, and M. Mendelsohn, eds.), p. 3. Wiley, New York, 1979.

[8] L. A. Herzenberg, R. G. Sweet, and L. A. Herzenberg, *Sci. Am.* **234**, 108 (1976).

[9] P. J. Crosland-Taylor, *Nature (London)* **171**, 37 (1953).

[10] L. K. Kamentsky, M. R. Melamed, and H. Derman, *Science* **150**, 630 (1965).

[11] W. Gohde and W. Dittirch, *Acta Histochem., Suppl.* **10**, 42 (1971).

[12] M. A. Van Dilla, T. T. Trujillo, P. F. Mullaney, and J. R. Coulter, *Science* **163**, 1213 (1969).

[13] R. G. Sweet, *Rev. Sci. Instrum.* **36**, 131 (1965).

nique to sorting cells on the basis of a Coulter-type resistive volume measurement.[14] Our group at Stanford[15] developed a sorting system for fluorescent-stained cells, and Bonner et al.[16] in our laboratory introduced optical measurements in the sorting jet rather than in an enclosed flow cell. A laser light scatter channel that was added to that system to provide detection of nonfluorescent cells proved to be useful not only for cell sizing but also for live/dead cell discrimination.[17] The first commercial cell sorting instrument (Becton Dickinson "FACS I") was based on that system, and most commercial sorters still use the same basic design.

In the last several years both investigator-built and commercial cell sorters have proliferated and evolved with changes particularly toward multiparameter measurement and toward more sophisticated and powerful computer analysis.

Theory of Operation and Measurement

Description of the Basic Cell Sorter Components and Their Functions

A sketch of the mechanical and optical components of a "typical" cell sorter and a block diagram of the signal processing electronics are shown in Fig. 1 and described below. Fluorescent and scattered light are produced when a cell contained in the liquid jet passes through the focused laser beam. The jet consists primarily of cell-free sheath fluid. The cell suspension is injected into the center of the nozzle, and hydrodynamic focusing[18] ensures that the cells remain centered in the jet. This centering is necessary for making accurate measurements on the cells.

A forward light scatter detector receives light from a selected angular range beyond the coverage of the laser beam stop, commonly from angles in the range from 1 to 15°. Fluorescent light is collected by a lens and divided between two photomultiplier detectors (PMTs) by a dichroic beamsplitter. Each fluorescence detector has further optical filtering to exclude scattered laser light and to pass light in the desired wavelength region for that detector.

For sorting, the nozzle assembly is vibrated by an oscillator-driven piezoelectric crystal (a device which expands and contracts slightly as a

[14] M. J. Fulwyler, R. B. Glascock, R. D. Hiebert, and N. M. Johnson, *Rev. Sci. Instrum.* **40,** 42 (1969).

[15] H. R. Hulett, W. A. Bonner, J. Barrett, and L. A. Herzenberg, *Science* **166,** 747 (1969).

[16] W. A. Bonner, H. R. Hulett, R. G. Sweet, and L. A. Herzenberg, *Rev. Sci. Instrum.* **43,** 404 (1972).

[17] M. R. Loken and L. A. Herzenberg, *Ann. N.Y. Acad. Sci.* **254,** 163 (1975).

[18] V. Kachel and E. Menke, *in* "Flow Cytometry and Sorting" (M. Melamed, P. Mullaney, and M. Mendelsohn, eds.), p. 41. Wiley, New York, 1979.

function of the varying voltage applied across it) at a frequency near the natural drop breakup frequency of the jet. This stabilizes the drop formation at that frequency, resulting in uniform drop size and a well-defined time delay between detection of a cell at the laser beam and incorporation of the cell into a free drop. If a cell is to be sorted, a potential in the 100 V

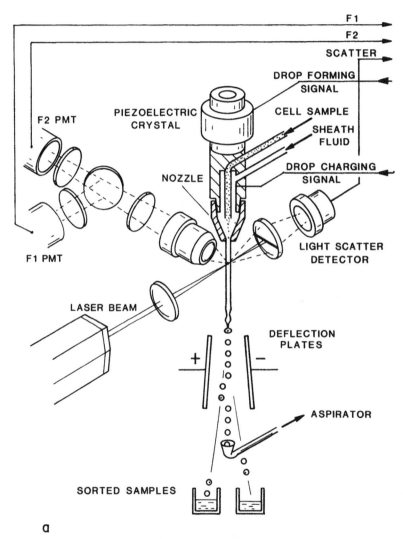

FIG. 1. Generalized cell sorter diagram. The functions of the components in (a) the mechanical and optical systems and (b) the signal processing and sorting electronics are described in the text.

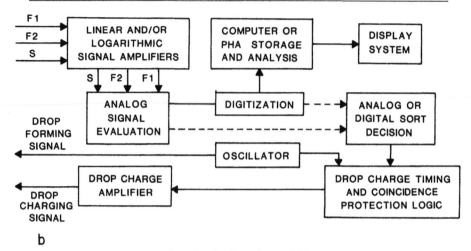

b

FIG. 1b. See legend on p. 201.

range is applied to the fluid inside the nozzle. The electrical conductivity of the fluid assures that any drops that break from the jet while the voltage is applied will carry a corresponding electric charge. The timing and duration of the applied voltage are chosen to charge one or more drops which will contain the desired cell. Two populations of cells can be sorted simultaneously by applying a positive charge to drops containing one population and a negative charge to drops containing the other. The train of drops passes between two plates charged at plus and minus several thousand volts, separating the charged drops from the uncharged ones. The undeflected drops are removed by an aspirator, and the deflected drops are collected in appropriate receptacles. The aspirator is very effective in preventing formation of aerosols by the undeflected drops.[19] This can be important when handling biohazard materials.

Proceeding to the signal processing electronics (Fig. 1b), the light scatter and fluorescence detector outputs are amplified and processed for evaluation. These operations may involve bandpass limitation, time gating, integration, peak detection, etc. with the usual objective of producing a processed signal for each detector that is the best possible estimator of the scattered light or fluorescence produced by each cell. These signals are digitized and fed to a computer or pulse height analyzer for storage and analysis. Cell frequency data in one or two of the scatter/fluorescence measurement dimensions can be displayed on a CRT screen or on a hard copy plotter to assist the investigator in visualizing and interpreting the cell population measurements.

[19] J. T. Merrill, *Cytometry* **1,** 342 (1981).

Sorting decisions can be made using either analog or digital data. Analog decision making is usually accomplished by combining independent "windows" on the signals so that a cell is selected for sorting if its light scatter signal falls within a selected range and each of its fluorescence signals falls within the range selected on that channel. In two dimensions (e.g., light scatter vs green fluorescence) the selection criteria can be visualized as a rectangular box such that cells whose signals correspond to a point in the box are to be sorted, and those outside the box are not. If the sort decision is made by a computer using digitized data, the selection may be defined on the basis of a more general combination of the signal values. In two dimensions useful results might be elliptical sorting regions or irregular polygon regions.

Once the decision about the preferred disposition of the cell has been made (i.e., left sort, right sort, or no sort) it is still necessary to produce the drop charging pulse at the appropriate time. It is also usually desirable to protect the purity of sorted fractions by examining situations in which a cell that fulfills the sorting criteria is close to a cell which does not ("coincidences") so that sorting of any drop that might contain the wrong cell can be suppressed. The system to carry this out could be hard-wired or computerized.

The fundamental limit on the complexity of computer decision making in cell sorter systems is the time between the detection of the cell and its incorporation into a free drop, which may be as short as 200 μsec. In addition, it may be necessary to process several cells during that interval.

Cell Sorter Elaborations and Variations

At this point we would like to mention a few additional features and ways in which some cell sorter systems differ from the "typical" one described above.

More Excitation Sources and Detectors. As the cell populations and questions being asked about them become more complex, it becomes important to increase the array of measurements made on each cell. A number of systems have been produced using two or three lasers to excite several dyes.[20-24] Use of such systems may require three or four fluorescence detection channels. Also light scatter measurements in angular ranges other than the moderate forward angles mentioned above have

[20] D. J. Arndt-Jovin, B. G. Grimwade, and T. M. Jovin, *Cytometry* **1,** 127 (1980).
[21] D. R. Parks, R. R. Hardy, and L. A. Herzenberg, *Immunol. Today* **4,** 145 (1983).
[22] D. R. Parks, R. R. Hardy, and L. A. Herzenberg, *Cytometry* **5,** 159 (1984).
[23] J. A. Steinkamp, C. C. Stewart, and H. A. Crissman, *Cytometry* **2,** 226 (1982).
[24] M. R. Loken and L. L. Lanier, *Cytometry* **5,** 151 (1984).

proved useful in analyzing complex populations. We discuss these additions in more detail below.

Sample Handling. Several years ago we developed a microprocessor-controlled, syringe-driven sample handling system as an aid in analyzing large numbers of cell samples. Samples may be taken directly from the tray in which they are stained and run without being transferred to individual tubes, allowing larger numbers of samples to be run conveniently.

Cloning. In work with tissue culture cells it is often convenient to do cloning directly with the cell sorter.[25] The sorting logic modifications that allow one and only one cell or a specified number of cells fulfilling the selection criteria to be sorted have been implemented on several commercial machines.

Flow Cell Interrogation. In some cell sorter systems the cell stream intersects the laser beam within a transparent flow chamber and then passes through the jet forming orifice. This system has some optical advantages in that the laser beam passes through flat windows thus avoiding the large amount of reflected and refracted light found in jet-in-air measurement systems. On the other hand the windows must be kept clean while the surface of a jet in air is constantly renewed and self cleaning. The extra time and the stream velocity changes occurring between cell detection and drop formation in a flow cell sorter lead to greater delay time uncertainties in sorting so that more drops may have to be allocated to each cell than with jet-in-air measurement.

Other Measurements in Flow

While light scatter and fluorescence are the primary measurements made in flow sorters, several other types of measurements can be made in flow systems, and some have been implemented in sorters. Coulter volume, based on the effect of a cell on the DC resistance across an orifice, involves a more complex geometry with separate measuring and jet forming orifices and also requires careful design to keep the sorter drop charging pulses from interfering with the resistive volume measurement.[26] Related AC measurements have been carried out,[27] but they do not seem to be in routine use.

Ultrasound measurements on cells in flow have been demonstrated,[28] but their utility in biological investigations has not yet been established.

[25] D. R. Parks, V. M. Bryan, V. T. Oi, and L. A. Herzenberg, *Proc. Natl. Acad. Sci. U.S.A.* **76,** 1962 (1979).

[26] V. Kachel, *in* "Flow Cytometry and Sorting" (M. Melamed, P. Mullaney, and M. Mendelsohn, eds.), p. 61. Wiley, New York, 1979.

[27] R. A. Hoffman and W. B. Britt, *J. Histochem. Cytochem.* **27,** 234 (1979).

[28] R. G. Sweet, M. J. Fulwyler, and L. A. Herzenberg, *Anal. Cytol. Cytometry, 9th* and *Int. Symp. Flow Cytometry, 6th* Abstract (1982).

Light extinction can be measured, particularly in flow cell geometries.[29] This is a measurement of the light removed from a laser beam by the passage of a cell. Theoretically, it should correspond to an all-angle integrated scatter measurement since light is removed from the laser beam by scatter and/or absorption, and absorption by cells stained with the usual dyes tends to be low. Strongly absorbing dyes have been used in some systems such as the Technicon Hemalog to distinguish different types of white blood cells by measuring light scatter and absorption.[30]

Light Scatter Analysis

Uses for Light Scatter Measurements

Light scatter measurements provide extremely valuable information in flow analysis and sorting. They (1) provide reliable detection of all cell-sized objects regardless of their fluorescence, (2) provide clear initiation signals for sort system timing, (3) give some information on relative cell size, (4) allow live-dead cell discrimination in some populations, and (5) provide useful cell type discrimination in mixed populations.

Light Scatter Theory

Light scatter measurements in flow systems have been reviewed by Salzman.[31,32] Because typical eukaryotic cells and their principal internal structures have dimensions in the range of a few times the wavelength of visible light, both large-object and small-object approximations fail to simplify light scatter calculations. As a result significant computations are required to predict light scatter angular distributions in even the most simplified model systems.[33] In practical applications light scatter theory provides some guidelines on what angular ranges may give sensitivity to features of interest, but, at present, it does not allow us to define specific properties and structures of a cell on the basis of light scatter measurements.

Some guidelines derived broadly from theoretical calculations and from measurements on real cells may be summarized as (1) light scatter

[29] J. A. Steinkamp, *Cytometry* **4**, 83 (1983).
[30] L. A. Cooper, *in* "Flow Cytometry and Sorting" (M. Melamed, P. Mullaney, and M. Mendelsohn, eds.), p. 679. Wiley, New York, 1979.
[31] G. C. Salzman, P. F. Mullaney, and B. J. Price, *in* "Flow Cytometry and Sorting" (M. Melamed, P. Mullaney, and M. Mendelsohn, eds.), p. 105. Wiley, New York, 1979.
[32] G. C. Salzman, *in* "Cell Analysis" (N. Catsimpoolas, ed.), p. 111. Plenum, New York, 1982.
[33] M. Kerker, "The Scattering of Light and Other Electromagnetic Radiation." Academic Press, New York, 1969.

intensities for cells are highest at very small angles falling off with oscillations in intensity to values several orders of magnitude lower at large angles, (2) light scatter is most proportional to overall cell size and least sensitive to internal structure at the smallest angles, (3) light scatter outside the smallest angles (e.g., in the range from 4–12°) shows more sensitivity to cell structure, (4) at large angles in the 90° range fine internal structure and granularity become important, and (5) back angles (near 180°) show weak signals that seem to be similar in information content to the stronger signals at corresponding forward angles (near 0°), although this may be an artifact caused by reflections of the bright small angle scattered light.

Light scatter from nonspherical cells is affected by cell orientation. One effect of the hydrodynamic focusing of the cell stream in a jet forming orifice is that nonspherical cells tend to be oriented with their long axes parallel to the direction of flow. Such aligned cells can still have different orientations with respect to the laser beam which can result in striking differences in light scatter measurements among essentially identical cells.[34]

Standard Measurements and Interpretation

The common light scatter measurements in cell sorter systems are made at forward angles and around 90°. Forward scatter provides a strong signal with considerable cell size dependence. In appropriate angular ranges it can discriminate live and dead lymphocytes quite well[17] and live and dead cells of less uniform cell types less well. It can also assist in discriminating among cell types. In our work we have found that forward scatter from the 3 to 12° range is better than scatter from smaller angles for live-dead discrimination and for distinguishing erythrocytes from lymphocytes, lymphocytes from monocytes, or lymphocytes from granulocytes.

We have found light scatter in the region near 90° to be a useful addition to forward scatter in making clear separations among cell types in mixed samples. Figure 2 shows how the two scatter signals taken together distinguish cell types in human blood. In some flow cytometers a fluorescence detector is well placed to measure 90° scatter with just a change from color filters to an appropriate neutral density filter to limit the light scatter signal level on the detector. Such a scatter detector may not work well for sorting since the drop forming oscillation modulates the laser light reflected from the jet resulting in "noise" on the scatter detec-

[34] M. R. Loken, D. R. Parks, and L. A. Herzenberg, *J. Histochem. Cytochem.* **25**, 790 (1977).

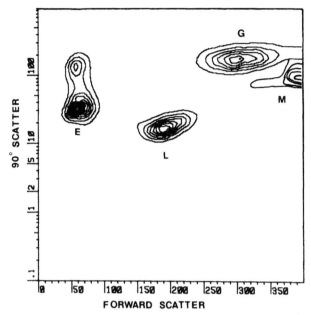

FIG. 2. Discrimination of cell types in human blood by two angles of light scatter. The cell populations are erythrocytes (E), lymphocytes (L), granulocytes (G), and monocytes (M). A linear measure of forward light scatter at 488 nm with an acceptance angle from about 2.5 to 12° is shown on the horizontal dimension. The vertical axis plots the logarithm of "90°" scatter with an acceptance range from about 70 to 110 °. The figure is a superposition of data from several subsamples enriched for different cell types.

tor. We now use a simple fiber optic collector leading to a separate photomultiplier tube for large angle scatter measurements comparable to 90° scatter. This collector is placed next to the fluorescence objective and below the plane of reflected/refracted laser light. Our system requires redesign to minimize background light, but it works equally well in sorting or nonsorting conditions.

Multiple laser systems open the possibility of measuring light scatter at two or more wavelengths. Little work has been done in this area, but there has been one report of useful information in dual wavelength measurements.[35] (Incidentally, this work was done not with two lasers but with one laser and special laser cavity mirrors.)

Light Scatter Standards

Since light scatter measurements tend to be quite useful without being measures of readily interpretable cell properties, it should not be surpris-

[35] G. R. Otten and M. R. Loken, *Cytometry* **3**, 182 (1982).

ing that light scatter standards are useful in monitoring system stability and in detecting problems in cell samples but not in making theoretically interesting comparisons. For example, we currently use 1.8-μm-diameter plastic microspheres for signal standardization (Polysciences, Inc. cat. no. 9719 or 9847). Hybridoma cells are nearly 1000 times as large in volume and scatter about 100 times as much light in our forward scatter measurement but are about equal to the microspheres in large angle scatter.

Cell Staining Reagents and Procedures

Fluorescent Reagents

Fluorescent reagents are used in most flow cytometric measurements because the low background possible with fluorescence signals allows high sensitivity and a large dynamic range in the measurements.

For our purposes the large array of fluorescent sources that are relevant in flow cytometry can be divided into three categories: (1) intrinsic cellular fluorescences, (2) inherently fluorescent probes, and (3) reagents whose specificity and fluorescent tag are independent. Intrinsic cellular fluorescences can be measured in the UV to give information on the content of particular molecules,[36,37] but usually they form part of the background against which other fluorescences must be measured. Inherently fluorescent probes have been used to measure cellular DNA content, protein content, membrane potential, membrane fluidity, intracellular pH, and a number of other properties. In general these reagents provide large fluorescence signals whose measurement is more-or-less straightforward.

In this chapter we are concerned primarily with the third group including antibodies, lectins, and antigens which bind specifically to appropriate cellular molecules and which are marked with a chromophore of the experimenter's choice. Our use of the term immunofluorescence will generally be applicable to staining with any reagents in this group. The chromophore may be attached directly to the binding molecule, or it may be carried on a second step reagent which binds to the first. The second step may be an antibody to the first step molecule, an antibody to a hapten which is coupled to the first step or avidin which binds to biotin coupled to the first step. Two step procedures often give brighter staining than direct coupling of dye to the primary reagent, but the high purity and specificity of monoclonal antibodies make directly labeled reagents adequate for most tasks.

[36] J. E. Aubin, *J. Histochem. Cytochem.* **27**, 36 (1979).
[37] B. Thorell, *Cytometry* **2**, 39 (1981).

Also, increased interest in multiple immunofluorescence measurements has led to more use of directly labeled reagents in order to minimize the possibilities for crossreactions among reagents.

Dyes for Immunofluorescence

An immunofluorescent dye must fulfill several stringent criteria. Since signal brightness is often the limiting factor in immunofluorescent measurements, the dye should absorb strongly at wavelengths for which a good excitation source is available and should have a high quantum efficiency for fluorescence emission. In addition, it must be possible to couple the dye to an antibody or other specificity-conferring molecule without disrupting that specificity or the dye fluorescence.

Figure 3A, B, C, D, and F shows excitation and emission spectra for a number of dyes that are useful for immunofluorescent-type cell labeling. Fluorescein and rhodamine have been used for many years in microscopy as immunofluorescent dyes. They are small molecules (MW 332 and 479, respectively) which are commonly attached to antibodies or other proteins by the reaction of their isothiocyanate derivatives with protein amino groups. Fluorescein is ideally suited to excitation by the argon-ion laser line at 488 nm. The 514.5 nm argon-ion laser line is commonly used to excite rhodamine, but it is not so well matched to the excitation spectrum of that dye. On the other hand, the mercury arc lamps commonly used in fluorescence microscopy are well matched to excite rhodamine and not nearly so good for fluorescein. Thus good reagents for one means of observation may not be optimal for another.

Texas Red (Molecular Probes, Inc., Johnson City, OR) is a sulfonyl chloride derivative of Sulforhodamine 101 which couples to proteins via their amino groups. It has been used most successfully as an avidin conjugate although some direct antibody reagents have also been used.

Phycobiliproteins are accessory photosynthetic pigments found in red algae and in cyanobacteria[38,39] that have recently been applied to immunofluorescent labeling.[40] They harvest solar energy at wavelengths where chlorophyll itself is not efficient and transfer the excitation nonradiatively to chlorophyll for use in photosynthesis. The principal types of phycobiliproteins are phycoerythrins, phycocyanins, and allophycocyanins. Different organisms have somewhat different forms of these molecules. Each whole phycobiliprotein molecule is a complex of several

[38] A. N. Glazer, in "The Evolution of Protein Structure and Function" (D. S. Sigman and M. A. B. Brazier, eds.), p. 221. Academic Press, New York, 1980.
[39] A. N. Glazer, Annu. Rev. Microbiol. **36,** 173 (1982).
[40] V. T. Oi, A. N. Glazer, and L. Stryer, J. Cell Biol. **93,** 981 (1982).

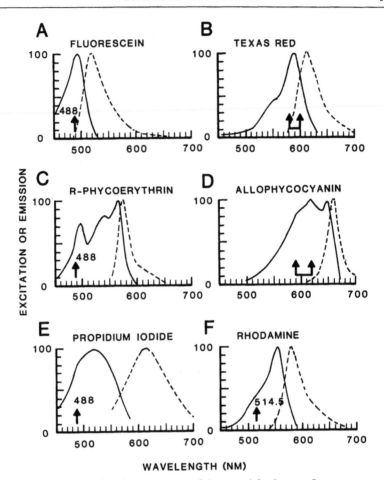

FIG. 3. Excitation and emission spectra of dyes used for immunofluorescence and of propidium iodide. Excitation spectra (solid lines) and emission spectra (broken lines) were taken with a SPEX Fluorolog instrument and are uncorrected. The vertical scale is in arbitrary units. The arrows in (A), (C), (E), and (F) mark the argon ion laser lines normally used to excite the dyes. The arrowed ranges in (B) and (D) mark the best range for excitation using a tunable dye laser. R-phycoerythrin (C) was from the red alga *Gastroclonium coulteri*, and the allophycocyanin (D) was from the cyanobacterium *Anabaena variabilis*.

polypeptides with open-chain tetrapyrrole chromophores attached. The polypeptide complexes range in form from $(\alpha/\beta)_3$ for phycocyanins and allophycocyanins to $(\alpha/\beta)_6\gamma$ for some phycoerythrins (where α, β, and γ are different polypeptide subunits in the different proteins). Allophycocyanins of molecular weight 105,000 carry 6 chromophores; phycoerythrins of MW 240,000 carry up to 40 chromophores. The spectral properties of the molecules derive both from the types of chromophores attached and from the protein structure. Some algae even vary the mix of chromophores in response to changes in the spectrum of the light they receive.

As immunofluorescent labels phycobiliproteins have several advantages. They can be excited efficiently over a fairly wide range of wavelengths. Each molecule carrys many chromophores, and the molecules, being naturally bioengineered, maintain higher quantum efficiencies for fluorescence emission than one would expect to obtain from an equivalent number of ordinary dye molecules on a comparably sized protein. At dye-to-antibody coupling ratios near one, phycobiliprotein reagents should exhibit intense fluorescence with minimal interference with binding specificities.

The procedures for coupling phycobiliproteins to specific reagents are just being explored, but information on some approaches that have been tried is contained in references 22, 24, and 40.

Dye-Reagent Considerations in Immunofluorescence

The details of reagent preparation and dye coupling are beyond the scope of this chapter and are the subject of an extensive literature.[41-43] The pitfalls and limitations of immunofluorescent reagents as used in flow cytometry must be mentioned, however.

The optimum dye to protein ratio for a flow cytometric reagent may be lower than for reagents to be used in visual microscopy since the high sensitivity of flow cytometric measurements means that the low levels of nonspecific staining often found with heavily dye-labeled reagents may be detected and misinterpreted. High coupling ratios may result in considerable quenching of the dye fluorescence so that fluorescence output is not as great as would be predicted from the optical absorbance of the reagent. This must be taken into account in making quantitative estimates of the reagent bound per cell.

Reagent and Cell Handling for Immunofluorescent Staining (This Volume [41])

Detailed procedures for immunofluorescent staining are available from various sources.[41,42] We would, however, like to emphasize several specific points. For optimal staining specificity it is important to remove aggregated material from reagents before use. This can be accomplished,

[41] D. M. Weir, ed., "Handbook of Experimental Immunology," Vol. 3. Blackwell, Oxford, 1978.
[42] B. B. Mishell and S. M. Shiigi, "Selected Methods in Cellular Immunology." Freeman, San Francisco, California, 1980.
[43] J. W. Goding, J. Immunol. Methods 13, 215 (1976).

for example, by centrifugation for 10 min at 20–50,000 g in a microfuge or airfuge. We normally incorporate sodium azide at a concentration of 0.1% in staining medium to prevent removal of surface stain by active cellular processes. When large numbers of samples are to be stained, it is convenient to do the cell handling in flexible 96-well trays.[5,44]

The use of propidium iodide (PI) to mark dead cells so that they can be excluded from analyses has been particularly useful in work involving rare cells or small subpopulations of cells,[45] but we find it worthwhile in almost all fluorescence analyses. PI stains DNA and RNA but it is excluded very effectively by cells with intact plasma membranes. As a last step before FACS analysis we wash the cells in medium containing about 1 μg PI per ml (a much lower concentration than that used in saturation staining for DNA quantitation). Several minutes exposure to the dye results in distinct labeling of the dead cells with no detectable effect on live cells. Excitation and emission spectra of PI labeled dead cells are shown in Fig. 3E. It excites well at 488 nm but emits fluorescence mostly above 600 nm. These characteristics make it possible to measure PI along with fluorescein and/or Texas Red as described below (Fluorescence Measurement Systems). Forward light scatter separates live cells from nonnucleated debris and from a fraction of dead cells which depends on the sample. Combining PI staining and forward light scatter criteria we can "gate" incoming data to assure that further analysis is carried out only on live cells.

Fluorescence Analysis

Fluorescence Excitation Sources, Filters, and Detectors

The function of the fluorescence excitation and detection system is to produce signals which accurately reflect the amount of each dye associated with each cell. For good sensitivity in immunofluorescent measurements we need to maximize the number of photoelectrons produced at the photomultiplier tube (PMT) photocathode by fluorescence of our dye and minimize interference due to light from other sources. The rate of photon emission by a cell is proportional to (1) the amount of dye associated with the cell, (2) the illumination intensity,[46,47] (3) the extinction coefficient of the dye at the exciting wavelength, and (4) the quantum efficiency of the

---

[44] K. Hayakawa, R. R. Hardy, D. R. Parks, and L. A. Herzenberg, *J. Exp. Med.* **157**, 202 (1983).

[45] J. L. Dangl, D. R. Parks, V. T. Oi, and L. A. Herzenberg, *Cytometry* **2**, 395 (1982).

[46] At very high excitation intensity dye saturation and photobleaching affect fluorescence output,[47] but these are small effects in most systems.

[47] M. F. Bartholdi, D. C. Sinclair, and L. S. Cram, *Cytometry* **3**, 395 (1983).

dye for fluorescence emission. The conversion of fluorescence emission to photoelectrons is governed by (1) the light collection efficiency of the fluorescence optics, (2) the transmission of the optical filters for the light emitted by the dye, and (3) the quantum efficiency of the PMT for producing photoelectrons from the filtered fluorescence. The dye related aspects of the process have been discussed above (Dyes for Immunofluorescence), and the other aspects are discussed in this section.

The principal excitation sources used in flow cytometry are argon-ion lasers, krypton-ion lasers, tunable dye lasers, and mercury arc lamps. Sensitive measurements of immunofluorescence require intense light sources well matched to the dyes being measured if they are to be made over the short time intervals required for high cell rate analysis and sorting. In sorting systems operating at several thousand cells per second the excitation source spot size should be small to minimize the frequency of situations in which more than one cell is in the beam at one time. The requirements for high intensity and a small spot size make arc lamp sources marginal or inadequate for sorting systems. Adequate amounts of fluorescence can be obtained in nonsorting arc lamp systems by illuminating a larger area and by passing the cells through the beam more slowly. The limited intensity of arc lamp-derived illumination is based on fundamental optical principles which limit secondary spot intensity to be no more than the original source intensity. While an arc lamp is an intense source by ordinary standards, its source intensity is much less than that of a laser of the types used in flow cytometry. Laser beams are not normally used with the beam focused to give the highest possible intensity but rather focused to give high intensity with adequate uniformity across the width of the cell stream.

Cell sorters are at a disadvantage compared to some other flow cytometers in terms of light collection. The sorting geometry generally allows only N.A. 0.6–0.75 light collecting lenses while analysis machines may employ N.A. 1.2 immersion lenses. The difference can be as much as a factor of 4 in collection efficiency.

The function of optical masking and filtering in fluorescence measurements is basically quite simple: to pass as much light as possible from a particular dye while passing as little light as possible from any other source. The sources to be discriminated against include scattered light from the exciting laser beam, fluorescence of the filters excited by scattered laser light, cell autofluorescence, raman scatter in the liquid jet, stray room light, and, of course, fluorescent light from other dyes and excitation sources in multifluorescence systems. Imaging the jet onto a mask with apertures passing light only from the region around each laser/jet intersection helps to minimize stray light acceptance (see Fig. 4).

Since laser light scattered by a cell is normally several orders of magnitude brighter than immunofluorescence, fluorescence filters must have an attenuation approaching 10^6 at the laser wavelength. Fluorescence of the filters themselves is much less of a problem with interference filters than with colored glass or other absorption filters, but even gel filters (e.g., Kodak Wratten series) can be used if they are kept well away from the detectors. The sensitivity of a detector to cell autofluorescence and to other dyes can be minimized by passing a narrow spectral band near the emission peak of the dye of interest, but narrowing the band decreases collection efficiency for that dye also. The optimum choice of filtering depends on the details of the system, but we have found that interference or interference-plus-absorption filters with a bandwidth of 30 to 40 nm are generally good for single dye measurements and also adaptable to multi-dye work. Such filters can have 50% transmission at a wavelength as close as 20 nm to the exciting laser and >80% transmission in the central region of the band. Depending on the spacing between the laser line and the dye emission peak we have obtained filters which transmit 35–55% of the total dye emission (Becton Dickinson FACS Division, Sunnyvale, CA).

Filters can be tested for blocking of scattered light from the laser by running nonfluorescent particles such as 2-μm-diameter polyvinyltoluene microspheres (Duke Scientific, Cat. No. 115). Comparing fluorescence detector signals with neutral-density filters and the color filters alternately in the filter holder gives an estimate of the blocking of the laser wavelength by the color filters.

Photomultiplier tubes (PMTs) are used for low level fluorescence measurements because they offer very high gain/very low noise amplification of the photoelectron signal produced by the fluorescent light. They also have good linearity over a wide range of signal levels. The spectral sensitivity of a PMT depends on the photocathode material. In the visible range, quantum efficiency usually decreases at increasing wavelengths, so it is important to use PMTs that have good red sensitivity if dyes emitting much beyond 600 nm are to be used. PMT dark current, the output due to thermal emission of electrons from the cathode and from the amplifying dynodes, is usually not an important source of noise signals in cell sorters; the main noise sources are invariably unwanted light reaching the detector.

Fluorescence Measurement Systems

To illustrate the optical systems, the diagrams in Fig. 4 show the layout of one, two, and three immunofluorescence systems using dyes shown in Fig. 3. The first part shows a system for measuring fluorescein

A FLUORESCEIN/PROPIDIUM IODIDE

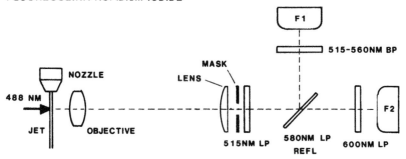

B FLUORESCEIN/TEXAS RED/PROPIDIUM IODIDE

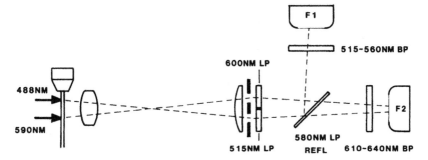

C FLUORESCEIN/PHYCOERYTHRIN/ALLOPHYCOCYANIN

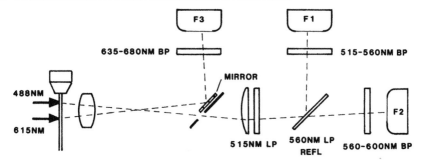

FIG. 4. Optical systems for one, two, and three immunofluorescence measurement systems. Schematic diagrams are shown for (A) fluorescein immunofluorescence plus propidium iodide, (B) fluorescein and Texas Red immunofluorescences plus propidium iodide, and (C) fluorescein, phycoerythrin, and allophycocyanin immunofluorescences. Optical filters are designated by their 50% transmission wavelength(s) and their type: LP, long wavelength passing filter; BP, bandpass filter; LP REFL, long wavelength passing dichroic reflector. Photomultiplier detectors are designated F1, F2, and F3. Some components in these diagrams have been rotated from their actual orientations to allow the drawing to be made in one plane.

immunofluorescence with PI viability gating. An argon-ion 488 nm laser beam excites both fluorescein and PI. Light collected by the objective is imaged onto an aperture which is adjacent to a field lens and a laser-blocking filter. The field lens ensures that all light coming from the objective and passing the mask and filters falls on the sensitive area of the PMTs. It also improves the spatial distribution of the signal on the PMT by imaging the back of the objective onto the PMT face. Such a lens may or may not be necessary depending on the geometry of the system. Light passing the laser blocking filter is divided by a dichroic beamsplitter into components above and below 580 nm, and each of these is filtered further and detected by a PMT. In this system fluorescein fluorescence falls primarily on detector F1 and PI fluorescence primarily on F2.

In the second part of Fig. 4 a dye laser beam at 590 nm is added to excite Texas Red. Light from the two laser spots is imaged onto two separate apertures each of which is adjacent to an appropriate laser-blocking filter. The rest of the system is like the previous one. The F1 detector is fluorescein sensitive while the F2 detector measures PI as the cell passes through the 488 nm laser beam and Texas Red as it passes through the 590 nm beam several microseconds later. (This is essentially the system described by Parks et al.[21])

In the third part of Fig. 4 two laser beams are used to excite three immunofluorescences. Light from the laser–jet intersections is imaged onto an aperture and 45° degree mirror assembly. Fluorescein and PE fluorescence pass through the aperture and laser blocking filter to a 560 nm dichroic beamsplitter and final signal defining filters. F1 is sensitive to fluorescein while F2 is sensitive to PE. APC fluorescence excited at 615 nm is reflected from the mirror passing through a laser-blocking and bandpass-defining filter to give a signal on detector F3.[22]

Spectral Overlap Correction When One Laser Excites Two Dyes

In each of the systems described in the previous section two dyes are excited by the 488 nm laser beam. The filters are selected to make each detector much more sensitive to one dye than to the other, but, in general, each detector has some sensitivity for both dyes. Since the relation between the signal components on each detector is linear (e.g., doubling the fluorescein emission doubles the fluorescein dependent signal on both F1 and F2), linear combinations of the two detector signals can be constructed that represent each of the dyes alone.[48] Stated simply, if we take the sensitivity of detector F1 for fluorescein to be 1 and for PE to be a and if we take the sensitivity of detector F2 for PE to be 1 and for fluorescein

[48] M. R. Loken, D. R. Parks, and L. A. Herzenberg, J. Histochem. Cytochem. 25, 899 (1977).

(FL) to be b, the signals from the F1 and F2 detectors are

$$F1 = FL + aPE \tag{1}$$
$$F2 = bFL + PE \tag{2}$$

Subtracting aF2 from F1 gives

$$F1 - aF2 = (1 - ab)FL \tag{3}$$

a signal proportional only to fluorescein. Similarly, subtracting bF1 from F2 gives

$$F2 - bF1 = (1 - ab)PE \tag{4}$$

This transformation can be performed simply by making analog combinations of the two detector signals in the fluorescence amplifiers. The amounts of each to be mixed are determined by running samples containing only one dye or the other and adjusting the output for the "wrong" dye to be zero. The result is two signals that can be processed like direct detector signals but which are each sensitive to only one dye. For analytical purposes the transformation described above could be performed after-the-fact by processing the recorded data, but we find analog real-time correction to be preferable since the signals we see and which are used to define sorting conditions correspond directly to the final recorded data.

One potential problem in multiple immunofluorescence measurements that we have not observed in practice is resonance energy transfer. Examining the excitation and emission spectra in Fig. 3 we see significant overlap between the emission spectrum of fluorescein and the excitation spectrum of Texas Red. We also see considerable spectral overlap between fluorescein and PE and between PE and APC. In such cases excitation of the first dye can result in radiation-less transfer of excitation to the second dye resulting in decreased fluorescence from the one and increased fluorescence output from the other compared to the actual amounts of the dyes. For energy transfer to occur with detectable efficiency the dye molecules must be in close proximity.

It seems that they are not usually so close in multiple immunofluorescent staining, but in particular cases, such as labeling two determinants on a single cell surface molecule, energy transfer could be a problem (or perhaps a solution depending on the questions being asked).[49]

Fluorescence Measurement Quality

The quality of fluorescence measurements is affected by a number of factors whose influence varies depending on the type of sample and the

[49] T. M. Jovin, *in* "Flow Cytometry and Sorting" (M. Melamed, P. Mullaney, and M. Mendelsohn, eds.), p. 137. Wiley, New York, 1979.

purpose of the measurements. The limiting cases, however, may be considered as one in which we want to measure brightly stained cells as accurately as possible and the other in which we want to distinguish minimally stained cells from unstained. The factors affecting the fluorescence signal are laser beam intensity, stability and uniformity, fluorescence collection, filter and PMT efficiency, background light sources, and cell autofluorescence. In all cases the accuracy of measurements is limited by the number of photoelectrons produced by a cell. Increases in laser beam power, light collection efficiency and PMT efficiency at the relevant wavelengths will all increase the absolute signal level. A tradeoff between laser beam intensity and illumination uniformity of the cells depends on the beam focusing and the size of the cell stream. Cells following different trajectories in the jet pass through different parts of the laser beam. If the laser beam is tightly focused to obtain higher beam intensity, the uniformity of illumination of different cells may suffer. Since the range of cell trajectories increases with the sample flow rate, observed laser beam uniformity will be best if cell samples are run at high cell concentration and low liquid flow rate. Laser beam stability refers to fluctuations in beam power or spatial distribution over the course of a set of measurements.

When low level fluorescence is to be measured, cell autofluorescence may set the real limit on our ability to estimate the reagent fluorescence. Autofluorescence is a function of cell size, cell type, excitation wavelength, and emission detection range. The fluorescent molecules are normal cell constituents like flavins[50] and cytochromes which tend to be found in greater quantity in larger cells. On the other hand cultured cells tend to be more autofluorescent than corresponding fresh cells and dead tissue culture cells can show enough increase in autofluorescence for this to be used as a criterion of nonviability.[25] Cell samples such as spleen and bone marrow include not only lymphocytes and other low autofluorescence cells but also a variable-sized fraction of much brighter cells which can interfere with efforts to define and characterize small subpopulations of cells by immunofluorescence.

What can we do about autofluorescence? Besides the obvious use of brighter reagents so that the relative effect of autofluorescence is diminished we may be able to use redder reagents and optimize filtering to decrease measured autofluorescence. While good spectra of the autofluorescence of ordinary cells are difficult to obtain, it seems that autofluorescence emission tends to be broad and that excitation at longer wave-

[50] R. C. Benson, R. A. Meyer, M. E. Zaruba, and G. M. McKhann, *J. Histochem. Cytochem.* **27,** 44 (1979).

lengths excites less autofluorescence. Taking advantage of this means, first, detecting dye fluorescence through proper band pass filters centered on the main emission range of the dye as described above and, second, bringing red dye systems into fully routine use. One limitation on signal improvements with redder dyes is the decrease in the quantum efficiency of normal PMTs at longer wavelengths so that more light is required to give a particular photoelectron signal.

Fluorescence Standards, Quantitation, and System Calibration

Tuning Samples. When aligning and tuning up a cell sorter it is very convenient to run a sample of uniform particles which give signals on each light scatter and fluorescence detector. Uniformity of the particles in light scatter properties and in fluorescence makes it easier to find optimum adjustment positions and helps to detect problems in the flow system or electronics that degrade measurement quality. In multiple fluorescence conditions we often use a mixture of two types of plastic microspheres as a tuning and calibration sample (e.g., Polysciences, Inc. Cat. No. 9719 and 7769).

Calibration. In order to make valid comparisons from one experiment to another it is useful to have standard calibration conditions so that a particular signal level corresponds to a constant amount of dye per cell. This can be accomplished by adjusting the signals from stable calibration particles to a standard output level. As long as the excitation wavelength and the emission filters and detectors are not changed, it is not necessary that the fluorescence spectrum of the calibration particles match that of the relevant cell labeling dye.

Standardization. True standardization that will allow valid comparisons of staining levels from one machine or laboratory to another is best accomplished with particles whose fluorescence spectrum is the same as that of the dye being standardized. At present there is no accepted set of stable reference particles with spectra matching immunofluorescence labels, but such a set would be quite useful.

Analytical Data and Data Processing

Signal Amplification

In a flow cytometer each cell generates electrical signals while crossing each laser beam (or other sensing area). These signals are amplified in linear or logarithmic fashion to yield output signals in the range of voltages that the rest of the electronics is designed to handle. The amplification usually includes some attenuation of frequency components outside the primary frequency range of the real signals thereby improving the signal-to-noise ratio. The choice between linear and logarithmic amplification depends on the range of signal levels to be measured and on the types of distributions we expect from the measurements. Logarithmic amplifiers are available for FACS signals which operate well over a 3.5–4.0 decade range (70–80 dB). We have found them to be very well suited to immunofluorescence measurements in which a single sample often includes cells with only their autofluorescence, stained cells a few times that bright and stained cells hundreds of times as bright. A logarithmic amplifier gives good resolution over the whole range. In addition, stained cell populations in logarithmic displays tend to have a symmetrical shape which is helpful in choosing break points between populations for analysis and sorting. What these more "normal" distributions may mean biologically is discussed in the next section.

Logarithmic amplifiers require occasional adjustment to set the output characteristics to give a chosen number of channels between signals that are, for example, a factor of 10 apart. On the other hand, in day-to-day and sample-to-sample operation they require no gain changes or other manipulation.

Linear amplifiers are appropriate for signals that vary over a relatively small range and/or represent biologically linear processes. Most measurements of light scatter or cellular DNA content fall in this category. Figure 5 shows a comparison of logarithmic and linear displays of some dual immunofluorescence data. Dot displays as illustrated in the left panels of the figure represent each cell as a dot located at a point whose x coordinate corresponds to one of the signals derived from the cell and whose y coordinate corresponds to the value of a second measurement on the same cell. Such displays are commonly used to provide immediate visualization of two parameter data and are quite valuable for monitoring cell sorter operation. The contour plots illustrated in the right panels of Fig. 5 show the same data in the form of topographic maps of the two dimensional histograms of the data. In our experience it is usually easier to distinguish subpopulations of cells and to evaluate and interpret their characteristics when immunofluorescences are displayed logarithmically.

Biological Implications of Log-Normal Data

While we originally began to use logarithmic amplifiers for immunofluorescence data just to handle the wide range of observed signals without continual changes in amplifier gain, we have concluded that logarithmic processing is in fact more appropriate biologically in many cases. Statistically speaking (based on the Central Limit Theorem), a signal that results from the sum of many randomly varying quantities tends to give a normal

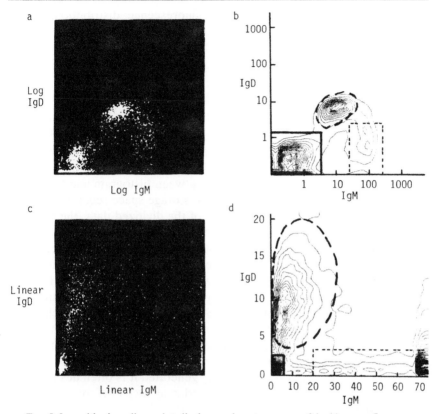

FIG. 5. Logarithmic vs linear dot displays and contour maps of dual immunofluorescence data. Mouse spleen cells were stained with fluorescein-conjugated antibody to IgM and biotin-conjugated antibody to IgD which was revealed by a second step of Texas Red avidin. Measurements were made with a two laser FACS. All of the fluorescence data were light scatter gated on the main lymphocyte population. In (a) and (c) each dot represents the measurements on one cell. (b) and (d) show contour plots of the logarithmic and linear data, respectively, with constant "vertical" spacing between contours. The rectangular boxes and broken lines enclose corresponding regions (i.e., the same cell populations) in the two displays.

distribution while a signal resulting from a product of random variables tends to give a log-normal distribution (which looks like a normal distribution in a logarithmic display). For example, if a cell surface antigen were produced by a number of different pathways whose contribution varied from cell to cell, the distribution of antigen per cell should tend to a normal distribution. If the amount of the antigen on the surface were controlled by the multiplicative effects of a series of synthesis and transport rates, a log-normal distribution should result. The latter is much closer to our understanding and expectations for most cell surface antigens.

Our observations are that immunofluorescence distributions often look approximately "normal" in logarithmic display. In such cases any statistical tests that assume normal distributions are better performed on the logarithmic data.[51]

Data Storage Format and Preliminary Data Reduction

In flow cytometric work the usual way to record information on a sample is to compile digitized measurements on a statistically adequate number of cells into a data file representing that sample. Cell measurements are stored as histograms or as lists with a digital value for each measurement on each cell. The choice between histogram and list mode storage is based mostly on the amount of storage space required. This in turn depends on the channel resolution of the digitized data, the number of measurements per cell, and the number of cells recorded per sample. Histograms are preferable when the data space is densely populated (i.e., when the number of cells in the data set is large compared to the number of elements required for a histogram), while listing is better when the data space is sparsely populated. Under typical flow cytometry conditions of 5000 to 100,000 cells per sample and 8 to 10 bit resolution in each measurement, histograms give more efficient storage of one-dimensional data while list mode storage is best for three or more dimensional data. The efficiencies are roughly comparable for two-dimensional data.

When multiparameter data are being generated it is often desirable or necessary to perform preliminary data reducing steps before storing the remaining data. For example, if data storage is in list mode we may exclude objects whose forward light scatter signal is too low for a live cell. Or we may set a live cell window on a PI measurement and not store data from dead cells or even the actual PI data from live cells thus saving data storage space.

One-dimensional histogram storage means that only one parameter of the data can be stored. Other measurements can only be used as windowing parameters which control storage of data on each cell but which are not stored themselves.

List mode storage has several advantages over immediate reduction to one dimension for histogram storage. Cell analysis operation of a FACS can be made more efficient since immediate data analysis decisions are minimized and data can be taken as long as the monitoring displays indicate that the machine is running properly so that "good" data will be taken. The ability to "recreate" the experiment from list mode data means that acceptance ranges set on some parameters for data analysis can be adjusted and optimized later.

[51] D. F. Heath, *Nature (London)* **213,** 1159 (1967).

Data Recovery and Auxiliary Information

The long-term usefulness of stored data is enhanced considerably by storage of "notebook" type information on each sample. This can be simple machine information on amplifier gains, etc., but full protocol information on the cells, stains, etc. used to generate the sample is part of the retrievable data in at least one recently implemented system (W. Moore, personal communication).

Data Displays in One and Two Dimensions

Often the most helpful tool in the analysis of complex data is a good visual display. One-dimensional data are conventionally displayed as a histogram of cell frequency in each signal brightness interval. Linear histograms of immunofluorescence data tend to have large numbers of cells concentrated in the lowest channels and/or falling off the top of the distribution. Logarithmic displays tend to be easier to estimate visually since it is normally possible to have both unstained cells and brightly stained cells well on scale in a single display. It is also easy to define reasonable separation points between the relatively symmetrical peaks corresponding to different populations.

Two-dimensional data can be displayed in several ways including dot displays as illustrated (Fig. 5) and "fishnet" drawings (perspective views of transects) which look like a net draped over a two-dimensional histogram of the data. Fishnet drawings viewed in perspective make it easy to visualize the two-dimensional histogram, but we have not found them to be very useful in defining conditions for further analysis. Contour maps as illustrated in Fig. 5 are not so easy to visualize at first, but they can be used more easily to compare different data sets and to define regions for numerical analysis. The spacing between the contour levels in Fig. 5 is a constant, but other contour spacings have been used with some success. Figure 6 shows a single data set displayed with four different contour level definitions. We are finding good results with the method illustrated in Fig. 6B which places contour planes on the two-dimensional histogram so that a constant fraction of the total cells is found between adjacent contours. This assures that minor cell populations will be marked by at least a few contours while major populations are not obscured by an unreadable number of contours.

Three-or-More Dimensional Data

Data with more than two measurements per cell is difficult to display or visualize in its entirety, but it can be dealt with effectively, especially in cases where the measurements can be divided into relatively independent

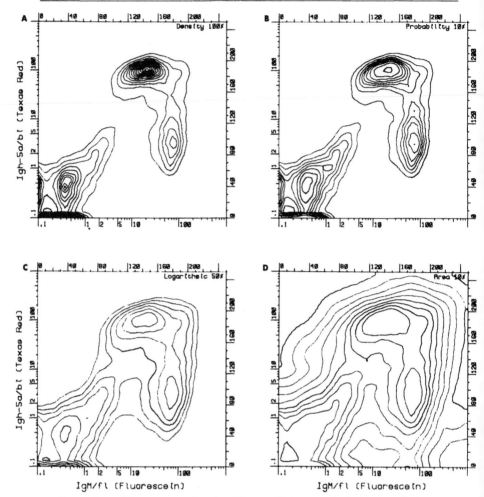

FIG. 6. Four different contouring algorithms applied to a single dual immunofluorescence data set. Mouse spleen cells were stained for IgM and IgD and measured on a two laser FACS. The first method gives the ordinary uniform contour spacing (A). In probability contouring (B) the region between adjacent contour levels includes a constant fraction of the cells (10% in this case). In the logarithmic spacing display (C) each contour level is twice as high as the previous one. The equal area contours (D) are defined to give equal parts of the area of the final two dimensional display (10% in this case) to the region between each successive pair of contour lines.

sets. For example, a 5 measurement system might include forward light scatter, large-angle light scatter, propidium iodide fluorescence (marking dead cells), fluorescein immunofluorescence, and Texas Red immunofluorescence. First, we would delineate the cell populations of interest on a two-dimensional display of forward scatter vs large angle scatter (like that illustrated in Fig. 2), eliminating most dead cells, debris, and aggregated cells. Then, one-dimensional gating to eliminate PI positive cells removes the remaining dead cells from the selected populations. We could then display the immunofluorescence measurements on each population of viable cells as a fluorescein vs Texas Red contour plot (as shown in Fig. 5). Analysis of these displays would then lead to numerical results and/or definition of sorting windows, depending on the experiment.

When groups of measurements are not conveniently divided into one- or two-dimensional subgroups, such as in three color immunofluorescence work, several iterations of analyses among the dimensions may be necessary. With such data "smarter" computer analysis to help reveal and delineate cell populations would be desirable, but as yet we know of no system that is generally applicable to cell sorter data.

Extraction of Numerical Results

The common numerical values used to characterize immunofluorescent populations are a fraction of the total cells in the population, bounds of the population in various dimensions, and signal level statistics for the population (including mean, median, variance, and coefficient of variation). Defining the division between "positive" and "negative" cells in a sample is often difficult, so that the specific criterion used should be specified in any publication quoting percentage positive cells, etc.

Means calculated on logarithmic data are geometric means with respect to the original data. In making certain comparisons, such as average cell fluorescence by FACS and bulk fluorometry, a mean of linear data is appropriate, but for characterizing a cell population that appears relatively symmetrical in logarithmic display the mean of the logarithmic data is probably more representative of the typical cell in the population than the linear mean.

Sorting

Sort Decision Making

The transition from cell analysis to sorting is conceptually simple, but actually carrying it out involves further considerations which are the subject of this section. Understanding these considerations is important in making the best use of cell sorting in particular applications.

Sort decisions can be based on any combination of measurements that can be made in a system designed for sorting. The simplest and most common way to select cells for sorting is to set an acceptance range on each measurement channel and to require each signal from a cell to fall within the acceptance range on the appropriate channel. This defines a rectangle in two dimensions, a rectangular box in three dimensions, etc. Since discrete cell populations tend to have shapes other than rectangles in two parameter displays, sort definitions involving arbitrary polygons, ellipses, or more general shapes would often be useful. In fully computerized systems very general sorting criteria can be defined (with the limitation that a sort decision must be reached before the cell traverses the length of the jet and is incorporated into a drop).

Drop Forming Processes and Effects

Stabilization of the breakup of the liquid jet into droplets is accomplished by imposing vibrations at a frequency near the natural drop formation frequency of the jet. For typical jet velocities of 8–10 m/sec the drop formation frequency is about 40 kHz for a 50-μm-diameter jet and 20 kHz for a 100-μm jet. The jet velocity and drop formation frequency can be increased by increasing the pressure inside the nozzle, but the increase cannot be too large or cell viability will be impaired. The vibration frequency defines the number of drops produced per second setting an upper limit on the number of theoretically separable events. In practice the number of cells per second that can be sorted with good efficiency is only a fraction of the drop formation rate.

Number of Drops per Sort

Initial sorting decisions are made on a cell-by-cell basis, but execution of sorting requires drop-by-drop charging. In the normal sorting mode any drop which might contain a desired cell and cannot contain an unwanted cell is sorted. In practice several factors make it difficult to predict precisely which drop a cell will be found in so that more than one drop must be assigned as a possible location for each cell. Loss of desired cells due to proximity of unwanted cells is minimized by assigning the smallest number of drops that will definitely contain a particular cell.

The factors limiting how well cells can be assigned to drops are (1) measurement uncertainty in setting the average delay between detection of a cell and breakoff of the drop containing it, (2) changes in the jet and breakoff conditions over time, and (3) the effect of the cells themselves on the drop forming conditions. In passing through the jet forming orifice, cells behave differently from the surrounding fluid and induce a modulation on the jet which interacts with the cyclic modulation produced by the piezoelectric transducer. This changes the drop formation conditions adding uncertainty to the cell-to-drop assignment.[52] The amount of uncertainty produced increases for larger cells and decreases for larger jet diameter or higher piezoelectric vibration amplitude.

To maintain good recovery of sorted cells it is necessary that a desired cell always be found within the range of drops assigned to it, and conversely high purity of sorted fractions requires that unwanted cells not be found outside their assigned drops. In practice, when we sort lymphocytes with a 60- to 80-μm-diameter jet driven at 25–35 kHz, we find that as few as 1.5 drops can be assigned per cell. This gives only ±0.25 drop cycle latitude for error in the cell-to-drop assignments, but with care proper conditions can be maintained for hours at a time. (Assigning 1.5 drops means that cells detected during half of the oscillator cycle are safely assigned to a single drop while cells in the other half cycle might appear in either of two drops both of which must be assigned to the cell. Thus an average of 1.5 drops is assigned.)

Cell Flow Rate and Coincidence Losses in Sorting

In large scale sorting experiments a high cell flow rate is desirable to minimize the time needed to obtain the necessary number of sorted cells. This is particularly desirable in live cell work where cell viability and staining uniformity may decrease over time. The problem is that at higher cell flow rates the fraction of desired cells lost due to proximity of unwanted cells ("coincidences") increases, thus diminishing the returns.

The theoretical recovery rate for given sorting conditions can be calculated as follows[53]:

A desired cell will be found in one of the particular drops assigned to it. If that drop is not in the assigned proximity of an unwanted cell, the

---

[52] R. T. Stovel, *J. Histochem. Cytochem.* **25**, 813 (1977).

[53] This calculation assumes that all cells are properly analyzed and that there is no cell clumping. Actual analysis electronics have some "dead" time following the appearance of a cell during which any new cells will not be analyzed. Real cell samples are likely to show some degree of clumping which results in a larger fraction of cells occurring close to other cells than would otherwise be expected. The result is to decrease actual recoveries somewhat from the theoretical expectation. The sorting mode modeled in this calculation has been called coincidence "out" or "normal" by Becton Dickinson, "anticoincidence on" by Coulter, and "charge gate" by Ortho. For the Becton Dickinson coincidence "in" or "full deflection envelope," or Ortho "anticoincidence," the calculation is the same except that $2n$ replaces n in the equations.

desired cell will be sorted and recovered. Assigning n drops per cell this condition is equivalent to having a region of n drops centered on the drop in question which does not contain the expected center position of any unwanted cells. At a drop frequency f and running R cells per second with a fraction a of desired cells, the rate of unwanted cells is $(1 - a)R$. Assuming that cells appear randomly, the probability that any drop contains the expected center position of no unwanted cells is calculated from the Poisson distribution for zero events

$$P(0) = \exp(-m) \qquad (5)$$

where m is the mean rate of events per drop which is in this case the ratio of the unwanted cell rate to the drop rate. Thus

$$m = (1 - a)R/f \qquad (6)$$

The probability that the n drop region around a wanted cell contains no unwanted cell center positions is $P(0)^n$ so that the expected fraction of desired cells actually sorted and recovered E is

$$E = P(0)^n = \exp(-mn) = \exp[-(1 - a)Rn/f] \qquad (7)$$

Thus, the cell flow rate giving any particular sorting efficiency is proportional to the drop formation rate and inversely proportional to the number of drops assigned per cell. The sorting rate S in cells per second is

$$S = aRE \qquad (8)$$

The recovery efficiency (expressed as percent) is illustrated in Fig. 7A as a function of cell flow rate for a particular choice of drop rate and fraction of desired cells. Note, for example, that the cell rate giving a particular recovery rate is twice as high for 1.5 drop sorting as for 3 drop sorting. Figure 7B shows the calculated absolute sorting rate for desired cells as a function of cell flow rate. The curves in Fig. 7C show the recovery efficiency and the recovery rate in generalized form as a function of M ($= mn$ above), the mean number of unwanted cells per sorting interval.

The choice of a cell flow rate depends somewhat on the availability of extra cells to be sorted and on the analysis dead time of the system. Actual recoveries may be significantly lower than estimated here if the electronic dead time in the anslysis system is long, or if there is considerable clumping of cells in the sample. In general, we tend to use cell flow rates giving mean M of approximately 0.3 which gives an estimated 74% recovery and a cell collection rate 60% of the theoretical maximum. To cover all cell losses including cells remaining in the tubes, cells used to set up the initial sorting conditions, etc., we try to start with a sample twice

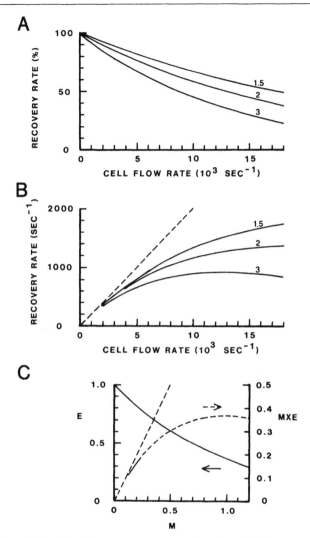

FIG. 7. Theoretical cell sorting recovery rates as a function of cell flow rate. (A) and (B) illustrate the expected sorting results for the particular situation of drop frequency 30,000 per second and a 20% population to be sorted. (A) shows the fraction of the input cells from the selected population that would be sorted with 1.5, 2, and 3 drops assigned to a cell. Note that for any particular percentage recovery rate the allowable cell flow rate is twice as high with 1.5 drop sorting as with 3 drop sorting. (B) shows the sorting rate for selected cells in total cells per second. (C) shows the sorting efficiency (E) and a parameter proportional to the total sorted cell rate ($M \times E$) in generalized form as a function of M, the mean number of unwanted cells per sorting interval. The straight broken lines in (B) and (C) show the total input rate for the selected cell population

as large as would be needed to supply the required sorted cells at 100% efficiency.

Sorting Procedure Details

The following suggestions and comments are designed to help prevent some common problems in sorting and to clarify procedures.

Maintaining Cell Viability. If the system sheath fluid is just normal saline, cells can become unhappy during the time they are held in the sort collecting vessel. In such cases the collecting vessel should be prestocked with buffered medium containing extra protein (e.g., serum) so that good conditions can be maintained as cells and saline are added. It is also helpful to mix the liquid in the collection vessel occasionally. For more sensitive cells we use RPMI (without pH indicator dye) as sheath fluid. Some cell types, such as mouse splenocytes, retain best viability when both the cell supply tube and the sorted fractions are kept ice cold.

Drop Drive Interference with Optical Measurements. In jet-in-air interrogation systems the sort drop forming vibrations can produce ripples on the jet surface which give oscillations in the reflected/refracted band of light. The oscillations can appear as periodic noise on the detectors. This causes the most trouble when a fluorescence type detector is used to measure 90° light scatter, but fluorescence or forward light scatter detectors may also be affected. In our experience the disturbance of the jet at the laser interrogation point tends to be greater for large diameter jets than for small. Such interference can be decreased by lowering the amplitude of the drop forming vibration, but this may result in greater uncertainty in assigning cells to drops.

Charge Buildup in Collection Vessels. Since the deflected drops are electrically charged, a significant potential can build up on insulated collectors to the point of repelling newly sorted drops away from the vessel and even attracting oppositely charged drops from the other side in two fraction sorting. Normally only plastic vessels are good enough insulators to hold much charge. A grounded wire in contact with the liquid in the vessel will solve the problem.

Selecting the Optimum Sort Delay Setting. In order to sort at 1.5 drops per cell it is necessary to set the delay between cell detection and drop charging quite accurately. The appropriate delay can be estimated by measuring the drop-forming frequency and the distance between free drops to obtain a velocity and using that with a measurement of the distance from the laser–jet intersection to the drop breakoff point to give a delay time. However, because of electronic delays in the system, differences between the free drop velocity and the cell-in-jet velocity, etc. a correction must be added or subtracted to obtain the best delay estimate. Once established, the correction is a constant if the drop frequency, sheath fluid pressure, and laser-to-drop-breakoff distance are not changed too much. We measure the optimum delay by sorting easily identified microspheres at several delay settings near an initial estimate and counting the number of microspheres actually recovered. Sorting about 50 single drop deflections into a pool on a slide allows a good estimate of the recovery rate. The optimum delay can be found by fitting the counts at various delay settings using the model that the recovery should be 100% at the best delay setting, and it should fall linearly to 0% at delays one drop cycle longer or shorter than the optimum.

Sterile (Aseptic) Sorting. Sterile sorting can be carried out routinely with no great difficulty. In our laboratory there are usually 2 days of sterile sorting per week. Since collection vessels may be open to the air for many minutes during a sort, it is best to keep the surrounding areas clean and avoid air turbulence in the vicinity.

Reanalysis of Sorted Fractions. It is often desirable to evaluate the purity of sorted fractions by reanalyzing a small portion of each fraction. Special care is required in cleaning the input lines leading to the nozzle since cells tend to remain after long sorts, and the sample for reanalysis is usually at low concentration. A good test is to run a "sample" of cell-free medium to see that very few extraneous cells will be included in the reanalysis.

Rare Cell Selection and Isolation

The FACS can be a very powerful tool in the identification and isolation of rare cells. The applications include enrichment of fetal cells from maternal blood,[54] selection of hybridoma variants producing different classes of immunoglobulin from the parent line,[45,55] isolation of cells incorporating and expressing a new gene in DNA transformation,[56] and selection of cells with increased gene copy number.[57,58]

Multiparameter gating is very useful in eliminating artifacts due to dead cells, debris, etc. which would otherwise greatly outnumber the rare cells being sought. In selection of cells from tissue culture populations most of the artifacts would not grow in subsequent culture, but their elimination in the analysis makes it possible to identify the regions in

[54] G. M. Iverson, D. W. Bianchi, H. M. Cann, and L. A. Herzenberg, *J. Prenatal Diagn.* **1**, 61 (1981).

[55] J. L. Dangl and L. A. Herzenberg, *J. Immunol. Methods* **52**, 1 (1982).

[56] P. Kavathas and L. A. Herzenberg, *Proc. Natl. Acad. Sci. U.S.A.* **80**, 524 (1983).

[57] R. N. Johnston, S. M. Beverley, and R. T. Schimke, *Proc. Natl. Acad. Sci. U.S.A.* **80**, 3711 (1983).

[58] P. Kavathas, and L. A. Herzenberg, *Nature (London)* **306**, 385 (1983).

which rare cells should be sought more accurately and to make better estimates of the frequency of rare cells.

With cell populations growing in culture the selection process can include several cycles of sorting for enrichment with intervening growth periods. FACS direct cloning of selected rare cells is usually effective when they have been enriched to a frequency in the range of 0.1–0.01% of the total.

Estimates of the frequency of rare cell types can sometimes be extended beyond what can be derived reliably from FACS analytical data alone by sorting a known number of selected cells into each well of a culture tray. For example, if secreted antibody from one hybridoma variant can be detected in a well that was stocked with 25 cells, then a sort of 1 cell per 1000 with 25 cells in each of 40 wells would give a frequency estimate among 1,000,000 (i.e., 1000 × 25 × 40) cells examined.[59]

Special sorting modes are sometimes useful in selecting extremely rare cells. These include "coincidence override" mode, in which the drops assigned to a desired cell are sorted without regard to what other cells may be carried along, and fluorescence triggering mode in which low fluorescence cells are ignored by the electronics. Fluorescence triggering is useful only if the cells being sought are part of a small proportion of brighter fluorescent cells. These modes allow the use of high cell flow rates that would lead to unacceptable analysis and/or sorting coincidence losses in normal sorting. In appropriate cases we have obtained useful enrichment of cells present at frequencies of a few in 10^8 by sorting at a flow rate of 30,000 cells per second. In such cases pure samples can be obtained by resorting the cells obtained in the first sort using normal sorting conditions.

Applications in Lymphoid Cell Analysis

Characterization of Lymphoid Cell Subpopulations

For all their diversity of function most lymphocytes look pretty much alike. FACS analysis and sorting using a variety of monoclonal antibodies to cell surface molecules of lymphocytes are now making it possible for us to define and study functional lymphoid subpopulations with a precision that is not available from other techniques. In particular, quantitative multiparameter measurements of light scatter and two or more immunofluorescences are revealing subpopulations that are not uniquely defined by any single antigen/monoclonal antibody.

[59] T. J. Kipps and L. A. Herzenberg, *in* "The Handbook of Experimental Immunology" (D. M. Weir, L. A. Herzenberg, and C. C. Blackwell, eds.), 4th ed. In press.

Here we will discuss primarily work in our laboratory on mouse lymphoid cell subpopulations, but similar work has been done on human lymphoid cells, often with strikingly parallel results.[60] The original distinction between B and T cells using heterologous antisera to immunoglobulin or the Thy-1 antigen led to FACS sorting of each type and initial functional studies. T cell subsets including helper or suppressor/cytotoxic functions were found using cytotoxic antisera, but monoclonal antibodies have facilitated more detailed study.

In our laboratory FACS analysis was used to screen hundreds of unselected hybridomas for production of antibodies binding to lymphocyte subpopulations.[61] This allowed concentrated efforts on the most promising hybridomas which turned out to include cell lines producing antibodies to Thy-1, Ly-1(Lyt-1), and Lyt-2, among others. FACS studies with these antibodies[62,63] showed that Lyt-2 is represented on a fraction of Thy-1 positive spleen and lymph node cells, and that the amount per cell is rather uniform. Antibody to Ly-1 stained a number of cells about equal to the total Thy-1 positive population, although the range of staining brightness among Ly-1 positive cells was large. The implication from these results that Lyt-2 positive cells in spleen or lymph node are also Ly-1 positive was in conflict with expectations from previous cytotoxicity-based studies. Studies were undertaken using cytotoxicity plus FACS analysis, additive staining with FACS analysis and two color FACS analysis to resolve the question.[63]

Cytotoxic depletion with various concentrations of antibody to Ly-1 plus complement followed by staining with a different antibody to Ly-1 and FACS analysis showed that, in general, the cells with larger amounts of Ly-1 were more susceptible to lytic attack with anti-Ly-1. Samples stained with anti-Ly-1 alone were compared with samples additively stained with anti-Ly-1 plus anti-Lyt-2 plus anti-Lyt-3. (Anti-Lyt-3 was added just to increase staining brightness since it stains the same cell population as anti-Lyt-2.) The other antibodies did not add any cells to the number stained by anti-Ly-1 alone, but they did remove cells from the dull staining portion of the Ly-1 staining histogram, resulting in a new brightly stained peak in the distribution. This indicates that the Lyt-2 positive cells

[60] J. A. Ledbetter, R. L. Evans, M. Lipinski, C. Cunningham-Rundles, R. A. Good, and L. A. Herzenberg, *J. Exp. Med.* **153**, 310 (1981).

[61] J. A. Ledbetter and L. A. Herzenberg, *Immunol. Rev.* **47**, 63 (1979).

[62] H. S. Micklem, J. A. Ledbetter, L. A. Eckhardt, and L. A. Herzenberg, in "Regulatory T Lymphocytes" (B. Pernis and H. J. Vogel, eds.), p. 119. Academic Press, New York, 1980.

[63] J. A. Ledbetter, R. V. Rouse, H. Spedding Micklem, and L. A. Herzenberg, *J. Exp. Med.* **152**, 280 (1980).

are Ly-1 positive but tend to have low levels of Ly-1. Two color staining with fluorescein-labeled anti-Lyt-2 and rhodamine-labeled anti-Ly-1 also indicated that the Lyt-2 positive cells carry low amounts of Ly-1. At that time we had only a single laser fluorescein/rhodamine system for two color immunofluorescence.[47] The two color staining and additive staining were both used since two color staining, although easily interpreted, gave suboptimal data quality, while additive staining gave good measurements whose interpretation was more difficult.

As described above (Fluorescence Measurement Systems), systems are now available which give uncompromised dual immunofluorescence measurements. Our two laser dual immunofluorescence system has been used to define several mouse B cell subpopulations which do not have known unique antigenic markers. The original subsets were defined on the basis of quantitative dual immunofluorescence for IgM and IgD. Comparing staining patterns among cell samples derived from different lymphoid tissues, different mouse strains and individuals of different ages, it was possible to discern three different but antigenically overlapping populations.[64,65] These were designated populations I (IgD high, IgM low), II (IgD high, IgM high), and III (IgD low, IgM high).

The dual immunofluorescence system has also allowed us to identify a Ly-1 positive B cell population whose relatively bright staining for IgM separates it from the IgM negative cells with similar Ly-1 staining.[44] As illustrated in Fig. 8 real staining for Ly-1 is present even though the population is not fully resolved from IgM positive, Ly-1 negative cells. It has been known for some time that spleen cells from NZB "autoimmune" mice secrete IgM *in vitro*. FACS sorting and culture of Ly-1 B cells from such mice has shown that this small fraction of the B cells probably accounts for all of the "spontaneous" IgM secretion.[44]

In some cases even dual immunofluorescence cannot readily resolve antigen expression patterns. In our efforts to characterize expression of the recently defined antigen BLA-1 (recognized by the monoclonal antibody 53-10.1), two color analysis of BLA-1, IgM, and IgD by pairs did not yield an unequivocal determination of the relative expression of the three antigens. To obtain full information cells were stained for all three antigens, and measurements were taken on the system diagrammed in Fig. 4C.[22] CBA spleen cells were stained with fluorescein anti-IgM, biotin anti-IgD revealed by phycoerythrin avidin, and allophycocyanin anti-BLA-1. Figure 9 illustrates the analysis.

[64] R. R. Hardy, K. Hayakawa, J. Haaijman, and L. A. Herzenberg, *Nature (London)* **297**, 589 (1982).

[65] R. R. Hardy, K. Hayakawa, J. Haaijman, and L. A. Herzenberg, *in* "Immunoglobulin D: Structure and Function" (G. J. Thorbecke and G. A. Leslie, eds.), p. 112. New York Acad. Sci., New York, 1982.

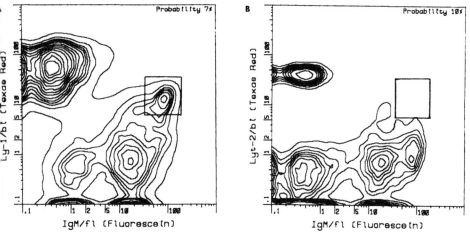

Fig. 8. Demonstration of Ly-1 B cells by two-color FACS analysis. (A) shows NZB spleen cells stained with fluorescein anti-IgM and biotin anti-Ly-1 revealed by Texas Red avidin. The box encloses IgM bright, Ly-1 positive cells. (B) shows a control staining in which the biotin reagent is anti-Lyt-2. In this case very few cells are found in the boxed region. NZB mice tend to have more Ly-1 B cells than most other mouse strains.

Two color analyses had shown that all IgM-bearing (B) cells in the spleens of young (2 week) mice express the BLA-1 antigen whereas the majority of B cells in adult (>2 month) spleen do not express this antigen. Three color analyses with IgM/IgD and BLA-1 demonstrated that population I B cells (the late-arising B cell population) constitute this BLA-1 negative group. Interestingly, the IgM/IgD pattern of the BLA-1 positive cells in adult spleen is very similar to the IgM/IgD pattern of B cells from spleens of young mice. This result suggests to us that these BLA-1 positive B cells found in adult spleen are less "mature" compared to BLA-1 negative B cells.

Phenotyping of Leukemias Using Cell Surface Markers

Identifying the type of proliferating cell in lymphoid neoplasms is important in making an accurate diagnosis, in predicting clinical characteristics, and in understanding the biology of the disease.[66] Patient groups based on cell morphology and enzyme histochemistry have not been adequate to guide staging and treatment. Staining with a panel of monoclonal antibodies offers promise of more consistent grouping. FACS measurements can be particularly valuable in some cases since light scatter mea-

[66] R. A. Warnke and M. P. Link, *Annu. Rev. Med.* **34,** 117 (1983).

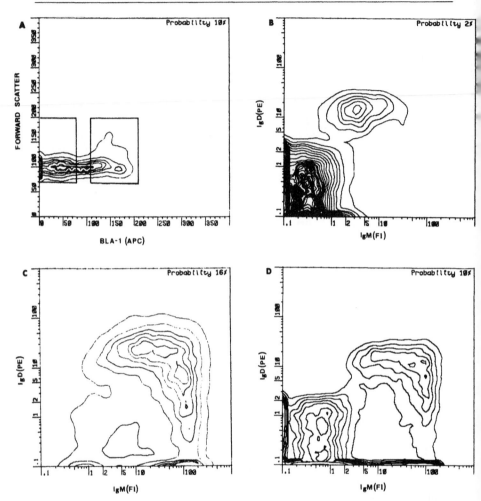

Fig. 9. Three-color immunofluorescence analysis for IgM, IgD and BLA-1 on CBA spleen cells. (A) shows an ungated display of BLA-1 vs forward light scatter with boxes enclosing BLA-1 negative and BLA-1 positive lymphocytes. (B) and (C) show IgM vs IgD displays of the BLA-1 negative and BLA-1 positive lymphocyte groups, respectively, as defined by the boxes in (A). (D) shows IgM vs IgD staining of spleen cells from a 2-week old CBA mouse. Note the similarity in pattern between the BLA-1 positive cells in the adult (C) and the total Ig positive cells in the juvenile.

surements and quantitative fluorescence can be used to identify a leuke-mic cell population and define its phenotype even though it is not the major cell population.

The basic results of phenotyping with various monoclonal antibodies have been (1) that each leukemia usually shows a single cell surface phenotype which is the same in initial presentation and in relapses, implying a lack of clonal evolution, and (2) that lymphomas and leukemias generally represent normal but arrested differentiation states, but some seem to have nonnormal phenotypes.

Such phenotyping has been useful diagnostically in distinguishing T CLLs, which express mature T cell markers of helper or cytotoxic/suppressor types and do not express thymic T cell antigens, from T ALLs which do express thymic markers. So far, typing for myeloid vs lymphoid cells and for T cell or non-T cell type among lymphoid cells has been shown to be important, but the clinical value of more detailed phenotyping has not yet been proven (R. A. Miller, personal communication).

Detection of Small Numbers of Monoclonal B Lymphocytes in Human Blood

Ault[67] has used FACS analysis of staining for κ and λ immunoglobulin light chains on aliquots from a single blood sample to detect monoclonal B lymphocyte populations. Knowledge of such populations may be an important clinical indicator in lymphoma patients. The basis for the test is that in normal human blood the distributions of staining for κ and λ light chains are broad and quite similar. A monoclonal B cell population typically has a narrower distribution of staining and expresses only one light chain leading to a difference in the shapes of κ and λ staining histograms.

Ault compared the histograms by normalizing them to the same number of stained cells and calculating the Kolmogorov–Smirnov D statistic from the difference between the two cumulative frequency distributions. The statistic should be linearly related to the fraction of monoclonal cells in the sample. The test was found to detect monoclonal populations in the 10% range routinely and populations below 1% in some cases. In tests on patients with lymphoid tumors, circulating monoclonal cells were found in a number of persons whose circulating cells were not detectably abnormal in the usual assays.

Other Applications

Amplification of the Gene for Dihydrofolate Reductase (dhfr)

FACS selection has been used to demonstrate and investigate spontaneous changes in copy number of the *dhfr* gene. Similar techniques could be used to amplify genes in order to facilitate gene cloning, analyze gene products, etc.

The basic prerequisite is having a fluorescent probe which marks a product of the gene within or on the surface of viable cells. The FACS application involves sterile sorting and single probe analysis.

[67] K. A. Ault, *N. Engl. J. Med.* **300**, 1401 (1979).

The drug methotrexate inactivates the enzyme *dhfr* which is necessary for normal cell proliferation. Methotrexate-resistant cell lines derived by exposure to gradually increasing concentrations of the drug show increased *dhfr* gene copy number and correspondingly increased enzyme production.[68] Using fluorescein conjugated methotrexate as a probe for FACS selection Johnston et al.[57] have shown that gene copy number variants occur spontaneously and are not just a product of drug-induced metabolic stress. Chinese hamster ovary (CHO) cells were grown in medium supplemented to prevent toxic effects of the probe, and 10 cycles of FACS selection for the brightest 2–5% of the cells were carried out with 7- to 12-day growth periods between FACS sorts. The resulting cell populations were 50 times as brightly staining as the parent and showed 40-fold amplification of the *dhfr* gene. Also cell lines previously drug selected for increased *dhfr* were FACS selected for increased or decreased *dhfr*. The FACS analysis data at each stage in the various selections made it possible to estimate the increase or decrease in gene copies per cell division giving values of 0.02 for one drug-amplified population and 0.001 for a recent subclone from a nonamplified line.

Selection of DNA Transfected Cells

Kavathas and Herzenberg[56] have used the FACS for efficient isolation and cloning of DNA transformed cells expressing cell surface molecules not expressed by the parent cell line. Thymidine kinase-deficient (TK⁻) mouse L cells were cotransformed with human DNA and herpes simplex virus *TK* gene. Culture in a selective medium (HAT) was used to eliminate cells which had not incorporated the *TK* gene. Cotransformation and growth in selective medium enriches for cells that have incorporated human DNA since cells that accept any DNA tend to incorporate a lot including both human and *TK* genes. Aliquots of the surviving cells were stained with monoclonal antibodies to several human T cell differentiation antigens, and the brightest 0.1–0.5% of live cells were sorted and cultured. After growth the cultures were restained and bright staining cells were FACS cloned. This procedure yielded stable transformants for several T cell differentiation antigens including Leu-1 and Leu-2.

Several Leu-2 transformant lines have subsequently been selected for increased Leu-2 expression using the original selecting reagent and several cycles of FACS sorting. The resulting gene amplified lines contain many copies per cell of the Leu-2 gene and express up to 40 times as much Leu-2 protein as the original transformant.[58]

The cell lines produced in this way are expected to be useful in gene cloning and in analysis of the gene products, and the selection techniques

---

[68] P. C. Brown, R. N. Johnston, and R. T. Schimke, *Symp. Soc. Dev. Biol.* **41**, 197 (1983).

are applicable to any cell surface molecule or other gene product that can be fluorescent labeled in viable cells. It should be possible to carry out transformations without the drug marker and TK cotransformation, but in that case one or two more cycles of FACS selection and growth would probably be required.

Evaluation of Inducible Enzyme Activity in Single Cells by Measurement of Decreased Substrate Fluorescence

Flow cytometric analysis has been used by Watson[69] to monitor the kinetics of enzymatic production of fluorescent product from a fluorogenic substrate. He was able to discriminate between cells from different cell lines at appropriate time points in a kinetic observation. Miller and Whitlock[70] have taken a similar approach following the disappearance of an already fluorescent substrate to investigate factors affecting inducible enzyme activity in the aryl hydrocarbon hydroxylase (AHH) system which is involved in the metabolism of a variety of environmental contaminants and carcinogens. They showed that after cells are exposed to the fluorescent substrate benzo[a]pyrene (BP), the rate of decrease in fluorescence is directly related to aryl hydrocarbon hydroxylase activity in the cells. FACS measurement of the residual benzo[a]pyrene fluorescence (excitation 351 nm + 364 nm; emission near 405 nm) allowed them to select cells with high or low enzyme activity compared to the general population.

Using the Hepa1c1c7 mouse hepatoma cell line, which can be induced to increase AHH activity by exposure to chemicals such as benz[a]anthracene or 2,3,7,8-tetrachlorodibenzo-p-dioxin (TCDD), they labeled both basal and induced cell samples with BP, and, after appropriate incubation to allow the substrate to be partially metabolized, they sorted fractions with high and low AHH activity. After 3–5 rounds of selection and growth populations were obtained with AHH activity significantly different from those of the parent populations.

Miller and Whitlock[71] have also investigated the induction of AHH by TCDD as a function of inducer concentration and time of exposure. At a TCDD concentration giving maximal induction of Hepa1c1c7 cells the population was found to be homogeneous with respect to the rate of induction, but at very low TCDD concentrations heterogeneity was observed in the rate of induction. In low activity variant clones (isolated after a combination of FACS sorting for low activity and selection by exposure to toxic concentrations of BP) induction heterogeneity was

[69] J. V. Watson, *Cytometry* **1**, 143 (1980).
[70] A. G. Miller and J. P. Whitlock, Jr., *J. Biol. Chem.* **256**, 2433 (1981).
[71] A. G. Miller and J. P. Whitlock, Jr., *Mol. Cell. Biol.* **2**, 625 (1982).

found even with maximal inducing concentrations of TCDD. Sorting high activity cells from the induced heterogeneous populations resulted in populations which, after several weeks of culture, were again heterogeneous and indistinguishable from the original population implying that the heterogeneity in induction sensitivity is transient and nonheritable. The authors suggest that the differences may be due to cell cycle phase.

Human Placental Cell Analysis and HLA mRNA in Trophoblast Cells

Work in our laboratory originating from efforts to isolate human fetal cells from maternal blood[54,72] has led to analysis of the cell populations found in the placenta[73] and to an investigation of the lack of cell surface HLA on the major cell population found in placental cell suspensions. Cell suspensions obtained by trypsin digestion of placental tissue were stained with Texas Red-labeled antibody to Trop-2[74] and with fluorescein-labeled antibody to a common framework determinant of HLA-A, B, and C. Two laser FACS analysis revealed 5 cell populations (Fig. 10). Only population A was actually positive for Trop-2; population B consisted of large, highly autofluorescent cells. Microscopy on sorted fractions showed population C to be erythrocytes while populations D and E included at least some white blood cells. Two-color FACS analysis with the HLA-A,B,C antibody and an antibody to a particular HLA type, in a case where the mother and fetus had an appropriate difference in that type, showed that populations D and E were of mixed fetal and maternal origin while population B was only fetal. Quinacrine Y-body analysis of sorted cell fractions (in a case where the fetus was male) indicated that populations A and B were of fetal origin.

The most interesting observation was the lack of HLA antigen expression on cells of population A, the major fetal-derived population. This may be related to maternal acceptance of the fetal allograft. The origin of this lack of HLA expression was traced by sorting population A cells with special care to maintain high purity and preparing messenger RNA from the cells.[75] In a Northern blot analysis with probes for HLA and β_2-microglobulin the cells were shown to be extremely low in HLA mRNA compared to other cells and somewhat low in β_2-microgobulin mRNA.

[72] L. A. Herzenberg, D. W. Bianchi, J. Schroder, H. M. Cann, and G. M. Iverson, *Proc. Natl. Acad. Sci. U.S.A.* **76,** 1453 (1979).

[73] M. Kawata and L. A. Herzenberg, *J. Exp. Med.* (in press).

[74] M. Lipinski, D. R. Parks, R. V. Rouse, and L. A. Herzenberg, *Proc. Natl. Acad. Sci. U.S.A.* **78,** 5147 (1981).

[75] M. Kawata, K. Sizer, S. Sekiya, J. R. Parnes, and L. A. Herzenberg, *Cancer Res.* (in press).

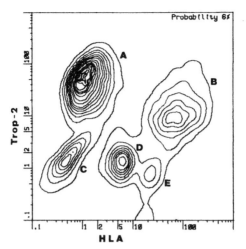

FIG. 10. Placental cell populations revealed by dual immunofluorescence measurements of Trop-2 and HLA. The antibody to HLA was directly fluoresceinated while the antibody to Trop-2 was biotin labeled and revealed with Texas Red avidin.

Thus the lack of HLA expression on these cells is primarily due to transcriptional regulation of HLA chain message rather than to regulation of transport to the cell surface or other posttranscriptional control.

Conclusions

Over the last several years the power of FACS analysis and sorting has been increasing rapidly in areas that are useful in work on lymphoid cell subpopulations and in the other areas discussed in this article. This has been due largely to the availability of an expanding array of monoclonal antibodies, to the use of new dyes and measurement systems for immunofluorescence and to improvements in data handling systems to take advantage of the greater information generating capacity.

We look forward to continued expansion of analytical capabilities, including some additional measurement parameters, and to modest improvements in cell sorting rates. Eventually, detailed division of lymphoid cell populations into truly specific functional units should be possible with appropriate combinations of monoclonal antibodies.

Acknowledgments

We thank Mr. Wayne Moore for mathematical and computer discussions, Richard Stovel for his illustrations, Randy Hardy for discussions and figures, and Richard Sweet for his criticism of the manuscript. This work was supported by N.I.H. Grant GM-17367.

[20] Isolation of Human Mononuclear Leukocyte Subsets by Countercurrent Centrifugal Elutriation

By HENRY C. STEVENSON

Introduction

Countercurrent centrifugal elutriation (CCE) is a continuous flow centrifugation technique utilizing a specialized chamber that has recently been applied to the isolation of purified subsets of human mononuclear leukocytes. These include B lymphocytes,[1] natural killer cells,[2,3] and human monocytes.[4,5] Isolation of small numbers of human monocytes can be achieved by CCE with the Sanderson elutriation chamber[6] or large numbers of monocytes can be obtained with the Beckman elutriation chamber.[4,5,7] The method described below utilizes the Beckman elutriation chamber and has been found to be reliable and efficient for isolating large numbers (8×10^8) of purified monocytes and monocyte-depleted purified lymphocytes (5×10^9). The monocytes thus obtained are greater than 99% viable and greater than 90% pure. The most common contaminating leukocytes are from the granulocyte series.[4] The lymphocytes that are separated at the same time are greater than 99% pure,[4] and are so effectively depleted of monocytes that numerous monocyte-dependent lymphocyte activation processes will not occur without the addition of autologous monocytes.[8] The lymphocytes and monocytes obtained by this procedure are sterile despite the fact that no antibiotics are used in any step of the isolation procedure; they are suitable for long-term suspension culture.[9]

[1] O. M. Griffith, *Anal. Biochem.* **87**, 97 (1978).
[2] E. Lotzová, *in* "Natural Cell-Mediated Immunity Against Tumors" (R. B. Herberman, ed.), p. 131. Academic Press, New York, 1980.
[3] T. Yasaka, R. J. Wells, N. M. Mantich, L. A. Boxer, and R. L. Baehnor, *Immunology* **46**, 613 (1982).
[4] T. J. Contreras, J. F. Jemionek, H. C. Stevenson, V. M. Hartwig, and A. S. Fauci, *Cell. Immunol.* **54**, 215 (1980).
[5] R. S. Weiner and V. O. Shah, *J. Immunol. Methods* **36**, 89 (1980).
[6] R. J. Sanderson, F. T. Shepperdson, A. E. Vatter, and D. W. Talmadge, *J. Immunol.* **118**, 1409 (1977).
[7] C. G. Figdor, W. S. Bont, J. de Vries, and W. van Es, *J. Immunol. Methods* **40**, 225 (1981).
[8] H. C. Stevenson, *in* "The Reticuloendothelial System: A Comprehensive Treatise" (J. A. Bellanti and H. B. Herscowitz, eds.) Vol. 6, p. 113. Plenum, New York, 1984.
[9] H. C. Stevenson, P. Katz, D. G. Wright, T. J. Contreras, J. F. Jemionek, V. M. Hartwig, W. J. Flor, and A. S. Fauci, *Scand. J. Immunol.* **14**, 243 (1981).

METHODS IN ENZYMOLOGY, VOL. 108

For smaller numbers of cells, one unit of peripheral blood or two buffy coats from manual cytapheresis collections can be utilized; for larger numbers of cells, machine cytapheresis collections can be used. Techniques have been devised to cryopreserve large numbers of monocytes and purified lymphocytes, with maintenance of most functions[10] (this volume [36]).

From a practical and theoretical standpoint, it should be noted that CCE allows for the collection of very large numbers of purified monocytes and lymphocytes (1) with high purity, (2) with excellent viability and functional status, (3) with high yield (80% of all monocytes in cytapheresis specimens and greater than 90% of all monocytes found in peripheral blood specimens[4]), and (4) obtained by a negative selection technique that leaves both the monocytes and lymphocyte preparations in a suspension state, thus reducing the likelihood of alteration of the native function of these circulating peripheral blood cells. The procedure takes about 3 hrs from the time that the white cells are obtained from the blood donor until the time that the purified monocytes and lymphocytes are obtained.

Materials

1. A Beckman elutriator chamber (Beckman Instruments, Palo Alto, CA).

2. A Beckman centrifuge with a porthole in the centrifuge lid and a strobe light apparatus assembly (must be capable of 1500 to 3500 rpm).

3. A cardiovascular pump (Sarns Medical Supply, Inc., Libertytown, IN) (or alternative pump with low pulsation properties).

4. A 2-liter sterile reservoir filled with elutriation medium (discussed below) connected by a 3-way stopcock to flask described below.

5. A pulse-suppressing 50-ml Erlenmeyer flask sealed with a rubber stoper through which both an 8-cm-long stainless-steel entry tube and a 10-cm-long stainless-steel exit tube penetrate. The pulse-suppressing flask should have a small stirring bar at the bottom and be placed on a magnetic stir plate (Fig. 1A).

6. A 20-ml syringe filled with the entry leukocytes suspended in elutriation media (see Fig. 1B).

7. Conical centrifuge tubes (50 ml) (Corning Plastics, Corning, NY).

8. A cell sizer system [Coulter Counter Model ZBI with a Channelizer H4 system (Hialeah, FL) or an Elzone Cell Sizer System (Chicago, ILL)].

9. Elutriation medium containing 0.05 M phosphate-buffered saline

[10] H. C. Stevenson, P. J. Miller, A. Akiyama, J. A. Beman, B. Stull, T. Favilla, G. Thurman, R. B. Herberman, A. Maluish, and R. K. Oldham, J. Immunol. Methods (in press).

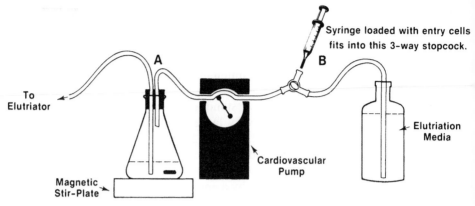

FIG. 1. Elutriator pumping system. (A) is the pulse-suppressor component of the system, consisting of an Erlenmeyer flask, entry and exit needles in a stopper, stir-bar, and a stirplate. (B) indicates the cell entry component of the elutriation system, consisting of a syringe with a sliding rubber plunger.

with 2% (v/v) human albumin (Cutter Laboratories, Berkeley, CA), filtered through a 0.45-μm filter unit just prior to use.

10. Ficoll–Hypaque mixture [prepared by Pharmacia (New York, NY). or from Litton Bionetics (Kensington, MD)].

11. RPMI 1640 (GIBCO, Grand Island, NY).

12. Sterile 21-gauge spinal needles.

13. All media should be documented to be endotoxin-free (by Limulus lysate assay, i.e., <0.1 ng/ml) prior to use.

Procedure

Basic Considerations

CCE operates on the principle that different sized cells held at the bottom of a centrifugation chamber can be differentially "washed away" by a current of medium flowing counter to the direction of the centrifugal force of the chamber. In a Ficoll–Hypaque gradient separation of human blood, three blood elements are encountered: (1) platelets, (2) small and large lymphocytes, and (3) monocytes. (There should be minimal numbers of red blood cells or granulocytes after Ficoll–Hypaque separation.) Figure 2 demonstrates a representative sizing analysis of the entering leukocytes encountered in a human mononuclear cell preparation from a normal donor. Lymphocytes and platelets will be the first cells to exit the elutriation chamber, are designated as the "lymphocyte" fraction, and should be maintained separate from the other mononuclear subset frac-

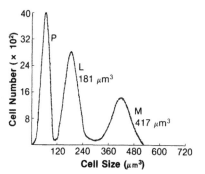

Fig. 2. Cell sizing analysis of Ficoll–Hypaque unfractionated mononuclear leukocytes entered into the elutriator system. P, Platelets; L, leukocytes; M, monocytes. Numbers next to each peak indicate the mean size of the peak (in cubic microns).

tions (see Fig. 3). Approximately two-thirds through the procedure, the largest of lymphocytes and the smallest of monocytes will exit the chamber at the same time (see Fig. 4); this set of fractions have been termed the "intermediate monocyte fraction" and are ordinarily separated from the "lymphocyte" and "pure monocyte" fractions. The last fractions collected from the chamber will be greater than 90% pure monocytes (see Fig. 5) and represent the largest cells in the mononuclear cell suspension. It should be noted that all of the cells placed in the CCE apparatus can be recovered in suspension and should be harvested in a laminar flow hood for sterile collection. If the CCE apparatus and the materials used to isolate the mononuclear cells are not sterile and free of possible monocyte activators (such as endotoxin, residual alcohol, or degranulated granulocytes), the entire procedure will have to be abandoned since the leukocytes will agglutinate in the chamber. Thus, sterile technique and endo-

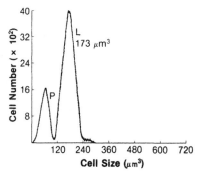

Fig. 3. Cell sizing analysis of the purified lymphocyte fraction consisting of the first 5 to 7 fractions that exit the elutriator chamber. Note that although there are virtually no monocytes in this preparation, platelet contamination is appreciable.

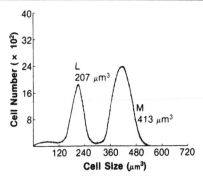

Fig. 4. Cell sizing analysis of the intermediate monocyte fraction consisting of a mixture of large lymphocytes and small monocytes.

toxin-free medium are absolutely essential for the successful completion of this procedure.

Preparation of Human Mononuclear Leukocyte Suspensions

If small numbers of human monocytes and lymphocytes are desired, one can utilize one unit of whole blood or two buffy coats from platelet-pheresis packs as the white cell substrate. Since it takes the same amount of effort to purify a large number of monocytes as it does a small number, we generally utilize a machine cytapheresis collection, containing nearly 10 times more mononuclear cells than the peripheral blood specimens.[11]

1. The peripheral blood or cytapheresis specimen is diluted to 10^7 cells/ml in a large flask with sterile phosphate-buffered saline, and 30 ml of this diluted blood is added to each of approximately 30 conical 50-ml centrifuge tubes.

2. Of the Ficoll–Hypaque solution 12 ml is layered under the diluted blood with a spinal needle; then all tubes are centrifuged at room temperature at 400 g for 35 min.

3. The bands of mononuclear cells at the Ficoll–Hypaque interface are harvested and combined, and then washed twice at 500 g for 10 min with cold RPMI 1640 (4°). The final mononuclear cell pellet thus obtained should be resuspended in 20 ml of elutriation medium (approximately 25 × 10^8 cells/ml of medium when cytapheresis cells are utilized).

4. Wright's and esterase stains[12] are performed on a small aliquot of this suspension to confirm the purity of the mononuclear cell prepara-

[11] H. C. Stevenson, J. A. Beman, and R. K. Oldham, *Plasma Ther.* **4,** 57 (1983).
[12] C. Y. Li, K. W. Lam, and L. T. Yam, *J. Histochem. Cytochem.* **21,** 1 (1973).

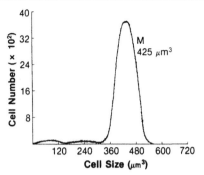

FIG. 5. Cell sizing analysis of the purified monocytes obtained from the CCE apparatus (last 2 to 4 fractions).

tions; trypan blue analysis of viability should also be performed. In addition, the unfractionated mononuclear leukocyte preparation should be sized on the cell sizer apparatus so that the total number of monocytes in the sample and recovery yields at the end of the CCE procedure can be calculated (this volume [6]).

Performance of CCE Procedure

1. The elutriator centrifuge temperature should be 18° and the elutriator rotor speed should be set at 2020 ± 10 rpm. The initial flow rate of elutriation medium through the rotor should be established at 6.0 ± 0.5 ml/min. (Assessed by the rate of effluent collection of elutriation medium into a 50-ml centrifuge tube.)

2. When the CCE system has been calibrated to the above settings, the unfractionated mononuclear cell suspension can be injected into the system with a syringe connected to the three-way stopcock. The three-way stopcock should then be turned, to stop the flow of elutriation medium and thus allow cells to enter the system. At the completion of cell entry, the three-way valve should be repositioned to resume elutriation medium flow from the reservoir. Extreme caution must be exercised to prevent any air bubbles from entering the elutriation chamber, since this will create a back pressure that will cause fluid flow through the elutriation chamber to cease.

3. Aliquots (50 ml) should be collected from the exit port of the elutriator centrifuge, and an aliquot of each should be examined in the cell sizing system to determine whether platelets, lymphocytes, or monocytes are exiting the chamber.

4. The countercurrent medium flow rate should be increased every 10 min by increments of 0.5 ml/min. Generally, the first five to eight 50-ml

fractions collected under these conditions will be pure lymphocytes (plus platelets) and can be designated the "lymphocyte" fraction.

5. At approximately 9 ml/min of medium flow rate, the smallest monocytes will be found to exit the chamber. Generally there will be two to four 50-ml fractions of mixtures of lymphocytes and monocytes. These two–four fractions may be pooled and termed "intermediate monocyte fraction." It should be noted that the total number of cells obtained in this fraction is generally only about 5% of the total number of mononuclear cells introduced into the system.

6. Finally, at a flow rate of approximately 11.0 ml/min, purified monocytes should exit the chamber. (There may be some donor-to-donor variation regarding the exact flow rate for monocyte exit.) The flow rate of the medium should then be increased until all of the monocytes can be seen to have exited the chamber; all of the monocyte fractions may be pooled and termed "purified monocyte fraction."

7. All of the 50-ml aliquots thus obtained (usually 10 to 14 tubes per individual run) should be centrifuged at 500 g for 10 min and the pellets pooled according to the above classification scheme. The cell pellets should be resuspended in the appropriate medium for the subsequent experimental procedures.

8. Aliquots from each of the three fractions (i.e., lymphocyte fraction, intermediate fraction, regular monocyte fraction) should be assessed for monocyte purity by Wright's and esterase staining coupled with an assay of phagocytosis (e.g., latex bead ingestion).

9. Further cell typing can be performed with a fluorescence-activated cell sorter (FACS) and appropriate antibody typing reagents (this volume [19]). These reagents can be used to assess the percentage of lymphocyte subsets, natural killer (NK) cells and monocytes in each fraction.

Maintenance of CCE Equipment

Sterility and preservation of function of monocyte and lymphocyte suspensions can be ensured by closely adhering to the aseptic technique detailed above. In addition, the entire CCE apparatus must be maintained in an aseptic fashion which includes the routine disassembly of the entire rotor system, cleaning it with disinfectant solutions (70% ethanol or dilute bleach), rinsing the component parts in distilled water, and ethylene dioxide gas sterilizing the entire system. The entire system should be wrapped in a gas sterilization package, to allow its maintenance in a sterile state for several days. The CCE apparatus should be assembled under sterile conditions, immediately prior to each CCE procedure. Each CCE experiment should be preceded by the passage of 500 ml of sterile phosphate-buffered saline through the CCE apparatus. The system is then ready for entry of elutriation medium and cell suspensions.

Conclusion

The CCE procedure described above is a very precisely controlled operation for purifying human monocytes and lymphocytes in suspension. Elutriation is the only technique currently available that will allow for the negative selection of large numbers of these cell types while meeting the additional criteria of (1) high purity, (2) high yield, and (3) excellent viability and functional status.

[21] Use of Antisera and Complement for the Purification of Lymphocyte Subpopulations

By FUNG-WIN SHEN

Introduction and Principles

In 1956 Gorer and O'Gorman[1] devised the complement-dependent cytotoxicity assay for detecting H-2 alloantigens on nucleated cells of the mouse. Ideally, when living cells in free suspension are exposed to antibody, together with complement, cells bearing the respective antigen on their surfaces are lysed, while antigen-negative cells remain viable. Dead cells admit Trypan Blue, and living cells do not. Thus counts of blue and unstained cells in a hemacytometer give the numbers and proportions of dead (antigen-positive) and live (antigen-negative) cells. The cytotoxicity assay is the starting-point of selective immunocytolysis with cell-discriminating antibody and complement. For instance, antibody made in H-2^b mice against cells of H-2^a mice will in this way eliminate H-2^a cells from a mixture of H-2^a and H-2^b cells, and so purify the H-2^b cell population.

Understanding the cytotoxicity assay is an aid to best use of immunocytolysis because the principles and ingredients are the same.

Later, it became known that certain genes and gene products are expressed only in thymus-processed (T) lymphocytes,[2–4] normal or malig-

[1] P. A. Gorer and P. O'Gorman, *Transplant. Bull.* **3**, 142 (1956).
[2] L. J. Old, E. A. Boyse, and E. Stockert, *J. Natl. Cancer Inst. (U.S.)* **31**, 977 (1963).
[3] E. A. Boyse, L. J. Old, and S. Luell, *J. Natl. Cancer Inst. (U.S.)* **31**, 987 (1963).
[4] E. A. Boyse, M. Miyazawa, T. Aoki, and L. J. Old, *Proc. R. Soc. London, Ser. B* **170**, 174 (1968).

METHODS IN ENZYMOLOGY, VOL. 108

nant, that expression of these genes and products on the cell surface (in the plasma membrane) is further restricted to particular stages and sublineages of T cell differentiation, and that the imprints of these molecules, forming distinctive cell surface phenotypes, are hallmarks of T cell sets and subsets programmed for different sorts of immune response, notably cell-mediated cytotoxicity (killing of foreign or altered cells), helper-inducer functions, and suppression.[5] Thus complement-dependent antibody-mediated cytolysis became the first procedure whereby T cell populations could be divided into their otherwise relatively inseparable and generally indistinguishable functional compartments. Much the same applies to the other (B) category of lymphocytes, and the advent of monoclonal antibodies has led to the description of homologous human lymphocyte compartments characterized immunogenetically by surface phenotype.[6,7]

Applications

The general purposes of using antibodies that distinguish T and B lymphocytes and that categorize compartments of these two lymphocyte classes, aside from theoretical extrapolations to differentiate development as a whole, are to enumerate and separate the severally programmed sets and subsets which lymphocytes comprise, to study their functional and other properties alone or in combination, and to deduce mechanisms and aberrations of immunity. The several presently known immunogenetic systems that distinguish T and B lymphocyte compartments of mice include the Lyb and Lyt series, Tla, Pca, Qa-1, and others.[8]

An antibody belonging to one of these discriminating immunogenetic systems could be used either as a negative selective agent, removing unwanted cells, or in positive selection, when the cells reacting with antibody are recovered by such means as release from a solid base to which the antibody has been bound. Since cells treated with antibody and complement are destroyed, this method is limited to negative selection. A further limitation is that only certain classes of antibody bind complement, which automatically precludes the use of monoclonal antibodies belonging to other subclasses. The effectiveness of a conventional antiserum must depend in part on its ratio of complement-fixing to non-comple-

[5] H. Cantor and E. A. Boyse, *Immunol. Rev.* **33,** 105 (1977).
[6] E. L. Reinherz and S. F. Schlossman, *Immunol. Today* **2,** 69 (1981).
[7] I. F. C. McKenzie and H. Zola, *Immunol. Today* **4,** 10 (1983).
[8] F. W. Shen, *in* "The Mouse in Biomedical Research" (H. L. Foster, J. D. Small, and J. G. Fox, eds.), Vol. 3, p. 381. Academic Press, New York, 1983.

ment-fixing antibody.[9] Further, commonly a given complement-fixing antibody is not lytic for particular cells, or only weakly so, although the presence of antigen on those cells is demonstrable by serological absorption; this is because there is too little antigen on the cells in question.

However, where these limitations do not apply, cytolytic elimination has the considerable advantages of simplicity and precision, and efficacy can readily be monitored by counts of live and dead cells.

Complementation

The choice and preparation of complement are the key to efficient cytolysis. Complementation is a vast topic, and relatively little is known about complement-mediated cytolysis of nucleated cells as compared with red cells. Certainly lysis of nucleated cells demands much more complement than red cells. With the latter, a single IgM antibody molecule may be sufficient for complement-mediated lysis of a single red cell. For nucleated cells, it is useful to view the cells of the target population as being at risk of a critical number of lytic lesions of the plasma membrane, each depending upon the formation of a lytic complex comprising antigen, antibody, and the several components of complement, in limited time, such that there is a given probability that each cell will be sufficiently damaged to be lysed. In strong lytic systems like H-2, lysis may approach 100% of the cells. But with the several systems that distinguish lymphocyte categories, the amounts of complement it is feasible to supply may be little more than marginal, and second cycles of cytolysis may need to be considered, perhaps with antibody of a second specificity if the target cell population carries more than one suitable surface antigenic marker. A main difficulty in supplying enough complement is that the most effective complements for lysis of nucleated cells come from foreign species, rabbit serum being the best source of complement for lysis of nucleated human cells[10] and mouse cells[11] and therefore the introduction of natural interspecies antibody is inevitable. Unfortunately such natural antibody is stable whereas complement is highly unstable, and therefore a prime aim is to avoid all sources of degradation in the preparation of complement, taking care also to eliminate from the cytolytic procedure all known anticomplementary factors such as cell debris and undue amounts of mouse serum, which itself is anticomplementary.[12]

Another important consideration is that efficiency of complementation

---

[9] F. W. Shen, E. A. Boyse, and H. Cantor, *Immunogenetics* **2**, 591 (1975).

[10] R. L. Walford, R. Gallagher, and G. M. Troup, *Transplantation* **3**, 387 (1965).

[11] E. A. Boyse, L. J. Old, and G. Thomas, *Transplant. Bull.* **29**, 64 (1962).

[12] E. A. Boyse, L. J. Old, and I. Chouroulinkov, *Methods Med. Res.* **10**, 39 (1964).

depends highly upon the *concentration* of complement, hence the benefit of adding the complement to cells that have been presensitized with antibody, washed, and pelleted, thereby favoring maximal complementation with minimal natural antibody.

For those without experience, there is no comparable substitute for obtaining an accredited sample of complement to serve as a standard for comparison with the complement pool that the investigator will prepare or acquire. In the author's experience, rabbit serum for use as complement is best prepared by selecting rabbits, no more than 48 hr beforehand, for low cytotoxicity combined with high complement titer, and then exsanguinating these rabbits and pooling their sera.[13] If this is skillfully accomplished, which requires good teamwork, it will be unnecessary to resort to attempts to remove natural antibody by absorption with cells or other materials, a procedure which even in the presence of EDTA[13] to minimize complement consumption, is suspect of lowering the natural toxicity of rabbit serum as much by removing complement as by removing natural antibody. There is no certain evidence that any rabbit of a particular age, sex, or strain is substantially better than another in regard to relative content of complement and unwanted antibody. It pays to make as large a pool as possible and to freeze in amounts suitable for a single day's work. Repeated freezing and thawing must be avoided.

Illustrative Procedures

The three ingredients are the cell suspension, the antibody, and rabbit serum as complement, maintained in an ice-bath until ready for incubation at 37°. The concentrations required are gauged initially from cytotoxicity assay[1,12,13] of the particular antiserum pool or monoclonal antibody of a given system. For making counts of live and dead cells, Trypan Blue is made up freshly each time by dilution of a stock aqueous solution of Trypan Blue with a suitable stock concentration of saline.[12] Release of ^{51}Cr is an alternative to Trypan Blue as the indicator of lysis.[14] Titration of antibody with a constant concentration of cells and complement typically gives a plateau indicating a range of antibody concentrations giving the same maximal percentage of lysed cells.[15]

Step 1: Cytotoxicity Assay. Assume for purposes of illustration that titration of a conventional antiserum, by standard cytotoxicity assay, with equal volumes of cells (5×10^6/ml), antibody in seral dilution, and rabbit serum as complement (say 1:15), shows a plateau of maximal lysis around a midpoint of (say) 1:160.

[13] E. A. Boyse, L. Hubbard, E. Stockert, and M. E. Lamm, *Transplantation* **10,** 446 (1970).
[14] H. Wigzell, *Transplantation* **3,** 423 (1965).
[15] E. A. Boyse, L. J. Old, and E. Stockert, *Ann. N.Y. Acad. Sci.* **99,** 574 (1962).

Step 2: Presensitization for Cytoelimination. Wash and pellet the cells, and resuspend at a concentration of 5×10^7/ml in antibody at a dilution of 1 : 16, for incubation on ice for 30–40 min with occasional shaking. The use of ice-bath temperature hinders dissociation and release of antibody from the sensitized cells.[16]

Step 3: Cytoelimination. Wash and resuspend the pelleted presensitized cells at a concentration of 5×10^7/ml in rabbit serum at 1 : 10. Note that there is less complement per cell than in the conventional cytotoxicity assay (Step 1) but that the concentration of complement is higher (see above under Complementation). Transfer directly from the ice-bath to a 37° water-bath because there is evidence that natural antibody is more effective and troublesome at intermediate temperatures.[13] Incubate, with gentle shaking, for 30–45 min. Check the viable and nonviable counts in the experimental preparation and controls. Dead cells can be removed, for example by suitable centrifugation or passage through a glass wool column.[17,18]

Control

The best control is substitution of cells of the same type from congenic mice of an allele and allotype different from that to which the antibody is directed,[9] in which case there should be no specific lysis and the nonspecific lysis that occurs is the best indicator of background lysis due to natural antibody and other causes. With conventional antisera, these other causes include autoantibody, which in the case of thymocytes as targets appears to be particularly prominent in immunizations between congenic mice or other mice that are closely similar genetically.[19] Where this control is not possible, as with human cells, and with monoclonal antibody from xenogeneic immunizations which recognize species determinants rather than allotypic determinants, controls in which medium is substituted for antibody will signify background toxicity due to the complement source, and substitution of a monoclonal antibody of the same class but to a different antigen is another criterion of specificity.

Acknowledgments

Supported in part by Grants CA-22131 and AI-00329 from the National Institutes of Health.

[16] S. Chang, E. Stockert, E. A. Boyse, U. Hammerling, and L. J. Old, *Immunology* **21**, 829 (1971).

[17] A. J. Feeney and U. Hammerling, *Immunogenetics* **3**, 369 (1976).

[18] T. H. Stanton, C. E. Calkins, J. Jandinski, D. J. Schendel, O. Stutman, H. Cantor, and E. A. Boyse, *J. Exp. Med.* **148**, 963 (1978).

[19] E. A. Boyse, E. Bressler, C. A. Iritani, and M. P. Lardis, *Transplantation* **9**, 339 (1970).

[22] Suicide of Lymphoid Cells

By A. Basten and P. Creswick

Introduction

Lymphocytes are unique among the cells of the immune system in that they recognize antigen specifically by way of surface receptors; once activated by antigen they undergo proliferation to produce a clone of effector and memory cells, each with specificity identical to their cell of origin.[1] The concept of specificity applies to all classes of lymphocytes including B cells which are antibody-forming cell precursors and the various T cell subsets which comprise lymphocytes responsible for mediating help (Th), suppression (Ts), cytotoxicity (Tc), and delayed type hypersensitivity (T_{DTH}).[2] Detection of antigen-specific lymphocytes has, however, proved difficult owing to their low frequency in lymphoid tissue (estimated to be of the order of 1 : 100 or less in the case of B cells) and to the fact that the density of surface receptors, particularly on T cells, is below the threshold of sensitivity of currently available methods.[3] The various "suicide" techniques provide a functional way of addressing this problem. For convenience they can be divided into two broad categories. The first involves exposure of lymphoid cell populations to radiolabeled (^{125}I) antigen followed by assessment of the capacity of that cell population to mount a specific response to the antigen in question.[4,5] The second category depends on the sensitivity of proliferating lymphocytes once activated by antigen to radioactive DNA precursors like tritiated thymidine (hot pulse technique)[6] or to fluorescent light after incorporation of bromodeoxyuridine.[7] The latter technique is described in this volume [23].

[1] F. M. Burnet, *in* "Theoretical Immunology" (G. I. Bell, A. S. Perelson, and G. H. Pimbly, eds.), p. 63. Dekker, New York, 1978.

[2] A. Basten, R. H. Loblay, R. J. Trent, and P. A. Gatenby, *Recent Adv. Clin. Immunol.* **2,** 33 (1980).

[3] C. G. Fathman and J. G. Frelinger, *Annu. Rev. Immunol.* **1,** 633 (1983).

[4] A. Basten, J. F. A. P. Miller, N. L. Warner, and J. Pye, *Nature (London), New Biol.* **231**(21), 104 (1971).

[5] G. L. Ada and P. Byrt, *Nature (London)* **222,** 1291 (1969).

[6] R. W. Dutton and R. I. Mishell, *J. Exp. Med.* **126,** 443 (1967).

[7] D. W. Thomas and E. M. Shevach, *Proc. Natl. Acad. Sci. U.S.A.* **74**(5), 2104 (1977).

METHODS IN ENZYMOLOGY, VOL. 108

Hot Antigen Suicide

Materials and Methods

Animals. Mice of any of the common inbred strains (e.g., CBA/CaH, C57BL/6J, BALB/cJ) can be used as a source of lymphocytes and for assay purposes.

Antigens. A soluble antigen containing tyrosine residues is required for iodination. Immunoglobulins such as human γ-globulin (HGG)[8] or fowl γ-globulin (FYG)[5] are suitable for this purpose. In the case of HGG the Cohn fraction II is chromatographed on a protein A-Sepharose CL-4B column (Pharmacia, Uppsala, Sweden) to yield purified IgG from which the F(ab')₂ fragments can be obtained by pepsin digestion.[9] FYG is prepared by ammonium sulfate fractionation of fowl serum followed by extensive absorption against murine lymphocytes to remove nonspecific anti-mouse lymphocyte activity.[5]

For assay purposes, horse erythrocytes (HRC) are used as a control antigen and sheep erythrocytes (SRC) for measurement of antibody forming cells. They are collected by venipuncture, stored in Alsever's solution, and washed three times in phosphate-buffered saline (PBS, pH 7.2) before use.

Antibody. Suicide of Ts and Th requires affinity purified antibody as well as antigen. High titer antisera to HGG, for example, can be raised in rabbits by immunization with three intramuscular injections of 0.5 mg HGG emulsified in Freund's complete adjuvant (total volume 2 ml) at two weekly intervals. Serum is then collected 14–21 days after the last injection and the anti-HGG titer measured by hemagglutination assay.[8] To obtain an affinity purified antibody the antiserum is initially fractionated on protein A-Sepharose CL-4B, following which the IgG fraction is loaded onto an HGG-Sepharose 4B column (Pharmacia, Uppsala, Sweden) and the specific antibody eluted under acid conditions with glycine–HCl (pH 2.8). The F(ab')₂ fragment is prepared by pepsin digestion.

Iodination. Antigen and affinity purified antibody are iodinated by means of the chloramine-T method as modified by Ada and Byrt.[5] In order to avoid denaturation of the proteins a concentration of at least 5 mg/ml is required. Fifteen micrograms protein and 1 mCi [¹²⁵I]iodide (Amersham International, England, Cat. No. IMS30, specific activity 100 mCi/ml NaI) are mixed together on ice in the presence of freshly prepared chloramine-T solution (5 μl, 1 mg/ml in degassed PBS, pH 7.2). The reaction is

[8] A. Basten, J. F. A. P. Miller, R. Loblay, P. Johnson, J. Gamble, E. Chia, H. Pritchard-Briscoe, R. Callard, and I. McKenise, *Eur. J. Immunol.* **8,** 360 (1978).
[9] L. Hudson and F. C. Hay, "Practical Immunology." Blackwell, Oxford, 1976.,

allowed to proceed for 10 min when it is stopped by addition of an equal volume of potassium metabisulfite (2.4 mg/ml in degassed PBS). To separate ^{125}I-labeled protein from free iodide, the reaction mixture is run through a Sephadex G-75 column (10 × 1 cm) which has been stabilized with 20 μl 5% BSA in degassed PBS. The effluent is collected in 0.2-ml aliquots and radioactivity measured on a gamma spectrometer. The specific activity of the second peak containing the labeled protein is then calculated (assuming a 100% recovery of protein). This should be of the order of 60–80 μCi/μg representing a substitution ratio of five iodine atoms/molecule since efficiency of iodination is usually 60–80%.

Cell Suspensions. Single cell suspensions are prepared by teasing donor lymphoid tissue (e.g., spleen) through 80-mesh stainless-steel sieves into cold HEPES buffered RPMI 1640 medium (Grand Island Biological Co, Grand Island, NY) containing 10% fetal calf serum (FCS). Cell populations enriched for T cells or B cells are obtained by passage through nylon wool columns[10] or treatment with anti-Thy-1 plus complement,[8] respectively.

Source of Cell Populations. Primed Th cells and B cells are obtained from animals immunised with antigen in immunogenic form at least 4–6 weeks previously, and Ts are generated by injecting animals with antigen in either immunogenic or tolerogenic form 7–10 days beforehand. For FYG and HGG, the precise methods have been described elsewhere.[4,5]

Suicide. The conditions for antigen specific suicide vary depending on the cell subset involved. For B cell suicide, cells are exposed directly to ^{125}I-labeled antigen in the cold. In the case of T cells, no killing occurs under these circumstances and a 37° step is required. To achieve suicide of Ts, cells are incubated with ^{125}I-labeled antibody since they already appear to be coated with antigen. Helper T cells, on the other hand, need to be incubated with cold antigen followed by hot antibody. Furthermore, it should be noted that although Th have been enriched by passage through nylon wool columns, a small number of antigen presenting cells still need to be present for suicide of help to occur. The technical details for each subset are summarized in the flow diagram (Fig. 1).

Functional Assay of Suicide and Its Applications

The efficacy of suicide is demonstrated by functional assay of the cells following treatment with ^{125}I-labeled ligands. To illustrate how the effects of suicide are measured, the results of an experiment involving suicide of Th with HGG–anti-HGG complexes are shown (Fig. 2). Purified Th obtained by passing spleen cells through nylon wool columns were incu-

[10] M. H. Julius, E. Simpson, and L. A. Herzenberg, *Eur. J. Immunol.* **3,** 645 (1973).

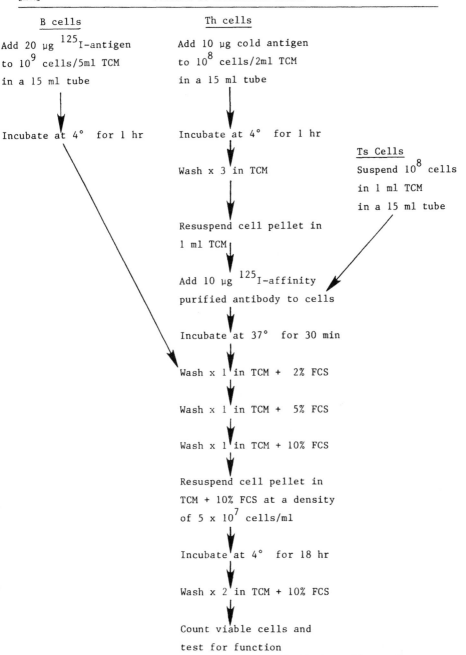

FIG. 1. Flow diagram of suicide of different lymphoid cell subsets. Tissue culture medium (TCM) is HEPES buffered RPMI 1640. FCS, Fetal calf serum. Cells are suspended in 15 ml conical centrifuge tubes (Falcon Plastics, No. 2087). Washing is performed by centrifugation at 1300 g for 5 min. Viability is determined by trypan blue dye exclusion. Note that the incubations with ^{125}I-labeled ligands are carried out in TCM lacking FCS.

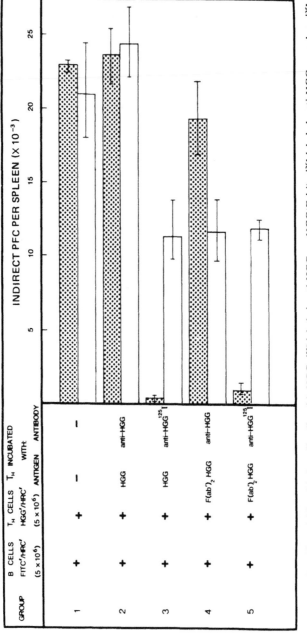

FIG. 2. Specific suicide of helper T cells with HGG—[125]I-labeled anti-HGG or HGG(Fab')$_2$—[125]I-labeled anti-HGG complex. ^{125}I-labeled HGG or ^{125}I-labeled anti-HGG alone are ineffective. Stippled areas depict anti-FITC antibody response; open areas depict control anti-HRC antibody response.

bated with HGG [or its F(ab')$_2$] fragment followed by ^{125}I-labeled anti-HGG antibodies as described previously. Control cells underwent the same treatment except that cold antibody was substituted for ^{125}I-labeled antibody. The use of F(ab')$_2$ fragments was designed to eliminate nonspecific binding via Fc receptors on the T cell membrane. Helper function was assayed in vivo by transferring Th with spleen cells primed to the hapten fluorescein (FITC) coupled to an irrelevant carrier (KLH) and a control antigen (HRC) into heavily irradiated syngeneic recipients which were then challenged with FITC-HGG and HRC. Seven days later the antibody responses to FITC and HRC were measured by plaque-forming cell assay. As shown in Fig. 2, the collaborative antihapten response was significantly reduced when Th had been exposed to HGG followed by hot anti-HGG antibody. The suicide effect was antigen specific since HRC antibody production by control and test cells was comparable. Similar results can be obtained with B cells[4] and Ts[11] when they are exposed to ^{125}I-labeled ligands under the appropriate conditions.

The suicide technique has a number of applications: first, it provides a system for demonstrating specific antigen binding to nondividing cells in a functional assay; second, it demonstrates unequivocally that both Ts and Th subsets can bind antigen specifically even though the number of receptor molecules is too low to be readily detected by other conventional methods[3]; third, use can be made of suicide conditions to enrich for specific antigen binding T cells which has application in production of T cell lines[3]; and, finally, it can be applied to the study of the nature of antigen specific receptor sites and their relationship to other surface determinants by preincubating cells with appropriate antibodies before exposure to ^{125}I-labeled ligands.

Suicide of Proliferating Cells by the Hot Pulse Technique

The aim of this form of suicide is to effect specific deletion of individual clones of cells which are undergoing proliferation in response to antigen.[6]

Materials and Methods

Cell Source and Media. In the original experiments[6] murine spleen cells were used but the source of lymphocytes is not critical and can be from any secondary lymphoid tissue of commonly available laboratory animals. Cell suspensions are prepared under aseptic conditions as de-

---

[11] A. Basten, R. Loblay, E. Chia, R. Callard, and H. Pritchard-Briscoe, *Cold Spring Harbor Symp. Quant. Biol.* **41**, 93 (1977).

scribed in the previous section except that Eagle's minimal essential medium (MEM) replaces RPMI 1640. MEM (100 ml) is supplemented with 1 ml, 200 mM glutamine, 1 ml, 100 × MEM nonessential amino acids, 1 ml, 100 mM sodium pyruvate, and 5% FCS. Conditioned medium was prepared by harvesting the supernatants from 24 hr mouse spleen cells.[6]

Remove spleen cells from normal animals aseptically

↓

Set up 1 ml cultures of 1.5×10^{7} spleen cells and

3×10^{6} SRC in 35 mm petri dishes containing TCM

↓

Incubate at 37°C in 10% CO_2 in air on a rocker

for 24 hours

↓

Add 10 μCi tritiated thymidine ± 100 μg cold thymidine

to each culture and incubate cells for a further 24 hours

↓

Wash cells x 2, resuspend in 'conditioned' medium containing

100 μg/ml cold thymidine and incubate for 48 hours

↓

At 96 hours harvest cells and assay for antibody production

to SRC by PFC assay

FIG. 3. Flow diagram of hot pulse technique for suicide of lymphocyte proliferation in response to antigen. TCM is MEM supplemented as described in the text.

Antigen. Soluble (e.g., PPD)[12] or particulate (e.g., SRC)[6] antigens can be used. In the protocol described below SRC is the example selected to illustrate the principles of the technique.

Suicide and Functional Assay

The precise steps involved in the hot pulse technique are summarized in Fig. 3. Briefly, murine spleen cells are cultured for a total of 96 hr together with 3×10^6 SRC in 35-ml Petri dishes (Falcon Plastics No. 3001) containing 1 ml MEM. Incubation is carried out at 37° at an atmosphere of 10% CO_2 in air on a Rocker (e.g., Bellco Glass Inc., New Jersey). Between 24 and 48 hr later each culture is pulsed with 10 μCi (0.16 μg) of tritiated thymidine while control cultures receive 100 μg/ml cold thymidine. All cultures are fed daily with 100 μl of "feeding" medium plus 30 μl FCS. Twenty-four hours after the thymidine pulse cells are washed twice, resuspended in 1 ml conditioned medium containing 100 μg/ml cold thymidine, and incubated for a further 48 hr. Cells are then harvested and tested for functional activity in a plaque-forming cell (PFC) assay, designed to measure the anti-SRC antibody response. Successful suicide is demonstrated by the abrogation of anti-SRC-PFC in cultures exposed to hot but not excess cold thymidine.

The technique can also be applied to other systems involving lymphocyte transformation to soluble antigens.[12] Although the readout is different (i.e., tritiated thymidine incorporation) the end result is the same with deletion of the clone of lymphocytes undergoing proliferation in response to antigen. In the case of antibody responses the cells involved include B and Th cells whereas in lymphocyte transformation T cells are predominantly affected.

Acknowledgments

The experimental studies from our laboratory were supported by grants from the National Health and Medical Research Council of Australia. We are indebted to Ms M. A. Stack for preparation of the manuscript.

[12] L. Cohen, R. S. Holzman, F. T. Valentine, and H. S. Lawrence, *J. Exp. Med.* **143**(4), 791 (1976).

[23] Elimination of Specific Immunoreactive T Lymphocytes with 5-Bromodeoxyuridine and Light

By ETHAN M. SHEVACH

General Considerations

The technology used to specifically delete a population of T lymphocytes reactive with a foreign protein antigen or cell bound alloantigen from a heterogeneous mixture of cells is derived from methods used in microbial[1] and somatic cell genetics[2] for the detection and isolation of nutritionally deficient mutants of established cell lines. A mixed cell population was placed in a limited medium in which nutritionally deficient forms could not grow and then an agent which would only kill dividing cells was applied. Two agents have been widely used for this purpose. Lethal amounts of [³H]thymidine ([³H]TdR) of high specific activity (>20 Ci/mmol) can be used to inactivate cells which are synthesizing DNA in response to a stimulating antigen. The technique to accomplish this deletion of response is based on the observation of Dutton and Mishell[3] that the plaque-forming antibody response of murine spleen cells to sheep cells could be abolished by exposing the spleen cells to [³H]TdR of high specific activity 24 hr after they had been exposed to the heterologous erythrocytes. Only the cells that were stimulated by the antigen participated in DNA synthesis and incorporated the compound; because of the high specific activity and short irradiation path of [³H]TdR only the antigen-stimulated cells destroyed themselves by incorporating a lethal dose of the compound in a form of "[³H]TdR suicide."

Although the [³H]TdR hot-pulse technique has been used by some investigators[4,5] to analyze the specificity of responding cell populations in the mixed leukocyte reaction (MLR), this method is subject to practical limitations in the leukocyte culture system where response is most easily measured by [³H]TdR incorporation. Furthermore, the large amounts of radioactivity needed for these studies render this methodology somewhat objectionable. High concentrations of radioactivity were also shown[4] to have a deleterious effect on the mitomycin C treated MLR stimulating population resulting in poor stimulation, lack of proliferation, and resul-

[1] B. Djordjević and W. Szybalski, *J. Exp. Med.* **112**, 509 (1960).
[2] T. T. Puck and F. Kao, *Proc. Natl. Acad. Sci. U.S.A.* **58**, 1227 (1967).
[3] R. W. Dutton and R. I. Mishell, *J. Exp. Med.* **126**, 443 (1967).
[4] H. Hirschberg and E. Thorsby, *J. Immunol. Methods* **3**, 251 (1973).
[5] S. E. Salmon, R. S. Krakauer, and W. F. Whitmore, *Science* **172**, 490 (1971).

tant poor elimination of the relevant histocompatibility antigen reactive cell population.

The incorporation of 5-bromodeoxyuridine (BUdR, a thymidine analog) into DNA of dividing mammalian cells in culture provides an alternative means of specifically inactivating a responding cell population. When BUdR-treated cells are exposed to visible or ultraviolet light they exhibit large numbers of both single- and multihit chromosomal aberrations.[2] It is highly likely that cell death results from inhibition of chromosomal repair mechanisms in BUdR-treated mammalian cells.

Deletion of Antigen- and Alloantigen-Specific T Lymphocytes

The BUdR and light suicide technique has been used for a number of distinct purposes in cellular immunology. The major question addressed in the original studies of Zoschke and Bach[6] was whether the cells that respond to different soluble antigens in cultures of primed human T lymphocytes were totipotential or whether they belong to distinct or partially distinct subpopulations of immunologically committed cells. A related question studied with this technique[7,8] is the relationship between cells responding to distinct populations of foreign histocompatibility antigens.

Elimination of Human T Lymphocytes Responsive to One Soluble Protein Antigen

1. Isolate mononuclear cells from the peripheral blood of an individual sensitive to two different protein antigens by centrifugation on Hypaque–Ficoll (this volume [9]).

2. Suspend responding cells (2×10^6/ml) in medium 199, Earle's base supplemented with penicillin 100 U/ml, streptomycin 100 μg/ml, and 10% cell free plasma.

3. Add the optimal stimulating concentration of the soluble protein antigens to be tested and leave some cultures unstimulated.

4. Maintain replicate 2 ml cultures in a humidified atmosphere of 5% CO_2 in air.

5. After 48 hr, add BUdR (Sigma Chemical Co., St. Louis, MO) to a final concentration of 10^{-5} M to the appropriate cultures.

6. Twenty-four hours later, all cultures should be illuminated for 90 min by placing the open bottom culture tube rack directly on or 3–5 cm above an inverted desk lamp containing two fluorescent lamps (Westinghouse Cool-Ray, Westinghouse Electric Corp., Bloomfield, NJ).

[6] D. C. Zoschke and F. H. Bach, *Science* **170,** 1404 (1970).
[7] D. C. Zoschke and F. H. Bach, *Science,* **172,** 1350 (1971).
[8] D. C. Zoschke and F. H. Bach, *J. Immunol. Methods* **1,** 55 (1971).

7. Remove excess BUdR by centrifugation followed by decantation of the culture medium.

8. Reculture cells (1×10^5/well) in microtiter plates in fresh medium containing either no antigen, the homologous antigen, or the heterologous antigen.

9. Continue incubation and assay [^{3}H]TdR incorporation either on day 7 or day 10. The day 7 assay demonstrates that the cell cultures were responding to the initial stimulants and that treatment with BUdR and light eliminated this response. Cultures that initially did not receive an initial stimulant but underwent BUdR and light treatment should respond well to either antigen when assayed on day 10. All cultures that initially received antigen A and were then treated with BUdR and light, should no longer respond to antigen A when tested on day 10, but should respond vigorously to antigen B. Similarly, cultures that initially responded to antigen B, followed by BUdR and light treatment, should no longer respond to antigen B on day 10, but should proliferate to antigen A.

Certain other parameters of the test system were also established by Zoschke and Bach.[8] Incubation of cell cultures with high concentrations of BUdR (10^{-3} or 10^{-4} M) eliminated the response without illumination; incubation of cell cultures with lower concentrations of BUdR (10^{-5} or 10^{-6} M) required subsequent exposure to light. Exposure to light for 90 min resulted in the maximal killing effect if the cell cultures had been treated with 10^{-5} M BUdR. Control cultures (no BUdR, light exposure only) were not significantly affected. Of critical importance was the observation that a considerable lag period existed after exposure of the responding population to soluble antigen before the responding cells became sensitive to BUdR and light. Thus, treatment with BUdR and light on days 1–2 of culture had no cytolethal effect. The same treatment on later days of culture, when the responding cells were actively proliferating, was effective in eliminating the response as measured on day 7.

Elimination of a Human T Lymphocyte Population Responsive to an Alloantigenic Stimulus[7,8]

1. Isolate mononuclear cells from peripheral blood as described above.

2. Stimulator populations are inactivated by treatment with 2500 R irradiation or mitomycin C (50 μg/ml for 60 min at 37°, followed by washing (Sigma Chemical Co.).

3. Culture responder cells (2×10^6) with stimulator cells (2×10^7) in 8 ml plastic petri dishes.

4. Add BUdR to a final concentration of 10^{-5} M on day 3 of culture and 24 hr later expose the cells to a fluorescent light source as described above.

5. Following washing, count the cells and resuspend (1×10^5/well) in microtiter plates with fresh stimulating cells (1×10^5/well).

6. Measure [^{3}H]TdR uptake on day 10 of culture.

The initial response of cells from individual A to stimulator populations from individuals B and C should be eliminated by treatment of the cultures with BUdR and light. In those cases in which the cells of A were initially stimulated with cells from individual B, there should be no subsequent restimulation with cells from B; however, restimulation with cells from individual C should result in a significant response. Conversely, initial stimulation of A with C eliminates the response to C, but leaves intact a significant response to B.

Deletion of Antigen-Specific Guinea Pig T Lymphocytes[9,10]

The BUdR and light technique has been used with guinea pig T lymphocytes from primed (2×13)F_1 donors to determine whether individual antigen-specific T lymphocytes displayed a selective capacity to be activated by antigen-pulsed stimulator cells from one or the other parental strain.[10]

1. Immunize animals with ovalbumin (OVA) emulsified in complete Freund's adjuvant (CFA) (Difco Laboratories, Detroit, MI).

2. Fourteen days later, isolate peritoneal exudate lymphocytes (PEL) from the immunized animals by passages over nylon wool columns as described in detail[11] (this volume [28]).

3. Obtain peritoneal exudate cells (PEC) from nonimmune donors which had received injections of 25 ml of sterile mineral oil (Marcol 52; Humble Oil and Refining Co., Houston, TX) 3 days previously. Incubate PEC with either no antigen, OVA (100 μg/ml) or purified protein derivative of tuberculin (PPD 100 μg/ml, Connaught Medical Research Laboratory, Willowdale, Ontario, Canada).

4. Mix PEL (6×10^6) with antigen-pulsed PEC (2×10^6) in 3 ml of medium RPMI-1640 supplemented with penicillin 100 U/ml, streptomycin 100 μg/ml, 2-mercaptoethanol (2ME, 5×10^{-5} M), and 10% normal guinea pig serum and culture in 17×100-mm plastic tubes.

[9] C. A. Janeway and W. E. Paul, *J. Exp. Med.* **144,** 1641 (1976).
[10] W. E. Paul, E. M. Shevach, S. Pickeral, D. W. Thomas, and A. S. Rosenthal, *J. Exp. Med.* **145,** 618 (1977).
[11] D. L. Rosenstreich, J. T. Blake, and A. S. Rosenthal, *J. Exp. Med.* **134,** 1170 (1971).

5. After 48 hr, add BUdR to a final concentration of 3×10^{-6} M and incubate for an additional 24 hr.

6. Illuminate cells with a fluorescent light source for 90 min.

7. Wash cells, count, and reculture in microtiter plates with a variety of stimulants. [³H]TdR incorporation is determined on day 6.

A representative experiment is shown in the table. When $(2 \times 13)F_1$ PEL from donors primed with OVA in CFA were initially cultured with OVA-pulsed strain 2 PEC, OVA-pulsed 13 PEC, PPD-pulsed 2 PEC, or PPD-pulsed 13 PEC, the response of the cells to the antigen-pulsed PEC used for the negative selection with BUdR and light was very low. In contrast, cells negatively selected with any given antigen-pulsed PEC responded well to antigen-pulsed PEC other than those used for the negative selection.

NEGATIVE SELECTION OF OVA AND PPD
IMMUNE $(2 \times 13)F_1$ T LYMPHOCYTES[a]

Stimulus first culture	Second culture [³H]TdR incorporation (cpm $\times 10^{-3}$)		
	Antigen	2 PEC	13 PEC
2-PPD	PPD	*4.1*	59.7
	OVA	125.5	52.6
13-PPD	PPD	91.0	*8.7*
	OVA	141.9	54.4
2-OVA	PPD	86.8	55.9
	OVA	*1.9*	25.4
13-OVA	PPD	95.3	67.7
	OVA	102.7	*4.5*

[a] T lymphocytes from OVA-CFA primed $(2 \times 13)F_1$ animals were cultured with PPD or OVA pulsed strain 2 or strain 13 PEC for 48 hr, at which time BUdR $(3 \times 10^{-6}$ $M)$ was added. After an additional 24 hr, the cultures were exposed to light for 90 min, washed, and recultured with 2 or 13 PEC which had been pulsed with OVA or PPD. Incorporation of [³H]TdR was measured 3 days later; values in italics indicate instances of specific negative selection. Adapted from Paul et al.[10]

Sensitization of T Lymphocytes to Soluble Protein Antigens in
    Association with Allogeneic Ia Antigens

It has been proposed that T lymphocytes do not recognize soluble protein antigens per se, but can only be sensitized to antigen-modified membrane components or the complexes of antigen combined with certain membrane molecules. One prediction of this hypothesis is that homology at the major histocompatibility complex (MHC) is not necessary for effective T lymphocyte macrophage interactions, but that T cells should recognize antigens associated only with the macrophage histocompatibility type used for initial sensitization. A direct test of this hypothesis would be to demonstrate T cell sensitization to antigen-pulsed allogeneic macrophages. However, experiments of this type have been difficult to interpret due to the magnitude of the MLR in such cultures of allogeneic macrophages and T cells. The use of the BUdR and light suicide technique has facilitated the removal of alloreactive T cells which respond in the MLR and has allowed subsequent priming and effective sensitization of T cells to protein antigens associated with allogeneic macrophages.

Generation of Allo-Ia Restricted, Antigen-Specific Guinea Pig T Lymphocytes[12,13]

1. Purify T lymphocytes by passing mesenteric lymph node cells from unprimed guinea pigs over nylon wool adherence columns (this volume [28]).
2. Obtain PEC from mineral oil injected animals and pulse expose them to antigen as described above.
3. Deplete normal T lymphocytes (5×10^6) of alloreactivity by culturing them with nonpulsed allogeneic PEC (1×10^6) in a final volume of 2 ml of complete medium (see above) in 24-well Costar vessels (Costar, Data Packaging, Cambridge, MA) for 3 days.
4. After 48 hr, add BUdR to a final concentration of 10^{-5} M.
5. Twenty-four hours later, illuminate the cultures by exposure to a fluorescent light source, wash, and count.
6. Mix antigen-pulsed allogeneic PEC (1×10^6) with the T lymphocytes (2×10^6) recovered from the BUdR and light treated cultures and maintain in 2-ml Costar vessels for an additional 7 days.
7. Recover the T cells from the priming cultures and culture them (1×10^5) with antigen-pulsed or unpulsed fresh allogenic PEC (1×10^5) in microtiter plates.
8. Determine [³H]TdR uptake after 72 hr.

12 D. W. Thomas and E. M. Shevach, *Proc. Natl. Acad. Sci. U.S.A.* **74,** 2104 (1977).
13 R. B. Clark and E. M. Shevach, *J. Exp. Med.* **155,** 635 (1982).

Generation of Murine Allo-Ia Restricted, Antigen-Specific T Cells[14]

1. Obtain T lymphocytes from normal unprimed spleens, remove red cells by osmotic lysis (this volume [6]), and pass over nylon wool adherence columns (this volume [28]).

2. Isolate macrophages by injecting 4–5 ml of media into killed mice, massaging the abdomen, and then removing the exudate with the same syringe and needle used for injection. Following washing, adhere the exudate cells (1 × 10^7/ml) to a 2-ml petri dish overnight. Nonadherent cells are then decanted and adherent cells (~20% of input) are harvested by the use of a rubber policeman (this volume [27]).

3. Culture T cells (4 × 10^7) with allogeneic PEC (1 × 10^7) in 20 ml of medium (RPMI-1640 supplemented with 5% horse serum, 2 ME, and antibiotics) in 200-ml plastic culture flasks.

4. After 48 hr, add BUdR to a final concentration of 6 × 10^{-5} M and 24 hr later expose the cultures to a fluorescent light source. Note that a higher concentration of BUdR is required to eliminate murine alloreactive T cells compared to the guinea pig system (see above).

5. Culture BUdR and light treated cells (1.5 × 10^7) with fresh allogeneic PEC (5 × 10^6) in 20 ml of medium in the presence of the optimal stimulating concentration of the priming antigen for 3 days.

6. After 3 days, wash cells, adjust to the initial cell concentration, and culture for 4 more days without the further addition of antigen. All cultures should be performed in 200 ml tissue culture flashes with the large surface area down.

7. At the end of 2 days of culture, carefully decant the nonadherent T cells from the flasks. The recovery is usually less than 5% of the input T cells.

8. Culture T cells (1 × 10^5) recovered from the priming culture with allogeneic PEC (1 × 10^5) with or without antigen for 3 days in microtiter plates.

9. Determine [^{3}H]TdR uptake during the last 18 hr of culture.

Elimination of Cytolytic T Lymphocytes with 33258 Hoechst Modified
BUdR and Light Treatment

Mammalian cells that have incorporated BUdR into DNA have been shown[15] to become extremely sensitive to photo-induced killing after treatment with the dye 33258 Hoechst. This method has been applied to enhance the killing of cytolytic T lymphocytes proliferating in secondary

[14] N. Ishii, C. N. Baxevanis, Z. Nagy, and J. Klein, *J. Exp. Med.* **154,** 978 (1981).
[15] G. Stetten, R. L. Davidson, and S. A. Latt, *Exp. Cell Res.* **108,** 447 (1977).

MLR cultures.[16] The use of 33258 Hoechst increased by at least 50 times the effectiveness of BUdR and light in reducing cytolytic activity generated in secondary MLR. The inclusion of the dye also allowed the use of lower, slightly less toxic doses of BUdR and a relatively short (30 min) light exposure. The modified BUdR and light protocol did not inhibit the cytolytic mechanism per se because it had no effect on cytolytic activity during the first 24 hr of culture.

1. Obtain spleen cells from animals that had been primed 2–6 months previously by ip injection of allogeneic cells (3×10^7).

2. Culture primed spleen cells (3×10^7) with irradiated (2000 R) allogeneic spleen cells (10^7) in 10 ml Dulbecco's modified medium supplemented with antibiotics, 10 mM HEPES, 5×10^{-5} M 2ME, and 5% FCS.

3. Maintain cultures in 25-cm^2 tissue culture flasks for 48 hr at 37°, 5% CO_2.

4. Add BUdR to a final concentration of 10^{-5} M.

5. Twenty-four hours later, add 33258 Hoechst (5 μg/ml, Bisbenzimid H 33258, Riedel-De-Haen AG, Seelze-Hannover, West Germany). Incubate for 2–3 hr at 37°.

6. Expose cultures to a fluorescent light source for 30–60 min, wash, count, and then continue culture for 24 hr.

7. Assay cytolytic activity of cells isolated from these cultures on ^{51}Cr-labeled targets.

Conclusions

Although the BUdR and light technique has been successfully used to conclusively demonstrate that T lymphocytes that respond to soluble protein antigens and foreign alloantigens are not totipotential, it is important to recognize that treatment with BUdR without exposure to light may have certain deleterious effects on cultured cells.[17] For example, BUdR may interfere with specific developmental programs and may prevent the subsequent appearance of functional effector cells. Alternatively, the formation of certain macromolecules associated with the phenotype of the fully differentiated state (for example, immunoglobulin) may be altered in the absence of overt killing of the cells.

Another difficulty encountered with BUdR and light treatment is the lack of an efficient colony assay to directly determine the effectiveness of the procedure in the elimination of a specific population of immunologi-

---

[16] J. L. Maryanski, J. C. Cerottini, and K. T. Brunner, *J. Immunol.* **124,** 839 (1980).

[17] D. Levitt and A. Dorfman, *Curr. Top. Dev. Biol.* **8,** 103 (1974).

cally committed cells. A limiting dilution analysis has recently[16] been used to address this question and has shown that the modified BUdR and light treatment with 33258 Hoechst resulted in a 98–99% effective depletion of cytotoxic T lymphocyte precursor cells present in an *in vivo* primed secondary MLR culture. On the other hand, it should be pointed out that complete depletion of a responding cell population may not be required in order to analyze the specificity of certain T lymphocyte responses.[6–10]

[24] Collection of Mouse Thoracic Duct Lymphocytes

By Robert Korngold and Jack R. Bennink

Introduction

A procedure for collecting thoracic duct lymph (TDL) from mice was first described by Shrewsbury[1] whose studies were carried out to determine lymph flow and lymphocyte output. Since those investigations others have made various modifications to the basic technique[2–4] and have used it for a wide range of experimental purposes. These applications have primarily included analysis of lymphocyte recirculation[5] and the obtaining of T lymphocyte populations which have been positively or negatively selected for specific antigens.[6–9] In this chapter we will describe in detail the basic cannulation technique for mice as well give a specific example of its usefulness for immunological studies.

Materials and Reagents

Collection Fluid. Lymph is collected in 15-ml Falcon tubes (Becton Dickinson, A. M. Thomas, Philadelphia, Pennsylvania) in 2.0 ml Dulbecco's phosphate-buffered saline (PBS) solution with 10% fetal bovine se-

[1] M. M. Shrewsbury, *Proc. Soc. Exp. Biol. Med.* **99,** 53 (1958).
[2] B. M. Gesner and J. L. Gowans, *Br. J. Exp. Pathol.* **43,** 424 (1962).
[3] J. L. Boak and M. F. A. Woodruff, *Nature (London)* **205,** 396 (1965).
[4] J. F. A. P. Miller and G. F. Mitchell, *J. Exp. Med.* **128,** 801 (1968).
[5] J. Sprent, *Cell. Immunol.* **7,** 10 (1973).
[6] J. Sprent and H. von Boehmer, *J. Exp. Med.* **144,** 617 (1976).
[7] D. B. Wilson, K. Fischer-Lindahl, D. H. Wilson, and J. Sprent, *J. Exp. Med.* **146,** 361 (1977).
[8] J. Sprent, *J. Exp. Med.* **147,** 1159 (1978).
[9] P. C. Doherty and J. R. Bennink, *J. Exp. Med.* **149,** 150 (1979).

rum (FBS) and 5 U/ml preservative-free heparin (Lipo-Hepin aqueous solution, Riker Laboratories, Northridge, California) on ice.

Infusion Solution. Mice are infused intravenously (iv) with balanced salt solution, Dulbecco's solution or normal saline at 0.5–1.0 ml/hr. This is important to minimize clot formation and optimize cell yields. Heparin can also be added at a concentration of 0.5 U/ml.

Avertin. The anesthesia of choice is "avertin" (tribromoethanol). Stock solutions of "avertin" are made by dissolving 10 g tribromethanol in 10 ml of *t*-amyl alcohol (Aldrich Chemical Co., Milwaukee, Wisconsin). This solution is stored at 4° in a light sealed container. A 1 : 50 dilution in normal saline is made from this stock and injected intraperitonally at approximately 0.015 ml/g body weight for anesthesia.

Oil. Mice can be given 0.2 ml olive oil orally shortly before cannulating. This is unnecessary but when learning the procedure enables one to visualize the duct with significantly greater ease, as the lymph will appear a milky white color.

Cannulas. Cannulas for thoracic duct insertion are made from PE-50 intramedic tubing (Clay Adams, A. H. Thomas, Philadelphia, Pennsylvania). The tubing is cut in 30- to 40-cm pieces. One end of each piece is bent and dipped in boiling water. The bend is finger formed so that its diameter is about 0.5 cm with the resultant short end of the cannula being 3–5 cm and parallel with the long end (25–35 cm). The short end of this cannula is cut with the tip of a scalpel blade just after the short end becomes parallel with the long end. It is best to cut the cannula tip twice so that it is "arrow-shaped."

Tail vein cannulas are made using PE-20 intramedic tubing (Clay Adams, A. H. Thomas, Philadelphia, Pennsylvania) and a 26-gauge needle broken off and inserted blunt end first into the PE-20 tubing. This needle is used to insert the cannula into the tail vein. Once in the vein the tail cannula is taped firmly in place. A saline-filled syringe with 26-gauge needle (Becton Dickinson, A. H. Thomas, Philadelphia, Pennsylvania) is used to infuse the mouse.

Equipment. The cannulation operation is facilitated by a good adjustable lamp. It is also worth noting that the operation is made easier by using a dissecting microscope or magnifying lenses but is in no way solely dependent upon such equipment. For overnight or longer collection, the tail vein cannula are connected to a peristaltic pump (Brinkman or Ismatic). A collection apparatus for suspending the mice is very important (Fig. 1). The ability of the mouse to run on the wheels greatly enhances lymph flow.

Cannulation Procedure. The left flank of an anesthetized mouse is shaved and the mouse is placed left flank up, head to the left of the

FIG. 1. Collection apparatus for suspending mice.

operator, on a styrofoam or cork dissection board. After swabbing with 70% ethanol, the skin is lifted with forceps or the index finger and thumb of the left hand and a thin piece of skin is cut away from the thigh to the ribs. This excision should leave an opening 2.0 cm long by 2.0 cm wide. Using scissors and forceps, a 1.5–2.0 cm incision is made in the abdominal muscle parallel with and approximately 0.75 cm posterior to the rib cage. Four retractors made from large paper clips are placed in the incision and pinned to the dissecting board using 18-gauge needles. The abdominal cavity should be exposed through a hole approximately 2.0 × 1.5 cm. A cotton tip swab (A. H. Thomas, Philadelphia, Pennsylvania) is taken in each hand. While using the swab in the left hand the kidney is gently drawn ventrally and the cotton swab in the right hand is placed behind the kidney and slowly twirled counterclockwise to dissect the fat holding the kidney from the lumbar muscles. The cysterna chyla (thoracic duct) should now be visible. If given oil, it will appear as a milky white color between the inferior vena cava and the lumbar muscles. A hole is made in the posterior abdominal muscles using a pair of extrafine microdissecting forceps (Clay Adams #6433, A. H. Thomas, Philadelphia, Pennsylvania). The long end of the cannula is inserted and drawn through the hole from the left. The short end with the beveled cut should now be over the duct. The cannula is filled with saline using a syringe and 26-gauge needle so that there is no air in the tubule. The cannula is placed in a position (horizontal) so that the saline does not flow out of either end. While using a cotton swab in the left hand to expose the duct, a small hole is made in the duct using extrafine microdissecting forceps held in the right hand. This is done by poking the closed tip forceps through the wall of the duct and then carefully spreading the tips. The hole produced should be about 1 mm in diameter.

The curved end of the cannula is grasped with the extrafine microdissecting forceps and the tip is placed into the hole in the duct. The lymph must enter freely into the cannula by lowering the long arm slightly below the horizontal. The cannula is placed back in the horizontal position and a tiny drop of Eastman 910 adhesive (VWR Scientific, Philadelphia, Pennsylvania) is placed over the area where the cannula enters the duct. At least 1 min should be allowed for the adhesive to set and then the system must be tested again to ensure free lymph flow. The retractors are removed and the wound is clipped with four or five autoclips (Clay Adams, A. H. Thomas, Philadelphia, Pennsylvania) through the skin. The abdominal muscles do not need to be sewn unless one plans to collect lymph over a prolonged period of time. Tape is then placed around the mouses' abdomen so it can be suspended from the crossbar of the collection rack. Care must be taken to let the mouse rest lightly on the wheel to allow mobility, while at the same time the tape must be tight enough to prevent escape. Food and water are provided *ad libitum* once the mice are fully awake. Cannulas must be checked constantly for the first 3–4 hr after cannulation and clots can be removed using a horse hair or nylon thread with a gentle twirling motion. A desirable lymph flow is indicated if the lymph flows back freely when the cannula is raised.

Experimental Application of Thoracic Duct Cannulation. The obvious use for thoracic duct cannulation is to study the blood to lymph recirculation of lymphocytes. Recirculation can be analyzed in at least two ways. In the first technique, variations in cell recirculation can be studied simply by determining the cell output in relation to the duration of collection and volume of the lymph.[1,2] The cell output for normal mice varies with the strain but is generally between 1 and 3 $\times$ 10^6/hr. This population is composed of approximately 65% T lymphocytes of which 30–50% are Lyt-2 positive. A second, more analytical method utilizes the tracer sample principle described by Gowans and Knight.[10] Generally, its adaptation for recirculation studies involves the labeling of a specific lymphoid cell population with radioactive sodium chromate and subsequent injection of these cells intravenously into the animal.[5,11] Radioactivity of the lymph and other organs is determined at various time points and analyzed in relation to the total number of cells and counts originally injected.

Aside from these straightforward recirculation studies, the technique of thoracic duct cannulation has been primarily used for experiments where depletion of specific antigen-reactive T lymphocytes is required. In the case of alloreactivity to the major histocompatibility complex (MHC)

---

[10] J. L. Gowans and E. J. Knight, *Proc. R. Soc. London, Ser. B* **159**, 257 (1964).
[11] W. L. Ford and M. E. Smith, *Adv. Exp. Med. Biol.* **149**, 139 (1982).

determinants, T cells injected iv into lethally irradiated (850–950 rad) MHC-incompatible rats[12] or mice[6] results in a sequence of negative and positive selection for the ability of donor origin TDL to recognize the alloantigen of the host. During the first 48 hr, after donor cell transfer, the specifically reactive T cells become sequestered in the lymphoid tissues where they encounter antigen and begin to proliferate as a consequence of stimulation. Therefore at this state of negative selection, the TDL is devoid of these host-reactive T cells. After another day, the reactive cells begin to reenter the circulation in ever-increasing numbers (blast cells) and reach a peak at about 5–6 days (positive selection stage).

As an example of this type of application, we have used the cannulation techniques to study whether T cells known to cause graft-versus-host disease (GVHD) across minor histocompatibility (H) barriers in mice required H-2-restricted recognition of these determinants.[13]

Acknowledgments

We are grateful to Dr. Jon Sprent for his efforts in teaching us the mouse cannulation procedures. This work was supported by Grants NS 11036 and AI 14162 from the National Institutes of Health.

[12] W. L. Ford and R. C. Atkins, *Nature (London), New Biol.* **243**, 178 (1971).
[13] R. Korngold and J. Sprent, *J. Exp. Med.* **151**, 1114 (1980).

[25] Methods for the Collection of Peritoneal and Alveolar Macrophages

By RICHARD M. MCCARRON, DIANA K. GOROFF, JORDAN E. LUHR, MARY A. MURPHY, and HERBERT B. HERSCOWITZ

Although the importance of macrophages in the immune response has been well established, their precise regulatory role has only recently begun to be elucidated. The procedures used to collect macrophages are, therefore, important if we are to increase our understanding of the mechanisms by which these cells can modulate the immune response.

In this chapter, techniques for the collection of macrophages from the lungs and peritoneal cavity of rabbits and mice are described. These basic techniques may be modified, with little effort, to obtain similar cell populations from other species. The collection of alveolar macrophages from rats,[1] dogs,[2] guinea pigs,[3] and humans,[4] as well as the collection of peritoneal macrophages from rats[5] and guinea pigs[6] have previously been described.

In order to increase the yield of macrophages obtained from a single animal, an eliciting agent (e.g., thioglycolate medium) may be injected into the animal prior to the collection of cells. The use of eliciting agents will also induce a change in the properties of the collected population (termed elicited) as compared to the cells obtained from an untreated animal (resident population). The techniques for obtaining an elicited population as well as some of the considerations involved in the use of these agents will be discussed in this chapter.

Collection of Peritoneal Cells

Materials

Equipment and Supplies

Animal clipper (Model 10, Oster; Milwaukee, WI) fitted with a No. 10 blade
Rabbit restraining board
Small animal dissecting board
Sterile surgical instruments: scalpel, scissors, toothed-forceps and hemostats
Sterile needles (18, 20, and 25 gauge)
Sterile syringes (5, 10, 30, and 50 ml volumes)
Sterile disposable centrifuge tubes (15 and 50 ml)
Sterile Pasteur pipets
70% alcohol
Suction apparatus, consisting of a sterile collection flask (500-ml Erlenmeyer) connected in series to a trap flask (250-ml Erlenmeyer) which is attached to a vacuum source (e.g., water aspirator or vacuum pump).

[1] Y. Kirkawa and R. Roneda, Lab. Invest. **30**, 76 (1974).
[2] M. J. Ansfield, H. B. Kaltreider, J. L. Caldwell, and F. N. Herskowitz, J. Immunol. **122**, 542 (1979).
[3] R. Oren, A. E. Farnham, R. Saito, E. Milofsky, and M. L. Ranowsky, J. Cell. Biol. **17**, 487 (1963).
[4] R. P. Daniele and J. H. Dauber, in "Manual of Macrophage Methodology" (H. B. Herscowitz, H. T. Holden, J. A. Bellanti, and A. Ghaffar, eds.), p. 23. Dekker, New York, 1981.
[5] J. D. Feldman and E. M. Pollock, J. Immunol. **113**, 329 (1974).
[6] J. D. Feldman, D. G. Tubergen, E. M. Pollock, and E. R. Unanue, Cell. Immunol. **5**, 325 (1972).

Collecting apparatus, consisting of a glass tube (o.d. 7 mm) 10–12 cm in length attached to rubber tubing connected to the suction apparatus. The glass collecting tube is surrounded by a 16 × 100-mm test tube with numerous perforations at the lower end (conveniently made from a cellulose nitrate or polypropylene tube which is perforated with a hot 16-gauge needle). The outer tube which prevents clogging of the collecting tube during aspiration of lavage fluid, is not autoclavable and should be gas-sterilized.

Ice bucket

Ice

Reagents

Brewer's thioglycolate medium (Cat. #0236-01, Difco Laboratories, Detroit, MI), prepared according to manufacturer's instructions, sterilized by autoclave, and stored in the dark at room temperature for one week prior to use.

Trypan blue stain [Grand Island Biological Company, Grand Island, N.Y. (GIBCO)], 0.04% in 0.15 M NaCl.

Collecting fluid: sterile phosphate-buffered saline without calcium and magnesium (PBS), pH 7.3, containing 10 U/ml of heparin (preservative-free; O'Neal, Jones and Feldman, St. Louis, MO).

Lethal solution (Diabutol; Diamond Laboratories, Des Moines, IO).

Tissue culture medium—the type to be used will be dependent upon the experiment to be performed [e.g., Eagle's minimum essential medium (Microbiological Associates; Walkerville, MD) or RPMI-1640 (GIBCO)].

Procedure

Injection of Eliciting Agent

Mice. Three to six days before harvesting the peritoneal cells, mice are given an intraperitoneal injection with the appropriate agent (e.g., thioglycolate medium) as follows.

1. Immobilize the mouse and dampen the abdominal area with 70% alcohol.

2. Using a 5/8-in., 25-gauge needle, inject 1–3 ml of thioglycolate medium into the peritoneal cavity at a 45° angle. (Note: aspirate the needle to ensure that it has not penetrated the viscera or a blood vessel.)

Rabbits. Three to five days before collection of the cells is desired, rabbits are injected with the appropriate agent (e.g., thioglycolate medium) as follows.

1. Using aseptic technique, fill each of four 50-ml syringes to capacity with thioglycolate medium and cover the tips to maintain sterility.

2. Restrain the rabbit on a board, ventral side up, by tying each of the four extremities.

3. Shave the fur from the lower abdominal region with the animal clipper and wet the area with 70% alcohol.

4. Attach a sterile 18-gauge needle to one of the syringes and insert the needle into the peritoneal cavity at a 45° angle.

5. Aspirate the syringe to ensure that the needle has not penetrated the viscera or a blood vessel. Inject the entire contents of the syringe and, with the needle still in the peritoneal cavity, replace the empty syringe with one of the remaining filled syringes and repeat.

Recovery of Cells from the Peritoneal Cavity

Mice

1. Sacrifice the mouse by rapid cervical dislocation and place the mouse, ventral side up, on the dissecting board; dampen the anterior abdominal region with 70% alcohol.

2. Retract the skin of the anterior abdominal region with forceps, and with sterile scissors make a longitudinal incision along the midline (inguinal region) (approximately 1/2 in.).

3. With thumb and forefinger (or toothed-forceps), pinch the fur above and below the incision and gently pull apart until the abdominal wall is exposed from the neck to the pelvic girdle (cautiously avoiding contact with the exposed abdominal wall). Wash the exposed abdomen with 70% alcohol to remove any loose fur.

4. Using a 20-gauge needle, carefully inject 3–5 ml of collecting fluid into the side of the abdominal wall, cautiously avoiding puncture of the viscera. (Note: leave approximately 0.2 ml of collecting fluid in the syringe to alleviate clogging of the needle which may occur during harvest.)

5. Once the peritoneal cavity becomes distended, gently massage (with the needle in place) with thumb and forefinger to ensure maximum suspension of cells (avoid punctures with the needle).

6. Using the needle, apply slight lateral pressure to lift the abdominal wall away from the mesentery to form a "pocket" of fluid.

7. Aspirate the fluid gently to avoid clogging the needle by viscera.

8. Upon completion, remove the needle from the syringe and transfer the solution to a sterile tube kept on ice.

Rabbits

1. Sacrifice the rabbit by intravenous injection of 5 ml of Diabutol or 20 ml of air.

2. Place the rabbit, ventral side up, on the restraining board, shave the entire abdominal area, and wet with 70% alcohol.

3. Open the peritoneal cavity by a midline incision approximately 3 in. long through the skin. Cut back skin flaps at right angles to the initial incision so that the underlying muscle layer is exposed.

4. Using a forceps, raise the muscle layer away from the abdominal viscera and make a midline incision about 2-1/2 in. long through the muscle layer.

5. Clamp hemostats to the opposing edges of the incision and lift to form a pocket.

6. Add 50–75 ml of the collecting fluid to the peritoneal cavity and while lifting, hold the hemostats together to close the opening and vigorously massage the abdominal area.

7. Insert the collection apparatus (described in the Materials section) into the peritoneal cavity and remove the fluid by suction into the collection flask which should be kept on ice.

8. Washing of the peritoneal cavity should be repeated as described above until a total of 250 ml of collecting fluid has been used. About 85–95% of the collecting fluid should be recovered.

9. Transfer the contents of the collection flask into sterile 50-ml tubes kept on ice.

Collection of Alveolar Macrophages

Materials

The materials are the same as described in the previous section with the following additions:
    Aluminum foil
    Gauze pads
    Suture 4-0 silk

Lavage Apparatus for Mice (Fig. 1)

A sterile 3-way stopcock (American Pharmaseal Inc.; Toa Alto, Puerto Rico) is fitted with a sterile 10-ml syringe which is in a vertical position and a 5-ml syringe which is in a horizontal position. The remaining outlet of the stopcock (opposite the 10-ml syringe) is fitted with a sterile 20-gauge needle which is attached to a 5-in. piece of gas-sterilized polyethylene tubing (Clay Adams; Parsippany, NJ) (o.d. 0.05 in.). The two syringes are clamped to a ringstand and the height is adjusted so that 1 in. of the tubing lies on the dissecting board. (Note: the bevel of the 20-gauge needle may be cut off in order to decrease the risk of damaging the

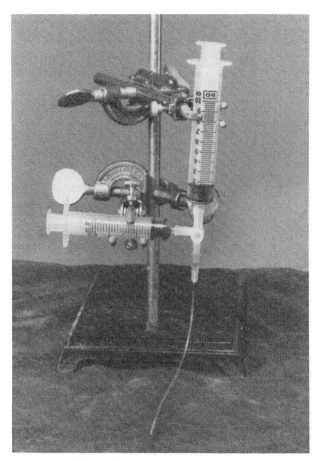

FIG. 1. Lavage apparatus for mice.

polyethylene tubing. A bevel cut at the end of the polyethylene tubing will facilitate its entry into the trachea.)

Lavage Apparatus for Rabbits (Fig. 2)

A sterile 30-ml syringe is attached to the middle outlet of a sterile 3-way stopcock (American Pharmaseal, Inc.) and is clamped in a vertical position to a ringstand. A solution administration set (cat. no. 2C0002, Travenol Laboratories; Deerfield, IL) and an anesthesia extension set (cat. no. 2C0050, Travenol Laboratories) are then attached to the remaining outlets of the stopcock. The connector at the drip chamber end of the intravenous tubing and a 5-in. length of glass tubing are inserted into a two-holed rubber stopper which is then fitted to a 250-ml Erlenmeyer flask

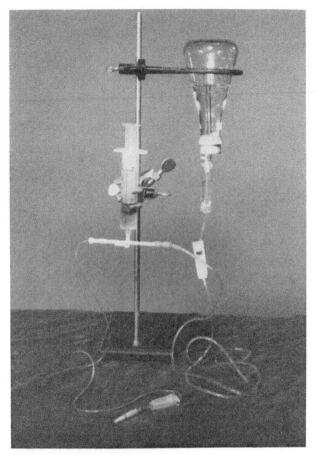

FIG. 2. Lavage apparatus for rabbits.

filled with PBS. The flask is then inverted in a ring on a stand. (Note: the intravenous tubing is gas sterilized and the flask and stopper are autoclaved before use.)

Procedure

Recovery of Alveolar Cells

Mice

1. Sacrifice the mouse by rapid cervical dislocation and place the mouse ventral side up on the dissecting board; dampen the anterior neck and thorax with 70% alcohol.

2. Retract the skin and cut the underlying tissue from the top of the rib cage up to the jaw. Then carefully remove the underlying tissue taking care not to sever arteries or veins in proximity to the trachea. Once the trachea is exposed, the sheath surrounding the trachea should be cut away.

3. Place a 20-gauge needle behind the trachea at a right angle to add stability. Lift up *gently* on the needle, pulling the trachea taut, and make a small lateral incision between two of the cartilage rings in the trachea. While still holding the support needle, place the lavage tubing into the incision and slide it down the trachea until resistance is met. Withdraw the tubing slightly and gently release the tubing and 20-gauge needle.

4. Fill the 10-ml syringe with warm (37°) PBS and fit it back onto the 3-way stopcock. Direct the flow through the stopcock from the 10-ml syringe to the lavage tubing. Fill the lungs with approximately 0.5 ml of PBS and turn the valve of the stopcock so that flow is now between the 5-ml syringe and the lavage tubing. Massage the chest and then gently withdraw the plunger of the syringe until the flow stops. Turn the valve so that the flow is in the original direction. Repeat this process until the lungs have been lavaged with 5.0 ml of PBS.

5. After the process is complete, empty the contents of the 5-ml syringe into a 50-ml conical tube which is kept on ice.

Rabbits

1. Sacrifice the rabbit by an intravenous injection of 20 ml of air.

2. The rabbit is tied to a restraining board, in the supine position, and the anterior neck and thorax are shaved and cleansed with 70% alcohol.

3. A midline incision is made in the neck to expose the trachea which is then clamped with a hemostat.

4. An incision is made in the chest wall along the sternum to the manubrium and the skin of the chest is flapped back.

5. The diaphragm is cut open and the rib cage cut back on one side to expose the lungs.

6. The remaining connective tissue holding the lungs in place is carefully dissected and the lungs are gently removed from the chest cavity with the trachea and heart attached.

7. The dissected lungs are washed with sterile PBS and placed on a piece of aluminum foil which is covered with sterile gauze pads moistened with PBS.

8. A small lateral incision is made between two of the cartilage rings in the trachea and the free end of the lavage tubing is inserted and tied in place with a silk suture.

9. The syringe is filled with 10–15 ml of PBS which is slowly instilled into the lungs. The lungs are massaged gently and the lavage fluid is withdrawn into the syringe and collected into 50-ml centrifuge tubes which are kept on ice. This procedure is repeated until all of the PBS is used.

Comments

Cell Yield

The collected cells are washed 2–3× in PBS by centrifugation (225 g, 10 min) and resuspended in appropriate solution (buffer or media) by gentle agitation. If cell preparations are contaminated with red blood cells, the latter may be lysed by either hypotonic shock or with Tris-buffered ammonium chloride.

The average yield of resident cells from the peritoneal cavity of mice is $2–5 \times 10^6$ cells of which 65–85% are macrophages. Depending upon the eliciting agent used, the cell yield may be $1–4 \times 10^7$ cells of which 75–95% are macrophages. In an unstimulated rabbit, the peritoneal cavity contains $1–5 \times 10^7$ cells of which 85–95% are macrophages. The elicited population, in rabbits, ranges from 5 to 25×10^8 cells of which 75–95% are macrophages.

The average yield of alveolar cells from the mouse is $2–3 \times 10^5$ cells of which 80–95% are macrophages. The average yield of rabbit alveolar cells is $1–5 \times 10^7$ of which 90–95% are macrophages.

If properly handled, all cultures should be >90% viable as determined by trypan blue exclusion.

Eliciting Agents

The number of macrophages in the peritoneal cavity differs among the various species[7] and, in some, is rather low. To increase the yield of macrophages a sterile peritoneal exudate can be induced by the intraperitoneal injection of an eliciting agent. It should be understood that the macrophages obtained from a stimulated animal differ from those found in a resident population in that many of the cells are recently immigrated from the circulation and are likely to be morphologically, biochemically, and functionally different. After injection of an eliciting agent both the resident and the new cells in the exudate may be activated with respect to their phagocytic and bactericidal capabilities. Furthermore, differences in function can also be associated with differences in the eliciting agent used

---

[7] J. Padawer and A. S. Gordon, *Anat. Rec.* **124**, 209 (1956).

TABLE I

AGENTS USED TO INCREASE MOUSE PERITONEAL MACROPHAGE (Mϕ) YIELD[a]

Eliciting agent	Volume[b] (ml)	Time[c] (days)	Relative Mϕ increase	Drawbacks
Horse serum	0.5	2–3	Moderate	Early exudates rich in PMN and lymphocytes
Glycogen	0.5	3–5	Moderate	Early exudate contains PMN
Proteose peptone	1–1.5	3–5	High	Mϕ are "sticky" and tend to clump
Thioglycolate	1–3	3–6	High	Mϕ tend to clump
Mineral oil	0.1–0.5	4–6	Very high	Must separate aqueous and oil phases; cells contain oil droplets
Peptone	0.1–0.5	5	Moderate	Mϕ may be "sticky"

[a] Adapted from A. E. Stuart, J. A. Habeshaw, and A. E. Davidson, in "Handbook of Experimental Immunology" (D. W. Weir, ed.), 3rd ed., p. 312. Blackwell, Oxford, 1978.

[b] Volume of eliciting agent injected intraperitoneally per mouse.

[c] Time between the injection of the eliciting agent and the harvesting of the cells.

in the study. Caution should be exercised to ensure removal of the eliciting agent from the harvested cell population. Several eliciting agents currently in use are not digestible and therefore remain in the macrophages. This is particularly true of mineral oil where the macrophages can be shown to contain oil-laden vacuoles which could possibly interfere with studies measuring phagocytic capabilities.

The time of harvesting after injection of an eliciting agent is also a critical factor. For example, it has been shown that the injection of thioglycolate medium 3 days prior to harvesting of rabbit peritoneal cells resulted in a 50-fold increase in recovered cells with 75% of the cells having characteristics of macrophages and the remainder of the population consisting of 20% lymphocytes and 5% granulocytes.[8] On the other hand, if the peritoneal exudate cells were harvested 5 days after injection of thioglycolate, the total cell recovery was only 30-fold enhanced with 80% of the cells having characteristics of macrophages and a shift in the remaining population to only 4% lymphocytes and 17% granulocytes. Examples of some of the currently used eliciting agents and their properties are listed in Tables I and II.

[8] R. E. Conrad, L. C. Yang, and H. B. Herscowitz, J. Reticuloendothel. Soc. 21, 103 (1977).

TABLE II
AGENTS USED TO INCREASE RABBIT PERITONEAL
MACROPHAGE (Mφ) YIELD[a]

Eliciting agent	Volume[b] (ml)	Time[c] (days)	Relative Mφ increase
Thioglycolate	200	3–5	Very high
Mineral oil	50	3–5	High
Glycogen	50	3–5	Moderate

[a] Adapted from Conrad et al.[8] Drawbacks for each agent are the same as described in Table I.
[b] Volume of eliciting agent injected intraperitoneally per rabbit.
[c] Time between the injection of the eliciting agent and the harvesting of the cells.

Induction of a granulotomous lesion by an intravenous injection of 150 μg of BCG (heat-inactivated bacillus Calmette-Guérin) suspension[9] or Fruend's adjuvant[10] 3–4 weeks before sacrifice will greatly increase the yield of alveolar macrophages from rabbits. A relatively moderate increase in mouse alveolar macrophage yield has been shown following an intravenous injection of 250 μg BCG, 96–120 hr before sacrifice.[11]

Collecting Fluid

An increase in the recovery of alveolar macrophages has been demonstrated when the collecting fluid contains EDTA[12] or is devoid of calcium and magnesium ions.[13] The addition of lidocaine to the collecting fluid has also been shown to increase the yield of alveolar cells.[14]

Attempts to increase the recovery of alveolar cells by increasing the number of washes or the volume used for each wash should be pursued cautiously, since such procedures may create leaks in the lung. These leaks will result in the decreased recovery of collecting fluid and a loss of alveolar cells. On the other hand, the use of smaller volumes for lavage may result in unexpanded areas of the lung leading to a decrease in the number of cells recovered from those areas.

[9] J. N. Myrvik, E. S. Leak, and S. Oshima, *J. Immunol.* **89**, 745 (1962).
[10] B. M. Gesner and J. G. Howard, *in* "Handbook of Experimental Immunology" (D. W. Weir, ed.), 3rd ed., p. 1009. Blackwell, Oxford, 1978.
[11] A. Blussé van Oud Alblas, B. van der Linden-Schrever, and R. van Furth, *J. Exp. Med.* **154**, 235 (1981).
[12] E. R. Pavillard, *Aust. J. Exp. Biol. Med. Sci.* **41**, 265 (1963).
[13] J. D. Brain, *Arch. Intern. Med.* **126**, 477 (1970).
[14] P. G. Holt, *Am. Rev. Respir. Dis.* **118**, 791 (1978).

[26] Separation and Collection of Rat Liver Macrophages

By Justine S. Garvey and Matthew F. Heil

Introduction

The isolation of Kupffer cells in high yield and purity is basically a two-step procedure: the first step is aimed at releasing these cells from the reticulum tissue (substratum) in which they reside; the second step is their isolation to purity by removal of other cells and debris. The first step involves enzyme treatment; this demands a decision concerning whether the parenchymal cells are to be obtained also. If so, the enzyme cannot be Pronase since it destroys the parenchymal cells. Additionally, Pronase destroys membrane receptors, thereby necessitating a period of culture for receptor renewal.[1] It may be assumed that following any enzyme treatment the membrane surface of the treated cell is changed to some degree and the degree of change in the cell receptors requires testing[2] under the precise conditions of the receptor assay. Collagenase is the enzyme of choice for isolation of the parenchymal cells; use of this enzyme does not interfere with the subsequent isolation of the nonparenchymal cells.

The second step in the procedure, the handling of the crude cell isolate, likewise requires an initial decision, this being whether the Kupffer cells are to be isolated by adhesion or by centrifugal elutriation. Adhesion is a simpler and less expensive method than elutriation but the yields are lower, less pure, and less reproducible than those provided by elutriation.

The method detailed here allows all the principal liver cell types to be isolated from one rat. This protocol dates from early work,[3] now improved to combine the recent perfusion[4] and elutriation[5] procedures pioneered in the laboratories of Seglen and of Knook, respectively, and subsequently modified by us to optimize the cell yield and purity for both nonparenchymal and parenchymal cell types.

[1] A. C. Munthe-Kaas, T. Berg, P. O. Seglen, and R. Seljelid, *J. Exp. Med.* **141,** 1 (1975).
[2] J. S. Garvey and T. J. Caperna, *in* "Sinusoidal Liver Cells" (D. L. Knook and E. Wisse, eds.), p. 421. Elsevier/North-Holland Biomedical Press, Amsterdam, 1982.
[3] J. S. Garvey, *Nature (London)* **191,** 972 (1961).
[4] P. O. Seglen, *in* "Methods in Cell Biology" (D. M. Prescott, ed.), p. 29. Academic Press, New York, 1976.
[5] D. L. Knook and E. Ch. Sleyster, *Exp. Cell Res.* **99,** 444 (1976).

METHODS IN ENZYMOLOGY, VOL. 108

Perfusion and Elutriation

Buffers and Media

These are prepared aseptically and in advance for the perfusion and elutriation, both of which are performed aseptically. *All* glassware should be thoroughly siliconized.

Buffer A. HBSS[5a] (Ca^{2+}-free, Mg^{2+}-free) + 0.5 mM EGTA, pH 7.4 (EGTA dissolves upon autoclaving)

Buffer B. HBSS (Ca^{2+}-free, Mg^{2+}-free), pH 7.4 (simply omit EGTA used in Buffer A)

Buffer C. RPMI 1640 + 70 U/ml collagenase Type I (Sigma), pH 7.4

GBSS. (for Metrizamide only)

Unit I (500 ml)		Unit II (500 ml)	
KCl	0.3728 g	$MgSO_4 \cdot 7H_2O$	0.0739 g
$CaCl_2 \cdot 2H_2O$	0.2352 g	$Na_2HPO_4 \cdot 7H_2O$	0.4557 g
$MgCl_2 \cdot 6H_2O$	0.1829 g	KH_2PO_4	0.0272 g
(Unit I can be autoclaved)		$NaHCO_3$	0.2268 g
		Glucose	0.99 g
		Sterile filter this solution; do not autoclave)	

For complete GBSS solution mix equal volumes of Unit I and Unit II; adjust pH to 7.2 with 1 N HCl.

Buffer D. HBSS

Buffer E. 2 × HBSS

Buffer F. L-15 media (GIBCO) + 10 mM HEPES

Buffer G. PBS (Dulbecco's) (Ca^{2+}-free, Mg^{2+}-free)

Buffer H. GBE (elutriation buffer, pH 7.0): NaCl, 16.0 g; KCl, 0.74 g; $Na_2HPO_4 \cdot 7H_2O$, 0.91 g; KH_2PO_4, 0.054 g; glucose, 2.0 g; phenol red, 0.02 g; HEPES, 4.25 g; EDTA, 0.0175 g; solid NaOH, 2 pellets.

Dissolve the above in distilled H_2O and bring volume to 500 ml. Freeze at 0 to −20° until used.

To use: Thaw GBE and add to 1 liter distilled H_2O. Add 500 ml of aqueous 0.3% gelatin (Difco) solution and 5 ml of 10% $NaHCO_3$.

30% Metrizamide. Prepared from centrifugation grade supplied by Accurate Chemical Co

---

[5a] Abbreviations: HBSS, Hanks' balanced salt solution; EGTA, Ethyleneglycol-bis-(β-aminoethyl ether) *N,N'*-tetraacetic acid, MW 380.4; RPMI 1640, Roswell Park Memorial Institute media 1640; GBSS, Gey's balanced salt solution; GBE, elutriation buffer; L-15, Leibowitz media 15; FCS, fetal calf serum; HEPES, *N*-2-hydroxyethylpiperazine-*N'*-2-ethanesulfonic acid, MW 283.3; NPC, nonparenchymal cells.

1. Mix 25 ml Unit I and 25 ml Unit II (from GBSS protocol) and adjust pH to 7.0 with 1 N HCl. Remove 20 ml and set aside.

2. Add 15 g Metrizamide to remaining 30 ml.

3. After Metrizamide is dissolved, adjust the volume to 50 ml using the 20 ml volume set aside in Step 1.

4. Store in 8-ml aliquots at $-20°$.

50% Percoll Gradient. Prepared from stock solution supplied by Pharmacia, Piscataway, NJ

1. Into each of two 50-ml round bottom polyacrylamide Oak Ridge tubes mix 16 ml 2 × HBSS and 16 ml Percoll.

2. Add 100 ml of 1 N HCl to obtain a pH of 7.2–7.4.

3. Invert tubes to mix; then centrifuge them at 31,000 g at 4° for 60 min in a Sorvall SS-34 rotor.

Material and Equipment

1. Surgical equipment including two 22-gauge indwelling catheters and surgical silk Type 00 (sterile except for items 13 through 16).

2. Graduated, capped, conical centrifuge tubes (Bellco #3048-00; 140 ml, 12).

3. Pyrex beakers (250 ml, 4).

4. Pipets (Bellco #1204; 10 ml, 8).

5. Conical centrifuge tubes (15 ml, 4).

6. Round bottom centrifuge tubes (Corex #8446, 25 ml screw cap; 30 ml, 4).

7. Erlenmeyer flasks (250 ml, 2).

8. Aseptic filling bells, 2.

9. Plastic disposable pipets (1 ml, 10 ml)

10. Polycarbonate Oak Ridge tubes (Fisher #05-529c; 50 ml, 2).

11. Media reservoir flask, stoppered and equipped with entry port for carbogen gas and exit port for media and buffers. (A spinner flask is useful if accommodated with a rubber stopper with 3 holes, one to accommodate an inlet tube from the carbogen tank, another for a tube leading to the perfusion pump, and the third for a funnel through which media are added to the flask; the side arms are stoppered prior to autoclaving of the assembled flask, 100 ml size.)

12. Centrifuge for elutriation (Beckman model J-21C) equipped with a JE-6 rotor plus sterile sample mixing chamber and connecting lines to rotor and from rotor into collecting tubes; see item 2.

13. Centrifuge for nonelutriation steps, e.g., IEC Centra -7R, rotor #216.

14. Masterflex pump, model 7014.20 or equivalent (two, one each for perfusion and elutriation if left assembled for each procedure).

15. Water bath (37°).

16. Carbogen gas (95% O_2, 5% CO_2).

Procedure (See Fig. 1 for Flow Chart of Composite Procedures)

Perfusion and Surgery. All buffers should be kept at 37° and the pH adjusted to 7.4 by bubbling carbogen gas through the buffer.

The rat is secured to a surgical board and anesthetized with anesthesia grade ether. The abdomen is swabbed with 70% EtOH and a midline incision from the sternum to the pubic symphysis is made to expose the body cavity. The intestines are reflected to the side to expose the vena cava and the right kidney. The right renal vein is clamped off (or ligated by suture). A loose ligature is placed around the vena cava superior to the entrance of the renal vein (this will be used to hold the indwelling catheter in place). Another loose ligature is placed around the hepatic portal vein (this will be used to secure a second indwelling catheter should collection of the perfusate be desired). At this time a 22-gauge 2-in. indwelling catheter is inserted into the vena cava and tied in place by tightening the ligature that was previously placed around the vein. The vena cava is then clamped superior to the diaphragm in the thoracic cavity. Buffer A is then perfused through the liver at 6 ml/min (100 ml total). The draining perfusate can be collected as mentioned above by inserting a second catheter into the hepatic portal vein and securing it. Alternatively, the portal vein can be severed and the perfusate allowed to drain into the abdominal cavity. Buffer B (50 ml) is then perfused at 8 ml/min followed by Buffer C (100–150 ml) at a flow rate of 10–12 ml/min.

Following perfusion the liver is excised, placed into a 250-ml beaker with approximately 100 ml of L-15 medium, minced with scissors, and forced through a 250-μm Nylon mesh (Tekco Inc., Elmford, N.Y.). Rinsing of the beaker and Nylon mesh with L-15 medium follows, bringing the volume of the cell suspension to approximately 140 ml.

Gradient Centrifugation of Cells. The cell suspension is then placed into a Bellco 140 ml graduated conical centrifuge tube and cold L-15 medium (4°) is used to rinse the beaker and to increase the cell suspension volume until the tube is filled. The suspension is then centrifuged for 5 min at approximately 25 g at 4°.

After centrifugation the supernatant is removed and collected into a 140-ml conical centrifuge tube. Removal of the supernatant is to a level above the pellet where no hepatocytes are removed. These washings of the cell pellet are repeated until four 140-ml Bellco tubes are filled with the

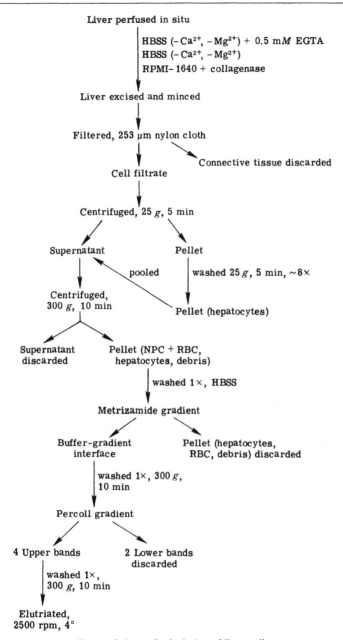

FIG. 1. Scheme for isolation of liver cells.

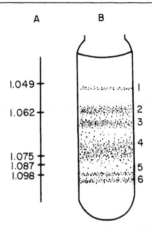

FIG. 2. Banding pattern of nonparenchymal cells on 50% Percoll gradient (see text). Numbers in column A indicate the corresponding density of regions (shown in B) at which various bands form and are designated 1–6. For elutriation of Kupffer cells and endothelial cells the bands 1–4 are pooled. Bands 5–6 are discarded. (Modified from Fig. 1 in Garvey and Caperna.[6])

supernatant. The supernatant containing the nonparenchymal cells is centrifuged at 300 g and 4° for 10 min. The recovered pellets are then pooled, washed 1× in HBSS (Buffer D) (300 g and 4° for 10 min), and are now ready for the Metrizamide gradient separation.

The cells are now resuspended in 10.0 ml HBSS (Buffer D). This suspension is then mixed thoroughly with 15 ml of 30% Metrizamide. Equal volumes of this mixture are carefully transferred to two 25-ml screw cap Corex tubes. A 1.0 ml overlay of HBSS (Buffer D) is carefully added to each tube and then the tubes are centrifuged at 1400 g and 22° for 15 min. (It is important to adhere to the specified times and temperatures for centrifugation of the gradient tubes in order to obtain the expected cell isolations.) After centrifugation the cell layer observed at the buffer–gradient interface is transferred to a 140-ml Bellco tube and the suspension is washed 1× in HBSS (Buffer D) 300 g at 4° for 10 min. The cells are resuspended with 6.0 ml HBSS (Buffer D) and 3.0 ml of this cell suspension is overlaid on each 50% Percoll gradient. The gradients are then centrifuged at 800 g and 4° for 30 min.

Centrifugal Elutriation. Cells are removed from appropriate bands within the gradient (see Fig. 2), pooled, washed 1× in HBSS (Buffer D), resuspended in a 10.0 ml volume of HBSS (Buffer D), and injected slowly (over a 2-min period) into the precooled (4°) mixing chamber.

Fractions are collected in 140-ml Bellco tubes through an aseptic filling

[6] J. S. Garvey and T. J. Caperna, *in* "Manual of Macrophage Methodology" (H. B. Herscowitz, H. T. Holden, J. A. Bellanti, and A. Ghaffar, eds.), p. 31. Dekker, New York, 1981.

bell attached to exit tubing from the elutriator. A filling bell is also attached to the media line entering the mixing chamber from the media reservoir. (Both filling bells may be Bellco #516 but the bell on the reservoir needs an extended stem in order to reach into the media.)

Four successive 140-ml fractions are collected in Bellco tubes by varying the pump flow rate and maintaining the rotor at 2500 rpm. These flow rates are Fraction I, 12.8 ml/min; Fraction II, 20 ml/min (endothelial cells); Fraction III, 25 ml/min; Fraction IV, 42 ml/min (Kupffer cells). Pump settings should be periodically rechecked to assure reproducible flow rates. Cells are recovered by centrifugation of the collected fractions at 300 g and 4° for 10 min.

Cell Characterization

The following procedures are common ones that are used routinely for characterization of nonparenchymal cells: both endothelial and Kupffer cells stain positively with esterase stain; peroxidatic activity is present only in Kupffer cells and phagocytosis is a reliable test for the functional integrity of Kupffer cells. Additional procedures are found in texts devoted to macrophage and monocytic cell procedures.[7,8]

Esterase Stain

Materials

Sorensen's phosphate buffer, 0.067 M, pH 7.4
   (*a*) Na$_2$HPO$_4$, 4.735 g/500 ml (8.942 g if Na$_2$HPO$_4$ · 7 H$_2$O)
   (*b*) KH$_2$PO$_4$, 4.54 g/500 ml (80.4 ml *a* and 19.6 ml *b* are mixed)
2-Methoxyethanol, Eastman No 2381, MW 76.10
NaNO$_2$, 4%
α-Naphthyl butyrate, Sigma N-8000, MW 214.3
Basic fuchsin or pararosaniline-HCl

Procedure for Preparation of Stain (to be prepared fresh immediately before use)

1. Stock basic fuchsin is prepared by adding 1 g basic fuchsin to 20 ml distilled H$_2$O + 5 ml concentrated HCl.

2. Two milliliters of 2-methoxyethanol and 0.04 ml of naphthyl butyrate are mixed and set aside.

[7] D. O. Adams, P. J. Edelson, and H. S. Koren, eds., "Methods for Studying Mononuclear Phagocytes." Academic Press, New York, 1981.
[8] H. B. Herscowitz, H. T. Holden, J. A. Bellanti, and A. Ghaffar, eds., "Manual of Macrophage Methodology." Dekker, New York, 1981.

3. 0.5 ml of 4% $NaNO_3$ and 0.5 ml stock basic fuchsin prepared in (1) above are mixed and allowed to react for 1 min; this mixture and also the solution prepared in (2) are added to 38 ml of Sorensen's buffer, pH 7.4.

4. The stain prepared in (3) is filtered with a qualitative filter and then with a 0.45-μm Millipore filter.

Staining Procedure

1. Slides containing cells prepared by centrifugation in a Cytospin centrifuge, or manually, are stained for 30–45 min in the esterase stain prepared above.

2. Slides are washed once in distilled H_2O.

3. Counterstaining in methyl green (2%; chloroform extracted) is optional.

Peroxidatic Stain

  Materials

  Tris buffer, 0.05 M, pH 7.4 at 37° (pH 7.7 at 25°), 10 ml
  Tris–sucrose, 7% sucrose dissolved in 0.05 M Tris buffer, 5 ml
  H_2O_2, 1/15 dilution of 30% H_2O_2, 50 μl
  Diaminobenzidine (DAB), 2.5 mg
  NaOH, 1 N 1 ml

  Staining Procedure (it is important that the stain be prepared *fresh*
    for each use)

1. Five milliliters of Tris–sucrose solution, 2.5 mg DAB, 50 μl 1/15 dilution of H_2O_2, and 20 μl 1 N NaOH are mixed.

2. To 10^6 cells in 100 μl is added 1 ml of the stain prepared in step 1. (The cells must not be fixed prior to staining.) The cell–stain mixture is incubated for 30 min at 37° in a shaking water bath.

3. The cells are pelleted and resuspended in PBS. A 50-μl sample of cell suspension is placed on a slide, air dried, and fixed for 2 min in 0.02% glutaraldehyde in PBS. Then the slide is rinsed well in double-distilled H_2O, air-dried, and mounted.

4. Counterstaining in 2% methyl green is optional.

Phagocytosis

  Materials

  Latex beads, Sigma, 5 μm
  RPMI 1640
  FCS, heat inactivated.

Procedure

1. A 1% suspension of latex beads is prepared in RPMI 1640 and 25% FCS.

2. The cells are pelleted by centrifugation and resuspended in 1.0 ml of latex bead suspension.

3. The mixture is then incubated in a 37° shaking water bath for 30 min.

4. The cells are resuspended and overlaid on a 2 ml FCS gradient and centrifuged for 3–4 min at 300 g.

5. The top debris is removed and the pellet resuspended (at this time step 4 above can be repeated if necessary). Observe the cells microscopically.

General Comments

The yield of Kupffer cells from male F-344 rats, averaging a body weight of 300 g and used in our studies with the described methodology, was $22.8 \pm 1.3 \times 10^6$ cells with viability of 95% and purity of 85%; the contaminating cell type is primarily endothelial cells. We have found it useful to thoroughly clean the rotor and to store it disassembled between uses even when usage is almost daily. This precautionary measure ensures good preventive maintenance and likewise sterility control. Some of the specified materials may be substituted with perhaps little change in the mentioned parameters; however, we have found the replacement of 40-ml conical centrifuge tubes with 140-ml Bellco tubes not only decreased the number of containers centrifuged but also increased the cell yield considerably (by as much as 58%).

The use of two types of gradients is necessary, one with Metrizamide to remove erythrocytes and cell debris followed by another with Percoll to remove hepatocytes which otherwise are a contaminant in elutriation that cause clumping. Whether or not either of these gradients contributes a negative effect in the subsequent use of the cells should be checked. From published findings the property of adherence may be decreased in cells isolated by Percoll[9] and we have found a high degree of chemiluminescence in Kupffer cells examined in the presence of luminol with an LS 7000 Beckman scintillation counter operated in the out-of coincidence mode (unpublished results) possibly indicating these cells have been activated.

For the reporting of extensive morphological and functional studies as

[9] J. S. Wakefield, J. S. Gale, M. V. Berridge, T. W. Jordan, and H. C. Ford, *Biochem. J.* **202**, 795 (1982).

well as recent efforts aimed at improved cell culture of Kupffer cells, there are two excellent sources that should prove helpful.[10,11]

Acknowledgment

Support to JSG from the National Institutes of Health (Grant AG 00111) is gratefully acknowledged.

[10] E. Wisse and D. L. Knook, eds., "Kupffer Cells and Other Sinusoidal Cells." Elsevier/North-Holland Biomedical Press, Amsterdam, 1977.
[11] D. L. Knook and E. Wisse, eds., "Sinusoidal Liver Cells." Elsevier, Amsterdam, 1982.

[27] Separation of Macrophages on Plastic and Glass Surfaces

By D. E. MOSIER

Introduction

Macrophages are a heterogeneous lineage of cells that appear to derive from bone marrow stem cells via circulating monocyte precursors.[1] Macrophages are widely distributed in both lymphoid and nonlymphoid tissues. They are an essential component in immunity, both directly by phagocytosis and killing of microorganisms and indirectly by providing a series of polypeptide growth factors for T and B lymphocytes.[2] Macrophages are large cells which will attach tenaciously to solid substrates during short periods of tissue culture and it is this property on which the following enrichment procedures are based. It is essential to note that macrophages are not the only cell type that adheres to plastic or glass, so adherence techniques alone cannot yield a 100% pure macrophage population.

Cell Preparation

Macrophages may be obtained from many tissue sources,[3] but the most convenient sources for murine macrophages are spleen or peritoneal

[1] R. van Furth and Z. Cohn, J. Exp. Med. 128, 415 (1968).
[2] S. B. Mizel, Immunol. Rev. 63, 51 (1982).
[3] See, for example, Chapters 1–17, in "Methods for Studying Mononuclear Phagocytes" (D. O. Adams, P. J. Edelson, and H. Koren, eds.). Academic Press, New York, 1981.

exudate cells. Peritoneal exudates are induced by the intraperitoneal injection of 1.0 ml thioglycolate broth 4–7 days before harvesting the cells. Peritoneal cells are collected after injection of 10 ml of Hanks' balanced salt solution (HBSS) containing 5% fetal bovine serum (FBS) and 10 U/ml sodium heparin immediately following cervical dislocation of the mice. The abdominal wall is exposed by a small midline incision (skin *only;* best performed by grasping the skin with toothed forceps and pulling it up and away from the abdominal muscles), the HBSS injected forcefully with a 10-ml syringe fitted with a 20-gauge needle (avoid puncturing the GI tract by using a shallow angle of entry and a high midline injection site), and the fluid slowly withdrawn. Old obese mice should not be used as their omental fat often clogs the needle at this point. The 8–9 ml recovered should be transferred to a sterile test tube and recovered by centrifugation at 300–400 g for 10 min at 4°. Between 20 and 40 $\times$ 10^6 cells per mouse should be obtained.

Spleen cell suspensions are prepared from spleens aseptically removed from normal mice killed by cervical dislocation. Spleens are trimmed free of any adherent fat or pancreatic tissue in a 60-mm Petri dish containing 3 ml cold HBSS. The intact spleens are then transferred to a second sterile dish containing 5 ml cold HBSS and a single cell suspension is made by teasing the spleens (after tearing the capsule) with curved toothed forceps. The cell suspension is transferred to a 15-ml centrifuge tube, allowed to stand for 5 min for large cell clumps to settle, and all but the settled cells transferred to a second centrifuge tube. The cells are collected by centrifugation at 300 g for 10 min at 4°. Red blood cells may be removed at this point by hypotonic lysis if desired. The cell pellet is resuspended in HBSS diluted 1 : 10 in sterile double-distilled, deionized water, vortexed for 10–15 sec. Immediately thereafter, a large excess of HBSS is added to restore isotonicity. Cells are washed in cold HBSS three times before resuspension in complete medium. Cell recovery should be 75–150 $\times$ 10^6 per spleen at this point.

Medium and Substrate Preparation

The adherence of macrophages to solid substrates is an energy-dependent process that is enhanced by high serum concentration.[4] Spleen or peritoneal cells are suspended at a concentration of 5–10 $\times$ 10^6/ml in RPMI 1640 medium supplemented with 20% fetal bovine serum (heat inactivation is unnecessary), 15 mM HEPES, 20 mM L-glutamine, 50 μM 2-mercaptoethanol, and 50 μg/ml gentamicin. Macrophages may be col-

---

[4] M. Rabinovitch and M. J. DeStefano, *Exp. Cell Res.* **77**, 323 (1973).

lected by adherence to plastic, glass, or collagen monolayers.[5] Only the first two methods will be described here.

Sterile 60-mm plastic culture dishes (sterilized by γ-irradiation, not ethylene oxide, e.g., Costar 3060) or *tissue culture washed* glass Petri dishes are preincubated with 5 ml complete medium (see above) for 30 min at 37° in a 5% CO_2–95% air atmosphere. This step precoats the surface with serum proteins and prewarms the dishes to hasten the initiation of adherence.

Adherence and Recovery of Adherent Cells

Three milliliters of cells suspended in RPMI 1640 + 20% FBS are added to each 60-mm dish. No more than 30×10^6 cells should be contained in this volume to avoid cell clumping. The contents of the dish is gently swirled to distribute the cells evenly, and the dishes are incubated for 1 hr at 37° in a humidified 5% CO_2–95% air atmosphere. Macrophages will attach quite firmly to plastic or glass during this incubation step. Nonadherent cells are removed by swirling the dish to resuspend settled or loosely adherent cells, aspirating these cells with a sterile Pasteur pipet, and then rinsing the dishes three times with cold HBSS to remove residual loosely adherent lymphoid cells. This procedure is best performed by tilting each dish about 30° from horizontal and directing a stream of HBSS across the surface of the dish with a Pasteur pipet. It is difficult to be too vigorous at this step. At this point, about 5% of spleen cells and 50–70% of peritoneal exudate cells remain firmly attached to the Petri dishes.

To harvest adherent cells, each dish is flooded with 3 ml of a prewarmed 1 : 5000 dilution of Versene in phosphate-buffered saline (Gibco). Dishes are incubated for 15 min at 37° and the cells are harvested by vigorously pipetting the Versene solution across the plate surface. The resuspended cells, are aspirated, added to a centrifuge tube containing an equal volume of RPMI 1640 + 20% FBS, and the cells are collected by centrifugation. In the meantime, the dishes are inspected with an inverted phase microscope to determine if residual adherent cells are present. If they are, 3 ml additional prewarmed 1 : 5000 Versene is added and the remaining cells are gently detached with a sterile rubber scraper (commonly called a "rubber policeman" for reasons I never understood). These cells may be aspirated, diluted in complete medium, and collected as above. The recovered macrophage-enriched cells are resuspended in RPMI 1640 + 10% FBS and a viable cell count taken.

[5] D. E. Moiser, *in* "Methods for Studying Mononuclear Phagocytes" (D. O. Adams, P. J. Edelson, and H. Koren, eds.), p. 179. Academic Press, New York, 1981.

If further enrichment of macrophages is desired, T lymphocytes (which may comprise 10–15% of recovered adherent cells) may be depleted by treatment with anti-Thy-1.2 antibody and complement. The cells are resuspended in an appropriate dilution of anti-Thy-1.2 antibody (predetermined in a microcytotoxicity assay) in HBSS, incubated 30 min at 4°, centrifuged, the cell pellet resuspended in prescreened rabbit complement (e.g., Low-Tox M, Accurate Sci. Corp.), and incubated at 37° for 45 min with frequent agitation. Finally, the cells are resuspended in RPMI 1640 + 10% FBS and recounted for viable cell recovery.

Evaluation of Macrophage Enrichment (This Volume [26,31])

For some purposes, it may be important to know what fraction of the cell population prepared by adherence is macrophages. In *no* instance will macrophages be 100% of the recovered cells. Several techniques are available for enumerating macrophages. Immunofluorescent staining with monoclonal antibodies specific for macrophages[6] may soon prove to be the most sensitive and reproducible way of assessing macrophage recovery. In the meantime, assays for phagocytosis or lysosomal enzymes are the simplest ways to enumerate macrophages. Cells may be incubated with dilute suspensions of latex beads in complete medium, washed, and the number of phagocytic cells counted using a hemocytometer and a phase microscope. The ingested beads appear as refractile granules within the cytoplasm. Histochemical staining for neutral esterases has been described elsewhere[7] and requires fixation of a small aliquot of the recovered adherent cells.

Final Comments

Macrophages are a heterogeneous group of cells and adherence techniques may not recover all subpopulations in equal proportions to their starting representation. The recovered adherent cells contain cell types other than macrophages in small but appreciable numbers. The metabolism of macrophages released from plastic or glass with Versene may be depressed, and cells should be allowed to recover from the separation procedure for one to several hours before physiological measurements are attempted. Finally, several well-characterized macrophage lines exist[8] and they may be well suited for some studies of macrophage function.

[6] E.g., R. F. Todd, L. M. Nadler, and S. F. Schlossman, *J. Immunol.* **126,** 1435 (1981).
[7] G. A. Miller and P. S. Monahan, *in* "Methods for Studying Mononuclear Phagocytes" (D. O. Adams, P. J. Edelson, and H. Koren, eds.), p. 367. Academic Press, New York, 1981.
[8] P. Ralph, *in* "Methods for Studying Mononuclear Phagocytes" (D. O. Adams, P. J. Edelson, and H. Koren, eds.), p. 155. Academic Press, New York, 1981.

[28] Separation of Lymphoid Cells on Nylon Wool Columns

By David A. Litvin and David L. Rosenstreich

Introduction

A variety of techniques for separating and purifying lymphoid populations have been employed to facilitate study of the cytology of cells and the analysis of immune responses involving distinctive cell types.

In this chapter, we will focus on separation of lymphoid cells on nylon wool columns. This technique employs fibers which have not been specifically coated with ligand, relying on a natural affinity of certain lymphoid cells to bind to this support. The advantages of this protocol is that it is an inexpensive, rapid single step procedure for the separation and purification of cells from heterogeneous mononuclear cell preparations that also produces cells with high viability for use in *in vitro* and *in vivo* experiments. This procedure may also avoid important but subtle chemical or physiological changes of membrane structures which result from incubation of cells with specific ligand coated fibers.

Basically, nonadherent T cells pass through the column material while B cells and monocytes are retained. With gentle agitation and differential centrifugation these adherent cells are easily collected. This method for the isolation of functional thymus derived murine lymphocytes was first popularized by Julius et al.[1] who noted that the effluent population derived from passage of spleen cells through nylon fiber columns was virtually devoid of B cell precursors and memory cell activity but contained cells with helper function and cytotoxic effector cell precursor activity when compared to unfractionated spleen cells. The utility of this fractionation technique was subsequently extended by Handwerger and Schwartz[2] who demonstrated that the adherent B cell population could be effectively recovered from the nylon wool with greater than 90% viability. This was accomplished with careful agitation, teasing, and compression of the nylon wool with the bound cells attached.

B cells recovered from the nylon fiber have been demonstrated to retain functional activity. This was first shown by Trizio and Cudkowicz[3] using B cells recovered from nylon fiber columns and measuring adoptive transfer of antibody-producing cells.

[1] M. H. Julius, E. Simpson, and L. A. Herzenberg, *Eur. J. Immunol.* **3**, 645 (1973).
[2] B. S. Handweger and R. H. Schwartz, *Transplantation* **18**, 544 (1974).
[3] D. Trizio and G. Cudkowicz, *J. Immunol.* **113**, 1093 (1974).

Thus, the nylon fiber column technique is useful for the rapid preparation of relatively pure, functionally active T and B cell populations.

Techniques

Preparation of Nylon Wool

Standard Technique

1. Place nylon wool (Fenwal Laboratories; Deerfield Ill. scrubbed nylon fiber, 3 denier, 3.81 cm, type 200) in beaker of deionized or distilled H_2O. Cover beaker with aluminum foil and boil nylon wool for 10 min.
2. Allow beaker to cool and drain nylon wool.
3. Repeat washing steps one and two, 3–5 times.
4. Wrap washed wool in clean cloth and squeeze out excess H_2O and place on clean absorbent cloth and dry in a 37° incubator for 2–3 days.

Alternative Technique. As an alternative, nylon wool may be wrapped in aluminum foil and dry heat sterilized at 120° for 18 hr. High temperature exposure of unwrapped fiber should be avoided or this will cause charring.

Differential centrifugation will be needed to remove nylon wool debris (see below).

Preparation of Columns

Nonsterile Technique

1. Tease apart 0.40 g of fiber under medium containing 1% glutamine, penicillin, streptomycin, and 20 mM HEPES.
2. Affix a 3-way plastic stopcock (Cobe Labs.; Cranbury, NJ) to a 5-ml plastic syringe (B-D. and Co.; Rutherford, NJ) and loosely pack the syringe with nylon wool, taking care that the nylon wool is saturated with medium at all times.
3. Wash the column with 50 ml of medium (above) containing 5% heat-inactivated fetal calf serum (FCS) and cover with parafilm (American Can Company; Greenwich, Connecticut).
4. Incubate the column in a 5% CO_2 incubator for 1 hr at 37°.
5. Following incubation, remove parafilm and wash column with 50 ml prewarmed (37°) medium (above) with 5% FCS.

Sterile Technique. For sterile procedures, nylon wool columns can be prepared in distilled H_2O, wrapped in aluminum foil, wet autoclaved for 15 min, then rinsed through with 50 ml of medium with FCS just prior to step 4 (above).

Addition of Cells to Columns

1. After the column is washed, 1.0 ml of a lymphoid cell suspension (containing $10-60 \times 10^6$ cells) in medium plus FCS is added to the top of the column and allowed to percolate into the nylon wool followed by the addition of 2 ml of the previously prepared medium containing 5% FCS.

2. The column is again sealed and allowed to incubate at 37° for 1 hr.

Cell Collection

Nonadherent Cells

1. After incubation, the nonadherent (T cells, some null cells) are removed dropwise by the slow addition of 10 ml of prewarmed medium containing 5% FCS, collecting 10 ml of the effluent.

2. The column is then rewashed with 50 ml of prewarmed medium with 5% FCS and the effluent discarded.

Adherent Cells. The adherent cells (B cells, monocytes, plasma cells, some accessory cells) are removed by gently tapping the column several times with a blunt object and then compressing the nylon wool with the syringe plunger. The nylon wool is then teased with forceps and fresh cold (4°) medium with 5% FCS is added to fill the column. This procedure is repeated 3 times until 30 ml of medium has been collected.

Column Capacity Considerations. More cells can be fractionated using proportionately larger columns, larger quantities of nylon wool, and appropriate amounts of reagents.

Cell Washing

In addition to platelet and granulocyte contamination found in both nonadherent and adherent populations, broken nylon wool particle debris is also found, which is particularly problematic in complement-dependent microcytotoxicity. This debris can be removed by differential centrifugation. First, after elution from the column, both populations of cells are washed 3 times (1000 g for 1 min) in a Fisher Centrifuge Model 59 (Fisher Scientific; St. Louis, Missouri) to eliminate platelets. The cells are then centrifuged at 5000 g for 1 min and the supernatant discarded. The cell pellet is very gently resuspended in fresh medium and spun at 1000 g for 4 sec. The supernatant which contains the purified cells is carefully removed and saved, and the pellet which contains granulocytes and fiber debris is discarded.

Alternative Fibers

Although most studies have been performed using nylon wool fractionated cells, glass wool and rayon can be used as well as cotton balls for lymphoid cell separation. Columns utilizing these materials should be packed to a moderate density (where additional fiber can no longer be packed in comfortably) and forcep teased to allow medium to flow through freely. As an approximation, four cotton balls in a volume of 35–45 ml would produce a good fiber density. If the fiber is packed too tightly, then cell yield will be poor. If the fiber is too loose, then cell purity will be diminished. All other steps outlined for nylon wool apply to those alternative materials.

Discussion

Technical Considerations

Cell Yield and Purity. Depending on the percentage of nonadherent cells in the starting population, nonspecific losses of cells on these columns of at least 25–40% are expected. Greater yields can be achieved at the sacrifice of purity by increasing elution flow rates. FCS, an important constituent in these isolations, should be kept at a concentration of 5–10%. A higher concentration would result in a decrease of adherent cells able to stick to fibers while a lower concentration would result in nonspecific adhesion. Variability in the percentages of isolated populations obtained depends heavily in the animal species and organ source of the starting lymphoid population. Weinblatt[4] reports that lymph node cells are easier to deplete of macrophages than spleen, peritoneal exudate, or peripheral blood lymphocyte populations.

Composition of the Nonadherent (T Cell) Population. Most investigators have found that the T cells isolated from nylon wool columns are representative of all T cell subpopulations found in the unfractionated population. Those T cell subsets reported to be contained in the effluent include unprimed and primed antigen specific helper cells, T cells that are precursors of cytotoxic effector cells, and T cells that proliferate in response to alloantigens and soluble antigens. However other studies suggest that certain T cell subpopulations are selectively retained on nylon

---

[4] A. C. Weinblatt, S. N. Vogel, and D. L. Rosenstreich, *in* "Manual of Macrophage Methodology: Collection, Characterization and Function" (H. Herscowitz, H. Holden, J. Bellanti, and A. Ghaffar, eds.). Dekker, New York, 1981.

wool.[5] In particular, there is some loss of Fc receptor-positive T cells.[6] These variations may be due to the degree to which the column is packed, the type of nylon wool used, and slight differences in both the preparative washing and cell removal techniques.

Composition of the Adherent Population. Depending on the starting tissue, the population that is adherent to nylon, and which can subsequently be recovered, usually contains varying amounts of B cells, monocytes/macrophages, and neutrophils. Cells collected in this manner have been used as stimulators in autologous mixed lymphocyte reactions, histocompatibility testing, cell surface marker studies, plaque assays for immunoglobulin secreting cells, antigen-specific proliferation studies, and for *in vivo* adoptive transfer studies.

Additional Purification Techniques. The T cell subpopulation can be additionally purified by passage over a second nylon wool column. A higher degree of T cell purity can be obtained by use of a second type of adherence column such as fine glass beads or Sephadex G-10 columns (this volume [29]). Additional purification will produce T cells that are so devoid of accessory cells that they will no longer proliferate in response to T cell mitogens unless accessory cells or factors are added.[7]

Since the adherent population contains both B lymphocytes and monocytes, it is more difficult to obtain purified B cells using this technique alone. Generally a second positive selection method such as rosetting or absorption to anti-Ig columns or a negative selection such as removal of the phagocytic cells is required to produce satisfactory monocyte-deficient B cell populations (this volume [14, 25–32]).

---

[5] R. D. Stout and L. A. Herzenberg, *J. Exp. Med.* **142,** 1041 (1975).

[6] M. P. Arala-Chaves, L. Hope, J. H. Korn, and H. Fundenberg, *Eur. J. Immunol.* **8,** 77 (1978).

[7] A. C. Weinblatt, J. J. Oppenheim, and D. L. Rosenstreich, *Cell. Immunol.* **68,** 332 (1982).

[29] Use of Sephadex G-10 to Separate Macrophages and Lymphocytes

By ROBERT I. MISHELL and BARBARA B. MISHELL

Introduction

The use of Sephadex G-10 as a column matrix for removing macrophages from suspensions of immunologically reactive cells was developed empirically.[1] The basis of separation is not fully understood, but probably involves both adherence and size. Anchorage-dependent cells including macrophages adhere to beads made of Sephadex (Cytodex)[2] and large cells may become trapped in the spaces between the relatively inelastic G-10 beads. Other standard chromatographic grades of Sephadex (G-25 and higher) do not permit efficient recovery of nonadherent lymphocytes because these beads are more elastic and thus pack more tightly, obstructing the flow of even small (5–7 μm diameter) cells.

Cytometric analysis of the effluent cells reveals that 95–100% of murine splenic cells with diameters greater than 12 μm are removed.[3] Although precursors of antibody formation, NK cells, and T cell-mediated cytolytic reactions and T cells that mediate both helper and suppressor activities from immune and nonimmune animals are quantitatively recovered in the effluent population, virtually all differentiated antibody-forming cells, some of the cytolytically active T effector cells, and large cycling cells such as mitogen- and antigen-induced blasts are retained with the macrophages.[1,4-6]

The technique has been widely employed to obtain macrophage-free lymphocyte-rich cells for various cellular immunological studies. In contrast to flow cytometry, the G-10 column cell separation procedure is a simple, inexpensive, rapid method for processing relatively large numbers of cells ($>10^8$) under sterile conditions. Thus, it is particularly suitable as

---

[1] A. Ly and R. I. Mishell, *J. Immunol. Methods* **5**, 239 (1974).
[2] E. C. Ren, *J. Immunol. Methods* **49**, 105 (1982).
[3] R. I. Mishell and B. B. Mishell, unpublished observations.
[4] R. H. Schwartz, A. R. Bianco, B. S. Handwerger, and C. R. Kahn, *Proc. Natl. Acad. Sci. U.S.A.* **72**, 474 (1975).
[5] A. Singer, C. Cowing, K. S. Hathcock, H. B. Dickler, and R. J. Hodes, *J. Exp. Med.* **147**, 1611 (1978).
[6] C. L. Miller and R. I. Mishell, *J. Immunol.* **114**, 692 (1975).

the first step in a more elaborate purification scheme if higher degrees of purity than those produced by G-10 alone are required.

Preparation of Reagents

Sephadex G-10

Sephadex G-10 (Pharmacia Fine Chemicals)
Sodium chloride, 0.15 M

Glass Beads

Sulfuric acid, concentrated
Nitric acid, concentrated
Sodium bicarbonate, 0.12 M
Hydrochloric acid (1 : 100 dilution of concentrated HCl)
Glass beads (Microbeads Subdivision of Cataphote Division, Ferro Corporation)
Class IV-A #456 Unisphere beads, 250–350 μm
Class IV-A #235.5 Unisphere beads, 500–710 μm

Procedure. Approximately 250 g of Sephadex G-10 is added to 1500–2000 ml of 0.15 M NaCl, gently stirred, and allowed to swell overnight at 4°. The liquid is removed by suction, and an additional quantity is added that is 3–4 times the bed volume. The Sephadex is resuspended and allowed to settle once again. The liquid is again removed by suction along with the fine Sephadex particles that have failed to settle. This procedure is repeated for 3–4 times until the fine particles are removed from the washed and swelled Sephadex. The ratio of settled Sephadex to 0.15 M NaCl is adjusted so that 40–45 ml of slurry contain 30–35 ml of packed Sephadex. The slurry is then distributed into 40- to 45-ml aliquots, and autoclaved for 40 min at 110° (slow exhaust). The sterilized Sephadex can be stored indefinitely at room temperature.

The glass beads, each size prepared separately, are soaked for 24 hr in a 50:50 mixture of concentrated sulfuric and nitric acid. They are then placed under gently running tap water for 8 hr, soaked in a large volume of 0.12 M sodium bicarbonate for 24 hr, and thoroughly rinsed in double-distilled water. The beads are then soaked in diluted HCl for 24 hr, and rinsed in double-distilled water until the pH of the rinse water is above 6, after which they are dried in a drying oven. The dried beads are then distributed into tubes in 5 g amounts (sizes are not combined) and auto-claved for 20 min at 121° (fast exhaust). The beads can be stored at room temperature indefinitely.

Separation Procedure

Reagents

Balanced salt solution[7] containing 50 units/ml penicillin, 50 μg/ml streptomycin, and 5–10% heat-inactivated (56°, 30 min) fetal calf serum (BSS-FCS)

Sephadex G-10

Glass beads, approximately 5 g of each size

Disposable syringe, 50 ml

Tissue culture dish, 60 mm

Three-way stopcock, No. K-75, Pharmaseal Laboratories

Ring stand and clamp

Cell suspension, 1.5×10^8 cells/ml in BSS

Procedure. The column separation is done at room temperature or 37°. A 50-ml plastic syringe is attached to a ring stand, the plunger discarded, and the top of the syringe is covered with a sterile lid of a 60-mm culture dish. The stopcock is attached to the syringe tip in closed position. The larger glass beads are poured into the syringe, followed by the smaller beads. These retain the Sephadex in the syringe. Approximately 10 ml of the BSS-FCS is added to the syringe to wash the beads from the sides. The Sephadex is gently pipetted on top of the glass beads, and allowed to settle in the syringe. A beaker is placed under the column and the stopcock opened to allow the BSS-FCS to pass through the column. The Sephadex is then washed with 100–150 ml of BSS-FCS and during the last wash the top of the Sephadex bed is gently stirred with a pipet tip to ensure an even, level top layer. All the fluid is allowed to penetrate the column. The cell suspension is then gently added to the column without disturbing the top of the column bed and allowed to penetrate the column. Small amounts of BSS-FCS are added to the syringe until the cells are approximately halfway down the column at which time 15–20 ml of BSS-FCS is added. The collection of the cells is begun when the cells reach the glass bead layer and is terminated after 10–20 ml is obtained. Any Sephadex contamination of the collected cells is allowed to settle for 2–3 min, the cells are decanted to another tube, and centrifuged at 200 g for 10 min.

A column of the size described above will accommodate a total of 4–6 $\times 10^8$ cells applied at a concentration of 1.5×10^8 cells/ml.

Comments. The recovery of separated cells is variable. In general the greater the recovery the poorer the separation. Cell recoveries typically

---

[7] B. B. Mishell and S. M. Shiigi, eds., "Selected Methods in Cellular Immunology," p. 447. Freeman, San Francisco, California, 1980.

range from 30 to 40% with murine spleen and lymph node cells.[1,2,8-10] Other investigators have reported somewhat higher recoveries with human peripheral blood leukocytes initially purified on Ficoll–Hypaque.[4,11] Depletion of macrophages as judged histologically and by phagocytosis ranges from 90 to 99%.[4,5,11,12] Approximately normal proportions of T and B lymphocytes are recovered in the effluent populations as judged by surface markers and by functional criteria.[1,4-6,12]

While the method is a convenient one for removing macrophages, data generated with this technique should be interpreted cautiously. Cells that develop functions similar to macrophages are often present in the filtered population, in the form of a functionally significant number of small, nonadherent accessory cells. For example, some preparations of Sephadex G-10-depleted cells can generate primary immune responses to sheep red blood cells if 2-mercaptoethanol is included as a medium constituent. We have observed that spleen cells from mice maintained under conditions that minimize microbial exposure are more difficult to deplete than lymphoid populations from mice maintained under conventional conditions. We suspect that cells of the macrophage lineage that have not been stimulated by environmental pathogens more easily pass through the columns. Macrophages provide two distinct types of required functions in the *in vitro* generation of immune responses. They produce mediators such as interleukin-1 that amplify responses and they process and present antigen to lymphocytes. Since usually fewer macrophages are required for antigen processing and presentation than for secreting adequate levels of interleukin-1, Sephadex G-10 separation alone may not be suitable for evaluating antigen processing.

In an attempt to obtain better separations with this technique, other investigators have used sequential passages of cells through two columns[5] and have conducted the separations at 42°.[13,14] Our experience is that these modifications do not significantly affect the functional depletion of cells when judged by the aforementioned criteria. Other approaches used to reduce the problem of residual contamination by functionally active accessory cells are to choose cell sources that contain relatively few macrophages, such as lymph nodes[9] or peripheral blood, to culture the

[8] S. B. Pollack, K. Nelson, and J. D. Grausz, *J. Immunol.* **116**, 944 (1976).

[9] J. I. Kurland, P. W. Kincade, and M. A. S. Moore, *J. Exp. Med.* **146**, 1420 (1977).

[10] T. M. Chused, S. S. Kassan, and D. E. Mosier, *J. Immunol.* **116**, 1579 (1976).

[11] N. T. Berlinger, C. Lopez, and R. A. Good, *Nature (London)* **260**, 145 (1976).

[12] C. Alonso, R. R. Beruabe, E. Moreno, and F. Diaz de Espada, *J. Immunol. Methods* **22**, 361 (1978).

[13] K. Pickel and M. K. Hoffmann, *J. Immunol.* **118**, 653 (1977).

[14] J.-P. Kolb, S. Arrian, and S. Zolla-Pazner, *J. Immunol.* **118**, 702 (1977).

depleted cells at low cell concentrations to dilute out the effects of contaminating cells, or to use antigens of limited complexity for immune induction since responses to such antigens apparently require more macrophages than do those to complex antigens.[5] We have had little experience in recovering and using the adherent population. A procedure for recovering this population has been described by Schwartz et al.[4] Recently Ren reported employing Cytodex beads in suspension to separate anchorage-dependent cells from lymphocytes.[2] His data showed good recovery of macrophages by treating the beads with trypsin following cell separation.

[30] Depletion of Macrophages from Heterogeneous Cell Populations by the Use of Carbonyl Iron

By DENNIS M. WONG and LUIGI VARESIO

Before the advent of more sophisticated methods for the separation of macrophages from lymphocytes and other nonlymphoid cells in heterogeneous cell populations, a common procedure was to expose the cells to iron particles and remove those cells which had phagocytized the iron with the aid of a magnet.[1] Macrophages, being highly phagocytic, are preferentially removed from the cell population treated with iron since lymphocytes do not phagocytize particulate materials.[2] Although any type of iron particle may be used in this procedure, the most common reagent is carbonyl iron, $Fe(CO)_3$. This technique has been used for the removal of monocytes from peripheral blood leukocyte cell suspensions in order to increase the yield of lymphocytes.[3-6] More recently, this technique has been applied to study the effects of removing macrophages from immune cell populations.[7-9]

[1] S. Levine, *Science* **123**, 185 (1956).
[2] M. F. Greaves, J. J. T. Owen, and M. C. Raff, "T and B Lymphocytes: Origins, Properties and Roles in Immune Responses," p. 57. Am. Elsevier, New York, 1973.
[3] S. Thierfelder, *Vox Sang.* **9**, 447 (1964).
[4] B. Lichtenstein, L. Palseltiner, R. Weingard, and R. Widmark, *Fed. Proc., Fed. Am. Soc. Exp. Biol.* **30**(1), 409 (1971).
[5] K. Tebbi, *Lancet* **1**, 1392 (1973).
[6] S. A. Shah and J. A. Dickson, *Nature (London)* **249**, 168 (1974).
[7] O. Sjoberg, J. Anderson, and G. Moller, *Eur. J. Immunol.* **2**, 123 (1972).
[8] C. Desaymard and M. Feldmann, *Cell. Immunol.* **2**, 106 (1975).
[9] K. C. Lee, C. Shiozawa, A. Shaw, and E. Diener, *Eur. J. Immunol.* **6**, 63 (1976).

METHODS IN ENZYMOLOGY, VOL. 108

Procedural Considerations

In most cases a small amount of carbonyl iron powder (G.A.F. Corporation, New York, NY or General Aniline and Film Corporation, Dyestuff and Chemical Division, Linden, NJ) is added to a cell suspension. The mixture is incubated at 37° for 30–60 min, and phagocytic cells which have ingested the iron are removed with the aid of a magnet. Critical in this technique appears to be the carbonyl iron. Some commercial lots of carbonyl iron may contain substances which are toxic to cells; therefore washing the iron in saline or tissue culture medium several times by repeated centrifugation is recommended. The washed preparation of carbonyl iron should be used within a short period of time since the effects of prolonged storage of this reagent in solutions containing electrolytes have not been studied. However, a commercial product known as lymphocyte separating reagent containing carbonyl iron suspended in an aqueous solution (Technicon Instruments Corp., Tarrytown, NY) can be stored for several months.

The next item to consider is the quantity of carbonyl iron that will eliminate the maximum number of macrophages. The exact amount of reagent to use must be determined by trial and error since the effects of adding carbonyl iron to a cell suspension will differ depending upon the types of cells present in the mixture. Excessive amounts of carbonyl iron will promote clumping of cells due to the adherence of cells to the iron particles. Peripheral blood lymphocytes (PBL) depleted of granulocytes by separation on Ficoll–Hypaque consist mostly of thymus-derived lymphocytes which are nonadherent.[10] Consequently, more carbonyl iron can be added to a suspension of PBL than to a suspension of spleen cells. The spleen contains macrophages, dendritic cells, and bone marrow-derived (B) lymphocytes which exhibit adherent properties. The possibility of clumping is therefore, increased by the presence of these cells. Thus, the optimal amount of carbonyl iron to use for each type of cell suspension must be determined by the addition of various concentrations of iron to the cell mixture. The concentration of carbonyl iron is then used which depletes the maximum number of macrophages without causing excessive cell clumping.

After incubation of the cells with carbonyl iron, macrophages which have ingested the iron particles are removed with the aid of a magnet. Although powerful electromagnets have been employed in this technique,[1,4] any large magnet may be used. Two Alnico-5 magnets (No. 36924-001, Fisher Scientific Co., Pittsburgh, PA) will provide enough magnetic force to achieve a complete removal of the iron-laden cells.

[10] A. Bøyum, *Scand. J. Clin. Lab. Invest.* **21,** Suppl. 97, 77 (1968).

Procedure

Two depletion methods will be described: one using plastic tissue culture dishes and the other using a magnetic column. In both cases, the cells are suspended in RPMI 1640 tissue culture medium (GIBCO Laboratories, Grand Island, NY) containing 20% fetal bovine serum (FBS) (Reheis Chemical Co., Chicago, IL), and mixed with an optimal concentration of carbonyl iron. The difference between the two methods occurs during the actual depletion of iron-laden macrophages.

Plastic Tissue Culture Dish Method

1. The cells are washed thoroughly by centrifuging several times at 200 g for 10 min and resuspending them in 9 ml of RPMI 1640 containing 20% FBS at a density of 1 × 10^7 cells/ml.

2. A carbonyl iron suspension is prepared by adding an optimal amount of iron (150–250 mg) to 1 ml of RPMI 1640 medium containing 20% FBS. The suspension is thoroughly mixed by vortexing and 1 ml of the carbonyl iron preparation is added to 9 ml of cell suspension.

3. The cell mixture is added to a 25 × 150-mm plastic tissue culture dish (No. 1013, Falcon Plastics, Oxnard, CA) which is placed on a 25 × 25-cm rocker platform (No. 7740-10010, Bell Glass, Inc., Vineland, NJ) with the rocking motion set at three-fourths maximum speed. The entire apparatus is placed into a 5% CO_2 incubator for 45 min at 37°.

4. After the incubation, the dish is placed on top of the poles of two magnets which have been fixed to a ringstand.

5. The cells are allowed to settle for 1 min. The supernatant is removed with a 10-ml pipet and transferred to another plastic tissue culture dish. Five milliliters of cold (4°) RPMI 1640 containing 20% FBS is added to the original dish. The dish is rocked gently for several seconds, and the supernatant is removed and pooled with supernatant from the original dish.

6. The original dish is replaced with the second dish containing the pooled supernatants and step 5 is repeated. Since a total volume of 20 ml of pooled supernatants is obtained after this step, two dishes containing 10 ml of pooled supernatants each may be used for the next step.

7. A third treatment of cell suspension with magnets may be necessary depending upon the type of cell population being treated. Depletion of macrophages from peripheral blood lymphocytes may only require two treatments of the cell mixture with magnets, while spleen cells may require three treatments. If a third treatment is required, then step 5 may be repeated.

8. After the final treatment, the cells are washed two times with RPMI 1640 containing 10% FBS.

Magnetic Column Method

1. The cells are treated and the carbonyl iron is added as described for the dish method (steps 1 and 2 above).

2. The cell suspension containing the carbonyl iron is then added to a 50-ml conical test tube. The tube is placed on a rocker platform for 1 hr at 37°.

3. A magnetic column is prepared as follows. (a) Add $1/2 \times 5/16$ in. magnetic stirring bars to a 35-ml syringe up to the 20 ml mark. (b) Wash with 50 ml of Hanks' balanced salt solution. (c) Put a three-way stopcock and a 21-gauge needle on the syringe.

4. After the incubation, the cell mixture is transferred from the 50-ml test tube to the magnetic column and incubated for 10 min at room temperature.

5. The cells are then eluted from the column.

6. If desired, the column can be washed by the addition of 10 ml of RPMI 1640, followed by incubation for 10 min at room temperature and elution.

7. The magnetic column can be recycled by filling it with 1 N HCl and incubating overnight at room temperature. This treatment dissolves the carbonyl iron and the iron-ingesting cells. The recycled column can either be autoclaved or filled with 70% alcohol and stored.

Results

The degree of macrophage depletion achieved by the use of carbonyl iron will vary depending on the composition of the untreated cell population. In a mouse spleen cell population, up to 80% of the splenic macrophages, as determined by positive esterase staining and ingestion of latex particles, can be removed.[11] This level of macrophage depletion is sufficient to suppress the antibody-forming immune response to polymeric flagellar protein.[9] With human peripheral blood samples, the purity of lymphocytes obtained after the removal of monocytes with carbonyl iron can exceed 96% with less than 0.5% contamination with monocytes.[4,5]

The table shows the results obtained when macrophages are depleted from mixed cell populations. Spleen cell suspensions were obtained from normal C57BL/6 mice (NSC) or mice injected intramuscularly (TB-SC) 14 days before with the Moloney strain of murine sarcoma virus (MSV). Injection of MSV causes the development of tumors and an increase in

---

[11] D. M. Wong, in "Manual of Macrophage Methodology" (H. B. Herscowitz, H. T. Holden, J. A. Bellanti, and A. Ghaffar, eds.), p. 105. Dekker, New York, 1981.

DEPLETION OF MACROPHAGES BY THE MAGNETIC COLUMN METHOD

| Cell source[a] | Before depletion: macrophages (%)[b] | After depletion | |
		Recovery (%)	Macrophages (%)[b]
NSC	7 ± 2^c	80 ± 8	0.7 ± 0.3
TB-SC	11 ± 4	75 ± 11	0.9 ± 0.4
CFT	61 ± 9	16 ± 8	2.6 ± 1.8
PEC	58^d	35	2

[a] NSC, Normal spleen cells from C57BL/6 mice; TB-SC, spleen cells from C57BL/6 mice bearing MSV-induced tumors, harvested 14 days after injection of virus; CFT, cells from enzymatically dissociated MSV-induced tumors 14 days after injection of virus; PEC, peritoneal exudate cells from C57BL/6 mice.

[b] Measured by evaluating the number of cells ingesting latex beads.

[c] Average ± standard error of the results obtained in four independent experiments.

[d] Results of one experiment.

the percentage of splenic macrophages.[12] Tumors removed 14 days after the injection of MSV contain a high infiltrate of macrophages as shown by cellular analyses of tumors which have been dissociated with collagenase (CFT).[13] Regardless of the composition of the initial cell population, high levels of macrophage depletion are obtained with the carbonyl iron technique. The efficiency of this method is shown by the depletion of the CFT from 61 to 2.6% macrophages with a single treatment with carbonyl iron. It has been shown that subsequent passage of this macrophage-depleted cell population on a nylon wool column reduces the macrophage contamination to less than 0.5%[13] (this volume [28]). It should be noted that not only the extent of depletion but also the recovery of depleted cells is quite constant. Depletion of macrophages by the use of carbonyl iron restores the ability of spleen cells from tumor-bearing mice to produce lymphokine in response to stimulation with tumor antigens or mitogens.[14,15]

Discussion

The presence of plasma cells and B cell blasts in the spleen cell population can cause considerable clumping or aggregation of cells during treatment with carbonyl iron, since these two types of cells exhibit adher-

[12] S. Landolfo, R. B. Herberman, and H. T. Holden, *J. Immunol.* **118**, 1244 (1977).

[13] L. Varesio, R. B. Herberman, J. M. Gerson, and H. T. Holden, *Int. J. Cancer* **24**, 97 (1979).

[14] L. Varesio and H. T. Holden, *Cell. Immunol.* **56**, 16 (1980).

[15] L. Varesio, M. Giovarelli, S. Landolfo, and G. Forni, *Cancer Res.* **39**, 4983 (1979).

ent qualities.[7,16,17] The nonspecific adherence of cells at high concentrations of carbonyl iron can reduce the number of cells recovered after treatment to as low as 50% of the original spleen cell suspension.[9] Hence, the need for using the minimal amount of carbonyl iron which will cause a maximum degree of macrophage depletion. Cell aggregation can also be minimized by using an increased concentration of FBS (e.g., 15–20%) in the tissue culture medium, applying more agitation during the incubation period to keep the cells in motion, or decreasing the cell density in the mixture from $1 \times 10^7/\text{ml}$ to $1 \times 10^6/\text{ml}$.

Removal of macrophages from a heterogeneous cell population by the use of carbonyl iron appears to be quite efficient when compared to other physical depletion methods such as adherence to Petri dishes and columns composed of glass beads or Sephadex G-10, which may result in the nonspecific removal of other types of cells. The carbonyl iron technique also has the advantages of being highly reproducible in terms of recovery of macrophage-depleted cells and extent of depletion. Thus, it is recommended for routine use where a reproducible and rapid fractionation of cells is important. It should be stressed that no one technique ensures the total removal of macrophages from a mixed cell population. Where a contamination of 1% or less of macrophages is critical (e.g., in the evaluation of accessory cell functions), at least two depletion procedures should be used, preferentially one based on the phagocytic and the other on the adherence properties of the macrophages.

A novel method employing a fluorescence-activated cell sorter to remove macrophages which have ingested fluorescein-conjugated latex particles appears promising.[18] However, this method is time consuming and the degree of macrophage depletion may be comparable to that obtained with carbonyl iron treatment. A better method may be the use of monoclonal antibodies. With the development of monoclonal antibody against specific cell determinants, the ideal antimacrophage serum may become available as the most specific reagent against macrophages. Since there is enough evidence to suggest that macrophages may be heterogeneous in structure (e.g., have different antigenic markers) and function,[19] the monoclonal antibody must be directed at an antigenic determinant shared

[16] K. Shortman, W. Byrd, N. Williams, K. T. Brunner, and J. D. Cerottini, *Aust. J. Exp. Biol. Med. Sci.* **50**, 323 (1972).

[17] M. H. Julius, E. Simpson, and L. A. Herzenberg, *Eur. J. Immunol.* **3**, 645 (1973).

[18] H. S. Boswell, S. O. Sharrow, and A. Singer, *J. Immunol.* **124**, 989 (1980).

[19] O. Forster and M. Landy, eds., "Heterogeneity of Mononuclear Phagocytes." Academic Press, New York, 1981.

by all macrophages for the complete removal of this cell type. Once such a reagent against macrophages becomes available, then the physical depletion of the cells through rosetting techniques, cellular immunoabsorbent columns, or complement-dependent lysis will be more effective than any of the methods available at the present time.

[31] Preparation and Use of Monoclonal Antimacrophage Antibodies

By MAY-KIN HO and TIMOTHY A. SPRINGER

Introduction

Macrophages are a diverse family of cells originating from pluripotent stem cells in the bone marrow. Outside the marrow, macrophages can take the form of blood monocytes, Kupffer cells, alveolar macrophages, peritoneal macrophages, Langerhans cells, and in almost every organ, as "fixed tissue macrophages." In addition to variation in anatomical localizations, macrophages can exhibit heterogeneity in function and state of differentiation. The advent of hybridoma technology has allowed the identification of over 40 macrophage antigens, some of which are useful for distinguishing macrophages from other cells whereas others are associated with distinct subsets of macrophages.

In this section, we have chosen a number of monoclonal antibodies (MAb) with relatively restricted specificities for macrophages and summarized their characteristics. Possible applications of these antibodies are also described.

Preparation of Monoclonal Anti-Macrophage Antibodies

Procedures for the production of monoclonal antibodies have already been described in detail elsewhere.[1] Most of the antibodies summarized here were prepared by using macrophages, monocytes, or macrophage cell lines as immunogens.[2-4]

[1] G. Galfre and C. Milstein, this series, Vol. 73, p. 1.

[2] J. Unkeless, *J. Exp. Med.* **150**, 580 (1979).

[3] S. Maruyama, T. Naito, H. Kakita, S. Kishimoto, Y. Yamamura, and T. Kishimoto, *J. Clin. Immunol.* **3**, 57 (1983).

[4] S. N. S. Hanjan, J. F. Kearney, and M. D. Cooper, *Clin. Immunol. Immunopathol.* **23**, 172 (1982).

Certain antigens, such as Mac-1, and histocompatibility antigens, are more immunodominant. Hence, MAb to these antigens may be repeatedly isolated from different fusions. To allow the production of MAb to less immunodominant molecules, Springer depleted several previously defined antigens from macrophage glycoproteins by affinity chromatography and used the effluent as immunogen. Several MAb recognising previously unidentified antigens were produced by this "cascade" procedure.[5]

In general, hybrids are screened for their ability to bind macrophages or to inhibit certain macrophage functions. Binding of MAb to cells can be detected by a secondary antibody labeled by a radioisotope, fluorescent dye, or enzyme conjugate. To avoid reactivity with the target cells, the second-stage antibody should be prepared in the same specie as the target cells, or absorbed with immunoglobulins from the appropriate species. A more elegant approach to decrease cross-reactivity is to use MAb as the secondary antibody as well. Several MAb against mouse[6] or rat[7] immunoglobulin subclasses and kappa chains are now available through the American Type Culture Collection (ATCC). Hybrids selected by the primary screen can be further characterized by (1) their reactivity pattern on a large number of normal cells, leukemic cells, and cell lines; (2) immunoprecipitation of the antigens recognized; and (3) *in situ* staining of tissue sections. Immunoprecipitation studies are especially useful because they provide an early indication of whether the newly isolated MAb have the same specificity as existing antibodies.

Macrophage Antigens Defined by Monoclonal Antibodies

Mouse Antigens

Antigens identified by some monoclonal antimacrophage antibodies are listed in Table I. Mac-1, the first macrophage antigen defined by a MAb, M1/70, is a general marker for distinguishing macrophages from lymphocytes. It is found on peritoneal macrophages elicited by a variety of agents, blood monocytes, alveolar macrophages, and free-lying macrophages in spleen, lymph node, and thymus.[8–10] Lymphocytes, Kupffer cells, and dendritic cells are Mac-1⁻.[10] Mac-1 is not entirely macrophage-

[5] T. Springer, *J. Biol. Chem.* **256**, 3833 (1981).
[6] D. E. Yelton, C. Desaymard, and M. D. Scharff, *Hybridoma* **1**, 5 (1981).
[7] T. A. Springer, A. Bhattacharya, J. T. Cardoza, and F. Sanchez-Madrid, *Hybridoma* **1**, 257 (1982).
[8] T. Springer, G. Galfre, D. S. Secher, and C. Milstein, *Eur. J. Immunol.* **9**, 301 (1979).
[9] M.-K. Ho and T. A. Springer, *J. Immunol.* **128**, 2281 (1982).
[10] T. J. Flotte, K. A. Haines, K. Perkman, T. A. Springer, I. Gigli, and G. J. Thorbecke, *in* "Mononuclear Phagocyte Biology" (A. Volkman, ed.). Dekker, New York (in press).

specific because granulocytes, 50% of bone marrow cells and natural killer cells also bear this antigen.[11] 1.21J, an independently isolated MAb, also recognizes Mac-1.[12]

Several antigens are expressed on only subpopulations of macrophages. They include AcM.1,[13] 54-2[14,15] M43, M57, M102, and M143.[16] AcM.1 is only detected on activated macrophages induced by pyran and *Corynebacterium parvum*, but not on monocytes, resident macrophages, and macrophages elicited by thioglycolate, peptone, and mineral oil. The 54-2 antigen is strongly expressed on surfaces of thioglycolate-induced, but not resident peritoneal macrophages. Mac-2 was originally found to have a similar distribution by immunofluorescence and immunoprecipitation.[17] However, recent immunohistochemical studies show that 5% of resident macrophages are strongly Mac-2$^+$ in the cytoplasm.[10] The remaining 95% show much weaker, but definite staining. It seems that only cells with large amounts of Mac-2 internally express this antigen on their surfaces. In addition to peritoneal macrophages, *in situ* staining showed that Mac-2 is found on macrophages of all lymphoid and nonlymphoid tissues examined thus far. Interdigitating dendritic cells and Langerhans cells are also Mac-2$^+$.[10] Another antigen identified from the same fusion, Mac-3, is present in all Mac-2$^+$ cells.[18] In addition, megakaryocytes and endothelial cells of postcapillary venules are Mac-2$^-$ but Mac-3$^+$.[10]

For the identification of macrophages in cell suspensions, F4/80 can be used. The F4/80 antigen is present on macrophages from the peritoneal cavity, spleen, thymus, lymph node, and lung. Blood monocytes and macrophages derived from *in vitro* bone marrow cultures also express this antigen. Lymphocytes, granulocytes, and fibroblasts are negative.[19] It seems that the amount of F4/80 expressed increases with the maturity of macrophages.[20]

Human Antigens

A summary of antigens on human monocytes/macrophages is presented in Table II. These antigens can be roughly categorized into three

[11] K. A. Ault and T. A. Springer, *J. Immunol.* **126**, 359 (1981).
[12] I. S. Mellman, R. M. Steinman, J. C. Unkeless, and Z. A. Cohn, *J. Cell Biol.* **86**, 712 (1980).
[13] T. Taniyama and T. Watanabe, *J. Exp. Med.* **156**, 1286 (1982).
[14] P. A. LeBlanc, H. R. Katz, and S. W. Russell, *Infect. Immun.* **8**, 520 (1980).
[15] H. R. Katz, P. A. LeBlanc, and S. W. Russell, *J. Reticuloendothel. Soc.* **30**, 439 (1981).
[16] D. Sun and M.-L. Lohmann-Matthes, *Eur. J. Immunol.* **12**, 134 (1982).
[17] M.-K. Ho and T. A. Springer, *J. Immunol.* **128**, 1221 (1982).
[18] M.-K. Ho and T. A. Springer, *J. Biol. Chem.* **258**, 636 (1983).
[19] J. M. Austyn and S. Gordon, *Eur. J. Immunol.* **11**, 805 (1981).
[20] S. Hirsch, J. M. Austyn, and S. Gordon, *J. Exp. Med.* **154**, 713 (1981).

TABLE I

RAT MONOCLONAL ANTIBODIES DEFINING MOUSE MACROPHAGE ANTIGENS

Antibody		Antigen		Distribution[a]		Functional significance[a]	Lysis	References
Designation	Subclass	Designation	Polypeptide chains	Fluid cells	Solid tissue			
M1/70[b]	IgG$_{2b}$	Mac-1	190,000 105,000	PEM, PM, M, 50% bone marrow, mature Mac lines	Mac in splenic red pulp, LN sinuses; alveolar Mac	Inhibits binding to Mac CR$_3$; blocks induction of Mac-3 expression on myeloblast; ADCC and NK effectors	Weak	8–11, 42–45
1-21J	NR	Mac-1	180,000 94,000	J774 Line	NR[c]	NR		2, 12
M3/31[b] M3/38[b]	IgM IgG$_{2a}$	Mac-2	32,000	TG-PEM, mature Mac lines	Mac in GC, red pulp, LN sinuses, and thymus; DC in red pulp PALS, and LN; alveolar Mac, Kupffer cells; Langerhans cells, epithelial cells	NR	NR	5, 10, 17
M3/84[b]	IgG$_1$	Mac-3	110,000	PEM, PM, Mac lines, some myeloid and B lymphoid lines	Same as for Mac-2, plus megakaryocytes and PCV endothelial cells	NR	NR	5, 10, 18

Antibody	Ig class	Cell line	M_r	Distribution		Comments	ADCC	Ref.
54-2[d]	IgG_{2a}	Mac-4	NR	BM-Mac, TG-PEM, mast cells	NR	NR	NR	14, 15
M3/37	NR	Mac-4	180,000	TG-PEM	NR	NR	NR	5
F4/80	IgG_{2b}	F4/80	160,000	PEM, PM, M, splenic Mac, thymic Mac, 8% BM, Mac lines, alveolar Mac	NR	Increases with maturity of Mac	–	19, 20
2.4G2	IgG	FcRII	47–70,000	Mac, B lymphocytes, Mac lines, PMN	NR	Inhibits binding to FcRII	NR	2
M43	IgM	M43	NR	30–40% PEM, BM-Mac	NR	Cytotoxic for lymphokine-activated Mac	+	16
M57	IgG_{2b}	M57	NR	20–30% PEM, BM-Mac, PMN	NR	Cytotoxic for ADCC effectors	+	16
M102	IgG_1	M102	NR	20–30% PEM, BM-Mac	NR	Cytotoxic for ADCC effectors	+	16
M143	IgG_{2a}	M143	NR	10–30% PEM, BM-Mac	NR	NR	+	16
AcM.1	IgG_{2c}	AcM.1	NR	Pyran and C. parvum-PEM	NR	Cytotoxic for tumorcidal macrophages	+	13

[a] ADCC, Antibody-dependent cellular cytotoxicity; BM, bone marrow, CR_3, Type III complement receptor; DC, dendritic cells; GC, germinal center; LN, lymph node; M, monocytes; Mac, macrophages; NK, natural killer; PALS, periarteriolar lymphatic sheath; PCV, postcapillary venule; PEM, peritoneal exudate macrophages; PM, resident peritoneal macrophages; PMN, polymorphonuclear cells; TG, thioglycolate.

[b] Cell lines available through American Type Culture Collection (ATCC).

[c] NR, Not reported.

[d] 54-2 and M3-37 precipitate polypeptides which coelectrophorese (M. K. Ho, unpublished).

TABLE II

MONOCLONAL ANTIBODIES DEFINING HUMAN MONOCYTE ANTIGENS[a]

| Antibody | | Antigen | | Distribution[b] | Functional significance[b] | Lysis | References |
Designation	Subclass	Designation	Polypeptide chains				
OKM1	IgG$_{2b}$	OKM1	NR[c]	PMN, null cells, blood monocytes	NR	+	21
M1/70[d]	IgG$_{2b}$	Mac-1	NR	PMN, NK, and ADCC effectors, blood monocytes	Inhibits binding to CR$_3$ on PMN; blocks induction of Mac-3 expression on NK and ADCC effectors	NR	11, 42, 44
Anti-Mo-1	IgM, IgG$_{2a}$	Mo-1	155,000 94,000	Monocytes, PMN, null cells, PM, immature BM. HL-60, U937, and KG-1 cells. Mac and monocytes in splenic red pulp, LN sinusoids, histiocytes	Inhibits binding to CR$_3$ on monocytes	+	22, 32
Anti-Mo-2	IgM, IgG$_{2b}$	Mo-2	55,000	Monocytes, 7% BM, PM, HL-60, cultured monocytes, Mac and monocytes in splenic red pulp, LN sinusoid and GC	"Pan-monocyte antigen"	+	22, 32
Anti-Mo-3	IgM	Mo-3	NR	Cultured monocytes, 6% BM. HL-60, and U937 lines	NR	NR	25
Mac-120	NR	Mac-120	120,000	37% blood monocytes	Lymphokine production, accessory cells for T cell proliferation to Con A and antigens	+	26

Antibody	Class	MW	Cellular distribution	Function	ADCC[b]	Reference
63D3[d]	IgG$_1$	200,000	Weak on PMN monocytes	Accessory cells for mitogen-induced proliferation	NR	23, 46
MφP-15	IgG$_1$	NR	70% monocytes, 79% pleural Mac or PM	NR	NR	24
MφP-9	IgG$_{2b}$	NR	Most monocytes, pleural Mac and PM; 20–40% PMN	NR	NR	24
MφS-1	IgG$_{2a}$	NR	PMN	NR	NR	24
MφS-39	IgG$_{2a}$	NR		NR	NR	24
C10H5	IgM	NR	Blood monocytes, <5% BM	Inhibits growth of myeloid/monocytic stem cell	+	27
D5D6	IgM	NR	Blood monocytes, <5% BM, HL60		+	27
MMA	IgM	NR	Blood monocytes, PMN, U937, HL60, and some T cell lines	Accessory cells for T cell proliferation to mitogen and antigen, cytotoxic for precursors of granulopoietic colony forming cells	+	4
UC45	IgM	66,000	Monocytes, neurons	NR	NR	28, 29
10-75-3	IgG$_{2a}$	135,000	Blood monocytes, platelets, HL60	Specific for α chains of fibrin	+	30, 31
M206	IgG$_{2a}$	180,000	Blood monocytes, U937, weak on PMN	Accessory cell for PWM-induced proliferation	+	3

[a] All antibodies listed, except M1/70, are mouse monoclonal antibodies.
[b] ADCC, Antibody-dependent cellular cytotoxicity; BM, bone marrow, Con A, concanavalin A; CR$_3$, Type 3 complement receptor with specificity for C3bi and its degradation products; GC, germinal center; LN, lymph node; Mac, macrophage; NK, natural killer; PMN, polymorphonuclear cells; PWM, pokeweed mitogen.
[c] NR, Not reported.
[d] Available through American Type Culture Collection.

groups. The first group comprises antigens found on blood monocytes/ macrophages and polymorphonuclear cells, but not on lymphocytes. They include OKM1,[21] Mac-1,[11] Mo-1,[22] 63D3,[23] MϕP-9, MϕS-1, MϕS-39,[24] and M206.[3] This group of markers serves to distinguish macrophages and monocytes from cells of the erythroid and lymphoid lineage.

The second group, including Mo-2,[22] Mo-3,[25] Mac-120,[26] MϕP-15,[24] C10H5, D5D6,[27] UC45,[28,29] and 10-75-3,[30,31] are on all or a subset of monocytes/macrophages, but not on lymphocytes or granulocytes. Therefore, they can be used to define monocytes from other hematopoietic cells.

The last group represents monocyte antigens which are expressed on unrelated cell types. For example, MMA[4] is found on abnormal T cells whereas UC45[28] is expressed on neurons.

Most of the data on distribution presented in Table II are from studies on cell suspensions. The only two antigens examined on tissue sections are Mo-1 and Mo-2.[32] Anti-Mo-1 stains polymorphonuclear cells and a few large mononuclear cells in splenic red pulp and lymph node sinusoids. Mo-2 is not found on granulocytes, but on large mononuclear cells in the red pulp, lymph node sinusoids, and germinal centers.

Applications

Identification of Macrophages in Vitro

Immunofluorescence. The most routine procedure for identifying macrophages is by indirect immunofluorescence. The cells to be assayed are

[21] J. Breard, E. L. Reinherz, P. C. Kung, G. Goldstein, and S. F. Schlossman, *J. Immunol.* **124**, 1943 (1980).

[22] R. F. Todd, III, L. M. Nadler, and S. F. Schlossman, *J. Immunol.* **126**, 1435 (1981).

[23] V. Ugolini, G. Nuñez, R. Graham Smith, P. Stastny, and J.D. Capra, *Proc. Natl. Acad. Sci. U.S.A.* **77**, 6764 (1980).

[24] A. Dimitriu-Bona, G. R. Burmester, S. J. Waters, and R. J. Winchester, *J. Immunol.* **130**, 145 (1983).

[25] R. F. Todd, III and S. F. Schlossman, *Blood* **59**, 775 (1982).

[26] H. V. Raff, L. J. Picker, and J. D. Stobo, *J. Exp. Med.* **152**, 581 (1980).

[27] M. Linker-Israeli, R. J. Billing, K. A. Foon, and P. I. Terasaki, *J. Immunol.* **127**, 2473 (1981).

[28] N. Hogg and M. Shusarenko, *Cell* **24**, 875 (1981).

[29] N. Hogg, *J. Exp. Med.* **157**, 473 (1983).

[30] W. H. K. Anderson, J. J. Burckhardt, J. F. Kearney, and M. D. Cooper, *Fed. Proc., Fed. Am. Soc. Exp. Biol.* **40**, 986 (1981).

[31] J. J. Burckhardt, W. H. K. Anderson, J. F. Kearney, and M. D. Cooper, *Blood* **60**, 767 (1982).

[32] R. F. Todd, III, A. K. Bhan, S. E. Kabawat, and S. F. Schlossman, in "Leukocyte Typing: Human Leukocyte Differentiation Antigens Detected by Monoclonal Antibodies" (A. Bernard and L. Boumsell, eds.), p. 424. Springer-Verlag, Berlin and New York (in press).

incubated with saturating amounts of an antimacrophage MAb, washed, and incubated further with an excess of fluoresceinated second-step antibody. The secondary antibody should be specific for the primary Mab, but not cross-react with the target cells. Because the Fc receptors on macrophages can bind antibodies, several precautions should be taken: (1) F(ab')$_2$ fragments of the second stage antibody should be used; (2) avoid repeated freezing and thawing of the antibody reagents; (3) store antibodies at 4° after deaggregation by ultracentrifugation at 100,000 g for 1 hr; and (4) include a control MAb of the same isotype as the primary MAb, but with an irrelevant specificity.

Some antigens are modulated and possibly internalized after interaction with antibody, hence leading to erroneous results. This problem can be minimized by carrying out procedures at 4° and by including 10 mM sodium azide in the incubation and washing buffers.

The stained cells can be centrifuged onto glass slides, fixed, and examined under an UV microscope. Alternately, they can be analyzed by the fluorescence-activated cell sorter (FACS)[33] For this purpose, the stained cells should be filtered through nytex (Small Parts, Miami, Fl #CMN-37) to remove cell clumps. The anti-Mac-1 MAb has been found in some cases to cause agglutination of macrophages. This can be prevented by lowering the antibody concentration. In addition to the number of positive cells, FACS analysis can also provide information on the relative fluorescent intensity per cell. Details on this procedure are discussed in this volume [19].

Immunofluorescent staining has also been performed on adherent cell monolayers.[34] Briefly, macrophages are allowed to adhere onto glass cover slips. After fixation in 1% paraformaldehyde (to prevent antigen redistribution), the cells are stained by placing the cover slip, cell-side-down, onto 15 μl of antibodies.

Autoradiography. This procedure allows the visual localization of bound antibody on individual cells. Both cell suspensions or adherent cell monolayers can be used. The method of labeling is essentially the same as that for immunofluorescence except that the second antibody is radiolabeled. Labeled cells on glass slides or cover slips are dried, fixed, and coated with photographic emulsion.[19,35] After an empirically determined time, the slides are developed, and counterstained with Giemsa. When examined by light microscopy, the bound antibodies can be localized to areas with black silver grains. Both the morphology and relative quantity

---

[33] L. A. Herzenberg and Lenore A. Herzenberg, in "Handbook of Experimental Immunology" (D. M. Weir, ed.), 3rd ed., Chapter 22. Blackwell, Oxford, 1978.

[34] D. I. Beller, J.-M. Kiely, and E. R. Unanue, *J. Immunol.* **124,** 1426 (1980).

[35] J. Watson, in "Selected Methods in Cellular Immunology" (B. B. Mishell and S. M. Shiigi, eds.), p. 166. Freeman, San Francisco, California, 1980.

of antibody bound to cells can be determined simultaneously. However, this procedure is tedious and time consuming.

Rosetting. Cells sensitized with an antimacrophage MAb can be detected by binding to erythrocytes which have been coated with $F(ab')_2$ fragments of a secondary antibody by chromic chloride.[36,37] Usually, the sensitized cells and erythrocytes are mixed in a ratio of 1:40 or 50, centrifuged at 200 g for 5 min, and incubated for 1–5 hr at 4°. The pellet is then gently dispersed and cells stained by acridine orange or toluidine blue. The number of positive cells can be determined by counting the cells with five or more attached erythrocytes.

Enrichment or Depletion

Immunofluorescence. Fluorescent cells prepared as indicated above can be separated according to their fluorescence intensity on the FACS.[33] This method offers the advantage that almost pure populations of positive as well as negative cells can be recovered at relatively high yield. If only the positive (or negative) fraction is needed, the time of separation can be reduced by starting with cells that are enriched for this fraction by another procedure which gives less purity.

Complement-Mediated Lysis. There are both IgM and IgG MAb which can lyse macrophages in the presence of complement (see Tables I and II). Conditions leading to lysis are more stringent than those for binding. For each MAb used, a titration has to be performed to determine the optimal concentrations of MAb and complement. It is advisable to screen a number of complement batches from different animal sources to obtain the highest specific lysis but the lowest level of toxicity (lysis in the absence of MAb). Some MAb, such as M57 and M102, are indirectly cytotoxic.[16] They require the presence of a facilitating (second step) antibody to mediate lysis. Again, a MAb of the same isotype, but with an irrelevent specifity, should be included as control.

Depending on the antigen density, not all cells which show binding with a MAb will be lysed. Sometimes, several rounds of treatment are required to lyse most of the positive cells. Another approach is to use a mixture of MAb which recognize different antigenic determinants on the same molecule or cell.

Rosetting. Rosetted cells can be separated on Percoll[36] or Ficoll–Hypaque[37] density gradients. The bound erythrocytes can be dissociated from cells by either hypotonic lysis or by incubation of excess second-stage antibody followed by density gradient centrifugation. Relatively large numbers of cells (up to 10^9) can be separated by this procedure.

[36] E. P. Rieber, J. Lohmeyer, D. Schendel, and G. Riethmuller, *Hybridoma* **1**, 59 (1981).
[37] E. T. Dayton, B. Perussia, and G. Trinchieri, *J. Immunol.* **130**, 1120 (1983).

Localization of Macrophages in Situ

In addition to studies of cell suspensions, *in situ* staining of macrophages is important for the characterization of antimacrophage MAb because of the following reasons: (1) macrophages in tissue sections can be localized and examined in their "natural" state; (2) cell suspensions prepared from lymphoid or nonlymphoid organs may not be representative of the cellular composition of these tissue;[9,38] (3) immunohistochemistry allows the study of cells and structures, such as Kupffer cells and postcapillary venules, which are difficult to isolate.

Depending on the nature of the antigen, the tissue of interest can be frozen in OCT compound (Miles laboratories, Elkhart, IN) or fixed and embedded before sectioning. Optimal conditions for the localization of different antigens can vary greatly. Mac-1 is destroyed by fixation in alcohol, requiring study in frozen sections.[9] In contrast, Mac-3 and Ia antigens are readily detected in alcohol-fixed tissue embedded in polyester wax. Mac-2 is even more stable. It is routinely localized in tissue processed in Carnoy's fixative and embedded in paraffin.[10] Before staining, the embedding material has to be removed. Paraffin is soluble in xylene whereas OCT is water soluble.

Tissue sections are generally labeled by immunoperoxidase[39] or immunofluorescence. The former involves overlaying the sections sequentially with the chosen MAb, excess bridging antibody, and a peroxidase antiperoxidase (PAP) complex. In the case of rat MAb, rabbit anti-rat Ig and rat PAP have been used. Endogenous peroxidase can be quenched by incubating with 0.3–0.6% H_2O_2 before staining. A modification of this technique employs a primary MAb, biotinylated secondary antibody, and a preformed avidin–biotinylated peroxidase complex (ABC).[40,41] The ABC system is more sensitive than the PAP technique. Furthermore, the antibody in the PAP complex has to be made in the same specie as the MAb, whereas an universal ABC can be used for MAb prepared in different species. Sometimes, staining can be increased by using four layers of antibodies. Peroxidase activity is detected by incubating with diaminobenzidine (DAB) and H_2O_2. To allow identification of cells and structures, the sections can be counterstained with hematoxylin or methyl green.

Immunohistochemical studies on Mac-1, Mac-2, and Mac-3 have allowed the classification of the macrophage-dendritic cell family into two

[38] K. A. Haines, T. J. Flotte, T. A. Springer, I. Gigli, and G. J. Thorbecke, *Proc. Natl. Acad. Sci. U.S.A.* **80**, 3448 (1983).
[39] L. A. Sternberger, "Immunocytochemistry". Wiley, New York, 1979.
[40] S. M. Hsu, L. Raine, and H. Fanger, *Histochemie* **29**, 577 (1981).
[41] J. D. Minna, F. Cuttita, S. Rosen, P. A. Bunn, D. N. Carney, A. F. Gazdar, and S. Kransnow, *In Vitro* **17**, 1058 (1981).

groups. Free-lying, round macrophages are positive for all three antigens whereas cells with dendritic morphology (excluding follicular dendritic cells) are Mac-1$^-$, but Mac-2$^+$ and Mac-$^+$.[10]

Functional Studies

The functional significance of antigens can be studies by performing the biological assay of interest in the presence of the appropriate MAb or after depletion of cells bearing the corresponding antigen. Control MAb reacting against other cell surface structures should be included in the assay.

Functional assays were used to select for and study an antibody to macrophage Fc receptor.[2] The M1/70 MAb inhibits the binding of C3bi to type III complement receptors (CR$_3$) on mouse macrophages and human polymorphonuclear cells. Binding to CR$_1$ and FcR are unaffected.[42] Anti-Mo-1 also has the same activity on human monocytes.[32] Therefore, it appears that these antigens are associated with the function of or identical to the CR$_3$. Anti-Mac-1 also inhibits the expression of Mac-3 on a mouse myelomonocytic line, M1, and a human monoblast line, U937, during induction by various agents.[43,44] This suggests that Mac-1 plays a regulatory role in macrophage differentiation. In addition, M1/70 defines the cells in human peripheral blood and mouse peritoneal exudate which have natural-killing activity.[11,45]

Some of the other MAb examined for functional significance include Mac-120, 63D3,[46] MMA, M206, M43, C10H5, and D5D6. The first four MAb are associated with accessory cells required for mitogen- or antigen-induced lymphocyte proliferation. M43 is cytotoxic for killer macrophages activated by lymphokines. MMA is cytotoxic for precursors of granulopoietic colony-forming cells whereas C10H5 and D5D6 inhibit the growth of myeloid/monocoyte stem cells.

Acknowledgments

This work was supported by USPHS Grant CA 31799 and Council of Tobacco Research grant 1307.

[42] D. I. Beller, T. A. Springer, and R. D. Schreiber, *J. Exp. Med.* **156,** 1000 (1982).
[43] P. Ralph, M.-K. Ho, P. B. Litcofsky, and T. A. Springer, *J. Immunol.* **130,** 108 (1983).
[44] P. Ralph, P. E. Harris, C. J. Punjabi, K. Welte, P. B. Litcofsky, M.-K. Ho, B. Y. Rubin, M. A. S. Moore, and T. A. Springer, *Blood* **62,** 1169 (1983).
[45] L. A. Holmberg, T. A. Springer, and K. A. Ault, *J. Immunol.* **127,** 1792 (1981).
[46] S. A. Rosenberg, F. S. Ligler, V. Ugolini, and P. E. Lipsky, *J. Immunol.* **126,** 1473 (1981).

[32] Elimination of Macrophages with Silica and Asbestos

By ELLIOTT KAGAN and DAN-PAUL HARTMANN

Silica and asbestos are naturally occurring, silicon-containing compounds which are widely distributed within geologic deposits in the earth's crust. Mechanical comminution of these mineral species produces particles which exhibit a diverse spectrum of size distributions. Asbestos differs from silica in that, unlike silica, asbestos particulates have a fibrous character (i.e., a length : width aspect ratio of greater than 3 : 1).

Certain types of asbestos and silica are toxic to macrophages, both *in vitro* and *in vivo*.[1-3] When macrophages are exposed *in vitro* to these toxic agents, they exhibit extreme cytoplasmic vacuolation, karyopyknosis, disappearance of pseudopodia, and effacement of plasma membranes.[3] Similar morphologic abnormalities have been observed in alveolar macrophages after experimental silica or asbestos inhalation in animals.[4,5] The cytoplasmic enzyme, lactate dehydrogenase, has proved to be a useful marker for assessing macrophage cytotoxicity induced by these mineral dusts.[6] Thus, the release of this enzyme into macrophage culture supernatants provides confirmatory evidence of macrophage lysis. The lethal action of silica is believed to result from a hydrogen bonding interaction of silicic acid with membrane protein and phospholipid moieties.[6] Chrysotile asbestos, on the other hand, appears to exert its cytotoxic effect through an electrostatic interaction of its surface magnesium groups with ionized carboxyl groups of membrane glycoprotein sialic acid residues.[6]

Silica does not appear to be directly cytotoxic to either T or B cells *in vitro*.[7] One study has, however, suggested that some silica preparations may suppress lymphoproliferative responses to concanavalin A and lipopolysaccharide mitogens in a macrophage-independent fashion.[8] Chrysotile asbestos, on the other hand, does not affect the viability of lympho-

---

[1] A. C. Allison, J. S. Harington, and M. Birbeck, *J. Exp. Med.* **124**, 141 (1966).
[2] K. Miller, *CRC Crit. Rev. Toxicol.* **5**, 319 (1978).
[3] E. Bey and J. S. Harington, *J. Exp. Med.* **133**, 1149 (1971).
[4] K. Miller and E. Kagan, *J. Reticuloendothel. Soc.* **21**, 307 (1977).
[5] E. Kagan, Y. Oghiso, and D. P. Hartmann, *Environ. Res.* **32**, 382 (1983).
[6] A. C. Allison and D. M. L. Morgan, *in* "Lysosomes in Applied Biology and Therapeutics" (J. T. Dingle, ed.), p. 14. North-Holland Publ., Amsterdam, 1979.
[7] E. J. O'Rurke, S. B. Halstead, A. C. Allison, and T. A. E. Platts-Mills, *J. Immunol. Methods* **19**, 137 (1978).
[8] J. J. Wirth, W. P. Carney, and E. F. Wheelock, *J. Immunol. Methods* **32**, 357 (1980).

cytes *in vitro*.[9,10] Indeed chrysotile has been shown to enhance the mitogen-induced blastogenic response of monocyte-depleted peripheral blood mononuclear cells.[10]

The selective lethal action of these silicon-containing compounds thus provides a useful tool for the elimination of monocytes and macrophages from cell cultures and body fluids. It is the purpose of this review to discuss the practical use of some of these agents as selective toxins for mononuclear phagocytes.

Requirements

Types of Silica

Silica (SiO_2, silicon dioxide) occurs naturally within geologic deposits in 5 structurally polymorphic crystalline forms: quartz, cristobalite, coesite, tridymite, and stishovite. Only quartz preparations are readily obtainable from commercial sources for biological studies. Noncrystalline (amorphous) silica samples are also available. These represent forms of colloidal silica which are processed by precipitation and subsequent desiccation.

A major problem experienced by investigators using different silica preparations is the great variation in biologic potency of samples from different commercial sources, since no international reference standard exists for silica samples. We have used high-quality preparations of Dowson and Dobson crystalline quartz particles. These samples were obtained from a vein of pure quartz, mined at the Witkop Quarry, Pietersburg, South Africa, and were provided to us by Mr. R. E. G. Rendall, National Centre of Occupational Health, Johannesburg, South Africa. We have also used commercial amorphous silica, obtainable from British Drug Houses, Poole, England. Many other commercial silica preparations are, however, available. Dörentrup D Q 12 quartz samples of <5 μm particle size can be obtained from Steinkohlenberg-Bauverein, Essen-Krei, West Germany, and Min-U-Sil quartz particles of various size specifications are obtainable from Pennsylvania Glass Sand Corporation, Pittsburgh, PA. Fransil amorphous silica particles, ranging in size from 0.1 to 0.5 μm, may be obtained from the Safety in Mines Research Establishment, Sheffield, England. In our experience, particle size (<2–10 μm) has not been a

[9] S. Kagamimori, M. P. Scott, D. G. Brown, R. E. Edwards, and M. M. F. Wagner, *Br. J. Exp. Pathol.* **61**, 55 (1980).
[10] B. E. Bozelka, H. R. Gaumer, J. Nordberg, and J. E. Salvaggio, *Environ. Res.* **30**, 281 (1983).

major determinant of biological potency with respect to most silica preparations.[11]

Types of Asbestos

The term "asbestos" refers to a group of naturally occurring fibrous silicates which occur in two main structural forms: serpentine and amphibole. The amphiboles have a double chainlike structure composed of linked SiO_4 tetrahedra, whereas serpentine asbestos is a sheet silicate with a curved, cylindrical, scroll-like configuration.[12] Chrysotile, the most common type of asbestos used commercially in the United States, is the only serpentine form of asbestos available. There are 5 varieties of amphiboles, however: crocidolite, amosite, anthophyllite, tremolite, and actinolite. Of these, only amosite and crocidolite were previously used extensively on a commercial basis.

International (U.I.C.C.) standard reference samples of asbestos are obtainable from Mr. R. E. G. Rendall, National Centre of Occupational Health, Johannesburg, South Africa, or from Dr. V. Timbrell Medical Research Council, Pneumoconiosis Unit, Llandough Hospital, Penarth, Glamorgan, Wales. Standard U.I.C.C. reference samples have been prepared for South African crocidolite, South African amosite, Canadian and Rhodesian chrysotile, and Finnish anthophyllite. Although the physical dimensions of these reference preparations have been well characterized,[13] there is considerable heterogeneity in each sample with respect to fiber length. We have also used less well-characterized asbestos samples obtained from the National Institute of Environmental Health Sciences, Research Triangle Park, NC (crocidolite and chrysotile), from the Manville Corporation, Denver, CO. (Jeffrey Mine Canadian chrysotile lot # AX3738-M), and from Union Carbide Corporation, New York, NY (RG 144, a naturally occurring short fiber chrysotile from the New Idria deposit in CA).

A toxicity gradient is noted for various types of asbestos. Thus, chrysotile is the most lethal for macrophages, whereas crocidolite exhibits minimal toxicity.[2,3] Amosite is intermediate in its cytotoxic potential.[3,14] For this reason, it is recommended that chrysotile be employed for *in vitro* macrophage elimination studies, if asbestos is chosen as the lethal agent.

[11] E. Kagan and K. Miller, *in* "Manual of Macrophage Methodology: Collection, Characterization and Function" (H. B. Herscowitz, H. T. Holden, J. A. Bellanti, and A. Ghaffar, eds.), p. 137. Dekker, New York, 1981.

[12] F. D. Pooley, *Semin. Oncol.* **8**, 243 (1981).

[13] R. E. G. Rendall, *IARC Sci. Publ.* **30**, 87 (1980).

[14] T. McLemore, M. Corson, M. Mace, M. Arnott, T. Jenkins, D. Snodgrass, R. Martin, N. Wray, and B. R. Brinkley, *Cancer Lett.* **6**, 183 (1979).

TABLE I
HETEROGENEITY OF FIBER LENGTH IN MILLED AND
UNMILLED ASBESTOS SAMPLES[a]

	Percentage of particles or fibers					
Materials used	<0.11 μm	0.11– 0.5 μm	0.51– 1.0 μm	1.1– 2.0 μm	2.1– 5.0 μm	>5 μm
UICC A (Rhodesian) chrysotile	0.0	4.9	13.4	19.8	36.5	25.4
RG 144 (Calidria) chrysotile, unmilled	20.3	49.2	12.4	8.5	7.5	2.1
RG 144 chrysotile, 50-sec milled	12.7	58.6	16.7	8.6	3.2	0.2
RG 144 chrysotile, 1200-sec milled	58.2	39.4	2.3	0.1	0.0	0.0
Quartz (Min-U-Sil 15)[b]	0.0	0.0	0.0	67.0	22.0	11.0

[a] Adapted from Yeager et al.,[16] with permission from the publisher.
[b] Quartz particles were used for comparative purposes.

There is considerable debate concerning the issue of asbestos fiber length as a determinant of biologic potency. In our experience, asbestos samples containing predominantly shorter fibers (<1 μm in length) are considerably more toxic to macrophages than samples of longer size characteristics.[15] Mechanical comminution of asbestos samples by ball-milling considerably accentuates their killing potency. Milling can be accomplished, by placing 50-mg aliquots of chrysotile in a 2.5 × 1.0-cm steel cylinder chamber containing a small lead pellet. The cylinder is attached to a vibrator operating at about 10 cycles per second. The chrysotile is then ball-milled for periods ranging from 1 to 6 min. Table I shows the variation in fiber length distributions in 2 asbestos preparations and the effect of ball-milling on these characteristics. A Min-U-Sil 15 quartz sample is used as comparison.[16] The *in vitro* effect of these different preparations on the viability of alveolar macrophages is shown in Table II.[16] The choice of preparation to be used by the investigator is thus critical.

Other Requirements

HEPES-buffered RPMI 1640 medium (available from GIBCO, Grand Island, NY). The medium is supplemented with penicillin (200 U/ml),

[15] A. M. Langer, M. S. Wolff, A. N. Rohl, and I. J. Selikoff, *J. Toxicol. Environ. Health* **4**, 173 (1978).
[16] H. Yeager, Jr., D. A. Russo, M. Yañez, D. Gerardi, R. P. Nolan, E. Kagan, and A. M. Langer, *Environ. Res.* **30**, 224 (1983).

TABLE II

COMPARISON OF THE LETHAL EFFECTS OF QUARTZ AND
VARIOUS MILLED AND UNMILLED ASBESTOS SAMPLES ON
HUMAN ALVEOLAR MACROPHAGES[a,b]

Mineral sample tested[c]	Macrophage viability (%)	
	50 μg/culture	100 μg/culture
UICC chrysotile A[d]	67.2	62.7
RG 144 chrysotile[d]		
Unmilled	47.4	42.0
60-sec milled[e]	39.7	27.4
1200-sec milled[e]	23.7	14.4
Quartz (Min-U-Sil 15)[d]	29.3	9.3
None[d]	88.3	

[a] Modified from Yeager et al.,[16] with permission from the publisher.

[b] Macrophages (1×10^5) in RPMI 1640, containing 10% fetal calf serum, were exposed to 2 concentrations of silica or various asbestos samples for 1 hr at 37°. After washing off excess particulates, fresh serum-supplemented medium was added, and the cultures were incubated for a further 24 hr at 37°.

[c] The mineral samples used were the same as those characterized in Table I.

[d] Mean of 10 experiments.

[e] Mean of 6 experiments.

gentamicin (10 μg/ml), streptomycin (10 μg/ml), and fresh L-glutamine (0.3 mg/ml).

Joklik's Ca^{2+}- and Mg^{2+}-free medium (GIBCO)

Trypan blue dye

Bouin's fixative

Hematoxylin and eosin stains

Fluorescein diacetate (Sigma Chemical Company, St. Louis, MO)

Lactate dehydrogenase test kit reagents (as described in
    Sigma Technical Bulletin No. 226-UV)

Microscope slide/tissue-culture chambers (obtained from
    Lab-Tek Division, Miles Laboratories, Napierville, IL)

CO_2 incubator

Polarizer and analyzer filters

Ultraviolet spectrophotometer and cuvettes

Fluorescence and phase-contrast microscopes

Ultrasonicator

Precautions

Both silica and asbestos samples, when in the dry, powdered state, are potentially hazardous if inhaled as aerosols. These agents are both capable of causing pneumoconiosis (a form of interstitial pulmonary fibrosis) in the unprotected industrial environment. All varieties of asbestos are also carcinogenic and are capable of causing lung cancer and malignant mesotheliomas of the pleura and peritoneum. Although the risk to laboratory personnel is considered small, we advocate the use of gloves, mask, and a fume hood when handling or weighing powdered forms of these toxic agents. For these procedures, we personally use a Comfo II custom respirator with type "F" particulate filter. This device, which is approved by both NIOSH and OSHA, is obtainable from Mine Safety Appliances Company, Pittsburgh, PA.

Preparation of Silica and Asbestos Samples

Fresh suspensions of silica and asbestos should be prepared immediately prior to use. The samples should be presterilized in a dry hot-air oven at 160° for 2 hr, before suspension in RPMI 1640. In order to promote even dispersion of silica particles or asbestos fibers, the suspensions should be ultrasonicated for 1–2 min immediately before use. It is advisable to determine the optimal dosage of administration for different sample batches, by preparing a range of test concentrations, (generally between 50 and 300 μg/ml). For *in vitro* studies, a concentration of 200 μg/ml of either silica or asbestos is generally sufficient for maximal macrophage destruction. As illustrated in Table III, approximately 87–98% of macrophages were eliminated when various silica samples were used at this concentration.[11]

In Vitro Procedures

Preparation of Cell Monolayers

Resident populations of alveolar and peritoneal macrophages can be obtained from rodents by bronchoalveolar[4] or peritoneal lavage, respectively, using cold Joklik medium (this volume [25]). In order to obtain blood monocytes, peripheral blood mononuclear cells are prepared from heparinized blood on Ficoll–Hypaque gradients[17] (this volume [9]). The various cell populations are then washed 3 times in cold Joklik medium, resuspended in RPMI 1640, and subsequently applied to microscope slide/

---

[17] J. J. Oppenheim and B. Schecter, *in* "Manual of Clinical Immunology" (N. R. Rose and H. Friedman, eds.), 2nd ed., p. 233. Am. Soc. Microbiol., Washington, D.C., 1980.

TABLE III
EFFECT OF SILICA CONCENTRATION AND PARTICLE SIZE ON GUINEA PIG ALVEOLAR
MACROPHAGE MONOLAYERS[a]

Silica source[b]	Silica concentration (μg/ml)	Particle size (μm)	Number of adherent cells per culture chamber	Macrophage depletion[c] (%)
Quartz	50	<2	280	97.4
	100		296	97.2
	200		210	98.0
Amorphous silica	50	1–2	2167	79.6
	100		1588	85.1
	200		1392	86.9
Quartz	200	2–5	310	97.1
		5–10	823	92.3
None[d]	—	—	10,624	0

[a] Adapted from Kagan and Miller,[11] with permission from the publisher.

[b] Macrophage monolayers were incubated with silica for 1 hr, washed, and incubated in serum-free medium for a further 3 hr.

[c] Expressed as

$$\left[1 - \frac{\text{Number of adherent cells in cultures containing silica}}{\text{Number of adherent cells in cultures without silica}} \right] \times 100.$$

[d] Macrophage monolayers were incubated in serum-free medium for 4 hr.

tissue-culture chambers at a concentration of 2.5×10^5 cells/chamber. The cells are allowed to attach to the underlying substrate at 37° for 45 min. During this period, the cultures are maintained in a humidified environment of 5% CO_2 : 95% air.

Treatment of Macrophages with Silica or Asbestos

In order to remove nonadherent cells, the culture supernatants are removed with a Pasteur pipet, whereupon the monolayers are washed 3 times with RPMI 1640. The culture chambers are then filled with the freshly prepared asbestos or silica suspension in RPMI 1640. The monolayers are subsequently incubated for 1 hr at 37° in a humidified atmosphere containing 5% CO_2 : 95% air. After removal of the culture supernatant again, the monolayers are vigorously washed 6 times more, to remove any superfluous silica or asbestos. It is usually not possible, however, to completely remove these particulates in the washing process. The culture chambers are again filled with fresh RPMI 1640, whereupon the monolayers are incubated for a further 3–24 hr at 37° in a humidified 5% CO_2 : 95% air milieu. Cultures, which have not been treated with silica or asbestos, are run in parallel as controls.

Evaluation of Macrophage Viability

This can be assessed in various ways. The simplest method is to quantitate the proportion of cells which exclude 0.1% trypan blue dye. A more reliable method of estimating cell viability is a fluorochromasia technique with fluorescein diacetate.[18] Only living cells can take up and hydrolyze the fluorochrome ester. Since dead cells are unable to hydrolyze the ester, they remain unstained. The percentage of living cells is thus readily determined by counting those cells which exhibit a bright apple-green fluorescence (due to liberated fluorescein). A stock solution of fluorescein diacetate can be prepared by dissolving the fluorochrome ester in acetone, at a concentration of 5 mg/ml. The stock solution can be stored at −20°. For the evaluation of cell viability, the stock solution is diluted, immediately prior to use, to a final concentration of 10 μg/ml in RPMI 1640. The dye-containing medium is then added to the cell cultures, which are now incubated at 37° for 10 min. Fluorochromasia is observed in the dark with an inverted fluorescence microscope.

The efficacy of macrophage or monocyte destruction by asbestos and silica preparations can be confirmed, by quantitation of lactate dehydrogenase levels in the culture supernatants.[6] We have used the manufacturer's kit supplied by Sigma Chemical Company.[5] The assay is based on the catalytic interconversion of lactate and pyruvate, with NAD as the hydrogen acceptor:

$$\text{Lactate} + \text{NAD} \xleftrightarrow{\text{lactate dehydrogenase}} \text{pyruvate} + \text{NADH}$$

The reaction mixture is maintained at 30° and absorbance is read over 3 min at 340 nm in a UV spectrophotometer. The complete procedural details are described in the manufacturer's technical bulletin (Sigma Technical Bulletin #226-UV, Revised August, 1982). Supernatants from cultures maintained in RPMI 1640 cannot, however, be assayed for lactate dehydrogenase activity, owing to the presence of phenol red indicator in the medium, which can produce interference. Medium 199 without phenol red (obtainable from GIBCO) is used instead as a substitute culture medium for enzymatic assays.

Assessment of Macrophage Detachment from Monolayers

After the desired period of incubation with either silica or asbestos preparations, the monolayers are examined under an inverted phase-contrast microscope, and the number of adherent mononuclear phagocytes

---

[18] K. I. Welsh and J. R. Batchelor, *in* "Handbook of Experimental Immunology" (D. M. Weir, ed.), 3rd ed., Vol. 2, p. 35.1. Blackwell, Oxford, 1978.

are quantitated. A permanent microscopic record can be obtained, by removing the culture chambers from the microscope slides, fixing the monolayers in Bouin's fixative for 1 hr, and staining the cells with hematoxylin and eosin.

Visualization of Ingested Silica and Asbestos

The use of a polarizer–analyzer filter system allows the microscopic identification of birefringent silica particles which are ingested by mononuclear phagocytes. Asbestos fibers longer than 5 μm in size can be visualized under phase-contrast microscopy. Shorter asbestos fibers may require transmission electron microscopy for more accurate visual identification. After 1 hr in vitro exposure to either agent, usually more than 95% of macrophages contain ingested particulate material.

In Vivo Applications

When silica particles are injected intravenously into experimental animals, they are rapidly sequestered by the sinusoidal mononuclear phagocytes in the liver, spleen, and other organs. Although there is no evidence that intravenous silica administration can effectively diminish the numbers of circulating monocytes or peritoneal macrophages in experimental animals, this procedure has been employed by a number of investigators to abrogate various macrophage-dependent immunologic functions. Thus, silica given intravenously has been shown to ablate the primary immune response to sheep erythrocytes,[8,19] to impair the generation of allogeneic killer cells,[19] and to promote the survival of skin allografts.[20] Intravenous silica can also abrogate the induction of natural killer cells,[21] and can impair genetic resistance to hybrid, allogeneic, and xenogeneic grafts.[22,23] This mode of silica administration has also been demonstrated to enhance susceptibility to tumor growth[24-26] and to parasitic infestation.[27]

[19] M. H. Levy and E. F. Wheelock, J. Immunol. 115, 41 (1975).
[20] N. N. Pearsall and R. S. Weiser, J. Reticuloendothel. Soc. 5, 107 (1968).
[21] R. Kiessling, P. S. Hochman, O. Haller, G. M. Shearer, H. Wigzell, and G. Cudkowicz, Eur. J. Immunol. 7, 655 (1977).
[22] E. Lotzova and G. Cudkowicz, J. Immunol. 113, 798 (1974).
[23] E. Lotzova, M. T. Gallagher, and J. J. Trentin, Biomedicine 22, 387 (1975).
[24] R. Keller, J. Natl. Cancer Inst. (U.S.) 57, 1355 (1976).
[25] M. A. Palladino and G. J. Thorbecke, Cell. Immunol. 38, 350 (1978).
[26] J. M. Zarling and S. S. Tevethia, J. Natl. Cancer Inst. (U.S.) 50, 149 (1973).
[27] F. Kierszenbaum, E. Knecht, D. Budzko, and M. C. Pizzimenti, J. Immunol. 112, 1839 (1974).

Only one study has demonstrated that the intraperitoneal inoculation of silica can almost completely eliminate macrophages from the peritoneal cavity.[28] In that study, mice were injected daily with silica for 5 days. Only about 5% of the normal number of peritoneal macrophages (many of which were functionally defective) were recoverable from silica-treated animals 6–20 days after completion of the treatment regimen. Silica given by the intraperitoneal route can also impair natural killer cell function.[29]

The dosage of silica used should be well tolerated by the animal, yet sufficient for the desired immunosuppressive effect. Mice tolerate intravenous injections containing 3 mg of quartz well, whereas rats can tolerate at least 10 mg by this route.[30] Macrophage depletion can be achieved in mice with intraperitoneal injections of 0.1–5 mg of silica/day.[28]

When asbestos samples are administered via the intraperitoneal route, they evoke a granulomatous response, and increased numbers of macrophages are recoverable from peritoneal exudates in these animals.[31] No studies have been performed to determine the lethal action of intravenous asbestos on macrophage function. Asbestos thus appears to be unsuitable for most *in vivo* studies of this nature. Prolonged asbestos inhalation can, however, reduce the number of recoverable alveolar macrophages by as much as 50%, an effect which persists as long as 1 year after asbestos exposure ceases.[5]

Variability of Results

Mention has already been made of the gradient in cytotoxic potential expressed by different types of asbestos and by variation in fiber length. Quartz samples from different suppliers also vary in their lethal action. All quartz preparations have a biologically active surface, or Beilby, layer.[32] Removal of this layer enhances the biological activity of quartz,[32] an effect which may vary with the mode of mechanical processing of different quartz samples. Surface contaminants, especially iron oxide, have been identified in some commercial silica preparations and these impurities have been shown to inhibit the biological effectiveness of these agents.[30] These contaminants can be removed, by boiling the quartz samples in 1 N HCl, a process which leaches off the iron as a green precipi-

---

[28] H. van Loveren, M. Snoek, and W. den Otter, *J. Reticuloendothel. Soc.* **22,** 523 (1977).

[29] J. R. Oehler and R. B. Herberman, *Int. J. Cancer* **21,** 221 (1978).

[30] A. C. Allison, in "In-Vitro Methods in Cell-Mediated and Tumor Immunity" (B. R. Bloom and J. R. David, eds.), p. 395. Academic Press, New York, 1976.

[31] K. Donaldson and J. M. G. Davis, *Environ. Res.* **29,** 414 (1982).

[32] A. M. Langer, *Q. Rev. Biophys.* **11,** 543 (1978).

tate.[30] After no further green color is observed, the quartz is extensively washed in distilled water and then dried for use.

The presence of serum in the culture medium considerably retards the cytotoxic action of silica[1,6] and asbestos. Potent quartz samples can eliminate greater than 95% of macrophages from monolayers after only a 3-hr incubation in serum-free medium (Table III).[11] If, however, the culture medium contains 10–20% fetal calf serum (obtainable from GIBCO) at the time the silica suspension is prepared, significant macrophage toxicity may only develop after 24–48 hr in culture. This delayed cytotoxic effect results from retardation of damage to the macrophage phagolysosomal membrane, due to serum coating of the silica particles. On the other hand, when serum is added after the quartz particles have already been ingested, toxicity will then be comparable with that attainable under serum-free conditions.

The effects of *in vivo* silica administration are extremely complex and are unlikely to be explained purely on the basis of a selective lethal action on macrophages. Silica treatment of macrophages *in vitro* enhances the production of interleukin 1,[33] a T cell mitogen which also augments antibody production by B cells.[34,35] Asbestos inhalation is also associated with enhanced antigen-directed interleukin 1 release.[36] The elaboration of the lymphokine, interleukin 2, is also augmented after asbestos inhalation.[36] These findings may be related to the increased expression of Ia antigens on alveolar macrophages following *in vivo* exposure to asbestos.[37]

The *in vivo* effects of silica tend to be of relatively short duration, generally lasting 1–3 days after intravenous administration and somewhat longer after intraperitoneal inoculation. Thus, the timing and mode of silica administration (particularly in relation to antigen administration) can profoundly influence its biological action. Heterogeneity in sample size has also been shown to affect the *in vivo* action of silica.[8]

Acknowledgment

This work was supported by U.S. Environmental Protection Agency Grant R80825201.

[33] I. Gery, P. Davies, J. Derr, N. Krett, and J. A. Barranger, *Cell. Immunol.* **64,** 293 (1981).
[34] J. J. Oppenheim, B. M. Stadler, R. P. Siraganian, M. Mage, and B. Mathieson, *Fed. Proc., Fed. Am. Soc. Exp. Biol.* **41,** 257 (1982).
[35] J. J. Farrar and M. L. Hilfiker, *Fed. Proc., Fed. Am. Soc. Exp. Biol.* **41,** 263 (1982).
[36] D. P. Hartmann, M. M. Georgian, Y. Oghiso, and E. Kagan, *Clin. Exp. Immunol.* **55,** 643 (1984).
[37] D. P. Hartmann, M. M. Georgian, and E. Kagan, *J. Immunol.* **132,** 2693 (1984).

[33] Use of Horseradish Peroxidase and Fluorescent Dextrans to Study Fluid Pinocytosis in Leukocytes

By JANET M. OLIVER, RICHARD D. BERLIN, and BRUCE H. DAVIS

Introduction

Pinocytosis, the uptake of medium and its dissolved or suspended contents by enclosure in small membrane vesicles that bud from the cell surface, occurs in essentially all cells except for mature erythrocytes. Two categories of pinocytosis have been distinguished.[1,2] Fluid pinocytosis refers to the internalization of medium constituents by their simple inclusion in droplets of medium engulfed by the cell. This process is nonselective, nonsaturable, and involves no specific adsorption of constituents to the plasma membrane. Kinetic measurements show that uptake is usually linear with time and with the concentration of marker in the medium. Among leukocytes, cells of the monocytic/macrophage lineage show the most active fluid pinocytosis, while lymphocytes and granulocytes are more weakly pinocytic. Adsorptive pinocytosis refers to the uptake of extracellular substances, and incidentally of medium, following their binding to membrane receptors. This uptake is always saturable and leads to selective uptake and usually concentration of the ligand. Again monocytes and macrophages show maximal activity. Granulocytes are also capable of extensive adsorptive pinocytosis and lymphocytes show the lowest activity.

This chapter is specifically concerned with methods to observe and quantify fluid pinocytosis. The procedures have been developed and applied in macrophages and granulocytes. However the same protocols may be used with other leukocytes and in a range of tissue culture cells.

Choice of Substrates

The ideal substrate for studies of fluid pinocytosis is a readily soluble, membrane-impermeable, nonmetabolized reagent that does not modify cellular activities, has no binding affinity for the plasma membrane, is readily visualized in both the light and electron microscope, and can be quantified with accuracy and high sensitivity at the level of single cells and cell populations.

[1] S. C. Silverstein, R. M. Steinman, and Z. A. Cohn, *Annu. Rev. Biochem.* **46**, 669 (1977).
[2] S. C. Silverstein, ed., "Transport of Macromolecules in Cellular Systems." Dahlem Konferenzen, Berlin, 1978.

METHODS IN ENZYMOLOGY, VOL. 108

COMPARATIVE PROPERTIES OF HORSERADISH PEROXIDASE AND
FLUORESCENT DEXTRANS RELATED TO FLUID PINOCYTOSIS

Property	HRP	F-Dextran
Resistance to lysosomal degradation	3+	4+
General applicability	2+	3+
Sensitivity of detection	3+	4+
Quantitation	2+	4+
Utility in observation of living cells	0	4+
Utility for light microscopy of fixed specimens	4+	4+
Utility for electron microscopy	4+	1+

Many substances satisfy some of these criteria. Radiolabeled proteins, sucrose, inulin, and polyvinylpyrrolidone have all been used to quantify fluid pinocytosis.[1,2] However, these cannot be observed morphologically; most proteins are degraded fairly rapidly within lysosomes; and sucrose causes the osmotic swelling of secondary lysosomes. Conversely electron dense markers such as ferritin, Thorotrast, and colloidal gold are useful as ultrastructural markers but their uptake is difficult to quantify and it may also be difficult to prevent their nonspecific adsorption to the membrane. This article therefore concentrates on the advantages, disadvantages, and protocols for use of two markers that enable both qualitative and quantitative observation of fluid pinocytosis: horseradish peroxidase (HRP) and fluorescent dextrans. Steinman, Cohn, and colleagues[1,3,4] have worked most extensively with HRP while we[5-10] have emphasized the versatility of fluorescent dextrans as pinocytic markers.

Relative Merits of HRP and Fluorescent Dextrans

The merits of HRP and fluorescent dextrans as pinocytic markers are compared in the table and discussed below.

1. Both HRP and fluorescent dextrans are stable within the intracellular (lysosomal) milieu. In macrophages $t_{1/2}$ for HRP degradation is 10–12

[3] R. M. Steinman and Z. A. Cohn, *J. Cell Biol.* **55**, 186 (1972).
[4] R. M. Steinman, S. E. Brodie, and Z. A. Cohn, *J. Cell Biol.* **68**, 665 (1976).
[5] R. D. Berlin, J. M. Oliver, and R. J. Walter, *Cell* **15**, 327 (1978).
[6] R. D. Berlin and J. M. Oliver, *J. Cell Biol.* **85**, 660 (1980).
[7] R. J. Walter, R. D. Berlin, J. R. Pfeiffer, and J. M. Oliver, *J. Cell Biol.* **86**, 199 (1980).
[8] B. H. Davis, R. J. Walter, C. B. Pearson, E. L. Becker, and J. M. Oliver, *Am. J. Pathol.* **108**, 207 (1982).
[9] B. H. Davis, *J. Cell Biol.* **97**, 103a (1983).
[10] B. H. Davis, *Am. J. Pathol.* (in press).

hr.[3,4] Dextrans show no detectable degradation during 24–48 hr of incubation.[6]

2. Fluorescent dextrans may be more widely applicable than HRP. The enzyme cannot be used to measure fluid pinocytosis in alveolar macrophages that bind HRP via mannan receptors[11] or in neutrophils that contain endogenous peroxidase and catalase activities. In contrast, fluorescent dextrans have not been found to bind or adsorb to membranes. Their use is compromised if the cells contain high levels of endogenous autofluorescent substances. However most mammalian cells show minimal autofluorescence.

3. Both HRP and fluorescent dextrans are readily soluble in aqueous media and have no known adverse effect on cell function. Hence they can be used at relatively high concentrations in weakly pinocytic cells. Furthermore, both reagents can be detected with high sensitivity. HRP is detected by means of its catalysis of reactions which yield either colored or electron-dense products.[3,4] This catalysis allows for large signal amplification [with some limitations due to the peroxidase activities of tissues (both specific and nonspecific) that raise background; and the loss of ultrastructural detail if postfixation is unduly delayed]. Fluorescent dextrans are detected directly by fluorescence microscopy, by (micro)spectrofluorimetry, by flow cytometry, and/or by photon counting techniques. These inherently sensitive methods can be enhanced, if necessary, by synthesizing very highly substituted fluorochrome-dextran conjugates.

4. Fluorescent dextrans may be preferable to HRP for quantitative studies of fluid pinocytosis. Steinman and colleagues[3,4] have recommended use of prolonged rinsing procedures to elute residual HRP from cell surfaces and from culture and incubation vessels prior to measurement of enzyme uptake. Recent evidence suggests that efflux of pinocytized material occurs fairly rapidly.[12,13] Thus the HRP method may yield erroneously low estimates of fluid pinocytosis by cell populations. In contrast, extracellular contaminating fluorescein-dextran can be rapidly washed away from cell suspensions or cell monolayers in preparation for quantitative analysis of ingested dextrans. Furthermore, fluorescent dextran uptake may be quantified in single cells by microspectrofluorimetry[5–8] and flow cytometry (FCM)[9,10] whereas no methods to quantify HRP uptake in single cells have been developed.

5. Obviously HRP cannot be observed in living cells. On the other hand cells incubated with fluorescent dextrans and washed free of excess

[11] P. D. Stahl and P. H. Schlesinger, *Trends Biochem. Sci.* **5**, 194 (1980).

[12] J. M. Besterman, J. A. Airhart, R. C. Woodworth, and R. B. Law, *J. Cell Biol.* **91**, 716 (1981).

[13] C. J. Adams, K. M. Maury, and B. Storrie, *J. Cell Biol.* **93**, 632 (1982).

substrate can be followed for changes in pH,[14] granule morphology, etc. during the intracellular processing of ingested substrate. Sequential incubation of cells with dextrans conjugated with optically isolated fluorochromes also allows pulse-chase studies of the fate of pinocytized substrates in vivo.[7]

6. Both substrates are useful for light microscopy of fixed cells. Fluorescent dextrans permit immediate observation of pinocytic vesicles. The slides are wet mounted and their quality decreases after several days of storage at 4°. HRP reaction product must be developed. However, the resulting slides can be mounted and stored indefinitely. Furthermore, HRP-labeled cells can be embedded and thick sectioned for light microscopic observation.

7. Where applicable, HRP remains the reagent of choice for ultrastructural observations of fluid pinocytosis. Metal contrasting techniques make polysaccharides, including dextrans, visible at the ultrastructural level.[15] Unfortunately these methods are not suitable for cells like polymorphonuclear leukocytes that contain deposits of another polysaccharide, glycogen. Dextrans substituted with a highly electron-dense iron core (iron dextrans) may serve as useful pinocytic markers in mammalian cells.[16–18]

Materials

Crystalline, highly soluble HRP (Type II) is purchased from Sigma Chemical Co., St. Louis, MO and other manufacturers. Stock solutions may be stored frozen at 10–20 mg/ml in phosphate-buffered saline (PBS). Aggregates are eliminated by passage of enzyme solution through a 0.22-μm filter (Millipore) before use. The HRP substrates, diaminobenzidine and o-dianisidine may be purchased from Sigma.

Fluorescein-conjugated dextrans of various molecular weight are available from Sigma. We routinely employ FD-70, a conjugate with average molecular weight of 70,000. Fluorescein conjugates may also be prepared according to Glabe et al.[19] Rhodamine dextran is simply prepared

[14] S. Ohkuma and B. Poole, Proc. Natl. Acad. Sci. U.S.A. 75, 3327 (1978).
[15] V. Herzog and M. G. Farquhar, Proc. Natl. Acad. Sci. U.S.A. 74, 5073 (1977).
[16] C. R. Ricketts, J. S. G. Fox, C. Fitzmaurice, and G. F. Moss, Nature (London) 208, 237 (1965).
[17] B. J. Martin and S. S. Spicer. J. Histochem. Cytochem. 22, 206 (1974).
[18] A. H. Dutton, K. T. Tokuyasu, and S. J. Singer, Proc. Natl. Acad. Sci. U.S.A. 76, 3392 (1979).
[19] C. G. Glabe, P. K. Harty, and S. D. Rosen, Anat. Biochem. 310, 287 (1983).

by incubation of dextran (Sigma or Pharmacia Fine Chemicals, Pharmacia Inc., Piscataway, NJ) with rhodamine isothiocyanate (Sigma) (approx. 100:1, w/w) followed by either three cycles of ethanol precipitation and resolubilization in water or exhaustive dialysis against Dulbecco's phosphate-buffered saline, pH 7.4 (PBS). Iron dextrans are manufactured by Fisons Pharmaceuticals Ltd., Loughborough, UK as pharmaceutical preparations for the treatment of iron deficiency anemias. Imposil, sold in the United States as Nonemic, is distributed for use in animals by Burns Biotec, Oakland, CA. Imferon is distributed for human use by Merrell-Dow, Cincinnati, Ohio. These preparations consist of 2- to 3-nm-diameter particles.[16] Prior to use, they are dialyzed against 0.15 M NaCl to remove phenol and are freed of uncomplexed dextran by 3 cycles of centrifugation for 6 hr at 60,000 rpm.[18]

HRP Uptake Viewed by Light and Electron Microscopy

The cytochemical procedure of Graham and Karnovsky[20] underlies all procedures for the observation of HRP.[4,5,7,13,21,22]

In this laboratory cell monolayers (grown on 13-mm-diameter glass coverslips) or cell suspensions are incubated at 37° with 0.2–5.0 mg/ml HRP in their normal tissue culture medium. The HRP concentration is selected on the basis of cell type (higher concentrations for weakly pinocytic cells) and on the duration of incubation (2 to 60 min). In our experience, cells pinocytize more extensively in culture medium with serum than in simple medium like PBS. After incubation, cell monolayers are rinsed through four changes of PBS (5 sec each); and cell suspensions (about 10^6 cells/ml) are washed four times by centrifugation for 3 sec in an Eppendorf microcentrifuge, and resuspension in PBS. The cells are then fixed for 5 min with 1% glutaraldehyde in 0.1 M sodium cacodylate, pH 7.4 at room temperature.

HRP is demonstrated by incubation of cells at room temperature with freshly prepared diaminobenzidine (DAB; 0.5 mg/ml) and hydrogen peroxide (0.01%) in 0.05 M Tris buffer pH 7.4. Incubation is usually for 15–60 min. Its progress can be estimated by the gradual browning of the coverslips or cell pellets. If longer incubation is needed, the substrate is changed every 30 min to maintain adequate levels of H_2O_2. Controls are run identically except for the omission of HRP or H_2O_2.

For light microscopy, cells may simply be rinsed in PBS or 0.05 M Tris buffer pH 7.4, mounted in 50% glycerol in PBS, and the brown granules

[20] R. C. Graham and M. J. Karnovsky, *J. Histochem. Cytochem.* **14,** 291 (1966).
[21] J. Gonatas, A. Stieber, S. Olsnes, and N. K. Gonatas, *J. Cell Biol.* **87,** 579 (1980).
[22] J. Thyberg and K. Stenseth, *Eur. J. Cell Biol.* **25,** 308 (1981).

observed by transmitted light illumination.[4] However, the DAB reaction product is greatly enhanced by 10 min extraction of the cells with acetone at freezer temperature followed by 30 min incubation with 2% osmium tetroxide (OsO_4) in 0.1 M sodium cacodylate pH 7.4. The acetone removes lipid that would otherwise be stained by OsO_4. The DAB product appears black after OsO_4.[5] Figure 1A shows a typical result of this procedure.

For electron microscopy, cell pellets or monolayers are rinsed as above, postfixed in 2% OsO_4, embedded in Epon, thin sectioned with a diamond knife, and sections are examined without further staining. Figure 1B shows the typical appearance of HRP reaction product in a cultured macrophage.

Quantification of HRP Uptake in Cell Populations

HRP uptake is quantified from the H_2O_2-dependent oxidation of *o*-dianisidine by cell extracts. Cells are incubated for various times with HRP in the range from 0.2 to 10 mg/ml. Depending on the protocol being followed, they are then washed extensively (5 times with intermediate soak periods for elution of dish and cell bound enzyme for peritoneal macrophages[3]) or rapidly (3 times by centrifugation at 200 g for 5 min in the cold, with a tube change each time, for Chinese hamster ovary (CHO) cells[13]). In the absence of consensus among investigators, we recommend preliminary measurement of the elution of HRP with time from the cell type being studied in order to select the briefest wash protocol that can remove all cell and dish adherent enzyme.

Washed cells (around 2 × 10[6] cells/assay) are lysed in 0.5–1.0 ml of 0.05% Triton X-100 in water and stored frozen for HRP assay. HRP content is estimated spectrophometrically from the rate of development of colored product at 460 nm in a mixture containing 0.1 ml of cell supernatant and 0.9 ml of substrate prepared freshly from stock solutions as follows: 6.0 ml of 0.05 M phosphate buffer pH 5.0 plus 0.06 ml of 0.3% (v/v) H_2O_2 plus 0.05 ml of 1.0g% (w/v) *o*-dianisidine in absolute methanol. Control samples are prepared using cells not exposed to HRP or exposed to HRP at 4°. Within each experiment, a standard curve is plotted relating the initial linear period of color development (1–3 min) to the amount by weight of Sigma Type II HRP (1–20 mg/ml stock enzyme solution in PBS + 0.1% bovine serum albumin). Prior to analysis of experimental samples, additional standard curves are generated to demonstrate that HRP uptake in the cells of interest is linear with time and with concentration of enzyme added to the incubation medium.[3,12]

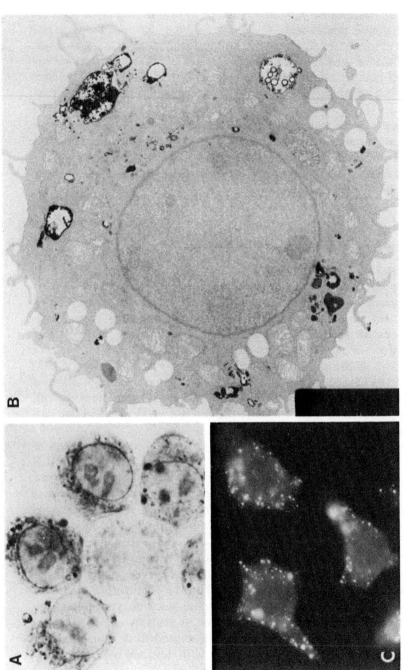

FIG. 1. HRP and fluorescein-dextran uptake view by light and electron microscopy. (A) Transmitted light micrograph of a group of J774.2 macrophages incubated for 15 min with 2 mg/ml HRP. The interphase cells contain DAB reaction product. The large central cell is in mitosis (anaphase) when all endocytic activities cease.[5] (B) Electron micrograph of an unstained thin section through a J774.2 macrophage incubated for 10 min 0.2 mg/ml HRP. (C) Fluorescence micrograph of a group of fluorescein dextran-labeled macrophages (FD-70, 10 mg/ml for 5 min) showing abundant pinocytic vesicles. (A, C) × 1200. (B) × 7100.

Dextran Uptake Viewed by Fluorescence Microscopy

Cell monolayers on glass coverslips or cell suspensions are incubated at 37° in complete medium supplemented with fluorescent dextran (usually Sigma FD-70, 1–20 mg/ml). After various periods of dextran uptake, cell monolayers are rinsed through 4 changes of PBS containing 1% bovine serum albumin (PBS-BSA) in 20-ml beakers (approx. 5 sec rinse per beaker) and fixed with 2% paraformaldehyde in PBS for 10 min. Cells labeled with fluorescein-dextran in suspension are washed by three or four successive 3-sec centrifugations in an Eppendorf microcentrifuge with resuspension of the cell pellets in 1 ml portions of PBS-BSA, then fixed as above. Fixed cells are washed once in PBS, mounted in 50% glycerol in PBS, and the distribution of dextran is observed by fluorescence microscopy[5,6] (Fig. 1C). Glutaraldehyde fixation should be avoided due to its induction of autofluorescence.

Dextran as an Ultrastructural Marker for Pinocytosis

Dextrans can be detected in the electron microscope by use of metal contrasting techniques.[15,23] Cells or tissues are incubated for various times with dextrans, then washed and fixed in 2% glutaraldehyde in 0.1 M phosphate buffer pH 7.4 (or any preferred fixative). The cells are postfixed in a mixture of equal volumes of 2% OsO_4 in 0.1 M phosphate buffer and a saturated solution of lead citrate in the same buffer. Alternatively, a one-step fixation staining is recommended by Simionescu et al.[23] For this the following stock solutions are used. A = 3% formaldehyde + 5% glutaraldehyde; B = 2% OsO_4; C = saturated lead citrate in 0.1 M phosphate or 0.1 M cacodylate buffer pH 7.4. These reagents are mixed in the ratio 3:2:1 prior to use and are kept cold during the fixation period (1–2 hr). After fixation, samples are dehydrated, embedded, thin sectioned and observed in the electron microscope.

Iron dextrans, which are electron dense, require no special fixation to be easily visible in the electron microscope.[17,18]

Fluorescent Dextran Uptake Measured in Single Cells

The pinocytic activity of single cells incubated with fluorescent dextrans can be measured by use of a commercial microspectrofluorimeter. We employ the Zeiss photometric system attached to a Zeiss Photomicroscope III.[6] Prior to the analysis of cell-associated fluorescence, the photometer is standardized by recording the intensity of a fluorescent stan-

[23] N. Simionescu, M. Simionescu, and G. E. Palade, J. Cell Biol. 53, 365 (1972).

dard (Zeiss) during its progressive attenuation with a set of neutral density filters. A linear curve of relative intensity vs percentage transmission ensures that the amplifier setting is within a linear range. The cells of interest are then incubated for various times and with various concentrations of fluorescent dextran in complete medium. Incubation is terminated by washing in PBS-BSA and fixation in 2% paraformaldehyde as described above. After a final wash, cells are plated onto slides (by inversion of cell monolayers on coverslips onto 10 μl of 50% glycerol-PBS pH 7.4, or by pipetting 10 μl of cell suspension onto slides and sealing under glass coverslips) and relative fluorescence intensities of the cells are measured. Fluorescence intensity $\pm$ SD is determined for at least 50 cells per point. Intensities measured on cells not exposed to fluorescent dextrans provide background measurements (and are usually negligible). These measurements should yield plots that show linear increases in fluorescence intensities with time and with dextran concentration as in intensities with time and with dextran concentration as in Fig. 2. Once linear standard curves are achieved, the experimental samples are analyzed.

We have found that measurements are best done in a single sitting, using the same diaphragm to frame the selected cells, blanking on cell-free regions of the same slide, and maintaining the same time interval between selecting a cell and taking a photometer reading (to minimize error due to fluorescence fading). Inclusion of n-propyl gallate (0.1–0.25 M) or phenylenediamine (100 mg%, w/v) in the buffered glycerol mounting medium is

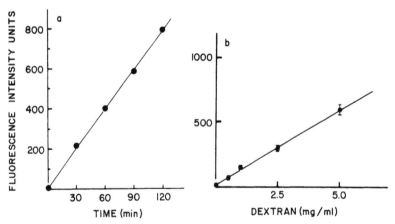

FIG. 2. Kinetics of fluorescein-dextran uptake measured in single cells. Suspensions of J774.2 macrophages were incubated with FD-70 for various times (a) or for 30 min with varying dextran concentrations (b), then fixed and washed. Fluorescence intensities of single cells were measured using a Zeiss microscope photometer. Each point represent the average fluorescence intensity measured in 50 replicate cells. Adapted from Berlin and Oliver,[6] with permission from *J. Cell Biol.*

an additional safeguard against error due to fluorescence photo-bleaching.[24,25]

If the standard curves fail to show linear uptake with concentration, the most likely explanation is the quenching of fluorescence emission due to the high fluorochrome concentration within vesicles. This is corrected by reducing the dextran concentration or the fluorochrome:dextran ratio. Nonlinear uptake kinetics with time observed in unfixed cells may be an artifact, reflecting the progressive increase in vesicle acidity and the strong, pH dependence of fluorescein emission intensity.[14] Paraformalde-hyde fixation eliminates this problem by normalizing granule pH to buffer pH. Nonlinear uptake with time in fixed cells may also occur as a cellular response to stimulation, e.g., the transient increased uptake of fluorescein-dextran in human granulocytes following exposure to the chemo-tactic peptide formylmethionyl-leucyl-phenylalanine.[8–10]

Fluorescent Dextran Uptake Measured in Cell Populations by Fluorescence Spectrophotometry

The uptake of fluorescein dextran by fluid pinocytosis is rapidly and sensitively measured in cell populations by fluorescence spectrophotome-try.[7,8] Cell suspensions or monolayers are incubated with fluorescein-dextran, and washed as above. Instead of fixation, the washed cells are lysed by vortexing in 0.1% Triton X-100 in PBS, pH 7.4. Supernatants are clarified by centrifugation and fluorescence intensities measured using a spectrofluorimeter with excitation and emission wavelengths of 486 and 515 nm, respectively. Extracts of cells not exposed to the pinocytic sub-strate or exposed at 4° serve as background (autofluorescence) controls. Results are expressed as relative fluorescence intensities per cell number or per pellet protein concentration. If samples of the medium are also measured, results can be converted to μl (volume of medium) pino-cytized.[7] Dilution of fluorescein dextran into PBS avoids quantification errors associated with fluorescence quenching or with the persistence of intracellular pH gradients. Preliminary kinetic experiments normally con-firm that the dextran uptake increases linearly with fluorochrome concen-tration and time under the conditions of the particular experiment (Fig. 3).

Fluorescent Dextran Uptake by Cells Measured by Flow Cytometry (This Volume [19])

Obviously, fluorescence spectrophotometry provides an average mea-sure of fluorescein dextran uptake over a whole cell population. However

[24] G. D. Johnson and G. M. de C. Nogueiran Araujo, *J. Immunol. Methods* **43**, 349 (1981).
[25] H. Giloh and J. W. Sedat, *Science* **217**, 1252 (1982).

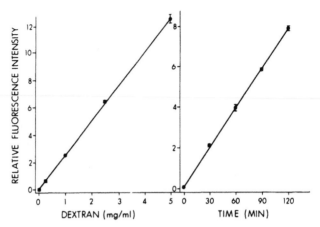

FIG. 3. Kinetics of fluorescein-dextran uptake by cell populations measured by fluorescence spectrophotometry. J774.2 cell suspensions (2×10^6 macrophages per ml) in complete medium were incubated for 30 min at 37° with a range of concentrations of fluorescein-dextran (left panel) or for various times with 1 mg/ml fluorescein-dextran (right panel). One-milliliter cell suspensions were subsequently washed three times by centrifugation with cold PBS, extracted in PBS that contained 0.1% Triton X-100, and dextran uptake was measured fluorimetrically. The data are the average ± SD of triplicate determinations. From Walter *et al.*,[7] with permission from *J. Cell Biol.*

cell preparations may contain several cell types or subpopulations of the same cell types that show different rates of pinocytosis. In these cases flow cytometry is the method of choice to quantify the pinocytic response.

Cell suspensions (around 10^6 cells/ml) are incubated with fluorescein dextran in PBS-BSA or other suitable medium for various times, then uptake is stopped by addition of an equal volume of 4% paraformaldehyde and collection on ice. Fixed cells are washed twice to remove excess dextran (extensive washing is unnecessary since the flow cytometer is relatively insensitive to extracellular fluorescence), stored at 4°, and analyzed within 24 hr for cell-associated fluorescence. Fixation preserves the cells and eliminates problems of quantification due to the pH sensitivity of fluorescein emission. Preliminary experiments are required to show that uptake is linear with concentration, indicating that fluorescence quenching at high substrate concentrations within vesicles is not compromising quantification.

Flow cytometric (FCM) analysis of samples is performed on a Coulter Epics V (or equivalent) flow cytometer with a 256 channel analyzer interfaced with a Multiparameter Data Acquisition and Displays computer system (MDADS). The argon ion laser (Coherent, Inc.) emits at a 488 nm line and is operated at 500 mW power. Fluorescence readings are ob-

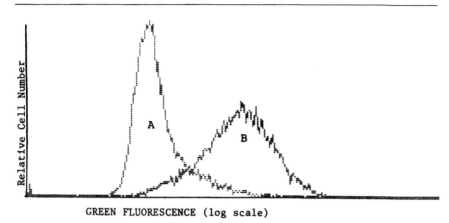

GREEN FLUORESCENCE (log scale)

Fig. 4. Fluorescein-dextran uptake by chemotactic factor stimulated human neutrophils measured by flow cytometry. PMNs were incubated for 10 min at 37° with 3 mg/ml fluorescein-dextran in buffer in the absence (A) and presence (B) of 10^{-6} M formylmethionyl-leucylphenylalanine. The relatively broad peak in (B) indicates that the increased uptake of fluorescein-dextran measured by FCM of stimulated cells is a heterogeneous response. In this experiment MCF values were 79 (A) and 129 (B).

tained through 515 nm long pass interference and 515 nm long pass absorbance filters; 10,000 cells are analyzed per histogram. Results may be expressed as mean channel of fluorescence (MCF), which is derived using the MDADS and represents the fluorescence channel at which 50% of the cells are less fluorescent and 50% are more fluorescent. Figure 4 shows a typical (FCM) analysis of fluorescent-dextran uptake by human blood neutrophils. The heterogeneity of uptake within these cell populations is evident.

With FCM analysis, appropriate gating may be used to eliminate uptake by irrelevent cell types, for example monocytes remaining in blood neutrophil preparations. With the more sophisticated flow cytometers now available at Los Alamos National Laboratories[26] and elsewhere, multiparameter sorting can be used to determine fluid pinocytosis as a function of cell cycle, presence of surface ligand-receptor complexes, cell volume, or other activities. Thus FCM represents a powerful tool for future studies of the activity and significance of fluid pinocytosis by leukocytes.

Acknowledgments

This work was supported in part by NIH Grants AI18564 and ES-01106 and by Grant BC179 from the American Cancer Society.

[26] J. A. Steinkamp, C. C. Stewart, and J. A. Crissman, *Cytometry* **2,** 226 (1982).

[34] Use of the Nude Mouse in Immunological Research

By Beppino C. Giovanella and James T. Hunter

Introduction

Since the discovery of the nude mouse in 1966[1] and the subsequent finding that the thymus gland was congenitally absent from mice homozygous for the autosomal recessive nude gene,[2] this animal has been used extensively in immunological research. Along with a multiplicity of reports in periodicals, the characterization and use of the nude mouse has been the subject of three major books,[3-5] one international symposium,[6] and three international workshops.[7-9] The nude mouse has been used in studies on immunoparasitology, allergy, autoimmunity, immunity to infectious diseases, transplantation immunity, immunological surveillance, tumor immunobiology, and has also been used to expand various tumor cell lines as well as to maintain hybridomas for monoclonal antibody production.

The nude mouse, although athymic, is not completely immunologically incompetent. Functional T lymphocytes are practically absent, but the B cell system, while not normal, is present. Indeed, some aspects of the defense system may be more developed in nude mice than in nonnude mice. For example, the nude mouse generally possesses higher titers of immunoglobulin M and greater NK activity than normal mice. Despite the presence of some degree of immunity, nude mice are subject to bacterial, parasitic, and viral infections (Table I) and have a very short life span if raised in a conventional environment. Even when the mice are raised

[1] S. P. Flanagan, *Genet. Res.* **8**, 295 (1966).
[2] E. M. Pantelouris, *Nature (London)* **217**, 370 (1968).
[3] J. Fogh and B. C. Giovanella, eds., "The Nude Mouse in Experimental and Clinical Research." Academic Press, New York, 1978.
[4] J. Fogh and B. C. Giovanella, eds., "The Nude Mouse in Experimental and Clinical Research," Vol. 2. Academic Press, New York, 1982.
[5] J. Rygaard, "Thymus and Self-Immunobiology of the Mouse Mutant Nude." F.A.D.L., Copenhagen, 1973.
[6] D. P. Houchens and A. A. Ovejera, eds., "Use of Athymic (Nude) Mice in Cancer Research." Fischer, New York, 1978.
[7] J. Rygaard and C. O. Povlsen, eds., "Proceedings of the First International Workshop on Nude Mice." Fischer, Stuttgart, 1974.
[8] T. Nomura, N. Ohsawa, N. Tamaoki, and K. Fujiwara, eds., "Proceedings on the Second International Workshop on Nude Mice." Univ. of Tokyo Press, Tokyo, 1977.
[9] N. D. Reed, ed., "Proceedings of the Third International Workshop on Nude Mice." Fischer, New York, 1982.

METHODS IN ENZYMOLOGY, VOL. 108

TABLE I
MOST IMPORTANT INFECTIOUS AGENTS CAUSING
REDUCED LIFE SPAN IN NUDE MICE

Agent	Syndrome
Murine hepatitis virus	Wasting
Sendai virus	Pneumonia
Murine adenovirus	Wasting
Toxoplasma	Wasting
Giardia and *Hexamita*	Wasting
Staphylococci	Facial abscesses and wasting

under specific pathogen-free conditions, the "normal flora" of the mice has a major impact and may cause dramatic variations in immunological parameters or even death of animals subjected to the stress of experimentation.

Breeding and Maintenance of Nude Mice

The nude mouse is an extremely useful tool for immunological experimentation, but good animal husbandry and strict precautionary measures are mandatory if meaningful experimental results are to be obtained.

Housing Systems

It is essential that nude mice be isolated from other animals, particularly other mice. Animals reared under standard conditions carry bacteria and viruses that are either species specific or adapted for growth in animals. Pathological changes due to these agents may not be manifest in fully immunocompetent mice but can be rapidly fatal for nude mice. Prime examples of this are the murine hepatitis virus and the Sendai virus. Both are capable of causing inapparent or mild infections in immunocompetent mice, but are a major cause of death in nude mice.[10]

Housing systems have been developed for maintaining nude mice by us[11] and are described in more general terms in the "Guide for the Care and Use of the Nude (Thymus-Deficient) Mouse in Biomedical Research."[12] We have used a system of plastic cages with sealed air filters

[10] K. Fujiwara, *in* "The Nude Mouse in Experimental and Clinical Research" (J. Fogh and B. C. Giovanella, eds.), Vol. 2. Academic Press, New York, 1982.

[11] B. C. Giovanella and J. S. Stehlin, *J. Natl. Cancer Inst.* **51**, 615 (1973).

[12] Institute of Laboratory Animal Resources, "Guide for the Care and Use of the Nude (Thymus-Deficient) Mouse in Biomedical Research." Natl. Acad. Sci., Washington, D.C., 1976.

successfully for the past 12 years. Standard transparent plastic cages and a lid covered with air filtering material are autoclaved. The assembly of the unit and all other procedures that require entering the cage are conducted under a hood with sterile positive laminar air flow by personnel in sterile surgical scrub clothing complete with double gloves, hat, and mask using strict aseptic technique. The junction of the cage and the filter are sealed with tape. Gloves are washed with antiseptic solution prior to changing cages. Thus each cage constitutes an independent unit, and the possibility of spreading diseases throughout the colony is minimized. As an added precaution, antibiotics are added to the sterile drinking water of the mice. Although expensive (due to requirements for large numbers of trained technicians), this system is the least costly and simplest method for successfully housing and conducting experiments with nude mice. The best system from the standpoint of isolation from the environment for maintenance of germ-free or pathogen-free nude mice is the use of isolator units. The isolators are sterilizable flexible or rigid chambers with filtered air supply and exhaust. All feed and supplies must be sterilized. Constant monitoring of the sterility of germ-free units or the microflora in specific pathogen-free units must be conducted. The principal disadvantage of the isolators is the extremely high-skilled labor requirement necessary to maintain even limited numbers of animals. Other secure but more expensive systems are provided by laminar air flow cage units and laminar air flow rooms. The former require careful animal husbandry practices in order to ensure continued sterility, while the latter require higher installation and maintenance costs and are more appropriate to large scale animals breeding facilities.

Breeding Systems

Systems which are standard for breeding mice have been described by many authors,[13–15] and specific articles on the breeding of nude mice have been published.[16–18] Female nude mice lactate poorly and are often infer-

[13] E. L. Green, in "Biology of the Laboratory Mouse" (E.L. Green, ed.), 2nd ed., p. 11. McGraw-Hill, New York, 1966.
[14] E. S. Russell, in "Biology of the Laboratory Mouse" (E. L. Green, ed.), 2nd ed., p. 11. McGraw-Hill, New York, 1966.
[15] G. L. Wolff, in "UFAW Handbook on the Care and Management of Laboratory Animals," p. 97. Livingstone, Edinburgh, 1967.
[16] B. C. Giovanella and J. S. Stehlin, J. Natl. Cancer Inst. 51, 615 (1973).
[17] J. Rygaard and C. O. Povlsen, Acta Pathol. Microbiol. Scand. 82, 48 (1974).
[18] K. Artzt, Transplantation 13, 547 (1972).

tile. Thus, mating schemes generally involve mating heterozygous ($nu/+$) females with homozygous recessive *(nu/nu)* males.

Three breeding systems are used: outbreeding, inbreeding, and cross-breeding. Outbreeding is the preferred system when mice of rigidly defined genetic background are not an absolute requirement. Outbred mice exhibit greater stamina than inbred mice. This is a considerable advantage in experiments with nude mice requiring long observation periods. In addition, the fertility of the mice is higher and larger litters are produced. Accidental inbreeding must be avoided or loss of fertility and stamina will eventually result. Mating systems for minimal inbreeding have been described.[19] Inbreeding (generally using brother × sister pairs set up at the time of weaning) is used if mice of specific and uniform genetic background are required. Such mice are more difficult to maintain than outbred animals. Crossbreeding of two different inbred nude mouse strains can be used in order to obtain genetically defined offspring with hybrid vigor.

The Immunobiology of the Nude Mouse

T Cell Function

The nude mouse has provided a model for the immunological situation of an animal deprived of its thymus. Anatomically, all the T cell compartments of the nudes lymphoid system appear empty and no or very few functional T cells have been detected in such mice. However, T cell precursors are present in the bone marrow. Reconstitution of the nude mouse can be achieved by thymus transplantation, but it is sustained and complete only if the transplant is made within a syngeneic system. Reconstitution by means of thymus extracts, or more or less purified thymic factors such as thymosin, has been at best partial and erratic.

Immunoglobulin Levels in Nude Mice

The main immunological lesion of the nude mouse is accompanied by numerous other abnormalities, such as deficiencies in the production of immunoglobulins. Unfortunately, whereas the major immunological defect of the nude (the lack or low level of functional T cells) is universally agreed upon, the secondary defects are much less well defined (Table II). The levels of immunoglobulins in the nudes are far from being well established. This is in part due to the fact that age and strain affect the levels of

---

[19] Institute of Laboratory Animal Resources, "Gnotobiotes: Standards and Guidelines for the Breeding, Care and Mangagement of Laboratory Animals." Natl. Acad. Sci., Washington, D.C., 1970.

TABLE II
EFFECT OF HOMOZYGOSITY FOR THE NUDE GENE
ON THE IMMUNOLOGICAL STATUS OF MICE

Parameter	Status
Thymus	Nonfunctional rudiment
Mature T cells	Virtually absent
Precursor T cells	Present
B cells	Functionally impaired
NK cells	Activity increased
Macrophages	Activity increased
Immunoglobulin levels	Variable
Interferon levels	Variable

immunoglobulins in the blood of the nudes (as they do, although to different degrees, in other mice). More significantly, different levels of antigenic stimulation provoke different levels of immunoglobulins. This is, perhaps, the most difficult variable to control. Different colonies of nudes are kept with different degrees of isolation and the bacterial flora of the mice varies considerably. The health status of these animals—especially for respiratory and gastrointestinal tract infections—is variable in the extreme. This variability reflects itself in the differences in longevity observed in different colonies of nudes of the same strain. The same can be said in regard to infestation with parasites. All of these pathological conditions translate themselves immunologically into different immunogenic stimuli which produce macroscopic variations in the levels of the different immunoglobulins forming the antibodies specific for the said antigens. Considering that different pathogens may provoke antibodies of the same or of a different class of immunoglobulins, there are enormous numbers of possible variations and permutations. This class of variables is so vast that it can account by itself for all the differences in immunoglobulin levels so far reported in nude mice.

Growth of Transplants of Normal and Neoplastic Tissues

As soon as the nude mouse became available, attempts were made to transplant fragments of normal and neoplastic tissues from mice or from other species into nude recipients.

Normal Tissues. The majority of the work in this field has been done with transplants of skin.[20] The methods used have been the standard full

[20] N. D. Reed and D. D. Manning, *in* "The Nude Mouse in Experimental and Clinical Research" (J. Fogh and B. C. Giovanella, eds.), p. 167. Academic Press, New York, 1978.

thickness skin transplant. No rejection has been observed. Skin from mice, rats, cats, chickens, lizards, and frogs have been grafted successfully and have remained viable indefinitely, with the exception of lizard and frog skin, which survived only for a limited time, owing to the difference in optimal temperature between the mice and these animals. All other skin grafts survived and continued to perform their normal functions, such as hair and feather growth, production of sebum, renewal of epithelium, etc. Good vascular connections were established between the host and the dermal stratum of the transplanted skin. Human fetal organs were also transplanted subcutaneously and most of them seemed to take, as did fetal rat tissues.

Neoplastic Tissues. Tumors developing in animals or other species, especially man, have been inoculated into nude mice; the first reports were in 1969.[21] The main route of inoculation has been subcutaneously by trocar, tissue mince, cell suspension, etc. All the other routes of inoculation commonly used for animal tumors have also been employed (intradermal, intramuscular, intraperitoneal, intracerebral, etc.).[22] Some less frequently utilized sites have been of considerable interest in the heterotransplantation of tumor in nudes, for example, the renal capsule.[23] Certain tumors, such as melanomas and colon carcinomas have a high take rate wherever they are inoculated. Others, such as lymphomas (with the exception of Burkitt lymphomas and of Hodgkins lymphomas) grow only or preferentially in selected sites such as the brain[24] (Burkitt lymphomas grow well subcutaneously,[25] and Hodgkins do not take in any location). For both human and animal tumors, there is no connection between the blood vessels of the host and the vascular bed of the neoplasm. The last degenerates and disappears completely after grafting.

The tumor tissues are vascularized by neoformed blood vessels of the host which invade the transplant (angiogenesis). After transplantation, the tumor has to survive by imbibition until vascularized. For this reason, it is useless to transplant tumor fragments which are thicker than 2 mm. The central part of any tissue thicker than 2 mm will die of anoxia and lack of nutrients before the implant receives its neovascular supply. Human tumors grow in nudes at a rate generally proportional to the growth rate in

[21] J. Rygaard and C. O. Povlsen, *Acta Pathol. Microbiol. Scand.* **77,** 1969 (1969).

[22] B. C. Giovanella and J. Fogh, *in* "The Nude Mouse in Experimental and Clinical Research" (J. Fogh and B.C. Giovanella, eds.), p. 282. Academic Press, New York, 1978.

[23] A. E. Bogden, D. E. Kelton, W. R. Cobb, and H. J. Esber, *in* "The Use of Athymic (Nude) Mice in Cancer Research" (D. P. Houchens and A. Ovejera, Eds.), p. 231. Fischer, New York, 1978.

[24] A. L. Epstein, M. M. Herman, H. Kim, R. Dorfman, and R. S. Kaplan, *Cancer* **37,** 2158 (1976).

[25] K. Nilsson, B. C. Giovanella, J. S. Stehlin, and G. Klein, *Int. J. Cancer* **19,** 337 (1977).

the host of origin. Anatomical structure at both the macro- and microscopic level is well preserved and the grafted tumor is generally indistinguishable from the structure of the tumor in the patient. Secretion of hormones in secretory tumors and the production of other biological substances (carcinoembryonic antigen, mucin, lipids, etc.) is maintained. Human tumors growing in nude mice produce metastases,[26] although not frequently. It has to be considered, however, that such tumors are growing in conditions that are far from ideal for metastasis. In humans, the primary tumor is generally removed or at least debulked. In the nude mouse the primary is left in place and in many cases is the cause of the animal's death after a few weeks. Further, the site of the primary tumor in the nude mouse is generally the subcutaneous space and not one of the internal organs as in man. From the point of view of the vasculature involved, this makes tumor dissemination by the lymphatic or hematological route much more difficult.

Some work has been done on the experimental production of metastases by intravenous inoculation of tumor cells. Not many tumors have been tested, but in the few cases studied, profound differences have been found between tumors of the same type and of similar biological characteristics in regard to their ability to produce tumors in the lungs when cell suspensions were inoculated intravenously into the tail vein. Even with this very preliminary result, it is clear that the ability to metastasize to the lungs is a biological characteristic of individual tumors, probably modulated by immunological factors. This is suggested by the fact that when nonmetastasizing tumors were inoculated, the lungs were free of metastases but heavily infiltrated by lymphocytes. Such infiltrates were totally lacking in the lungs where metastases developed.[27] Subcutaneous and renal transplants of human tumors have been extensively used for the study of experimental chemotherapy and of other therapeutic modalities (X-rays, hyperthermia, etc.). In such cases, the parameters studied are the delay of growth of the treated tumors compared with the growth of untreated controls. In both cases, these parameters are easily measured (the volume of subcutaneously growing tumors by calipers and the area of renal tumors by optical measurements with a micrometric ocular applied to a dissection microscope). For tumors growing in the brain, the effects of various treatments on the growth of the tumor is measured by inoculating known numbers of neoplastic cells into silent areas of the brain where animal death is caused exclusively by tumor compression of the brain— which of course is a measure of tumor volume. Animals receiving inocula

[26] B. C. Giovanella, S. O. Yim, A. C. Morgan, J. S. Stehlin, and L. J. Williams, *J. Natl. Cancer Inst.* **48,** 1531 (1972).
[27] B. C. Giovanella, unpublished results (1982).

of a given number of neoplastic cells all die within a very narrow time interval.[28]

Treatments effective in blocking tumor growth or causing extensive tumor cell death increase the survival of the treated animals. These increases are proportional to the effectiveness of the agent tested. Using panels of tumors of the same histological type, treated with various anticancer agents, it has been possible to demonstrate that the response of a given human tumor type to chemotherapy is essentially the same in nude mice and in the patient.[29] Work is now in progress to extend these studies (already completed for melanomas, breast carcinomas, and colon carcinomas) to other classes of human neoplasms. More important, from the practical standpoint, are the attempts being made to speed up the time frame of such studies. Today, in order to establish the effectiveness or lack of effectiveness of a drug on a specific tumor requires a time period of a month or more. Even under ideal conditions, this excessive time lag severely limits the effectiveness of the system. Attempts are now being made, by the use of prelabeled tumor cells, to drastically reduce such waiting periods.[30]

Infectious Diseases

The lack of T cells has rendered the nude mouse a good model for the study of certain bacterial diseases. For example, inoculation of *Mycobacterium leprae* into the paws of nude mice has produced a disease very similar to human leprosy in many of its symptoms.[31] The incubation period is quite long, more than 6 months, which makes it imperative for such studies to have available healthy nude mice, capable of surviving long enough to permit the experiment's completion. The evolution of the disease is similar to that observed in man. Inoculation of *M. leprae* in the foot pad even in small amounts (1×10^2 organisms) is followed after a long incubation period by multiplication of the bacteria in loco and by spread of the infection to other low temperature areas of the body. Bacteria can also be isolated in lower numbers from the liver and the spleen. After 18–22 months, macroscopic lepromatoid lesions are evident. The treatment with chemotherapeutic agents of nude mice affected by leprosy has given essentially the same results obtained with the same agents in man. Ac-

[28] B. C. Giovanella, unpublished results (1982).

[29] B. C. Giovanella, J. S. Stehlin, R. C. Shepard, and L. J. Williams, *Cancer* **152**, 1146 (1983).

[30] A. Lockshin, B. C. Giovanella, D. M. Vardeman, C. Quian, J. T. Mendoza, and J. S. Stehlin, *Proc. Am. Assoc. Cancer Res.* **24**, 307 (1983).

[31] J. M. Colston and K. Kohsaka, *in* "The Nude Mouse in Experimental and Clinical Research" (J. Fogh and B. C. Giovanella, eds.), Vol. 2, p. 247. Academic Press, New York, 1982.

cordingly, the nude mouse has proved to be a valuable model for the study of experimental chemotherapy of leprosy in man.

Immunoparasitologists have extensively used the nude mouse to study the effects of the lack of functional T lymphocytes and related functions on the development of parasitic infestations.[32] The results, although variable, have been interesting. In general, nude mice have a tendency to eliminate parasites, both macro and micro, more slowly than their normal counterparts. This results in higher and longer parasitemias with both protozoa and worms. However, such higher and more prolonged parasitemias do not always produce a more pronounced disease or a shortening of the life span. In some cases, the nude mice actually survive longer than normal mice although carrying a higher number of parasites. This paradox is explained with the assumptions that in such cases, death in the normal mice is caused by an allergic reaction to the parasite and that the allergic reaction is weaker in nude mice.

Use in Studies of Autoimmunity and Allergy

The nude mouse provides a model for studying autoimmunity, and the mechanisms involved in the maintenance of self-tolerance. The unique inability of the nude mouse to reject foreign cell transplants permits the study of the effector cells involved in or suspected to be involved in many types of autoimmune disease. For instance, injection of lymphocytes from diabetic patients intraperitoneally into nude mice induces a diabetic state in the mice.[33] Most of the studies to date have been concerned with the nature of the effector cell involved in autoimmune attack, but the increasing availability of nude mice of various genotypes may permit the precise analysis of the mechanisms involved in the control of the autoimmune response. The nude mouse has been used as a tool for studying the various components of allergic reactions. The fact that nude mice apparently lack the ability to generate specific anaphylactic type antibody responses[34] provides the possibility for experimentation to characterize the helper factors required for the expression of such responses in reconstituted nude mice.

Future studies in the nude mouse will undoubtedly utilize mice that are doubly or triply deficient in immunological function. Such mice may

[32] G. F. Mitchell, in "The Nude Mouse in Experimental and Clinical Research" (J. Fogh and B. C. Giovanella, eds.), Vol. 2, p. 268. Academic Press, New York, 1982.

[33] K. Buschard and J. Rygaard, in "The Nude Mouse in Experimental and Clinical Research" (J. Fogh and B. C. Giovanella, eds.), Vol. 2, p. 291. Academic Press, New York, 1982.

[34] N. R. Lynch, in "The Nude Mouse in Experimental and Clinical Research" (J. Fogh and B. C. Giovanella, eds.), Vol. 2, p. 345. Academic Press, New York, 1982.

provide revealing insights into the details of regulatory mechanisms controlling the immune system and the influence of the combined immunological deficiencies on tumor growth and other disease states. For example, nude mice have been produced that carry the beige mutation. These mice are not only nude, but are also profoundly deficient in NK activity. Such mice, although the product of prolonged breeding schemes, are now becoming more commonly available. A novel approach we are using to more rapidly create doubly deficient animals is parabiosis. Parabiosis between histoincompatible strains is impractical except in the case of the nude mouse. We have been able to produce parabionts between strains of nude mice with different secondary immunodeficiencies that survive as parabionts for well over 1 year.[35] Finally, the nude mutation in the rat has been observed,[36,37] and has been introduced into a number of different inbred strains of rats. The effects of the nude gene in rats have not been as fully characterized as those in mice but the advantage of the size alone will enable more complex surgical manipulations.

[35] B. C. Giovanella, J. T. Hunter, D. W. Gary, J. S. Stehlin, and J. Ledbetter, *Proc. Int. Workshop Immunodef. Anim. 4th, 1984*, p. 114.

[36] N. W. Festing, D. May, T. A. Connors, D. Lovell, and S. Sparrow, *Nature (London)* **247**, 365 (1978).

[37] L. J. McNeilage, B. F. Heslop, and M. V. Berridge, *Proc. Univ. Otago Med. Sch.* **57**, 47 (1979).

[35] Isolation of Epidermal Langerhans Cells

By Vera B. Morhenn and Eva A. Pfendt

Introduction

The human epidermis consists of a heterogeneous population of cells including keratinocytes in various stages of differentiation, Langerhans cells (LC),[1] melanocytes and Merkel cells. In particular, LC, which represent only 2–6% of the total epidermal population have proven difficult to separate as a relatively pure cell population.[1a] LC are of bone marrow

[1] Abbreviations: EDTA, ethylenediamine tetraacetic acid; FACS, fluorescence-activated cell sorter; FCS, fetal calf serum; R/M-FITC, fluorescein-isothiocyanate conjugated rabbit anti-mouse Ig; HLA-DR antigen, the human equivalent of the murine Ia or immune response-associated antigen; LC, Langerhans cells; PBML, peripheral blood mononuclear leukocytes; PBS, phosphate-buffered saline; SLR, skin cell lymphocyte reaction.

[1a] P. Langerhans, *Virchows Arch. Pathol. Anat. Physiol.* **44**, 325 (1868).

origin and are thought to play a major role in contact dermatitis.[2,3] LC express an antigen which binds antibody OKT6 and Ia (or HLA-DR) antigens, bear Fc and C3 receptors and present antigen to sensitized thymus-derived (T) cells.[4-7] Currently, LC cells are separated using a variety of techniques including Percoll and Ficoll–Hypaque gradient centrifugation and flow cytometry sorting.[7-10] However, all of these methods result in either low yield of LC and/or low purity. To investigate the mechanisms which regulate antigen presentation by LC, to study the role of LC in growth and differentiation of the epidermis, and to understand the functional interrelationships between the various cellular components of the skin, an antibody-coated Petri dish technique termed "panning" previously described for lymphocytes has been modified to separate LC and other epidermal cell subpopulations.[11] To attach LC to a goat anti-mouse IgG-coated plastic surface, antibody OKT6, a murine monoclonal antibody which binds specifically to LC in the epidermis was used.[4] With this method, populations of LC that are over 90% pure can be separated.

Materials

  Trypsin (M. A. Bioproducts)
  Antibody OKT6 (Ortho Pharmaceuticals)
  Goat anti-mouse immunoglobulin (IgG) (Tago, Inc.)
  Fluorescein isothiocyanate conjugated rabbit anti-mouse Ig (R/M-FITC) (Miles Laboratory), preadsorbed on dispersed, normal human epidermal cells
  100 × 25-mm Lab Tek Petri dish (Scientific Products)
  Castroviejo keratotome (Storz Instrument Co., St. Louis, MO)
  Dulbecco's phosphate-buffered saline (GIBCO, Grand Island, NY)

[2] J. G. Frelinger, L. Hood, S. Hill, and J. A. Frelinger, *Nature* (*London*) **282**, 321 (1979).
[3] J. Silberberg-Sinakin, *Cell. Immunol.* **25**, 137 (1976).
[4] E. Fithian, P. Kung, G. Goldstein, M. Rubenfeld, C. Fenoglio, and R. Edelson, *Proc. Natl. Acad. Sci. U.S.A.* **78**, 2541 (1981).
[5] G. Stingl, S. I. Katz, E. M. Shevach, E. C. Wolff-Schreiner, and I. Green, *J. Immunol.* **120**, 570 (1978).
[6] G. Stingl, E. C. Wolff-Schreiner, W. Pichler, F. Gschnait, W. Knapp, and K. Wolff, *Nature* (*London*) **268**, 245 (1977).
[7] G. Stingl, S. I. Katz, L. Clement, I. Green, and E. M. Shevach, *J. Immunol.* **121**, 2005 (1978).
[8] M. M. Brysk, J. M. Snider, and E. B. Smith, *J. Invest. Dermatol.* **77**, 205 (1981).
[9] A. Scheynius, L. Klareskog, U. Forsum, P. Matson, L. Karlsson, P. A. Peterson, and C. Sundstrom, *J. Invest. Dermatol.* **70**, 452 (1982).
[10] V. B. Morhenn, C. J. Benike, D. J. Charron, A. Cox, G. Mahrle, G. S. Wood, and E. G. Engleman, *J. Invest. Dermatol.* **79**, 277 (1982).
[11] L. J. Wysocki and V. L. Sato, *Proc. Natl. Acad. Sci. U.S.A.* **75**, 2844 (1978).

Procedure

Preparation of Dispersed Skin Cells. Single cell suspensions of normal skin are prepared from skin obtained at surgery.[12] Trimmed skin is cut into 1 × 5-cm strips and split-cut with a Castroviejo keratotome set at 0.1 mm. The resulting slices are treated for 25 min at 37° with 0.3% trypsin plus 0.1% EDTA in GNK (0.8% NaCl, 0.04% KCl, 0.1% glucose, 0.084% NaHCO$_3$, pH 7.3). Dispersed cells are suspended in RPMI 1640 medium supplemented with 10% heat-inactivated pooled human serum, 100 units/ml penicillin, 100 μg/ml gentamicin, 25 mM HEPES buffer, and 2 mM L-glutamine (complete RPMI medium). Viability, as determined by trypan blue exclusion immediately after trypsinization, is 90% or better.

Panning. A 100 × 25-mm Lab Tek Petri dish is coated with goat anti-mouse immunoglobulin (IgG) at 10 μg/ml in 0.05 M Tris buffer, pH 9.5 and incubated for 40 min at room temperature and then washed 3× with Dulbecco's phosphate-buffered saline (PBS) and 1× with 5% FCS/PBS.[11] The epidermal cells are washed 1× with PBS and are incubated with monoclonal antibody for 20 min at room temperature, washed with PBS 2×, resuspended in 5% FCS/PBS, layered on the Petri dish surface, incubated at 4° for 40 min, swirled, and incubated for another 30 min. The supernatant containing unattached cells is poured off gently, the Petri dish surface is washed 5× with 1% FCS/PBS. The cells obtained from the first 3 washes are pooled with the supernatant cells and the cells in the last 2 washes are discarded. Attached cells are scraped off with a rubber policeman in 1% FCS/PBS, centrifuged, and resuspended in medium.

Immunofluorescence (IF) Staining and Fluorescence-Activated Cell Sorter (FACS) (This Volume [19])

Analysis. To determine the number of cells which bind antibody, IF is applied. Attached and unattached cells are washed 2× with PBS and stained by suspending 2 × 10^6 cells in 0.2 ml R/M-FITC plus 0.02% sodium azide for 30 min, washed, and resuspended in 5% FCS/PBS. In preliminary experiments, it was determined that enough of the antibody used for panning remains on the cells which bind the antibody for subsequent IF. Cells are either examined with a Zeiss fluorescence microscope using excitation filters LP 455 plus SP 490 and a barrier filter of LP 520 or are processed in a FACS III (Becton Dickinson Electronic Laboratories, Mountain View, CA) at 1000 cells per second.[10] Background fluorescence is determined by analyzing unseparated (not preincubated with monoclonal antibody) skin cells labeled with R/M-FITC alone.

[12] S.-C. Liu and M. A. Karasek, *J. Invest. Dermatol.* **71,** 157 (1978).

Panning Using Antibody OKT6. Dispersed epidermal cells are incubated with antibody OKT6 and panned.[13] The yield of viable OKT6 positive cells is 1.3% of the starting population. To determine the purity of the cell population enriched for OKT6 positive cells (OKT6 enriched), cells attached to the Petri dish surface are scraped off, labeled with R/M-FITC, and the percentage of fluorescent cells determined using the fluorescence microscope. Over 90% of the cells are fluorescent and these cells are round and of uniform size. By contrast, the cell population depleted of OKT6 positive cells (OKT6 depleted) contains less than 1% OKT6 positive cells. An aliquot of the OKT6 enriched cells is processed for electron microscopy. The majority of these cells show nuclear features of mononuclear cells and on thin sections Birbeck granules characteristic of LC can be demonstrated in their cytoplasm (Fig. 1). The remaining cells in the OKT6 enriched fraction consist mainly of keratinocytes.

The stained adherent cell fraction may be also analyzed using the fluorescence-activated cell sorter. Ninety-seven percent of the cells demonstrate binding of OKT6 antibody.[14] For the initial experiments using the panning method reported in our original paper, 1% FCS in PBS is used to wash the Petri dish surface after coating with goat anti-mouse IgG.[13] By using 5% FCS in PBS to wash the antibody-coated Petri dish before incubating with antibody coated epidermal cells, the purity of the LC preparations is in excess of 90%.

Since LC express HLA-DR antigen and are, therefore, capable of stimulating allogeneic lymphocytes to proliferate in the skin cell lymphocyte reaction (SLR), this reaction may be used as an independent assay for the presence and viability of LC in the adherent, vis-à-vis nonadherent, cell fractions. Various concentrations of OKT6 enriched or depleted cells are coincubated with allogeneic PBML from 3 unrelated individuals in 2 separate experiments (see the table). In every instance, OKT6 enriched cells are much better stimulators than OKT6 depleted cells. In general, the stimulation of allogeneic lymphocytes is dependent on the number of stimulator cells present.

A number of techniques for separating epidermal cells into individual subpopulations have been described recently.[8–10,15] We have found that by far the best results are achieved with panning which, in our hands, achieves an enrichment of 15 to 20-fold for LC. The panning technique, while achieving at least the same purity as FACS sorting has the addi-

[13] V. B. Morhenn, G. S. Wood, E. G. Engleman, and A. R. Oseroff, *J. Invest. Dermatol.* **81,** 127s (1983).

[14] D. N. Saunder, C. A. Dinarello, and V. B. Morhenn, *J. Invest. Dermatol.* **82,** 605 (1984).

[15] V. B. Morhenn, E. Starr, C. Terrell, A. J. Cox, and E. G. Engleman, *J. Invest. Dermatol.* **78,** 319 (1982).

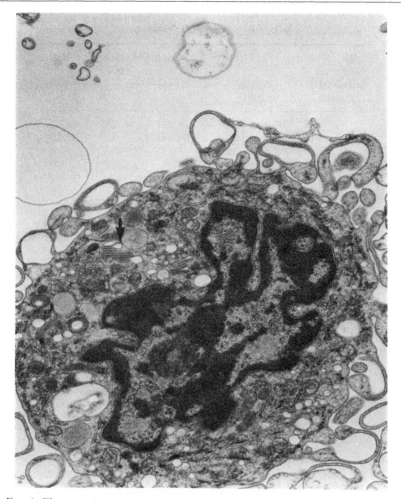

FIG. 1. Electron micrograph of a human LC (×13,600). OKT6 enriched skin cells are aliquoted and centrifuged. An aliquot is fixed for 30 min in Karnovsky's fixative and routinely processed for electron microscopy. Tennis racket shaped (Birbeck) granules (arrow) characteristic of LC are present in the cytoplasm.

tional advantages of higher yield of viable cells, shorter time, and lower cost. Moreover, using panning, it is very easy to maintain sterile technique.

Using other murine monoclonal antibodies such as VM-1 and VM-2, both of which are directed against an antigen found on human basal cells, it is possible to separate basal cells from the other epidermal cells.[13] Moreover, with this method, we have separated basal cells using serum

COMPARISON OF THE CAPACITY OF OKT-6-ENRICHED AND
OKT-6-DEPLETED CELLS TO STIMULATE ALLOGENEIC LYMPHOCYTES

Number of stimulator cells/well	Type of cell[a]	Response of allogeneic PBML (cpm)		
		Donor A	Donor B	Donor C
5×10^4	OKT6 enriched	69,014[b]	14,126[b]	49,800
	OKT6 depleted	14,960	571	1,153
10^4	OKT6 enriched	47,620	3,869	4,416
	OKT6 depleted	4,549	489	199
2×10^3	OKT6 enriched	67,015	733	819
	OKT6 depleted	2,059	620	118

[a] Unseparated skin cells (50,000 cells/well) coincubated with 50,000 PBML from donors A–C stimulated incorporation of 75,205, 5,916, and 1,153 cpm, respectively.

[b] Mean of triplicate wells; standard errors of the mean were less than 20%. The SLR is performed as described previously [V. B. Morhenn, C. J. Benike, and E. G. Engleman, *J. Invest. Dermatol.* **75,** 249 (1980)]. Peripheral blood mononuclear leukocytes (PBML) are obtained from healthy volunteers by Ficoll–Hypaque gradient centrifugation of fresh defibrinated blood [A. Bøyum, *Scand. J. Clin. Invest.* *(Suppl.)* **21,** 77 (1968)]. Fifty thousand PBML are coincubated in microtiter plates with varying numbers of allogeneic skin cells in complete RPMI medium in a final volume of 0.2 ml. The cultures are maintained in air/10% CO_2 for 6 days at 37°. [³H]Thymidine is then added (1 μCi/well) and the plates harvested 18 hr later.

obtained from a patient with bullous pemphigoid and have shown that most of these cells bind antibody VM-1.[16] Furthermore, using the monoclonal antibody T-200, it is possible to separate LC from the other cell types in mouse epidermis.[17]

Thus, with the appropriate antibody, it appears that the panning technique can be applied to enrich for most epidermal cell subpopulations including melanocytes and Merkel cells to a high degree of purity. This type of cell separation can be performed using human as well as mouse epidermal cells and therefore probably can be used to separate skin subpopulations of most mammalian species.

[16] V. B. Morhenn, unpublished data.
[17] G. S. Wood, J. Kosek, E. Butcher, and V. B. Morhenn, *J. Invest. Dermatol.*, in press.
[18] V. B. Morhenn, C. J. Benike, and E. G. Engleman, *J. Invest. Dermatol.* **75,** 249 (1980).
[19] A. Bøyum, *Scand. J. Clin. Lab. Invest.*, Suppl. **21,** 77 (1968).

[36] Cryopreservation of Lymphoid Cells

By J. Dixon Gray *and* S. H. Golub

Introduction

In vitro assessment of the cellular immune response can be influenced by day to day test variation, differences in responses by different individuals, and variation in reagents employed. Clinical samples provide additional problems as these often arrive at unpredictable times producing logistic difficulties when assays must be set up in advance. Cryopreserved lymphocytes have been used as a means to reduce these problems. Cryopreservation permits an analysis of the characteristics or functions of several lymphocyte samples on the same day under the same assay conditions. Furthermore, the use of cryopreserved lymphocytes facilitates the standardization of assay procedures to be employed.

The cryopreservation technique described in this contribution is one that we have found to be very effective with regard to cell recovery, viability, and function. This procedure is routinely used to cryopreserve lymphoid cells derived from human peripheral blood, lymph nodes, and thymus specimens as well as cultured cell lines.

Cryopreservation Technique

Equipment

Cryo-Med model 900 programmable freezing apparatus or similar apparatus
Liquid nitrogen tank
Liquid nitrogen freezer

Reagents

Tissue culture medium, e.g., RPMI 1640
Serum (one can use fetal calf serum, human AB serum, or human serum depleted of γ-globulin)
Dimethyl sulfoxide (DMSO) reagent grade

Procedure. Human blood leukocytes are purified on Ficoll–Isopaque gradients to remove red cells and polymorphonuclear cells (this volume [9]). Further purification of lymphocyte subpopulations can also be performed without affecting the cryopreservation efficiency. We have found that nylon wool column passage is particularly useful for cryopreserving

METHODS IN ENZYMOLOGY, VOL. 108

thymocytes which tend to be more fragile than other lymphocytes (this volume [28]).

Lymphocytes are resuspended just prior to freezing in an ice cold cryopreservative solution consisting of 50% v/v tissue culture media, 40% v/v serum, and 10% v/v DMSO. When preparing the cryopreservative solution DMSO is the final solution to be added. Once the lymphocytes have been resuspended in the cryopreservative solution, 1-ml aliquots are added to screw-cap plastic vials and immediately placed in ice. The vials are then placed in the freezing chamber of the programmed freezer which has been precooled to 4°. One sample vial is required to monitor the freezing rate. This is done by securely inserting the temperature probe into the vial. Following the instructions of the freezing apparatus used, liquid nitrogen is introduced at a rate to give a decline in temperature of 1°/min. There will be a sudden increase of 1–5° at the phase transition (usually between −10° and −20°). It is important to introduce more liquid nitrogen to prevent excess warming of the samples. This can be accomplished by a manual override of the machine program, although many of the newer instruments allow one to preprogram in influx of liquid N_2 to counteract the heat of fusion. The exact duration of override must be determined by experimentation with each instrument. It is important not to add an excessive amount of liquid N_2 as a sudden decline in temperature at this point can reduce viability.

Once the temperature of the samples reaches −50°, the manual override can again be used to rapidly bring the temperature down to −100 to −150°. This final cooling is necessary as there is always some warming of the samples in the transfer to the storage liquid nitrogen freezer. Samples should then be stored in the vapor phase of a liquid nitrogen freezer.

Procedure for Thawing of Cryopreserved Samples

Samples to be thawed are placed in a 37° water bath. Once the last ice crystal has melted, the sample is diluted with tissue culture medium containing 5% serum at room temperature. Ice cold diluent will significantly reduce viability. Slow dilution is crucial for optimal recovery of viable cells and the diluent should be added dropwise (0.1 ml) with frequent mixing until 10 volumes of diluent to 1 volume of cryopreserved cells have been added. This will result in >90% viable cell recovery.

Comments on Technique

1. It is important to use optimal concentrations of DMSO (7.5–10%) in the freezing solution to insure maximal cell recovery. With optimal

concentrations of DMSO, the inclusion of serum is not essential but is generally preferred.[1,2]

2. The number of cells which can effectively be frozen is variable. Cell concentrations between 1×10^6 and 2×10^8 have been frozen without any difference in percent recovery.[3,4]

3. Optimal freezing rate is between -1 and $-2°/min$ although with hardier cells such as some cultured cell lines, "crash" freezing (placing sample directly into vapor phase of liquid nitrogen freezer) can be performed.

4. Functional activity can still be recovered from lymphocytes which have been stored in liquid nitrogen for as long as 3 years.[5]

5. With fragile lymphoid samples, e.g., thymocytes, some clumping of cells may be detected. Passage of thymocytes over a nylon wool column before freezing can overcome this problem. However, if after thawing and resuspension the sample contains cell clumps, then treatment with 0.1 ml of DNase at 30 mg/ml for 30 min will result in a single cell suspension.[6]

Discussion

Although lymphocytes can successfully be frozen, stored, and recovered, it is essential that the lymphocytes recovered from cryopreservation retain the properties and functions of fresh lymphocytes.

One approach to determine lymphocyte functional activity is to measure their response to specific mitogens. Frozen lymphocytes do not differ from fresh lymphocytes in their ability to respond to lymphocyte mitogens such as PHA, Con A, or pokeweed mitogen.[7-9] However, Birkeland[9] observed that the response of frozen lymphocytes to PHA was delayed in that the peak response occurred on day 7 whereas fresh lymphocytes peak response appeared on day 4.

[1] S. A. Birkeland, *Cryobiology* **13,** 442 (1976).

[2] R. S. Weiner, *J. Immunol. Methods* **10,** 49 (1976).

[3] S. H. Golub, *in* "In Vitro Methods in Cell Mediated and Tumor Immunity" (B. R. Bloom and J. R. David, eds.), p. 731. Academic Press, New York, 1976.

[4] H. T. Holden, R. K. Oldham, J. R. Ortaldo, and R. B. Herberman, *in* "In Vitro Methods in Cell Mediated and Tumor Immunity" (B. R. Bloom and J. R. David, eds.), p. 723. Academic Press, New York, 1976.

[5] D. M. Strong, J. N. Woody, M. A. Factor, A. Ahmed, and K. W. Sell, *Clin. Exp. Immunol.* **21,** 442 (1975).

[6] M. Torten, N. Sidell, and S. H. Golub, *J. Exp. Med.* **156,** 1545 (1982).

[7] S. H. Golub, H. L. Sulit, and D. L. Morton, *Transplantation* **19,** 195 (1975).

[8] R. K. Oldham, J. H. Dean, G. B. Cannon, J. R. Ortaldo, G. Dunston, F. Applebaum, J. L. McCoy, J. Djeu, and R. B. Herberman, *Int. J. Cancer* **18,** 145 (1976).

[9] S. A. Birkeland, *Cryobiology* **13,** 433 (1976).

In mixed lymphocyte cultures frozen lymphocytes are equally effective as fresh lymphocytes in their ability to function as responders or stimulators in the assay.[5,7-9]

Human T lymphocytes express receptors for sheep erythrocytes (E receptor). Some authors report no difference betweeh fresh and frozen lymphocytes in the expression of the E receptor,[5] others detect an increase in the number of sheep erythrocytes with frozen lymphocytes,[10] and we found a decrease in E binding.[11] These differing findings may reflect differences in lymphocyte storage times since Oldham et al.[8] found that the number of frozen lymphocytes binding sheep erythrocytes increased with increasing time of storage in liquid nitrogen. Other lymphocyte markers such as EAC receptors[5,10] are found to be similarly expressed on frozen and fresh lymphocytes, although T cells bearing $Fc\gamma$ or $Fc\mu$ receptors have been reported to be sensitive to cryopreservation.[11] T cell differentiation antigens, such as those detected with monoclonal antibodies, can be effectively studied on cryopreserved lymphocytes.[12]

The expression of surface immunoglobulin on B lymphocytes is comparable on frozen and fresh lymphocytes.[5] Furthermore, analysis by a fluorescent-activated cell sorter revealed no difference between frozen and fresh lymphocytes in their expression of surface IgM, IgD, IgG, and IgA.[13]

It would therefore appear that cryopreserved lymphocytes retain most of the characteristics of fresh lymphocytes. However, frozen lymphocytes are less efficient than fresh lymphocytes in mediating cytotoxic activity as detected in cytotoxicity assays of 4–16 hr duration.[8,10,11] Cytotoxicity assays of 18 hr or longer in duration show no differences in cytolytic activity indicating that the depressed activity is temporary.[7,14,15] Decreased activity in NK and antibody-dependent cellular cytotoxicity assays (ADCC) may reflect impairment of the $Fc\gamma$ receptor.[11] We have compared fresh and frozen lymphocytes in their ability to mediate natural killer cell activity using both 4-hr ^{51}Cr release and 3-hr single cell assay. Cytotoxic activity in both assays is reduced with frozen lymphocytes as

[10] X. L. Kaprovitch, E. Rosenkovitch, H. Ben-Busset, and G. Izak, Cryobiology 17, 12 (1980).

[11] C. D. Callery, M. Golightly, N. Sidell, and S. H. Golub, J. Immunol. Methods 35, 213 (1980).

[12] L. M. Karavodin, A. E. Giuliano, and S. H. Golub, Cancer Immunol. Immunother. 11, 251 (1981).

[13] R. B. Slease, D. M. Strong, R. Wistar, Jr., R. E. Budd, and I. Scher, Cryobiology 17, 523 (1980).

[14] C. O'Toole, Natl. Cancer Inst. Monogr. 37, 19 (1975).

[15] M. A. Factor, D. M. Strong, J. L. Miller, and K. W. Sell, Cryobiology 12, 521 (1975).

effector cells. However, there is no difference between frozen and fresh lymphocytes in their ability to form conjugates indicating that the target recognition structures remain fully intact on frozen lymphocytes.

In summary, frozen lymphocytes retain most of the characteristics of fresh lymphocytes. This procedure allows several samples obtained at different times to be tested under identical assay conditions. Cryopreserved lymphocytes can provide a source of reference lymphocytes for standardization of assays where daily variation is a significant problem.[16]

[16] S. H. Golub, F. Dorey, D. Hara, D. L. Morton, and M. W. Burk, *JNCI, J. Natl. Cancer Inst.* **68,** 703 (1982).

Section III

Methods for the Study of Surface Immunoglobulin

[37] Ligand-Induced Patching and Capping of Surface Immunoglobulins

By F. LOOR

Patching and capping phenomena have been described for a variety of membrane components in a number of different cell types.[1] The basic model remains, however, the patching and capping of membrane immunoglobulin (mIg)[1a] of B lymphocytes.[2,3] With other membrane components, however, the conditions for obtaining such redistribution phenomena may be significantly different. It is not possible to describe here in detail all the various methods which have been used successfully for inducing patching and capping of mIg and other membrane components. Therefore, anybody planning to redistribute a novel membrane component should first work with mIg, then with other membrane components whose redistribution has already been obtained by others. Knowledge of the mechanisms of these phenomena is also mandatory. A number of reviews have appeared which cover several thousand references up to 1980 and deal with various aspects of redistribution phenomena.[1,4–7] The relevance of patching to cell activation has been discussed.[4,6] A detailed model for the mechanisms of patching and capping, and related membrane phenomena such as pinosome formation, microvilli formation, endocytosis, shedding, etc. has also been developed.[7]

Briefly, the formation of large clusters of a given membrane antigen upon ligand binding is a passive redistribution process, as spotting/patching does not require the cell to be alive or metabolically active. Clustering is actually best observed on cells whose capping machinery is inhibited either by specific drugs or as a consequence of a general inhibition of the cellular metabolism. The clustering of membrane components is an immunoprecipitation reaction or a lectin-mediated agglutination occurring in

[1] F. Loor, *Prog. Allergy* **23**, 1 (1977).

[1a] Abbreviations: BSA, bovine serum albumin; FCS, fetal calf serum; PBS, phosphate-buffered saline; F-PBS, paraformaldehyde-phosphate-buffered saline; DMSO, dimethyl sulfoxide; mIg, membrane immunoglobulin; Con A, concanavalin A.

[2] R. B. Taylor, P. H. Duffus, M. C. Raff, and S. De Petris, *Nature (London)* **233**, 225 (1971).

[3] F. Loor, L. Forni, and B. Pernis, *Eur. J. Immunol.* **2**, 203 (1972).

[4] G. F. Schreiner and E. R. Unanue, *Adv. Immunol.* **24**, 37 (1976).

[5] S. De Petris, *Cell Surf. Rev.* **3**, 643 (1977).

[6] F. Loor, *Adv. Immunol.* **30**, 1 (1980).

[7] F. Loor, *Cell Surf. Rev.* **7**, 253 (1981).

a two-dimensional pattern at the level of the plasma membrane. It will, therefore, be affected not only by all factors that commonly control such phenomena occurring in the fluid aqueous phase in three dimensions, but also by additional factors intrinsic to the plasma membrane, such as its viscosity, organization in discrete domains or in a continuous fluid mosaic. The outcome of the cluster formation will not only depend on the biophysical and biochemical parameters of the membrane itself but will also be influenced by biological and physiological interactions of the membrane components with other components of the plasma membrane and/ or the cytoplasmic cortex of the cell.

The general characteristics of the capping phenomenon, particularly the energy requirements, temperature dependence, inhibition or reversion by microfilament-directed drugs, and the accumulation of microfilaments under the cap, have led to the suggestion that capping is probably a cellular contractile phenomenon. Contractile microfilaments drag anchored patches of aggregated membrane components into an area of the cell where they can be endocytosed and digested, or shed from the surface. The role of microtubules in the capping phenomenon is more complex and they do not appear to enhance microfilament mobility but rather to inhibit it. Capping of any membrane component probably occurs as a consequence of microprecipitation resulting in entrapment within the lattice formed in the plasma membrane, of a microfilament-associated membrane component: the ''membrane capping component''[8] or ''integral protein X.''[9] This may then selectively sweep any *aggregated* membrane components or adsorbed material toward the cap area. The hypothesis has been formulated, however, that not only clustered components but also whole domains of membrane might be brought into the cap.[7] This concept puts severe restriction on the use of capping as a means to selectively remove given membrane components from the plasma membrane.

Classification of Fluorescence Patterns

The microscopic appearance of fluorescent ligand-coated cells may fall into three main categories: ring (Fig. 1), spots/patches (Fig. 2), and caps (Fig. 3).[2,3] Ring fluorescence is observed when focusing on the edge of the cell, at the level of its equatorial plane. It appears as a smooth circular ring of fluorescence. Spots or patches appear as an interrupted ring of fluorescence when focusing at the equatorial plane level; they should disappear (and new ones should appear) when focusing up and down, as they go out of (and into) the depth of the focus of the optical system. Caps appear as a smooth or speckled fluorescence, on only one-

---

[8] F. Loor, N. Block, and J. R. Little, *Cell. Immunol.* **17**, 351 (1975).
[9] L. Y. W. Bourguignon and S. J. Singer, *Proc. Natl. Acad. Sci. U.S.A.* **74**, 5031 (1977).

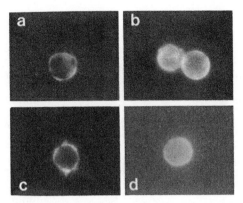

FIG. 1. Fluorescence patterns classified as "rings" (Loor *et al.*[3]). Mouse spleen cells. (a) Fluorescence-labeled rabbit Fab anti-mouse T cells (perfectly smooth fluorescence); (b) fluorescence-labeled Con A (the fluorescence is not quite homogeneous); (c) and (d) Tobacco mosaic virus binding cells coated in antigen excess and detected by fluorescence-labeled rabbit Fab anti-TMV; (c) ring with a fuzzy appearance (microvilli); (d) ring with a smooth gradient of fluorescence.

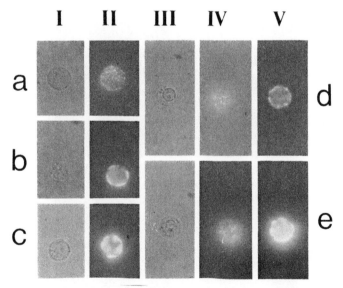

FIG. 2. Fluorescence patterns classified as "spots or patches" (from experiments reported in Loor *et al.*[8]). Mouse embryonic thymus, early blasts labeled with fluorescent Tla antigens (a) and fluorescent Thy-1 antigens (b–e). (I) and (II) Three different cells photographed under visible (I) and fluorescent (II) light showing clustered fluorescence: typical small spots (a), intermediate-size clusters (b), and large confluent patches (c). (III–V) Two different cells (d and e) photographed each under visible (III) and fluorescent (IV, V) light with two different levels of focus [superior pole (IV) and equatorial plane (V)] showing the different aspects of fluorescence clusters on different cells and on the same cell depending on the focus.

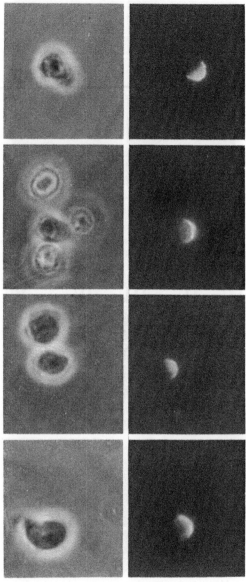

FIG. 3. Fluorescence patterns classified as "caps" (from experiments reported in Loor[12]). Cells are from the same mouse spleen preparation coated with 1 μg/ml fluorescent Con A followed by anti-Con A rabbit IgG.

half or less of the cell surface, frequently on a prominent uropod. These fluorescence patterns however are not well defined and there are a number of intermediate forms. Furthermore, instead of discrete clusters or caps one may find only an area of fluorescence brighter than the background. Thus, the classification of the fluorescence patterns remains rather subjective.

It is also semantic and subjective to distinguish between spots and patches, the latter being larger than the former. What at the electron microscope appears as large clusters,[5] may appear as very small spots or even as a ring when examined with the optical microscope. However, clusters of very small size might be detected as small fluorescent spots if a large amount of fluorescence is emitted from the cluster and little or none from the adjacent membrane area.

Endocytotic vesicles (pinosomes) containing the fluorescent ligand may be distinguished from spots by focusing up and down. This also may establish whether they reside in the cytoplasmic or on the membrane, although this distinction may be subjective and difficult for pinosomes close to the cell periphery. The best way to achieve a clear distinction between pinosomes and spots of ligand present on the cell surface is to fix the cell following the redistribution of the ligand by the first fluorescent antibody (see below) and to expose it to a second antibody directed against the first one and coupled to a different fluorochrome. Pinosomes will be labeled by the first fluorochrome only, whereas all types of surface clusters will display both fluorochromes.

The most sensitive method (indirect immunofluorescence) for revealing only the spots or patches on the cell surface consists of binding an unlabeled antibody to the cell. The surface-bound antibody is then revealed on the fixed cells by a fluorescent antibody directed against it.

Quantitation and Kinetics of Redistribution Processes

Before studying quantitatively the capping of membrane components and/or the possible interference of various treatments, it is advisable to follow carefully the kinetics of the whole process under various conditions. One must take into consideration all the cells of the suspension, fluorescent, and nonfluorescent. Since not all membrane components redistribute as fast as mIg on the cell surface, we recommend an initial examination of the cells in the fluorescence microscope at various times (0, 5, 15, 30, 60, and 120 min) under typical capping conditions, and then determination of the relative percentages of the different staining patterns (none, ring, clusters, cap) at each time. One should also perform controls in order to distinguish membrane aggregates from endocytotic vesicles (pinosomes). Under certain conditions, cells with microprojections (hairy cells) may appear as cells with spots or caps, depending on the distribution of the microvilli all around or on the uropod.[10]

One also has to keep in mind that the various treatments undergone by the cells before being put under "capping conditions" may affect not only the degree of capping but also the nature of the clusters which are formed and the patterns of redistribution which are obtained. Thus, for example, lymphocytes kept at room temperature will not cap with Con A at 37°, while they will if their microtubules have been disrupted by cold or drugs.[10-12] B lymphocytes treated in the cold with anti-Ig show a number of typical spots and caps, while when treated at room temperature with anti-Ig, they show a diffuse membrane fluorescence with brighter areas that may be pinocytotic vesicles or microvilli (depending on the medium). The few caps that are present may be only gradients of fluorescence brightness toward a pole of the cells or a polar accumulation of endocytotic vesicles or microvilli on an uropod.[10]

Basic Procedure for the Study of the Redistribution of Membrane Immunoglobulins

One milliliter of the various media to be assayed [including 0.5% BSA in PBS (which allows capping) and the same medium containing 0.01 M NaN$_3$ (which allows spotting only)] and containing 2.5×10^6 cells is dispensed in each of a series of round bottom tubes (Falcon Plastic No. 2054). The cells will be kept in their respective media at 0–4° except for two incubations at 37°, the first before coating the cells with antibody and the second after washing away the excess antibody and before fixing the cells. The cells are first equilibrated for 15 min, in a water bath at 37°; they are then pelleted by centrifugation (10 min at 100–150 g), and the pellets are resuspended in 0.1 ml of their respective media. The fluorescent antibody reagent [F(ab')$_2$ fragments, from whole-antiserum IgG, at 500 μg/ml in 1% BSA-PBS, see Appendix] is added in a minimal volume (25–50 μl) and the mixture is incubated on ice for 15 min. The excess unbound antibody is removed by four washes of the cells in 2.0 ml of medium (100–150 g, 10 min, refrigerated centrifuge). The last cell pellet is then resuspended in 0.1 ml of medium and incubated 10–15 min in a water bath at 37° to allow spotting and capping. The cells are then fixed with 1 ml of 3.7% paraformaldehyde in PBS (F-PBS), for 15 min at room temperature, pelleted, resuspended in 50–100 μl of a 9:1 (v/v) mixture of glycerol and

[10] F. Loor and L.-B. Hägg, Eur. J. Immunol. **5**, 854 (1975).
[11] G. M. Edelman, I. Yahara, and J. L. Wang, Proc. Natl. Acad. Sci. U.S.A. **70**, 1442 (1973).
[12] F. Loor, Eur. J. Immunol. **4**, 210 (1974).

0.01 *M* barbital buffer, pH 8.6, and mounted between slides sealed with nail polish.

Comments

1. Alkalinization of the glycerol is recommended particularly when fluorescein-labeled antibody is used, because this fluorochrome displays poor fluorescence in acidic or neutral conditions. F-PBS-fixed cells can be kept for weeks at 4° without alteration of the fluorescence. When cells show autofluorescence (visible through all filters), it is better to discard the preparation because the resolution of specifically fluorescent structures becomes very poor.

2. It is possible to omit the fixation step and put the stained, washed cells on microscope slides either directly after removing the unbound antibody or after the second incubation at 37°. By so doing, the development of some redistribution phenomena may be observed. These phenomena, however, occur under conditions difficult to reproduce (e.g., temperature of the room and the microscope plate, effect of the microscope light on the cell suspension, "living conditions" in the microscopic chamber between slide, coverslip and nail polish, etc.).

3. Considering that the volume of medium in the cell pellet is less than 0.1 ml, the washing procedure corresponds to about a 10^4-fold dilution of the unbound fluorescent antibody left in the final pellet before the cells are brought at a temperature which allows redistribution phenomena to take place. This dilution is usually adequate for direct immunofluorescence staining. In the indirect immunofluorescence method, however, a higher dilution of the unbound antibody may be necessary in order to avoid significant formation of soluble complexes between the first and the second antibody (which may be nonspecifically taken up through Fc). Unbound antibody may be removed in a single step by centrifuging the antibody-treated cell suspension at 4° through a "washing medium" of high protein content. This medium can be FCS or 5% (w/v) BSA in PBS. These media may be used either as such or as discontinuous gradients, prepared as follows: 3 ml of FCS or 5% BSA-PBS, 1.5 ml of 75% FCS or 3.75% BSA-PBS, and 1.5 ml of 50% FCS or 2.5% BSA-PBS. The 0.1–0.2 ml cell pellet is taken up in 0.5 ml volume of PBS and carefully layered on top of the washing medium. Small lymphocytes will sediment in 15 min at 200 *g*. The supernatant is removed by gentle suction through a Pasteur pipet connected to a vacuum pump. Since the pellet may be loosely packed, it is advisable to leave about 0.5 ml of medium above it, to avoid aspirating cells. A further wash with 2.0 ml 0.5% BSA-PBS is needed when further processing of live cells is required (e.g., treatment by an-

other antibody). This washing procedure considerably reduces the length of time for which the cells are kept at 0–4°. However, it is recommended to check the cell recovery, particularly when the cell suspension contains cells of types other than small lymphocytes, since they may not be sedimented under these conditions.

4. For each individual sample, the proportion of cells showing caps, spots, rings, or no fluorescence is determined by examining 250–1000 cells per sample when 25–80% of the cells are fluorescence labeled. When lower percentages of cells are labeled, one first counts fluorescent cells in a sample of appropriate size to give a statistically meaningful value of the percentage of these cells, then 250–500 cells for each different type of fluorescence patterns are counted. Although in preparations from lymphoid organs a high percentage of cells has mIg, we routinely use lymphocytes from lymph nodes of nude mice (this volume [34]) in studies of capping inhibitors. These preparations contain 85–95% B cells and their capping efficiency is about 60%. Thus, any effect of capping inhibitors is actually tested on a large proportion of cells rather than on a subpopulation.

Evaluation of Capping Capacity

The capping capacity of a cell suspension can be defined as the percentage of cells which are able to cap. Capping capacity should not be determined by taking into account only the fluorescent cells (i.e., those showing caps, spots, or rings). Indeed, in the absence of inhibitors (positive control for capping) there is a low percentage of cells which do not form caps and show spots only. Other cells, however, cap so fast that they rapidly become nonfluorescent. This phenomenon may cause an underestimate of the capping capacity. There are cells, on the other hand, which can show preformed polar caps of some components ("spontaneous" caps). For instance, a fast capping of mIg is obtained on B cells of 6- to 8-week-old mouse treated with anti-Ig in the cold (age affects the capping rate[6]). The cells start capping within minutes at 37° and many cells have lost their cap within 30 min. Loss of caps may occur by shedding in the medium or by endocytosis, presumably as a result of pinosome–lysosome fusion and degradation of the fluorescent ligand. Loss of caps can also occur by a reversion of cap to spots, which may be caused by some drugs or occur as a spontaneous process after the polar redistribution of the surface clusters of membrane components but before their endocytosis. Sometimes, following endocytosis of the clusters at one pole of the cell, the pinosomes redistribute all around the cell and give a patchy appearance, presumably due to some interference at the level of pinosome–lysosome fusion.

The best conditions for evaluating capping ability are when a majority of cells have left the spot stage but only a minority have already lost their caps. To this end, we usually incubate anti-Ig coated B cells for 10–15 min at 37°. The extent of capping is then best evaluated by scoring as positive the cells which still have spots (with methods described hereafter).

The capping capacity of a system is determined by comparing the percentage of noncapped cells in the absence of inhibitor (C^+) in 0.5% BSA-PBS, or 5 mM glucose-PBS, with or without 0.5% DMSO (which may be used to improve the solubilization of the material) and the percentage of noncapped cells in the presence of 0.01 M NaN$_3$ (C^-). The difference $C^- - C^+$ is the capping capacity of the system. It should be noted that the capping capacity of a system varies from one experiment to another, due mainly to C^+. For example, the percentage of cells in a nude mouse lymph node suspension (>85% B cells) capable of forming mIg caps within 10 min at 37° following coating with antibody at 4° and whose capping is 100% inhibited by 0.01 M NaN$_3$ was 62 ± 10% (SD) (average of about 100 experiments).[13,14]

Evaluation of Inhibitors of Capping

The ability of a compound to inhibit capping is best expressed relative to the inhibition given by 0.01 M NaN$_3$, according to the formula

$$(X - C^+/C^- - C^+) \times 100 = \text{percentage of inhibition of the process}$$

where X, C^+, and C^- are the percentage of noncapped cells in the sample, the positive control, and the negative control, respectively. This formula allows a comparison of the degree of capping inhibition obtained with various compounds.

Azide is not the best inhibitor of capping. Difficulties may be encountered with the inhibition of capping of healthy cells. Particularly in glucose-containing media, NaN$_3$ may not be very effective and many cells may cap. A more complete inhibition of capping may be achieved by other inhibitors such as sulfhydryl reagents, cytochalasins, or even fixatives.[13,14]

Evaluation of Enhancement of Capping

Enhancement of capping may be determined in the same medium used for the study of inhibition except for the absence of inhibitor and the

[13] F. Loor and L. Ångman, *Exp. Cell Res.* **129**, 289 (1980).
[14] F. Loor, C. Martin-Pélissard, and L. Ångman, *Cell. Immunol.* **57**, 73 (1981).

presence of 5 mM glucose. The final incubation is limited to 5 min at 37°. The reaction is then stopped by fixation. This procedure has been applied with success to mIg and Con A-binding glycoconjugates. By examining the cells soon after coating with the ligand, early enhancement of capping may be shown. Control experiments are needed to show that capping is complete and that the drug does not cause abortive capping.

If the enhancement of slow-capping membrane components is studied, the phenomenon should be followed for an appropriate length of time. In general, experimental protocols must be adapted to the expected effects. For instance, when a substance is expected to decrease membrane viscosity, suitable conditions are created by working at a temperature below that permitting capping to occur on control cells.

Reversibility of Inhibition of Capping

Cells are processed as for the study of inhibition of capping up to the stage at which the ligand-coated cells are allowed to redistribute their membrane components (i.e., incubation at 37° for 10–15 min). The excess inhibitor is removed by washing the cells several times with the control medium (0.5% BSA-PBS or 5 mM glucose-PBS, with or without DMSO, as appropriate) or on a FCS or BSA gradient. The cells are then incubated for 10 min at 37°. Capping is stopped by fixation with F-PBS as above.

Mouse B cells incubated at 37° for 10 min, in medium giving 100% inhibition of capping (0.01 M NaN$_3$), show[14] only partial recovery of their ability to develop mIg caps, after washing with BSA-PBS medium and reincubation for 10 min at 37°. Only about 70% of the cells which normally cap when put directly in such conditions actually cap. Furthermore, when anti-Ig coated control cells are incubated for 10 min at 37°, washed, and then reincubated for 10 min at 37°, significantly more (5–10%) capped cells are found at the end of the second incubation than at the end of the first. Thus, it is not clear whether one has to take into account the entire period spent by the cells at 37°, or only the time spent at 37° under capping conditions.

Differential Redistribution or Coredistribution of Membrane Components or Membrane and Cell Cortex Elements

The method of differential redistribution of spots and caps has been used to study the possible physical and/or physiological relationships of various components of the cell membrane.

If two components A and B are stably associated in the membrane, the clustering and capping of A should bring B in the cap too (copatching and cocapping). If, on the other hand, A and B are not associated, the clustering of one should not affect the distribution of the other. For instance, since mIg and histocompatibility antigens do not cocap, it may be concluded that these are different and independent molecules.

In practice, the easiest way to study a possible differential redistribution of two membrane components is to study their capping by using two different antibodies, one labeled with fluorescein (anti-A) and the other labeled with rhodamin (anti-B). First the cells are coated with anti-A antibody in the cold, washed, and allowed to cap under appropriate experimental conditions. Then, a relatively large volume of medium at 37° containing 0.01 M NaN$_3$ is added (e.g., 5 ml to a 0.1 ml pellet of capped cells). The cells are incubated for 10 min at 37°, pelleted by centrifugation, and the pellet is brought to 0°. The cells are resuspended in 0.1 ml of the anti-B antibody, incubated with it in ice for 15 min, and washed in the cold in NaN$_3$-containing medium. They can then be fixed with F-PBS either directly or following incubation for 10 min at 37° to allow the second antibody to redistribute the corresponding membrane components. Comparison of cells treated in such a way with cells which have been treated with only one of the two antibodies may reveal independent redistribution or, conversely, cocapping of the two membrane components. If the second antibody does not induce redistribution (e.g., a fluorescent monovalent Fab), any alteration of its distribution on the cells is the result of the redistribution induced by the first antibody. Obviously, one has to avoid a reaction of the two antibodies with each other. Furthermore, a number of membrane components require two layers of bivalent antibody in order to cluster or cap, thus increasing the risk of unwanted cross-reactions. In some cases, antibodies with distinct allotypic markers may be used in the first layer and antiallotype antibodies coupled with different fluorochromes in the second layer. In other cases, it may be possible to use unlabeled antibodies raised in different non-cross-reacting species in the first layer and antispecies, differently labeled, antibodies in the second layer. An elegant method consists in using antibodies labeled with a hapten in the first layer and an antihapten antibody in the second layer.[15]

Possible Pitfalls of the Differential Redistribution Method

In addition to the possibility of artifactual coredistribution phenomena due to cross-reactivity of antibodies, the differential redistribution method has some limitations regarding its intrinsic informative value and one should be careful not to overestimate its potential and significance. The following considerations apply not only to the interactions among

[15] L. Wofsy, P. C. Baker, K. Thompson, J. Goodman, J. Kimura, and C. Henry, *J. Exp. Med.* **140**, 523 (1974).

different membrane components, but also to the interactions of membrane components with components of the cell cortex such as actin, tubulin, and myosin.

1. A and B are associated within the membrane. The binding of anti-A, however, may cause the dissociation of A from B, due, for instance, to conformational change induced in A by anti-A binding. In this case, A will not redistribute B. If, on the other hand, the binding of B does not cause dissociation of the A–B complex, B will redistribute A.

2. A and B are in reversible equilibrium. This equilibrium may be displaced in either direction (association or dissociation) by binding of antibody to A and/or B. For instance, anti-A may bind only to the dissociated form of A. If the affinity of A for anti-A is higher than its affinity for B, the capping of A by anti-A will leave B unredistributed, although A and B interact.

3. A and B are independent, but the binding of anti-A to A causes its interaction with B (e.g., through quaternary structure rearrangements). In this case anti-A will cause partial or total cocapping of B. Anti-B alternatively may or may not cause interaction between A and B.

4. There is a stable association (e.g., covalent) between A and B, but some excess of free B. The binding of one antibody (e.g., anti-A) interferes with the binding of the other antibody (anti-B). Capping of A-associated B by anti-A will not be detectable with anti-B, unless the capping of A removes a large proportion of the total B from the cell surface.

5. If there is a large excess of A over B, the redistribution of A by anti-A may nonspecifically entrap B components and leave no B components on the surface following capping, even if A and B are not associated. On the contrary, the capping of B by anti-B will leave essentially all of A on the surface.

Examples of some of these cases have been shown or inferred.[1,11,12]

Artifactual redistribution may also occur when indirect immunofluorescence methods are used, with two or three successive layers of antibodies. It is probably due to the clustering (and, perhaps, activation) of the Fc portion of the antibodies and their possible interaction with Fc receptors present on the cell surface. This may lead to altered redistribution patterns. This type of problem may be avoided by working with fluorochrome conjugates of $F(ab')_2$ fragments of antibodies.

Detection of Changes of Surface and Cytoplasmic Components on
    Single Cells

In some cases a change of the distribution of membrane components is concomitant with the redistribution of cytoplasmic structures (e.g., fibrillar structures interacting with clusters or caps present in the membrane). In order to detect these changes, one has to proceed in two steps: first the cell surface determinant is stained and redistributed with the ligand; then, the cells are fixed and their membrane is made permeable (by treatment with organic solvents) to the second reagent specific for the cytoplasmic component under study (see an example of such a procedure below). While the first step is performed on live cells, the second step is usually performed on smears of surface stained cells.

Smears are prepared by using a cytocentrifuge, although this may alter the cell architecture.[16,17] F-PBS-fixed cells are resuspended in a minimum volume of 3% BSA-PBS. With a capillary Pasteur pipet, small drops of cell suspension are layered along one or several lines on the microscope slides. The slides are air-dried, fixed for 5–10 min in ethanol, washed several times with PBS, once with 3% BSA-PBS, and, finally, stained for the cytoplasmic components. To avoid alterations of cell architecture often occurring during this procedure, it is advisable to work with cell suspensions. In this case, the staining and redistribution of the membrane components are carried out under the usual conditions including the fixation of the cells with F-PBS. The cells (typically $3–10 \times 10^6$ cells/ml) are pelleted in conical glass tubes (each time for 10 min at 100–150 g at 4°) and resuspended (2 ml for 5 min at 0°) in sequence in acetone–water (1 : 1), acetone, acetone–water (1 : 1), PBS, and PBS containing 0.5% BSA. The pellets are finally resuspended in PBS containing 3% BSA and incubated for 15 min at 37°. This procedure markedly reduces the nonspecific background of cytoplasmic fluorescence after treatment with fluorescent antibodies. The pelleted cells are then incubated with 50–100 μl of fluorescent antibody fragments $F(ab')_2$ (not whole IgG) for 5 min at 0°, washed repeatedly in PBS containing 1% BSA, so that at least a 10,000-fold dilution of the fluorescent ligand is attained, washed once more in PBS, and finally resuspended in buffered glycerol (as described below). The antigen–antibody reaction is fast. Optimal detection can be reached after 1–2 min of staining time. In this procedure it is important to keep the background fluorescence (usually due to insufficient diffusion of unbound reagent outside of the cell) low. This may be achieved by keeping the concentration of the fluorescent reagent as low as possible. A high background fluorescence may be removed with prolonged washing or incubation of the cells in BSA-PBS. When two antibodies are used (indirect method), if the first antibody has not been completely washed away at the time of the addition

[16] L. Forni, in "Immunological Methods" (Y. Lefkovitz and B. Pernis, eds.), p. 151. Academic Press, New York, 1979.

[17] D. De Luca, in "Antibody as a Tool" (J. J. Marchalonis and G. W. Warr, eds.), p. 189. Wiley, New York, 1982.

of the second antibody, precipitates may form inside the cell. In this case, the high background fluorescence cannot be lowered.

Appendix: Preparation of Fluorescent Antibody Reagents

This topic will be limited to a few comments about the choice and preparation of fluorochromes. Detailed procedures may be found elsewhere[16,17] (this volume [41]).

Choice of antibody. Immunofluorescence may be performed with pure antibody (affinity purified) rather than whole-antiserum immunoglobulin. Pure antibody has the advantage that lower Ig concentration may be used, thus reducing the level of nonspecific fluorescence background. Possible artifacts due to the presence of unwanted antibody specificities (e.g., natural, autoantibodies) are also minimized. Compared to polyclonal antibodies, monoclonal antibodies have the additional advantages of the exquisite recognition of restricted epitopes and the possibility of adapting the conjugation procedure to the precise antibody subclass. This is particularly useful when $F(ab')_2$ are used, in view of the different susceptibility of mouse IgG subclasses to proteolytic attack. It should be kept in mind, however, that direct immunofluorescence with monoclonals will give rings only, since no cross-linking can occur if the relevant epitope occurs only once per membrane component and exists as an independent monomer on the cell surface. The lack of interactions between antibody and different epitopes of the membrane component (the "bonus effect") also weakens the stability of the microprecipitates. In order to have the advantages of the restricted specificity of monoclonal antibodies together with the clustering efficiency of polyclonal antibodies, it may be advisable to use a cocktail of monoclonal antibodies directed toward different epitopes of the relevant membrane component.

We also recommend the use of $F(ab')_2$ instead of IgG. This avoids nonspecific uptake of fluorescent antibody through the Fc receptors of the cells. Moreover, unlike most IgG of rat or mouse origin, the $F(ab')_2$ fragments can be recovered from the anion exchanger column used for the preparation of the fluorescent antibody. The preparation of IgG and their $F(ab')_2$ is carried out following the procedures appropriate for that animal species, and in the case of monoclonal antibodies for the given IgG subclass.

Fluorochrome coupling and reagent selection. IgG from rabbit antimouse Ig serum is prepared by two or three successive precipitations with $(NH_4)_2SO_4$ (1.6 M, 0°). The material is then dialyzed vs 0.0175 M phosphate buffer, pH 6.3, and further purified by anion exchange chromatography on column or by batch adsorption. Most contaminating proteins bind to the column whereas the rabbit IgG is concentrated to 10–30 mg/ml by ultrafiltration and dialyzed vs 0.1 M sodium acetate buffer, pH 4.5. The IgG (including precipitated material, when present) is then digested with 1% (w/w) crystallized pepsin for 20 hr at 37° with gentle stirring. The material is dialyzed against a continuous flow of 0.06 M sodium acetate buffer, pH 5.5, and the precipitate is discarded. The soluble protein is then applied to a cation exchanger column equilibrated with the same buffer. Undigested IgG and Fc fragments bind to the column, whereas F(ab')$_2$ is eluted. The IgG or the F(ab')$_2$ is then dialyzed vs 0.01 M phosphate buffer, pH 7.5. Antibody fractions of limited isoelectric point heterogeneity are obtained by stepwise elution from an anion exchange column with increasing concentrations of NaCl (0, 0.05, 0.1, 0.2, and 0.3 M) in 0.01 M sodium phosphate, pH 7.5. The protein fractions at the peaks (at 5–20 mg/ml) are dialyzed vs 0.15 M NaCl. Conjugation with either fluorescein isothiocyanate (12.5 μg fluorochrome/mg protein) or tetramethylrhodamine isothiocyanate (30 μg fluorochrome/mg protein) is carried out in the dark at 0° for 2 hr in 0.01–0.1 M carbonate-bicarbonate buffer, pH 9.0–9.5, adjusted with 0.01 M NaOH. The material is kept overnight at 4°. Unbound and hydrolyzed fluorochrome may be removed on a Sephadex G-50 column equilibrated with 0.01 M phosphate buffer, pH 7.5. Fractions of conjugated antibodies having a narrow range of isoelectric points can thus be obtained. It should be noted that, since binding of the fluorochrome causes a decrease in the isoelectric point of the antibody, if one starts with homogeneous isoelectric fractions of antibody, the second anion exchange chromatography selects for antibodies having similar numbers of fluorochrome molecules bound per molecule. This permits the exclusion of undercoupled antibody, which binds to the antigenic sites but cannot be detected, and overcoupled antibody, which adsorbs nonspecifically to cell membranes. An estimate of the degree of conjugation can be made by measuring the absorption ratio 280 nm/495 nm or 280 nm/515 nm (for fluorescein or rhodamine, respectively). Fluorescent antibody fractions with a ratio between 2 and 3 are usually adequate for immunofluorescence (for methods for determination of the number of fluorochrome groups bound per protein molecule, see Forni[16] and De Luca[17] (this volume [41]). The solutions of fluorescent antibody are made isotonic by dialysis vs PBS. The protein concentration is adjusted at 0.5 mg/ml. BSA (10 mg/ml) is added and the material is stored at 4° in the dark (with or without 10 mM NaN$_3$). Some of our fluorescent antibodies are still active nearly 10 years after conjugation, with the greatest loss of staining efficiency occurring within the first 6–12 months of storage. Before using conjugates which have been stored for long periods of time it is advisable to remove both precipitated and dialyzable material to avoid nonspecific binding.

[38] Direct and Mixed Antiglobulin Rosetting Reaction

By DAVID G. HAEGERT

Introduction

B cells may be defined as lymphocytes that synthesize surface membrane immunoglobulin (smIg) molecules. For many years the standard method for detecting smIg has been direct immunofluorescence (DIF).[1] However, recently both the mixed antiglobulin rosetting reaction (MARR) and the direct antiglobulin rosetting reaction (DARR) have been advocated as alternatives to DIF. This is because in direct comparative studies these Ig-rosette tests have been shown to be more sensitive than DIF[2,3]; in addition to typical DIF$^+$ B cells these tests have delineated a B lymphocyte population in several species which is smIg$^-$ by DIF[2-4] and includes (in humans at least) some killer (K)[5] and natural killer (NK) cells.[6] Because of their sensitivity it has been suggested that more widespread use of these Ig-rosette tests may expand our knowledge of B cell populations and provide further insight into B cell immunodeficiencies.[1] The objectives of this chapter, therefore, are to review the technology of the MARR and the DARR and to discuss the methods we have used to detect human B cell smIg but the same principles apply to detection of B cells in other species.

Principles

The DARR is a direct binding technique in which anti-Ig is linked to indicator particles. Usually these particles have consisted of erythrocytes (E) to which anti-Ig is covalently bound by chromic chloride ($CrCl_3$) (see below for $CrCl_3$ coupling procedure) but anti-Ig may also be bound to polyacrylamide[7] or latex beads.[8] Coombs and co-workers have exten-

[1] D. G. Haegert, *J. Immunol. Methods* **41,** 1 (1981).

[2] D. G. Haegert, *J. Immunol.* **120,** 124 (1978).

[3] D. G. Haegert, C. Hurd, and R. R. A. Coombs, *Immunology* **34,** 533 (1978).

[4] R. M. Binns, S. T. Licence, D. B. A. Symons, B. W. Gurner, R. R. A. Coombs, and D. E. Walters, *Immunology* **36,** 549 (1979).

[5] O. Eremin, D. Kraft, R. R. A. Coombs, J. Ashby, and D. Plumb, *Int. Arch. Allergy Appl. Immunol.* **55,** 112 (1977).

[6] O. Eremin, R. R. A. Coombs, D. Plumb, and J. Ashby, *Int. J. Cancer* **21,** 42 (1978).

[7] A. J. Amman, D. Borg, L. Kondo, and D. W. Wara, *J. Immunol. Methods* **17,** 365 (1977).

[8] C. R. Parish and T. J. Higgins, *J. Immunol. Methods* **53,** 367 (1982).

METHODS IN ENZYMOLOGY, VOL. 108

sively investigated the parameters which influence the sensitivity of anti-Ig coated E[9] and at the present time it appears that anti-Ig coated E are the most suitable indicator particles for the DARR. Direct antiglobulin rosettes are formed by centrifugation of unsensitized lymphocytes together with anti-Ig coated E. Binding of anti-Ig to corresponding determinants in B cell smIg induces rosette formation.

In contrast to the DARR, the MARR is not a true direct binding technique. To detect B cell smIg, lymphocytes are sensitized with free anti-Ig then centrifuged together with E to which non-antiglobulin Ig, corresponding to the specificity of the anti-Ig, is linked. Mixed antiglobulin rosettes form when the anti-Ig bound to smIg also binds to corresponding determinants on the indicator E.

With DIF a major concern is that Ig adsorbed onto lymphocyte IgG(Fc) receptors may be falsely interpreted as intrinsic smIg.[1] In this regard the use of insolubilized anti-Ig for the DARR renders the Fc portion of the anti-Ig unavailable for binding to IgG(Fc) receptors.[3] With the MARR free anti-Ig may bind to IgG(Fc) receptors but does not also bind to Ig coated indicator E to form rosettes.[2] Consequently with both the MARR and the DARR IgG anti-Ig reagents may be used routinely; this contrasts with DIF where it may be necessary to pepsin digest anti-Ig reagents or to preincubate lymphocytes in serum-free medium or at pH 4.0 to avoid detection of Ig bound to IgG(Fc) receptors (reviewed in Haegert[1]).

Reagents

Immunoglobulins

We prepare human IgG from normal or myeloma sera by DEAE-cellulose chromatography in 0.03 M phosphate buffer pH 7.3; the fall through fraction is IgG. IgM is prepared by repeated euglobulin precipitation followed by Sephadex G-200 chromatography. IgA is readily available from commercial sources (e.g., Cappel Laboratories, Cochranville, PA). Contaminating IgG may be removed when necessary by absorption on protein A Sepharose CL-4B. The purity of each Ig preparation is established by passive hemagglutination (see below for details).[2]

Anti-Ig Reagents

In contrast with DIF, high titer anti-Ig reagents are unnecessary with the MARR and the DARR to detect those B cells which express a high density of smIg[9]; we have termed these typical B cells the B major population.[10] However, high titer reagents are needed to detect these B cells which express a low density of smIg and are DIF$^-$; we have termed these cells the B minor population.[10] For practical purposes then all smIg$^+$ B cells may be detected by high titer polyvalent anti-Ig reagents. Such reagents may consist of pooled anti-heavy chain or anti-light chain reagents,[2] anti-Fab,[9] or anti-F(ab')$_2$ reagents.[2] B cell isotypic determinants may be detected by anti-heavy chain or anti-light chain reagents of appropriate specificities. In our experience commercially available anti-Igs may be used for these purposes but an important caveat is that the specificity of each reagent must be carefully established.

Reagent specificity is investigated in a number of ways.

Reverse Passive Hemagglutination (RPH).[2] This assay is based on the ability of anti-Ig coated E to agglutinate in the presence of corresponding antigen. Its advantages is its extreme sensitivity detecting antigen at 10 to 20 ng/ml or lower[2,3]; test sensitivity depends upon the amount of anti-Ig coupled to E.[9]

Each anti-Ig reagent is dialyzed against 0.9% NaCl then made to approximately 1 mg/ml in 0.9% NaCl. Fifty microliters is added to each of four 3 ml round bottom tubes containing 50 μl of packed, fresh sheep E (less than 1 week old) which have been washed three times with 0.9% NaCl. While mixing the tubes on a vortex mixer, 50 μl serial dilutions (usually 1/10, 1/20, 1/40, and 1/80), made in 0.9% NaCl, of a 1% mature chromic chloride (CrCl$_3$) solution (mature means prepared more than 3 weeks previously[11]) are added to each tube. After 5 min incubation at room temperature, the CrCl$_3$ E–anti-Ig preparations are washed three times in 0.9% NaCl or PBS. It has been reported that overnight incubation of the E–anti-Ig–CrCl$_3$ mixture may enhance the strength of anti-Ig binding to E.[11] We have not investigated this point but our indicator cells prepared after 5 min of incubation appear to be as sensitive as indicator cells prepared by the overnight incubation procedure.[1]

RPH is then performed in U bottom microtiter plates using 0.9% NaCl or PBS as the diluent. Fifty microliters of Ig corresponding to the putative specificity of the anti-Ig coupled to E is added to the first well of each of four rows (each row corresponds to a different dilution of the CrCl$_3$ solution used to couple anti-Ig). The Ig is then serially diluted and 50 μl of 1% CrCl$_3$ E–anti-Ig is added to each well. The plate is manually agitated to mix the preparation then read after 2 hr incubation on the bench top.

[9] R. M. Binns, S. T. Licence, B. W. Gurner, and R. R. A. Coombs, *Immunology* **47**, 717 (1982).

[10] D. G. Haegert and R. R. A. Coombs, *Lancet* **2**, 1051 (1979).

[11] N. R. Ling and P. R. Richardson, *J. Immunol. Methods* **47**, 265 (1981).

Agglutination is indicated by failure to settle into a tight button. The $CrCl_3$ E–anti-Ig preparation having the highest hemagglutination titer is then tested against a battery of Igs including IgG, IgA, IgM, and commercially available Bence-Jones proteins (e.g., Behringwerke, A. G. Marburg Germany). This procedure helps to exclude undesirable antiisotypic specificities.

Passive Hemagglutination. In a second procedure to exclude contaminating antiisotypic specificities monoclonal human Ig preparations are $CrCl_3$ coupled to fresh sheep E. Each Ig preparation is dialyzed against 0.9% NaCl then made to approximately 1 mg/ml. The $CrCl_3$ coupling is as described above: 50 μl of Ig, 50 μl of sheep E, and 50 μl of 1/10, 1/20, 1/40, and 1/80 dilutions of 1% $CrCl_3$ are mixed, incubated for 5 min, and then washed. Hemagglutination is then performed using the principles described for RPH.

Ig-Rosette Tests on T-Enriched Human Peripheral Lymphocytes. Human T cells are prepared by double centrifugation of sheep E rosetting peripheral blood mononuclear cells over Ficoll–Hypaque with collection of the pelleted cells.[12] The MARR and/or the DARR is then performed (see below for procedural details) on aliquots of these T cell preparations. The number of MARR $^+$ or DARR$^+$ cells should not exceed the number of non-T cells contaminating the T-enriched preparations; that is the percentage of sheep E rosetting cells (a human T cell marker[12]) plus the number of Ig-rosette$^+$ cells should not exceed 100%. This procedure excludes contaminating specificities directed against antigens (e.g., HLA) shared by T and B cells and excludes cross-reacting specificities shared between Ig and certain non-Ig lymphocyte molecules (e.g., β_2-microglobulin[13]). This procedure also excludes undesirable anti-Ig specificities which could bind to the Ig-related T cell receptor[14] and produce erroneously high values for B cell numbers.

Inhibition Experiments. Monoclonal IgG, IgA, and IgM at 5 mg/ml are heat aggregated at 63° for 15 min then serial dilutions are made. The MARR is then performed as follows. Aliquots of anti-Ig sensitized lymphocytes (50 μl) are mixed with serial dilutions of various monoclonal Ig preparations (50 μl), incubated together for 5 min, then centrifuged together with indicator E (100 μl) coated with Ig corresponding to the putative specificity of the anti-Ig. We have found that lymphocyte reactivity in the MARR after sensitization with a particular antiisotypic reagent is totally inhibitable by the corresponding Ig at 5 mg/ml but is not inhibited by other Ig molecules.[2] This procedure therefore excludes reactivity in

[12] D. G. Haegert and R. R. A. Coombs, *J. Immunol.* **116**, 1426 (1976).
[13] A. B. Gottlieb, E. Englehard, and H. G. Kunkel, *J. Immunol.* **119**, 2001, (1977).
[14] J. J. Marchalonis, *Immunol. Today* **3**, 10 (1982).

the MARR due to spurious rosetting or unusual antimembrane antibodies.

Comments on Specificity Tests. The anti-Ig coated E prepared for RPH studies are used as indicator cells for the DARR. However, it has been reported that indicator cells of low sensitivity may show no agglutination with antigen by RPH but still form rosettes with lymphocyte smIg in the DARR.[9] Presumably, contaminating specificities could be responsible for false positive rosettes but be at such a low concentration as to be undetectable by RPH. For this reason we also test anti-Ig specificity against Ig coated E by passive hemagglutination.

Testing of reagent specificity is a less crucial issue for the MARR. This is because the MARR has two advantages over all other test systems for demonstrating smIg. First, if an anti-Ig contains contaminating specificities (e.g., if anti-μ contains anti-γ and anti-λ) each antibody will bind to corresponding lymphocyte determinants but only μ will induce rosette formation because only anti-μ will also bind IgM-coated indicator cells; this is provided the IgM on the indicator cells is IgMκ but not IgMλ. Second cross-reacting specificities between Ig and non-Ig membrane components (e.g., between κ and β_2-microglobulin) will be detected only if the membrane component is also cross-reactive with the particular Ig present on the indicator cells.

Experimental Procedure

Preparation of Indicator E

The MARR. Isotypic determinants are most easily demonstrated using sheep E to which myeloma proteins are coupled by CrCl$_3$.[2] The coupling procedure is exactly as described for passive hemagglutination. The sensitivity of the MARR correlates closely with the passive hemagglutination titer and the preparation with the highest titer and a negative control (diluent alone) is selected; usually the titer is 1/1600 or higher with a conventional rabbit anti-Ig reagent (1 mg/ml).

To detect IgG, IgA, and IgM determinants, sheep E are coated respectively with IgG or Cohn II, IgA, and IgM molecules. Light chain determinants may be detected by coupling myeloma Ig of the corresponding light chain type to E.[2]

The DARR. Two variables are important in preparing sensitive indicator E.[1] The first is selection of conditions for CrCl$_3$ coupling which promote binding of optimal amounts of anti-Ig to E. In this regard Ling *et al.*[15] have emphasized several points: careful washing of E with 0.9% NaCl to remove anions reactive with Cr; fresh E and mature $CrCl_3$ (see above). The $CrCl_3$ coupling procedure is exactly as described for RPH and indicator E with the highest RPH titer are used for the DARR.

The second variable involves preparation of indicator E in such a way that the aggregation of lymphocytes and indicator E is maximal during the rosetting procedure. As reviewed recently,[1] native rabbit E, which has a low sialic acid surface content, ox E treated with trypsin and sheep E treated with neuraminidase to remove sialic acid form indicator E of a similar order of sensitivity and in several species detect apparently optimal numbers of $smIg^+$ B cells[1,4,9]; removal of sialic acid reduces the net negative surface charge on E. Indicator cells consisting of native sheep E or ox E coated with anti-Ig are less sensitive.

We routinely treat sheep E with *Vibrio cholerae* neuraminidase (VCN) (Behringwerke) as follows. Sheep E are washed in PBS, then 0.4 ml of packed cells is suspended in a 2 ml PBS/0.2% BSA containing 400 units of VCN for 30 min at 37°. The cells are then washed in 0.9% NaCl and anti-Ig is then $CrCl_3$ coupled.

Indicator E for the MARR and the DARR may be stored in MEM, 0.9% NaCl, or PBS for 3 days without loss of activity. The cells should be washed once after storage to remove any Ig or anti-Ig spontaneously eluted from the indicator E.

Lymphocyte Preparation

Mononuclear cells are isolated from heparinized blood over Ficoll–Hypaque then incubated in MEM/20% fetal calf serum with 20–50 μl of 1.01 μm size polystyrene particles (Dow Diagnostics) for 30–60 min at 37° to facilitate identification of phagocytic monocytes.[2] For the DARR, if sheep E are not incubated with VCN before coupling of anti-Ig then the lymphocytes are VCN-treated as follows to enhance test sensitivity to optimal levels.[16] Mononuclear cells (10^7) are suspended in 1 ml MEM/2% BSA containing 50 units VCN, incubated for 30 min at 37°, then washed three times in MEM/0.2% BSA before rosetting with indicator E.

Sensitization of Lymphocytes for the MARR

Anti-Ig reagents are centrifuged at 100,000 *g* for 30–60 min before use to remove aggregates. Lymphocytes (2×10^6) are sensitized with twice

[15] N. R. Ling, S. Bishop, and R. Jefferis, *J. Immunol. Methods* **15**, 279 (1977).
[16] D. G. Haegert, *J. Immunol. Methods* **22**, 73 (1978).

the concentration of anti-Ig found to give a plateau of percentage-positive lymphocytes in the MARR.[2]

Controls for the MARR and the DARR

When using VCN-treated lymphocytes and rabbit anti-Ig reagents for the DARR, it is essential to rosette the lymphocytes with $CrCl_3E$ coated with non-antiglobulin rabbit Ig. When the DARR is performed with VCN-treated E, lymphocytes are rosetted with VCN-treated $CrCl_3E$ coated with non-antiglobulin Ig. For the MARR a requisite control is rosette formation of unsensitized lymphocytes with Ig coated indicator E. Using sheep E as the indicator system up to 2% of lymphocytes bind these various control preparations. It has been found however that these false positive rosettes are readily abolished by rosette formation in 0.5% azide. Thus all Ig-rosette tests using sheep E are carried out in the presence of azide. In this regard an important advantage of using rabbit E or trypsin treated ox E is that these cells do not form spontaneous rosettes with lymphocytes in control tests.[1]

The DARR and the MARR

Indicator cells are made to 1% and mononuclear cells to 2000/ml.[3] Because we use sheep E, the DARR is performed in MEM/0.2% BSA/ 0.5% azide: 50–100 μl of unsensitized mononuclear cells and an equal volume of indicator cells are mixed in a 3 ml centrifuge tube then centrifuged at 4° for 3 min at 200 g. To recapitulate, the DARR is optimal when using either VCN-treated lymphocytes together with $CrCl_3E$–anti-Ig or VCN-treated $CrCl_3E$–anti-Ig together with freshly prepared lymphocytes.

The MARR is optimal when performed in MEM/5% BSA/0.5% azide; the high protein concentration reduces cell surface zeta potential, enhances the lymphocyte-indicator E aggregation phase, and obviates the need for VCN treatment of E or of lymphocytes.[2] Anti-Ig sensitized lymphocytes (50–100 μl) are centrifuged with equal volumes of indicator cells as described for the DARR.

After Ig-rosette formation, the cells are immediately resuspended with a Pasteur pipet and counted on a glass slide. A lymphocyte binding 3 or more E is regarded as a rosetting cell. Monocytes are recognized by their phagocytosis of polystyrene beads (this volume [26]) and by morphology and granulocytes are recognized by morphology.[2] A useful adjunctive procedure is to resuspend rosetted cells, centrifuge them in a cytocentrifuge, and stain the cells with Wright's stain.[17]

[17] J. L. Smith and D. Haegert, Clin. Exp. Immunol. **17,** 547 (1974).

Preparation of T-Enriched Preparations

The DARR may be used in a one stage procedure to prepare T-enriched lymphocytes.[2,6] Peripheral blood lymphocytes are rosetted in the DARR then resuspended immediately and centrifuged over Ficoll–Hypaque (this volume [7]). The interface contains more than 90% T cells and functional studies may be performed on the T-enriched preparations.

The MARR has not been adapted for this procedure.

Comparisons of the MARR and the DARR

The DARR is a more rapid and simpler procedure than the MARR and can be used to detect idiotypic, allotypic,[18] and isotypic B cell determinants. With the MARR it may be difficult or impossible to prepare certain indicator cells, e.g., those bearing idiotypes or δ and ε isotypes. Thus, the MARR is severely limited in its applications. However, as discussed above, the MARR has the particular advantage that it is not always essential to exclude contaminating specificities in an anti-Ig reagent before use. It is predicted that the more widespread use of monoclonal antibodies will avoid the problem of contaminating specificities and the DARR will then become the preferred Ig-rosette test. It should play an increasingly important role in demonstrating smIg because of its sensitivity.

[18] V. Cabana, M. Teodorescu, and S. Dray, *J. Immunol.* **125**, 2355 (1980).

[39] Ultrastructural Localization of Surface and Intracellular Immunoglobulin in Human Lymphoid Cell Suspensions

By D. G. NEWELL

Introduction

The differentiation of normal B lymphocytes, from the pre-B cell to the plasma cell, involves progressive morphological change accompanied by differences in the synthesis and surface expression of immunoglobulin (Ig). This pathway may be studied using heterogeneous normal or homogeneous neoplastic B cell populations. Unfortunately, biochemical analysis of immunoglobulin production and expression in normal and neoplastic cell preparations encompasses the whole population, obscuring cellular variations. Such variations must be analyzed in individual cells and the ultrastructural visualization of Ig is one approach to this problem.

METHODS IN ENZYMOLOGY, VOL. 108

Surface and intracellular Ig may be ultrastructurally localized using antiimmunoglobulin antibodies either directly or indirectly labeled with electron-dense markers. Many such markers are available, including ferritin and gold, but peroxidase is the most frequently used for qualitative analysis because the conjugates are readily available and suitable for high resolution.

Although immunocytochemical methods have been widely used to identify Ig-containing cells at the light microscope level, the extension of these techniques to the ultrastructural level has been restricted by a number of technical difficulties. The problem arises from the apparently incompatible requirements of preservation of ultrastructural morphology, retention of Ig antigenicity, and penetration of the antibody marker through the cell membrane. In contrast, the labeling of surface Ig is not dependent upon membrane permeability and is therefore easier.

Glutaraldehyde fixation provides the best preservation of ultrastructural morphology but the cross linkage of proteins may result in loss of Ig antigenicity[1] and severely restrict conjugate penetration into fixed cells.[2] Low concentrations (0.1%) of glutaraldehyde, however, do not appear to affect this antibody–antigen interaction.[1] Conjugate penetration of these fixed cells may be enhanced in several ways including the use of smaller conjugates, for example, microperoxidase and Fab fragments.[3] Recently, detergents, such as digitonin[4] and saponin,[5] have been used to solubilize the fixed cell membranes and render them permeable to immunological reagents.

Materials and Methods

Preparation of Peroxidase-Antibody Conjugates

Sheep IgG antibodies directed against human heavy or light chains, purified by affinity chromatography, are conjugated to horseradish peroxidase by the 2-step glutaraldehyde technique.[6]

Ten milligrams of horseradish peroxidase (Grade VI, Sigma Chemical Co., Poole) in 0.2 ml 1.25% glutaraldehyde (Grade I, Sigma Chemical Co.) in 0.1 M phosphate buffer, pH 6.8, is incubated at 20° for 18 hr with gentle

[1] W. Van Ewijk, R. C. Coffman, and I. L. Weissman, *Histochem. J.* **12**, 349 (1980).
[2] D. G. Newell, C. Bohane, S. Payne, and J. L. Smith, *J. Immunol. Methods* **37**, 275 (1980).
[3] J. P. Krachenbuhl and J. D. Jamieson, *Int. Rev. Exp. Pathol.* **13**, 1 (1974).
[4] S. Dales, P. J. Gomatos, and K. C. Hsu, *Virology* **25**, 193 (1965).
[5] J. G. Hall, M. S. C. Birkbeck, D. Robertson, J. Peppard, and E. Orlands, *J. Immunol. Methods* **19**, 351 (1978).
[6] S. Avrameas and T. Ternynck, *Immunochemistry* **8**, 1175 (1971).

stirring. The glutaraldehyde is removed from the reaction mixture on a Sephadex G-25 (coarse) column (1×30 cm) eluted with 0.15 M NaCl. The peroxidase-containing fractions are determined by optical density at 403 nm, pooled, and concentrated to 1 ml by ultrafiltration.

The activated peroxidase is incubated with either 2.5 mg F(ab')$_2$ or 5.0 mg IgG in 1.0 ml 0.15 M NaCl plus 100 μl 0.1 M carbonate buffer, pH 9.5 for 24 hr at 4°, with stirring. The reaction is terminated by the addition of 100 μl 0.2 M lysine. The conjugate is dialyzed against 0.02 M Tris–HCl buffer, pH 7.6, containing 0.001 M EDTA and 0.1 M NaCl overnight at 4°.

The immunoglobulin-peroxidase conjugate is purified on a Sephacryl S-200 superfine column (90×1.6 cm) eluted with 0.02 M Tris, 0.1 M NaCl buffer. The fractions are monitored at 403 and 280 nm and those containing conjugate are pooled and may be stored at 1.0 mg ml^{-1} protein concentration in 0.01 M phosphate-buffered saline (PBS) pH 7.2 containing 0.1 mg ml^{-1} Thimersal (Sigma Chemical Co.) for up to 6 months without detectable loss of activity. The molar ratio of peroxidase:IgG usually varies between 0.43 and 0.96. A control conjugate may be produced with normal sheep IgG.

The Isolation of Human Lymphoid Cells

Peripheral blood or aspirated bone marrow is collected in the presence of 10 units ml^{-1} of preservative-free heparin. Cell suspensions are harvested from spleen, lymph nodes, and tonsils by gentle teasing of the tissue and filtration of the cell suspension through a metal gauze. Hematopoietic cell lines, grown in RPMI 1640 (Gibco Europe, Paisley) containing 10% fetal calf serum (Sera Lab., Crawley Down) with 50 units ml^{-1} benzyl penicillin and 10 mg ml^{-1} streptomycin, are harvested at the log phase of growth.

Lymphoid cells are isolated from blood and tissue cell suspensions and the nonviable cells removed from the cell cultures by Ficoll/Triosil gradient centrifugation.[7] Up to 10 ml of cell suspension or blood is layered onto 10 ml of Histopaque-1077 (Sigma Chemical Co.) and centrifuged at 600 g for 30 min. The lymphoid cells are removed from the interface and washed three times in minimal essential medium (MEM), (Earle's salts) containing 25 mM HEPES buffer (Gibco Europe, Paisley). Since storage of cell suspensions at 4° or lower temperatures before fixation appears to reduce permeability of the treated cell membrane to the conjugate, fresh cell preparations are used whenever possible.

---

[7] A. Boyum, *Scand. J. Clin. Lab. Invest.* **21**, Suppl. 97 (1968).

Cell Prefixation for Labeling of Surface Immunoglobulin

All fixatives and subsequent washing fluids are filtered (0.45-μm cellulose acetate filter) to eliminate the accumulation of particulate contaminants, during the centrifugation steps, which may damage glass and diamond knives during thin section cutting. Similarly, plastic pipets and plastic tubes are used to prevent the accumulation of glass particles.

Washed cells (1×10^6–1×10^7) are resuspended in 20 ml 0.1% glutaraldehyde in 0.1 M sodium phosphate buffer pH 7.2, at 20° for 15 min. These prefixed cells are washed three times in the same buffer and stored at 4° until needed. All centrifugation steps after fixation may be performed in a microcentrifuge (Eppendorf, Hamburg) for 30 sec, in siliconized tubes to reduce cell loss.

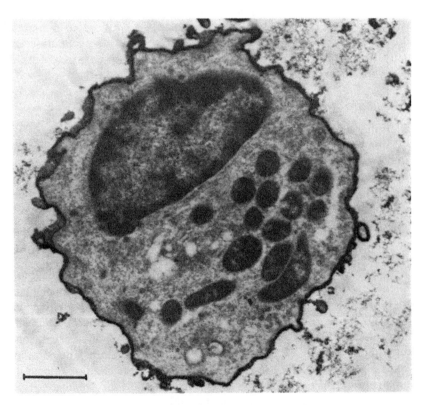

FIG. 1. A tonsil lymphocyte stained for surface immunoglobulin with anti-Fab γ conjugated to peroxidase. Bar equals 1 μm.

Cell Prefixation and Saponin Treatment for Labeling of Intracellular Immunoglobulin

Washed cells (1×10^6–1×10^7) are fixed in 20 ml 0.1% glutaraldehyde in 0.1 M sodium phosphate buffer, pH 7.2 at 20° for 15 min, washed three times in the same buffer, and stored at 4° until needed. Prefixed cells are then incubated with freshly made and prewarmed 1.0% saponin (Sigma Chemical Co.) at 55° for 10 min and then washed three times in PBS.

Immunostaining of Prefixed Cells

Cell pellets are resuspended in 100 μl of antibody-peroxidase conjugate (1 mg ml^{-1}) and incubated for 1 hr at 20° and then washed three times in PBS.

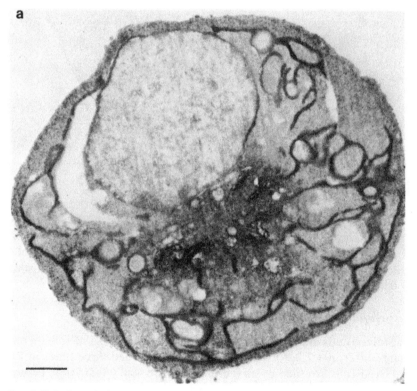

Fig. 2. Cells from the lymphoblastic cell line EB2 stained for (a) γ heavy chains and (b) λ light chains. Bar equals 1 μm. From Newell *et al.*[12]

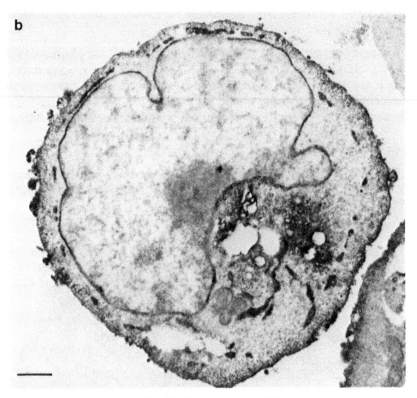

FIG. 2b. See legend on p. 397.

Peroxidase activity is detected using the diaminobenzidine substrate of Graham and Karnovsky.[8] 3,3'-Diaminobenzidine tetrahydrochloride (10 mg) (Grade II, Sigma Chemical Co.) is freshly dissolved in 20 ml 0.05 M Tris–HCl buffer, pH 7.6. The pH is readjusted to 7.6 and 0.2 ml 1% hydrogen peroxide is added. The cells are incubated with the substrate for 30 min at 20° in the dark and then washed three times in PBS.

Postfixation and Embedding

Stained cells are postfixed in 1% glutaraldehyde in 0.1 M sodium phosphate buffer, pH 7.2, for 15 min at 20°, then washed three times in PBS. The fixed cells are then postfixed in 1% osmic acid (Agar Aids, Bishop's Stortford) in veronal-acetate buffer, pH 7.2, for 30 min at 20°, followed by three washes in distilled water.

[8] R. C. Graham and M. J. Karnovsky, *J. Histochem Cytochem.* **14**, 291 (1966).

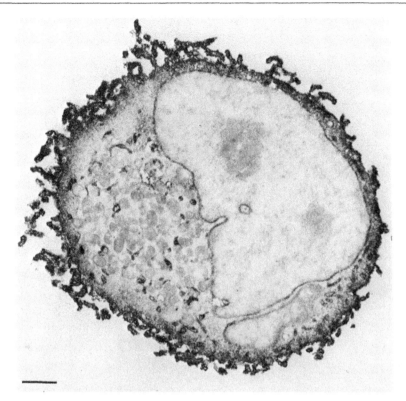

FIG. 3. Cell from the Jijoye cell line stained for κ light chains showing staining in the PNS and RER and nonspecific staining of the cell periphery which was also observed in the control conjugate. Bar equals 1 μm.

Before embedding in agar, all the excess liquid is removed from the cell pellet and the microfuge tube, containing the pellet, is placed in a water bath at 60°. The cell pellet is mixed with 10–20 μl of 2% agar which is then allowed to solidify. The agar is prepared as follows: 2% molten Noble Agar (Difco Labs., Detroit) is centrifuged at 1200 g and allowed to solidify. Any contaminating debris is removed from the bottom of the agar block. The agar is then sterilized by autoclaving and stored at 4°. The solidified agar–cell mixture is removed from the microfuge tube, trimmed with a razor blade, and stored in 70% ethanol. The cell-agar block is dehydrated in acetone before embedding in Spurr Resin (Agar Aids, Bishop's Stortford). Thin silver sections are cut with diamond knives and viewed without further staining at 40 kV in a Phillips 201 transmission electron microscope.

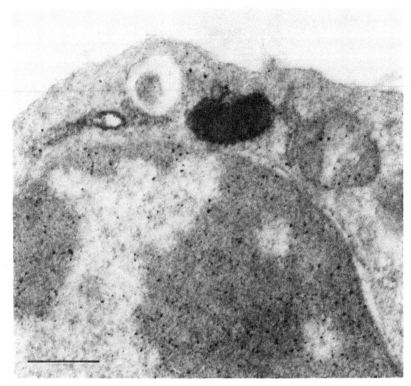

FIG. 4. A chronic lymphocytic leukemic cell stained with anti-μ conjugate showing some stain on the membranes of the RER and an immunoglobulin inclusion. There is no staining of the PNS. Bar equals 0.5 μm.

Critical Comments

The immunoperoxidase technique may be used to visualize the ultrastructural distribution of surface Ig on normal and leukemic human B lymphocytes.[9,10] As penetration of the conjugate is not a problem with the labeling of surface Ig, the larger conjugates, produced by the periodate conjugation technique,[11] and indirect sandwich labeling techniques are suitable. Furthermore higher concentrations of glutaraldehyde, up to 1.25%, may be used providing antigenicity is preserved.[9]

[9] F. Reyes, J. L. Lejonc, M. F. Gourdin, P. Mannoni, and B. Dreyfus, *J. Exp. Med* **141**, 392 (1975).

[10] M. F. Gourdin, F. Reyes, J. L. Lejonc, P. Mannoni, and B. Dreyfus, *Haematol. Blut-transfus.* **19**, 207 (1976).

[11] M. B. Wilson, and P. K. Nakane, *in* "Immunofluorescent and Related Staining Techniques" (W. Knapp, K. Holubar, and G. Wicks, eds), p. 215. Elsevier, Amsterdam, 1978.

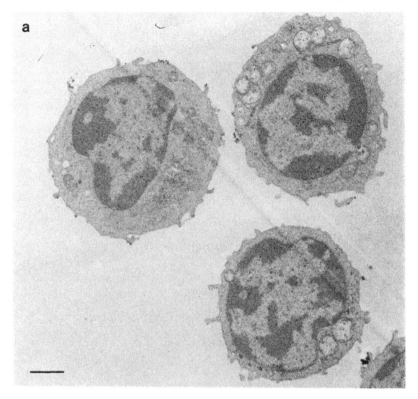

FIG. 5. Chronic lymphocytic leukaemic cells stained for (a) μ heavy chains and (b) λ light chains showing the differential distribution of immunoglobulin heavy and light chains in immature lymphoid cells. After culturing for 6 days with pokeweed mitogen the intensity and distribution of staining for (c) μ heavy chains becomes similar to that for (d) λ light chains reflecting a more mature lymphoid cell. Bar equals 1 μm. From Newell *et al.*[13]

In the majority of prefixed B lymphocyte preparations, surface Ig is randomly distributed (Fig. 1). Similar techniques are applicable to unfixed cells for observing the redistribution of surface Ig accompanying incubation with bivalent antiimmunoglobulin. Some leukemic B cells exhibit a nonrandom distribution of surface Ig, i.e., preferentially localized over a polar cap of microvilli.[10]

Although intracellular Ig may be visualized after glutaraldehyde prefixation alone, the staining is unpredictable in monoclonal B cell populations, presumably due to poor penetration of the conjugate.[2] The treatment of lightly glutaraldehyde-prefixed cells with saponin allows reproducible staining of intracellular Ig. However, conjugate penetration may be restricted for a number of reasons, including the use of the large

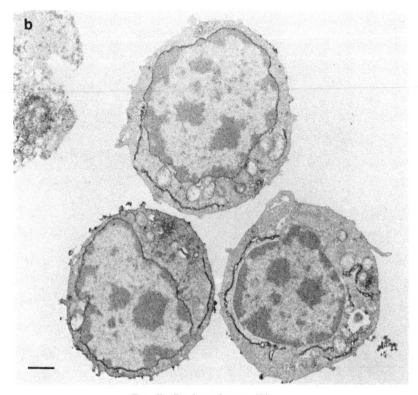

FIG. 5b. See legend on p. 401.

conjugates produced by the periodate conjugation technique[11] and the cooling of unfixed cells.

The ultrastructural immunoperoxidase technique may also be employed to study the distribution of Ig heavy and light chains in a number of heterogeneous and homogeneous lymphoid populations.[2,12,13] In each case a specificity of staining is observed which correlates with the monoclonality seen by immunofluorescence and [³H]leucine incorporation. Additionally, there is an increased sensitivity over immunofluorescence techniques and possibly over the [³H]leucine incorporation technique.[12]

With this method, intracellular Ig is found to be localized in the cisternae of the perinuclear space (PNS), rough endoplasmic reticulum (RER),

[12] D. G. Newell, A. Hannan-Harris, A. Karpas, and J. L. Smith, *Br. J. Haematol.* **50**, 445 (1982).
[13] D. G. Newell, A. Hannan-Harris, and J. L. Smith. *Blood* **61**, 511 (1983).

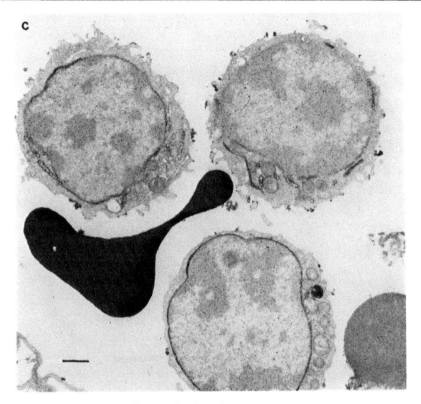

FIG. 5c. See legend on p. 401.

and Golgi apparatus (Fig. 2). Diffuse cytoplasmic staining is rarely observed and is usually nonspecific in that it is also found with the normal sheep IgG peroxidase control conjugate (Fig. 3). Such nonspecific staining is generally associated with poor cell viability, restricted conjugate penetration, or the presence of unconjugated peroxidase. However, Yasudu et al.[14] consistently found diffuse cytoplasmic staining in most of the cells from several patients with chronic lymphoid leukemia (CLL), although, in the majority of cases tested the Ig was localized in the PNS, RER, and Golgi complexes.

The treatment of cells with saponin induces, even after glutaraldehyde prefixation, loss of membrane and cytoplasmic integrity and swelling of the mitochondria and endoplasmic reticulum. This damage may be significantly reduced by limiting the exposure period of the cells to saponin to

[14] N. Yasuda, T. Kanoh, S. Shirakawa, and H. Uchino, Leuk. Res. **6,** 659 (1982).

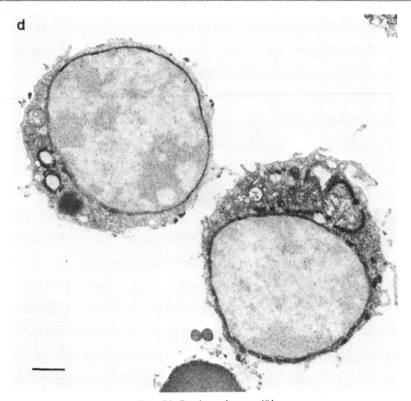

FIG. 5d. See legend on p. 401.

5–10 min. The susceptibility of cells to this treatment varies with the cell source. For example, cells from lymphoid tissues and hematopoietic cell lines are significantly more susceptible to saponin digestion than normal and leukemic peripheral blood lymphocytes. It is, therefore, essential to monitor the morphological damage in each cell type and balance morphological preservation against permeability to the conjugate. There is no evidence to suggest that saponin treatment results in cytoplasmic redistribution of Ig.[2] However, this treatment causes a loss of surface Ig which ensures that surface and intracellular Ig are not visualized at the same time.

The visualization of the differential distribution of Ig heavy and light chains in monoclonal populations has been useful in the diagnosis of B cell neoplasms[15] and has provided considerable data on the mechanisms

[15] D. G. Newell, M. Sattar, A. C. Hannan-Harris, M. I. D. Cawley, U. Jayaswal, and J. L. Smith, *Scand. J. Haematol.* **28**, 441 (1982).

of Ig synthesis, cytoplasmic transport, and secretion.[12,13] In immature B cells, such as CLL cells, the light chain is synthesized and secreted in excess and is localized in the PNS and RER, while staining for heavy chains is negative or poor and restricted to the membranes of the RER, except in those cases where a block in Ig secretion results in the formation of Ig inclusions (Fig. 4). Conversely, more mature B cells (e.g., mitogen stimulated B cells) have a balanced synthesis and secretion of Ig which is reflected in an increased intensity of staining for heavy chain localized in the cisterns of the PNS and RER (Fig. 5).

Acknowledgment

Financial support for this project was kindly provided by the Cancer Research Campaign.

[40] Use of Protein A in the Detection and Quantitation of Immunoglobulin G on the Surface of Lymphocytes

By VICTOR GHETIE and JOHN SJÖQUIST

Cell-associated IgG molecules either produced by, or attached to, lymphoid cells can be detected and quantitated using labeled protein A (SpA)[1-11] (see the table).

There are two main methods for detection and/or quantitation of surface IgG. (1) Cytochemical methods using SpA labeled with fluorescent dyes detected by fluorescent microscopy[2,9]; radionuclides detected by

[1] Abbreviations: ES, erythrocytes coated with SpA; PBS, phosphate-buffered saline; RBC, red blood cells; SpA, protein A of *S. aureus*.

[2] V. Ghetie, H. A. Fabricius, K. Nilsson, and J. Sjöquist, *Immunology* **26**, 1081 (1974).

[3] V. Ghetie, K. Nilsson, and J. Sjöquist, *Eur. J. Immunol.* **4**, 500 (1974).

[4] V. Ghetie, K. Nilsson, and J. Sjöquist, *Scand. J. Immunol.* **3**, 397 (1974).

[5] G. Dorval, K. I. Welsh, and H. Wigzell, *Scand. J. Immunol.* **3**, 405 (1974).

[6] P. Biberfeld, V. Ghetie, and J. Sjöquist, *J. Immunol. Methods* **6**, 249 (1975).

[7] K. I. Welsh, G. Dorval, and H. Wigzell, *Nature (London)* **254**, 67 (1975).

[8] G. Dorval, K. I. Welsh, and H. Wigzell, *J. Immunol. Methods* **7**, 237 (1975).

[9] G. Dorval, H. Wigzell, W. Leibold, and D. Killander, *J. Immunol. Methods* **9**, 251 (1976).

[10] A. C. Bancu and I. Moraru, *Rev. Roum. Biochim.* **14**, 235 (1977).

[11] M. A. Dobre, I. Moraru, and E. Mandache, *Rev. Roum. Biochim.* **18**, 258 (1981).

DERIVATIVES OF SpA USED IN THE DETECTION AND QUANTITATION OF
SURFACE-ASSOCIATED IgG

SpA derivative	Label	Method of detection
Fluorescent SpA	Fluorescein isothiocyanate	Fluorescent microscopy, cytophotometry
Radiolabeled SpA	^{125}I	Autoradiography and cell surface radioassay
	^{131}I	
	^{3}H, tritium	
Electron-dense SpA	Horseradish peroxidase	Electron microscopy
	Ferritin	
	Hemocyanin	
	Colloidal gold	
Microparticle-bound SpA	Fluorescent staphylococci	Fluorescence microscopy and scanning electron microscopy
	Red blood cells	Light microscopy
	Polyacrylamide microbeads	Light and electron microscopy
	Iron-containing albumin	Scanning electron microscopy
	Microspheres	

autoradiography[12]; electron-dense materials detected by electron microscopy,[13–16] or microparticles detected by light,[3] fluorescent,[4] or scanning[4] microscopy. (2) Radiochemical methods using SpA labeled with ^{125}I,[6,8] ^{131}I,[17] or ^{3}H,[18] followed by determination of radioactivity in a gamma or scintillation counter.

Preparation of Labeled SpA

The labeled SpA preparations should exhibit three main features: (1) the ability to react with IgG, (2) low content (if any) of free, unbound tracer, and (3) low nonspecific binding to the cell surface. This last property of labeled SpA preparations can be easily checked by omitting treatment of cells with the specific antibody when antigenic markers are tested.

[12] J. W. Goding, *J. Immunol. Methods* **20**, 241 (1978).
[13] M. Dubois-Dalcq, H. McFarland, and D. McFarlin, *J. Histochem. Cytochem.* **25**, 1201 (1977).
[14] T. Bächi, G. Dorval, H. Wigzell, and H. Binz, *Scand. J. Immunol.* **6**, 241 (1977).
[15] E. L. Romano and M. Romano, *Immunochemistry* **14**, 711 (1977).
[16] M. M. Miller, C. D. Stroder, and J. P. Revel, *Scanning Electron Microsc.* **2**, 125 (1980).
[17] P. M. Zeltzer and R. C. Seeger, *J. Immunol. Methods* **17**, 163 (1977).
[18] R. L. Wilder, C. C. Yuen, B. Subbarao, V. L. Woods, C. B. Alexander, and R. C. Mage, *J. Immunol. Methods* **28**, 255 (1979).

Fluorescent SpA[2,6,9]

Fluorescein isothiocyanate (Isomer I, Sigma) is coupled to SpA (10 mg/ml) in 0.25 M carbonate-bicarbonate buffer, pH 9.0 by adding 0.05 mg fluorochrome/mg SpA. The mixture is incubated overnight at 4°. Excess of fluorescein isothiocyanate is removed by gel filtration on a Sephadex G-25 (Pharamacia, Uppsala) column (1 × 20 cm) equilibrated with phosphate-buffered saline (PBS), pH 7.2. The fluorescent fraction is collected, concentrated and the F/P ratio is estimated according to the formula: F/P = 0.01[(μg F/ml)/(μg SpA/ml)]. The concentration of fluorochrome (in μg/ml) of the conjugate is calculated using the formula: F (μg/ml) = $(A_{495\,nm} - 0.5A_{320\,nm})$ × dilution/0.2. The SpA concentration (in mg/ml) may be determined using pure SpA $A_{280\,nm}^{10\,mg/ml}$ = 1.6 as standard. The F/P ratio should range between 2.5 and 3.0. Higher F/P ratios are not recommended since high background staining of cells is obtained.

Radiolabeled SpA

SpA can be labeled with ^{125}I using the lactoperoxidase method,[6] chloramine-T method,[8] or the Bolton–Hunter reagent.[19] The iodination procedure using the lactoperoxidase technique is described below[6]:

Ten microliters of lactoperoxidase (Sigma) solution in PBS (0.8 mg/ml prepared immediately before use), 10 μl Na[125]I (Amersham, England) (about 1 mCi), 75 μl PBS, and 5 μl perhydrol solution in PBS (0.03%) are added rapidly in sequence with stirring to 5 mg of SpA in 0.4 ml of PBS. The mixture is incubated at room temperature for 10 min with occasional stirring, then 10 μl of sodium azide (2.5 M) is added. To remove free iodine, the iodinated SpA is passed through a Sephadex G-25 column eluted with PBS. Alternatively, the material can be applied to a human IgG-Sepharose 4B column, washed with PBS to remove free radioiodine. The iodinated SpA is eluted with 0.1 M glycine buffer, pH 3.0. The effluent is neutralized with 0.5 M Tris–HCl buffer, pH 8.0 and dialyzed against PBS. These two procedures give similar results. The free radioiodine can be determined by passing an aliquot (50 μl) of ^{125}I-labeled SpA through a small column of IgG-Sepharose 4B and measuring the radioactivity eluted from the column with PBS. A suitable preparation of ^{125}I-labeled SpA should contain no more than 5% free iodine.

If a higher specific activity is needed, the amount of SpA must be lowered up to 100-fold (50 μg). Free iodine is then removed by dialysis or three-four successive ultrafiltration on Diaflo membrane PM-10.

[19] J. J. Langone, M. D. P. Boyle, and T. Borsos, *J. Immunol. Methods* **18**, 281 (1977).

Electron-Dense SpA

SpA can be covalently linked to a number of electron-dense proteins such as peroxidase,[13] ferritin,[14,20] or hemocyanin[16] or adsorbed on colloidal gold particles.[15] Excess of protein is removed by chromatography on IgG-Sepharose 4B. The coupling of SpA with peroxidase is described below[21]:

Glutaraldehyde to a final concentration of 1.25% is added to 10 mg of horseradish peroxidase (type VI, Sigma) dissolved in 0.20–0.25 ml 0.1 M phosphate buffer, pH 6.8. The mixture is kept overnight at room temperature and then applied to a Sephadex G-25 column (1 × 50 cm) equilibrated with 0.1 M phosphate buffer, pH 6.8. The colored fractions are pooled and concentrated to 1 ml by ultrafiltration on Diaflo PM-10. Five milligrams of SpA in 1 ml of 0.5 M carbonate-bicarbonate buffer, pH 9.5 is added to 1 ml of glutaraldehyde-activated peroxidase. The mixture is kept overnight at 4°. Two tenths of one milliliter of 0.2 M lysine in carbonate-bicarbonate buffer is then added and the mixture is incubated for 2 hr at room temperature. The mixture is then applied to a IgG-Sepharose 4B column (1 × 10 cm) equilibrated with PBS. The column is washed until the colored material is completely removed and then eluted with 0.1 M glycine buffer, pH 3.0 as previously described. The neutralized peroxidase-SpA conjugate is dialyzed, concentrated, and kept at 4° (sterile or with sodium azide, 1 mg/ml).

Immobilized SpA

SpA can be bound to various types of microparticles which may be used for the localization of IgG on cell membrane. SpA is a component of the cell wall of *Staphlococcus aureus* Cowan-1 strain and the bacteria themselves (1 μm diameter) can be used as an immobilized SpA reagent for the detection of surface IgG. Since the visualization of the bacteria adherent to the cell surface is difficult with the light microscope, the use of fluorescent bacteria in conjunction with the fluorescent microscope is recommended. SpA can also be bound to the surface of erythrocytes[3,22] allowing the detection of surface IgG by a rosette technique.[22] Polyacrylamide microbeads with entrapped SpA can also be used for identification of IgG-bearing cells.[23]

[20] C. L. Templeton and R. J. Douglas, *FEBS Lett.* **85**, 95 (1978).

[21] V. Ghetie and J. Sjöquist, unpublished technique.

[22] V. Ghetie, G. Stålenheim, and J. Sjöquist, *Scand. J. Immunol.* **4**, 471 (1975).

[23] R. Lindmark, E. Larsson, K. Nilsson, and J. Sjöquist, *Acta Pathol. Microbiol. Immunol. Scand., Sect. C* **90**, 117 (1982).

The preparation of fluorescent staphylococci and SpA-coated red blood cells (RBC) is presented below.

Fluorescent Staphylococci[4] (This Volume [45])

In the direct method, strain Cowan-1 staphylococci are fixed in 0.5% formaldehyde in PBS for 3 hr at room temperature and then inactivated by heating at 80° for 5 min. Fifty milligrams of wet bacteria is resuspended in 10 ml of 0.5 M carbonate–bicarbonate buffer, pH 9.0 and 1 mg fluorescein isothiocyanate in 0.2 ml carbonate–bicarbonate buffer is added. The labeling reaction is carried out at room temperature for no more than 10 min. The preparation is centrifuged and repeatedly washed with cold PBS until the supernatant is free of fluorescence. The fluorescent bacteria are then resuspended in PBS containing sodium azide (1 mg/ml) at a concentration of $\sim 10^9$ bacteria/ml. The suspension should contain about 5 μg fluorochrome/ml which is estimated by reading the optical density of a lysostaphin[24] (Sigma) digest at 495 and 320 nm and applying the formula: $(A_{495\ nm} - 0.5A_{320\ nm})/0.2$. The suspension of bacteria is stable for 2 weeks at 4° or for 1 month at −20°. It can be freeze-dried. In this case, however, addition of 5 mg proteose-peptone (Difco)/ml suspension to avoid aggregation of fluorescent bacteria is necessary.

SpA-Coated Erythrocytes (ES)[3,22]

RBC (sheep, ox, or human) are coated with SpA by using the chromium chloride technique[25]: 0.25 ml of packed RBC is suspended in 0.5 ml of 0.15 M NaCl containing 2.5 mg SpA. Five milliliters of 0.5 mM CrCl₃ in saline is then added. After 1 hr incubation at 30° the cells are washed twice in 0.15 M NaCl or veronal buffered saline and resuspended in PBS or other tissue culture medium to give a 3% ES suspension. ES suspensions containing sodium azide (1 mg/ml) can be stored at 4° for 2 weeks. Alternatively, RBC can be coated with SpA by using glutaraldehyde-treated RBC.[26] ES cells prepared with glutaraldehyde can be freeze-dried and stored indefinitely.

Detection of Surface IgG by SpA

Direct and indirect immunofluorescence staining can be carried out on living cells. The cells are washed with tissue culture medium (e.g., Ham's F-10 medium) without serum and suspended at 5 × 10⁶–10⁷ cells/ml.

[24] J. Sjöquist, B. Meloun, and H. Hjelm, *Eur. J. Biochem.* **29**, 572 (1972).
[25] G. Stalenheim, O. Götze, N. R. Cooper, J. Sjöquist, and H. J. Müller-Eberhard, *Immunochemistry* **10**, 501 (1973).
[26] G. Stålenheim and J. Sjöquist, *J. Immunol.* **105**, 944 (1970).

In the *direct* method, 50 μl of this suspension is incubated with 50 μl of fluorescent SpA (1–2 mg/ml) at 4°. After 30 min incubation the cells are washed three times with cold medium. The material is then resuspended in 50 μl of medium containing sodium azide (1 mg/ml). One drop is placed on a glass slide under a coverslip sealed with paraffin oil. The slides are immediately examined under a fluorescence microscope equipped with a HBO-200 mercury lamp. Photomicrographs can be taken using high speed Ektachrom films. The percentage of fluorescent cells is calculated by examining and scoring 200–300 cells in each preparation.

In the *indirect* method, 50 μl of cell suspension is incubated for 15 min at 4° with 50 μl of an appropriate dilution (usually 1/10–1/50) of an antiserum containing antibody against a specific membrane antigen. The cells are then washed three times and suspended in 50 μl medium containing 1 mg/ml sodium azide. Fluorescent SpA is then added and the procedure detailed above is followed. A control with antibody-untreated cells should be included to determine nonspecific binding of SpA.

For staining with fluorescent Staphylococci the following procedure is recommended: 0.1 ml of the cell suspension (10^6 cells), treated with specific antibody (see above), is mixed with 0.1 ml of the fluorescent bacteria (10^2 bacteria) in a small plastic tube. The mixture is centrifuged at 4° for 10 min at 300 g and then incubated in ice for 1 hr. The pellet is resuspended and one drop of it is placed on a slide under a coverslip and examined with a fluorescent microscope.

Quantitation of Surface IgG with Radiolabeled SpA

[125]I-labeled SpA is frequently used in conjunction with specific antibodies as a reagent for quantitation of cell surface antigens. The advantage of using ^{125}I-labeled SpA as a substitute for radiolabeled antibodies resides in its ability to react with a wide range of antibodies.[12,19] However, when it is used in the indirect method, radiolabeled SpA is a measure of antibody bound to the corresponding surface antigen rather than of the membrane antigens itself. If a 1 : 1 ratio of antibody to the membrane antigen and of SpA to the antibody is assumed,[3,19] the number of SpA molecules bound per cell at saturation level corresponds approximately to the number of specific antigen molecules present on the surface of a cell.

The radioassay is carried out as follows[6]: 100 μl of the cell suspension (containing 2×10^6 cells) is added to 100 μl of serial dilution of the specific antiserum (starting from 1/10). After a 30 min incubation at 4°, the cells are washed with PBS and resuspended in 100 μl of medium containing sodium azide (1 mg/ml). Ten microliters of radiolabeled SpA (1 mg/ml: specific activity 10^5 cpm/μg) is then added to each tube. After a 1 min incubation, the tubes are centrifuged and the resulting pellet is washed four times with cold medium. The radioactivity of the pellet is then determined. The binding of SpA can be expressed as percentage of bound radioactivity or, by using the Avogadro number, as number of SpA molecules bound per cells. Under optimal conditions a linear relationship is found between the cpm bound to cells (or percentage of SpA bound) and the concentration of IgG antibody used to sensitize the cells. To quantitate cell-bound IgG antibody, ^{125}I-labeled SpA must be added in excess to the antibody.

Detection of Surface IgG by Electro-Dense SpA Derivatives[13,14,16]

Fifty microliters of SpA conjugated with peroxidase, ferritin, or hemocyanine (about 2 mg/ml) is added to 50 μl of cell suspension (2 × 10^6 cells) previously treated with specific antibody (see above). The protein is allowed to react with the cells for 15 min at room temperature. The cells are then centrifuged and washed three times with medium. The pellet is finally resuspended and incubated for 30 min in 1 ml 2% glutaraldehyde in 0.1 M phosphate or sodium cacodylate buffer, pH 7.2. The cells are then washed twice with buffer and postfixed for 60 min at 4° with 2% osmic acid in buffer. The fixed cells are washed with distilled water and incubated overnight in 2% uranyl acetate in 25% ethanol. The cell pellet is dehydrated in graded ethanol (25, 50, 80, 95, and 100%) and embedded in plastic for the preparation of thin sections. The sections are mounted on grids and examined at the electron microscope. When peroxidase-labeled SpA is used, the cells (after glutaraldehyde fixation) are incubated for 15 min at room temperature in a medium containing 0.05% 3,3'-diaminobenzidine and 0.02% perhydrol in 0.05 M Tris–HCl buffer, pH 7.2.[27] The cells are then washed and postfixed with osmic acid as detailed above.

Detection and Quantitation of Surface IgG with SpA-Coated RBC[3,22]

ES form rosettes with Ig-bearing lymphoid cells. Inhibition of rosette formation by SpA can, therefore, be used for quantitation of the antigens on the cell surface.

A cell suspension (2 × 10^6–5 × 10^6 cells) in 0.5 ml medium is treated with 25 μl of the appropriate antiserum for 1 hr in an ice bath. The cells are then centrifuged and, after several washings with cold medium, the pellet is resuspended in 0.45 ml of medium. Fifty microliters of 3% ES-indicator cells is then added. The mixture is centrifuged at 100 g for 10

---

[27] F. Mandache, E. Moldoveanu, G. Mota, I. Moraru, and V. Ghetie, *J. Immunol. Methods* **35**, 33 (1980).

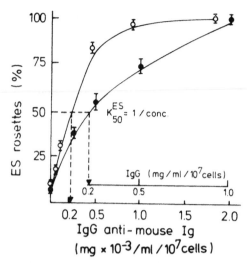

FIG. 1. Ability of mouse lymphocytes treated with various amounts of rabbit IgG and rabbit IgG anti-mouse Ig to form rosettes with ES. ○, Rabbit IgG; ●, rabbit IgG anti-mouse Ig. The maximal percentage of rosettes formed was 49% for rabbit IgG and 46% for rabbit IgG anti-mouse Ig. These figures were taken as 100% rosette formation. The relative affinity constant for rabbit IgG was $K_{ES/50} = 1/0.22 = 4.5$ and for rabbit IgG anti-mouse Ig was $K_{ES/50} = 1/0.23 \times 10^{-3} = 4350$. Thus, the affinity constant of rabbit IgG for Fc receptor was 1000-fold lower than that of a rabbit IgG that reacts through its antigen combining sites with surface-associated Ig.

min at 4° and incubated further for 1 hr or overnight at 4°. The sediment is gently resuspended and 0.05 ml 0.2% toluidine blue in PBS is added. After 10 min, 200–300 cells are counted. Cells binding more than three ES cells are considered to be forming rosettes. The percentage of ES-positive cells (bearing the specific antigen) is then calculated. ES rosette formation of cells treated with IgG antibody can be inhibited by free SpA. The minimum number of antigenic sites on the cell surface can be approximately calculated from the amount of added SpA that inhibits rosette formation by 50%.[3,28]

A relative affinity constant ($K_{ES/50}$) of IgG antibody or IgG ligands (for Fc receptor sites) can be calculated from the curves of ES binding to cells at different IgG concentration (Fig. 1).[28] $K_{ES/50}$ is expressed as the reciprocal of the IgG concentration giving half of the maximal percentage of ES rosettes ($K_{ES/50} = 1/IgG$).

A variant of the ES rosette assay which makes use of the ability of bovine RBC coated with SpA to bind IgG antibody without agglutinating

[28] V. Ghetie, C. Medesan, and J. Sjöquist, *Scand. J. Immunol.* **5,** 1199 (1976).

can also be used. RBC-linked IgG antibody can be prepared in this way and the reagent can be used directly for identification of surface antigenic markers.[29]

Applications

Both radiolabeled SpA and SpA-coated RBC (ES) are used for the detection and quantitation of various antigenic markers on the cell surface (in conjunction with specific antibody) or for the identification of antibodies belonging to the IgG class in normal and pathologic sera directed against some surface antigens of various types of cells. A comprehensive reference list of these applications has been published recently.[30]

Acknowledgments

Our investigation on the use of SpA in detection and quantitation of IgG on cell surfaces was supported by grants from the Swedish Medical Research Council and the Romanian Academy of Medical Sciences and was done in cooperation with Drs. K. Nilsson, H. A. Fabricius (Uppsala), C. Medesan, and S. Mihaescu (Bucharest).

[29] S. Mihaescu, J. Sjöquist, and V. Ghetie, *J. Immunol. Methods* **29,** 79 (1979).
[30] J. J. Langone, *Adv. Immunol.* **32,** 158 (1981).

[41] Use of Fluorescent Antibodies in the Study of Lymphoid Cell Membrane Molecules

By LUCIANA FORNI and STEFANELLO DE PETRIS

Introduction

Visualization of antigen–antibody reactions at the level of light microscopy is made possible by covalent coupling of fluorescent substances to antibody molecules, a method originally described by Coons *et al.*[1] The large variety of reagents, built in the antibody system and made recently practical by the hybridoma technique,[2] the specificity of the reactions, and the simplicity of the technique, make immunofluorescence the method of choice for the localization of macromolecules in cells and tissues. Its application to the study of membrane-bound components of

[1] A. H. Coons, H. J. Creech, and R. N. Jones, *Proc. Soc. Exp. Biol. Med.* **47,** 200 (1941).
[2] G. Köhler and C. Milstein, *Eur. J. Immunol.* **6,** 511 (1976).

living cells, which provided in the last decade relevant information on basic processes of cell physiology[3,4] and on the organization of biological membranes,[5,6] together with the possibility of detecting two antigens at the time by using antibodies of different specificity coupled to different fluorochromes, makes it possible to study association, independence, and interactions of various membrane structures at the single cell level.

Materials and Equipment

Antibodies

Source of antibodies can be either sera of hyperimmunized animals or products of hybridoma cell lines. Detailed indications for preparation of antisera and monoclonal antibodies, as well as immunochemical tests for antibody activity and specificity, are dealt with elsewhere in this series.[6a]

Fluorochromes and Fluorescence Microscopes

A property of fluorescent substances is the emission of light of a given wavelength upon absorption of light in a shorter wavelength range (excitation). The two fluorochromes most widely used in immunofluorescence are the isothiocyanate derivatives of fluorescein (FITC) and rhodamine (TRITC). The emission characteristics of the two substances, in the green range (517 nm) for fluorescein and in the orange-red range (580 nm) for rhodamine, make them suitable for double-staining experiments. Covalent coupling to proteins occurs through the thiocyanate group to the ε-amino groups of lysine. Fluorescein and tetramethylrhodamine isothicyanate are available from many chemical manufacturers (Sigma).

A water-soluble fluorescein derivative, dichlorotriazinylaminofluorescein (DTAF),[7] with fluorescence and conjugation characteristics very similar to that of FITC, is also available (Sigma).

Fluorescence microscopes for membrane studies are preferentially set for epiillumination, an adaptation of the vertical illumination used in reflection microscopy, where the light beam passes through the objective that acts as a condenser. The light emitted by a high pressure mercury or xenon lamp is selected and directed to the object by a dichroic mirror, that reflects and transmits light of nonoverlapping wavelength ranges.

[3] L. D. Frye and M. Edidin, *J. Cell Sci.* **7**, 319 (1970).
[4] S. de Petris, *J. Cell Biol.* **65**, 123 (1975).
[5] S. J. Singer and G. L. Nicolson, *Science* **175**, 720 (1972).
[6] S. de Petris and M. C. Raff, *Ciba Found. Symp.* **14**, 27 (1979).
[6a] See this series, Vol. 73, Part B.
[7] D. Blakeslee and M. G. Baines, *J. Immunol. Methods* **13**, 305 (1976).

Dichroic mirrors are especially designed to match the characteristics of excitation and emission of the various fluorochromes, so that only light of the required wavelength band is focused on the object, and only the emitted fluorescence, with little need of additional suppression, enters the eyepiece. The epiillumination provides therefore sufficient fluorescence intensity at the high magnifications required for membrane studies of single cells. The microscope is usually also equipped with transmitted phase-contrast illumination for the morphological observation of the cell sample.

Fluorescence microscopes with epiillumination are manufactured by several companies including Wild-Leitz, Zeiss, Vickers, and Reichert.

Detailed information on the principle of fluorescence and fluorescence microscopy with transmitted and incident illumination can be found elsewhere.[8-11]

Methods

Preparation of Conjugates

The original conjugation method of Coons et al.[1] has undergone substantial improvements aimed mostly at preventing nonspecific staining caused by hyperconjugated antibody or contaminant molecules. Purity and limited heterogeneity as regards the charge of the antibody to be used for conjugation is an important prerequisite for subsequent preparation of suitable conjugates. For this reason, the method of choice for purification of immunoglobulins from antisera is still ion-exchange chromatography, since other methods, such as affinity chromatography on antigen or on protein A from *Staphylococcus aureus* (SpA), while ensuring purity of the immunoglobulin preparation, do not guarantee any charge homogeneity. On the other hand, affinity chromatography procedures are perfectly suitable for purification of monoclonal antibodies from supernatants of hybridoma cultures or ascites.

Purification of Immunoglobulins for Conjugation. Conventional Antisera. Immunoglobulin-enriched fractions are prepared by salting-out with ammonium sulfate at a final concentration of 1.6 M. A 3.2 M solution of ammonium sulfate is added dropwise to the serum under stirring; the

[8] R. C. Nairn, ed., "Fluorescent Protein Tracing." Churchill-Livingstone, Edinburgh and London, 1976.
[9] E. J. Holborow, ed., "Standardization of Immunofluorescence." Blackwell, Oxford, 1970.
[10] J. S. Ploem, *Z. Wiss. Mikrosk. Mikrask. Tech.* **68**, 129 (1967).
[11] J. S. Ploem, *Ann. N.Y. Acad. Sci.* **254**, 5 (1975).

precipitate collected by centrifugation (20–30 min at 6000 g) is dissolved in phosphate-buffered saline (PBS) (0.01 M phosphate pH 7.5, 0.15 M NaCl) to a volume of one-third that of the original serum, and dialyzed in the cold against three changes of 100 volumes of PBS, to remove sulfate ions. Finally, the protein solution is equilibrated by dialysis against the starting buffer of ion exchange chromatography (0.01 M pH 7.5 sodium phosphate), and the insoluble fraction is removed by centrifugation (10 min at 5000 g). Alternatively, the dialysis against PBS can be substituted by extensive dialysis against large volumes of 0.01 M pH 7.5 sodium phosphate buffer.

Diethylaminoethylcellulose (DEAE-cellulose) (4 g/10 ml of original serum) is washed twice for a few hours with 0.01 M pH 7.5 phosphate buffer, and packed in a chromatography column. The dialyzed protein solution is applied to the ion exchanger and elution is started with the equilibration buffer. The more basic population of IgG molecules is not retained, and elutes with the void volume. For sheep and goat sera, part of the IgG is retarded, and elutes as a second peak with the same buffer: the two fractions can be pooled for conjugation with FITC, but must be kept separate for TRITC coupling. For all sera, an additional fraction of more negatively charged IgG is removed from DEAE-cellulose with 0.05 M NaCl/0.01 M pH 7.5 phosphate buffer: this can be satisfactorily labeled with TRITC, but poorly with FITC.[12]

It is advisable to test the antibody concentrations in the various chromatographic fractions, since the specific antibody activity might not be, and often is not uniformly distributed.

The protein fractions are made isotonic by addition of appropriate volumes of 1.5 M NaCl, concentrated by pressure dialysis to 5–7 mg/ml, and kept frozen at $-20°$.

Monoclonal Antibodies. Mouse monoclonal IgG antibodies can be conveniently purified by affinity chromatography on insoluble protein A from *Staphylococcus aureus*. The method outlined here is basically the one described by Ey *et al.*[13] for isolation of IgG subclasses from mouse serum on the basis of differential binding affinity on protein A. A preliminary determination of the protein A binding characteristics of the antibody to be isolated is suggested, remembering that a limited heterogeneity exists in this respect within a given IgG subclass. In general, binding of IgG_1 occurs only at pH 8.0, while IgG_{2a}, IgG_{2b}, and IgG_3 can bind at pH 7. Conversely, IgG_1 are eluted at pH 7 to 6, IgG_{2a} and IgG_3 at pH 4.5, and IgG_{2b} at pH 3. However, a minor proportion of IgG_1 is eluted only at pH

[12] J. J. Cebra and G. Goldstein, *J. Immunol.* **95**, 230 (1965).
[13] P. L. Ey, S. J. Prowse, and C. A. Jenkin, *Immunochemistry* **15**, 429 (1978).

4.5, and part of IgG_{2a} is eluted only at pH 3. One must remember that the capacity of the absorbent for different proteins also varies, and is rather low for IgG_1. Therefore, some care should be taken to avoid overloading of the column and possible loss of antibody.

Protein A-Sepharose CL-4B (Pharmacia, Uppsala) is swollen in 0.1 M pH 8 phosphate buffer, and the smaller particles removed by suction. The gel is packed in a small chromatography column, or in a syringe with a glass-wool support. Supernatants from hybridoma cultures (a 5- to 6-fold concentration by pressure dialysis is suggested purely for time-saving purposes), or ascitic fluid is brought to pH 8 by addition of 0.5 M phosphate buffer, and filtered on the absorbent column at a flow rate of 15–20 ml/hr. The column is then washed with 0.1 M pH 8 phosphate buffer until no proteins are detectable by UV absorption at 280 nm, and elution started. The requirement of pH 8 is strict for IgG_1 antibodies. For proteins binding to protein A at physiological pH, the gel can be equilibrated at pH 7–7.2 (with PBS or 0.1 M phosphate), and no adjustment of pH of the supernatant of ascites fluid is required. Eluent buffers are 0.1 M phosphate or citrate buffers pH 6, 0.1 M KH_2PO_4 or 0.1 M acetate or citrate pH 4.5, and 0.1 M acetate or citrate buffer pH 3, according to the characteristics of the protein. One step elution is performed for supernatants, while for ascitic fluids it is convenient to remove all proteins that elute at pH higher than the one specific for the relevant protein in each case. In this way, the monoclonal antibody will be contaminated only by normal immunoglobulins with the same protein A binding properties. The absorbent can be used many times with no loss of efficiency. It has to be washed with 1 M acetic acid, reequilibrated in 0.1 M pH 8 phosphate buffer containing 20 mM sodium azide, and stored at $+4°$.

The eluted antibody is dialyzed against PBS, concentrated to 2–3 mg/ml, and kept frozen at $-20°$.

Binding of rat IgG to protein A occurs only in rather critical and stringent conditions.[14] Therefore, purification of rat monoclonal IgG antibodies by this procedure does not offer any advantage over ion exchange chromatography. In the authors' experience, convenient conditions for elution of rat IgG from DEAE-cellulose is 0.025 M sodium phosphate buffer pH 8. The IgG fraction is preliminarily enriched from concentrated supernatant by precipitation with ammonium sulfate (final concentration 1.8 M) before chromatography (see preceding section).

Fluorochrome–Protein Coupling. Covalent coupling of FITC and TRITC to proteins occurs in aqueous solution at pH 9.0–9.5. At lower pH the reaction is considerably slower. Several procedures of conjugation

---

[14] A. Nilsson, E. Myhre, G. Kronvall, and H. O. Sjögren, *Mol. Immunol.* **19**, 119 (1982).

have been described.[12,14–16] The coupling method reported here presents in our experience some advantages: it can be carried out at fairly low protein concentration (as low as 1 mg/ml); the short time of exposure of the protein to the high pH required for coupling results in very little protein denaturation, and higher yield of suitable conjugate; and the shift in the relative magnitude of the absorption of TRITC at 515 and 555 nm is minimized,[17] resulting in TRITC conjugates of higher efficiency.

Conjugation Procedure. The protein solution, at a concentration of 1–4 mg/ml, is brought to pH 9.1–9.2 either by dialysis against 0.05 M carbonate–bicarbonate buffer, 0.10 M NaCl, or by addition of concentrated (0.5 M) buffer. The pH can also be adjusted by addition of 0.1 M NaOH. In any case, buffers containing amines, such as Tris buffers, must be avoided.

The fluorochrome (FITC isomer I or TRITC isomer R) is dissolved at 1.5 mg/ml in dimethyl sulfoxide (DMSO), and immediately added to the protein solution in the proportion of 50–100 μl/ml, regardless of the protein concentration (final concentration of the fluorochrome ~75–150 μg/ml). The mixture is gently stirred at room temperature for 40–45 min, protected from light.

The same method can also be used with higher protein concentrations (up to 10 mg/ml or more) provided the fluorochrome concentration is also increased so as to maintain a suitable fluorochrome/protein molar ratio (of the order of 3, but depending on the characteristics of the particular fluorochrome).

Since efficiency of coupling of fluorochromes to IgG varies from species to species, in general lower concentrations of fluorochromes are required for optimal conjugation of goat and sheep immunoglobulins than of rabbit immunoglobulins; even lower concentrations must be used for mouse and rat antibodies. Although some variability between antibodies even in the same species is observed, this is not necessarily related to the IgG subclass.

Some attention should be paid to the possibility that conjugation with fluorochromes could result in loss of antibody activity, due either to denaturation of proteins by exposure to high pH, or to interference of the bound fluorochrome at the binding site. With conventional antisera, this can happen, in the authors' experience, with mouse alloantisera of different specificities, while no significant changes in the average binding efficiency is usually observed after conjugation of heterologous antiimmunoglobulin antisera. This is not the case for monoclonal antibodies,

[15] B. T. Wood, S. H. Thompson, and G. Goldstein, *J. Immunol.* **95**, 225 (1965).
[16] L. Amante, A. Ancona, and L. Forni, *J. Immunol. Methods* **1**, 289 (1972).
[17] P. Brandtzaeg, *Scand. J. Immunol.* **2**, 273 (1973).

where fluorochrome labeling can grossly impair binding activity. This inconvenience is unpredictable, and applies also to other kinds of labeling (haptens in the hapten-sandwich technique,[18] or biotin in the biotin–avidin system[19]).

Purification of Conjugates. Labeled proteins and unbound fluorochrome are separated by gel filtration on Sephadex G-25 or G-50, equilibrated with PBS when only the gel filtration step is performed, or with 0.01 M pH 7.5 phosphate buffer if the conjugate has to be further purified by DEAE-cellulose chromatography. A column of 1 cm diameter and 60 cm length is required for filtration of 2–4 ml of conjugation mixture.

The effluent protein fraction is applied to a small column (1.0–1.5 g) of DEAE-cellulose equilibrated with 0.01 M pH 7.5 sodium phosphate buffer. The proteins that are not retained by the ion-exchanger and elute with the void volume are usually hypoconjugated. Fractions with increasing coupling ratios are eluted with increasing sodium chloride concentrations in the same buffer (0.05, 0.1, 0.2, 0.3 M NaCl in 0.01 M pH 7.5 sodium phosphate).

For FITC conjugates suitable conjugates are eluted with 0.1 and 0.2 M NaCl with some variability, depending on the species. However, the fraction eluted with 0.05 M NaCl can be used in some cases although it is advisable to test them for staining efficiency. Fractions eluted with higher NaCl concentrations are often hyperconjugated, and tend to give nonspecific staining.

For TRITC conjugates, in the majority of cases the whole conjugated protein elutes in a single fraction, usually with 0.05 M NaCl when the conjugate is made with a "basic" chromatographic fraction (originally eluted with 0.01 M phosphate alone) and with 0.1 M NaCl when the more "acidic" fraction (originally eluted with 0.05 M NaCl) is used for labeling. This makes in most cases the DEAE-cellulose step unnecessary for TRITC conjugates.

When only small amounts of proteins are available, to avoid excessive dilution of the conjugate the ion-exchange step can be performed with smaller amounts of DEAE-cellulose in a Pasteur pipet with a glass-wool plug.

Basically the same conjugation procedure can be applied to proteins other than antibodies, such as lectins or avidin. In these cases, only the gel-filtration step (which for some lectins can act as affinity chromatography[20]) is performed for the purification of the conjugate, and it can be substituted by extensive dialysis.

---

[18] S. Cammisuli and L. Wofsy, *J. Immunol.* **117**, 1695 (1976).
[19] H. Heitzmann and F. M. Richards, *Proc. Natl. Acad. Sci. U.S.A.* **71**, 3537 (1974).
[20] S. de Petris, *J. Cell Biol.* **65**, 123 (1975).

Characterization of the Conjugates. The most important parameter characterizing a conjugate is the degree of coupling, expressed as number of moles of fluorochrome bound per mole of protein, and is therefore dependent on the protein and fluorochrome concentration.

The *protein concentration* can be conveniently measured on the basis of the absorption at 280 nm: the measurement is approximate, but still adequate for practical purposes. The absorption value contributed by the fluorochrome at 280 nm, which represent a constant proportion of the absorption at the specific wavelength, must be subtracted from the total absorption at 280 nm before deriving the protein concentration in mg/ml according to the following formula:

FITC conjugates: $(OD_{280\ nm} - 0.35\ OD_{495\ nm})/1.4^{20a}$
TRITC conjugates:
  amorphous TRITC: $(OD_{280\ nm} - 0.56\ OD_{515\ nm})/1.4^{20a}$
  crystalline TRITC: $(OD_{280\ nm} - 0.65\ OD_{555\ nm})/1.4^{20a}$

The *fluorochrome concentration* is measured as absorption value at a wavelength in the visible range corresponding to an absorption peak, in general 495 nm for FITC and 515 or 555 nm for TRITC. In regard to the latter fluorochrome, pure isomer R ("crystalline" TRITC) has an absorption peak at 554–555 nm, whereas "amorphous" TRITC contains an additional component with a major peak (after conjugation) at about 515 nm (although the 555 nm component partially contributes to the absorption at that wavelength). The component with the 515 nm peak however absorbs light, but does not fluoresce.[21] The OD at 555 nm in both "crystalline" and "amorphous" TRITC therefore gives a better estimate of the TRITC groups actually capable of emitting fluorescent light. More detailed information on factors affecting the precision of measurement of the fluorochrome concentration can be found elsewhere.[22]

The fluorochrome/protein (F/P) ratio is derived from the absorption data on the basis of the molar extinction coefficient of the protein at 280 nm and of the fluorochrome at the specific wavelength. Due to the approximation of values for both protein and fluorochrome concentration obtained by the spectrophotometric measurement, only approximate F/P ratios can be calculated. Formulas based on slightly different constants are suggested by various authors[22-24] for the evaluation of the F/P ratio of

[20a] The value 1.4 is an approximate extinction coefficient for immunoglobulin molecules, that can be applied for practical purposes to all immunoglobulin isotypes from most species.
[21] P. Ravdin and D. Axelrod, *Anal. Biochem.* **80,** 585 (1977).
[22] S. de Petris, *Methods Membr. Biol.* **9,** 1 (1978).
[23] A. F. Wells, C. E. Miller, and M. C. Nadel, *Appl. Microbiol.* **14,** 271 (1966).
[24] A. Kawamura, Jr., "Fluorescent Antibody Techniques and their Application." University Park Press, Baltimore, Maryland, 1969.

FITC/IgG conjugates: although resulting in somewhat different F/P ratios for the same absorption values, they can all be considered adequate for most practical purposes.

The F/P ratio of TRITC conjugates can be evaluated in the same way only when the crystalline substance is used. For the amorphous fluorochromes, the ratio of the absorption values ($OD_{280 nm}/OD_{515 nm}$) is routinely used.

For the "crystalline" fluorochromes, we suggest the following formulas, that apply to IgG conjugates at pH 7.2:

FITC isomer I: $3.3\ OD_{495}/(OD_{280} - 0.35\ OD_{495})$
TRITC isomer R: $8.0\ OD_{555}/(OD_{280} - 0.65\ OD_{555})$

With some variability from one antibody preparation to another, suitable FITC conjugates have a F/P ratio of 2–4, and TRITC conjugates of 1.5–2.5. For "amorphous" TRITC, suitable conjugates have a ratio OD_{280}/OD_{515} of 3 to 6.

Storage of Conjugates. The conjugates are made isotonic or slightly hypertonic (0.2 M NaCl) by addition of appropriate volumes of 1.5 M NaCl and adjusted at a protein concentration of 0.5 mg/ml. They are sterilized by filtration through Millipore filter (0.45 μm) and stored at 4°. Alternatively, they may be stored at higher concentration in the presence of 5–10 mM sodium azide and diluted just before use.

Millipore filtration is strongly recommended: in addition to sterilization, it substitutes efficiently high speed centrifugation (30 min at 50,000 g) for elimination of aggregates that could bind to Fc receptors (although in our personal experience, fluorochrome coupling interferes with sites involved in Fc receptor binding) and at the same time removes denatured "sticky" molecules that could give nonspecific staining.

Staining Procedures

Two similar methods, differing only in the temperature at which the incubation of cells with antibodies occurs, and in the presence of metabolic inhibitors, are used for simple detection of membrane antigens, or for studies of mobility and interdependence of membrane components. In either case, the staining can be *direct,* using a conjugated antibody, or *indirect* (sandwich, piggyback) using an unlabeled antibody followed by a fluorescent antiimmunoglobulin antibody (or a hapten-derivatized antibody followed by a conjugated antihapten antibody,[18] or a biotin-labeled antibody followed by fluorescent avidin[19]). Two antigens can be detected at the same time by *double staining* using two reagents labeled with FITC and TRITC, respectively.

For simple detection of membrane antigens, small cell samples (25 to

100 μl containing 5×10^5–2×10^6 cells) are incubated with the antibody at a final concentration of 10–30 μg/ml (100–200 μg/ml for IgG fractions of conventional antisera). The staining is performed preferentially in small round bottom disposable tubes (plastic or glass). The incubation is carried out on ice for 15–30 min: the samples are then washed three times by centrifugation (5 min at 100 g) with 2 ml of cold PBS or any balanced salt solution containing 5% fetal calf serum (FCS) or 0.3% bovine serum albumin (BSA) and 10 mM sodium azide. The staining procedure is repeated under the same conditions for indirect staining.

For studies on ligand-induced redistribution of membrane antigens, the cell samples are first incubated with the antibody on ice for 15 min, then to higher temperature (30–37°) for variable periods of time according to the membrane molecules under study. Usually, 10–15 min at 37° is sufficient for redistribution (capping) of membrane immunoglobulins, while redistribution of other antigens (e.g., histocompatibility antigens) requires longer periods of incubation (up to 1 hr). No metabolic inhibitors (such as sodium azide) should be present in the medium during this step.

Studies on the *interdependence of membrane components* are performed by a double staining technique and a combination of the two above procedures. First, the cell samples are exposed to one conjugated antibody under "capping conditions" (37° and absence of metabolic inhibitors), so as to allow for the redistribution of the corresponding antigen. The cell samples are then chilled, washed with cold medium containing sodium azide, and subsequently exposed to a second reagent, labeled with a different fluorochrome, strictly in the cold and in the presence of sodium azide. Indirect stainings can be performed in both steps (this volume [37]).

After the last washing, the cell samples are ready for the microscopic observation. For quick counting of positive cells, a drop of the cell suspension is placed on a slide and covered with a coverslip which is sealed with nail polish. For more careful observations, and mostly for studies on the interdependence of membrane markers, any uncontrolled redistribution occurring at room temperature during the microscopic examination should be prevented. In these cases, the cell samples are fixed either in suspension (2% formaldehyde) or more conveniently on the slide. In the latter case, the cells are washed in a concentrated protein solution (50% FCS or 3% BSA) to ensure attachment to the glass. Smears can easily be made by drawing parallel lines with a pen nib, a fine Pasteur pipet, or a plastic micropipet tip. The cells are then air-dried, fixed for 5 min in 95% ethanol, washed 2–3 times in PBS, and mounted under a coverslip in 60% glycerol in PBS or other mounting media such as polyvinyl alcohol.[25]

[25] B. M. Thomason and G. S. Gowart, *J. Bacteriol.* **93,** 768 (1967).

Addition of 10^{-2} M p-phenylendiamine[26] or 10^{-1} M n-propyl gallate[27] to the glycerol prevents very effectively the fading of fluorescence during the microscopic observation.

Microscopic Observation

As discussed before, a fluorescence microscope equipped with a high-pressure mercury source, epiilumination, and phase contrast transmitted illumination is required for membrane immunofluorescence. This combination makes it possible to determine the frequency of cells positive for a given membrane antigen, as well as to observe the overall morphology of the positive cells.

Here we describe briefly the fluorescence pattern one observes for the three staining procedures outlined above.

The staining on cells exposed to divalent (or multivalent) ligands in the cold and in the presence of metabolic inhibitors appears in most cases as a pattern of discrete patches of variable size separated by empty negative areas. The patches are distributed on the whole cell surface, as it can be seen by changing the plane of focus. When the equitorial plane of the cell is in focus, a rim of bright fluorescent spots is observed. With monovalent ligands and with some lectins, the staining is diffuse, appearing as a continuous ring in the equatorial plane.

Cells incubated with divalent ligands at 30°–37° show the stain concentrated on one part of the cell membrane only, with the appearance of a cap. In some cases, after prolonged incubation at 37°, the antigen–antibody complexes are pinocytosed and appear as bright spots inside the cell, concentrated in an area corresponding to the Golgi zone. The phenomenon of redistribution and the conditions determining it, in the context of the organization of the cell membrane, have been extensively reviewed.[28,29]

General criteria to evaluate interdependence of membrane antigens by ligand-induced redistribution are as follows (this volume [37]):

1. Molecules that are stably connected always coredistribute regardless of which antigen is experimentally induced to redistribute. Their appearance under the fluorescence microscope is that of doubly-stained polar caps. Any nonredistributed positivity will also show overlapping

[26] G. D. Johnson, R. S. Davidson, K. C. McNamee, G. Russell, D. Goodwin, and E. J. Holborow, *J. Immunol. Methods* **55,** 231 (1982).
[27] H. Giloh and J. W. Sedat, *Science* **217,** 1252 (1982).
[28] E. R. Unanue and M. J. Karnowsky, *Transplant. Rev.* **14,** 184 (1973).
[29] S. de Petris, *in* "Dynamic Aspects of Cell Surface Organization" (G. Poste and G. Nicolson, eds.), p. 643. North-Holland Publ., Amsterdam, 1977.

patterns for the two fluorochromes. This is the case for polypeptides that are covalently linked, or strongly associated, such as H and L chains of immunoglobulins, H chain of H-2 class I antigens and β_2-microglobulin, α and β chain of H-2 class II antigens, and for different antigenic determinants present on the same polypeptide chain.

2. Conversely, molecules that are independent, as a rule do not coredistribute: at the microscope, one antigen (one fluorochrome) will be concentrated in a polar cap and the other is spread all over the cell membrane in patches (or in a diffuse pattern). Examples are membrane IgM and IgD and membrane immunoglobulins and H-2 class I antigens.

3. Many molecules on the cell membrane are neither permanently connected nor independent, but undergo interactions that are probably reversible in the unperturbed membrane. These interactions can be visualized as coredistribution when they are unidirectional, that is when the crosslinking and capping of one antigen causes the redistribution of another antigen, but the reciprocal does not occur. Bidirectional interactions cannot be distinguished from permanent association by these criteria.

Comments on the Sensitivity and Specificity of the Method

The absolute *sensitivity* of immunofluorescence techniques is difficult to evaluate since it depends on several factors such as the intrinsic properties (affinity) of the antibody, the efficiency of the conjugate, and the characteristics of the microscopic equipment. The sensitivity of visual detection is higher in conditions when the membrane molecules are crosslinked into discrete patches or redistributed into polar caps. Indirect (sandwich) techniques are 4–10 times more sensitive than the direct techniques.

As for the *specificity* of the detection, two separate cases must be considered, at the level of the original antibody and at the level of the fluorescent reagent, respectively. In either case, controls of specificity have to be performed directly by staining of positive and negative control cells of defined characteristics. It is rather easy to distinguish under the microscope whether an unexpected positivity is due to cross-reactive antibodies or to contaminating antibodies of unwanted specificity, or to lack of specificity of the conjugate. In the first case one observes a typical positive staining in rings and/or discrete patches confined to the cell membrane, while in the second case the staining is dull and diffuse and present as a colored background also on cells displaying a characteristic positive staining pattern, and it is often indistinguishable from autofluorescence.

Spurious positive staining can arise in double-staining experiments, especially when sandwich techniques are used, because of the possible cross-reactivity of anti-Ig antibodies directed against immunoglobulins of one species for Ig molecules of different species. Thus, for example, a (say, goat) anti-mouse Ig antibody may recognize some antigenic determinants on rabbit Ig, and conversely, an anti-rabbit Ig antibody may recognize some determinants on mouse Ig. This not infrequent possibility, that applies also to monoclonal antibodies, must be kept in mind. Therefore, antiimmunoglobulin antibodies to be used in double-staining experiments must be tested for possible reactivity with other Ig molecules and any unwanted cross-reactive antibody carefully removed with suitable immunoabsorbents. This should be done not only for the anti-Ig antibodies to be used as a second reagent, but also for the ones to be used as a first reagent. In fact, because of the reversibility of antigen–antibody reactions, already bound anti-Ig antibody may still be able to react with immunoglobulin added subsequently.

Another source of nonspecific staining, independent from either specificity of the antibody and of the conjugate, is the binding of antibody aggregates to cells expressing Fc receptors. This inconvenience can be prevented by ultracentrifugation or Millipore filtration of the conjugates. Only for special experimental purposes is it advisable to use pepsin $F(ab')_2$ fragments of antibodies, prepared according to standard methods.[30,31] Needless to say, since Fc receptors bind antigen–antibody complexes with much higher affinity than free IgG molecules, any absorption the antibody preparation may require, before or after conjugation, must be performed exclusively with insoluble antigens.

Acknowledgment

The Basel Institute for Immunology was founded and is supported by Hoffmann-La Roche Ltd. Basel, Switzerland.

[30] A. Nisonoff, L. N. Wissler, L. N. Lipman, and D. L. Koernley, *Arch. Biochem. Biophys.* **89,** 230 (1960).
[31] E. Lamoyi and A. Nisonoff, *J. Immunol. Methods* **56,** 235 (1983).

[42] Biosynthetic, Surface Labeling, and Isolation of Membrane Immunoglobulin

By DOROTHY YUAN and ELLEN S. VITETTA

Labeling of Ig by Surface Iodination

Cell surface iodination is a powerful tool for the study of cell surface immunoglobulins (Ig): (1) The surface Ig status of viable cells can be determined[1]; (2) molecules which are present in low numbers or have a slow biosynthetic rate can be studied biochemically[2]; and (3) the effect of various agents which change the surface Ig phenotype can be monitored.[3,4]

A number of methods for cell surface labeling have been reported. The most popular is lactoperoxidase-catalyzed iodination[5] using H_2O_2 directly or H_2O_2 generated by the glucose-oxidase system.[6] If properly used, this procedure achieves the vectorial labeling of membrane but not cytoplasmic proteins and is not toxic to cells.[6] The choloramine-T method[7] utilizes a powerful oxidant followed by strong reducing conditions to terminate the reaction and is often deleterious to cells. Recently, 1,3,4,6-tetrachloro-3α, 6α-diphenylglycoluril (Iodo-gen) has been used successfully as an iodinating reagent.[8]

Procedure of Lactoperoxidase-Catalyzed Iodination

Cell suspensions should be washed extensively to eliminate proteins which adhere to the cells from media or sera. The cells are then resuspended at 1 to 10 × 10^7 cells per ml of phosphate-buffered saline (PBS) pH 7.3 containing 400 μg/ml lactoperoxidase (Sigma Chemical Co., St. Louis, MO) in a 50-ml conical tube. In general, cells with a larger surface area are suspended at a lower concentration. Neutralized $Na^{125}I$ (2 to 4 mCi) is then added followed by two additions of 25 μl of 0.03% H_2O_2 at 5 min intervals. The reaction is terminated at 10 min by the addition of a large excess of cold PBS. The cells are then pelleted by centrifugation. If the

[1] E. S. Vitetta, S. Baur, and J. W. Uhr, J. Exp. Med. 134, 242 (1971).
[2] D. Yuan and E. S. Vitetta, J. Mol. Immunol. 20, 367 (1983).
[3] E. S. Vitetta and J. W. Uhr, J. Immunol. 117, 1579 (1976).
[4] D. Yuan, J. Mol. Immunol. 19, 1149 (1982).
[5] D. R. Phillips and M. Morrison, Biochem. Biophys. Res. Commun. 40, 284 (1970).
[6] A. L. Hubbard and Z. Cohn, J. Cell Biol. 55, 390 (1972).
[7] C. Jone and L. P. Hager, Biochem. Biophys. Res. Commun. 68, 16 (1976).
[8] M. A. K. Markwell and C. F. Fox, Biochemistry 17, 4807 (1978).

total number of cells labeled is less than 1×10^7, it may be necessary to add carrier cells immediately before centrifugation to ensure complete recovery. The cells are subsequently washed two more times with large volumes of PBS and lysed according to methods described below.

If a gamma radiation monitor is available, it is convenient to measure the total amount of $Na^{125}I$ added before starting the reaction since commercial vials of $Na^{125}I$ may be quite variable in content. Subsequent to labeling, the amount of radioactivity associated with the cell pellet should also be measured to ascertain the efficiency of labeling. In general, following two washes, incorporation of more then 20% of the initial radioactivity is indicative of good labeling efficiency. If a monitor is not available, an aliquot of the resuspended cells should be precipitated in 5% trichloroacetic acid (TCA) in order to estimate the efficiency of labeling. If the cells are not adequately labeled, they can, at this stage, be reiodinated. However, cell viability should be determined at the end of the reaction to ascertain that there is no decrease in viability.

The optimal density for labeling normal murine spleen cells, LPS-stimulated B lymphocytes, as well as large tumor cells such as BCL_1[9] is 1 to 10×10^7 cells/ml. At lower cell densities, the total amount of radioactivity incorporated into proteins may be higher. However, the amount of radioactivity specifically associated with cell surface Ig is reduced. At very high cell densities, the labeling efficiency is greatly reduced.

Cell Surface Labeling with Iodo-gen

Lymphocytes have also been surface labeled using Iodo-gen[8] (Pierce Chemical Co.) although the extent of intracellular labeling using this method has not been as carefully documented as in the lactoperoxidase method.

Glass scintillation vials or tubes are coated with 10 to 100 μg of Iodo-gen dissolved at 1 mg/ml in chloroform and dried under nitrogen. Cells to be labeled are washed and resuspended at 10^7 cells/ml in the glass vials or tubes and are incubated for 10 min at 25° in the presence of 0.5 mCi $Na^{125}I$. The reaction is terminated by adding a large volume of PBS followed by centrifugation.

Biosynthetic Labeling of Ig

Biosynthetic labeling of Ig is carried out to determine (1) the rate of biosynthesis and secretion of the Ig molecules[10,11] and (2) the intracellular

---

[9] S. Slavin and S. Strober, *Nature (London)* **272**, 624 (1977).
[10] J. W. Uhr and E. S. Vitetta, *Fed. Proc., Fed. Am. Soc. Exp. Biol.* **32**, 35 (1973).
[11] E. S. Vitetta and J. W. Uhr, *J. Exp. Med.* **139**, 1599 (1974).

fate of the labeled molecules.[12,13] It can also be used to prepare labeled molecules for peptide mapping, amino acid analysis or sequencing.[14]

Labeling Medium

A good medium for biosynthetic labeling of Ig is minimum essential medium (MEM) supplemented with antibiotics, glutamine, 5% fetal calf serum (FCS), and amino acids, except the one used for labeling. MEM and all of the supplements can be obtained from GIBCO, Grand Island, NY. The FCS can be dialyzed overnight against PBS to remove amino acids which may compete with the labeled amino acid.

For biosynthetic labeling of the carbohydrate portion of Ig molecules, MEM lacking glucose should be used. However, glucose at concentrations varying from 10 to 50% of that used in complete medium may be necessary in some cases in order to maintain optimum cell viability.

Cell Concentration

The optimum concentration of cells which should be used depends on the length of the labeling period. The incorporation of amino acids into protein is more efficient when cells are cultured at high densities. However under these conditions, optimal cell viability cannot be maintained for long periods of time. Murine splenocytes, for example, can remain viable for as long as 12 hr if the cell concentration used for labeling is below 1×10^7 per ml. Cells can be labeled at concentrations as high as 2×10^8 per ml without loss of viability if the labeling period is less than 30 min. This method is very useful if it is necessary to label cells to high specific activity for short term kinetics or for pulse chase experiments.

Isotopes

The radiolabeled amino acid chosen for biosynthetic labeling depends on the nature of the experiment. ^{3}H- or ^{14}C-labeled amino acids have lower specific activities but their long half-lives permit prolonged storage of the labeled material. [^{35}S]Methionine which has a higher specific activity can be used to label proteins which are synthesized at low rates. The incorporation efficiency is directly related to the amount of radioactive isotope added, although incorporation of more than 30% of the total radioactivity cannot be achieved. The radioactive medium can be reused albeit with a somewhat reduced rate of incorporation. It is helpful to add fresh glutamine to the medium prior to reusing it.

[12] A. Tartakoff and P. Vassalli, *J. Cell Biol.* **83,** 284 (1979).
[13] C. Sidman, *Cell* **23,** 379 (1981).
[14] D. Yuan, J. W. Uhr, and E. S. Vitetta, *J. Immunol.* **125,** 40 (1980).

Isolation of Ig Molecules

Cell Lysis

Subsequent to labeling, cells are washed at least two times with a 20- to 50-fold volume of PBS to reduce the amount of nonspecifically bound label. The cells can then be lysed with nonionic detergents such as Triton X-100 or Nonidet P40 (NP-40). Generally, a 0.5% solution in PBS will result in maximum lysis of cells without causing disruption of the nuclear membrane. Nuclei and other debris are removed by centrifugation. In order to minimize proteolytic degradation of labeled molecules, protease inhibitors such as phenylmethylsulfonyl fluoride (1 mM) can be added to the lysate. Addition of 5% fetal calf serum (FCS) increases the substrate concentration for other nonspecific proteases and hence minimizes degradation of the labeled proteins. If the cells are adequately washed prior to lysis, dialysis of the lysate is not necessary. In the presence of FCS, lysates can be stored at $-20°$ for up to 6 months without detectable breakdown of the labeled Ig molecules.

Immunoprecipitation of Ig

Labeled Ig can be isolated by immunoprecipitation using coprecipitation with cold carrier[1,2] or indirect precipitation with a secondary antibody.[1] Direct precipitation of labeled lysates is carried out by adding 25 μg of carrier immunoglobulin of the same species along with an excess of antiimmunoglobulin. Control precipitations of a parallel aliquot should be carried out using a *heterologous* immunoglobulin and antiimmunoglobulin to monitor the extent of nonspecific radioactivity trapped in the precipitate. The reaction mixtures are incubated at $37°$ for 30 min and overnight at $4°$.

An alternate and more rapid method of precipitation is the indirect, or "sandwich" method. An optimum amount of antiimmunoglobulin is incubated with the lysate for 15 min at $37°$ and an excess of secondary antisera reactive with the primary anti-Ig is added. The incubation is continued for 30 min at $37°$ followed by incubation at $4°$ until flocculation occurs (1–24 hr). In this case, the control precipitation should contain normal or hyperimmune serum to an unrelated antigen but the same secondary anti-Ig antibody.

For both methods, the precipitates should be washed exhaustively with PBS by repeated centrifugation and resuspension (usually 4×). Prior to dissolving precipitates, they should be transferred to clean tubes and pelleted.

Isolation of Immune Complexes by Binding to S. aureus
(*This Volume* [40])

Since the Fc region of the IgG of most species binds to protein A or fixed intact *Staphylococcus aureus*,[15] a convenient method for isolating Ig is to add a suitable anti-Ig reagent followed by binding the immune complexes to *S. aureus*. *S. aureus* is commercially available (Pansorbin, Carbiochem-Behring Co., La Jolla, CA). Each lot of *S. aureus* must be titrated to determine the optimal amount to be used for each antiserum since Ig of different species vary in their affinity for *S. aureus*. Different lots must also be screened for nonspecific binding. Prior to use, *S. aureus* is washed at least twice with PBS to remove debris. It is often desirable to "preclear" labeled lysates by adding *S. aureus* alone. The bacteria are removed by centrifugation and the supernatant is used for subsequent precipitation. Optimal binding of labeled Ig is performed by adding antibody to the lysate and incubating for 15 min at 37°. Immune complexes are bound to *S. aureus* by incubation for 30 min at 4°. The bacteria are then pelleted and washed extensively with 0.5 *M* NaCl, 0.05 *M* Tris pH 8.0, 0.5% NP40, 0.1% SDS, and 0.2% DOC to remove nonspecifically bound material.

If the antiserum used for immunoprecipitation is affinity purified or of monoclonal origin, it can be conjugated to a solid support such as Sepharose 6B beads and be used to bind Ig molecules in cell lysates. Alternatively, purified protein A from *S. aureus* can be conjugated to Sepharose 6B and used as an immunoabsorbant.[16]

Biochemical Analysis of Immunoprecipitates

Subsequent to immunoprecipitation or isolation of immune complexes by binding to *S. aureus*, the labeled Ig can be analyzed by sodium dodecyl sulfate–polyacrylamide gel electrophoresis (SDS–PAGE).[17] The immunoprecipitates are dissolved and the *S. aureus*-bound radioactivity is eluted by boiling the pellet in 2% SDS and removing insoluble material by centrifugation. Eluates can be reduced with 2-mercaptoethanol. Under reducing conditions, the heavy and light chains of IgM, IgG, IgA, and IgD are all easily resolved by 10% SDS–PAGE. Slab gels[18] are most convenient in that many samples can be compared. After fixation and dehydration, gels containing Ig labeled with ^{125}I can usually be visualized by

[15] S. W. Kessler, *J. Immunol.* **117**, 1482 (1976).
[16] S. Fuchs and M. Sila, *in* "Handbook of Experimental Immunology and Immunochemistry" (D. M. Weir, ed.), p. 10. Blackwell, Oxford, 1978.
[17] A. L. Shapiro, E. Vinuela, and J. V. Maizel, Jr., *Biochem. Biophys. Res. Commun.* **28**, 815 (1967).
[18] U. K. Laemmli, *Nature (London)* **227**, 680 (1970).

exposure to X-ray film for 12–16 hr. Autoradiograms of Ig from biosynthetically labeled cells must be enhanced by fluorography[18a] before exposure. In addition, if the amount of radioactivity associated with each Ig band is to be quantitated, the bands can be cut out of the gels and the radioactivity determined in a liquid scintillation or gamma counter.

Fluorography cannot be used, however, if the Ig fractionated by SDS–PAGE must undergo further biochemical analysis such as peptide mapping or amino acid sequencing. For these purposes, the dissolved immunoprecipitates are subjected to disc gel electrophoresis. The gels are cut with a gel slicer and the labeled material is eluted from the gel slices by overnight incubation in 2% SDS. On an aliquot of each fraction the radioactivity is determined, while the remaining peak fractions can be dialyzed, lyophilized, and processed.

In addition to one-dimensional PAGE, Ig can be analyzed by two types of two dimensional gel electrophoresis. (1) Nonreducing–reducing PAGE. This is performed by first electrophoresing immunoprecipitates in one dimension without reduction followed by electrophoresis in a second dimension under reducing conditions. Patterns of disulfide linkages can be disclosed in this manner[19]: (2) Isoelectrofocusing PAGE.[20] The charge properties of Ig molecules can be analyzed in this manner and the carbohydrate side chain characteristics of many Igs have been thus resolved.[21]

Isolation of Biosynthetically Labeled Cell Surface Ig

In order to determine which class of biosynthetically labeled Ig is present on the cell surface, one can isolate the biosynthetically labeled Ig by first incubating cells with anti-Ig serum prior to lysis. Immune complexes are isolated after lysis either by indirect precipitation or with *S. aureus*. Using this procedure, care must be taken to saturate (prior to cell lysis) remaining unbound sites on the primary antibody used to treat the intact cells so that these cell-bound antibodies do not bind additional intracellular molecules following cell lysis. These methods have been successfully used to determine the species of biosynthetically labeled Ig residing on the cell surface[11] as well as to determine the rate of synthesis of cell surface Ig.

Use of ^{125}I-Labeled Anti-Ig Reagents

[125]I-labeled anti-Ig reagents are invaluable for the quantification and characterization of Ig in both cellular secretions and on the cell surface. In

[18a] R. A. Laskey and A. D. Mills, *Eur. J. Biochem.* **56**, 335 (1975).
[19] J. W. Goding, *J. Immunol.* **128**, 2416 (1982).
[20] P. P. Jones, *J. Exp. Med.* **146**, 1261 (1977).
[21] J. W. Goding and L. A. Herzenberg, *J. Immunol.* **124**, 2540 (1980).

addition, they can be used to estimate the total amount of Ig in cell lysates.

Preparation of ^{125}I-Labeled Anti-Ig Reagents

Anti-Ig reagents cannot be used effectively for radioimmunoassay unless they have been purified extensively. Heterologous, polyspecific antisera should be purified by affinity chromatography while with monoclonal antibodies ammonium sulfate precipitation is usually sufficient. Subsequent to purification, the binding characteristics of each preparation must be ascertained by testing its ability to bind to a panel of myeloma proteins. It should be noted that specificities not apparent in the original serum are often amplified by affinity purification.

The anti-Ig reagents can be ^{125}I labeled by lactoperoxidase or Iodo-gen. We have found the following conditions of labeling of purified proteins by lactoperoxidase to be most effective. The protein to be labeled should be dialyzed against PBS and resuspended at approximately 1 mg/ml. Fifty micrograms is placed in a 5-cm Falcon plastic tube. Na^{125}I (0.5 mCi) is added, followed by 5 μl of lactoperoxidase (used at 1 mg/ml) and 5 μl of 0.03% H_2O_2. The tube is vortexed briefly and incubated at room temperature for 2 min. PBS (0.5 ml) is added to the mixture which is then immediately dialyzed extensively against large volumes of PBS. The dialysis buffer should be changed at least 5 times in a course of 2 days. After the first change of buffer, 0.5 ml of FCS is added to the sample to stabilize the labeled protein. Anti-Ig labeled in this manner and stored at $-20°$ is stable for 2–5 weeks.

Radioimmunoassay of Secreted Ig

Radioimmunoassay of Ig in cell secretions is most conveniently performed by an indirect binding procedure[22] in which heterologous anti-Ig preparations from a species different from the one measured is used. For example, in order to measure mouse IgM, wells of a 96-well microtiter plate are first coated with rabbit anti-IgM followed by the addition of the material to be quantitated. The amount of mouse IgM bound to the rabbit antibody is then determined by adding a third layer of ^{125}I-labeled antibodies specific for mouse Ig. Binding of anti-Ig to polyvinyl plates is carried out by incubation at room temperature for 2 hr or at 4° overnight. Subsequently, unbound antibody is washed away with distilled water followed by saturation of the unbound sites by incubation with a 50% solution of serum in PBS. The binding of each subsequent layer is performed in a

[22] N. R. Klinman, *J. Exp. Med.* **136**, 241 (1972).

similar manner. We have found that background binding can be minimized by a final wash with 0.15 M NaCl 0.01 M Tris (pH 8.0) containing 0.5% NP40.

For quantitative measurement of Ig, standard curves using myeloma proteins of the same class and subclass must be determined in each experiment. Similarly, the optimal amount of anti-Ig to be used in the first layer coated onto the plate should be determined.

Radioimmunoassay of Ig in cell lysates can be performed in a similar manner except that titration curves should be constructed with lysates containing known amounts of Ig. Quantification of cell bound Ig using a radioimmunoassay is more complicated due to the possibility of steric hindrance. However, relative measurements can be performed. The assay is carried out in a manner similar to that described above except that the ^{125}I-labeled anti-Ig is bound directly to cells which are then aliquoted in equal numbers into wells of polyvinyl plates.

Acknowledgments

We thank Ms. Tam Dang and Ms. Earlene Carlton for expert technical assistance and Ms. G. A. Cheek for cheerful secretarial assistance. We are indebted to Dr. J. W. Uhr for his participation in many of the studies described in this chapter. Supported by Grants AI-11851 and AI-12789.

Section IV

Methods for the Detection, Purification, and Biochemical
Characterization of Lymphoid Cell Surface Antigens

[43] Molecules Encoded within the Murine Major Histocompatibility Complex: An Overview

By W. LEE MALOY and JOHN E. COLIGAN

Introduction

Loci mapping to the murine seventeenth chromosome control a variety of traits (Fig. 1), including embryonic differentiation (e.g., *T, Fu,* and *gk*), hair growth (e.g., tk and thf), and various isozymes (e.g., *PgK-2, Ce-2,* and *Glo-1*).[1] Occurring approximately in the middle of chromosome 17 is a group of genes which encode antigens that are intimately involved in immunological and possibly developmental processes; they form the major histocompatibility complex (MHC), referred to as *H-2* in the mouse, and the *Tla* complex of genes. The demarcation between the *H-2* and *Tla* genetic regions is somewhat artificial in that recent studies have shown that loci in the *Tla* region encode molecules that are functionally and biochemically related to certain *H-2* loci (see below). Thus, the murine MHC in its extended version includes both the traditional *H-2* and *Tla* complexes of genes.[2]

Based on their structural characteristics, MHC gene products can be grouped into at least three classes. Class I molecules are defined as non-covalently associated dimers of a 45,000-dalton MHC encoded glycoprotein and β_2-microglobulin (β_2m), a 12,000-dalton, non-MHC encoded protein. As shown in Fig. 1, these molecules are encoded in the *K* and *D* regions of the *H-2* complex and are the only class of molecules known to be encoded in the *Tla* complex. The class II molecules, encoded in the *H-2I* region, consist of two noncovalently associated, membrane-bound glycoproteins with molecular weights of 31,000–35,000 (α chain) and 25,000–29,000 (β chain). Under certain conditions, a third chain (I_i, 31,000 daltons) can be coisolated with *I* region molecules. Functionally, the class I gene products differ from class II gene products in that the former guide cytolytic and the latter regulatory (helper) T lymphocytes.[3]

The *H-2S* region encodes two molecules, Ss (serum serological) and Slp (sex-limited protein), which are termed class III molecules. Both of these molecules correspond structurally to the C4 component of the com-

[1] J. Klein, F. Figueroa, and D. Klein, *Immunogenetics* **16**, 285 (1982).
[2] J. Klein, *Science* **203**, 516 (1979).
[3] J. Klein, A. Juretič, C. N. Baxevanis, and Z. A. Nagy, *Nature (London)* **291**, 455 (1981).

METHODS IN ENZYMOLOGY, VOL. 108

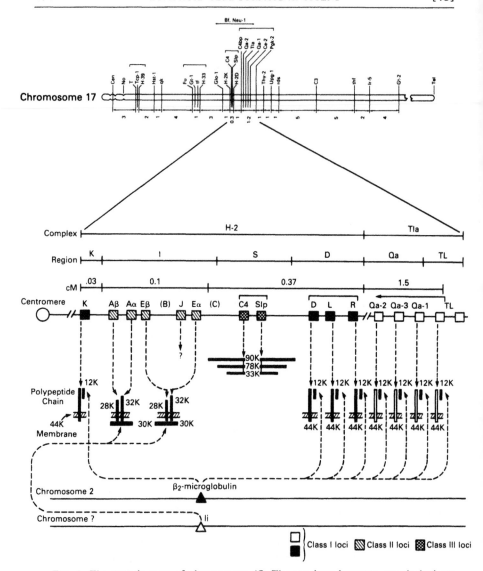

FIG. 1. The genetic map of chromosome 17. The numbers between genetic loci are approximate distances in centimorgans; brackets and arrows indicate uncertainty with regard to the relative order of the encompassed loci. MHC loci designated by the same type of squares belong to the same class. Class I molecules are associated with β_2m and class II molecules with invariant chain (I_i). Based on information provided by Klein et al.[1] and Hood et al.[8]

plement cascade, but only Ss has been shown to have activity in functional tests.[1]

In addition to the three classes of molecules mentioned above, other molecules have been reported to be encoded by genes located within the boundaries of the MHC. These include a series of intracellular proteins having molecular weights of 15,000–30,000 which map between the K and I regions[4]; a neuraminidase which is encoded in the S region[5]; and complement factor B, a 93,000 molecular weight protein which is a key component of the alternative complement pathway.[6]

Class I Molecules

The class I molecules encoded in the H-2, K, and D regions are the classical transplantation antigens which are primarily responsible for acute allograft rejection. They appear to be present on all nucleated somatic cells. Serological analyses of class I molecules have revealed an unparalleled degree of genetic polymorphism, which is reflected in two ways. First, there are multiple loci which encode H-2 and Tla class I molecules. For example, serological analyses have revealed that d haplotype mice[6a] (e.g., BALB/c) apparently express at least six H-2 alloantigens[7]—K1^d, K2^d, D^d, L^d, R^d, and M^d. This number of H-2 class I molecules agrees reasonably well with studies on the genetic fine structure of the H-2 complex which have identified 2 K-region loci and 3 D-region loci in BALB/c mice.[8] Other strains of mice, such as those of the b haplotype, appear to have fewer H-2 class I loci and consequently may express fewer class I molecules as indicated by the absence of a D or L molecule in this strain.[7] Second, in addition to the polymorphism provided by multiple class I loci, each H-2 locus can apparently encode one of many alleles, e.g., it is estimated that there may be upward of 100 alleles for each of the H-2, K, and D loci.[9] Thus, a completely heterozygous mouse may express 10–12 different K/D region alloantigens.

While the function(s) of class I antigens have not been clearly defined, it would seem likely that the extensive polymorphism of class I H-2 mole-

[4] J. J. Monaco and H. O. McDevitt, *Proc. Natl. Acad. Sci. U.S.A.* **79**, 3001 (1982).

[5] F. Figueroa, D. Klein, S. Tewarson, and J. Klein, *J. Immunol.* **129**, 2089 (1982).

[6] M. H. Roos and P. Démant, *Immunogenetics* **15**, 23 (1982).

[6a] Individual inbred strains of mice have distinct combinations of alleles, each of which is termed a haplotype. For example, the BALB/c mouse MHC is of the d haplotype and the corresponding gene products are generally denoted by a capital letter with a superscript for the haplotype designation, e.g., K^d, D^d, A$_\alpha$^{d}, A$_\beta$^{d}, E$_\alpha$^{d}, and E$_\beta$^{d}.

[7] E. S. Kimball and J. E. Coligan, *Contemp. Top. Mol. Immunol.* **9**, 1 (1983).

[8] L. Hood, M. Steinmetz, and B. Malissen, *Annu. Rev. Immunol.* **1**, 529 (1983).

[9] W. R. Duncan, E. K. Wakeland, and J. Klein, *Immunogenetics* **9**, 261 (1979).

cules is important for species survival. Research by many investigators[10] indicates that foreign molecules, such as viral antigens expressed on the host cell surface, are recognized by cytotoxic T lymphocytes (CTL) in the context of the host's own class I H-2 molecules and suggests a possible function for these molecules and the role of polymorphism. Accordingly, the recognition of viral antigens in the context of host class I antigens might serve to focus effector immune cells on the source of the virus, the infected cell, rather than on free viral particles, as a more effective means of combating infection. Polymorphism and heterozygousity may serve to maximize the repertoire of associative recognition, i.e., certain allelic products might be more suitable to present a particular antigen than other allelic products of that same or another locus.[7]

Analysis at the DNA level has demonstrated that there are 36 (recently revised to 32) class I genes in the BALB/c mouse genome.[8] Five of these map to the K/D region of the H-2 complex with the remainder mapping to the Tla complex. Although Tla encoded class I molecules appear to be structurally indistinguishable from H-2 encoded class I molecules, they differ in that they have a more restricted tissue distribution and presumably a different function.[11] Of the TL encoded molecules, some are hematopoietic differentiation antigens which are found on thymocytes at early stages of development, as well as on leukemic cells, whereas others appear to be expressed only on leukemic cells. The Qa molecules are found on many different types of lymphoid cells. Thus, in general, Tla molecules are thought to be hematopoietic differentiation antigens.

The determination of the complete primary structure of H-2K^{b}[12] and HLA-B7[13] (a human transplantation antigen) led to the derivation of the model for class I molecules shown in Fig. 2. As depicted, such molecules consist of three regions: an extracellular region (residues 1–283); a hydrophobic, transmembrane region (~residues 284–307); and a hydrophilic, cytoplasmic region (~308 to the carboxy terminal). The extracellular region can be further subdivided into three domains (N, C1, and C2) of about 90 residues each. Postulation of the existence of such domains was prompted by evidence of primary structural homologies of class I heavy chains and β_2m to IgG constant region domains.[14] Further data supporting

[10] R. M. Zinkernagel and P. C. Doherty, *Adv. Immunol.* **27,** 51 (1979).

[11] J. Michaelson, E. A. Boyse, M. Chorney, L. Flaherty, I. Fleisner, U. Hammerling, C. Reinisch, R. Rosenson, and F.-W. Shen, *Transplant. Proc.* **15,** 2033 (1983).

[12] J. E. Coligan, T. J. Kindt, H. Uehara, J. Martinko, and S. G. Nathenson, *Nature (London)* **291,** 35 (1981).

[13] H. L. Ploegh, H. T. Orr, and J. L. Strominger, *Cell* **24,** 287 (1981).

[14] H. T. Orr, J. A. Lopez de Castro, P. Parham, H. L. Ploegh, and J. L. Strominger, *Proc. Natl. Acad. Sci. U.S.A.* **76,** 4395 (1979).

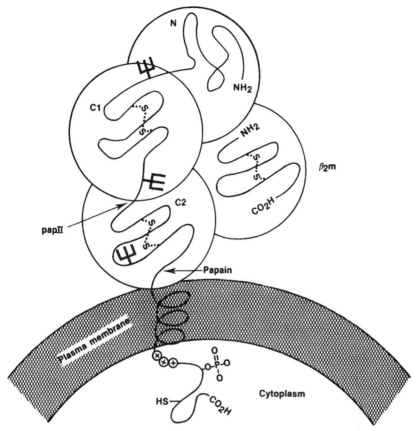

FIG. 2. A schematic representation of an H-2 class I molecule in the plasma membrane.

the organization of class I heavy chains into domains come from studies on the genetic organization of class I genes which show that they are organized into exons which essentially coincide with the postulated domain structure.[8]

Figure 2 shows three carbohydrate moieties on class I heavy chains. All H-2 class I molecules, which have been analyzed, have glycosyl groups attached to both the N and C1 domains at Asn-86 and Asn-176, respectively. Some H-2 class I molecules, e.g., H-2K^d and H-2D^b, are known to carry a third carbohydrate group that is attached to Asn-256 in the C2 domain.[15] There also is variability in the length of the cytoplasmic domain[7] with K^b and K^d being 9 residues longer than D^b and possibly L^d.

[15] W. L. Maloy and J. E. Coligan, *Immunogenetics* **16**, 11 (1982).

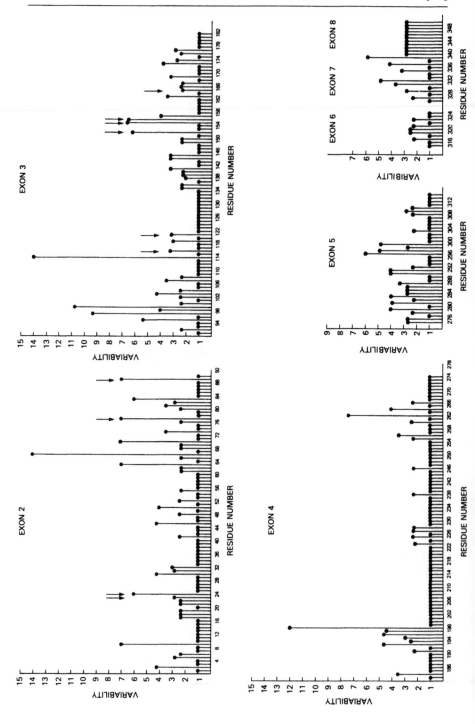

Whether this feature distinguishes between *K* and *D* region molecules remains to be seen. The role of the phosphate moiety detected in the C-terminal portion of class I molecules remains to be determined.

A major goal of primary structural studies on histocompatibility antigens is to determine, by means of sequence comparisons, regions of diversity and similarity in the hope of relating certain portions of the molecules to various biological reactivities. For such comparisons, extensive primary protein structural information is now available for 6 *H-2* encoded class I molecules.[7] In addition, extensive amino acid sequence data have been deduced from cDNA and genomic clones of mouse MHC class I antigens. The homologies among these molecules range from 82 to 94%. The variability in protein sequence is most easily examined in the form of a variability plot (Fig. 3). The greatest variability primarily occurs in three regions, residues 62–83, 95–121, and 193–198. The greatest homology in the sequence occurs between residues 199 and 254. Studies have indicated that variable regions most likely include sites of alloantigenicity, as well as T cell recognition sites,[16] while the highly conserved region on the C2 domain (199–254) is most likely responsible for binding β_2m.[17]

There are only sparse structural data available for *Tla* complex encoded molecules. Peptide mapping data indicate that *Tla* molecules are less diverse than *H-2* class I molecules. Comparison of different *H-2* class I molecules to one another yields 35–50% peptide homology while comparison of *Qa* molecules to each other yields 70–80% peptide homology, as does comparison of *TL* molecules to each other.[18] (80% peptide homology indicates at least 95% amino acid sequence homology.) Comparisons of TL and Qa molecules to each other or to H-2 molecules indicate the same levels of homology as when *H-2* class I molecules are compared to each other, i.e., about 80% amino acid sequence homology.[19] Partial N-terminal amino acid sequence data of Qa-2[20] and amino acid sequence data derived from the Qa region pseudogene (27.1)[21] support the conclu-

[16] M. R. van Schravendijk, W. L. Maloy, and J. E. Coligan, *Surv. Immunol. Res.* **2,** 199 (1983).

[17] K. Yokoyama and S. G. Nathenson, *J. Immunol.* **130,** 1419 (1983).

[18] K. Yokoyama, E. Stockert, L. J. Old, and S. G. Nathenson, *Proc. Natl. Acad. Sci. U.S.A.* **78,** 7078 (1981).

[19] M. J. Soloski, K. R. McIntrye, J. W. Uhr, and E. S. Vitetta, *Mol. Immunol.* **19,** 1193 (1982).

[20] M. J. Soloski, J. W. Uhr, and E. S. Vitetta, *Nature (London)* **296,** 759 (1982).

[21] M. Steinmetz, K. W. Moore, J. G. Frelinger, B. T. Sher, F.-W. Shen, E. A. Boyse, and L. Hood, *Cell* **25,** 683 (1981).

FIG. 3. Variability in the amino acid residues of murine class I molecules arranged according to the exon structure. Variability = number of different residues at a given position divided by the frequency of the most common residues.

sions drawn from the peptide mapping data about the high homology between *H-2* and *Tla* region class I molecules. However, there are apparently minor structural features which may distinguish Qa molecules from H-2 class I molecules such as the insertion of additional amino acid residues and the substitution of amino acid residues which have been invariant in H-2 molecules.[7]

($\beta_2 m$)

β_2m is an integral component of all class I molecules. The fact that there is only one β_2m gene per haploid genome indicates that the same β_2m gene encodes the light chain of all class I molecules.[22] There are two alleles of the murine β_2m gene, $\beta_2 m^a$ and $\beta_2 m^{b\,23}$, which encode molecules differing by a single amino acid[24] (single base change[22]).

Although the function of β_2m, as an integral component of all class I molecules is still undefined, a number of studies with human β_2m suggest that it may serve a major role in stabilizing the conformation of the class I heavy chain. These include (1) circular dichroism studies, which show that isolated heavy chains lose some of their β-pleated sheet structure and assume more of a random coil configuration in the absence of β_2m; (2) immunochemical studies, which demonstrate that HLA alloantisera are less able to recognize the class I molecule if β_2m has become dissociated from the heavy chain and that antigenic activity can be restored if the heavy chain–β_2m complex is carefully renatured; and (3) studies with Daudi cells, which fail to synthesize β_2m, reveal that cytoplasmic HLA molecules, not associated with β_2M, are not recognized by alloantisera but can be recognized by xenoantisera raised against denatured HLA heavy chains.[25]

The β_2m controlled heavy chain conformation also appears to be important as a recognition element involved in the intracellular processing and transport of class I heavy chains. Daudi cells are only able to express cytoplasmic HLA and a mutant murine cell line (R1(β_2-), which, like Daudi cells, does not synthesize β_2m, fails to express cell surface H-2 or Tla class I molecules.[26] Furthermore, it has been shown[27] that in a human mutant cell line, which expresses a form of HLA-A2 that does not bind

[22] J. R. Parnes and J. G. Seidman, *Cell* **29**, 661 (1982).

[23] J. Michaelson, E. Rothenberg, and E. A. Boyse, *Immunogenetics* **11**, 93 (1980).

[24] F. T. Gates, III, J. E. Coligan, and T. J. Kindt, *Proc. Natl. Acad. Sci. U.S.A.* **78**, 554 (1981).

[25] M. S. Krangel, H. T. Orr, and J. L. Strominger, *Scand. J. Immunol.* **11**, 561 (1980).

[26] R. Hyman and V. Stalling, *Immunogenetics* **4**, 171 (1977).

[27] M. S. Krangel, D. Pious, and J. L. Strominger, *J. Biol. Chem.* **257**, 5296 (1982).

β_2m, the HLA-A2 fails to undergo oligosaccharide processing and does not migrate to the cell surface. (It has previously been shown that glycosylation is not necessary for heavy chain expression.[28]) Thus, it is tempting to speculate that processing of class I molecules and translocation to the cell surface are dependent upon a particular conformation imparted by association with β_2m.

Class II Molecules

Class II molecules which are involved in the regulation of the immune response,[29] are encoded in the *I* region of the MHC. Unlike the class I molecules, the class II molecules are not found on all cells but primarily on B cells and macrophages, and to a lesser extent on T cells, Langerhans cells, dendritic cells, and epithelial cells.[30] Until recently, the *I* region was divided into five subregions (*A*, *B*, *J*, *E*, and *C*). However, critical reevaluation of the data leading to the proposal of the *B* and *C* subregions, including the fact that no serologically detectable products map to these subregions, suggests that these subregions are genetic artifacts and should be removed from maps of the H-2 complex.[31] In this same re-evaluation of the *H-2I* region, it was also suggested that the molecules bearing I-J determinants, which correlate with suppressor T cell activity, may also not be encoded within the *I* region. Analysis of 200 kb of DNA from the *I* region revealed genes for the *I-A* and *I-E* subregion molecules and left only 2.0 kb of DNA remaining for the entire *I-J* subregion.[32,33] Since most structural genes require significantly greater than 3.5 kb of DNA, it is unlikely that this remaining 2.0 kb of DNA encodes any structural genes.

The class II molecules encoded by the *I-A* and *I-E* subregions are composed of two glycoprotein chains, α and β. Prior to cell surface expression, the α, β complex is associated with an invariant glycoprotein, I_i. The α and β chains of the I-A antigen are both encoded in the *I-A* subregion and therefore are designated I-A$_\alpha$ and I-A$_\beta$. For the I-E subregion antigen, the α chain is encoded in the *I-E* subregion, but, the β chain is encoded in the *I-A* subregion and therefore is designated as either I-E$_\beta$

[28] H. L. Ploegh, H. T. Orr, and J. L. Strominger, *J. Immunol.* **126**, 270 (1981).

[29] L. T. Clement and E. M. Shevach, *Contemp. Top. Mol. Immunol.* **8**, 149 (1981).

[30] J. H. Freed, *in* "Ia Antigens and Their Analogs in Man and Other Animals" (S. Ferrone and C. David, eds.), Vol. 1, p. 1. CRC Press, Boca Raton, Florida, 1982.

[31] C. N. Baxevanis, Z. A. Nagy, and J. Klein, *Proc. Natl. Acad. Sci. U.S.A.* **78**, 3809 (1981).

[32] J. A. Kobori, A. Winoto, J. McNicholas, and L. Hood, *J. Mol. Cell. Immunol.*, **1**, 125 (1984).

[33] M. Steinmetz and L. Hood, *Nature (London)* **222**, 727 (1983).

or I-A$_e$. The invariant chain,[33a] which is common to the I-A and I-E region antigens, is not encoded on chromosome 17.[34,35]

All of the *I* region encoded α and β chains are polymorphic; however, peptide mapping and N-terminal amino acid sequence analyses and now several complete gene sequences indicate that there are different degrees of polymorphism, with the *I-A* subregion encoded α and β chains (A$_\alpha$, A$_\beta$, and E$_\beta$) being more polymorphic than the *I-E* subregion encoded I-E$_\alpha$ chain. For example, the recently determined sequences of the *I-E$_\alpha^d$* and *I-E$_\alpha^k$* genes[36,37] indicate only three amino acid differences in the entire sequences (99% homology) whereas gene sequences of six different haplotypes of I-A$_\alpha$ chains show homology from 92 to 97%,[38] three different I-A$_\beta$ sequences show homologies between 91 and 95%[39] and two I-E$_\beta$ sequences show 91% homology.[40,41] Previous peptide mapping studies and N-terminal sequence analyses[30] revealed no significant homology among the different I-A and I-E chains, but a recent comparison of the *I-E$_\alpha^k$* and *I-A$_\alpha^k$* genes shows a 50% amino acid sequence homology[37] and comparison between I-E$_\beta$ and I-A$_\beta$ shows a 63–65% homology.

The I-E molecule is not expressed in mice bearing the *b*, *s*, *q*, and *f* haplotypes in the *I-E* region. This can be traced to their inability to produce an E$_\alpha$ chain.[36,42] The I-E$_\beta$ chains in these mouse strains can be "rescued" in heterozygous mice by trans complementation with the I-E$_\alpha$ chain from a different haplotype.[30] The phenomenon of trans complementation appears to be general for both the I-A and I-E region antigens and therefore a mouse that is heterozygous in both the *I-A* and *I-E* subregions can express four different I-A and four different I-E molecules. However,

[33a] A recent publication indicates that the association of invariant chains with class 1 molecules may be artifactual [F. P. Thinnes, N. Hilschmann, H. Kayser, and H. Götz, *Hoppe-Seyler's Z. Physiol. Chem.* **364**, 1805 (1983)].

[34] N. Koch, G. J. Hämmerling, J. Szymura, and M. R. Wabl, *Immunogenetics*, **16**, 603 (1982).

[35] C. E. Day and P. P. Jones, *Nature (London)* **302**, 157 (1983).

[36] J. J. Hyldig-Nielsen, L. Schenning, U. Hämmerling, E. Widmark, E. Heldin, P. Lind, B. Servenius, T. Lund, R. Flavell, J. S. Lee, J. Trowsdale, P. H. Schreier, F. Zablitzsky, D. Larhammar, P. A. Peterson, and L. Rask, *Nucl. Acids Res.* **11**, 5055 (1983).

[37] C. O. Benoist, D. J. Mathis, M. R. Kanter, V. E. Williams II, and H. O. McDevitt, *Proc. Natl. Acad. Sci. U.S.A.* **80**, 534 (1983).

[38] C. O. Benoist, D. J. Mathis, M. R. Kanter, V. E. Williams II, and H. O. McDevitt, *Cell* **34**, 169 (1983).

[39] E. Choi, K. McIntyre, R. N. Germain, and J. G. Seidman, *Science* **221**, 283 (1983).

[40] H. Saito, R. A. Maki, L. K. Clayton, and S. Tonegawa, *Proc. Natl. Acad. Sci. U.S.A.* **80**, 5520 (1983).

[41] L. Mengle-Gaw and H. O. McDevitt, *Proc. Natl. Acad. Sci. U.S.A.* **80**, 7621 (1983).

[42] D. J. Mathis, C. Benoist, V. E. Williams II, M. Kanter, and H. O. McDevitt, *Proc. Natl. Acad. Sci. U.S.A.* **80**, 273 (1983).

no evidence has been found for the association of I-A α and β chains with I-E chains.

As with class I molecules, the class II molecules provide the context in which antigens are recognized by T lymphocytes.[2,29] However, in the case of class II molecules, the regulated T lymphocytes are helper T cells[43] rather than cytotoxic T cells. The regulation occurs by I region restriction of B cell, T cell, and macrophage collaboration. Specifically, in order for an antigen to be "presented" by one cell to another to stimulate a helper response, the antigen must be presented with an I region encoded molecule. Therefore, the extensive polymorphism seen for I region encoded molecules may be necessary to allow association with many different protein antigens and the inability to respond to certain antigens could be due to lack of an appropriate I region molecule for association. This concept is supported by studies on the I-region mutant B6.C-H-2[bm12]. This mutant has an α chain which is identical to, but a β chain which is structurally different from, the parent B6 (b haplotype) I-A molecule.[44] This structural alteration correlates with the loss in ability of this mutant mouse strain to respond to certain antigens (e.g., beef insulin).[45] The fact that the immune response to other antigens dependent on the I-A[b] molecule is not affected indicates that at least two interaction sites on the I-A[b] molecule function in antigen presentation.

Although a complete protein structure has not been determined for a murine class II antigen, there are several complete gene sequences; six for I-A$_\alpha$, two for I-E$_\alpha$, three nearly complete sequences for I-A$_\beta$, and two sequence for I-E$_\beta$. The overall structure of the murine I-A and I-E antigens appears to be very similar, allowing for the construction of a general model for class II molecules as shown in Fig. 4. The molecules can be divided into extracellular, transmembrane, and cytoplasmic regions. The extracellular region can be further divided into two domains of about 90 amino acids. For the β chain, the two domains (β_1 and β_2) each contain a disulfide loop while the α chains contain an N-terminal domain (α_1) lacking a disulfide loop followed by a disulfide loop containing domain (α_2). As in class I molecules, the transmembrane region is composed of a sequence of about 25 hydrophobic amino acids which terminates in one or more basic amino acids, this is followed by the intracellular region which is about 15 residues in length. Virtually all of the allelic variability in the I-A and I-E α and β chains is found in the α_1 and β_1 domains and therefore it is

[43] J. S. Bromberg, A. Tominaga, M. Takaoki, and M. I. Greene, *Surv. Immunol. Res.* **1**, 67 (1982).

[44] D. J. McKean, R. W. Melvold, and C. David, *Immunogenetics* **14**, 41 (1981).

[45] C-C. S. Lin, A. S. Rosenthal, H. C. Passmore, and T. H. Hansen, *Proc. Natl. Acad. Sci. U.S.A.* **78**, 6406 (1981).

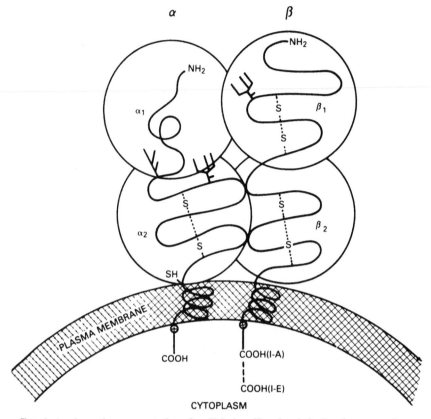

FIG. 4. A schematic representation of an H-2 class II molecule in the plasma membrane.

assumed that these domains contain the alloantigenic determinants. For the molecules sequenced, the α chains of both I-A and I-E appear to have two N-linked glycosyl units whereas the β chains have only one. The type of glycosyl units appears to vary with the cell type upon which the antigen is found; macrophages have a different glycosylation pattern from that of B cells.[46]

The structure of the I-J molecule, which may not be encoded within the MHC, is still a matter of controversy. The molecule has been detected on the surface of suppressor T cell hybridomas and also in solution as part of a suppressor factor. The I-J determinants are reported to be associated with immunoglobulin (Ig) determinants and antigen binding activity, ei-

[46] S. E. Cullen, C. S. Kindle, D. C. Shreffler, and C. Cowing, J. Immunol. 127, 1478 (1981).

ther as a single protein molecule or as a complex of two protein chains.[47] Recently, evidence has been presented for suppressor factors which have both antigen binding activity and also possess either I-A or I-E determinants.[48] Finally, H. Cantor and co-workers[49] have shown that a gene on chromosome 4 is involved in the expression of I-J in addition to a gene in the MHC most likely from the *I-E* region.

Homology among MHC Molecules and Other Molecules

A much discussed issue has been the evolutionary relationship between MHC molecules and Ig. Several years ago β_2m was found to be about 30% homologous to Ig constant region domains and was thus referred to as a free Ig domain. Similarities of the heavy chains of class I molecules to Ig and β_2m are suggested by the presence of Ig-like domain subregions in H-2K^b (C2 domain) and HLA-B7 (α3 domain).[7] A recent analysis, which compared the cDNA sequence of the H-2^d clone pH-2II to the DNA sequences of the mouse Ig μ constant region domains, supported the homology discerned by protein sequence analyses.[50] More importantly, however, this study revealed that the third base positions in the codons of these molecules are highly conserved. This strongly suggests that transplantation antigens and Ig are related through divergent rather than convergent evolution. Thus, Ig and transplantation antigens appear to be members of the same supergene family.

Recent data indicate that not only class I but also class II molecules are members of the Ig supergene family (Fig. 5). The domains in each chain nearest the cell membrane display homology with the Ig-like (α3) domain of HLA heavy chains, β_2m and Ig light and heavy chain constant domains. The levels of homology are at least as strong as the homology among other members of this supergene family.

These homology comparisons indicate that the humoral and cellular branches of the immune system have diverged from a common ancestral gene. However, recent evidence showing that the Thy-1 antigen is a member of the same supergene family suggests that the conserved structure among these molecules is not directly involved in the functioning of the immune system since Thy-1 is found in tissues unrelated to the immune response, e.g., the brain.[51] Also there is evidence for the presence of a

---

[47] J. Kapp, C. Pierce, S. Cullen, D. Shreffler, and B. Schwartz, *Immunol. Today* **4**, 1 (1983).

[48] Z. Ikezawa, C. N. Baxevanis, B. Arden, T. Tada, C. R. Waltenbaugh, A. Z. Nagy, and J. Klein, *Proc. Natl. Acad. Sci. U.S.A.* **80**, 6637 (1983).

[49] C. E. Hayes, K. K. Klyczek, D. P. Krum, R. M. Whitcomb, D. A. Hullett, and H. Cantor, *Science* **223**, 559 (1984).

[50] M. Steinmetz, A. Winoto, K. Minard, and L. Hood, *Cell* **28**, 489 (1982).

[51] A. F. Williams, and J. Gagnon, *Science* **216**, 696 (1982).

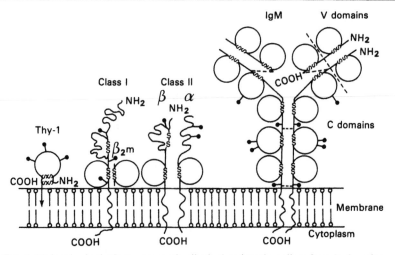

FIG. 5. Molecules in the Ig supergene family depicted on the cell surface, Ig domains, and their homologous regions in the other molecules are represented by circles. Intrachain disulfide bonds are shown by S-S symbols and interchain bonds in IgM by dashed lines. N-linked carbohydrate structures are shown by the clubs. It is believed that the class I, class II, and IgM molecules are integrated into the membrane by a hydrophobic protein sequence; but for Thy-1 the evidence indicates that a nonprotein structure, perhaps a fatty acid, is responsible for membrane integration.[51]

Thy-1-like molecule in the brain of relatively primitive animals (squids). Consequently, it has been proposed that Thy-1 is one of a set of Ig-related structures that mediate cell recognition during morphogenesis.[51] In this hypothesis, the Ig-like domain common to these molecules is considered a stable platform for the display of determinants and thus the immunoglobulin domain would serve as a basic recognition unit for cell interactions (Fig. 5).

Class III Antigens

Both the Ss (C4) and Slp proteins have a molecular weight of 200,000 and consist of three covalently linked polypeptide chains, α (87,000), β (78,000), and γ (33,000). The three chains are derived by cleavage from a single precursor chain and are thus presumably encoded by a single gene.[52] In the classical complement pathway, the activated C1s component cleaves a low molecular weight fragment (C4a) from the N-terminus of the C4 α chain. The remainder of the C4 molecule (C4b) then combines with activated C2, forming the $\overline{C4b\overline{C2}a}$ complex. Ss refers to an antigenic

[52] M. H. Roos, J. P. Atkinson, and D. C. Shreffler, *J. Immunol.* **121**, 1106 (1978).

determinant on the γ chain of the C4 protein which is detected by rabbit antisera. Using this reagent, sera from H-2^k strains of mice have been shown to express relatively low levels of C4 and are thus designated Ss^l, whereas all other strains express much higher levels and are thus Ss^h.

The Slp protein is structurally very similar to the Ss (C4) protein. The Slp protein was originally thought to be limited to males of positive strains (hence Slp for sex-limited protein) but recent findings indicate that females of Slp-positive strains do express low levels of the protein. The alleles controlling presence or absence of the protein are designated Slp^a and Slp^0, respectively.[1] Although no recombinants between Ss and Slp have been detected, the loci are very likely distinct because the Ss and Slp proteins differ in their affinity for the C4-binding protein, in their recognition by antihuman C4 sera and slightly in their respective molecular weights. Most significantly, only the Ss protein has C4 activity,[53] while the function, if any, of the Slp protein is unknown.

Other MHC Encoded Antigens

A series of proteins which are biochemically and genetically distinct from previously described murine MHC antigens have been detected by congenic anti-H-2^d sera.[4] Sixteen such proteins have been defined which exhibit a range of molecular weights (15,000–30,000) and isoelectric points (pI 4–9). Their lack of glycosylation and large charge separation suggest that they are not related to each other by posttranslational modifications. These proteins appear to be only present intracellularly and are most strongly expressed in macrophages, although they are readily detectable in fibroblasts, B, T, and null cell lines. The genes controlling the expression of these proteins have been tentatively mapped between the K and I-A subregions. Three alleles have been defined. Mice of the b and q haplotypes have not been demonstrated to express the proteins and hence, are termed to have a "null" allele. These molecules from mice of the d haplotype can be distinguished by their two-dimensional gel pattern from those of mice of all other positive H-2 types tested thus far (a, k, f, and s).

N-Acetylneuraminic acid hydrolase or neuraminidase is an enzyme that cleaves N-acetylneuraminic acid from complex oligosaccharide moieties; $Neu-1$, which maps between the I-E_α and D regions, most likely contains the structural gene for this enzyme.[54] There are three alleles at this locus, $Neu-1^\alpha$, which codes for a deficient enzyme present in SM/J mice, $Neu-1^b$, which codes for a normally active enzyme present in most

[53] A. Ferreira, V. Nussenzweig, and I. Gigli, *J. Exp. Med.* **148**, 1186 (1978).
[54] J. E. Womack, D. L. S. Yan, and M. Potier, *Science* **212**, 63 (1981).

strains of mice, and *Neu-1ᶜ*, which codes for a weakly active enzyme present in A.SW and a few other strains of mice. Previously described variations in the posttranslational processing of acid phosphatase, α-mannosidase, arylsulfatase-β, and α-glucosidase are attributed to pleiotropic effects of this gene.[5]

Complement factor B (*Bf*; Fig. 1) also known as C3 proactivator, either maps within the MHC or is closely linked to it.[1] This protein, which is a key component of the alternative complement pathway, consists of a single 93,000 molecular weight polypeptide chain. Its role is analogous to that of C2 in the classical pathway and in fact, these two components appear to be evolutionarily related.

Concluding Remarks

The murine MHC encodes a variety of molecules many of which are intimately involved in immune defense mechanisms. The fact that about 60 different traits[55] have been mapped to this region which contains 1000–2000 kb of DNA[33] suggests that other still to be identified molecules are encoded here. Many of these are likely to have no direct involvement with the immune response.

[55] J. Klein, *Adv. Immunol.* **26**, 55 (1978).

[44] Analysis of H-2 and Ia Antigens by Two-Dimensional Polyacrylamide Gel Electrophoresis

By Patricia P. Jones

A variety of serological and biochemical approaches have been used in the analysis of the structure and genetics of murine H-2 and Ia antigens. Two-dimensional polyacrylamide gel electrophoresis (2-D PAGE) has proved to be an especially valuable technique; the combined charge and size separation resolves individual molecular forms of H-2 and Ia antigens from complex mixtures, providing molecular fingerprints that are characteristic of each distinct protein. From the 2-D gel patterns generated by immunoprecipitated H-2 and Ia antigens, information can be obtained on the number and identity of gene products and their genetic polymorphism and biosynthetic processing. The major focus of this chapter is on the techniques used to produce high resolution 2-D PAGE separations of H-2 and Ia antigens. Some consideration also is given to special approaches

METHODS IN ENZYMOLOGY, VOL. 108

that can be used to identify distinct gene products within mixtures of immunoprecipitated proteins, one of the most valuable uses of 2-D PAGE. For discussion of the basic methodology readers are encouraged to read the original papers on 2-D PAGE by O'Farrell[1] and O'Farrell et al.[2] from which most of the procedures described below were derived. The use of 2-D PAGE for the analysis of H-2 and Ia antigens as well as other lymphocyte proteins has been described in detail.[3-5]

Methods

Radiolabeling of Cells (This Volume [42])

For most purposes, cells should be labeled biosynthetically with radioactive amino acids. While ^{125}I can be used to specifically label cell surface forms of H-2 and Ia antigens, for several reasons this is not normally the method of choice. First, not all proteins can be labeled by this technique (e.g., the Ia E_α chain[3]). Second, the spots in the autoradiograms are fuzzier than with ^{35}S, ^3H, or ^{14}C, due to the higher energy of the radiation emitted. Finally, since the cytoplasmic forms of the protein are not labeled, the spot patterns generated are less complex and therefore less complete as molecular fingerprints of each gene product.

For biosynthetic labeling, ^{35}S is the preferred isotope because amino acids containing ^{35}S are available at much higher specific activities than are those containing ^3H or ^{14}C. [^{35}S]Methionine is most commonly used; it can be supplemented with [^{35}S]cysteine to obtain higher specific activity labeling. To label normal lymphocytes, such as from the spleen, 5×10^7 viable lymphocytes are suspended in 2 ml of RPMI 1640 medium lacking methionine, supplemented with 5% fetal or newborn calf serum (which can be dialyzed to remove free methionine and improve incorporation of the [^{35}S]methionine) and containing 250 μCi/ml [^{35}S]methionine (>1000 Ci/mmol). The suspension is incubated in a 60-mm plastic tissue culture plate for 4 hr in a 37° incubator at 5% CO_2. After the incubation the cells are transferred to a 15-ml conical centrifuge tube. Residual cells are removed by rinsing the dish with 3 ml of cold washing buffer [phosphate-buffered saline (PBS: 10 mM sodium phosphate, 0.15 M NaCl, pH 7.2) containing 2 mg/ml methionine and 0.02% NaN_3] and added to the centri-

[1] P. H. O'Farrell, J. Biol. Chem. 250, 4007 (1975).
[2] P. Z. O'Farrell, H. M. Goodman, and P. H. O'Farrell, Cell 12, 1133 (1977).
[3] P. P. Jones, J. Exp. Med. 146, 1261 (1977).
[4] P. P. Jones, D. B. Murphy, D. Hewgill, and H. O. McDevitt, Mol. Immunol. 16, 51 (1979).
[5] P. P. Jones, in "Selected Methods in Cellular Immunology" (B. B. Mishell and S. M. Shiigi, eds), p. 398. Freeman, San Francisco, California, 1980.

fuge tube. The cells are then centrifuged at 200 g for 10 min at 4° and washed twice with 5 ml cold washing buffer. After removing all the supernatant fluids the cell membranes are solubilized by adding to the pellet 0.5 ml of cold extraction buffer (10 mM Tris–HCl, 0.15 M NaCl, 0.5% Nonidet P40 (NP-40), 0.02% NaN$_3$), vortexing, and then incubating the mixture for 15 min on ice. The nuclei and other insoluble materials are removed by centrifugation at high speed (e.g., 15 min at 27,000 g in a Sorvall centrifuge using the SS-34 rotor or 10 min at top speed in a microfuge). The supernate is then transferred to a fresh tube and stored at −70° until use.

This procedure routinely gives 70,000–120,000 cpm of TCA-insoluble [^{35}S]methionine radioactivity per 1 μl of extract. If shorter labeling times are required, for example, for pulse-chase experiments, the amount of [^{35}S]methionine added can be increased provided that the [^{35}S]methionine is concentrated by freeze-drying. Proteolysis in the extracts is not usually a problem as long as the extracts are kept at 4° during preparation and handling. For added precaution the proteolytic enzyme inhibitors phenylmethyl sulfonyl fluoride (1 mM) and aprotinin (Sigma; 1%) can be added to the extract.

Immunoprecipitation of H-2 and Ia Antigens

For most purposes it is advisable to use an "immunoprecipitation" method involving recovery of the soluble immune complexes by binding to an insoluble adsorbent. The more classical approach of immunoprecipitating the complexes with a second-step antiimmunoglobulin antibody results in high levels of IgG being present in the pellet, which can distort the gel pattern. An effective, simple-to-use, and reasonably inexpensive adsorbent is *Staphylococcus aureus,* Cowan I strain (SaC); these bacteria have an IgG-binding protein, protein A, on their surfaces.[6] Heat-killed, formalin-fixed SaC for this purpose are available commercially (e.g., IgG-sorb, from the Enzyme Center, Boston, MA).

The general immunoprecipitation procedure is as follows; all steps should be carried out on ice. A volume of SaC (as a 10% suspension) that is enough to provide 300 μl per immunoprecipitate is washed three times at 700 g with 10 volumes of SaC buffer [PBS containing 0.5% (w/v) NP-40, 2 mM methionine, and 0.02% NaN$_3$] and resuspended to the initial volume (as a 10% suspension) in SaC buffer containing 1 mg/ml ovalbumin. Each extract should be precleared with SaC to reduce the nonspecific sticking of extract proteins to the SaC. One hundred microliters of the

[6] S. W. Kessler, *J. Immunol.* **115**, 1617 (1975).

washed and resuspended SaC per immunoprecipitate is placed in a 12 × 75-mm plastic tube and centrifuged at 700 g for 5 min. After the supernate is removed, 100 μl of extract is added to the pellet, vortexed, and incubated 15 min on ice; the SaC are then centrifuged as before, and the precleared extract is transferred to a clean tube. This preclearing step can be carried out batch-wise on a large volume of extract and the adsorbed extract then pipetted as 100-μl aliquots into separate tubes for immunoprecipitation. To each, a saturating amount of anti-H-2 or Ia antibody is added. The appropriate amount depends on the titer of each individual antibody preparation. In general, 30 μl of an alloantiserum or 10 μg of a monoclonal antibody is adequate. After 30 min on ice, 200 μl of the washed and resuspended SaC is added and the mixture is incubated for 15 more min with occasional mixing. The SaC–antigen–antibody complexes are then diluted with 2 ml SaC buffer, centrifuged 10 min at 700 g, and washed twice with 2.5 ml SaC buffer. The SaC is then resuspended in 0.5 ml, transferred to a 6 × 50-mm glass tube or a 1.5-ml microfuge tube, and centrifuged into a tight pellet. All detectable supernatant fluid is then removed with a drawn-out Pasteur pipet. To elute and dissociate the antigen–antibody complexes, the pellet is resuspended by vigorous vortexing at room temperature in 50 μl isoelectric focusing (IEF) sample buffer[1]: 9.5 M urea (Schwartz/Mann Ultra-Pure), 2% (w/v) NP-40, 1.6% pH 5–7 Ampholine (LKB), 0.4% pH 3.5–10 Ampholine (LKB), and 5% 2-mercaptoethanol (this sample buffer is stored in tightly-capped 0.5-ml aliquots at −70°). The SaC is pelleted by centrifugation at 700 g for 10 min at room temperature. The supernate is removed with a drawn-out Pasteur pipet and stored in a tightly capped microfuge tube until needed. Twenty-five microliters is generally sufficient as a sample for 2-D PAGE separation.

Usually, with this procedure 0.2–0.5% of the TCA-precipitable [35S]methionine-labeled radioactivity is precipitated specifically with antibodies to single H-2 or Ia antigens. The specific counts are determined by comparison to counts in a control sample always prepared in parallel, using an inappropriate (nonreactive) antibody. The major disadvantage to using the SaC immunoprecipitation technique is that some antibodies (e.g., mouse IgM and IgG$_1$) are not bound by the protein A. The SaC may be adapted for use with antibodies of these isotypes by precoating the bacteria with antibodies that recognize mouse immunoglobulins, such as rabbit anti-mouse Ig. In practice this approach has proven somewhat difficult. Purified (or monoclonal) anti-Ig antibodies are generally required. Moreover, it is essential that conditions be worked out to ensure that all of the antigen–antibody complexes be bound to the anti-Ig-coated SaC.

Preparation of Samples Consisting of Total Lymphocyte Proteins

For some purposes it is valuable to examine H-2 and Ia proteins in cell extracts without relying on antibodies for immunoprecipitation. Whole cell extracts can be prepared by solubilizing cells directly in IEF sample buffer at 80 μl for 2 $\times$ 10^6 cells; the small amount of insoluble material remaining may be removed by centrifugation. Alternatively, NP-40 extracts prepared as described above can be processed for direct analysis by 2-D PAGE. To 10 μl of NP-40 extract 10 mg crystalline urea and 10 μl IEF sample buffer are added. The urea is dissolved by vortexing. For either type of extract, samples for 2-D PAGE should contain 300,000 cpm of TCA-insoluble radioactivity.

First Dimension: Charge Separation

Two alternative methods can be used for the separation of H-2 and Ia antigens according to their isoelectric point in thin polyacrylamide tube gels. The first, IEF, was the method initially described by O'Farrell.[1] Proteins are electrophoresed to their equilibrium positions in a pH gradient ranging from 4.5 to 7. The more basic proteins are not resolved with this method, however. Because of this deficiency a second, modified electrophoretic system was developed. In the initially described procedures for nonequilibrium pH-gradient electrophoresis (NEPHGE)[2] a basic pH gradient is created using basic ampholytes. To resolve both basic and acidic proteins on the same gel, important for the analysis of Ia antigens, NEPHGE utilizing a broader pH gradient obtained with pH 3.5–10 ampholytes has proven to be most effective.[4,5] Aside from the pH gradient, the major differences between IEF and NEPHGE are the direction in which the samples are electrophoresed (from basic to acidic for IEF; from acidic to basic for NEPHGE) and the length of the run (5600 V-hr for IEF, 2750 V-hr for NEPHGE).

The choice of which first dimension charge separation method should be used depends on the type of information one hopes to obtain. To resolve mouse Ia β chains, I$_i$ chains, and β_2-microglobulin, which are very basic, NEPHGE is required. Good sharp spot patterns are obtained for all H-2 and Ia antigens with broad pH-range NEPHGE. It is also a shorter and somewhat easier technique than IEF, so this is the most commonly used method. However, if one is interested in obtaining the best spot separation for the more acidic Ia α chains and H-2 antigens, IEF gives better results. A comparison of results obtained with the two methods will be given below.

Preparation of Cylindrical Glass Gel Tubes. Both types of first dimension gels are poured in 13.0 cm lengths of soft glass tubing, 2.5 mm inside diameter. Prior to their initial use, the gel tubes should be seasoned by polymerizing acrylamide in them. After this seasoning, and after each use, the tubes are rinsed in water and given a two-step treatment to ensure that the gels stick appropriately to the tubes. The tubes are soaked for 2–24 hr in dichromate cleaning solution, rinsed thoroughly in deionized water, and soaked an additional 2–24 hr in KOH-saturated ethanol (188 g KOH dissolved in 500 ml 95% ethanol). The tubes are then rinsed vigorously in 7 changes of deionized water and dried. Prior to use, a 12.5 cm length is marked off on each tube, and the bottom of the tube is sealed with 4–5 layers of Parafilm. Up to 24 gels can be handled comfortably in a single experiment.

Isoelectric Focusing (IEF). The cathode electrode solution, 20 mM NaOH, is prepared at least several hours prior to the planned start of electrophoresis by dissolving 0.80 g of NaOH in 5 ml of deionized water. This is then added to 1 liter of deionized water that has been boiling for 10 min, and the solution is boiled an additional 5 min, evacuated, and stored under vacuum. For preparing the IEF gel mixture, the following procedure gives 8 gels containing 9.2 M urea, 4% acrylamide, 2% NP-40, and 2% Ampholines. The recipe can be scaled up as needed. Add to 2.75 g urea (Schwartz/Mann Ultra-Pure) in a 125-ml Erlenmeyer flask, 0.665 ml 30% IEF-gel acrylamide stock (28.38% acrylamide and 1.62% bisacrylamide, both from Bio-Rad), 1.0 ml 10% (w/v) NP-40, 0.98 ml deionized water, 200 μl pH 5–7 Ampholine, and 50 μl pH 3.5–10 Ampholine (both from LKB). The urea is dissolved by gentle swirling in a 37° water bath, but the mixture is not warmed above room temperature. To initiate polymerization 5.0 μl 10% ammonium persulfate (stored at 4° up to 2 weeks) and 3.5 μl TEMED are added in sequence. The gel tubes are then filled with the mixture from bottom up, using a 10 cm^3 syringe with an 8-in. 20-gauge needle (BOLAB, Inc, Derry, NH). The tubes should be tapped to release trapped air bubbles and overlayed with 20 μl deionized water. After 1–2 hr, the water and unpolymerized gel mixture are removed and replaced with 20 μl IEF sample buffer, overlayed with 10 μl deionized water. After 1–2 hr these overlay solutions are removed and the gels are placed in a tube-gel electrophoresis tank containing the anode electrode solution (10 mM H$_3$PO$_4$). The gels are overlayed with 20 μl IEF buffer followed by the degassed NaOH cathode electrode solution. The top chamber is then filled up with cathode solution and the gels are prerun for 15 min at 200 V, 30 min at 300 V, and 30 min at 400 V. At the end of the prerun, the solution on top of the gels is removed. The samples, generally one-half of each H-2 or Ia immunoprecipitate as described above or a volume of total cell extract containing 300,000 cpm of [^{35}S]methionine, are applied and overlayed with 10 μl of sample overlay solution (9 M urea,

0.8% pH 5–7 Ampholine, 0.2% pH 3.5–10 Ampholine; stored at −70°), followed by cathode electrode solution. The upper chamber is filled with the same solution. The samples are electrophoresed at 300–400 V for a total of 4800 V-hr (i.e., 12–16 hr). To sharpen the bands, the voltage is increased to 800 V for a final hour.

After removing each gel tube the liquid at the ends of the tube should be removed. The gel is extruded using air pressure from a 5 cm³ syringe attached by a 1-in. length of 3/8-in. Tygon tubing to the basic (upper) end of the gel tube. The gel should be extruded directly into a 16 × 125-mm screw-cap glass tube containing 5 ml SDS sample buffer (0.0625 M Tris base, 10% (w/v) glycerol, 5% 2-mercaptoethanol, 2.3% SDS (BDH), adjusted to pH 6.8 with HCl). The gels need to be equilibrated in the SDS sample buffer with gentle rocking for a total of 2 hr prior to the second dimension. They can be quick-frozen in a dry ice-ethanol bath and stored at −70° at any stage after the initial 30 min of equilibration.

Nonequilibrium pH-Gradient Electrophoresis (NEPHGE)

While NEPHGE gels are very similar to IEF gels, the method in general is shorter and simpler than IEF. The gels can be poured and the run completed on the same day; alternatively, the gels can be poured the day before. The following procedure gives 8 NEPHGE gels: to 2.75 g urea is added 0.665 ml IEF-gel 30% acrylamide stock (see IEF gel recipe), 1.0 ml 10% NP-40, 0.98 ml deionized water, 250 μl pH 3.5–10 Ampholine (LKB), 7.0 μl 10% ammonium persulfate, and 4.0 μl TEMED. Following the procedures described above for IEF gels, the gels are poured and overlayed initially with water and then with IEF sample buffer and water. In NEPHGE the direction of electrophoresis is from anode to cathode, therefore the 20 mM NaOH cathode electrode solution (freshly made, but no boiling or evacuation is necessary) is placed in the lower chamber of the electrophoresis tank. No prerunning is necessary; the samples can be loaded directly, followed by 10 μl sample overlay solution and the 10 mM H$_3$PO$_4$ anode electrode solution which is also used to fill the upper chamber. The samples are electrophoresed using *reverse polarity* (anode at the top, cathode at the bottom), for 5 hr at 550 V. For extruding the gels the syringe and tubing again should be attached to the basic end (this time the bottom of the tube). The gels are equilibrated in SDS sample buffer and stored frozen as described before.

Second-Dimension (Size) Separation

The size separation is accomplished by electrophoresing the proteins out of the first-dimension gel into a discontinuous SDS polyacrylamide slab gel. Good resolution of all of the component chains of H-2 and Ia antigens except β_2-microglobulin is obtained with 10% acrylamide gels; to resolve β_2-microglobulin, 14% acrylamide gels should be used. Twelve-cm-long separating gels generally provide adequate resolution of these proteins. Slab gel plates are cut from 1/8-in. window glass; the back and front plate are $6\frac{1}{2} \times 6\frac{1}{2}$ in. The front plate has a $5\frac{1}{4} \times 3/4$-in. notch cut in the top. The glass at the base of the notch should be beveled at about a 45° angle to enable the tube gel to be anchored more firmly to the top of the slab gel. Most vertical slab gel electrophoresis tanks can be used for the second dimension of 2-D PAGE. To facilitate running multiple slab gels at the same time (12 gels can be processed simultaneously without difficulty), it is advisable that each tank have two terminals each for the upper and lower reservoirs so that adjoining tanks can be connected together; the gels will in effect be in parallel with each other. Detailed plans for an appropriate slab gel tank have been published recently.[4]

The slab gels should be poured 4–24 hr before use. Any standard method can be used for assembling the gel plates. The following recipe gives enough mixture for four 0.030 inch (0.75 mm) 10% acrylamide gels (it can be scaled up as needed.): 16 ml lower gel buffer [1.5 M Tris base, 0.4% SDS (BDH); adjusted to pH 8.8 with HCl], 26.8 ml deionized water, 21.2 ml 30% SDS-gel acrylamide stock (29.2% acrylamide and 0.8% bis-acrylamide, both Bio-Rad), 100 μl 10% ammonium persulfate, and 50 μl TEMED. The mixture should be poured to a line about 7/8 in. below the notch to give a 12-cm gel and then overlayed; 2 ml water-saturated *sec*-butanol is highly recommended for the overlay. After about 2 hr the overlay and any unpolymerized gel mixture should be removed and replaced with a 1:4 dilution of lower gel buffer containing ammonium persulfate and TEMED at the same concentration as found in the stacking gel (i.e., 3 μl 10% ammonium persulfate and 1 μl of TEMED per ml of 1:4 diluted lower gel buffer). The gels should be covered with Saran Wrap if not used immediately.

Prior to pouring the stacking gel the overlay is removed, stirring up any particles that might have settled on top of the gel. The following recipe gives enough stacking gel mixture for four gels: 5.0 ml upper gel buffer (0.5 M Tris base, 0.4% SDS; adjusted to pH 6.8 with HCl), 11.83 ml deionized water, 3.17 ml 30% SDS-gel acrylamide, 60 μl 10% ammonium persulfate, and 20 μl TEMED. The stacking gel mixture is poured up to the notch. To form a flat surface and reduce contact with air, a flat spacer of 0.030-in. Teflon (1×5.125 in.) is inserted to a depth of 2 mm below the notch, avoiding trapping air bubbles. Any gel mixture that may be pooled on the beveled edge of the notch in front of the spacer should be removed. After 20 min the Teflon spacers are removed from the gels, any debris is cleaned from the back plates, and spacers and gaskets used for assembling the gel plates are removed.

The first dimension tube gels can now be readied. After the SDS sample buffer has been poured off from the thawed, fully equilibrated IEF or NEPHGE gels, each gel is placed on a piece of Parafilm and aligned parallel to and about 1/4 in. from the edge of the Parafilm. To load the tube gel onto a slab gel, the slab gel is leaned against a test tube rack at about a 45° angle, 1 ml of melted 1% agarose in SDS sample buffer (maintained in a boiling water bath) is quickly pipetted on top of the stacking gel, and the tube gel is slid from the Parafilm into the agarose (Fig. 1). Care should be taken not to trap any bubbles under the tube gel. After several minutes the gels can be loaded into the tanks. Two drops of 0.1% bromophenol blue are placed in the upper reservoir, which is then filled with SDS running buffer (0.025 M Tris base, 0.192 M glycine, 0.1% SDS; can be made as a 10× stock). The pH of the buffer should be 8.2–8.4; if it is off, it should not be readjusted. The lower reservoir may be filled with used running buffer. Any air bubbles that may be trapped between the gel plates below the bottoms of the gels should be removed. The gels are electrophoresed at 20 mA/gel (constant current) until the bromophenol blue dye marker reaches the bottom of the gel (3–4 hr). The tube gel and stacking gel are discarded and the lower gel placed in fixative with or without stain (e.g., 50% TCA containing 0.1% Coomassie blue for 20 min). The gels can then be destained in 7% acetic acid.

Fluorography. Although [^{35}S]methionine produces relatively high-energy β particles, making it quite efficient for autoradiography, exposure times can be reduced severalfold by processing the gels for fluorography. A variety of chemicals can be used for fluorography, including PPO-

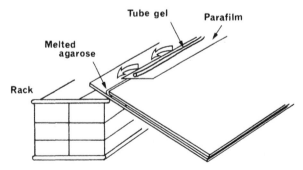

Fig. 1. Loading the first-dimension tube gel on the SDS–PAGE slab gel. Note that the slab gel plates must still be held together by clamps at this stage. To simplify the drawing the clamps have not been shown.

DMSO as originally described[7,8] and EN³HANCE (New England Nuclear). A rapid, less expensive, and less noxious alternative to these fluorogenic substances is sodium salicylate,[9] which gives virtually indistinguishable results. After fixation in acid, the gels need to be soaked in 20 volumes of water for 30 min to prevent precipitation of the salicylate; they can then be transferred to 10 volumes of 1 M sodium salicylate for 30 min. The gels are then placed on Whatman 3MM filter paper, dried down, and exposed to preflashed Kodak XAR-5 X-ray film at −70°. Standard exposure times are 2–3 days for total cell extract gels and 5–7 days for immunoprecipitate gels.

Results and Discussion

2-D PAGE with either IEF or NEPHGE as the first-dimension charge separation generates spot patterns characteristic of individual H-2 and Ia polypeptide chains. Figure 2 shows a comparison of the patterns obtained for identical immunoprecipitates run in parallel on IEF and NEPHGE gels. The very basic A_β and I_i chains are included in the NEPHGE but not in the IEF 2-D gel. Figure 3 shows the patterns generated by total cell extracts using the two different charge separation systems. Also demonstrated in Fig. 3 is the feasibility of recognizing H-2 and Ia antigen spots in gels of total extract proteins (compare to Fig. 2).

Elucidation of General Features of H-2 and Ia Antigens (This Volume [43])

An important use of 2-D PAGE is in the detection of conserved and polymorphic features of H-2 and Ia polypeptide chains. Each of the four I-region products migrates in a characteristic region of the gel; the order of chains in the charge dimension, from acidic to basic, is always A_α, E_α, E_β, and A_β. While the pattern generated for each chain is similar from haplotype to haplotype, variation in the pattern due to genetic polymorphism is readily apparent (compare the patterns of A_β^k in Fig. 2b and A_β^b in Fig. 4b). The ability to distinguish between allelic forms of individual Ia polypeptide chains has been extremely useful in genetic mapping studies.[10-13] For H-2 antigens the situation is quite different. All H-2 antigens

[7] W. M. Bonner and R. A. Laskey, *Eur. J. Biochem.* **46,** 83 (1974).
[8] R. A. Laskey and A. D. Mills, *Eur. J. Biochem.* **56,** 335 (1975).
[9] J. P. Chamberlain, *Anal. Biochem.* **98,** 132 (1979).
[10] P. P. Jones, D. B. Murphy, and H. O. McDevitt, *in* "Ir Genes and Ia Antigens" (H. O. McDevitt, ed.), p. 203. Academic Press, New York, 1978.
[11] P. P. Jones, D. B. Murphy, and H. O. McDevitt, *J. Exp. Med.* **148,** 925 (1978).
[12] P. P. Jones, *J. Exp. Med.* **152,** 1453 (1980).
[13] C. E. Day and P. P. Jones, *Nature (London)* **320,** 157 (1983).

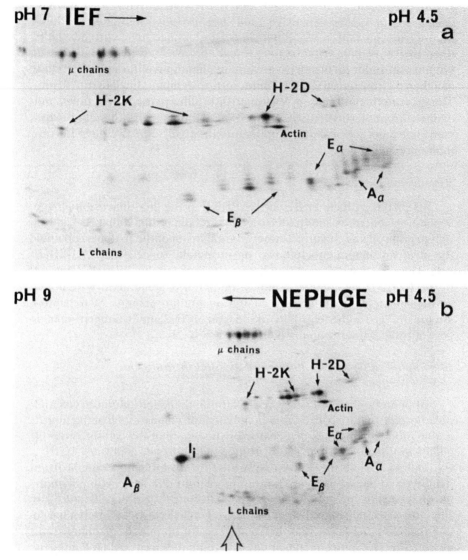

FIG. 2. Comparison of H-2 and Ia antigens separated by IEF and NEPHGE. H-2 and Ia antigens from C3H (*H-2^k*) spleen cells were immunoprecipitated with a C3H.SW (*H-2^b*) anti-C3H antiserum as described in the Methods section. Half of the immunoprecipitate was run on an IEF first dimension gel (a) and half on a broad-range NEPHGE gel (b). The second dimension consisted of a 10% acrylamide SDS slab gel. Spots corresponding to actin and to immunoglobulin μ and light chains, which bind to the SaC, are indicated. Sets of spots generated by individual H-2 and Ia polypeptide chains are labeled. The vertical arrow at the bottom of b indicates the position in the NEPHGE gel corresponding to the basic end of the IEF gel. Adapted from Jones *et al.*[4]

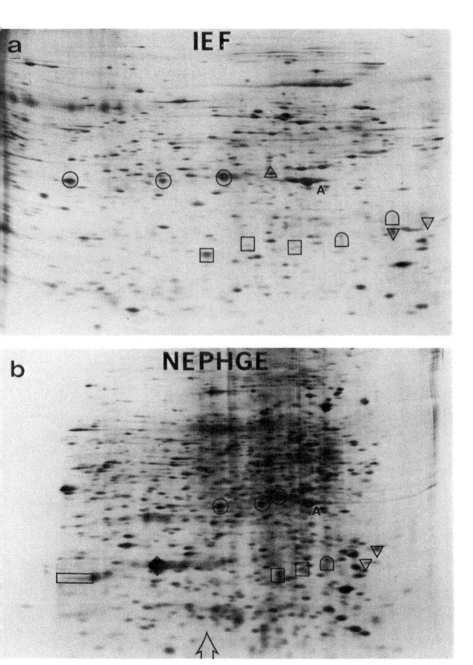

Fig. 3. Comparison of total C3H lymphocyte proteins separated by IEF (a) and NEPHGE (b): 300,000 cpm of TCA-insoluble [^{35}S]methionine-labeled proteins were separated by the two first dimension methods. Spots corresponding to several of the H-2 and Ia polypeptide chains can be recognized in these gels: H-2K, $\bigcirc$; H-2D, $\triangle$; A$_\alpha$, ∇; A$_\beta$, $\square$; E$_\alpha$, $\bigcap$; E$_\beta$, $\square$; I$_i$, $\diamond$. The vertical arrow at the bottom of b indicates the position in the NEPHGE gel corresponding to the basic end of the IEF gel. A, Actin

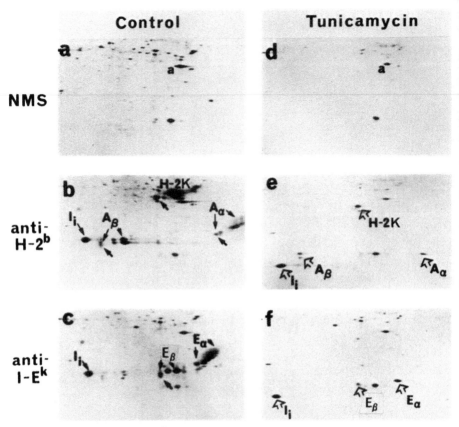

FIG. 4. Detection of nonglycosylated H-2 and Ia polypeptide chains following tunicamy-cin treatment: 5×10^7 B10.A(3R) (I-A^b, I-E^k) cells were precultured for 1 hr in 2 ml methionine-free RPMI-1640 containing 5% newborn calf serum and 5 μg/ml tunicamycin prior to the addition of 250 μCi/ml [^{35}S]methionine. A control culture was precultured in medium without tunicamycin. Sera used for immuniprecipitation were normal mouse serum (NMS) (a,d), C3H anti-C3H.SW (recognizes H-2K^b, I-A^b) (b,e), and C3H.SW anti-C3H (recognizes I-E^k) (c,f). The proteins were separated on NEPHGE first-dimension gels and 12% acrylamide second-dimension SDS gels. The closed upward arrows (b,c) indicate the partially glycosylated cytoplasmic precursors; the open upward arrows (e,f) indicate the nonglycosylated proteins. Adapted from Sung and Jones.[14]

migrate in the region of the gel near actin; however, H-2K and H-2D patterns do not have distinguishing features (see Figs. 2b and 4b), in agreement with data obtained at the amino acid sequence level.

Identification of Distinct Gene Products

A second important application of 2-D PAGE in the analysis of H-2 and Ia antigens is for the identification of individual gene products within complex mixtures of proteins. As shown in Fig. 2a, the patterns generated by individual H-2 and Ia proteins are complex, resulting primarily from the varying carbohydrate content of different biosynthetic forms. The most basic and lowest molecular weight form of each H-2 or Ia chain consists of a cytoplasmic precursor containing only high-mannose oligosaccharides; the other forms are all on the cell surface and have complex oligosaccharides including sialic acid.[3,14,15] Using several approaches it is possible to recognize the partially glycosylated precursors and the more mature cell surface forms of individual chains within complex spot patterns; thus 2-D PAGE can be used to determine the number of different polypeptide chains present in an immunoprecipitate. These techniques will be discussed briefly below.

One set of methods permits the identification of the partially glycosylated cytoplasmic precursor forms of the antigens. The simplest approach utilizes the antibiotic tunicamycin during the [^{35}S]methionine labeling to block the addition of the N-linked oligosaccharide core residues to H-2 and Ia antigens. The proteins synthesized in the presence of tunicamycin lack all of their N-linked carbohydrate residues, the only type of carbohydrate found on H-2 and Ia gene products. As shown in Fig. 4e and f, single spots are immunoprecipitated for each of the H-2 and Ia chains; these spots have the identical charge but are lower in molecular weight than the partially glycosylated cytoplasmic precursor forms.[14] Using this approach, we recently were able to recognize a novel H-2 protein controlled by the *H-2K*^{d} region.[16] A related method is to treat immunoprecipitated H-2 and Ia antigens with the enzyme endoglycosidase H, which strips high-mannose groups off glycoproteins but does not affect the processed complex oligosaccharides. The 2-D gel patterns obtained from endogylcosidase H-treated samples have the normal cell surface forms of H-2 and Ia, but instead of having spots corresponding to the high mannose-bearing

[14] E. Sung and P. P. Jones, *Mol. Immunol.* **18**, 899 (1981).

[15] P. P. Jones, C. E. Day, D. King, J. McNicholas, and E. Sung, *in* "Ir Genes: Past, Present, and Future" (C. W. Pierce, S. E. Cullen, J. A. Kapp, B. D. Schwartz, and D. C. Shreffler, eds.), p. 135. Humana Press, Clifton, New Jersey, 1983.

[16] M. Tryphonas, D. P. King, and P. P. Jones, *Proc. Natl. Acad. Sci. U.S.A.* **80**, 1445 (1983).

cytoplasmic precursors they have the lower molecular weight spots generated by the nonglycosylated proteins (identical to the spots obtained following tunicamycin treatment).[15] Due to the difficulty in obtaining good preparations of endoglycosidase H, tunicamycin is the method of choice; tunicamycin currently is available from Calbiochem. A final procedure for recognition of biosynthetic precursor forms is through pulse-chase experiments. Five to fifteen minute pulses of [35S]methionine label only the partially glycosylated precursors of H-2 and Ia antigens in mouse lymphocytes.[3] By 1 hr of chase significant label can be found in the cell surface forms.

Three additional procedures useful in the identification of distinct H-2 and Ia gene products involve the selective isolation of the cell surface forms of these proteins. The first is "cell surface immunoprecipitation,"[14] in which the intact radiolabeled cells are exposed to the antibodies prior to detergent extraction; the solubilized antigen–antibody complexes are recovered by addition of SaC to the extracts. This method is simple and gives excellent results provided that the cells are extracted in the presence of nonradiolabeled H-2 and Ia antigens to block free antibody combining sites.[14] The second method for obtaining cell surface forms consists of the isolation of plasma membranes from radiolabeled cells.[14] This procedure is somewhat tedious and requires 2×10^8 radiolabeled cells as starting material to provide sufficient purified plasma membranes for immunoprecipitation analysis. The final method utilizes a unique property of the detergent Lubrol WX. This detergent selectively solubilizes cell surface forms of H-2 and Ia antigens; the cytoplasmic precursor forms can be extracted from the Lubrol-insoluble material with NP-40 or Triton X-100.[17]

All of the approaches described above can facilitate the identification of sets of spots corresponding to individual gene products within complex patterns. This capability is especially useful when antibodies are not available for the selective immunoprecipitation of distinct proteins. In addition, these methods provide information on the biosynthetic processing of these complex membrane proteins. As the study of the structure, function, and expression of H-2 and Ia antigens turns more toward the use of genetic engineering technologies, 2-D PAGE will continue to be an important tool for the molecular fingerprinting of these gene products.

[17] J. P. Moosic, E. Sung, A. Nilson, P. P. Jones, and D. J. McKean, *J. Biol. Chem.* **257**, 9084 (1982).

[45] Isolation of Antigenic Proteins from Lymphocytes by Binding Immune Complexes to Staphylococcal Protein A

By SUSAN E. CULLEN

The ability of staphylococcal protein A (SpA) to bind numerous types of immunoglobulin (Ig) with high affinity was first noted by Verwey in 1940.[1] This protein has been studied extensively in the intervening years with two purposes in mind. First, better definition of the structural feature(s) shared by Ig molecules with which SpA reacts was sought, and second, information permitting exploitation of SpA as an immunological developing reagent was collected. The latter types of investigation have led to the use of SpA, either *in situ* on the bacterial surface or in isolated form, in a myriad of applications which previously required the use of a specific second antibody (anti-Ig).

Excellent descriptions of staphylococcal protein A-mediated isolation of immune complexes have appeared in recent years. Goding[2] presented a general description of the uses of staphylococcal protein A in various immunological contexts, including its use as an immunoadsorbent for isolation of antigen–antibody complexes. More recently, Kessler[3] has reviewed the specific topic of SpA-mediated isolation of immune complexes containing antigens derived from cells, and MacSween and Eastwood[4] have discussed methods for recovering the antigenic materials once they are bound to SpA. In addition, Langone[5] has examined the types of interactions that have been found to exist between protein A and immunoglobulins of various classes and subclasses, and from various species. The methods for growing bacteria with high protein A content, and of preparation of fixed bacteria have also been extensively discussed.[3,5]

The scope of this chapter will therefore be restricted to the most recent findings on practical matters related to the use of SpA to bind polyclonal and monoclonal reagents, and through this Ig binding capacity, to isolate the celular antigens with which these antibodies react.

---

[1] E. F. Verwey, *J. Exp. Med.* **71,** 635 (1940).
[2] J. W. Goding, *J. Immunol. Methods* **20,** 241 (1978).
[3] S. W. Kessler, Vol. 73, p. 442.
[4] J. M. MacSween and S. L. Eastwood, Vol. 73, p. 459.
[5] J. J. Langone, *Adv. Immunol.* **32,** 157 (1982).

General Considerations

The procedure generally used for isolation of cell-derived antigens (usually proteins or glycoproteins), which may be present in very small quantities, involves a series of steps that are by now quite standardized in outline. The antigen of interest is radiolabeled either biosynthetically or postsynthetically, and is then solubilized using a nondenaturing detergent. The preparation containing the antigen is mixed with specific antibody, the immune complexes which form are isolated, and the specifically bound antigen is released for further study. The size of the immune complexes generated in this procedure is generally small, probably because of the relatively low concentration of antigen, and the complexes do not precipitate. Protein A is an extremely useful tool for isolation of these immune complexes.[6-10]

The most common applications of protein A in the isolation of lymphocyte derived antigens have been examinations of immunoglobulins and of antigens of the major histocompatibility complex (MHC). These topics are discussed in this volume [40]. Some other leukocyte antigens isolated using SpA are listed in Table I. The isolation procedures used in these studies vary considerably in detail, employing different radiolabeling methods, different detergents, and different types of antibody preparations, but all are quite similar in their basic features.

If a suitable antibody is available, the only serious limitations on ability to isolate the antigen with which it reacts appear to be sufficient density of that antigen on the cells from which it is to be isolated, and ease of radiolabeling the material of interest. Adequate biosynthetic labeling requires a reasonably high rate of synthesis, and may be more difficult if the antigen of interest is synthesized only during a limited portion of the cell cycle. This type of labeling may also be difficult when the cell of interest rapidly loses capacity to synthesize the antigen *in vitro*. Postsynthetic labeling may obviate some of these difficulties, but is subject to other problems. Postsynthetic labeling (usually lactoperoxidase-catalyzed iodination) requires that the molecule have the appropriate functional groups accessible to the labeling reagents. This procedure may also label components adsorbed to, rather than synthesized by, the cells. It is also possible that postsynthetic radiolabeling will destroy some antigenic de-

[6] S. Jonsson and G. Kronvall, *Eur. J. Immunol.* **4**, 29 (1974).
[7] S. W. Kessler, *J. Immunol.* **115**, 1617 (1975).
[8] S. E. Cullen and B. D. Schwartz, *J. Immunol.* **117**, 136 (1976).
[9] S. W. Kessler, *J. Immunol.* **117**, 1482 (1976).
[10] M. J. Brunda, P. Minden, T. R. Sharpton, J. K. McClatchy, and R. S. Farr, *J. Immunol.* **119**, 193 (1977).

TABLE I
Some Leukocyte Antigens Isolated with Antibody and Protein A-Bearing Staphylococci

Antigen	Isolating antibody	Investigation
Thy-1	Rabbit antiserum	I. S. Trowbridge, P. Ralph, and M. J. Bevan, *Proc. Natl. Acad. Sci. U.S.A.* **72**, 157 (1975)
Ly-1	Rabbit alloantiserum	P. J. Durda, C. Shapiro, and P. D. Gottlieb *J. Immunol.* **120**, 53 (1978)
Ly-2,3	Mouse alloantiserum and monoclonal Ab	E. B. Reilly, K. Auditore-Hargreaves, U. Hammerling, and P. D. Gottlieb, *J. Immunol.* **125**, 2245 (1980)
	Monoclonal Ab	J. A. Ledbetter, W. E. Seaman, T. T. Tsu, and L. A. Herzenberg, *J. Exp. Med.* **153**, 1503 (1981)
	Monoclonal Ab	J. W. Goding and A. W. Harris, *Proc. Natl. Acad Sci. U.S.A.* **78**, 4530 (1981)
T200 (Ly-5)	Rat monoclonal Ab	R. Hyman and I. Trowbridge, *Immunogenetics* **12**, 511 (1981)
Thymocyte marker proteins	Rabbit xenoantiserum	D. C. Hoessli, P. Vassali, and J. R. L. Pink, *Eur. J. Immunol.* **10**, 814 (1980)
TLA	Mouse alloantiserum	K. D. Pischel and J. R. Little, *Mol. Immunol.* **17**, 305 (1980)
Qa-1	Mouse Alloantiserum	T. H. Stanton and L. Hood, *Immunogenetics* **11**, 309 (1980)
Tsu^d	Mouse alloantiserum	G. W. Spurll and F. L. Owen, *Nature*
Tind^d	Monoclonal Ab	*(London)* **293**, 742 (1981)
LFA-1	Rat monoclonal Ab	D. Davignon, E. Martz, T. Reynolds, K. Kurzinger, and T. A. Springer, *Proc. Natl. Acad. Sci. U.S.A.* **78**, 4535 (1981)
Mac-1	Mouse monoclonal Ab	I. S. Trowbridge and M. B. Omary, *J. Exp. Med.* **154**, 1517 (1981)
E rosette receptor of human T cells	Monoclonal Ab	M. Kamoun, P. J. Martin, J. A. Hansen, M. A. Brown, A. W. Siadak, and R. C. Nowinski, *J. Exp. Med.* **153**, 207 (1981)
	Monoclonal Ab	F. D. Howard, J. A. Ledbetter, J. Wong, C. P. Bieber, E. B. Stinson, and L. A. Herzenberg, *J. Immunol.* **126**, 2117 (1981)
Human TCGF (IL-2) receptor	Mouse monoclonal Ab	W. J. Leonard, J. M. Depper, T. Uchiyama, K. A. Smith, T. A. Waldmann, and W. C. Greene, *Nature (London)* **300**, 267 (1982)
LMP "low molecular weight proteins"	Mouse alloantiserum	J. J. Monaco and H. O. McDevitt, *Proc. Natl. Acad. Sci. U.S.A.* **79**, 3001 (1982)

terminants, a particular problem when a monoclonal reagent reactive with a single antigenic determinant is to be used for isolation. Since each type of radiolabeling has its strengths and weaknesses, both should be attempted when an antigen is first sought.

Protein A–Immunoglobulin Interactions

Initially it was thought that protein A bound only IgG molecules,[11] with some variation in affinity for different IgG subclasses.[12] IgM, IgA, and IgE were considered to be nonbinding. IgG binding is certainly the most prominent reaction of protein A with immunoglobulins. The major type of interaction that occurs between protein A and IgG is binding through the Fcγ region.[11-15] It has been shown that Fcγ fragments bind to SpA, and that they inhibit the binding of intact IgG to SpA. Protein A has different affinities for the Fc fragments of various IgG subclasses within a single species, and it also varies in its affinity for IgG from different species. Of the species that are more frequently used as antibody sources for the types of experiments discussed here, SpA affinity for rabbit ~ human > mouse > rat > goat IgG. Relative affinity for subclasses within these species is noted in Table II.

An alternative binding site in the F(ab')₂ regions of human Ig molecules has also been documented. Substantial binding of IgA and IgM from human serum or myelomas had been noted,[16-21] and at first it was assumed that this reaction was entirely Fc mediated.[17,18] Later, when an interaction between some polyclonal human IgE molecules and protein A was observed, it was found that the binding of IgE was through its F(ab')₂ε portion.[22] This binding reaction is restricted to the F(ab')₂ε segment of about 6–9% of polyclonal IgE molecules, and no interaction whatever occurs between Fcε and SpA. The interaction of protein A with IgM and IgA classes and subclasses was then reinvestigated and it was found that

[11] A. Forsgren and T. Sjöquist, J. Immunol. 97, 822 (1966).
[12] G. Kronvall and R. C. Williams, J. Immunol. 103, 828 (1969).
[13] A. Forsgren and U. Forsum, Infect. Immun. 2, 387 (1970).
[14] G. A. Stewart, R. Varro, and D. R. Stanworth, Immunology, 35, 785 (1978).
[15] C. Endresen and A. Grov, Acta Pathol. Microbiol. Scand., Sect. C 86C, 193 (1978).
[16] G. McDowell, A. Grov, and P. Oeding, Acta Pathol. Microbiol. Scand., Sect B: Microbiol. Immunol. 79B, 801 (1971).
[17] M. Harboe and I. Folling, Scand. J. Immunol. 3, 471 (1974).
[18] A. Grov, Acta Pathol. Microbiol. Scand., Sect C 83C, 173 (1975).
[19] A. Grov, Acta Pathol. Microbiol. Scand., Sect C 83C, 325 (1975).
[20] E. Saltvedt and M. Harboe, Scand. J. Immunol. 5, 1103 (1976).
[21] A. Grov, Acta Pathol. Microbiol. Scand., Sect. C 84C, 71 (1976).
[22] S. G. O. Johannson and M. Inganas, Immunol. Rev. 41, 248 (1978).

TABLE II

SUMMARY OF BINDING ACTIVITY OF STAPHYLOCOCCAL PROTEIN A FOR SELECTED
IMMUNOGLOBULINS

Immunoglobulin	Comments on binding activity	Investigation
Human		
IgG	IgG_1, IgG_2, IgG_4 strongly bound	G. Kronvall and R. C. Williams, *J. Immunol.* **103**, 828 (1969)
	IgG_3 variably bound: myelomas, nil	G. Kronvall and R. C. Williams, *J. Immunol.* **103**, 828 (1969)
	Polyclonal, 35%	F. Skvaril, *Immunochemistry* **13**, 871 (1976)
	G3m(u⁻) allotypes bound	E. Van Loghem, B. Frangione, B. Recht, and E. C. Franklin, *Scand. J. Immunol.* **15**, 275 (1982)
	G3m (u⁺) allotypes not bound	
	$F(ab')_2$ binding selective	M. Inganas, S. G. O. Johannson, and H. H. Bennick, *Scand. J. Immunol.* **12**, 23 (1980)
IgM	Variably bound; some myelomas	M. Harboe and I. Folling, *Scand. J. Immunol.* **3**, 471 (1974)
	Some polyclonal	E. Saltvedt and M. Harboe, *Scand. J. Immunol.* **5**, 1103 (1976)
	IgM "subclasses" based on SpA binding	A. Grov, *Acta Pathol. Microbiol. Scand., Sect. C* **83C**, 325 (1975)
	$F(ab')_2$ binding selective	M. Inganas, *Scand. J. Immunol.* **13**, 343 (1981)
IgA	Variably bound 20–30% polyclonal	G. McDowell, A. Grov, and P. Oeding, *Acta Pathol. Microbiol. Scand., Sect. B* **79B**, 801 (1971)
	IgA_2 commonly IgA_1 sometimes	G. J. Van Kamp, *J. Immunol. Methods* **27**, 301 (1979)
		M. J. Brunda, P. Minden, and H. Grey, *J. Immunol.* **123**, 1457 (1979)
	$F(ab')_2$ binding selective	M. Inganas, *Scand. J. Immunol.* **13**, 343 (1981)
IgE	$F(ab')_2$ binding selective, 6–9% of polyclonal IgE, no Fc mediated binding	S. G. O. Johansson and M. Inganas, *Immunol. Rev.* **41**, 248 (1978)
Mouse		
IgG	IgG_{2a}, IgG_{2b}, IgG_3 strongly bound, IgG_1 more weakly bound	G. Kronvall, H. Grey, and R. C. Williams, *J. Immunol.* **105**, 116 (1970)
		G. F. Mitchell, J. W. Goding, and M. D. Richard, *Aust. J. Exp. Sci.* **55**, 165 (1977)

(*continued*)

TABLE II (*continued*)

Immunoglobulin	Comments on binding activity	Investigation
		M. R. MacKenzie, N. L. Warner, and G. F. Mitchell, *J. Immunol.* **120,** 1493 (1978)
		P. L. Ey, S. J. Prowse, and C. R. Jenkin, *Immunochemistry* **15,** 429 (1978)
		M. P. Chalon, R. W. Milne, and J.-P. Vaerman, *Scand. J. Immunol.* **9,** 359 (1979)
	F(ab')₂ binding; probably none	M. Inganas, S. G. O. Johannson, and H. H. Bennich, *Scand. J. Immunol.* **12,** 23 (1980)
IgM	Polyclonal, no detectable binding	G. F. Mitchell, J. W. Goding, and M. D. Richard, *Aust J. Exp. Sci.* **55,** 165 (1977)
	Myelomas occasionally bound	M. R. MacKenzie, N. L. Warner, and G. F. Mitchell, *J. Immunol.* **120,** 1493 (1978)
IgA	Polyclonal antibody tested and no binding found	G. Kronvall, H. Grey, and R. C. Williams, *J. Immunol.* **105,** 116 (1970)
Rat		
IgG	IgG₁ and IgG₂c strongly bound, IgG₂B more weakly bound	G. A. Medgyesi, G. Fust, J. Gergely, and H. Bazin, *Immunochemistry* **15,** 125 (1978)
		J. Rousseaux, M. T. Picque, H. Bazin, and G. Biserte, *Mol. Immunol.* **18,** 639 (1981)
	F(ab')₂ binding: probably none	M. Inganas, S. G. O. Johannson, and H. H. Bennich, *Scand. J. Immunol.* **12,** 23 (1980)
IgM	15–20% of polyclonal bound, monoclonal IgM variably bound	G. A. Medgyesi, G. Fust, J. Gergely, and H. Bazin, *Immunochemistry* **15,** 125 (1978)
IgA, IgE	A few monoclonals were not bound but data are sparse	G.A. Medgyesi, G. Fust, J. Gergel, and H. Bazin, *Immunochemistry* **15,** 125 (1978)

these molecules or their $F(ab')_2$ fragments inhibited the binding of radiolabeled $F(ab')_2\varepsilon$ to SpA but did not inhibit binding of radiolabeled $Fc\gamma$.[23] This indicates that for the antibodies tested, binding was at least partially mediated by an $F(ab')_2\varepsilon$-equivalent site. Other authors have evidence for

[23] M. Inganas, *Scand. J. Immunol.* **13,** 343 (1981).

Fc mediated binding of human IgA to SpA,[21,24–26] and it appears to be likely that both Fc and F(ab')$_2$ binding exists, although perhaps not always on the same IgA molecules. It has been noted that some human IgG molecules have an F(ab')$_2$ binding site in addition to their Fc binding site.[27] The binding of Ig molecules which have a low affinity Fc interaction may be positively influenced by presence of the F(ab')$_2$ interaction.

This alternate F(ab')$_2$ binding site appears to exist in Ig from some species other than humans. It has been found that relatively crude polyclonal IgG fractions from Rhesus monkey, guinea pig, pig, and dog have considerable capacity to inhibit the binding of radiolabeled human IgE to protein A.[27] Pig F(ab')$_2\alpha$ fragments bind to protein A.[28,29] Species apparently lacking the F(ab')$_2\varepsilon$-equivalent binding site in polyclonal IgG include mouse, rat, rabbit, and goat.[27] The occasionally observed binding between rodent IgM, or IgA and SpA[30–32] has not been intensively examined to determine whether this binding is via Fc or F(ab')$_2$ moieties.

It has also been noted that SpA appears to have greater affinity for antibody in the form of immune complexes than for free antibody.[7] When SpA is limiting, immune complexes may compete favorably with free antibody. It has even been observed that Ig molecules which fail completely to bind to protein A when in the free state may be quite readily bound when they are complexed to antigen.[33,34] Because of the varied modes and affinities of its interaction with Ig molecules, the binding of SpA to myeloma proteins or monoclonal antibodies, either free or complexed to antigen, is not totally predictable, and should be checked empirically.

Isolation of Antigen-Antibody Complexes with SpA

Insoluble Particles. The particle most commonly used for protein A-mediated binding of immune complexes is the bacterium which synthesizes it, *Staphylococcus aureus*. The Cowan I strain (ATCC 10832, NCTC

[24] A. Dalen, A. Grov, R. Matre, and O. L. Myking, *Clin. Exp. Immunol.* **27,** 421 (1977).
[25] M. J. Brunda, P. Minden, and H. M. Grey, *J. Immunol.* **123,** 1457 (1979).
[26] E. Van Loghem, B. Frangione, B. Recht, and E. C. Franklin, *Scand. J. Immunol.* **15,** 275 (1982).
[27] M. Inganas, S. G. O. Johansson, and H. H. Bennich, *Scand. J. Immunol.* **12,** 23 (1980).
[28] A. Milon, M. Houdayer, and J. J. Metzger, *Dev. Comp. Immunol.* **2,** 699 (1978).
[29] C. Endersen, *Acta Pathol. Microbiol. Scand., Sect. C* **87C,** 177 (1979).
[30] S. Kessler, *J. Immunol.* **117,** 1482 (1976).
[31] M. R. MacKenzie, N. L. Warner, and G. F. Mitchell, *J. Immunol.* **120,** 1493 (1978).
[32] G. A. Medgyesi, G. Fust. J. Gergely, and H. Bazin, *Immunochemistry* **15,** 125 (1978).
[33] T. Barkas and C. M. J. Watson, *Immunology* **36,** 557 (1979).
[34] G. Mota, I. Moraru, J. Sjöquist, and V. Ghetie, *Mol. Immunol.* **18,** 373 (1981).

10344) is a good producer of protein A, and is universally used for this purpose. Heat-treated, formaldehyde-fixed protein A-bearing *S. aureus* Cowan I (SaC) with bound immune complexes can be easily washed free of nonbound material by low speed centrifugation. Although this procedure is certainly not immune "precipitation" in the immunochemical sense, it functionally resembles precipitation and may be casually referred to in this way even in formal articles.

Purified protein A can also be attached to Sepharose to form a particle useful for isolation of immune complexes. This method is more expensive, but may eliminate some undesirable nonspecific binding which can occur with SaC. It is especially valuable to try this method if the antigen of interest binds nonspecifically (without using antibody) to SaC, although Sepharose itself is not free of nonspecific binding effects. Protein A-Sepharose can be used in batch-type isolations, or can be packed in a column and used in affinity chromatography.[35,36]

In practice, investigators have generally carried out batch-type isolation (i.e., washing in tubes) of immune complexes using SaC, and if background effects were troublesome, if scaling-up was required, or if other problems arose, they have usually directly coupled antibodies to Sepharose and isolated antigen by antibody affinity chromatography. The major application of protein A-Sepharose has been isolation of Ig molecules.

Preclearance of Antigen Preparations. A preclearance step involving exposure of antigen preparation to SaC alone or SaC plus control Ig-containing reagents is usually required in order to remove radiolabeled material which binds nonspecifically to the bacterial surface. Radiolabeled Ig molecules, when not the object of the study at hand, are also easily removed by the preclearance step. Obviously if the antigen itself binds to SaC during this step, another approach to isolating it must be sought.

Additional radiolabeled components from the antigen preparation may adsorb to SaC in the presence of normal (nonimmune) serum components or components of normal (nonimmune) ascitic fluid. If the experiment calls for the use of normal serum (or ascites) controls, it may be preferable to perform the preclearance step in the presence of control serum. This step will reduce background in the specific isolation step to be performed subsequently.

If the specific isolation step contemplated employs a non-SpA binding monoclonal reagent and therefore requires an anti-Ig developing reagent, then a preclearance step including the anti-Ig developing reagent

[35] F. S. Walsh and M. J. Crumpton, *Nature (London)* **269,** 307 (1977).
[36] S. Werner and W. Machleidt, *Eur. J. Biochem.* **90,** 99 (1978).

may be useful. This procedure would remove radiolabeled non—SpA binding Ig before the specific isolation step is performed.

If the experimental design does not permit preclearance and the antigen preparation contains radiolabeled Ig, this material may interfere with the subsequent examination of the antigen under study (e.g., in SDS–PAGE). Bearing in mind that many membrane proteins are not disulfide linked oligomers,[37,38] it may be possible to selectively omit or employ reduction of the sample to allow separation of the antigen from intact Ig (no reduction) or its subcomponents (reduction).

Specific Isolation Procedure. A generalized protocol for antigen isolation includes the following steps.

1. Prewash of bacteria. The fixed bacteria are kept suspended at 10% v/v in buffered saline with 0.2% sodium azide for short-term storage. SaC to be used in the preclearance step and in subsequent experimental steps should be prewashed in the same detergent-containing buffer used for preparation of the cell extract. This washing procedure removes fine, nonsedimenting bacterial particles, and releases some bacterial proteins. After washing, the bacteria may be resuspended at 10% v/v in appropriate buffer, or the quantity required can be left as a pellet to which antigen will be added.

2. Preclearance of cell extract. For preclearance, a cell extract (for example, an extract from 1×10^8 mouse B cells, volume 4 ml) is added to a pellet of washed SaC (this example, 400 μl). The SaC are resuspended, and after a 15 min incubation at 4° the suspension is centrifuged and the supernatant is collected. The SaC used in preclearance may be washed, and the bound components, which in this example would include IgG and other labeled molecules, may be analyzed.

A second preclearing step might be carried out at this point. If, for example, the specific antibody to be used is a monoclonal mouse reagent which does not directly bind to SpA, it would be appropriate to mix the SaC-precleared preparation with a small quantity of a rabbit anti-mouse Ig antiserum. After 30 min at 4°, the mixture is transferred to a new SaC pellet sufficient to completely bind the rabbit Ig. The bacteria are resuspended and, after a 15 min incubation at 4°, removed by centrifugation. The rabbit anti-mouse Ig reagent will have reacted with mouse Ig not removed by the first SaC preclearance. This step thus prevents endogenously radiolabeled Ig from interfering with subsequent analysis.

3. Isolation of antigen. Specific analytical precipitation is accomplished by mixing an aliquot ($\sim 10^6$–10^7 cell equivalents) of the precleared

[37] R. J. Allore and B. H. Barber, *Mol. Immunol.* **20**, 383 (1983).
[38] J. W. Goding and A. W. Harris, *Proc. Natl. Acad. Sci. U.S.A.* **78**, 4530 (1981).

antigen preparation with antibody, and incubating at 4° for 2–4 hr in 10 × 75-mm round bottom disposable glass tubes. To continue with our example, the antibody reagent might be 10 μl of a monoclonal mouse anti-"B cell" antibody obtained from a hybridoma grown in tissue culture. The control for this specific isolation step is to mix an aliquot (same number of cell equivalents) of antigen with an equivalent quantity of an irrelevant mouse monoclonal antibody. If necessary, the developing anti-mouse Ig serum is then added to specific and control mixtures, and allowed to react for 15 min. At this time, an appropriate excess quantity of a 10% v/v suspension of washed SaC is added for 15 min.

4. Washing of immune complex-bearing SaC. After the incubations with antigen and antibody, the SaC, now bearing absorbed immune complexes (as well as uncomplexed Ig) must be washed free of nonbound materials. Buffered saline containing detergent is usually used as a washing reagent. One milliliter washing buffer can be added to multiple tubes with a repeating (e.g., Cornwall) syringe. The tubes should be centrifuged at 3000 g in a refrigerated centrifuge. The supernatant is decanted and additional buffer added for the second washing step. Decanting of supernatants may not be successful if plastic tubes have been used, because the bacterial pellet may not adhere well to the bottom of the tube. Use of conical tubes results in formation of pellets which are difficult to resuspend adequately. The smaller bacterial pellets generated in RIA type applications or with potent monoclonal reagents may be more rapidly processed using small conical tubes and small high speed centrifuges such as the Eppendorf. Bacteria should be resuspended using a vibrating mixer (e.g., Vortex), centrifuged as before, and the supernatants decanted.

After addition of buffer for the third washing, a tube transfer should be performed. The bacterial suspension is transferred to a fresh tube using a Pasteur pipet. The transfer is necessary because there is a substantial amount of nonspecific adsorption of materials in the antigen preparation to the wall of the glass tube. These radiolabeled materials may be released from the tube in the step in which antigen is released from the SaC, and the tube transfer averts this problem. After the final centrifugation, the supernatants are decanted and the tubes allowed to drain briefly onto absorbent paper. These antigen-containing or control bacterial pellets can either be treated immediately to release radiolabeled material, or stored frozen until needed.

The quantities of reagents mentioned in the foregoing example have been arbitrary, and appropriate quantities of the reagents to be used must be decided on a case-by-case basis. Clearly, the titers of antibody in serum, ascitic fluid, and hybridoma tissue culture supernatants will vary quite widely. Also present will be somewhat variable quantities of Ig other than the specific antibody, and these free Ig molecules will influence binding of complexes to SaC. The amount of SaC used should be in sufficient excess to bind free Ig as well as immune complexes. This will provide the safety factor required to avoid binding perturbations caused by such factors as aggregated Ig (see below). If antibody reagents are in such short supply that titration of SpA binding is not practical, then the quantity of SaC required can be set by the following *very* crude approximation of the volume ratios of reagent to 10% SaC suspension:

Whole serum: SaC suspension (1 : 10 v/v)

Whole ascites: SaC suspension (1 : 5 v/v)

Hybridoma supernatant: SaC suspension (1 : 2 v/v).

When antibody has been partially purified, the purification factor must also be considered.

5. Release of antigen from SaC. Dissociating conditions compatible with subsequent experimental steps should be chosen to release the antigen bound to SaC. Release can be achieved using ionic detergents, dissociating agents such as urea, guanidine, or KSCN, or even by modification of pH.[3,4] The antibodies and a considerable amount of staphylococcal constituents are also released by these treatments,[8,39] and although these components are not radiolabeled they may cause protein overloading problems in subsequent separation steps.

Properties of Antibody-Containing Reagents

The types of antibody-containing reagents currently available for use have different properties in SpA-binding reactions. Some of these will be discussed below.

Antibodies in Serum or Ascitic Fluid. Antigen–antibody complexes formed using polyclonal antibody in whole serum are highly variable in size and composition, containing several classes and subclasses of immunoglobulin. If the antiserum used comes from a species which has some immunoglobulins that are bound by SpA, then it is virtually certain that the complexes formed by antibodies in that serum all will contain SpA-binding Ig and thus will be bound, provided that some simple precautions are taken.

Antiserum should be treated at 56° for 30 min to inactivate complement. Binding of C1q to immune complexes has been found to impede the binding of protein A to the Fc portion of the immunoglobulins.[40–43] The

[39] P. G. Natali, L. Walker, and M. A. Pellegrino, *Clin. Immunol. Immunopathol.* **15,** 76 (1980).

[40] J. Sjöquist and G. Stahlenheim, *J. Immunol.* **103,** 467 (1969).

[41] G. Kronvall and H. Gewurz, *Clin. Exp. Immunol.* **7,** 211 (1970).

[42] G. Stahlenheim, O. Götze, N. R. Cooper, J. Sjöquist, and H. J. Müller-Eberhard, *Immunochemistry* **10,** 501 (1973).

[43] J. J. Langone, M. D. P. Boyle, and T. Borsos, *J. Immunol.* **121,** 327 (1978).

effect is not complete, but could reduce the efficiency of binding. Another serum component, α_2-macroglobulin, has also been reported to bind to the Fc portion of some immunoglobulins, and thus to inhibit SpA binding.[24] No simple method for disrupting this interaction has been suggested. It is possible to dissociate α_2-macroglobulin from immunoglobulin by gel filtration in 8 M urea and to recover the SpA-binding immunoglobulin,[24] but such a procedure is not practical on a routine basis. The extent to which Ig-bound α_2-macroglobulin affects the typical SpA binding reaction is not known, but the problem could be more troublesome when one uses serum obtained from individuals with conditions leading to increased levels of α_2-macroglobulin.

The problems caused by non-Ig components of serum also include cleavage of antigen by endogenous serum enzymes, or by enzymes released by tissues into the serum during absorption steps.

Partial purification of the Ig from serum or ascitic fluid may reduce some of the problems caused by proteases, complement, and other serum components, and may reduce nonspecific adsorption of serum components or components from the antigen preparation. Partial Ig purification can be accomplished by salt precipitation,[44] ion exchange chromatography and gel filtration,[45] or affinity to protein A (see Table II). The first method can have some limited selectivity, but usually is not used to fractionate Ig classes. It is a frequent preliminary step used to concentrate Ig and remove some unwanted protein. The second two methods have Ig class selectivity and can separate IgG, IgA, and IgM, but they are generally not subclass selective. The last method is the most selective, providing Ig free of detectable contaminants, but it does fail to bind some antibodies which may be of value. Conditions can be adjusted to permit selective isolation of IgG subclasses because the affinity of protein A for different IgG subclasses varies (see Table II).

The unwanted effects of non-Ig serum components may also be avoided by modifying the isolation procedure outlined previously. Whole serum can be premixed with SaC and washed to remove nonbinding components. The SaC with bound antibody can then be mixed with the antigen preparation and immune complexes will form on the bacterial surface.[3,46] With this method a lower binding efficiency is achieved, and frequent mixing is required to facilitate contact between free antigen and

[44] K. Heide and H. G. Schwick, in "Handbook of Experimental Immunology" (D. M. Weir, ed.), Sect. 7. Blackwell, Oxford, 1978.

[45] J. L. Fahey and E. W. Terry, in "Handbook of Experimental Immunology" (D. M. Weir, ed.), Sect. 8. Blackwell, Oxford, 1978.

[46] P. G. Natali, M. A. Pellegrino, L. Walker, S. Ferrone, and R. A. Reisfeld, J. Immunol. Methods 25, 255 (1979).

the particulate Ab-SaC. In a variation of this technique, antibodies can be cross-linked to the SaC so that subsequently, antigen alone can be dissociated from the particles.

Serum to be used in SpA binding should be frozen in small aliquots so that repeated freezing and thawing is avoided. Aggregates of immunoglobulins formed during freeze–thaw cycles not only reduce the titer of specific antibody, but also compete with specific immune complexes for SpA binding sites.[7] Aggregates can be removed by centrifugation, but this separation is imperfect because the minimal sedimentation rate of IgG aggregates (dimer = 11 S) is lower than the sedimentation rate of IgM pentamer and IgA dimer and trimer, and because some IgG monomer is inevitably lost in the separation process. The volume of serum to be freed of aggregates is usually intentionally small, and centrifuge tubes of low capacity are preferable. The Beckman Airfuge provides the most convenient means of performing this separation. Aggregate-free serum can be obtained by centrifuging at 100,000 g for the appropriate length of time, and then carefully collecting an appropriate volume from the upper portion of the column of serum in the tube. Table III lists some centrifugation conditions suitable for specific quantities of serum. It is obvious that the centrifugation procedure results in considerable loss of antibodies from all classes, including most of the IgM and IgA and up to half of the monomer IgG, depending on the starting volume. Except in special experimental circumstances, it is preferable to compensate for the presence of aggregates by using an excess of SpA.

The cautionary notes which apply to the use of whole serum as a source of antibody are equally applicable to the use of ascitic fluid har-

TABLE III

CONDITIONS REQUIRED TO OBTAIN Ig
AGGREGATE-FREE SERUM

Starting volume (μl)	Time[a] (min)	Volume[b] to be collected (μl)
175	33.0	125
175	27.0	100
150	25.8	100
150	19.8	75
125	19.2	75
100	12.0	50

[a] Duration of centrifugation at 100,000 g in a Beckman Airfuge.

[b] Only the upper portion of the serum is free of aggregates of 11 S or greater.

vested from animals making polyclonal antibody or bearing monoclonal antibody-secreting hybridomas. The ascitic fluid contains Ig and other "serum" constituents from the host animal, and these materials may interfere with antibody–SpA interaction.

Antibodies from Hybridoma Culture Fluid. When the source of antibody to be used in antigen isolation is tissue culture medium in which a monoclonal antibody-secreting hybridoma has been grown, the quantity of nonantibody Ig present is much lower than it is in serum or ascitic fluid. If the fusion partner used to produce the hybridoma was an immunoglobulin nonsecretor, then the cloned hybridoma should produce only the antibody of interest. The only other source of Ig in the preparation is the serum supplement which is used in the culture medium. Hybridomas are usually cultured in fetal bovine serum or agamma horse serum to diminish the contribution of Ig from external sources. When the presence of this small quantity of Ig is undesirable, passage of the serum to be used in the medium over protein A-Sepharose prior to its inclusion in the medium will remove residual Ig.[47] This step can provide a hybridoma culture supernatant in which the only SpA-binding Ig is from the hybridoma.

The quantity of antibody produced by hybridomas in culture is much lower than the quantity produced in ascitic fluid of mice bearing the hybridoma. Concentration of antibody from culture supernatants may therefore be desirable. Methods generally used for increasing Ig concentration include salt precipitation or binding to protein A-Sepharose.

The affinity of monoclonal antibodies for antigen is frequently quite low, and immune complexes may not be stable enough to permit antigen isolation. Combination of two monoclonal reagents reactive with the same molecule may significantly improve capacity to isolate antigen.[48-50] In systems which are not well developed, it may be difficult to determine whether moncolonal reagents are detecting epitopes on the same molecule, or separate molecules of similar type.

[47] P. A. Underwood, J. F. Kelly, D. F. Harman, and H. M. MacMillan, *J. Immunol. Methods* **60**, 33 (1983).
[48] R. Tosi, N. Tanigaki, R. Sorrentino, R. Accolla, and G. Corte, *Eur. J. Immunol.* **11**, 721 (1981).
[49] P. H. Ehrlich, W. R. Moyle, Z. A. Moustafa, and R. E. Canfield, *J. Immunol.* **128**, 2709 (1982).
[50] W. R. Moyle, C. Lin, R. L. Corson, and P. H. Ehrlich, *Mol. Immunol.* **20**, 439 (1983).

[46] Use of Monoclonal Antibodies in the Study of the Fine Specificity of Antigens Coded for by the Major Histocompatibility Complex in Mice

By LORRAINE FLAHERTY and MICHAEL A. LYNES

The major histocompatibility complex (MHC) of the mouse, known as the *H-2* complex, is composed of a series of linked genes with related characteristics.[1–3] Genes of this region control a wide variety of cell-surface antigens and lymphoid functions. Four different classes of molecules are encoded by the *H-2* region. Class I genes encode cell surface glycoproteins having molecular weights of approximately 40,000 to 45,000, and are associated with a β_2-microglobulin subcomponent. This class can be further subdivided into two categories.[4] The first category, class I(a), consists of molecules bearing the classical transplantation antigens found on a majority of cells in the mouse. These molecules play an important role in associative recognition. Examples include H-2K, H-2D, and H-2L. The second category, class I (b), consists of molecules which are limited to cells of the hematopoietic system and which have no established function.[5] Examples include Qa-2 and TL. Class II molecules encode a set of cell-surface molecules called Ia which are composed of two subunits (35,000 and 28,000 daltons) and are involved in cell–cell interactions of the immune system. Class III genes encode molecules which belong to the complement system. Recently, Monaco and McDevitt[6] have described a fourth class of MHC molecules that have a low molecular weight and are detectable by gel electrophoresis techniques.

Class I and II molecules are most commonly recognized by use of alloantibodies directed against polymorphic antigenic determinants. Before the advent of monoclonal antibodies, greater than 50 specificities were known to exist for class I molecules. These specificities were divided into two classes—public, those unique to a single haplotype, and

---

[1] J. Klein, "The Biology of the Mouse Histocompatibility-2 Complex." Springer-Verlag, Berlin and New York, 1975.

[2] G. Snell, J. Dausset, and S. Nathanson, "Histocompatibility." Academic Press, New York, 1976.

[3] L. Flaherty and N. Cohen, *in* "Immune Regulation" (L. N. Ruben and M. E. Gershwin, eds.), p. 1. Dekker, New York, 1982.

[4] M. A. Lynes, S. Tonkonogy, and L. Flaherty, *J. Immunol.* **129**, 928 (1982).

[5] J. Forman, J. Trial, S. Tonkonogy, and L. Flaherty, *J. Exp. Med.* **155**, 749 (1982).

[6] J. J. Monaco and H. O. McDevitt, *Proc. Natl. Acad. Sci. U.S.A.* **79**, 3001 (1982).

private, those shared by at least two unrelated haplotypes.[1-3] Since most of these alloantisera were made between strains differing at more than one class I or class II locus, extensive genetic, serologic, and biochemical studies were often necessary to identify these specificities. The use of monoclonal reagents has improved our understanding of these determinants and has reinforced the notion that cross-reactive antigens are present on diverse class I and class II molecules. Generally, the same procedures used to detect and characterize alloantiserum reactivity are applicable to monoclonal antibody reagents. Where exceptions exist, they will be noted.

Detection Systems

Hemagglutination Assay

The hemagglutination assay was first used by Gorer[7] to demonstrate class I molecules on erythrocytes. Two methods are currently used—the dextran–fetal calf serum (FCS) method[8] and the polyvinylpyrrolidone method.[9] Both methods are designed to enhance the agglutinability of mouse erythrocytes by reducing their electrostatic repulsion. A modified dextran–FCS procedure is used in our laboratory and described in detail below.

Reagents

10% dextran (M_r 100,000–200,000, Schwartz-Mann) in 6% dextrose solution (sterilize and keep at 4°; good for several months). Since the lot and brand of dextran is important, several batches should be tested for their ability to enhance agglutination.

Heat-inactivated FCS.

Dextran working solution (2.0 to 2.3%) made by diluting 10% dextran with saline (keep frozen; good for several months). The exact percentage of dextran should be determined in preliminary assays. Choose a concentration which gives efficient agglutination with antibody but no agglutination with dextran and red blood cells alone.

50% v/v of 3.8% sodium citrate in saline.

Procedure

Bleed mice and immediately dilute blood in 50% sodium citrate solution.

[7] P. A. Gorer, *Br. J. Exp. Pathol.* **17**, 42 (1936).
[8] P. A. Gorer and Z. B. Mikulska, *Proc. R. Soc. London, Ser. B* **151**, 57 (1959).
[9] J. H. Stimpfling, *Transplant. Bull.* **27**, 109 (1961).

Wash cells once in saline and resuspend. Store on ice until needed.

Serially dilute antibody in dextran working solution. Put one drop of each dilution into wells of a Linbro Multi-well Disposo-tray (round bottom wells). Include a dextran control.

Spin down cells a final time and aspirate off supernatant. Pipet one drop of packed red blood cells into 15 drops of FCS, making sure that no saline is taken with the packed red blood cells.

Add one drop of red blood cell suspension to wells containing dextran or diluted antibody. Mix very well. Seal top with plastic wrap and incubate at 37° for 1–1.5 hr.

A positive reaction will be indicated by an enlarged and irregular pellet at the bottom of the well.

These assays are suitable for preliminary screenings of H-2 antibodies or when large numbers of mice are to be tested. However, they are relatively insensitive and will not detect antibodies against certain class I(a) antigens. [The entire set of class I(b) and class II antigens is not detectable on erythrocytes.] Since this assay depends on the capacity of antibodies to cross-link erythrocytes, only certain classes of monoclonal antibodies will be detected. IgM monoclonals, however, are particularly suited for hemagglutination assays.

Cytotoxicity Assay (This Volume [21])

The cytotoxicity assay is a sensitive and accurate technique for the detection of class I and class II antigens.[10,11] This assay is applicable to complement-fixing antibodies, usually IgM and IgG, in the mouse. Briefly, cells bearing the class I or class II antigens are incubated for 45–60 min with antibody and complement in either a one- or two-stage test. The cytolytic effect of the antibody on the cells is then determined either by vital dye exclusion or by ^{51}Cr release.

Since the cytotoxic test is commonly used in assays for class I and II antigens, the procedure currently being used in our laboratory will be described in detail. Antibody is diluted in M-199 containing 1% γ-globulin-free FCS and kept at 4°. Target cells are prepared by mincing lymph nodes or thymus in approximately 1 ml of cold media. The cells are then resuspended in 15 mls of cold media and centrifuged at 700 g for 10 min at 4°. The supernatant is discarded and the cells are gently suspended in approximately 0.5 ml of media and kept on ice. An aliquot is then adjusted to a concentration of 5×10^6 cells/ml immediately before use. (The rest of the cells can be used in later tests. They remain viable for at least 2–4 hr.)

[10] P. A. Gorer and P. O'Gorman, *Transplant. Bull.* **3**, 142 (1956).
[11] E. A. Boyse, L. J. Old, and I. Chouroulinkov, *Methods Med. Res.* **10**, 39 (1964).

Equal parts (usually 25 μl) of cells and diluted rabbit serum as a source of complement (usually at a 1/5 to 1/15 dilution) are mixed in glass tubes. (The complement should be freshly thawed and diluted no more than 5 min before addition to the diluted antibody.) Immediately after the addition of the diluted complement, 25 μl of lymphoid cell suspension is added to each tube. The tubes are shaken and placed in a 37° shaking water bath for 45 min. After incubation the tubes are immediately placed on ice, and 50 μl of trypan blue solution is added. The trypan solution is made by combining 4 ml of 0.2% trypan blue in water with 1 ml of 4.25% NaCl. This solution should be prepared daily. The live and dead (stained) cells are then counted in a hemacytometer. Two complement controls (with no antibody) should be included and read at the beginning and end of the counting procedure to ensure that the cells have not lost viability during this time period. Serum controls (with no complement) should also be included.

The selection of a complement source is critical. Nontoxic rabbit serum has largely replaced guinea pig serum as a source of complement because of its higher activity in the cytotoxicity test.[12] Most class I(b) and class II antigens are detectable only with rabbit complement.

The correct selection of the individual rabbits as a serum source is also crucial. Certain rabbit sera contain high amounts of heteroantibody against mouse cells. Therefore, it is often necessary to screen a number of rabbits and select only those which have low heteroantibody and high complement activity in their sera. To screen effectively for heteroantibody, target cells (usually BALB/c thymocytes) should be chosen which are particularly sensitive to complement lysis. If the individual rabbit serum is nontoxic at a 1/10 starting dilution, then further testing of the sera for complement activity should be conducted. It is advisable to test for activity on the target cells and with the antibody that will be used in the assay. It is our experience that complement sources which are active with TL antibody–antigen complexes on thymocytes are generally not very active with Qa antibody–antigen complexes on lymphocytes and vice versa. Therefore, a useful general purpose complement source is usually made from pools of several individually tested rabbit sera. Rabbit sera should be stored at −70° in small aliquots since freezing and thawing of the sera result in a large loss of activity. It is generally advised that the complement source should be thawed only twice before being discarded. If heteroantibody is still a problem, absorption procedures such as those described by Boyse et al.[12] can be used.

[12] E. A. Boyse, L. Hubbard, E. Stockert, and M. E. Lamm, Transplantation **10**, 446 (1970).

Antibody Binding Assays

There are several different types of binding assays that have been used successfully for the detection of class I and II molecules. In general, these assays use radioactive or fluorescent markers to determine the degree of binding. Since these methods do not depend on a particular class of immunoglobulin (e.g., complement-fixing) they are more versatile in their applications. Two types of binding assays are commonly used—the direct method and the sandwich technique. In the direct binding assay the mouse alloantibody is labeled, whereas in the sandwich technique labeled antiimmunoglobulin or labeled Protein A, in addition to an unlabeled alloantibody, is used for detection.

Monoclonal antibodies can be radiolabeled in a variety of ways. Since they are readily prepared in large quantities, many laboratories choose to label these proteins by standard biochemical methods such as radioiodination. For many purposes these procedures provide ample sensitivity and stability. One must be aware, however of the fact that these biochemical techniques can alter the binding capacity of the antibody, either by nonspecific chemical action on the antibody or by the attachment of the radioactive moiety in the binding site. The use of shorter lived radioisotopes such as ^{125}I also means that the radiolabeled preparations have a short-lived usefulness.

To avoid these shortcomings, internally labeled antibody can be used.[13] These antibodies are prepared by including labeled amino acids in the culture media. Antibodies prepared with ^3H-labeled amino acids are stable and retain their original specificity. The following procedure is generally applicable to internal labeling of monoclonal antibodies.

1. Collect hybridoma cells at the end of log-phase growth.
2. Wash the cells in sterile media and resuspend in RPMI 1640 plus 15% FCS at 5×10^6 cells/ml.
3. Aliquot 0.5 ml of this cell suspension in wells of Costar 24 well cluster plate (Cambridge, MA).
4. Add 0.1 ml of 1:2 dilution of L-[^3H]lysine (New England Nuclear, Boston, MA, 60–80 Ci/mmol) in RPMI 1640 (pH adjusted to 7.0).
5. Incubate 16 to 20 hr at 37° in 5% CO_2.
6. Collect culture media. Centrifuge at 700 g for 10 min at 4°.
7. Harvest supernatant.
8. Remove excess L-[^3H]lysine by passing supernatant over a Sephadex G-25 column and collecting the first protein peak.

[13] G. Galfré and C. Milstein, this series, Vol. 73, p. 3.

Radiolabeled Antibody Binding

1. Wash viable cells twice in M199 plus 10% FCS (newborn bovine serum or horse serum can be used in place of FCS).
2. Aliquot cells (3–5 × 10^6/tube) into 6 × 50-mm glass tubes.
3. Centrifuge at 700 g for 10 min at 4°, and aspirate media, taking care not to aspirate cells.
4. Add 0.1 ml of radiolabeled antibody and resuspend cells.
5. Incubate on ice with intermittent mixing for 30 min.
6. Add 0.5 ml of M199 plus 10% FCS and mix. Underlayer mixture with 0.1 ml of FCS.
7. Centrifuge at 700 g for 10 min at 4°. Wash cells twice in M199 plus 10% FCS.
8. After the second wash resuspend cells in 0.1 ml of phosphate-buffered saline (PBS). Add 0.5 ml of NCS tissue solubilizer (Amersham, Arlington Heights, IL); add scintillation fluid and count.

Characterization

Since many class I and II molecules are specified by the MHC, each monoclonal antibody should be characterized to determine the molecule or molecules that are reactive. This examination includes genetic, biochemical, epitope, and distributional analyses.

Genetic Analysis

Three major genetic studies are necessary for defining the reactivity of a MHC monoclonal antibody. First, a preliminary strain distribution pattern is necessary to determine the congenic strains that can be used for further genetic mapping. It is also desirable to note whether levels of activity vary from strain to strain since this information will give some indication of the effects of background influences and/or cross-reactive determinants. MHC congenics can then be selected which react either strongly or not at all with the monoclonal. If the immunization used for the production of the hybridoma is an MHC congenic combination, then selection of the donor and recipient strains is desirable for this purpose.

Next, a survey of H-2 congenics should be performed. This survey will confirm that only genes inside the MHC influence monoclonal reactivity. The testing of more than one MHC congenic combination is essential since mutations and/or background influences could complicate interpretations.[14] Finally, the subregion control of this reactivity should be ascertained. This task is most easily accomplished by the use of recombinant congenic strains. In this case, however, one must be extremely careful about the choice of strains and interpretations of results, since cross-reactivities between different class I molecules are common. It is desirable to select recombinant congenics whose parents clearly type strongly positive and completely negative.

In addition, several recombinant congenic series should be tested since genes at some distance proximal or distal to the H-2K and H-2D genes could be influencing monoclonal reactivity. The recent data of Steinmetz et al.[15] indicate that there may be as many as 36 functional class I genes. Their exact positions on the seventeenth chromosome are unknown. Since the appearance of these genes was not monitored in the development of these congenics (they could be either from the donor or background parent), misinterpretations are possible. This point is exemplified by the typings of enzyme markers distal to the MHC. As shown in Table I, a number of recombinant congenic strains are the result of two crossover events on chromosome 17.

If specific class I or class II mutants are available, these strains should also be examined. Here, more ready identification of the reactive molecule may be made since most of the available mutant strains have only one affected molecule. These typings, however, may be unrewarding since most of these mutations affect only a single antigenic specificity.

Biochemical Analysis

Biochemical analysis can aid in determining the reactivity of monoclonals. First, class I and II molecules can be distinguished from one another by molecular weight determinations. In addition, the Qa-2 and Lq molecules can be distinguished from class I(a) molecules by their lower molecular weight (40,000 vs ~45,000).[16,17]

Moreover, coprecipitation studies can be used to determine if any two monoclonals or serological reagents react with the same molecule. Since reagents to a number of class I and class II molecules are available, preclearing of the membrane lysate with one reagent, and reactivity of

---

[14] L. Flaherty, in "The Mouse in Biomedical Research" (H. L. Foster, J. D. Small, and J. G. Fox, eds.), p. 215. Academic Press, New York, 1981.

[15] M. Steinmetz, A. Winoto, K. Minard, and L. Hood, Cell 28, 489 (1982).

[16] J. Michaelson, L. Flaherty, Y. Bushkin, and H. Yudkowitz, Immunogenetics 14, 129 (1981).

[17] P. Démant and M. H. Roos, Immunogenetics 15, 461 (1982).

TABLE I

SEVENTEENTH CHROMOSOME ORIGINS OF SOME COMMONLY USED CONGENIC AND RECOMBINANT CONGENIC STRAINS[a,b]

Strain	Centromeric region	H-2K	S	H-2D	Qa2	TL	Pgk-2	Upg-1
B10.A	B10 or A	A	A	A	A	A	B10 or A	B10
B10.A(2R)	B10 or A	A	A	B10	B10	B10	B10	B10
B10.A(3R)	B10 or A	B10	A	A	A	A	B10 or A	B10
B10.A(4R)	B10 or A	A	B10	B10	B10	B10	B10	B10
B10.A(5R)	B10 or A	B10	A	A	A	A	B10 or A	B10
B10.D2	B10 or DBA/2	DBA/2	DBA/2	DBA/2	DBA/2	DBA/2	DBA/2 or B10	B10
B10.S(7R)	A.SW, B10, or A	A.SW	A.SW	A	A	A	B10 or A	B10
B10.S(8R)	A.SW, B10, or A	A	A.SW	A.SW	A.SW	A.SW	B10 or A	B10
B10.S(9R)	A.SW, B10, or A	A.SW	A	A	A	A	B10 or A	B10
A-Tla[b]	A or B6	A	A	A	A	B6	B6 or A	A
B6-Tla[a]	A or B6	B6	B6	B6	B6	A	B6 or A	B6
B6.K1	AKR, B6, or A	B6	B6	B6	AKR	AKR	AKR	AKR
B6.K2	AKR, B6, or A	B6	B6	B6	B6	AKR	AKR	AKR

[a] J. Klein, F. Figueroa, and D. Klein, *Immunogenetics* **16**, 285 (1982).
[b] D. Klein, S. Tewarson, F. Figueroa, and J. Klein, *Immunogenetics* **16**, 319 (1982).

this precleared lysate with a second reagent would confirm the nonidentity of the target molecules.

Unfortunately, not all monoclonal reagents give successful immunoprecipitation reactions—an essential feature for these biochemical studies. This is especially true of murine IgM antibodies. Other characterization techniques are therefore preferable for these reagents.

Epitope Studies

Because of the fine specificity of monoclonal reagents, it is often possible to use these antibodies for identifying different domains on a particular MHC molecule. Generally, these studies are performed by determining whether preincubation of cells with one monoclonal will subsequently interfere with the binding of a second, labeled monoclonal. Such studies have confirmed the identity of at least three polymorphic domains on the IA class II molecule[18] as well as a number of domains on class I molecules.[19,20]

Three complications, however, may arise in this type of experiment. First, preincubations with certain class I monoclonal antibodies can increase the subsequent binding of other monoclonals directed against the same molecule.[20] These results suggest that the antibody binding reaction can induce conformational changes in the molecule thus facilitating the binding of a second monoclonal. In addition, monoclonals against one molecule can often sterically hinder the binding of antibodies to a separate and distinct molecule.[21] Here, coprecipitation studies and careful analysis of cross-reactive determinants are necessary to confirm that the monoclonals are reacting with the same molecule. Finally, certain epitopes are the result of combinatorial molecules and exist only on a small proportion of molecules in the total population. This situation is especially true of class II molecules of F_1 mice which express new antigens resulting from the combination of one subunit from one parent with another subunit from the other parent.[22] Moreover, recent results of new Qa antigens recognized by monoclonal reagents indicate that the association of different β_2-microglobulin allotypes influences antigenicity.[23]

There are several interesting applications to epitope studies that will be valuable to our understanding of the nature of class I molecules. For

[18] M. Pierres, C. Devaux, M. Dosseto, and S. Marchetto, *Immunogenetics*, **14,** 481 (1981).

[19] H. Lemke, G. H. Hammerling, and U. Hammerling, *Immunol. Rev.* **47,** 175 (1979).

[20] K. Ozato, P. Henkart, C. Jensen, and D. H. Sachs, *J. Immunol.* **126,** 1780 (1981).

[21] M. A. Lynes, M. Karl, K. DiBiase, and L. Flaherty, *J. Immunogenet.* **9,** 475 (1982).

[22] W. P. Lafuse, J. F. McCormick, and C. S. David, *J. Exp. Med.* **151,** 709 (1980).

[23] V. R. Sutton, P. M. Hogarth, and I. F. C. McKenzie, *J. Immunol.* (in press).

example, Evans *et al.*[24] have recently used monoclonal reagents to identify different domains on the H-2L molecule. That information was used, in turn, to determine whether the expression of hybrid class I molecules is possible. By recombinant DNA techniques, they showed that these hybrid molecules retain the domain antigenicity of their genetic origin. These experiments can be further exploited to determine the functional and active sites of the class I molecule in associative recognition as well as other roles of the domains on the class I molecule.

Distributional Analysis

Identification of the tissue and cell distribution of class I and class II antigens can serve two purposes. First, they can be used for preliminary classification of the target molecule into one of the established classes. It has previously been determined that class I(a) antigens are present on most adult mouse tissues including nonhematopoietic ones, while Qa antigens are represented only on cells of the hematopoietic system. Class II antigens are further restricted to B cells, some T cells, macrophages, and other antigen-presenting cells. TL antigens (class I(b)) are the most restricted being present only on thymus cells.

Second, the tissue and cell distribution of these antigens can contribute to our understanding of the functions of these molecules as well as be useful in studies of the hematopoietic and lymphoid systems. For example, certain class I(b) and class II molecules have been successfully used to study T and B cell ontogeny.

Special Applications of Monoclonals

Cross-reaction Studies

Even though cross-reactive antigens have been clearly demonstrated through the use of polyvalent alloantisera, monoclonal antibodies are far more useful for this purpose. With alloantisera it is often unclear whether there are multiple antibodies reactive to more than one antigen or whether a single antigenic specificity is involved. Monoclonal antibodies are a more useful tool in these studies because of their specificity. These reagents have confirmed the results of biochemical and molecular genetic studies showing a large degree of homology among murine class I mole-

---

[24] G. A. Evans, D. Margulies, B. Shykind, J. G. Seidman, and K. Ozato, *Nature (London)* **300**, 755 (1982).

cules.[25] In addition, phylogenetic studies have also shown conservation of antigenic sites between mouse and man.[26]

Furthermore, lack of cross-reactivity of certain monoclonals with other molecules has been informative. So far, no monoclonal antibody has been isolated which cross-reacts between the TL molecule and any other class I molecule. This observation confirms the biochemical results of McIntyre et al.[27] that TL is the most unique class I molecule so far described.

Alloantisera vs Monoclonal Reagents

With monoclonal antibodies new antigenic specificities have been discovered. In fact, the reverse statement is probably even more applicable—antigens detected with monoclonal reagents generally do not correlate with the strain distributions of alloantiserum-defined specificities. Several explanations could account for this discrepancy: (1) alloantisera detect more than one specificity; (2) monoclonal antibodies have very high titers and activity and may therefore be more sensitive in detecting cross-reacting specificities; (3) since monoclonal antibodies are more specific, slight changes in conformational or molecular structures, such as the presence of carbohydrate residues or presentation on different cell types, or background genetic regulatory influences may alter the antigenicity to such an extent that changes in expression are readily evident; and (4) in alloantisera some antibodies are present in large amounts, while others are present in small amounts. Hybridoma technology, however, can be used to expand these minor antibodies and reveal their antigenic specificities.

Tumor and Cell-Line Typings

In the past, tumor cells, mitogen-stimulated cells, and tissue culture lines have been difficult to type directly for class I and II antigens. This difficulty stems from the contamination of MHC alloantisera with antiviral components, class I(b) antibodies, and other normal mouse serum components. Extensive absorption tests were often needed to confirm the cell-surface phenotypes of these cells. Monoclonal reagents have now revolutionized these typings. Since these reagents can be made *in vitro*

[25] J. Michaelson, L. Flaherty, E. Vitetta, and M. D. Poulik, *J. Exp. Med.* **145,** 1066 (1977).

[26] N. Rebai, P. Mercier, T. Kristensen, C. Devaux, B. Malissen, C. Mawas, and M. Pierres, *Immunogenetics* **17,** 357 (1983).

[27] K. McIntyre, J. Uhr, E. Vitetta, U. Hammerling, J. Michaelson, L. Flaherty, and E. A. Boyse, *J. Immunol.* **125,** 601 (1980).

with defined media components and since these reagents are monospecific, they have largely eliminated these problems.

Caution, however, should be recommended in the use of monoclonal antibodies obtained from ascites fluid since these may have contaminating activities. In addition, controls which verify the possibility that nonspecific binding of particular classes of antibody to certain types of cells takes place, should be included. Absorption tests should also be included to ensure that the detected antigenic specificity of the cell line is the same as that detected on normal resting cells and is not a cross-reactive one.

Quantitative and Tissue-Specific Expression

The use of monoclonal antibodies has also illustrated the possible influence of outside genetic and environmental effects on class I expression. In studies of two monoclonals against a public specificity correlating

TABLE II
MONOCLONAL ANTIBODIES AVAILABLE THROUGH ATCC

Specificity	Ig class	ATCC number	Reference
H-2 (all haplotypes)	IgG_{2a}, HLK	TIB 126	a
I-A^b	IgM	HB 38	b
I-A^{b,d}	IgM, K	HB 35	b
I-A^{b,d}	IgG_{2a}, K	HB 26	b
I-A^{b,d}	IgG_1	TIB 154	c
I-A^d	IgG_{2a}, K	HB 3	d
I-A^k	IgM	HB 15	—
I-A^k	IgM	HB 42	—
I-A^k (i haplotype only)	IgG_{2b}	TIB 94	e
I-A^k	IgG_{2a}	TIB 92	e
I-A^k	IgG_{2b}	TIB 93	e
I-A^s	IgG_{2b}	HB 4	d
I-E^k, C^k	IgG_{2a}, K	HB 32	f
I-E^k	IgG_{2a}, K	HB 6	f
K^b	IgG_1, K	TIB 139	g
K^{b}D^b	IgG_{2a}, K	HB 11	b
K^{b}D^b	IgG_{2a}, K	HB 51	b
K^k	IgG_{2a}, K	HB 5	f
K^k	IgG_{2a}, K	HB 16	f
K^k	IgG_{2a}, K	HB 25^l	f
K^{k}D^k	IgG_{2a}, K	HB 20	f
K^{k}D^k	IgM, K	HB 50	f
K^{k}D^k	IgG_{2b}, K	HB 53	f
K^{k}D^k	IgG_{2a}, K	HB 13	f
K^{k}D^k	IgG_{2a}, K	HB 14	f
K^k	IgG_{2a}	TIB 95	e

TABLE II (*continued*)

Specificity	Ig class	ATCC number	Reference
L^d molecules in D^d and D^q region	IgG_{2a}, K	HB 27[l]	h
D^b	IgM, K	HB 36	b
TL.3	IgG_{2a}, K	HB 7[l]	i
D^b	IgM, K	HB 19	b
D^k	IgG_{2a}, K	HB 24	f
$\beta_2 M$	IgG_{2b}	HB 28	j
L^d, D^q, $L^{q,b}$	IgG_{2a}, K	HB 31	h
Ia^b	IgM	HB 37	f
$L^d D^q$ and/or L^q	?	HB 52	h
$I-E^d$	?	HB 74	k
D^d	?	HB 75	k
D^d	?	HB 76	k
K^d	?	HB 77	k
K^d, D^d, $K^{b,s,r,p,q}$	?	HB 79	k
$I-A^d$	?	HB 85	k
D^d	?	HB 87	k

[a] T. A. Springer, in "Monoclonal Antibodies" (R. H. Kennett, T. J. McKearn, and K. B. Bechtol, eds.), p. 185. Plenum, New York, 1980.

[b] K. Ozato and D. H. Sachs, *J. Immunol.* **126**, 317 (1981).

[c] F. W. Symington and J. Sprent, *Immunogenetics* **14**, 53 (1981).

[d] J. W. Kappler, B. Skidmore, J. White, and P. Marrack, *J. Exp. Med.* **153**, 1198 (1981).

[e] V. T. Oi, P. P. Jones, J. W. Goding, L. A. Herzenberg, and L. A. Herzenberg, *Curr. Top. Microbiol. Immunol.* **81**, 115 (1978).

[f] K. Ozato, N. Mager, and D. H. Sachs, *J. Immunol.* **124**, 533 (1980).

[g] G. Kohler, K. Fisher-Lindahl, and C. Heusser, in "The Immune System, Volumes I and II, Festschrift in Honor of Niels Kaj Jerne, on the Occasion of his 70th Birthday" (C. M. Steinberg and I. Lefkovits, ed.), p. 202. Karger, Basel, 1981.

[h] K. Ozato, T. H. Hansen, and D. H. Sachs, *J. Immunol.* **125**, 2473 (1980).

[i] H. Lemke, G. H. Hammerling, and U. Hammerling, *Immunol. Rev.* **47**, 175 (1979).

[j] F. W. Brodsky, W. F. Bodmer, and P. Parham, *Eur. J. Immunol.* **9**, 536 (1979).

[k] D. H. Sachs, *Transplantation* (in press).

[l] Contaminated with mycoplasma.

with the strain distribution of H-2.8, Tonkonogy and Amos[28] have shown that quantitative variations in the expression of H-2 antigens are evident, confirming the previous studies of O'Neill and McKenzie[29] with alloanti-

[28] S. Tonkonogy and D. B. Amos, *Immunogenetics* **14**, 497 (1981).

[29] H. C. O'Neill and I. F. C. McKenzie, *Immunogenetics* **11**, 225 (1980).

sera. This quantitative control is perhaps most striking when Qa antigens are examined. Each new Qa monoclonal appears to have a unique tissue distribution (and usually a unique strain distribution).[3,4,23,30] Each also exhibits a different level of activity on the various haplotypes. Mapping studies indicate that all these monoclonals react to an antigen or antigens encoded in the *Qa2* subregion. Currently, it is unclear whether each of these monoclonals recognizes a unique antigen or whether conformational or biochemical modifications of the same antigen are responsible for the altered activity.

Monoclonal Listing

Many laboratories have successfully produced monoclonals against class I and II antibodies. To list all of the monoclonals produced by various investigators is outside the scope of this chapter. However, a listing of hybridoma cell lines available through a National Institutes of Health contract and distributed by the American Type Culture Collection is given in Table II.

[30] L. Flaherty, *in* "The Role of the MHC in Immunobiology" (M. Dorf, ed.), p. 33. Garland Press, New York, 1981.

[47] β_2-Microglobulin

By ETTORE APPELLA and JANET A. SAWICKI

β_2-Microglobulin (β_2M) is a protein having a molecular weight of 12,000 that is found in various body fluids and on cell surfaces. It was originally found in urine.[1] Indeed, urine from patients with Wilson's disease, chronic cadmium poisoning or renal failure is a good source of β_2M. Tissue culture supernatants of human lymphoid cells have also been used as a source for protein purification.[2] Smithies and Poulik[3] were the first to isolate β_2M from urine of dogs with kidney damage caused by surgery and X-ray treatment. Similar proteins were also isolated from the urine of

[1] I. Berggard and A. G. Bearn, *J. Biol. Chem.* **243**, 4095 (1968).
[2] K. Nakamuro, N. Tanigaki, and D. Pressman, *Proc. Natl. Acad. Sci. U.S.A.* **70**, 2863 (1973).
[3] O. Smithies and M. D. Poulik, *Proc. Natl. Acad. Sci. U.S.A.* **69**, 2914 (1972).

rabbits[4] and guinea pigs[5] treated with sodium chromate which is known to produce tubular damage and increase urinary excretion of β_2M. However, attempts to isolate mouse β_2M from urine have been unsuccessful. It was originally purified from livers following 3 M NaSCN extraction.[6] Improvements in yield and reproducibility on this purification procedure have recently been published.[7]

β_2M is homologous in amino acid sequence to the sequences of the constant regions of both light and heavy chains of immunoglobulins.[8,9] Physicochemical studies show that β_2M and immunoglobulin domains are also similar in three-dimensional structure.[10] These similarities have suggested that the gene for β_2M may have evolved from a precursor gene that, by duplication, also gave rise to immunoglobulin light and heavy chains. Unlike the immunoglobulins, however, β_2M is synthesized not only by lymphoid cells but also by a variety of other cell types.[11] It has been demonstrated that β_2M is in close association with several cell-surface glycoproteins encoded within the major histocompatibility complex (MHC) of man (HLA) and mouse (H-2).[2,12–14] This association may represent one aspect of the important link between the immune system and the histocompatibility system. The nature of the MHC–β_2M interaction has not yet been well characterized, and the function of β_2M in this association remains unknown. One possible reason for this association may be to stabilize the conformation of the MHC antigens in the cell membrane to allow appropriate restricted recognition.

Biochemical Purification of β_2M

Mouse β_2M has been purified from liver using the following procedure.[7]

Crude Membrane Preparation. Livers from 500 mice (wet weight about 500 g) are homogenized at 4° in Littlefield's medium (0.05 M Tris–

[4] I. Berggard, *Biochem. Biophys. Res. Commun.* **57,** 1159 (1974).
[5] R. Cigen, J. A. Ziffer, B. Berggard, B. A. Cunningham, and I. Berggard, *Biochemistry* **17,** 947 (1978).
[6] T. Natori, M. Katagiri, N. Tanigaki, and D. Pressman, *Transplantation* **18,** 550 (1974).
[7] L. Ramanathan, G. C. DuBois, E. A. Robinson, and E. Appella, *Mol. Immunol.* **19,** 435 (1982).
[8] B. A. Cunningham, J. L. Wang, I. Beggard, and P. A. Peterson, *Biochemistry* **12,** 4811 (1973).
[9] O. Smithies and M. D. Poulik, *Science* **175,** 187 (1972).
[10] F. A. Karlsson, *Immunochemistry* **11,** 111 (1974).
[11] K. Nilsson, P. E. Evrin, I. Berggard, and J. Ponten, *Nature (London)* **244,** 44 (1973).
[12] P. A. Peterson, L. Rask, and J. B. Lindblom, *Proc. Natl. Acad. Sci. U.S.A.* **71,** 35 (1974).
[13] E. S. Vitetta, M. D. Poulik, J. Klein, and J. W. Uhr, *J. Exp. Med.* **144,** 179 (1976).
[14] T. H. Stanton and L. Hood, *Immunogenetics* **11,** 309 (1980).

HCl, pH 7.2, 0.25 M sucrose, 25 mM KCl, 5 mM MgCl$_2$) containing protease inhibitors (0.1% aprotinin, 1 mM each in TPCK, TLCK, and PMSF)[14a] first in a tissue press and then in a motor-driven tissue grinder with a Teflon pestle. The crude homogenate (1 liter) is clarified by centrifugation at 1000 g for 15 min to remove nuclei and unbroken cells and the resultant supernatant is centrifuged at 100,000 g for 60 min to obtain the crude membrane pellet (wet weight ~86 g).

3 M NaSCN Extraction. The membrane fraction is resuspended by homogenization in 300 ml of 3 M NaSCN, pH 7.4, containing protease inhibitors, and then stirred for 2.5 hr at 4°. The NaSCN extract is centrifuged at 100,000 g for 30 min. The clear, yellow supernatant is dialyzed versus 4 liters of 20 mM Tris–HCl buffer, pH 8.1, 0.15 M NaCl, 0.02% NaN$_3$ using membranes of a pore size that retains β_2M.

Acid Precipitation. The 3 M NaSCN supernatant is titrated to pH 4.5 with 1 N HCl at 4° and the precipitate, which forms immediately, is removed by centrifugation. The supernatant is neutralized with 1 N NaOH, and iodoacetamide is added to a concentration of 10 mM. It is then concentrated to about 20 ml by ultrafiltration on a UM2 membrane (Amicon Corp.)

Sephadex G-75 Column. The acid supernatant, containing 300,000 cpm of ^{125}I-labeled β_2M as a marker (see below for preparation), is chromatographed on a 5 × 87 cm column of Sephadex G-75 equilibrated in 20 mM Tris–HCl (pH 8.1), 0.15 M NaCl containing proteolytic inhibitors. Fractions (10 ml) are collected. The fractions containing the ^{125}I-labeled β_2M are diluted in 10 volumes of 0.2 glycine–HCl buffer, pH 2.3, containing 0.5% Triton X-100 and incubated for 1 hr at 4° to dissociate the β_2M from the heavy chain. Each fraction is then neutralized with 1 M Tris, pH 9.0 prior to testing for β_2M activity in radioimmunoassay. Acid dissociation of the two chains is necessary because it has been found that free β_2M has ~2.8 times the inhibiting effect of bound β_2M in the radioimmunoassay. Fractions containing the bulk of β_2M activity are pooled, concentrated to 10 ml by ultrafiltration, and dialyzed vs 1 liter of 20 mM Tris–HCl (pH 8.1), 0.02% NaN$_3$ with one change of buffer.

DEAE-Cellulose Column. Two preparations of 500 livers each are combined after the gel filtration step and subjected to ion-exchange chromatography on a 1.4 × 18.5 cm DEAE-cellulose (Whatman DE52) column equilibrated with the pH 8.1 dialysis buffer containing protease inhibitors. [β_2M-B, an allelic variant found in C57BL/6 mice (see below), is not absorbed at this pH, and so the initial pH must be raised to 8.65.] Frac-

---

[14a] Abbreviations: TPCK, L-1-tosylamido-2-phenylethyl chloromethyl ketone; TLCK, W-2-p-tosyl-L-lysine chloromethylketone; PMSF, phenylmethylsulfonyl fluoride; PBS, phosphate-buffered saline.

tions containing the bulk of β_2M activity are pooled, concentrated, and dialyzed against 1 liter of 0.01 M K phosphate buffer, pH 6.5, with four changes of buffer.

CM-Cellulose Column. The sample obtained from the preceding step is applied to a CM-cellulose (Whatman CM52) column (0.9 × 6.5 cm) equilibrated with 0.01 M K phosphate, pH 6.5. After the initial protein peak, the column is eluted with a linear phosphate gradient (0.01 M K phosphate, pH 6.5, to 0.1 M K phosphate, pH 7.4). The phosphate gradient is determined by measurement of pH and conductivity. Fractions containing β_2M are determined by radioimmunoassay, pooled, concentrated to ~2 ml, and washed with 0.01 M NH$_4$HCO$_3$ by ultrafiltration. The purified β_2M is stored at −70°. Characterization of β_2M by SDS–polyacrylamide gel electrophoresis, alkaline–urea polyacrylamide gel electrophoresis, high-performance liquid chromatography, and amino terminal sequence analysis indicated the presence of a homogeneous product.

Radiolabeled Preparations of β_2M. The immunochemical preparation of radiolabeled murine β_2M has been carried out by incorporating radiolabeled amino acids into murine tumor cell lines.[15] H2-K^b, H-2K^d, and H-2D^d alloantigens are isolated by specific immunoprecipitation with alloantisera and β_2M is separated from the heavy chain by chromatography on a column of Sephadex G-75 in 1 M formic acid. To iodinate β_2M, the protein is first immunoprecipitated and isolated as above and then iodinated using conventional methods.

Methods of Detecting β_2M

Synthesis: Radiolabeling of Spleen Cells with [^{35}S]Methionine and Immunoprecipitation of β_2M

Newly synthesized β_2M can be detected using the following procedure. Mouse spleen cells are teased into RPMI 1640 medium lacking methionine and supplemented with 2 mM glutamine. Cells are washed three times in this medium, suspended at a concentration of 10^8 cells/ml, and labeled for 1 to 4 hr at 37° in the same medium with 100 μCi of [^{35}S]methionine/ml. Proteins are then solubilized for 15 min in ice cold 20 mM Tris–HCl, pH 7.4/0.1 M NaCl/1% Triton X-100 and 1 mM phenyl methylsulfonyl fluoride. After centrifugation at 2000 g for 10 min to remove the nuclei, preabsorption with protein A-Sepharose, and centrifugation at 20,000 g for 15 min, the supernatants are ready for immunoprecipitation. The precipitation step can be carried out with a rabbit antiserum to

[15] J. E. Coligan, F. T. Gates, III, E. S. Kimball, and W. L. Maloy, Vol. 91, p. 413.

purified mouse β_2M^7 or an antibody to H-2. After incubation of the extract with antiserum, protein A-Sepharose or a 10% suspension of heat-killed, formalin-fixed *Staphylococcus aureus,* Cowan I strain, are added and the mixture incubated at 4° for 10 min to 3 hr. The immune complexes are removed by centrifugation and then solubilized by resuspending the pellets in O'Farrell's lysis buffer A [9.5 M urea, 2% w/v NP-40, 2% ampholines (1.6% pH 5–7 and 0.4% pH 3–10) and 5% 2-mercaptoethanol].[16] Proteins are separated by sodium dodecyl sulfate–polyacrylamide gel electrophoresis in 10–15% gradient gel slabs under reducing conditions.[17] To analyze precipitates prepared with the rabbit anti-mouse β_2M serum by two-dimensional gel electrophoresis, nonequilibrium pH gradient gels having an approximate pH gradient of 4.5–7.9 are used as the first dimension.[18] Separation in the second dimension is done using a 4.5% polyacrylamide stacking gel and a 14% polyacrylamide separation gel.

Cell Surface Expression: Immunofluorescence with Rabbit Anti-β_2M

β_2M can be detected on the surface of cells by indirect immunofluorescence. Poulik and Motwani[19] were the first to show that rabbit antiserum to human β_2M, when appropriately absorbed, allows the detection of β_2M on the lymphocyte surface. Further work, using the same antiserum followed by incubation with a fluorescein-labeled goat antiserum to rabbit immunoglobulin, showed that aggregation of β_2M was induced at the lymphocyte surface.[20] For this study of cap formation by immunofluorescence microscopy, cells were first sensitized with anti-β_2M and then mixed with rabbit antiserum to human Ig, in the presence of cycloheximide (100 μg/ml for 2 × 10⁶ cells). Cycloheximide, an inhibitor of protein synthesis, prevents resynthesis and "recovery" of antigens sequestered in the cap. After 2 hr at 20°, the cells were washed three times and then incubated for 2 hr at 37° with goat antiserum to rabbit Ig coupled with rhodamine in the presence of cycloheximide. Under these conditions, more than 50% of the cells showed typical caps where fluorescence was concentrated in less than one-third of the cell contour. When cells that had been treated with anti-β_2M were subjected to a second sensitization in the cold with human antibodies to HLA and then treated with antiserum to human Ig-labeled with fluorescein, complete overlapping of the two fluorochromes was observed. In contrast, capping of the HLA determinants with anti-HLA antiserum resulted in capping of only a portion of β_2M present on the surface.[20] These findings indicate that a physical link-

[16] P. H. O'Farrell, *J. Biol. Chem.* **250,** 4007 (1975).
[17] G. Blobel and B. Dobberstein, *J. Cell Biol.* **67,** 835 (1975).
[18] P. Z. O'Farrell, H. M. Goodman, and P. P. O'Farrell, *Cell* **12,** 1133 (1977).
[19] M. D. Poulik and N. Motwani, *Clin. Res.* **20,** 795 (1972).
[20] M. D. Poulik, M. Bernoco, D. Bernoco, and R. Ceppellini, *Science* **182,** 1352 (1973).

age exists between HLA and β_2M and suggest that there are some β_2M molecules that are not associated with HLA. A quantitation of β_2M on the surface of lymphocytes has also been performed by a competition of a radioimmunoassay using ^{125}I-labeled purified β_2M and an anti-β_2M antiserum.[21] The results show the presence of 1 to 3×10^5 molecules per cell on different populations of lymphocytes. Values for HLA antigens are of the order of 2 to 3×10^4 molecules per cell. These numbers are consistent with there being β_2M in excess on the lymphocyte surface.

Allelic Variants of β_2M

Despite numerous attempts to detect genetic polymorphism of β_2M in humans, no such differences have been demonstrated. This is not too surprising considering that β_2M is a single copy gene as opposed to a member of a multigene family. It also seems reasonable that β_2M's role as a common subunit for many proteins must restrict the variation that is functionally tolerable. Analysis of somatic cell hybrids between human and mouse cells has established that the gene for human β_2M is on chromosomes 15, unlinked to the HLA complex which is on chromosome 6.[22] Recently, structural variants of mouse β_2M have been reported.[23] Cells from mouse spleen or thymus were labeled with ^{125}I or with [^{35}S]methionine, solubilized with NP-40 and precipitated with either rabbit anti-β_2M or anti-H-2K, anti-TL, anti-Qa-2, or anti-H-2D. The β_2M of mouse strains B6 and B10.A migrates faster than β_2M of strain A on the SDS–PAGE system. These two variants have been named β_2M-A and β_2M-B because they were first identified in strains A and B6. Two-dimensional gel electrophoresis reveals that the isoelectric point of β_2M-B is more basic than that of β_2M-A (Fig. 1). In $(A \times B10)F_1$ mice, both β_2M variants are seen on SDS–PAGE and no quantitative differences are apparent. The strain distribution of β_2M-A/β_2M-B phenotypes shows that, as in man, mouse β_2M is not closely linked to the H-2 region. Classical genetic analyses employing the β_2M allelic variation, as well as the analysis of somatic cell hybrids between mouse and Chinese hamster cells, have located the β_2M structural gene on chromosome 2, closely linked to the *H-3, Ly-4, Ly-m11* loci.[24,25] Since these last loci have not

[21] R. T. Kubo, S. M. Colon, R. T. McCalmon, and H. M. Grey, *Fed. Proc., Fed. Am. Soc. Exp. Biol.* **35,** 1183 (1976).

[22] P. Goodfellow, E. A. Jones, V. Heyningen, E. Solomon, M. Bobrow, N. Miggiano, and W. F. Bodmer, *Nature (London)* **254,** 267 (1975).

[23] J. Michaelson, E. Rothenberg, and E. A. Boyse, *Immunogenetics* **11,** 93 (1980).

[24] J. Michaelson, *Immunogenetics* **13,** 167 (1981).

[25] D. R. Cox, J. A. Sawicki, D. Yee, E. Appella, and C. J. Epstein, *Proc. Natl. Acad. Sci. U.S.A.* **79,** 1930 (1982).

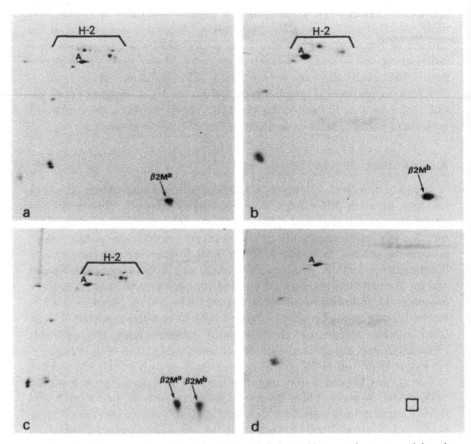

FIG. 1. Electrophoretic variants of mouse β_2M. Polypeptides were immunoprecipitated from [^{35}S]methionine-labeled splenic lymphocytes. (a) SWR/J lymphocytes, anti-β_2M; (b) C57BL/6, anti-β_2M; (c) SWR/J lymphocytes, anti-β_2M + C57BL/6 lymphocytes, anti-β_2M; (d) SWR/J lymphocytes, normal rabbit serum (negative control). The square in d indicates the site to which β_2M would migrate on the gel. "A" indicates actin (MW = 44,000). The basic side of each gel is on the right.

been separated from β_2M by either recombination or strain distribution, they are either closely linked loci or pleiotropic manifestations of a single locus.

The complete amino acid sequences of β_2M from BALB/c and C57BL/6 has been determined.[7,26] The two allelic forms differ at a single position (residue 85); BALB/c has aspartic acid at this position, C57BL/6 has alanine. This interchange is in good agreement with the differences ob-

[26] F. T. Gates, III, J. E. Coligan, and T. J. Kindt, *Proc. Natl. Acad. Sci. U.S.A.* **78**, 554 (1981).

served in amino acid composition and isoelectric point between the two forms of β_2M. Nucleic acid sequence analysis of different β_2M clones revealed a nucleotide base difference in agreement with the amino acid difference at position 85.

Molecular Cloning of β_2M

In order to understand the organization of the *B2m* gene, recombinant DNA technology has been applied through the synthesis of cDNA probes. To construct cDNA probes, total polysomal mRNA was extracted from livers of C57BL/6 mice according to the procedure of Efastratiatis and Kafatos.[27] The poly(A)$^+$ enriched fraction selected by oligo(T) cellulose was then centrifuged through a linear sucrose gradient and fractions containing different size mRNAs were assayed for β_2M mRNA using *in vitro* translation followed by immunoprecipitation with a rabbit antiserum to purified mouse β_2M. The fractions of 9 S–10 S containing β_2M mRNA were used for the synthesis of cDNA. The double-stranded cDNA was then inserted in the *Pst*I site of pBR322 by the use of poly(dG) and poly(dC) tails and transformation of *E. coli* strain LE392 was carried out as described by Dagert and Ehrlich.[28] When the 9 S–10 S mRNA was translated *in vitro*, less than 1% of the synthesized protein was immunoprecipitated. Enough double-stranded cDNA was obtained from 5 μg of poly(A)$^+$ mRNA to make more than 10,000 independent clones of which 95% were sensitive to ampicillin and thus contained inserts. The screening procedure for such clones was hybrid selection, a method which uses the ability of a cDNA clone immobilized on a nitrocellulose filter to hybridize to β_2M mRNA. The hybridized mRNA can be eluted from the filter at high temperature and its *in vitro* translation product can be assayed by immunoprecipitation with the rabbit anti-β_2M. The screening of 1400 cDNA clones gave three positives which contained sequences identical to the mouse β_2M protein sequence.[29] Using these clones, a study of the gene encoding β_2M in the mouse was undertaken. The results show that a single *B2m* gene exists per haploid genome. The genetic organization is similar to that of the evolutionarily related H-2 and immunoglobulin domains with four coding regions separated by three intervening sequences.[30] A comparison of the nucleotide sequences of the *B2m* gene, the *H-2L^d* gene and mouse immunoglobulin heavy and light chain genes revealed a 50–60% homology over stretches of 50–100 base pairs in the

[27] A. Efstratiadis and F. C. Kafatos, *Methods Mol. Biol.* **8**, 1 (1976).
[28] M. Dagert and S. D. Ehrlich, *Gene* **6**, 23 (1979).
[29] J. R. Parnes, B. Velan, A. Felsenfed, L. Ramanthan, V. Ferrini, E. Appella, and J. G. Seidman, *Proc. Natl. Acad. Sci. U.S.A.* **78**, 2253 (1981).
[30] J. R. Parnes and J. G. Seidman, *Cell* **29**, 661 (1982).

regions encoding the cysteine residues. This nucleotide sequence homology together with similarities in the genetic organization support the hypothesis that $B2m$, H-2, and immunoglobulin genes are derived from a common ancestral gene encoding a primitive domain.

Regulation of the Expression of β_2M

Despite the fact that the function of β_2M is still unknown, our relatively advanced understanding of the protein and the gene that encodes it make this an attractive system for studying the regulation of gene activity. Information about how the $B2m$ locus is regulated at the transcriptional and translational levels derives from two general lines of study. One approach is the study of cultured cell lines in which the expression of β_2M is abnormal. The other approach is based on the observation that the $B2m$ gene is differentially expressed during normal development.

The Daudi cell line, established from a Burkitt lymphoma,[31] has been extensively studied because neither HLA class I antigens[32,33] nor β_2M[34,35] are expressed on its surface. Chromosome analysis has revealed that one of the two chromosome 15s in these cells is deleted in the long arm region q14–q21, the region where the $B2m$ locus has been mapped.[36] These cells, therefore, have only one $B2m$ gene. Biosynthetic studies have demonstrated the synthesis and cytoplasmic localization of HLA antigens in Daudi cells, whereas no β_2M synthesis can be detected.[37] When Daudi cells are fused to cells of other lines, HLA-specific antigens associate with β_2M derived from the other line and are expressed on the surface of the hybrid cell.[38–40] Taken together, these observations demonstrate that the absence of expression of HLA antigens and β_2M on the surface of Daudi cells is the result of a mutation of the single $B2m$ gene. The explanation

[31] E. Klein, G. Klein, J. S. Nadkarni, J. J. Nadkarni, H. Wigzell, and P. Clifford, *Cancer Res.* **28**, 1300 (1968).

[32] E. A. Jones, P. N. Goodfellow, J. G. Bodmer, and W. F. Bodmer, *Nature (London)* **256**, 650 (1975).

[33] R. A. Reisfeld, E. D. Sevies, M. A. Pellegrino, S. Ferrone, and M. D. Poulik, *Immunogenetics* **2**, 184 (1975).

[34] K. Nilsson, P. E. Evrin, and K. I. Welsh, *Tranplant. Rev.* **21**, 53 (1974).

[35] M. Raff, *Nature (London)* **254**, 287 (1975).

[36] J. Zeuthen, U. Friedrich, A. Rosen, and E. Klein, *Immunogenetics* **4**, 567 (1977).

[37] H. L. Ploegh, L. E. Cannon, and J. L. Strominger, *Proc. Natl. Acad. Sci. U.S.A.* **76**, 2273 (1979).

[38] G. Klein, P. Terasaki, R. Billing, R. Honing, M. Jondal, A. Rosen, J. Zeuthen, and G. Clements, *Int. J. Cancer* **19**, 66 (1977).

[39] M. Fellous, M. Kamoun, J. Wiels, J. Dausset, G. Clements, J. Zeuthen, and G. Klein, *Immunogenetics* **5**, 423 (1977).

[40] J. Zeuthen, G. Klein, R. Ber, G. Masucci, S. Bisballe, S. Povey, P. Terasaki, and P. Ralph, *JNCI, J. Natl. Cancer Inst.* **68**, 179 (1982).

for this malfunction has recently been determined.[41,42] Northern blot hybridization, with cDNA plasmid clones as probes, reveals a β_2M mRNA in Daudi cells that apparently is comparable in size and intracellular concentration to β_2M mRNA in cells that express this polypeptide normally. Daudi β_2M mRNA, however, when purified by hybridization-selection with β_2M plasmid DNA, is unable to function as a messenger in protein synthesis and is therefore a translationally inactive mRNA. The molecular basis for the inability of this mRNA to serve as a template is not known.

Cells from a mutant mouse cell line, R1 (β_2^-), derived from a C58 thymoma, do not express surface β_2M, H-2, and thymic leukemia antigen.[43] Similarly to Daudi cells, R1 (β_2^-) cells do not synthesize β_2M.[30] This mutant was studied at the gene level by Southern blot analysis of restriction enzyme digests. The results of this analysis indicate that both *B2m* genes are defective in these cells. It is not clear, however, whether one of the *B2m* genes is deleted, as in Daudi cells. Further studies to characterize β_2M mRNA in R1 (β_2^-) cells should reveal the functional consequences of altered DNA sequences in this mutant.

An alternative to studying mutant cell types has been to examine the control of expression of β_2M in cells during differentiation. F9 teratocarcinoma stem cells are similar in many ways to cells derived from early mouse embryos. Neither β_2M nor H-2 histocompatibility antigens are expressed on the surface of F9 cells. Using cloned DNA probes specific for β_2M and H-2 mRNAs in blot hybridization analyses, no β_2M or H-2 specific mRNA was detected in stem cells.[44] These mRNA sequences were detected, however, in differentiated cells derived from F9. These cells also express both β_2M and H-2 antigens on their surface. Considering the level of sensitivity of the blot hybridization technique, it was concluded that F9 stem cells contain no more than one-tenth of the β_2M and H-2 mRNAs of F9 differentiated cells. These observations suggest that the regulation of the *H-2* and *B2m* genes in F9 cells and by inference, in mouse embryo cells, is under transcriptional control.

It is generally thought that H-2 class I antigens that are expressed on the surfaces of all cells in adult mice either are not expressed or are expressed at extremely low levels on cell surfaces in preimplantation mouse embryos. It was of interest, therefore, to determine whether β_2M was expressed at these same early stages of development. Using a rabbit anti-mouse β_2M serum in an indirect immunofluorescence assay, β_2M is

[41] C. De Préval and B. Mach, *Immunogenetics* **17**, 133 (1983).
[42] F. Rosa, M. Fellous, M. Dron, M. Tovey, and M. Revel, *Immunogenetics* **17**, 125 (1983).
[43] R. Hyman and V. Stallings, *Immunogenetics* **3**, 75 (1976).
[44] C. M. Croce, A. Linnenbach, K. Huebner, J. R. Parnes, D. H. Margulies, E. Appella, and J. G. Seidman, *Proc. Natl. Acad. Sci. U.S.A.* **78**, 5754 (1981).

first detected on the surface of trophoblast cells of blastocysts (J. Sawicki, T. Magnuson, and C. Epstein, unpublished results). β_2M had previously been observed on the surface of delayed implanting blastocysts.[45] Using appropriately absorbed alloantisera having reactivity to specific H-2 antigens, similar assays failed to detect H-2 antigens on the surface at this stage.

Embryos can also be radiolabeled with amino acid precursors to study newly synthesized proteins. The procedure for labeling preimplantation embryos with [^{35}S]methionine and preparing radiolabeled embryo extracts is as follows: 100–400 eggs or embryos are delivered in as small a volume as possible to a 30 μl drop of modified Whitten's medium[46,47] containing 1/120 μCi lyophilized [^{35}S]methionine (specific activity 1200 Ci/mmol) in a 35 × 10 mm tissue culture dish filled with paraffin oil equilibrated with Whitten's medium. The embryos are incubated in the labeling medium for 4–5 hr at 37° in an atmosphere of 95% air and 5% CO_2. After labeling, they are washed three times in cold PBS, transferred to 50 μl Tris-buffered saline, pH 7.4, and incubated for 15 min on ice. Nuclei and cell membranes are removed by centrifugation at 10,000 g for 5 min and the supernatant extract is stored at −70°.

When embryos are radiolabeled in this manner and immunoprecipitates are prepared and analyzed on two-dimensional gels, newly synthesized β_2M is detected at all stages of preimplantation development except the fertilized egg (J. Sawicki, T. Magnuson, and C. Epstein, unpublished results). A similar analysis fails to detect the synthesis of H-2 antigens prior to implantation. It appears that the activity of the *B2m* and *H-2* genes may not be coordinately regulated in early mouse embryos.

Advantage was taken of the fact that β_2M is polymorphic to determine the time at which the paternally derived *B2m* gene is first expressed in embryos. Embryos heterozygous at the *B2m* locus were derived from matings between SWR/J female (*B2m*^{a}) and C57BL/6 males (*B2m*^{b}).[48] These embryos were radiolabeled and immunoprecipitates were prepared using a rabbit anti-mouse β_2M serum. The immunoprecipitates were analyzed on 2D gels. β_2M-B was first detected from immunoprecipitates of 2-cell embryos thus demonstrating that the paternally-derived *B2m* gene is active at this early stage.

Acknowledgments

We wish to thank Sarah Butler for her assistance in the preparation of this manuscript.

[45] S. Hakansson and P. A. Peterson, *Transplantation* **21**, 358 (1976).

[46] C. J. Epstein, A. Wegienka, and C. W. Smith, *Biochem. Genet.* **3**, 271 (1969).

[47] M. S. Golbus and C. J. Epstein, *Differentiation* **2**, 143 (1974).

[48] J. A. Sawicki, T. Magnuson, and C. J. Epstein, *Nature (London)* **294**, 450 (1981).

[48] Structural Analysis of Murine Class I Major Histocompatibility Complex Antigens

By RODERICK NAIRN

Background

The murine major histocompatibility complex *(MHC)*, also known as *H-2*, is a cluster of loci coding for proteins involved in the processes of cell-mediated immunity. There is a considerable degree of homology among the various *MHC* genes, presumably indicating an origin from common ancestor genes. On the basis of this evolutionary homology the *MHC* loci have been divided into classes.[1] In this article the methods used to obtain structural information about the H-2 Class I antigens will be outlined.

There are two groups of Class I genes (and antigens). Class I genes located in the left-hand side of the *MHC* (denoted *K, D, L,* and in addition, at least in certain haplotypes, *M* and/or *R*) encode the cell-surface molecules that represent the "classical" transplantation antigens known to mediate graft rejection. These antigens are to be found on virtually all nucleated cells in the mouse. The Class I genes found to the right-hand portion of the *MHC*, known as the *Qa* and *Tla* region, encode cell surface antigens that are distributed in a tissue-specific fashion. The Class I genes to the left-end of the *MHC* are particularly notable for their extensive polymorphism (especially *K* and *D*), whereas those to the right-hand end are far less polymorphic. Additional background data on the genetic structure of the murine *MHC* can be found in the recent reviews by Klein *et al.*[2] and Hood *et al.*[3]

There are structural differences (e.g., chain length, extent of glycosylation) among the H-2 Class I antigens but they do have certain features in common. There is a membrane-bound, glycosylated polypeptide chain of MW. 45,000, noncovalently associated with β_2-microglobulin (β_2M), a polypeptide of 12,000 MW, which is not encoded in the *MHC*. Currently, the classical transplantation antigens are the best characterized of the

---

[1] J. Klein, *in* "The Major Histocompatibility System in Man and Animals" (D. Götze, ed.), p. 339. Springer-Verlag, Berlin and New York, 1977.

[2] J. Klein, F. Figueroa, and Z. A. Nagy, *Annu. Rev. Immunol.* **1,** 119 (1983).

[3] L. Hood, M. Steinmetz, and B. Malissen, *Annu. Rev. Immunol.* **1,** 529 (1983).

Class I antigens (see reviews by Nathenson *et al.*[4] and Kimball and Coligan[5]) although structural data on antigens of the *Qa* and the *Tla* regions can be expected to accumulate in the near future.

Isolation of Murine MHC (H-2) Class I Antigens for Structural Studies

These molecules have almost exclusively been extensively characterized by the use of radiochemical micromethods, since even small amounts of material (mg) are difficult to obtain. While it has been possible to exploit the radiochemical approach to obtain extensive amino acid sequence data, other studies (e.g., on three-dimensional structure, or functional studies) are not feasible unless larger quantities of material are available.

Source of Material

In order to obtain milligram quantities of material one can start either with large numbers of mice[6] or large volumes of cultured cells.[7,8] These earlier procedures were not very successful, but recent modifications of these methods employing immunoadsorbent columns prepared with monoclonal antibodies[9-11] offer some hope that adequate amounts of Class I antigens can be obtained.

For the radiochemical approach, cells from spleen and lymph node, as well as cell lines, have been used as a source of antigen. Cell-surface labeling techniques (e.g., lactoperoxidase-catalyzed iodination) are useful in identification of antigens and, although no longer widely utilized in studies of transplantation antigens, are still used to help identify the molecules recognized by antisera raised against products of the *Qa* and *Tla* regions.[12-14] By far the most widely applied and fruitful approach to obtaining starting material for structural studies of the H-2 Class I antigens has been biosynthetic labeling in cell culture of continuous cell lines or

[4] S. G. Nathenson, H. Uehara, B. M. Ewenstein, T. J. Kindt, and J. E. Coligan, *Annu. Rev. Biochem.* **50,** 1025 (1981).

[5] E. S. Kimball and J. E. Coligan, *Contemp. Top. Immunol.* **9,** 1 (1983).

[6] O. Henriksen, E. A. Robinson, and E. Appella, *J. Biol. Chem.* **254,** 7651 (1979).

[7] J. H. Freed, D. W. Sears, J. L. Brown, and S. G. Nathenson, *Mol. Immunol.* **16,** 9 (1979).

[8] M. J. Rogers, E. A. Robinson, and E. Appella, *J. Biol. Chem.* **254,** 1126 (1979).

[9] S. H. Herrmann and M. F. Mescher, *J. Biol. Chem.* **254,** 8713 (1979).

[10] K. C. Stallcup, T. A. Springer, and M. F. Mescher, *J. Immunol.* **127,** 923 (1981).

[11] J. E. Mole, F. Hunter, J. W. Paslay, A. S. Bhown, and J. C. Bennett, *Mol. Immunol.* **19,** 1 (1982).

[12] J. Michaelson, L. Flaherty, Y. Bushkin, and H. Yudkowitz, *Immunogenetics* **14,** 129 (1981).

[13] E. Rothenberg and D. Triglia, *Immunogenetics* **14,** 455 (1981).

[14] R. G. Cook, R. N. Jenkins, L. Flaherty, and R. R. Rich, *J. Immunol.* **130,** 1293 (1983).

cell suspensions from spleen or lymph node. This appoach will be dealt with in detail here.

Biosynthetic Labeling of Cells in Culture[15-17] (*This Volume* [49])

Reagents

Growth medium: Dulbecco's modified Eagle's medium (DME) for established cell lines and RPMI 1640 for spleen and/or lymph node cell suspensions. These media are supplemented with serum [usually 10% (v/v) heat-inactivated fetal calf serum] and may contain antibiotics (usually penicillin and streptomycin).

Labeling medium: as above, except minus all amino acids (GIBCO, Long Island, NY).

Radioactive amino acids: Amersham (Arlington Heights, IL) usually supplies amino acids in ethanol/water. If the final concentration of ethanol in the labeling medium is too high, the solution of labeled amino acids should be lyophilized and then redissolved in labelling medium. New England Nuclear (Boston, MA) usually supplies amino acids in dilute HCl. These solutions should be neutralized with $NaHCO_3$ (7.5% w/v) using phenol red as an indicator. Correct salt concentrations should be reestablished in both cases by the addition of $10\times$ Earle's balanced salt solution.

Procedure

In general, a total of $5 \times 10^7 - 4 \times 10^8$ cells are used at a concentration of $2 \times 10^6/ml - 2 \times 10^7/ml$. Cells are washed in labeling medium at room temperature and used at the desired density. Usually, since more spleen cells are required than tumor cells, the former are incubated at higher densities. Factors influencing the number of cells to be used are (1) ability to detect the antigen in question, and (2) amount of antibody needed to recover the antigen.

Cells should be incubated in labeling medium with radiolabeled amino acid(s) at a concentration of label not greater than 20% (v/v) of the total culture volume. Optimal incubation times vary between 6 and 12 hr. Eight hours is generally both convenient and efficient. Incubation is carried out at 37° in a humidified CO_2 incubator (5% CO_2, 95% air). At the end of the labeling period the cells are harvested by centrifugation. Cells and medium are sampled for radioac-

[15] R. Nairn, S. G. Nathenson, and J. E. Coligan, *Eur. J. Immunol.* **10**, 495 (1980).
[16] J. E. Coligan and T. J. Kindt, *J. Immunol. Methods* **47**, 1 (1981).
[17] J. E. Coligan, F. E. Gates, III, E. S. Kimball, and W. L. Maloy, this series, Vol. 91, p. 413.

tivity measurement to determine percentage incorporation of label. The radiolabeled cells may then be treated immediately with NP-40 or the cell pellet frozen.

Usually ³H-labeled amino acids are used for structural studies. They are cheaper and, in general, of higher specific activity. ^{14}C- and/or ^{35}S-labeled amino acids may be used for special purposes. For example, ^{14}C with ³H in a double-label comparison, [^{35}S]methionine to locate overlap tryptic peptides for BrCN fragments, and [^{35}S]cysteine to locate Cys residues.

A change in procedure that improved the percentage incorporation of radiolabel over that obtainable with the methods used previously was the use of a labeling medium lacking all amino acids (R. Nairn and S. G. Nathenson, unpublished). However, some amino acids were found to be incorporated more efficiently than others. For example, Tyr and Arg often show >80% uptake, whereas Glu and Asp usually show around 2% uptake. Gates et al. (see Coligan et al.[17]) have developed the use of the transaminase inhibitor, aminooxyacetic acid, to help increase incorporation of amino acids such as Asp by as much as 10-fold. In some situations it is possible to increase incorporation by stimulating the cells with mitogens such as LPS.[18] Alternatively, one might try to isolate cells expressing increased amounts of the antigen of interest. Such cells were very useful in studies of the HLA Class I antigens.[19] With respect to H-2 Class I antigens, the C14 (H-2^d) cell line expresses approximately 10-fold greater amounts of H-2D^d antigen than that found on H-2^d spleen cells.[15]

Coligan et al.[16,17] have discussed at some length the appropriate mixtures of amino acids to be used in sequence analysis of radiolabeled peptides. Of the greatest importance is that the members of a multilabel group should be chosen in such a way that they are incorporated with similar specific activities, thus avoiding problems of residue identification caused by having a residue of high specific activity adjacent to one of low specific activity. Those amino acids which do not incorporate well (Gly, Ala, Asp, Glu) should be used singly. Finally, biosynthetic interconversions are known to occur with Glu, Gln, Ser, Gly, Asn, and Asp. This fact should be taken into account when label groups are chosen.[17]

Antigen Solubilization (This Volume [49])

The first stage in the isolation of the radiolabeled Class I antigens involves their dissociation from the lipid bilayer. A major part of the

[18] D. J. McKean, A. J. Infante, A. Nilson, M. Kimoto, C. G. Fathman, E. Walker, and N. Warner, *J. Exp. Med.* **154,** 1419 (1981).

[19] J. N. McCune, R. E. Humphreys, R. R. Yocum, and J. L. Strominger, *Proc. Natl. Acad. Sci. U.S.A.* **72,** 3206 (1975).

molecule, retaining alloantigenic activity, can be obtained in a water-soluble form by treatment with proteolytic enzymes such as papain.[20] This approach, however, gives poor yields with murine Class I antigens. Detergent solubilization has been found to give greater yields and also produces an intact molecule.

When the material is to be further purified by immunological methods, the detergent Nonidet-P40 (BDH, England—available through Gallard Schlesinger, Carle Place, NY) is used to extract the membrane-bound Class I antigens.[21] The advantage of this detergent over other similar nonionic detergents is that it does not lyse the nuclear membrane and therefore potential DNA contamination is avoided. If the cells are stored frozen before treatment, however, nuclear damage may not be avoided.

Cells are suspended in "lysing buffer" (0.01 M Tris–HCl, pH 7.4, 0.15 M NaCl, 1.5 mM $MgCl_2$, and 10% (v/v) NP-40 added to give a final concentration of 0.5% (v/v)). The suspension is mixed at intervals by vigorous vortexing during incubation (1 hr at 4°). The number of cells/ml should be adjusted depending on cell size. For example, with spleen cells a concentration of 1×10^8 cells/ml is optimal whereas with most tumor cell lines, 5×10^7 cells/ml or even 1×10^7 cells/ml [as in the case of the C14 (H-2^d) cell line] can give more efficient solubilization.

To obtain the operationally defined "soluble" fraction, the suspension is centrifuged to obtain a 10^7 $g \times$ min supernatant. Given the volumes normally involved this can be readily carried out in 10 ml ultracentrifuge tubes (polycarbonate, Oak Ridge Type, from Fisher Scientific, Springfield, NJ) in a Beckman 50 Ti rotor for 100 min at 40,000 rpm and 4°. The supernatant can be stored at −70° before use. Addition of protease inhibitors has not been found to be necessary, at least for the classical transplantation antigen group of Class I antigens.

With a spleen cell preparation, if the Class I antigen is isolated using immunological methods, it is necessary to first remove any radiolabeled Ig from the detergent lysate. This can be accomplished by incubating the lysate with a heterologous antibody against mouse Ig (e.g., goat anti-mouse Ig). About 80 μl of antibody should be added for each 1 ml of NP-40 lysate and the mixture incubated for 30 min at 4°. An immunoprecipitate is formed by adding an "equivalent" amount of normal mouse serum and incubating at 4° for 2 hr. After centrifugation at 4° for 10 min at 1000 g the supernatant is removed and kept frozen at −70° until needed.

As a general rule, greater than 90% of the radioactivity found in the cell pellet should be recovered in the NP-40 soluble fraction.

[20] A. Shimada and S. G. Nathenson, *Biochemistry* **8**, 4048 (1969).
[21] B. D. Schwartz and S. G. Nathenson, *J. Immunol.* **107**, 1363 (1971).

Lectin Affinity Chromatography (*This Volume* [49])

Before immunoprecipitation of a Class I antigen, it is usually desirable to partially purify the NP-40 lysate by affinity chromatography on a column of the lectin from *Lens culinaris* (lentil lectin) coupled to BrCN-activated Sepharose 4B (Pharmacia, Piscataway, NJ).[22,23] "Cleaner" antigen preparations are obtained from lysates subjected to lectin affinity chromatography, particularly if alloantisera are used as immunoprecipitating reagents. Whether this step is essential with monoclonal antibodies remains to be reliably determined. Preliminary observations suggest that it may not be.

Lentil lectin can be prepared by the procedure of Howard *et al.*,[24] as modified by Hayman and Crumpton.[22] The common lentil obtainable from most supermarkets is generally suitable. Some manufacturers' treatments may affect the activity of the lectin (D. Snary and M. J. Crumpton, personal communication), and a number of different brands may have to be tried. Both the large green bean and the smaller orange bean have been used successfully.

Affinity columns prepared with 10 mg lectin coupled to 10 ml Sepharose 4B have proved to be more than adequate for purification, in good yield, of radiolabeled Class I transplantation antigens from up to 10^9 labeled cells.

The column is equilibrated at 4° with 0.25% (v/v) NP-40, 0.01 M Tris–HCl, pH 7.4, 0.15 M NaCl. The radioactive sample is applied to the column which is washed to background radioactivity (usually 5 column volumes) with the equilibration buffer. Elution of the "glycoprotein pool" containing the Class I antigen is accomplished with 0.1 M α-methyl mannoside (Sigma, St. Louis, MO) in the same buffer. Based on the presence of radioactivity determined in a small aliquot of each column fraction, pools of peak fractions are made.

Before applying a second lysate, the column is washed with 10 ml of a solution of 1% (w/v) bovin serum albumin, 1 M α-methyl mannoside, 1.5 mM MgCl$_2$ in 0.25% (v/v) NP-40, 0.15 M NaCl, 0.01 M Tris–HCl, pH 7.4, followed by 50 ml of 0.25% (v/v) NP-40, 0.15 M NaCl in 0.01 M Tris–HCl, pH 7.4. This washing procedure is repeated twice. It is also advisable to keep separate columns for ^{3}H- and ^{14}C-labeled preparations. The lectin columns are very stable at 4° and show no significant alteration in binding capacity over periods of up to 2 years.

[22] M. J. Hayman and M. J. Crumpton, *Biochem. Biophys. Res. Commun.* **47**, 923 (1972).
[23] J. L. Brown and S. G. Nathenson, *J. Immunol.* **118**, 98 (1977).
[24] I. K. Howard, H. J. Sage, M. D. Stein, N. M. Young, M. A. Leon, and D. F. Dykes, *J. Biol. Chem.* **246**, 1590 (1971).

In general 1–5% of the radioactivity applied to the *LcH* column as NP-40 lysate is recovered in the "bound" (i.e., glycoprotein) pool.

Immunoprecipitation/Immunoaffinity Chromatography (This Volume [49])

Indirect immunoprecipitation has been widely used for identification and characterization of Class I antigens.[25] This procedure originally involved the use of alloantisera and heterologous (usually goat) anti-mouse immunoglobulin as the coprecipitating reagent.[21,25] More recently the Cowan I strain of *Staphylococcus aureus* (SAC I) (American Type Culture Collection) has been used as an alternative coprecipitating reagent.[26] The Sepharose-linked protein-A reagent (Pharmacia, Piscataway, NJ), while more expensive, can also be used.

A recent advance has been the use of monoclonal antibodies as immunoprecipitating reagents[27] or, bound to a gel, as immunoaffinity reagents.[9] Not all monoclonal antibodies function well in indirect immunoprecipitation, or as immunoadsorbent reagents, however. Alloantisera generally function well in indirect immunoprecipitation but rarely as immunoadsorbent reagents. There is no general rule to be followed as to the best strategy for utilizing antisera to isolate a particular Class I antigen.

With immunoprecipitation protocols it is essential to ascertain the optimal amount of antiserum required in order to obtain a "clean" immunoprecipitate from the detergent lysate or glycoprotein pool. The radioimmunoassay described by Freed *et al.*[7] is useful in this regard. Even monoclonal antibodies can yield impure immunoprecipitates[27] and careful titrations and analysis by sodium dodecyl sulfate–polyacrylamide gel electrophoresis of the polypeptides recovered can save additional purification steps. Extensive washing of an immunoprecipitate is critical (at least 4 times with 2× the protein pellet volume of a solution of 0.15 M NaCl, 0.1 M Tris–HCl, pH 7.4) before proceeding.

If the antigen is going to be cleaved with cyanogen bromide (BrCN), the use of SAC I as the coprecipitating agent is essentially precluded, since SDS containing buffers are required to efficiently elute the antigen–antibody complexes from SAC I and the consequent oxidation of Met residues during this treatment results in poor yields of BrCN fragments. A second antibody is probably the method of choice as a coprecipitating agent if BrCN fragment preparation is planned after immunoprecipitation.

The radioactivity recovered with the immunoprecipitated antigen is in

---

[25] S. G. Nathenson and S. E. Cullen, *Biochim. Biophys. Acta* **344,** 1 (1974).

[26] S. E. Cullen and B. D. Schwartz, *J. Immunol.* **117,** 136 (1976).

[27] W. L. Maloy, G. Hämmerling, S. G. Nathenson, and J. E. Coligan, *J. Immunol. Methods* **37,** 287 (1980).

the range of 1–15% of the radioactivity present in the glycoprotein (i.e., "bound") pool.

Polyacrylamide Gel Electrophoresis in Sodium Dodecyl Sulfate (SDS–PAGE)

A widely used method for analyzing Class I H-2 antigens, and separating the subunit chains for further structural analysis, is SDS–PAGE. Most studies use the discontinuous buffer system of Laemmli.[28,29] SDS–PAGE is an essential part of establishing whether or not determinants are present on the same or different molecules. The method used is known as sequential immunoprecipitation and has been used most recently, with Class I antigens, to demonstrate the existence of at least three D region encoded molecules: D, L, and R,[30] to demonstrate that TL.4 and TL.1,2 determinants are on the same molecule[31] and that two Qa-1 molecules are expressed in the d haplotype.[14]

The principle of sequential immunoprecipitation is to take a soluble preparation of the radiolabeled antigen(s) and treat it with an excess of anti-"A" antiserum (where A is any antigenic specificity). Immune complexes and excess anti-"A" are then precipitated with anti-Ig or SAC I. The supernatant resulting from this treatment is tested for the presence of a second specificity (B) by the same procedure with anti-"B" antiserum. If A and B are on the same molecule, SDS–PAGE analysis of the two precipitates will show radioactive peaks only in the first anti-"A" precipitation. If A and B are on separate molecules, radioactive peaks will be present on SDS–PAGE of both the anti-"A" and anti-"B" immune precipitations.

SDS–PAGE has also been used preparatively to obtain material for tryptic peptide mapping and partial amino acid sequence analysis of H-2 Class I antigens.[32–34] Gel slices containing the radioactive material are eluted with SDS-containing buffer and the material is then analyzed further by the techniques described below.

[28] U. K. Laemmli, *Nature (London)* **227**, 680 (1970).

[29] U. K. Laemmli and M. Favre, *J. Mol. Biol.* **80**, 575 (1973).

[30] T. H. Hansen, K. Ozato, M. R. Melino, J. E. Coligan, T. J. Kindt, J. J. Jandinski, and D. H. Sachs, *J. Immunol.* **126**, 1713 (1981).

[31] K. Yokoyama, E. Stockert, L. J. Old, and S. G. Nathenson, *Proc. Natl. Acad. Sci. U.S.A.* **78**, 7078 (1981).

[32] R. M. Maizels, J. A. Frelinger, and L. Hood, *Immunogenetics* **7**, 425 (1978).

[33] L. Lafay, S. J. Ewald, M. McMillan, J. A. Frelinger, and L. Hood, *Immunogenetics* **12**, 21 (1981).

[34] K. R. McIntrye, U. Hämmerling, J. W. Uhr, and E. S. Vitetta, *J. Immunol.* **128**, 1712 (1982).

Amino Acid Sequence Analysis

A detailed discussion of the strategy of radiochemical sequence determination and the methods required for its performance has been given elsewhere.[16,17] The first step in the amino acid sequence analysis of Class I antigens is the separation of the two chains composing the molecule. The fact that there is no need to remove contaminating antibody, or other unlabeled protein present is an advantage of the radiochemical approach. The presence of unlabeled protein also minimizes losses of the minute amounts of radiolabeled material present.

Separation of β₂-Microglobulin

In structural studies of the H-2 antigens, chain separation has been achieved by SDS–PAGE or gel filtration. SDS–PAGE has been discussed in this volume [44]. In general, antigens prepared by this method have been sequenced only for the NH_2-terminal 30 residues (see, for example, Maizels *et al.*[32]). In order to serparate the two chains, immune precipitates can also be dissolved in 2 *M* HCOOH, and chromatographed on a Sephadex G-75 column (1.5 × 75 cm or 2.5 × 100 cm) equilibrated with 1 *M* HCOOH.[15,35] Under these conditions, most of the 45,000 MW material appears in the excluded volume of the column, while the β₂-microglobulin is eluted at approximately the same volume as horse heart cytochrome *c*. The larger column (2.5 × 100 cm) is recommended if an alloserum immunoprecipitate containing a substantial quantity of antibody must be fractioned. These columns can be run relatively quickly (2.5 ml/hr). Aliquots (0.5–5%) of the fractions are taken for determination of radioactivity and pools of peak fractions made.

Papain Digestion and Separation of β₂M

Sometimes there is an advantage in treating the immunoprecipitate with papain and preparing the water-soluble papain fragment (MW ~ 37,000). For example, this procedure reduced contamination of the immunoprecipitated H-2D^d molecule by unknown higher molecular weight proteins.[15]

The optimal amount of papain (Sigma Chemical Co., St. Louis, MO., suspension of 2× crystallized, 25 mg/ml, 30 U/mg), determined to be 40 µl papain for every 1600 µl of goat anti-mouse IgG antiserum, is preactivated by incubation at 37° for 5 min in digestion buffer (0.15 *M* NaCl, 0.054 *M* 2-mercaptoethanol, 0.01 *M* Tris–HCl, pH 7.4). This solution is then added

[35] B. M. Ewenstein, T. Nisizawa, H. Uehara, S. G. Nathenson, J. E. Coligan, and T. J. Kindt, *Proc. Natl. Acad. Sci. U.S.A.* **75**, 2909 (1978).

directly to the immune precipitate and digestion continued for 15 min at 37°. The resulting suspension is then chromatographed on a Sephadex G-75 column as described above. The papain digestion product (MW ~37,000) elutes just after the excluded volume. Sometimes a second cleavage takes place and the products of this reaction can also be separated on a Sephadex G-75 column (2 × 190 cm) equilibrated with 1 M HCOOH.[36]

BrCN Fragment Preparation and Separation

Met residues are highly conserved in H-2 Class I antigens.[4,5] This has led to a general strategy for radiochemical sequence analysis of these molecules based on initial cleavage at Met by CNBr. Ewenstein et al.[35] first applied cleavage with CNBr to the H-2 Class I antigen H-2K^b using the method of Gross.[37]

Preparation of BrCN Fragments. The lyophilized, radiolabeled preparation of Class I antigen containing carrier bovine serum albumin (Sigma, Fraction V) is dissolved in 70% HCOOH at 10 mg/ml and treated with a 25-fold excess (w/v) of BrCN (Eastman Chemical Co., Rochester, NY) for 1 hr at room temperature and then for 23 hr at 4°.[35] The reaction mixture is then diluted with a 10-fold (v/v) excess of distilled water and lyophilized. All these steps must be carried out in a well ventilated fume hood.

Separation of BrCN Fragments. The fragments thus produced are dissolved in 2–3 ml of 8 M guanidine–HCl (Ultrapure, Schwarz/Mann, Spring Valley, NY) by incubation at 37° for a few hours. Distilled water is then added to give a final concentration of 6 M guanidine–HCl and carrier cytochrome c is added as a molecular weight marker. Separation is achieved on a Sephacryl S-200 column (2 × 190 cm) in the presence of 6 M guanidine–HCl with a flow rate of 3 ml/hr and a fraction size of 1.3 ml. Guanidine–HCl is not compatible with all scintillation fluids. High efficiency counting can be achieved in Biofluor (New England Nuclear, Boston, MA), for example. BrCN fragments are recovered by lyophilization after desalting on a Sephadex G-15 column (2.5 × 40 cm) in 2 M HCOOH.

Reduction and Carboxamidomethylation of Disulfide-Linked Fragments. Disulfide bonds can be reduced by dissolving BrCN fragments (together with 2–5 mg horse heart cytochrome c as carrier protein) in 3 ml of 6 M guanidine–HCl, 0.3 M Tris–HCl, 1 mM EDTA, 0.1 M dithiothreitol under N$_2$ for 3 hr at 37°. Cys residues are alkylated in the

---

[36] J. E. Coligan, T. J. Kindt, R. Nairn, S. G. Nathenson, D. H. Sachs, and T. H. Hansen, *Proc. Natl. Acad. Sci. U.S.A.* **77,** 1134 (1980).

[37] E. Gross, Vol. 11, p. 238.

dark at 20° by adding solid iodoacetamide to a final concentration of 0.25 M and adjusting the pH to 8.5. After 20 min the reaction is halted by adding 2-mercaptoethanol to a final concentration of 0.75 M.[35] The mixture of reduced BrCN fragments is fractionated on a Sephacryl S-200 column in the presence of 6 M guanidine–HCl, as for the unreduced fragments.

Usually the BrCN fragments of K, D, and L molecules obtained by the procedures detailed above can be shown to be pure by NH2-terminal sequence determination.[15,35,36] Occasionally some additional steps are required.[15] For the first molecule to be characterized in this way (H-2K^b) Ewenstein et al.[35] aligned the BrCN fragments by means of NH2-terminal sequence analysis, identification of partial BrCN cleavage products, and localization of the site of papain cleavage. For subsequent H-2 Class I molecules encoded by different loci the data from H-2K^b were used to align fragments based on sequence homology.[15,36] The presence of certain radioactive compounds in the BrCN fragments can also provide useful structural insights and help in characterizing the BrCN fragments. For example, incorporation of [^{35}S]Cys suggests which fragments may be disulfide bonded. Incorporation of radioactive sugars (e.g., [^{3}H]fucose) helps to identify the fragments containing carbohydrate side-chains. The larger BrCN fragments must be digested with other enzymes in order to produce fragments of a size suitable for complete sequencing. Digestion with thrombin, trypsin, V8 protease, and chymotrypsin has been utilized to study the H-2 Class I antigens. Purification of the resultant peptides is achieved by gel filtration, ion-exchange chromatography, and reverse-phase HPLC.[38–45]

Sequencing of Radiolabeled Peptides

Extensive details of these methods have already been given, along with the criteria for residue assignments and some discussion of the advantages and disadvantages of radiochemical sequencing, in a companion

[38] H. Uehara, B. M. Ewenstein, J. M. Martinko, S. G. Nathenson, J. E. Coligan, and T. J. Kindt, Biochemistry 19, 306 (1980).
[39] H. Uehara, B. M. Ewenstein, J. M. Martinko, S. G. Nathenson, T. J. Kindt, and J. E. Coligan, Biochemistry 19, 6182 (1980).
[40] J. M. Martinko, H. Uehara, B. M. Ewenstein, T. J. Kindt, J. E. Coligan, and S. G. Nathenson, Biochemistry 19, 6188 (1980).
[41] H. Uehara, J. E. Coligan, and S. G. Nathenson, Biochemistry 20, 5936 (1981).
[42] H. Uehara, J. E. Coligan, and S. G. Nathenson, Biochemistry 20, 5940 (1981).
[43] R. Nairn, S. G. Nathenson, and J. E. Coligan, Biochemistry 20, 4739 (1981).
[44] E. S. Kimball, S. G. Nathenson, and J. E. Coligan, Biochemistry 20, 3301 (1981).
[45] W. L. Maloy, S. G. Nathenson, and J. E. Coligan, J. Biol. Chem. 256, 2863 (1981).

volume of this series.[17] Essentially, the only differences in the method of sequencing radiolabeled peptides, as compared to unlabeled peptides, reside in the analysis of the resulting PTH amino acid derivatives.

Sequencing of radiolabeled peptides is usually performed in the presence of unlabeled "carrier" protein, typically bovine serum albumin or cytochrome *c*. Polybrene (Aldrich Chemical Co., Milwaukee, WI) and glycylglycine may be added if the peptide to be sequenced is less than 70 residues. The automated sequencer (Beckman 890C) with cold trap modification and the Quadrol program has been used for most of the sequencing of radiolabeled H-2 Class I antigen peptides.[17]

For each sequence step about 25% of the butyl chloride extract is taken for determination of radioactivity. Those fractions showing radioactivity are used for the preparation of PTH derivatives and then mixed with PTH-amino acid standards. The PTH amino acids are separated by HPLC using the procedure of Gates *et al.*[46] and then the radioactivity determined.

Structural Analysis by Peptide Mapping

Peptide-mapping techniques, utilizing radiolabeled molecules, have been extensively used in structural comparisons of the products of different H-2 loci,[31,33,34,47] of allelic products,[31,33,47] and of standard ("wild-type") and mutant molecules.[48] This technique, first applied to the *H-2* system by Brown *et al.*,[47] has been very valuable and informative. Thus, whole-molecule comparisons have been useful in the study of the structural relatedness of molecules encoded by different *H-2* loci. Peptide mapping of fragments of a molecule has in addition allowed the assignment of structural difference(s) to a particular segment of the molecule. Using this technique preparatively, peptides can be isolated for amino acid sequence determination. This procedure has also been used to establish the nature of the sequence difference(s) between various standard and mutant H-2 molecules.

The Class I antigens to be compared are prepared by isolating one antigen labeled with ^{14}C-labeled amino acid and the other antigen labeled with ^{3}H-labeled amino acid. Double-label comparisons allow more reliable conclusions than comparisons of data from separate single-label chromatographic analyses. It is better to combine the two antigens (in appropriate amounts, such that one isotope is not in great excess over the

[46] F. E. Gates, III, J. E. Coligan, and T. J. Kindt, *Biochemistry*, **8,** 2267 (1979).
[47] J. L. Brown, K. Kato, J. Silver, and S. G. Nathenson, *Biochemistry* **13,** 3174 (1974).
[48] R. Nairn, K. Yamaga, and S. G. Nathenson, *Annu. Rev. Genet.* **14,** 241 (1980).

other) as early as possible, usually at the stage of immune precipitation. This minimizes the possibility of artifactual differences. After the Class I antigen material (whole-molecule or fragment) is purified, it is digested with trypsin. The tryptic peptides are resolved by ion-exchange chromatography in pyridine-acetate buffers and/or by reverse-phase HPLC. With intact molecules the use of radiolabeled Arg or Lys theoretically allows the identification of all the tryptic peptides except for the COOH-terminal peptide. In practice, however, insoluble peptides and "low-yield" peptides are not identified or are difficult to identify. Moreover, there may be other peptides that will not be visualized. For example, since the specificity of trypsin normally limits its action to peptide bonds associated with the carboxyl groups of lysine and arginine, the COOH-terminal homoserine lactone-containing peptide of a BrCN fragment will not, in general, be labeled by either Arg or Lys. To control for this possibility, a separate double-labeled [^{35}S]-, and [^{3}H]Met whole-molecule tryptic map comparison must be made. In addition, inappropriate trypsin cleavage at Tyr residues has been observed with the Class I antigens.[49] This type of cleavage releases non-Arg and non-Lys-containing peptides which can be visualized by utilizing Tyr-labeled material. Other limitations of the peptide-mapping technique can be overcome by not depending exclusively on one kind of chromatographic analysis to compare two antigens. For example, some sequence differences might occur which do not affect the charge of a peptide and these would not be detected by ion-exchange peptide mapping. They might be detected by reverse-phase HPLC, however. Despite these difficulties and limitations, peptide mapping remains a very valuable method of comparing two antigens. Often, a complete comparison can be accomplished.

Preparation and Comparative Analysis of Tryptic Peptides from Class I Antigens

For whole-molecule peptide mapping comparisons, the 45,000 MW chain of the Class I molecules has been purified by elution from gel slices after SDS–PAGE[33,34] or by chromatography in SDS on columns of BioGel A (Bio-Rad Labs, Richmond, CA).[47] Provided that the higher resolving capacity of SDS–PAGE is not a requirement (e.g., to separate the 45,000 MW chain from contaminating proteins of similar molecular weight in the immunoprecipitate), to carry out one column chromatographic separation is more convenient than to run a large number of polyacrylamide gels.

[49] B. M. Ewenstein, H. Uehara, T. Nisizawa, R. W. Melvold, H. I. Kohn, and S. G. Nathenson, *Immunogenetics* **11**, 383 (1980).

SDS/BioGel Purification

1. The washed immunoprecipitate is dissolved in 2 ml of 9% SDS in 0.6 M Tris–HCl, pH 8.5. This may require warming to 37° for a few minutes and may be easier with a frozen and thawed sample.

2. The SDS-treated material is then reduced and carboxamidomethylated by first adding dithiothreitol (Calbiochem, San Diego, CA) to a final concentration of 0.11 M (34 mg). The suspension is transferred to a boiling water-bath and incubated for 5 min to dissolve the proteins.

3. After saturation with nitrogen the solution is incubated at 37° for 30 min in a capped tube. Then iodoacetamide (Eastman Chem. Co., NY) is added to a final concentration of 0.25 M (92.5 mg) and the pH adjusted to 7.8–8.0 by addition of 1 M NaOH.

4. Following incubation at room temperature for 30 min *in the dark* the reaction is stopped by addition of 60 μl of 2-mercaptoethanol.

5. The sample is loaded onto a column (2.5 × 110 cm) of BioGel A-0.5M (200–400 mesh) equilibrated immediately before use with buffer containing 0.5% (w/v) SDS, 0.002 M dithiothreitol, in 0.05 M Tris–HCl, pH 7.5. Flow rate is 6 ml/hr with a fraction size of 1.5 ml. Fractions containing the 45,000 MW H-2 molecule are pooled after determining the radioactivity in a small aliquot of each fraction. Dithiothreitol should be replenished in the column reservoir daily.

6. The material is recovered from the fractions by precipitation with CCl_3COOH. The protein content of the pooled fractions is determined from the OD at 280 nm. If necessary, human IgG (Fraction II) is added to give 12 mg total protein. Then 18 μl CCl_3COOH is added for every 100 μl of column effluent (i.e., a final CCl_3COOH concentration of 15%) and the sample is incubated at 4° overnight. The resulting precipitate is washed with 5 ml of each of the following: (1) 15% CCl_3COOH, (2) a 1 : 1 mixture of ethanol and ether, and (3) anhydrous ether. The dried ether precipitate is warmed at 37° for a few minutes to remove the ether and then frozen, or used directly for digestion with trypsin.

Trypsin Digestion. A stock solution of TPCK-trypsin (Worthington, Millipore Corp., Bedford, MA) in 0.001 M HCl is prepared. The radiolabeled antigen preparation isolated, as above, in the presence of reduced and alkylated IgG as carrier protein, or in the presence of reduced and alkylated horse heart cytochrome c as carrier protein in the case of a BrCN fragment,[49] is suspended in 0.1 M ammonium bicarbonate, pH 8.55 to give a protein concentration of 10 mg/ml. TPCK-trypsin is added to give a protein : trypsin ratio of 20 : 1. After incubation at 37° for 1 hr, half as much trypsin as the first time is added and the incubation continued for 3 more hours. The reaction is stopped and the tryptic peptides are dissolved in the appropriate chromatography buffer. For example, for cation-exchange chromatography, the reaction is stopped by lowering the pH to 2.0 with acetic acid and the insoluble "core" peptides removed by centrifugation. The acid-soluble peptides are applied onto the column in 0.05 M pyridine–acetic acid, pH 3.1.

Chromatography of Tryptic Peptides. 1. *High-performance liquid chromatography with reverse-phase columns.* Initially, most HPLC peptide mapping used modifications of the method of Hancock *et al.*,[50] which employes a fatty acid analysis column developed with a gradient of acetonitrile in the presence of phosphoric acid. Thus, Rose *et al.*[51] used this system to compare H-2L molecules from different haplotypes and found peptide differences not detectable by ion-exchange peptide mapping. Lafay *et al.*[33] utilized a Zorbax O.D.S. column eluted with an acetone gradient in the present of phosphoric acid to examine allelic variation in K, and D locus gene products. Perhaps the trifluoroacetic acid based solvent systems described by Mahoney and colleagues[52–54] offer the best possibility for a universally applicable HPLC peptide-mapping system. HPLC methods, however, require rather expensive equipment. Fortunately, HPLC peptide-mapping is not an indispensable tool. Considerable success has been achieved using cation-exchange chromatography.

2. *Ion-exchange chromatography.* The methods used for analysis of Class I antigens are based on that first described by Brown *et al.*[47] The analyses carried out in Dr. S. G. Nathenson's laboratory have used the cation-exchange resin, Spherix type XX8-60-0 (Phoenix Co., Long Island, NY), described by Brown *et al.*[47] This resin has some useful hydrophobic properties in addition to its ion-exchange properties. Chromobeads Type P (Technicon, Tarrytown, NY) has also been used[33,34,55–57] and gives peptide patterns very similar to those obtained with Spherix. Unfortunately, these resins are no longer manufactured, although since they were in common use in amino acid analyzers it may be possible to obtain them from colleagues. Another possible substitute is Dionex DC-6A (Dionex Corp., Sunnyvale, CA), a sulfonated divinylbenzene-polystyrene (8% cross-linked) cation exchange resin which has been used in Dr. J. D.

[50] W. S. Hancock, C. A. Bishop, R. L. Prestridge, and M. T. W. Hearn, *Anal. Biochem.* **89,** 203 (1978).
[51] S. R. Rose, T. H. Hansen, and S. E. Cullen, *J. Immunol.* **125,** 2044 (1980).
[52] W. C. Mahoney and M. A. Hermodson, *J. Biol. Chem.* **255,** 11199 (1980).
[53] J. D. Pearson, W. C. Mahoney, M. A. Hermodson, and F. E. Regnier, *J. Chromatogr.* **207,** 325 (1981).
[54] W. C. Mahoney, *Biochim. Biophys. Acta* **704,** 284 (1982).
[55] D. W. Sears and C. M. Polizzi, *Immunogenetics* **10,** 67 (1980).
[56] B. Arden, E. K. Wakeland, and J. Klein, *J. Immunol.* **125,** 2424 (1980).
[57] S. J. Ewald, J. Klein, and L. Hood, *Immunogenetics* **8,** 551 (1979).

Freed's laboratory for peptide-mapping of murine Class II antigens (personal communication). These resins are packed in Altex medium-pressure (1000 psi) water-jacketed glass columns (0.9 × 25 cm, from Rainin Instruments, Woburn, MA). The chromatography is carried out at 50°. The elution buffers are pumped at 45 ml/hr with a high-pressure metering pump [e.g., Beckman "Accu-Flo" (Beckman, Berkeley, CA) or Milton-Roy Mini-Pump (Lab Data Control, Division of Milton Roy Co., Riviera Beach, FL]. The elution buffer gradient is formed in a gradient mixing chamber (a chamber similar to the now no longer manufactured Virtis Varigrad is obtainable from MRA Corp., Clearwater, FL). Of each of the following pyridine-acetic acid buffers 120 ml is used:

1. 0.05 M, pH 3.13, 4 ml pyridine, 80 ml glacial acetic acid, H_2O to 1000 ml
2. 0.10 M, pH 3.54, 4 ml pyridine, 32 ml glacial acetic acid, H_2O to 500 ml
3. 0.20 M, pH 4.02, 8 ml pyridine, 25 ml glacial acetic acid, H_2O to 500 ml
4. 0.50 M, pH 4.50, 20 ml pyridine, 26 ml glacial acetic acid, H_2O to 500 ml
5. 2.00 M, pH 5.05, 158 ml pyridine, 100 ml glacial acetic acid, H_2O to 1000 ml.

The tryptic digest is applied and washed into the column in Buffer 1. Five fractions of 60 drops each are collected. Then the gradient mixer is started and a total of 200 fractions of 60 drops each are collected. An LKB model 7000 fraction collector is particularly suitable for this use as the pump can be plugged into the "Live During Operation" outlet of the collector. This turns the pump off at the end of the run. Fractions are collected in scintillation vials, and the pH of every fifth fraction measured. For this purpose it is recommended to use the expanded scale of a pH meter such as the pH Meter 26, Radiometer, Copenhagen. After evaporating the fractions to dryness in a vented oven at 100°, 0.2 ml of water is added to each vial to redissolve the tryptic peptides, and liquid scintillation fluid added. These ion-exchange resins should be washed at the end of a run with 8 M pyridine buffer (316 ml pyridine, 5 ml glacial acetic acid, H_2O to 500 ml) and with 0.5 M NaOH to elute off any residual peptides. The column is then regenerated by extensive washing with Buffer 1.

Concluding Remarks

Application of the methods and approaches described above has led over the past 5 or 6 years to a complete characterization of certain of the

H-2 Class I antigens. This has been a remarkable achievement given the fact that a relatively short time before these methods began to be applied, the H-2 antigenic structure was a rather nebulous entity, thought at one time or another to be carbohydrate, nucleic acid, lipid, or protein.[58] Today we are in a period of greater and even faster advances as the tools of molecular genetics are applied to the *H-2* system.[2,3]

The impetus for biochemical studies of the Class I antigens came from a need to determine the overall molecular characteristics of these molecules and then to have a structural basis for understanding the complex serologic and genetic data that already existed. These goals have largely been realized for the transplantation antigens although additional complexity undoutedly remains to be discovered.

Based on protein data for 6 H-2 antigens and data deduced from cDNA and genomic clones,[4,5] the molecular characteristics of the Class I transplantation antigens are as follows. The average molecular weight of the *MHC* encoded chain is 45,000 but there is some size heterogeneity, due to different chain length, and different extent of glycosylation. Thus, for example, the K^d and D^b antigens have three carbohydrate side chains as opposed to the "usual" two; the K^b antigen is 10 residues longer than D^b or L^d. Structural features of the K, D, L molecules include a hydrophobic membrane-spanning segment (residues 284–304), and a disulfide-looped domain structure. Two disulfide loops, between residues 101–164 and 203–259, respectively, connect three domains situated outside of the cell membrane, each of approximately 90 amino acids. The predicted protein domains correlate very well with the known exon structure of the Class I genes.[3]

An important feature of the *H-2* system is its extensive serologically detected polymorphism.[58] Peptide mapping and amino acid sequence studies have led to the determination of regions of diversity and similarity between allelic products of the same loci and between products of different loci.[4,5] Amino acid sequence differences between H-2 Class I molecules tend to occur in clusters. A region of major diversity is between residues 61 and 83. Others are 95–99, 114–116, and 152–157. Amino acid residues in the second disulfide loop domain are highly conserved among K, D, L molecules (homology is around 95%). This domain which shares extensive sequence homology with immunoglobulins[5] may be the site of binding of $\beta_2 M$.[59]

In contrast to the predictions from data obtained in peptide mapping and NH$_2$-terminal sequence studies, comparisons of extensive sequence

[58] J. Klein, "Biology of the Major Histocompatibility Complex." Springer-Verlag, Berlin and New York, 1975.

[59] K. Yokoyama and S. G. Nathenson, *J. Immunol.* **130**, 1419 (1983).

data reveal possible amino acid sequences unique to particular allelic gene products.[5] However, the sequence differences observed between "allelic products" are much greater than would be expected for alleles.[4,5]

Peptide mapping and amino acid sequencing using the radiochemical methods outlined in this article have been especially fruitful in the study of the H-2 Class I gene products from mutant strains of mice.[48,60-62] In the case of the *H-2* mutant studies, application of the above methods allowed determination of the exact position of the mutations and, in many cases, the nature of the amino acid interchange. These biochemical studies have been instrumental in providing the ultimate proof for the placement of the mutations in the *H-2K* or *H-2D* loci. Furthermore, they helped to demonstrate that T cells are capable of recognizing very subtle (1 or 2 amino acids) changes in H-2 Class I antigen structure.

Looking to the future it seems probable that methods like those described above for determining the protein structure of H-2 Class I antigens will be most useful at the early stages of the investigation of a new cell-surface antigen. One might guess that they will never be used to their fullest extent again in the study of the *MHC* as progress can now be made more rapidly in this area by using recombinant DNA technology. Perhaps the greatest value of these methods in the future will be to complement DNA sequence studies of *MHC* genes, and to investigate less well defined antigen systems where identification, preliminary molecular characterization, and structure comparison are required before attempts are made to clone the genes and determined the complete amino acid sequence from the nucleotide sequence.

Acknowledgments

Along with those cited in the reference list, I would specifically like to acknowledge the contributions of the following people to the development of the methods I have outlined here: Lynne Brown, John Coligan, Bruce Ewenstein, John Freed, Frederick Gates, Ed Kimball, Tom Kindt, Lee Maloy, John Martinko, Stan Nathenson, Tosiki Nisizawa, Duane Sears, Hiroshi Uehara, and Karen Yamaga. It is a pleasure to acknowledge the expert secretarial and word processing capabilities of Morag Nairn who prepared this manuscript. Financial support of R. N. by NIH Grant AI 18556 is gratefully acknowledged.

[60] L. R. Pease, B. M. Ewenstein, D. McGovern, R. W. Melvold, T. Nisizawa, and S. G. Nathenson, *Immunogenetics* **17**, 7 (1983).
[61] K. M. Yamaga, G. M. Pfaffenbach, L. R. Pease, D. McGovern, T. Nisizawa, R. W. Melvoid, H. I. Kohn, and S. G. Nathenson, *Immunogenetics* **17**, 19 (1983).
[62] K. M. Yamaga, G. M. Pfaffenbach, L. R. Pease, D. McGovern, T. Nisizawa, R. W. Melvoid, H. I. Kohn, and S. G. Nathenson, *Immunogenetics* **17**, 31 (1983).

[49] Isolation and Analysis of the Murine Ia Molecular Complex

By SUSAN E. CULLEN

The Ia Molecular Complex

The murine Ia molecules are recognized to be participants in communication between cells of the immune system, and are considered to be the "immune response (Ir) gene products" because structural alteration in Ia molecules or binding of anti-Ia monoclonal antibodies alters immune response phenotype.[1-3] When structural studies on the murine Ia (Class II major histocompatibility complex molecules) began about 10 years ago,[4,5] the extent of their molecular heterogeneity was not appreciated, and it remains incompletely described to the present date. Since interest in examining Ia molecular structure continues, this chapter will describe the Ia molecular complex, and the means currently used to separate its components for further study.

The Ia molecular complex is composed of genetically polymorphic and nonpolymorphic components. It is generally agreed that the polymorphic components, the α and β chains, are those directly responsible for Ir gene variability and attendant phenotypic traits including capacity to stimulate mixed lymphocyte alloreactivity (MLR). The β chain appears to vary more extensively than the α. The functional role of the major nonpolymorphic component, I_i, and other specifically associated cellular compo-

[1] E. A. Lerner, L. A. Matis, C. A. Janeway, P. P. Jones, R. H. Schwartz, and D. B. Murphy, *J. Exp. Med.* **152,** 1085 (1980).

[2] C. N. Baxevanis, D. Wernet, Z. A. Nagy, P. H. Maurer, and J. Klein, *Immunogenetics* **11,** 617 (1980).

[3] C. S. Lin, A. S. Rosenthal, H. C. Passmore, and T. H. Hansen. *Proc. Natl. Acad. Sci. U.S.A.* **78,** 6406 (1981).

[4] S. E. Cullen, C. S. David, D. C. Shreffler, and S. G. Nathenson, *Proc. Natl. Acad. Sci. U.S.A.* **71,** 648 (1974).

[5] E. S. Vitetta, J. Klein, and J. W. Uhr, *Immunogenetics* **1,** 82 (1974).

METHODS IN ENZYMOLOGY, VOL. 108

nents is unclear at this time, but these components may have an effect either on the display of α–β at the cell surface, on the capacity of α–β to interact with its receptors on the surfaces of other cells (e.g., helper T cells), or on the transduction of the signal presumably generated by the interaction of α–β with its receptor. Thus, the study of these nonpolymorphic components may increase our understanding of the cellular machinery required to support response to a specific signal.

Polymorphic Components

There are two readily isolated and genetically distinct types of mouse Ia heterodimers, which are currently called I-A and I-E. The α and β glycopeptide components of these molecules are encoded within the *I* region of the major histocompatibility complex (MHC) (Fig. 1). The I-A molecule includes the A_β and A_α chains which are encoded entirely within the *I-A* subregion.[6,7] The I-E molecule includes the E_β and E_α chains which are encoded by genes which are separated by a site of high frequency recombination, in or just telomeric to the segment of DNA which encodes the E_β chain.[8] The E_β chain is at least partly encoded within the confines of what is genetically definable as the *I-A* subregion,[9,10] and for that reason the E_β gene and its product are sometimes called A_e. The E_α chain gene is located telomeric to the high frequency recombination site, in the *I-E* subregion. Peculiarly, there are several laboratory mouse strains with apparently adequate immune response repertoires in which I-E molecules are not produced, and absence of these molecules has been attributed either to partial deletion of the E_α gene or to an mRNA defect.[11] In strains which are I-E negative, the E_β chain is sometimes translated in the absence of E_α chains[12] and in an F_1 hybrid, E_β may be assembled into a fully normal Ia molecule if there is a normally expressed E_α gene in the trans position.[9] The formation of "hybrid" Ia molecules in F_1 animals has been documented for both I-A[13] and I-E[14,15] molecules, so that in a (Y $\times$

---

[6] R. G. Cook, J. D. Capra, J. L. Bednarcyzk, J. W. Uhr, and E. S. Vitetta, *J. Immunol.* **123**, 2799 (1979).

[7] J. Silver and W. A. Russell, *Immunogenetics* **8**, 339 (1979).

[8] L. Hood, M. Steinmetz, and B. Malissen, *Annu. Rev. Immunol.* **1**, 529 (1983).

[9] P. P. Jones, D. B. Murphy, and H. O. McDevitt, *J. Exp. Med.* **148**, 925 (1978).

[10] R. G. Cook, E. S. Vitetta, J. W. Uhr, and J. D. Capra, *J. Exp. Med.* **149**, 981 (1979).

[11] D. J. Mathis, C. Benoist, V. E. Williams, M. Kanter, and H. O. McDevitt, *Proc. Natl. Acad. Sci. U.S.A.* **80**, 273 (1983).

[12] P. P. Jones, D. B. Murphy, and H. O. McDevitt, *Immunogenetics* **12**, 321 (1981).

[13] J. Silver, S. Swain, and J. J. Hubert, *Nature (London)* **286**, 272 (1980).

[14] R. G. Cook, E. S. Vitetta, J. W. Uhr, and J. D. Capra, *J. Immunol.* **124**, 1594 (1980).

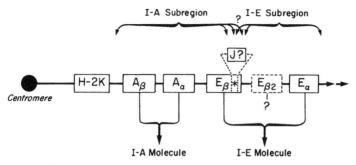

FIG. 1. Simplified map of the *I* region based on molecular genetic data (not drawn to scale). The order of genes represented is based on analysis of overlapping DNA segments isolated in cosmids. [From M. Steinmetz, K. Minard, S. Horvath, J. McNicholas, J. Frelinger, C. Wake, E. Long, B. Mach, and L. Hood, *Nature (London)* **300**, 35 (1982).] The *I-A* subregion includes the A_β, A_α and at least the 5′ end of E_β. In agreement with protein structural data, some of the E_β coding sequence is located in the *I-A* subregion, but the exact placement of the high frequency recombination site (*) relative to the E_β coding sequences is not known. The segment marked "J" is the small stretch of DNA which appears to be the candidate *I-J* segment if all the assumptions made are valid. (From Hood *et al.*[8]) The placement of the boundary between *I-A* and *I-E*, or between *I-A*, *I-J*, and *I-E* is indicated as ambiguous. The *I-B* subregion is not marked on this map. Its existence has been questioned, but the issue may be mainly semantic, since the "I-B mapped" Ir gene phenomena could result from α–β gene product interactions or from intragenic recombination. Initially, the gene order A_α, A_β was hypothesized because the I-A α chain in the *tl* haplotype was altered in a way that suggested an A^α intragenic recombination centromeric to A_β. The order A_β, A_α which was established by structural analysis of the DNA suggests that the event that occurred in *tl* was not a simple recombination, but perhaps a mutation, a double cross over, or a gene conversion. It is not known whether $E_{\beta2}$ is expressed or is a pseudogene.

Z)F_1 hybrid one may find molecules of the $A_\alpha^y A_\beta^z$, $A_\alpha^z A_\beta^y$, $E_\alpha^y E_\beta^z$, and $E_\alpha^z E_\beta^y$ types, as well as the expected "parental" type cis-paired heterodimers. No clear example of interregional mixing of the type $A_\alpha E_\beta$ or $E_\alpha A_\beta$ has been noted. The fraction of trans-paired and cis-paired α and β heterodimers in a heterozygous mouse varies with the haplotypes involved[16] and possibly with other parameters, and it is not always possible to demonstrate all types of pairing in a given F_1 hybrid.

Nonpolymorphic Components

Nonpolymorphic components are defined as being specifically associated with the Ia molecular complex because they are coisolated with the

[15] J. Silver, *J. Immunol.* **123**, 1423 (1979).
[16] J. McNicholas, D. B. Murphy, L. A. Matis, R. H. Schwartz, E. A. Lerner, C. A. Janeway, and P. P. Jones, *J. Exp. Med.* **155**, 490 (1982).

Ia α–β heterodimer when specific alloantisera, or more critically, when monoclonal anti-Ia reagents are used. They are not isolated when negative control reagents or reagents binding other lymphocyte antigens are employed. Moreover, once conditions for detecting these nonpolymorphic components are established, they are found to be associated with both I-A and I-E molecules, and are present in all MHC haplotypes, though perhaps to varying extents. They are not necessarily present in all cell types bearing Ia, nor are they necessarily absent from all cell types which lack Ia, and this information has not been sought in all cases.

The best studied of the nonpolymorphic components of the Ia molecular complex is invariant chain,[17] or I_i. Actually, two electrophoretically distinct and presumably allelic versions of I_i are now known to exist, and study of segregation of these two alleles has permitted the conclusion that the I_i structural gene maps outside the MHC.[18] All commonly used laboratory mouse strains bear the same I_i allele however, and so the "invariant" designation remains functionally correct for standard mice (*Mus musculus*). Although only a fraction of the total cellular I_i chains is found in the Ia molecular complex, the association of I_i with Ia α and β chains is specific, since it is not isolated with other lymphocyte markers such as H-2 molecules.

Several other proteins have been coisolated with Ia or I_i by some investigators, but they have not been studied as extensively. For example, a monoclonal anti-I_i antibody coprecipitates nonpolymorphic components of M_r 41,000, 27,000, 25,000, and 15,000[19] whose structural relationship to I_i is unknown. Some of the same nonpolymorphic components appear to have been identified in monoclonal anti-Ia precipitates, along with additional proteins, some of which may be I_i processing intermediates.[20]

Recently, a sulfated proteoglycan component has also been found in association with Ia and I_i.[21] This type of specific association is novel and its significance is not yet understood.

[17] P. P. Jones, D. B. Murphy, D. Hewgill, and H. O. McDevitt, *Mol. Immunol.* **16**, 51 (1979).

[18] C. E. Day and P. P. Jones, *Nature (London)* **302**, 157 (1983).

[19] N. Koch and G. J. Hammerling, *in* "Ir Genes: Past, Present and Future" (C. W. Pierce, S. E. Cullen, J. A. Kapp, B. D. Schwartz, and D. C. Shreffler, eds.), p. 159. Humana Press, Clifton, New Jersey, 1983.

[20] K. Reske, *in* "Ir Genes: Past, Present and Future" (C. W. Pierce, S. E. Cullen, J. A. Kapp, B. D. Schwartz, and D. C. Shreffler, eds.), p. 153. Humana Press, Clifton, New Jersey, 1983.

[21] A. J. Sant, B. D. Schwartz, and S. E. Cullen, *J. Exp. Med.* **158**, 1979 (1983).

Preparation for Structural Studies

The Use of Antibody as the Major Purification Tool

The most basic requirement for critical analysis of Ia glycoproteins in murine systems is the application of selective serological reagents to isolate the material to be studied. For analysis of standard homozygous strains it is no longer difficult to find such reagents, since alloantisera and monoclonal reagents are readily available,[22] at least for commonly used haplotypes.

On the other hand, appropriate use of these reagents to isolate Ia molecules from heterozygous animals, recombinants, or variant animals or cells can be complicated, because the reagents may be selective in a variety of ways. For example, reagents reacting with an epitope on a particular α or β chain may react with both cis-paired and trans-paired α and β heterodimers present in heterozygous cells and this feature may or may not be desirable depending on the experimental context. Conversely, alloantisera or monoclonal antibodies may selectively detect combinatorial determinants formed from the interaction of some but not all α and β chains, thus excluding isolation of particular α–β heterodimers. A determinant can be identified as combinatorial if it is present when one particular chain is paired with some, but not all allelic versions of its partner chain, Thus, the Ia.23 combinatorial determinant occurs when the E_β^d chain is paired with any of the E_α^d, E_α^k, or E_α^r chains, but not with E_α^u chains.[23] Thus, some Ia molecules found in particular F_1 hybrids, may not be detectable by reagents directed against combinatorial determinants.

If allele-specific selectivity in the expression of the determinant cannot be shown, then serological approaches cannot distinguish whether the determinant is combinatorial or is expressed on only one chain. The latter alternative can be confirmed if the separated chain reacts directly with the antibody. A few chain-specific determinants have been identified by the reaction of antibody with renatured separated chains isolated in propionic acid/urea.[24] The Western blotting procedure, in which two-dimensional gel-separated chains react in situ with tagged antibody, is another suitable approach which has been applied with some success to the study of human Ia homologues.[25]

[22] NIH Ia Alloantiserum and Hybridoma Banks. Contact Dr. John Ray, IAID Program, NIAID, Room 7A07, Westwood Bldg. NIH, 20205

[23] W. P. Lafuse, J. F. McCormick, P. S. Corser, and C. S. David, Transplantation 30, 341 (1980).

[24] J. M. Kupinski, M. L. Plunkett, and J. H. Freed, J. Immunol. 130, 2277 (1983).

[25] S. F. Radka, C. Machamer, D. N. Howell, and P. Cresswell, Fed. Proc., Fed. Am. Soc. Exp. Biol. 42, 1231 (1983).

Another complication in reagent selection is that certain anti-Ia antibodies may identify epitopes shared between I-A and I-E molecules and therefore their use may result in isolation of mixtures of Ia molecules. A-E cross-reaction rarely has been detected among alloantisera,[26] but cross-reaction is more commonly detected with monoclonal reagents.[27-29]

Selection of anti-Ia antibodies for tissue or cellular distribution studies, or to examine biosynthesis also requires care since the reagents may be more or less inclusive in their reactivity. The anti-Ia.w39 alloantiserum[30] is an example of a reagent which reacts with a subset of I-A^b molecules that may be a marker for a mature B cell population. Although the structural feature distinguishing Ia.w39$^+$ I-A molecules from other I-A molecules is not known, the general implication is that some reagents may be selective for Ia molecules predominantly expressed on particular cell types.

The reactivity of some anti-Ia reagents may be influenced by posttranslational modifications or by the interactions of the $\alpha-\beta$ heterodimer with other cellular components.[31] Two monoclonal antibodies, each reacting with I-A^b molecules, have been observed to interact with subsets of I-A^b molecules which differ in glycosylation. This kind of selectivity has obvious importance in studies of Ia biosynthesis.

Selectivity for subsets of Ia molecules present within one cell or in different cell types is clearly a valuable characteristic in an anti-Ia reagent only when it has been recognized. Undetected selectivity may be present in many reagents currently in use, and may account for differences in Ia molecules detected by reagents thought to be equivalent.

Cell Source

The most frequently used cell source for structural studies on Ia molecules is unseparated normal splenocytes. The Ia molecules derived from this source are predominantly from B cells,[32-34] since Ia is not easily detectable on resting T cells, and Ia$^+$ macrophages represent a very small

[26] P. A. Lowry and D. B. Murphy, *Immunogenetics* **14**, 189 (1981).

[27] A. Bhattacharya, M. Dorf, and T. Springer, *J. Immunol.* **127**, 2488 (1981).

[28] F. W. Symington and J. Sprent, *Immunogenetics* **14**, 53 (1981).

[29] M. Pierres, W. P. Lafuse, M. Dosseto, C. Devaux, D. Z. Birnbaum, and C. S. David, *Tissue Antigens* **20**, 208 (1982).

[30] B. T. Huber, *Proc. Natl. Acad. Sci. U.S.A.* **76**, 3460 (1979).

[31] J. P. Moosic, A. Nilson, G. J. Hammerling, and D. J. McKean, *J. Immunol.* **125**, 1463 (1980).

[32] D. H. Sachs and J. L. Cone, *J. Exp. Med.* **138**, 1289 (1973).

[33] V. Hauptfeld, M. Hauptfeld, and J. Klein, *J. Immunol.* **113**, 181 (1974).

[34] G. J. Hammerling, B. D. Deak, G. Mauve, U. Hammerling, and H. O. McDevitt, *Immunogenetics* **1**, 68 (1974).

fraction of the spleen cell population.[35] The other frequently used cell source is an Ia$^+$ B cell line with sufficient antigen density. Most of the available Ia$^+$ cell lines are of the *Ia^d* or *Ia^k* haplotypes (Table I), and this limits their application in the study of Ia polymorphism, although they are useful for general structural studies. This limitation has been partially relieved by the development of B cell hybridomas derived from fusion of an Ia$^+$ cell line with normal spleen cells of the desired haplotype.[36] Although the cell lines and hybridomas listed in Table I express relatively high levels of Ia, the quantitative yield from mouse lines cannot compare to that of the human B lymphoblastoid hyperproducers.

Normal macrophages, i.e., splenic adherent cells (SAC),[37] peritoneal exudate adherent cells (PEAC),[38] and Langerhans cells[39] have also been used as a source of Ia antigens. In untreated animals the yield of Ia$^+$ macrophages from the spleen is greater than from the peritoneal cavity. Peritoneal macrophages can be harvested in much greater quantity however, and it is therefore worth the effort to use one of the pretreatments recently found to increase the number of Ia$^+$ PEAC. These treatments include infection with *Listeria*,[40] administration of T cell supernatant factors,[41–43] and treatment with inhibitors of prostaglandin synthesis.[44] Among these treatments, use of prostaglandin synthesis inhibitors seems to be the most accessible method, and it gives easily reproducible results. Intraperitoneal injection of 50 μg of Indomethacin (Merck, Sharpe and Dohme) in 5% ethanol every 12 hr for 3 days results in about a 2- to 5-fold increased yield of PEAC and a 3- to 5-fold increase in the fraction of Ia$^+$ PEAC.[44] These cells are a remarkably good source of biosynthetically labeled Ia antigens.[45]

Although several murine cell lines with characteristics of macrophages have been shown to be Ia$^+$ by serological and functional methods, such lines generally have been a rather poor source of Ia molecules for structural studies. Reagents which can stimulate Ia$^+$ macrophages *in vivo* may

---

[35] C. Cowing, B. D. Schwartz, and H. B. Dickler, *J. Immunol.* **120**, 378 (1978).

[36] J. Kappler, J. White, D. Wegmann, J. E. Mustain, and P. Marrack, *Proc. Natl. Acad. Sci. U.S.A.* **79**, 3604 (1982).

[37] S. E. Cullen, C. S. Kindle, D. C. Shreffler, and C. Cowing, *J. Immunol.* **127**, 1478 (1981).

[38] R. H. Schwartz, H. B. Dickler, D. H. Sachs, and B. D. Schwartz, *Scand. J. Immunol.* **5**, 731 (1976).

[39] J. G. Frelinger, J. A. Frelinger, and L. Hood, *Immunogenetics* **12**, 569 (1981).

[40] D. I. Beller, J.-M. Kiely, and E. R. Unanue, *J. Immunol.* **124**, 1426 (1980).

[41] R. M. Steinman, N. Nogueira, M. D. Witmer, J. D. Tydings, and I. S. Mellman, *J. Exp. Med.* **152**, 1248 (1980).

[42] M. G. Scher, D. I. Beller, and E. R. Unanue, *J. Exp. Med.* **152**, 1684 (1980).

[43] P. S. Steeg, R. N. Moore, and J. J. Oppenheim, *J. Exp. Med.* **152**, 1734 (1980).

[44] D. S. Snyder, D. I. Beller, and E. R. Unanue, *Nature (London)* **299**, 163 (1982).

[45] T. Rumbarger and S. E. Cullen, submitted for publication.

TABLE I

B CELL LINES FROM WHICH Ia MOLECULES HAVE BEEN ISOLATED

Cell line	Passage	Origin	Haplotype	Original description	Users
A20-1.11	In vitro	BALB/c	$H\text{-}2^d$	a	b
A20.2J (formerly L10A.2J)	In vitro	BALB/c	$H\text{-}2^d$	a, c	d, b
LB-15.13	In vitro	BALB/c, C57BL/10 (fusion of A20.2J with spleen cells)	$H\text{-}2^d$, $H\text{-}2^b$	e	b
LK-35.2	In vitro	BALB/c, B10.BR (fusion of A20.2J with spleen cells)	$H\text{-}2^d$, $H\text{-}2^k$	e	b
WEHI-5	In vitro	(BALB/c × NZB)F$_1$	$H\text{-}2^d$	f	d
X16.C8.5 (X16.4.3)	In vitro	BALB/c	$H\text{-}2^d$	a	b
BCL1	In vivo	BALB/c	$H\text{-}2^d$	g	h, i, b
AKTB-1B	In vivo	AKR/J	$H\text{-}2^k$	j	k
CH1	In vivo	B10.H-2^a H-4^b p/Wts	$H\text{-}2^a$	l	m

[a] K. J. Kim, C. Kannelopoulos-Langevin, R. M. Merwin, D. H. Sachs, and R. Asofsky, *J. Immunol.* **122**, 549 (1979).

[b] M. Harris and S. E. Cullen, unpublished observations.

[c] D. J. McKean, A. J. Infante, A. Nilson, M. Kimoto, C. G. Fathman, E. Walker, and N. Warner, *J. Exp. Med.* **154**, 1419 (1982).

[d] D. J. McKean, *J. Immunol.* **130**, 1263 (1983).

[e] J. Kappler, J. White, D. Wegmann, E. Mustain, and P. Marrack, *Proc. Natl. Acad. Sci. U.S.A.* **79**, 3604 (1982).

[f] E. B. Walker, L. L. Lanier, and N. L. Warner, *J. Immunol.* **128**, 852 (1981).

[g] S. Strober, E. S. Gronowicz, M. R. Knapp, S. Slavin, E. S. Vitetta, R. A. Warnke, B. Kotzin, and J. Schroder, *Immunol. Rev.* **48**, 169 (1979).

[h] M. R. Knapp, P. P. Jones, S. J. Black, E. S. Vitetta, S. Slavin, and S. Strober, *J. Immunol.* **123**, 992 (1979).

[i] M. Letarte, G. Meghji, and G. Lorimer, *Mol. Immunol.* **19**, 557 (1982).

[j] M. M. Zatz, B. J. Mathieson, C. Kannelopoulos-Langevin, and S. O. Sharrow, *J. Immunol.* **126**, 608 (1981).

[k] J. M. Kupinski, M. L. Plunkett, and J. H. Freed, *J. Immunol.* **130**, 2277 (1983).

[l] M. A. Lynes, L. L. Lanier, G. F. Babcock, P. J. Wettstein, and G. Haughton, *J. Immunol.* **121**, 2352 (1978).

[m] M. F. Mescher, K. C. Stallcup, C. P. Sullivan, A. P. Turkewitz, and S. H. Herrman, Vol. 92, p. 86.

be suitable to stimulate macrophage lines *in vitro,* and thus may provide an answer to this problem.[46]

While Ia molecules are present on activated T cells by several functional and serological criteria, it has been exceedingly difficult to study murine T cell Ia by biochemical approaches. The quantity obtained in some early studies of thymocytes was insufficient for extensive analysis.[47] A controversy arose concerning the cellular origin of "T cell" Ia, which was thought by some authors to be acquired from other cell donors.[48–50] It is probable that both endogenously and exogenously synthesized Ia molecules can be found on T cells, though perhaps not on the same T cell subsets. Biosynthetically radiolabeled Ia molecules from T cell lines or hybridomas known to be free of other cell types have been isolated.[51,52] Thus far, the I-A molecules isolated from T cells show no structural differences from their B cell counterparts, although the analysis has not yet progressed to extremely fine structural detail.

The issue of the T cell specific *I* region determined I-J molecules continues to be examined. Evidence from several laboratories indicates that T cell regulatory factors bear unique *I*-region mapped antigenic determinants detected by monoclonal (anti I-J) antibodies. These factors are produced by some T cell hybridoma lines and can be isolated in sufficient quantity to permit structural analysis.[53–55] The genetic region responsible for the expression of these I-J determinants has been mapped between the *I-A* and *I-E* subregions by serological analysis of the Ia antigens of recombinant mice. Molecular genetic data have now shown that this mapped site for the *I-J* subregion is located within the very short DNA segment which already includes the E_β gene and the high frequency recombination site (see Fig. 1), leaving little room for the presumed *I-J* structural gene.[8]

[46] J. McNicholas, D. P. King, and P. P. Jones, *J. Immunol.* **130**, 449 (1983).

[47] B. D. Schwartz, A. M. Kask, S. O. Sharrow, C. S. David, and R. H. Schwartz, *Proc. Natl. Acad. Sci. U.S.A.* **74**, 1195 (1977).

[48] Z. A. Nagy, B. E. Elliott, and M. Nabholz, *J. Exp. Med.* **144**, 1545 (1976).

[49] S. O. Sharrow, B. J. Mathieson, and A. Singer, *J. Immunol.* **126**, 1327 (1981).

[50] M. I. Lorber, M. R. Loken, A. M. Stall, and F. W. Fitch, *J. Immunol.* **128**, 2798 (1982).

[51] N. Koch, B. Arnold, G. J. Hammerling, J. Heuer, and E. Kolsch, *Immunogenetics* **17**, 497 (1983).

[52] S. K. Singh, E. J. Abramson, C. J. Krco, and C. S. David, *Fed. Proc., Fed. Am. Soc. Exp. Biol.* **42**, 837 (1983).

[53] M. Taniguchi, T. Tokuhisa, M. Kanno, Y. Yaoita, A. Shimizu, and T. Honjo, *Nature (London)* **298**, 172 (1982).

[54] M. Taniguchi, *in* "Ir Genes: Past, Present and Future" (C. W. Pierce, S. E. Cullen, J. A. Kapp, B. D. Schwartz, and D. C. Shreffler, eds.), p. 575. Humana Press, Clifton, New Jersey, 1983.

[55] D. R. Webb, J. A. Kapp, and C. W. Pierce, *Annu. Rev. Immunol.* **1**, 423 (1983).

This puzzling problem of the product without a gene is being actively investigated.

Radiolabeling (This Volume [48])

The small quantity of Ia molecules which can be recovered from murine cells necessitates the use of radiolabeling techniques for most types of experiments. Radiolabeling may be biosynthetic or postsynthetic.

Biosynthetic Radiolabeling. Choice of Precursor. Biosynthetic labeling may employ a wide variety of precursors, most typically including ³H-, ¹⁴C-, or ³⁵S-labeled amino acids, or ³H-labeled monosaccharides. Amino acids frequently used singly include leucine, arginine, lysine, methionine, cysteine, and tyrosine. Mixtures of up to 10 amino acids have been used for limited amino acid sequence analysis in which the identity of each released radioactive amino acid derivative is subsequently determined.[56–60] [³H]Arg and [³H]Lys labeling should be carried out in separate preparations when tryptic peptide analysis is contemplated. Making two separate preparations is important because the specific activities of the [³H]Arg and [³H]Lys labeled molecules are often quite different and because the separation pattern of the peptide digest is usually overly complex if both precursors are used. Other precursors such as ³H-labeled monosaccharides or [³⁵S]cysteine may be used to identify particular peptides in proteolytic digestion experiments.

Comparative peptide mapping using trypsin or other proteolytic enzymes to generate the peptides has been a powerful tool for estimation of the extent of homology between molecules and for detection of small differences between related molecules, including Ia molecules.[15,61,62] Ideally, such comparisons should employ digestion of a mixture of independently prepared ¹⁴C- and ³H-labeled samples followed by separation of the cleavage products using the technique of choice. Such ¹⁴C/³H double label comparisons represent a gold standard with an equivalently high price tag. For many laboratories the use of ¹⁴C-labeled amino acid is prohibi-

[56] J. Silver and L. Hood, *Nature (London)* **256,** 63 (1975).

[57] J. Silver, W. A. Russell, B. L. Reis, and J. A. Frelinger, *Proc. Natl. Acad. Sci. U.S.A.* **74,** 5131 (1977).

[58] M. McMillan, J. M. Cecka, D. B. Murphy, H. O. McDevitt, and L. Hood, *Proc. Natl. Acad. Sci. U.S.A.* **74,** 5135 (1977).

[59] R. Cook, E. S. Vitetta, J. D. Capra, and J. W. Uhr, *Immunogenetics* **5,** 437 (1977).

[60] J. E. Coligan, F. T. Gates, E. S. Kimball, and W. L. Maloy, Vol. 91, p. 413.

[61] J. H. Freed, C. S. David, D. C. Shreffler, and S. G. Nathenson, *J. Immunol.* **121,** 91 (1978).

[62] M. McMillan, J. M. Cecka, L. Hood, D. B. Murphy, and H. O. McDevitt, *Nature (London)* **277,** 663 (1979).

tively expensive. However, samples labeled with ³H-labeled amino acid can be reliably compared if certain criteria are satisfied. First, differences noted in the comparison of two independently analyzed ³H-labeled preparations should be reproducible when new digests are analyzed. Second, unique peaks identified in the separate digests of two different ³H-labeled samples should also be detectable in an analysis of the digest of the combined ³H-labeled samples. Small shifts in peak position noted in the independent analysis of each sample should be verifiable during the analysis of combined samples if the shift is not artifactual.

[³⁵S]Methionine is a high specific activity amino acid which is frequently used to label Ia molecules. [³⁵S]Met is incorporated into I_i extremely well,[17] and [³⁵S]Met-labeled I_i may overshadow α and β chain in some experiments.

The monosaccharide best suited to study of N-linked glycosylation is D-[2-³H]mannose. Mannose is present in N-linked oligosaccharides of both the high mannose and complex types, and thus it is an inclusive labeling method. Radioactivity in [2-³H]mannose is found mainly in mannose and fucose after short-term labeling.[63] Use of the higher specific activity precursor [2,6-³H]mannose is undesirable for detailed study of N-linked oligosaccharides since the mannose conversion products will not all be labeled to the same specific activity, and their contribution to structure cannot be quantitatively assessed.

For the study of O-linked glycosylation, [1,6-³H]glucosamine is probably the precursor of choice. This monosaccharide is rapidly converted to galactosamine and sialic acid,[63] and these monosaccharides label O-linked oligosaccharides fairly heavily, and to higher specific activity than is achievable with [³H]galactose. Of course, [³H]glucosamine is also incorporated into N-linked oligosaccharides, but N-linked oligosaccharides can be recognized because their synthesis is blocked by tunicamycin,[64] while O-linked oligosaccharides are distinctive because they can be released by alkaline borohydride.[65]

In general the specific activity of the glycoprotein which can be achieved with monosaccharide precursors is approximately 1–10% of that obtained with amino acid precursors and therefore monosaccharide label should be employed only when it is specifically needed. Cells take up the monosaccharides very well but if these are not quickly incorporated into glycoconjugates they are rapidly metabolized. For example, radioactivity in labeling medium to which [³H]mannose has been added can be com-

[63] P. D. Yurchenco, C. Ceccarini, and P. H. Atkinson, this series, Vol. 50, p. 175.
[64] J. S. Tkacz and J. O. Lampén, *Biochem. Biophys. Res. Commun.* **65,** 248 (1975).
[65] K. Tanaka and W. Pigman, *J. Biol. Chem.* **240,** 1487 (1965).

pletely recovered as mannose at time zero, but after 4 hr of incubation with cells, 90% of the tritium in the medium is found in volatile catabolites.[66]

Labeling Procedure. In principle, radiolabeling should be carried out with the labeled compound at the highest concentration possible. It is not necessary to label large numbers of cells, and the cells labeled should be at the highest density compatible with good viability. For normal mouse cells, incubation at $2-5 \times 10^7$ cells/ml for periods of up to 5–6 hr is typical. For cell lines, the density must be individually adjusted. A 5 ml culture in which a total of $1-2 \times 10^8$ normal spleen cells are labelled with 2.5–5.0 mCi of a ^{3}H-labeled amino acid should provide adequate material for many analytical experiments or for chain isolation.

For amino acid labeling, the medium should lack the amino acid to be used as the radiolabeled precursor. Serum, if used, should be predialyzed against Hanks' balanced salt solution to reduce the concentration of free amino acids. Various media, including Dulbecco modified Eagle's medium (DMEM) and RPMI medium lacking the appropriate amino acid are suitable. For ^{3}H-labeled monosaccharide, glucose-depleted medium supplemented with pyruvate as an energy source is frequently used, e.g., 1 mCi/ml of [2-^{3}H]mannose in RPMI 1640 medium with glucose at 100 mg/liter, 1.0 mM pyruvate, 15 mM HEPES and glutamine supplementation.

Postsynthetic Radiolabeling. Postsynthetic labeling usually involves lactoperoxidase (LPO) catalyzed ^{125}I-iodination.[67] Because LPO remains outside of viable cells, this labeling method is vectorial, and results in selective labeling of surface components. Vectorial labeling using the Iodogen method may also be possible.[68] Ia molecules at the cell surface contain both α and β chains, but the β chain is much more efficiently iodinated.[67] The α chain contains multiple tyrosine residues which could potentially be derivatized by ^{125}I, so that the failure to iodinate α is probably secondary to lack of accessibility. When precipitated by anti-Ia reagents, I_i is found to be poorly labeled by LPO-catalyzed iodination. It is likely that this is due to the fact that I_i is only complexed with $\alpha-\beta$ heterodimers intracellularly, and not on the cell surface.[69,70] The non-Ia-associated I_i on the cell surface can be iodinated.[19]

Methods for iodinating α using membrane-seeking reagents have been

[66] E. P. Cowan and S. E. Cullen, unpublished observations.
[67] B. D. Schwartz, E. S. Vitetta, and S. E. Cullen, *J. Immunol.* **120,** 671 (1978).
[68] M. F. Mescher, K. C. Stallcup, C. P. Sullivan, A. P. Turkewitz, and S. H. Herrmann, Vol. 92, p. 86.
[69] E. Sung and P. P. Jones, *Mol. Immunol.* **18,** 899 (1981).
[70] J. P. Moosic, E. Sung, A. Nilson, P. P. Jones, and D. J. McKean, *J. Biol. Chem.* **257,** 9684 (1982).

successfully used,[71] but the reagents are not stable and must be generated immediately before use.

Solubilization (This Volume [48])

Solubilization of Ia antigens is readily accomplished with many detergents[4,5,70] including 0.5% Nonidet P-40 (NP-40), 0.5% Triton X-100 (or X-102), 0.5% Brij 98 (or Brij 56), 0.5% Lubrol-WX, 0.5% deoxycholate (DOC), and sodium dodecyl sulfate (SDS). The first two detergents are used most frequently, and do not alter the antigenic determinants as far as can be determined. They contain aryl moieties, and are not compatible with spectrophotometric monitoring of protein at 280 nm. Brij contains only alkyl groups, and permits monitoring at 280 nm. The special properties of Brij 56 and Lubrol-WX will be discussed in the section on selective solubilization. DOC forms very small micelles and is the only readily dialyzable detergent cited above. Detergents such as SDS solubilize Ia molecules completely but destroy their antigenic activity, although perhaps not irreversibly. With the exception of SDS, the detergent in which a sample is solubilized may be exchanged during affinity chromatography. Ia molecules in one detergent can be bound to a lectin or antibody affinity column, and eluted after the column is washed extensively in buffer containing the second detergent.

It is a fairly common practice to include inhibitors of proteolytic activity during detergent solubilization in order to prevent breakdown of Ia molecules, which are quite sensitive to endogenous enzymes. These enzymes may be especially troublesome if the cell source is a macrophage. Inhibitors are sometimes used in combinations. Some examples of inhibition cocktails are (1) 1% Aprotinin (same as Trasylol); (2) 0.14 mM L-p-tosyl-amino-2-phenylethyl chloromethyl ketone (TPCK), 0.14 mM Nα-p-tosyl-L-lysine chloromethyl ketone (TLCK), 1 mM phenylmethylsulfonyl fluoride (PMSF); (3) 3 mM sodium azide, and 1 mM PMSF (Sigma). Sodium azide inhibits lactoperoxidase and should be avoided if iodination of the extract is planned. Inhibitors are usually not used past the first isolation step performed on the detergent lysate.

Solubilization is performed on cells washed free of culture medium after labeling. Typically, for NP-40 solubilization cells are suspended at $0.5–1 \times 10^8$/ml in 0.15 M NaCl, 0.01 M Tris–HCl, pH 7.4 and mixed with an equal volume of 1% NP-40 in the same buffer, with inhibitors if appropriate. Lysis is allowed to proceed for 15–30 min at 4°, the lysate is

[71] C. L. Sidman, T. Bercovici, and C. Gitler, Mol. Immunol. 17, 1575 (1980).

centrifuged at 100,000 g for 60 min to pellet cellular debris, and the supernatant is retained.

Limited proteolysis to release Ia molecules from cell surfaces has been used,[72] but it is not a practical approach because the yield is poor. Murine Ia molecules are quite sensitive to proteolysis by various enzymes including papain. The procedure is difficult to control and cleavage products of consistent size and activity are not easily obtained.

Isolation of Ia Molecules Using Antibody

Immune Precipitation (This Volume [40, 48])

Indirect immune precipitation is a well-established procedure for isolation of antigens, including lymphocyte marker antigens in general and the Ia molecules in particular. Once appropriate reagents have been obtained, it is an easy and relatively trouble free method. In the procedure first designed, the solubilized antigen preparation was mixed with antibody, and the soluble antigen–antibody complexes were precipitated with anti-mouse immunoglobulin. For each antibody used, complete precipitation required determination of the amount of antiglobulin producing maximal precipitation (equivalence).[73] The mass of this isolation product was of concern because it provided a matrix to which labeled contaminants could nonspecifically absorb, and because the large quantity of unlabeled immunoglobulin sometimes interfered with subsequent analytical procedures.

Fixed protein A-bearing *Staphylococcus aureus,* Cowan I strain (SaCI) was later introduced as a better developing reagent. Protein A binds to the Fc portion of immunoglobulin (Ig) of many subclasses from animals of many species. Use of SaCI obviated the need for determining optimal proportions, and also had the effect of reducing the amount of unlabeled non-Ia protein in the isolation product.[73–75] Use of monoclonal antibodies is a further improvement, because it permits use of less SaCI for the same amount of binding activity, and thus further reduces the amount of matrix. The inability of protein A to bind certain classes of Ig is a minor drawback which can be overcome by the addition of very small (nonprecipitating) amounts of antiglobulin which create an antibody "sandwich" readily bound by SaCI. Some monoclonal reagents may not have sufficiently high affinity for Ia molecules to be useful for immune

[72] M. Hess, *Eur. J. Immunol.* **6,** 188 (1976).
[73] S. E. Cullen and B. D. Schwartz, *J. Immunol.* **117,** 136 (1976).
[74] S. W. Kessler, *J. Immunol.* **115,** 1617 (1975).
[75] S. E. Cullen, this volume [45].

precipitation. Two low-affinity reagents reactive with the same molecule, when combined, may be more successful in precipitation studies.[76]

Before immune precipitation is attempted, the lysate should be precleared of endogenous (radiolabeled) immunoglobulin by mixing it with SaCI alone, normal serum and SaCI, or, if anti-Ig is to be subsequently used as a sandwich reagent, with anti-Ig and SaCI.

In a typical analytical "immune precipitation" reaction, detergent lysate is mixed with antiserum or monoclonal antibody in glass tubes (10 × 75 mm). The lysate from 0.5–5 × 10⁶ cells labeled with an amino acid should suffice for analytical evaluation. The quantity of serum needed will vary, but for mouse alloantiserum of reasonable quality, the amount will be 10 μl or less. The quantity of monoclonal antibody required should be individually established. After incubation of the mixture at 4° for 1–4 hr, a 10% suspension of fixed SaCI is added. Generally the volume of the SaCI suspension required is about 10 times the volume of antiserum used, but much less can be used with monoclonal reagents, particularly if they have been harvested from culture supernatants. After 30 min on ice the bacteria are washed free of nonbound materials by low-speed centrifugation. Three washes in buffer should be performed, with a transfer to fresh tubes between the second and third wash.

Preparative precipitation requires larger amounts of reagents and it may be advantageous to perform two separate precipitation steps on the same aliquot of lysate to improve the final yield. The volume can be kept at reasonable levels by adding the lysate–antibody mixture to pelleted SaCI, rather than using the SaCI suspension. The Ia molecules in antibody complexes adsorbed to SaCI can be released by SDS, urea, or other dissociating agents, the choice of which depends on the subsequent analytical step.

Immune Affinity Chromatography (This Volume [48])

Isolation of Ia antigens can also be effectively carried out by a method in which antibody is attached to an insoluble matrix such as Sepharose, and used in affinity chromatography.[24,30,68,77] Elution conditions which preserve the capacity of the eluted Ia molecules to react with antibodies have been determined for some Ia antigens and include the following: 0.14 M NaCl, pH 11; 0.5 M NaCl, pH 11; 3 M KSCN, 100 mM Tris–HCl, 0.5% NP-40, pH 7.4; and 50 mM diethylamine, 0.15 M NaCl, 0.25% NP-40, pH 10.7. Conditions which are suitable for a particular antibody and Ia may

---

[76] R. Tosi, N. Tanigaki, R. Sorrentino, R. Accolla, and G. Corte, *Eur. J. Immunol.* **11**, 721 (1981).

[77] R. Zecher and K. Reske, *Mol. Immunol.* **19**, 1037 (1982).

not apply to all antibody-Ia combinations. In many cases, the activity of the conjugated antibody is retained well enough to allow repeated use of the same column. Since the antibody remains attached to the column, the antigen-containing eluate is relatively free of contaminating immunoglobulin. This method is extremely useful for the isolation of large amounts of Ia molecules. The released Ia molecules must be kept in nonionic detergent or in dissociating solvents to retain solubility.

Optional Preisolation Fractionation Steps

Differential Solubilization

The detergent Lubrol-WX can be used to solubilize a subpopulation of Ia molecules.[69,70] It has been hypothesized that this detergent is selective for the plasma membrane, leaving intracellular membranes intact. Another possibility is that it is less effective at solubilizing molecules associated with the cytoskeletal matrix, as has been seen with other detergents using other cell types. Lubrol solubilizes predominantly Ia molecules which are not associated with I_i, and which contain a higher proportion of complex carbohydrates. The Ia molecules which are predominantly insoluble in Lubrol-WX but soluble in NP40 have associated I_i, and have a greater proportion of high mannose oligosaccharides. The conclusion from studies employing Lubrol is that the Ia molecules which it solubilizes are from a predominantly mature, plasma membrane associated subpopulation. Selective use of detergents therefore provides a means for crudely separating mature and immature molecules and thus for examining the kinetics of biosynthesis of Ia α, β and associated components. The Lubrol-WX detergent is no longer commercially available, but Brij 56 or polyoxyethylene ether W-1 (Sigma) appear to have similar properties.[70,78]

Lectin Affinity (This Volume [48])

The Ia molecules are glycoproteins and it has been quite common to employ lectin affinity chromatography to prepare a partially purified glycoprotein pool from which the Ia molecules can be antibody precipitated with less nonspecific contamination than from whole detergent extract. We have found it preferable to conjugate lectin to cyanogen bromide-activated Sepharose at high concentration (e.g., 10 mg/ml of packed beads). This reduces the size of the column needed for binding (1.0 ml columns can be used for preparations from 10^9 spleen cells or more), and thereby reduces the amount of matrix potentially contributing to nonspe-

[78] C. S. Kindle, unpublished observation.

cific binding. Lectin should be conjugated in the presence of its competing sugar to protect the active site during the conjugation procedure. Freshly made lectin-Sepharose should be pretreated with a protein solution such as 1% BSA before use. This treatment reduces the amount of nonspecific adsorption by the Sepharose. Lentil lectin (LcH; *Lens culinaris* hemagglutinin) was the first lectin used for affinity purification of Ia antigens[79] and it is still very useful in this regard. Contaminants which may obscure Ia components during analysis of immune precipitates are much reduced by this approach. Ia molecules are recovered from LcH-Sepharose by a very gentle elution procedure which involves competition of free sugar (0.2 M α-methyl glucoside or 0.2 M α-methyl mannoside) for lectin binding sites. Other lectins suitable for this purpose include concanavalin A (competing sugar; 0.5 M α-methyl mannoside), and *Ricinus communis* (competing disaccharide; 0.2 M lactose). Con A has considerably greater affinity for Ia molecules than does LcH, and therefore recovery from lentil lectin is easier, and yields are in the range of 30–50%. Lentil lectin and ricin do not bind exactly the same population of Ia molecules.

The binding selectivity of lectins for specific sugars makes lectin affinity purification incompatible with protocols in which information about Ia glycosylation is sought, or dealing with biosynthesis in which Ia molecules at different processing steps must be analyzed. In addition, loss of associated materials such as I_i may occur during the affinity purification step. Overall, lectin affinity purification is inferior in yield to antibody affinity chromatography, and the latter is the preferable procedure when a suitable antibody is available.

Oligosaccharide structure itself can be studied by a technique employing sequential lectin affinity,[80] but for this purpose Ia antigens must be purified initially by antibody. The isolated Ia glycoprotein chains are then cleaved into glycopeptide fragments for lectin affinity analysis.

Release of I_i

For some experimental purposes it is desirable to release I_i from the isolated Ia molecular complex. This has been accomplished in several different ways. Solubilization with Lubrol-WX (see above) provides an Ia-containing preparation which has much reduced Ia-associated I_i, and thus isolation of Ia molecules from Lubrol-solubilized preparations minimizes the amount of I_i coisolated. Brief pretreatment of detergent lysates with 3 M KSCN and subsequent dialysis before precipitation also mini-

---

[79] S. E. Cullen, J. H. Freed, and S. G. Nathenson, *Transplant. Rev.* **30**, 236.
[80] E. P. Cowan, R. D. Cummings, B. D. Schwartz, and S. E. Cullen, *J. Biol. Chem.* **257**, 11241 (1982).

mizes the amount of I_i in precipitates.[77] The Ia–I_i association is also sensitive to solvent conditions after precipitation, and washing of immune precipitates with a buffer containing 20% pyrrolidinone[31] may reduce the I_i. The association is also sensitive to low concentrations of ionic detergent, so that introduction of 0.02% SDS, 0.02% DOC into the final wash buffer for immune precipitates elutes more I_i than α and β. It has also been suggested that freeze thawing of detergent extracts or lectin affinity chromatography reduces the I_i contribution to immune precipitates.[31] In experiments which employ isoelectric focusing (IEF) or nonequilibrium pH gradient electrophoresis (NEPHGE), I_i is easily distinguished from α and β and it may not be necessary to dissociate I_i from Ia prior to analysis.

Separating Components of the Ia Molecular Complex

Once the Ia molecular complex has been isolated by binding to antibody, its components must be separated. This is easier to do for analytical than for preparative purposes.

Analytical Methods (This Volume [44])

Analytical experiments may employ SDS–polyacrylamide tube or slab gels (SDS–PAGE), two-dimensional nonreducing/reducing slab gels (NRR–PAGE), IEF, NEPHGE, and two-dimensional IEF/SDS–PAGE or NEPHGE/SDS–PAGE. The one-dimensional methods are easier, and, in the slab format, present the opportunity to directly compare samples run side-by-side.

SDS–PAGE. The SDS–PAGE system is a stacking system[81] which employs buffers modified from those of Laemmli.[67] The sample is applied to a 2-layer gel composed of an upper (stacking) gel and a lower (running) gel with an appropriately selected acrylamide concentration (for Ia, usually 7.5 or 10%).

For 10% gels, the stock solutions needed are the following:
  A. Acrylamide; 40 g acrylamide, electrophoresis grade, 1.06 g methylene bisacrylamide (BIS), and water to make 100 ml
  B. TEMED; 1,2-bis(dimethylamino)ethane
  C. pH 8.8 buffer; 18.15 g Tris/100 ml, adjust pH with HCl
  D. pH 6.8 buffer; 6 g Tris/100 ml, adjust pH with HCl
  E. 10% SDS
  F. 10% ammonium persulfate (freshly made)
  G. Electrode buffer; 3 g Tris, 14.4 g glycine, 0.1% SDS/liter

For each 10 ml of running gel polymerization mixture, 6.1 ml water,

---

[81] L. Ornstein, *Ann. N.Y. Acad. Sci.* **121**, 321 (1964); B. J. Davis, *ibid.* p. 404.

1.25 ml solution C. 2.5 ml solution A, 0.1 ml solution E, 5 μl reagent B, and 0.1 ml solution F are added. The mixture is degassed and used to cast tube or slab gels. For each 10 ml of stacking gel polymerization mixture, 7.55 ml water, 1.25 ml solution D, 1.0 ml solution A, 0.1 ml solution E, 5 μl reagent B, and 0.1 ml solution F are added. The mixture is degassed and used to form the upper gel on tube or slab gels.

If the samples to be applied are analytical-size SaCI immune precipitates, the washed pellets are suspended in 100 μl sample buffer (pH 6.8 buffer 1 : 8, with 2% SDS, $\pm$ 2% 2-mercaptoethanol) and boiled for 1 min. The sample is then centrifuged at room temperature, and the clear supernatant is collected and mixed with an equal volume of 15% glycerol, 2% SDS, and 0.004% phenol red before application to the gel. After electrophoresis the separation is evaluated. For tube gels, this involves slicing the gel and determining radioactivity in each slice. For slab gels, the gel is impregnated with scintillation reagents, and after drying, it is exposed to X-ray film to record the separation pattern.[82]

Interpretation of the gel pattern is facilitated by coelectrophoresing molecular weight standards, and by comparing one sample with another. Identification of bands may sometimes be equivocal because the major components of Ia are close in molecular weight, and several forms of the α and β chains having slightly different apparent M_r are often present in biosynthetically labeled samples. These different molecular forms of α and β may include processing intermediates, or mature forms which are derived from minority cell populations which process the molecules differently. For this reason, citation of M_r values has been avoided whenever possible in this article, since the assignment of M_r by SDS–PAGE is not accurate enough to permit comparison of the values generated by different laboratories using slightly different methods.

α *Chain Separation by SDS–PAGE.* At least three subspecies of the α chain called α_1, α_2, and α_3 have been identified.[19,83] These three α subspecies are known to differ from one another in glycosylation, and may differ in other parameters. Despite the processing differences their tryptic peptide maps are the same,[83] and thus they are probably all variants of the same gene product. The "α_3" subspecies described by Koch and Hammerling[19] is also said to differ from α_1 and α_2 in cleavage by staphylococcal V8 protease, but the basis for this variability has not been clearly identified. Since the apparent molecular weights of these α subspecies are very similar, and close to that of other components of the Ia molecular complex, independent means of verifying the identity of the chains are important for interpretation of this SDS–PAGE data.

[82] W. M. Bonner and R. A. Laskey, *Eur. J. Biochem.* **46**, 83 (1974).
[83] E. P. Cowan, B. D. Schwartz, and S. E. Cullen, *J. Immunol.* **128**, 2019 (1982).

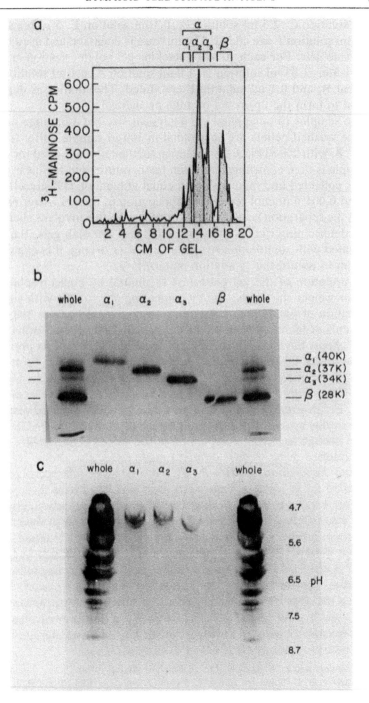

A suitable method for independent identification is examination of the chains on IEF, NEPHGE, or 2-D gels. Under these conditions α chains are acidic and can be easily distinguished from I_i. The three I-A α chain subspecies have isoelectric points in the 4.5–5.0 range.[84] The most basic of these species is α_3, which bears mainly high-mannose type oligosaccharides that are uncharged, and thus do not lower the pI as do the sialic acid residues in complex oligosaccharides. Figure 2 shows an example of preparative separation of subspecies of [^{3}H]mannose-labeled α chain by tube SDS–PAGE, followed by verification of the efficacy of the separation on an analytical slab gel, and confirmation by NEPHGE that all the putative α subspecies were actually α chains and did not contain I_i.

β *Chain Separation by SDS–PAGE.* It has been known for some time that β chains of Ia molecules have variable apparent M_r in SDS–PAGE depending on whether or not they are reduced with 2-mercaptoethanol.[85] Presumably this anomalous behavior results from the fact that β chains contain intrachain disulfide bonds[86] which affect their secondary structure and the extent to which they bind SDS. The change is most prominent among I-E molecules. Separation of α and β of the I-E^{d,p,u} molecules

---

[84] The I-E α chains have a consistently higher pI (5.0–5.5) than the I-A α chains, and one can virtually assign an Ia molecule to the I-A or I-E series by examining the pI of the α chains [S. E. Cullen, S. M. Rose, and C. S. Kindle *in* "Current Trends in Histocompatibility" (R. A. Reisfeld and S. Ferrone, eds.), p. 391. Plenum, New York, 1980].

[85] R. G. Cook, J. W. Uhr, J. D. Capra, and E. S. Vitetta, *J. Immunol.* **121**, 2205 (1978).

[86] D. R. Lee and S. E. Cullen, *Mol. Immunol.* **20**, 77 (1983).

FIG. 2. Separation of I-A^k α chain subspecies from spleen cells. (Adapted from Cowan *et al.*,[83] with permission.) (a) Preparative SDS–PAGE separation of [^{3}H]mannose-labeled I-A^k antigens. The I-A^k precipitates were washed in 0.5% NP-40, 0.2% DOC, 0.1% SDS in phosphate-buffered saline to release I_i. The profile shows the radioactivity present in 5% aliquots of SDS extracts from 2-mm gel slices of an 11 × 200 mm 10% polyacrylamide tube gel. Pools were made as indicated and their identity was checked by two methods (b and c). (b) Aliquots of the pools from a were run on an analytical 10% polyacrylamide SDS–PAGE slab gel. The outside lanes contained a whole immune precipitate of ^{3}H-labeled amino acid-labeled I-A^k antigens. These control lanes show that α_2 is the most prominent α subspecies, with smaller amounts of α_1 and α_3 present. The isolated [^{3}H]mannose-labeled α and β chains each correspond in mobility to one of the [^{3}H]leucine-labeled bands, and each appears to be free of cross-contamination. (c) Aliquots of the α chain pools from a were analyzed on NEPHGE gels to demonstrate that they were free of I_i. The outside lanes contained a whole immune precipitate of ^{3}H-labeled amino acid-labeled I-A^k antigens. The isolated α chains migrate to the low pH region of the gel, and basic bands are not present, indicating that I_i is absent. The α_1 and α_2 subspecies have a slightly lower pI than the α_3 subspecies. This difference is due to absence of sialic acid on α_3, which contains only high mannose oligosaccharides. (From Cowan *et al.*[80])

under reducing conditions is particularly difficult, and these chains will be better resolved if they are electrophoresed without reduction.[87]

There is also evidence that certain Ia molecules may contain β chains which are heterogeneous with respect to intrachain disulfide bonding.[88] The presence of β chains, only some of which have an intrachain bond, gives rise to the separation of two β chain subspecies with apparently different M_r because the presence or absence of the bond alters mobility in SDS–PAGE. In addition to these forms of β, electrophoretically separable β subspecies which differ in oligosaccharide processing have also been noted.[89]

A peculiar component migrating with an apparent M_r of 20,000 has been observed when haplotype d spleen cells or cell lines have been examined.[90] This polypeptide has structural features suggesting that it might be an aberrant fusion protein incorporating part of α and part of β. It has not been observed in any haplotype other than H-2^d.

NRR–PAGE. A variation of SDS–PAGE which has been recently applied is the 2D nonreducing/reducing slab (NRR–PAGE). This method permits analysis of interchain disulfide interactions,[91–93] although, as in one dimensional PAGE, chain identification can be equivocal. A nonreduced sample is run on a tube gel in the first dimension, and after completion of this separation, the tube gel is equilibrated in 2-mercaptoethanol and put on top of a slab gel containing dithiothreitol. Chains which did not have interchain disulfide bonds will migrate on a diagonal in the slab gel, while disulfide linked molecules will generate spots which migrate off the diagonal. To distinguish naturally occurring from artifactual disulfide bonds, acetylation (addition of 10 mM iodoacetamide) during the detergent lysis step is recommended.[85,93,94] Oxidants often present in detergent preparations can cause artifactual linkage of free sulfhydryl groups.[95] As has been pointed out, even artifactual linkages may be informative.[84,85] They present a way of determining that components are associated nonco-

[87] It is also possible to separate I-E^p α and β chains using the original Laemmli buffer system [J. M. Kupinski, P. A. Gamble, C. S. David, and J. H. Freed, *Immunogenetics* **16,** 393 (1982)]. The modified Laemmli system is still preferred for all other Ia separations.

[88] N. Koch and G. J. Hammerling, *Immunogenetics* **14,** 437 (1981).

[89] E. P. Cowan, R. D. Cummings, D. R. Lee, B. D. Schwartz, and S. E. Cullen, *Mol. Immunol.*, in press.

[90] D. J. McKean, *J. Immunol.* **130,** 1263 (1983).

[91] N. Koch and G. J. Hammerling, *J. Immunol.* **128,** 1158 (1982).

[92] N. Koch and D. Haustein, *Hoppe-Seyler's Z. Physiol. Chem.* **361,** 885 (1980).

[93] R. J. Allore and B. H. Barber, *Mol. Immunol.* **20,** 383 (1983).

[94] S. E. Cullen, C. S. Kindle, and D. R. Littman, *J. Immunol.* **122,** 855 (1979).

[95] H. W. Chang and E. Bock, *Anal. Biochem.* **104,** 112 (1980).

valently although not normally disulfide linked. Another approach to the problem of identifying associated components is intentional cross-linking.[96]

IEF and NEPHGE Gels. Direct analysis of isolated Ia molecules by IEF or nonequilibrium pH-gradient electrophoresis (NEPHGE) is not common, but it has been useful for certain analytical purposes. The regions of IEF or NEPHGE with a pH>6 contain the β chains, but the patterns are complicated by the presence of contaminants. Thus, the chief value of the method is for rapid, side-by-side comparisons of α chains. For example, the α chains of I-A and I-E antigens focus in two series of bands with characteristic relative positions, and therefore IEF can be used to rapidly distinguish I-A from I-E antigens.[84] Some of the processing variants of α have distinctive focusing patterns, and IEF and NEPHGE comparisons provided the first indication that the Ia from macrophages was processed differently from that of B cells.[37]

2D-PAGE. The most effective analytical separation method is two-dimensional gel electrophoresis (2D-PAGE) employing IEF[97] or NEPHGE[98] in the first dimension, and SDS gel electrophoresis in the second dimension. This procedure is discussed in detail in this volume [44]. The separation is based on the two parameters of charge and apparent size (or, more accurately, ability to bind SDS). This technique, developed for analysis of Ia primarily by Jones,[17,99,100] has permitted quite subtle analysis of Ia antigens. It was used to establish the existence of I_i and to analyze its relation to Ia.[17] It was used to determine that the E_β gene was encoded in the *I-A* subregion through a comparison of A/E recombinant strains.[9] Its use has permitted the discovery of an apparent intragenic recombinant in the E_β gene.[101] It continues to be useful in analysis of precursor product relationships and Ia processing, and in uncovering other components associated with Ia, such as the proteoglycan mentioned above.

The main drawback of the 2D method is difficulty in comparing samples to one another, particularly when differences are subtle. This difficulty can arise from variability in either dimension. In the first (IEF) dimension, the samples are run in separate tube gels which may differ slightly from one another in gel porosity or in the pH gradient established

[96] G. F. Dancey, J. Cutler, and B. D. Schwartz, *J. Immunol.* **123,** 870 (1979).

[97] P. H. O'Farrell, *J. Biol. Chem.* **250,** 4007 (1975).

[98] P. Z. O'Farrell, H. M. Goodman, and P. H. O'Farrell, *Cell* **12,** 1133 (1977).

[99] P. P. Jones, *J. Exp. Med.* **146,** 1261 (1977).

[100] M. McMillan, J. A. Frelinger, P. P. Jones, D. B. Murphy, H. O. McDevitt, and L. Hood, *J. Exp. Med.* **153,** 936 (1981).

[101] P. P. Jones, *J. Exp. Med.* **152,** 1453 (1980).

during the run. In the second dimension, the polyacrylamide slab gels may have local aberrations in gel porosity or in heating during the run which result in variation in mobility. The successful use of 2D-PAGE thus requires rigid standardization and adherence to protocol. Samples to be directly compared should be separated on gels cast at the same time and using the same reagent mixture.

There is an additional strategy which can be employed to verify subtle differences. Samples thought to differ can be run both separately and as a mixture to determine whether or not the separate patterns are additive when the mixture is run. Spots which have different positions in the separate runs should remain distinguishable in the mixed run.

Double-label 2D-PAGE is a theoretically applicable technique which requires higher specific activity than is usually available for studies of Ia antigens.

Preparative Methods

Preparative separation of the components of Ia molecules is difficult, and the variability in resolution of any of the methods employed requires that the purity of every preparation be analytically checked, ideally by both size and charge separation methods. Preparative methods include SDS–PAGE, IEF, hydroxylapatite chromatography, ion exchange chromatography in denaturing solvents, and reverse phase high-pressure liquid chromatography. The goal of any of these methods is to obtain purified chains in a single step, because all of the procedures employed are plagued by large losses of material.

Preparative SDS–PAGE. Electrophoresis of antibody-isolated Ia antigens can be performed in preparative size (11×200 mm) tube gels. The additional length permits better resolution. The upper gel should be about 2 cm long to ensure proper stacking. A sample containing as much as 500 mg protein can be applied to such a gel. The major species of unlabeled proteins in precipitated samples are immunoglobulin heavy and light chains. The migration of these bands does not usually distort the pattern in the Ia region of the gel, although it significantly alters the banding of H-2 antigens. Ia purified by antibody affinity columns is less subject to problems caused by protein overload because immunoglobulin is largely absent.

The α and β bands (and I_i, if present) can be eluted from the gel by mincing 2 mm gel slices in 500 μl water containing 0.01% SDS, and allowing the gel pieces to incubate for 1 to 2 days. The profile of radioactivity in the gel is determined by sampling aliquots of the eluted samples. Appropriate fractions can then be pooled and their purity rechecked on analytical SDS–PAGE and IEF or NEPHGE (see Fig. 2). The yield from this method is in the 30–50% range. Electrophoretic elution may improve yield somewhat.[102]

Preparative IEF. IEF has not been used extensively as a single step preparative tool in Ia separations. The radioactive profiles of Ia antigens separated by IEF are complicated by the presence of contaminants, since a number of different coprecipitating protein species migrate at the same pI as Ia components. In some instances, α and β chains overlap in IEF separations.

Hydroxylapatite Chromatography. Hydroxylapatite chromatography (HAP) in SDS–phosphate buffer has been successfully applied to the separation of lectin prepurified, antibody precipitated Ia α and β chains.[103] Thus far, the procedure has been performed on immunoprecipitated samples only after a preliminary gel filtration step on BioGel A-0.5m to reduce total protein applied to hydroxylapatite. Immune affinity-purified samples might be directly separable on HAP without the intervening gel filtration step. Samples to be applied to HAP are reduced and alkylated and dialyzed against 0.1% SDS, 0.01 M sodium phosphate, pH 6.4. The column is prewashed with the same buffer containing 1 mM dithiothreitol (DTT), and all operations are carried out at 37° to prevent precipitation of SDS. After sample application to the column, a linear gradient (0.01–0.6 M) of sodium phosphate buffer, pH 6.4 in 0.1% SDS, 1 mM DTT buffer is applied, and the status of the gradient is monitored by measuring refractive index. The elution occurs between 0.25 and 0.38 M sodium phosphate, with some variability from run to run. In this system the β chains are eluted before the α chains for both I-A and I-E molecules. Whether I_i would be separated from α and β on HAP is not known, since the samples analyzed thus far have been depleted of I_i. The yield of material from HAP is reported to be somewhat better than from SDS–PAGE, but it does not seem to resolve α or β subspecies from one another.

Ion Exchange Chromatography. Separation of α and β using CM-Sephadex ion exchange chromatography in propionic acid–urea has also been attempted.[24] I-A^k molecules purified by immune affinity chromatography are mixed with carrier protein and dissociated in 9 M urea, 25 mM KI, 1 mM EDTA, 1 M propionic acid, 0.25% NP-40, pH 4.5. The sample is then passed over a CM-Sephadex C-50 column which had been pretreated with 1 mg bovine γ-globulin in 1 M KCl to prevent nonspecific adsorption. The majority of the α chain is not bound to the column, and is recovered from the flow-through volume after sample application. The β chain is

[102] M. W. Hunkapiller, E. Lujan, F. Ostrander, and L. E. Hood, this series, Vol. 91, p. 227.
[103] J. H. Freed, *Mol. Immunol.* **17**, 453 (1980).

eluted with KI. A gradient of 25 to 350 mM KI is used, and the I-A β chains are eluted at about 175 mM. A shoulder eluted just prior to the main β chain peak contains an unidentified heterogeneous mixture of components which may contain α.

This technique has not yet been extended to different I-A or I-E molecules, so its general applicability is not known. The behavior of I_i in this system is also not known. The method has the advantage that it permits separation of α and β chains without using SDS. Thus, it may be the method of choice when subsequent analysis of the chains would be impeded by the presence of SDS. For example, successful and reproducible cleavage of the chains by cyanogen bromide requires the omission of SDS at all stages of preparation.

HPLC Methods. Separation of α and β chains by molecular sieving using HPLC columns may eventually be successful for Ia. Some attempts made using the Waters I-125 and the Altex TSK 2000 and 3000 have not been successful, but a reverse phase HPLC procedure for separating Ia components has been described.[104] Separation is achieved using a μBondapak C_{18} column with an initial mobile phase of 100% 0.2% Triton X-100, 1% triethylamine, and 1% trifluoroacetic acid, pH 3.0, which is altered by a step gradient of 40, 50, and 60% acetonitrile. Maintenance of low pH is critical to recovery from this column. The material separated in this way was an antibody affinity purified, Lubrol solubilized preparation, so that behavior of I_i in this system was not known initially. It has subsequently been found that I_i is not separated from α chain with this procedure, and that in scaling up to more preparative separations, the procedure becomes less reliable.[105]

Overall, despite its cumbersome nature and low yield, preparative Ia separation in SDS–PAGE tube gels seems to be the mainstay method in the field despite some intensive efforts in various labs to devise a better approach. Some of the procedures outlined are advantageous for certain purposes however, and should be considered when appropriate.

[104] D. J. McKean and M. Bell, *in* "Methods in Protein Sequence Analysis" (M. Elzinga, ed.), p. 417. Humana Press, Clifton, New Jersey, 1982.

[105] D. J. McKean, personal communication.

[50] Immunochemical Purification and Analysis of Qa and TL Antigens

By Mark J. Soloski and Ellen S. Vitetta

Introduction

Qa and TL antigens are encoded in the *Tla* region by genes located telomeric to the H-2 complex.[1] Immunochemical studies have demonstrated that at least three of these loci (*Qa-1*, *Qa-2*, and TL) encode molecules consisting of a 40,000–44,000 MW heavy chain noncovalently associated with β_2-microglobulin.[2-5] Thus, these cell surface structures bear a striking resemblance to the major self-recognition antigens H-2K, -D, and -L. Unlike H-2 antigens, however, *Tla* region antigens have a restricted tissue distribution and are expressed at levels severalfold below those of H-2 antigens.[1]

In this chapter, we will describe the methodology used to isolate and characterize murine Qa and TL antigens. These isolations rely heavily on radiolabeling and immunoprecipitation techniques reported previously, but have been modified to permit the isolation and characterization of cell surface molecules present in minute quantities. These techniques have allowed us to isolate and obtain NH_2-terminal sequence information of the Qa-2 antigen.[6]

Purification of Qa and TL Antigens

Cell Types

Since *Tla* region gene products are expressed at low levels and in a tissue-specific manner, selection of the cell population to be used for their isolation is critical. Approaches that are commonly utilized are to obtain cell populations enriched for high levels of expression of the antigens in question.

Activated T cells prepared by either concanavalin A or allogenic stimulation have been useful in isolating radiolabeled Qa-1 (R. Cook, personal

[1] L. Flaherty, *in* "The Role of the Major Histocompatibility Complex in Immunobiology" (M. E. Dorf, ed.), p. 33. Garland Press, New York, 1981.
[2] T. H. Stanton and L. Hood, *Immunogenetics* **11,** 309 (1980).
[3] R. G. Cook, R. R. Rich, and L. Flaherty, *J. Immunol.* **127,** 1894 (1981).
[4] J. Michaelson, L. Flaherty, E. Vitetta, and M. Poulik, *J. Exp. Med.* **145,** 1066 (1977).
[5] E. S. Vitetta, J. W. Uhr, and E. A. Boyse, *Cell. Immunol.* **4,** 187 (1972).
[6] M. J. Soloski, J. W. Uhr, and E. S. Vitetta, *Nature (London)* **296,** 759 (1982).

METHODS IN ENZYMOLOGY, VOL. 108

communication) and Qa-2 antigens.[7] Although there are several tumor cell lines which bear serologically detectable Qa-2 antigens (EL-4 and ERLD), immunochemical analysis has been difficult because of the small amounts of these antigens synthesized by such cells.[8]

The tumor cell line, ASL-1, has been used extensively to isolate radiolabeled TL antigens.[9,10] In addition, RADA-1, ERLD leukemias, and normal thymocytes have also been used to isolate TL antigens for structural analysis.[10,11]

Radiolabeling of Cells

Radiolabeling of the cells is performed in minimal essential Eagle's medium containing 5% heat-inactivated fetal calf serum (FCS) and lacking the appropriate amino acid(s).

Radiolabeled amino acids are obtained from New England Nuclear (Boston, MA) or Amersham (Chicago, IL). As a rule, the highest specific activity isotopes are utilized. ^{14}C-labeled amino acids are lyophilized and redissolved in medium prior to addition to cells. ^{3}H-labeled amino acids are diluted into medium and cells with no apparent adverse effect.

Cells are washed two times with Hanks' balanced salt solution (HBSS), buffered with 20 mM HEPES, pH 7.0. The washed cells are resuspended at 0.5 to 1.0 × 10^7 cells/ml in labeling medium containing ^{3}H-labeled (50–75 μCi/ml) or ^{14}C-labeled (8.5–12.5 μCi/μl) amino acids. When cultured cell lines are radiolabeled, lower cell densities (5 × 10^6/ml) are used. The cultures are gassed with 10% CO_2 and incubated for 8–10 hr at 37° on a rocking platform.

Preparation of Cytoplasmic Extracts and Isolation of Glycoproteins

The radiolabeled cells are centrifuged at 1000 rpm for 10–15 min and washed once with 20 mM Tris pH 7.4, 150 mM NaCl (TBS). The cell pellet is then resuspended at 5 × 10^7 cells/ml in TBS containing 0.5% Nonidet P-40 (NP40) and incubated on ice for 15–30 min. The nuclei are removed by centrifugation at 2000 g for 15 min. The supernatant (cytoplasmic extract) is then centrifuged at 10,000 g for 15 min to remove

[7] M. J. Soloski, J. W. Uhr, L. Flaherty, and E. S. Vitetta, *J. Exp. Med.* **153,** 1080 (1981).

[8] M. J. Soloski, unpublished observation (1979).

[9] K. R. McIntrye, U. Hammerling, J. W. Uhr, and E. S. Vitetta, *J. Immunol.* **128,** 1712 (1982).

[10] K. Yokoyama, E. Stockert, L. J. Old, and S. G. Nathenson, *Proc. Natl. Acad. Sci. U.S.A.* **78,** 7078 (1981).

[11] J. Michaelson, J.-S. Tung, L. Flaherty, U. Hammerling, and Y. Bushkin, *Eur. J. Immunol.* **12,** 257 (1982).

insoluble material. An aliquot of the supernatant is removed (10–20 μl) and the acid precipitable radioactivity is determined. In a typical experiment where Con A-activated spleen cells are radiolabeled with [³H]arginine in medium without arginine, acid precipitable radioactivity should range from 2 to 3 × 10^8 cpm/10^8 cells.

The centrifuged extract is then applied to a lentil lectin-Sepharose column. This column is prepared as described by Hayman and Crumpton.[12] For an extract from 3 to 4 × 10^8 cells, a column (130 × 5 cm) containing 2–3 mg lentil lectin/ml packed Sepharose is sufficient to remove all lentil lectin-binding material from the extract. The nonadherent proteins are allowed to fall through and the column is washed with three bed volumes of TBS–0.25% NP40. Adherent glycoproteins are eluted with 30–40 ml of TBS–0.25% NP40 containing 0.3 M α-methyl mannoside. Columns are extensively washed (100–200 ml) with TBS–0.25% NP40 prior to reuse. These columns can be used up to 30 times without any apparent loss in efficiency. It is recommended that the column be washed (occasionally) with TBS–0.25% NP40 containing 2 mM CaCl$_2$ and 2 mM MnCl$_2$. The lentil lectin-adherent fraction is then concentrated to a volume of 5–7 ml by negative pressure dialysis against TBS.

Serological Reagents

A number of serological reagents have been developed which have reactivities toward *Tla* region encoded cell surface determinants. Tables I (alloantibodies) and II (monoclonal antibodies) list those reagents useful in immunochemical isolation of *Tla* region encoded gene products. Also, listed in Table II are several monoclonal, TL-specific reagents which, although invaluable serologically, have no detectable immunoprecipitating activity. Consequently, in the case of the Qat-4 and Qat-5 specificities, we have no information concerning the molecules which bear these determinants.

It should be noted that a number of the alloantisera can have multiple reactivities. However, by using the appropriate cell population or monoclonal reagents, the isolation of a single molecular species can be achieved.

In addition to the anti-TL antibodies, a number of other reagents are needed for immunprecipitation. These are listed below.

Normal mouse serum (NMS): Purchased from Pel-Freeze Biological, Inc., Rogers, AZ. This reagent is a preparation of pooled serum from out-bred mice.

---

[12] J. J. Hayman and M. J. Crumpton, *Biochem. Biophys. Res. Commun.* **47**, 923 (1972).

TABLE I

Tla Region Specific Alloantiserum

Cell surface antigen system	Strain combination	Serological specificity recognized	Cell populations analyzed	Molecules recognized	Reference
Qa-1^l	(B6 × A.Tla^b)F$_1$αASL1 (A.TL × A.SW)F$_1$αA.TH A.Tla^{b}αA	Qa-1.1	Spleen, lymph node cells	46,000 + 12,000 55,000 – 75,000	a, b, c, d
			Thymocytes	45,000 + 12,000	b, c
	(B6·Tla^a × A)F$_1$ αA.Tla^b	Qa-1.2	Spleen, lymph node	46,000 + 12,000	b
Qa-2^m	B6.K1αB6 B6.K1αB6.K2	Qa-2.1	Spleen, lymph node Thymocytes	40,000 + 12,000	e, f
TL	(B6 × A.Tla^b)F$_1$αASL1 B6αASL1	TL-1,2,3,5 TL-1,2,3,5	Thymocytes and the ASL1, RARA1, ERLD tumor cell lines	45,000 + 12,000	g, h, i, j, k
	(B6·Tla^a × A)F$_1$αERLD (A.AC × B10A(2R)F$_1$ αB6.Tla^a thymocyte	TL-4 TL-5	ERLD, ASL1 Thymocyte, ERLD	45,000 + 12,000 45,000 + 12,000	j g, j, k

[a] T. H. Stanton and L. Hood, *Immunogenetics* **11**, 309 (1980).

[b] R. G. Cook, R. R. Rich, and L. Flaherty, *J. Immunol.* **127**, 1894 (1981).

[c] E. Rothenberg and D. Triglia, *Immunogenetics* **14**, 455 (1981).

[d] R. G. Cook, R. Jenkins, L. Flaherty, and R. R. Rich, *J. Immunol.* **130**, 1293 (1983).

[e] J. Michaelson, L. Flaherty, E. Vitetta, and M. Poulik, *J. Exp. Med.* **145**, 1066 (1977).

[f] J. Michaelson, L. Flaherty, Y. Bushkin, and H. Yerdkowitz, *Immunogenetics* **14**, 129 (1981).

[g] K. R. McIntyre, E. S. Vitetta, U. Hammerling, J. Michaelson, L. Flaherty, and J. W. Uhr, *J. Immunol.* **125**, 601 (1980).

[h] K. R. McIntyre, U. Hammerling, J. W. Uhr, and E. S. Vitetta, *J. Immunol.* **128**, 1712 (1982).

[i] E. Rothenberg and E. A. Boyse, *J. Exp. Med.* **150**, 777 (1979).

[j] K. Yokoyama, E. Stockert, L. J. Old, and S. G. Nathenson, *Proc. Natl. Acad. Sci. U.S.A.* **78**, 7078 (1981).

[k] J. Michaelson, J. S. Tung, L. Flaherty, U. Hammerling, and Y. Bushkin, *Eur. J. Immunol.* **12**, 257 (1982).

[l] As indicated below, alloantisera recognizing Qa-1 will also recognize TL on thymocytes.

[m] Only one allele, Qa-2^a, expresses a product. The alternative allele is null. Also, these reagents do not bind to Staph-A. The Qa-2 alloantiserum is serologically complex with potential reactivities toward Qa determinants −3, 4, and 5. Immunochemical analysis has revealed that the immunoprecipitating activity is directed entirely against the Qa-2 molecule.[f]

TABLE II
MONOCLONAL REAGENTS AGAINST *Tla* REGION GENE PRODUCTS

Cell surface antigen system	Hybridoma	Antibody class	Serological specificity recognized	Molecules immunoprecipitated	Reference
Qa-2	141-16.6	IgG$_{2b}$κ	Qam-2	40,000 + 12,000	a
	141-15.8	IgG$_{2b}$κ	Qam-2	40,000 + 12,000	a
	D3-262	IgMκ	Qa-2.1	None[c]	b
TL	18/20	IgG$_{2a}$κ	TL-m3	46,000 + 12,000	d, e, f
	18-407	IgG$_{2a}$κ	TL-m2	46,000 + 12,000	d, e, f
	12-120	IgG$_{2a}$κ	TL-m1	46,000 + 12,000	f
Qat-4	B16-146.R1	IgM/κ	Qat-4.2	Unknown	g
Qat-5	B16-167.R19	IgM/κ	Qat-5.2	Unknown	g

[a] P. M. Hogarth, P. E. Crewther, and I. F. C. McKenzie, *Eur. J. Immunol.* **12,** 374 (1982).
[b] J. Forman, J. Trial, S. Tonkongay, and L. Flaherty, *J. Exp. Med.* **155,** 749 (1982).
[c] M. J. Soloski, unpublished observation (1982).
[d] K. R. McIntrye, E. S. Vitetta, U. Hammerling, J. Michaelson, L. Flaherty, and J. W. Uhr, *J. Immunol.* **125,** 601 (1980).
[e] K. Yokoyama, E. Stockert, L. J. Old, and S. G. Nathenson, *Proc. Natl. Acad. Sci. U.S.A.* **78,** 7078 (1981).
[f] J. Michaelson, J. S. Tung, L. Flaherty, U. Hammerling, and Y. Bushkin, *Eur. J. Immunol.* **12,** 257 (1982).
[g] H. Lemke, G. H. Hammerling, and U. Hammerling, *Immunol. Rev.* **47,** 175 (1979).

Rabbit anti-mouse Ig (RAMIg): Prepared by the immunization of rabbits with purified MOPC-104E, TEPC-15, and serum IgG. This antiserum contains reactivities against μ, γ, α, κ, and λ chains.[13] It is used as a 1:1 dilution of whole serum in PBS.

Goat anti-mouse Ig (GAMIg): This antiserum is prepared against purified mouse IgG and is reactive with γ, κ, and λ chains.[14]

Cowan I strain of Staphylococcus aureus (S. aureus): This reagent is purchased from Calbiochem, La Jolla, CA. Each batch of *S. aureus* is obtained in 1-liter lots and individually tested for its ability to bind both rabbit and mouse Ig. In general, a 10:1 ratio of *S. aureus* (10 μl mouse serum + 0.1 ml of a 10% suspension of *S. aureus*) is sufficient for complete binding. Prior to use, *S. aureus* is washed 2× with TBS–0.25% NP40.

[13] E. S. Vitetta, K. Artzt, D. Bennett, E. A. Boyse, and F. Jacob, *Proc. Natl. Acad. Sci. U.S.A.* **72,** 3215 (1975).
[14] E. S. Vitetta, M. D. Poulik, J. Klein, and J. W. Uhr, *J. Exp. Med.* **144,** 179 (1976).

Immunoprecipitation

The vacuum-concentrated samples are centrifuged at 10,000 g for 15 min and the pellet is discarded. An aliquot (10–20 μl) is removed from the supernatant and the total acid precipitable cpm is determined. Typically, 4–8% of the labeled protein is recovered in the concentrated glycoprotein fraction.

The glycoproteins are rigorously "precleared" of immunoglobulin and "nonspecific" materials as follows. Immunoglobulin is immunoprecipitated from the concentrated glycoprotein pool by incubation with rabbit anti-mouse immunoglobulin (RAMIg) (300 μl/3 $\times$ 10^8 cells) for 45 min at 40°. The immune complexes are adsorbed to *S. aureus* by incubation for 45 min at 4° and the bacteria are pelleted. This preclearing step is necessary only when cell populations which contain B lymphocytes are used since immunoglobulin present in these extracts will coprecipitate when alloantibody-Ig-binding reagents [*S. aureus* or goat anti-mouse Ig (GAMIg)] are added. When *in vitro* lines of T cell tumors are used, this step can be omitted.[9] Immunoglobulin-depleted glycoproteins are further depleted of "nonspecific" material by adding normal mouse serum (NMS) (250 μl/3 $\times$ 10^8 cells) for 45 min at 4°, followed by two treatments with *S. aureus*. Precleared glycoprotein pools are then reacted with the appropriate alloantibody at 4° for 2–12 hr. All alloantibodies should be titrated so as to remove all reactive molecules. The immune complexes are removed by either absorption to *S. aureus* or by immunoprecipitation with GAMIg. The immunoprecipitates or *S. aureus*-bound complexes are washed three times with TBS containing 0.1% sodium dodecyl sulfate (SDS), 0.25% NP40, and 0.2% deoxycholate (DOC), followed by one wash with TBS.

Isolation of Tla-Region Encoded Molecules by SDS–PAGE

Immune complexes eluted from the *S. aureus* or immune precipitates are dissolved by boiling for 2 min in electrophoresis sample buffer (0.0625 M Tris, pH 6.8, 2.3% SDS, 10% glycerol, 5% 2-mercaptoethanol). All insoluble material is removed by centrifugation at 10,000 g for 5 min. Samples are then electrophoresed for 4–5 hr on 12.5% polyacrylamide tube gels using the Laemmli discontinuous system.[15] The stacker gel is discarded and the resolving gel is cut using a gel slicer (Bio-Rad, Richmond, CA, Model 195) equipped with a 1.5 mm slicing grid. Each slice is incubated overnight (minimum 12 hr) in 1.0 ml of 0.01% SDS to elute the labeled protein from the gel matrix. The radioactivity of aliquots (50 μl) of

[15] R. G. Cook, J. W. Uhr, J. D. Capra, and E. S. Vitetta, *J. Immunol.* **121,** 2205 (1978).

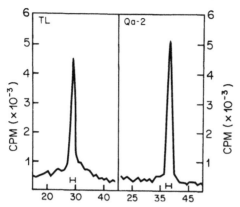

FIG. 1. An SDS–PAGE profile of Qa-2 and TL antigens. Precleared glycoprotein fractions from [³H]arginine-labeled C57BL/6 Con A-stimulated spleen cells or ASL-1 tumor cells were precipitated with anti-Qa-2 alloantiserum (K1αB6) or anti-TLm3 antibody, respectively. Immunoprecipitates are solubilized and analyzed on 12–5% SDS–PAGE tube gels as described above. Only fractions 20–50 are shown. The fractions to be pooled are indicated by the bars.

each fraction is determined in scintillation cocktail and the radioactivity in the gel fractions is calculated. A typical gel profile is shown in Fig. 1. The appropriate fractions are then pooled, the gel particles are removed by filtration through a 0.45-μm Millipore filter (Millipore Corp., Bedford, MA), and the samples are lyophilized.

Analysis of Isolated Qa and TL Antigens

The purification of *Tla* region encoded antigens, by the procedure described above, has yielded material suitable for structural analysis. Tryptic peptide map analysis[7,10,11] and NH$_2$-terminal sequence determinations[6] are the most common analytical techniques used and will be briefly described below.

For tryptic peptide map analysis, the lyophilized material is dissolved in 2–3 ml H$_2$O containing 1.0 mg human γ-globulin (HGG) as carrier and dialyzed against 0.1 M NH$_4$HCO$_3$ pH 8.0. The material is then relyophilized and dissolved in 1.0 ml of 0.1 M HN$_4$HCO$_3$, pH 8.0, with 20 μg TPCK-trypsin. After a 1 hr incubation at 37°, additional trypsin (100 μg) is added and the incubation is continued for 12–16 hr at 37°. The pH is then lowered by the addition of 2–3 drops of acetic acid and the sample is lyophilized. Before chromatography, the peptides are solubilized in 2.0 ml 0.05 M pyridine acetate (PA), pH 3.13. Acid-insoluble material is removed by centrifugation at 10,000 g for 5 min. Tryptic peptides are analyzed by cation exchange chromatography on Technicon Chromobeads

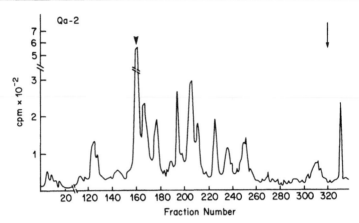

Fɪɢ. 2. Tryptic peptide map of the Qa-2 antigen isolated from C57BL/10 Con A-stimulated spleen cells. Qa-2 antigens metabolically labeled with [³H]arginine are isolated by immunoprecipitation and SDS–PAGE from a lysate of concanavalin A-stimulated spleen cells. The isolated molecules are digested with trypsin and the resulting peptides analyzed by cation-exchange chromatography on Technicon Chromobeads type P. The peak marked with an arrowhead has been shown to cochromatograph with [¹⁴C]arginine. The last peak, marked with an arrow, represents the peptides eluted with 2 N NaOH.

type P (Technicon T and T Corp, McFarland, WI) using a 3 × 150-mm microbore column maintained at 54° by a water jacket. The solubilized peptides are applied to the column in 0.05 M PA, pH 3.13. After washing the column with 5–6 ml of starting buffer, a pH-ionic strength gradient is applied using a Varigrad gradient maker (Phoenix Precision Instruments, Gardiner, NY). Thirty milliliters of each of the following PA buffers is used: 0.05 M, pH 3.13; 0.1 M pH 3.54; 0.2 M, pH 4.02; 0.5 M pH 4.50; 2.0 M pH 5.0. After completion of the gradient, 2–3 ml of 2 N NaOH is applied to remove all bound material. Fraction (12 drops) are collected into minivials and allowed to evaporate. Water (0.35 ml) and scintillation fluid are added and the radioactivity of the samples is determined. A typical tryptic peptide map of the Qa-2 molecule is shown in Fig. 2. The sensitivity of this technique, which can detect one single amino acid difference between two molecules, has been especially useful in comparative analysis of class I molecules.[16,17] For example, if a mixture of ³H-labeled Qa-2 antigen and another ¹⁴C-labeled protein is codigested with trypsin and cochromatographed, an assessment of the degree of relatedness can be made by the percentage of coeluted peptides.[7,9,10,16]

[16] J. L. Brown, K. Kato, J. Silver, and S. G. Nathenson, *Biochemistry* **13**, 3174 (1974).
[17] J. L. Brown and S. G. Nathenson, *J. Immunol.* **118**, 98 (1977).

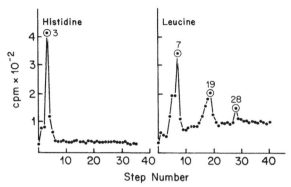

FIG. 3. Amino acid sequence determination of the leucine and histidine residues in the NH$_2$-terminus of the Qa-2 antigens. Qa-2 antigens radiolabeled with either [^{3}H]leucine or histidine are immunoprecipitated from C57BL/6 concanavalin A-stimulated spleen cells and isolated by SDS–PAGE. The purified antigens are dialyzed to remove SDS and loaded onto a Beckman 890C sequencer for analysis. The data are represented as the amount of radioactivity recovered after each Edman step. The circles note those steps where an amino acid assignment is indicated.

For NH$_2$-terminal sequence analysis, the radiolabeled antigen isolated by SDS–PAGE is lyophilized, dissolved in 1.0 ml H$_2$O with 1 mg human IgG (HGG) as carrier, and dialyzed against H$_2$O for 4 hr to remove free SDS. The dialyzed material is applied onto a Beckman 890C sequencer for analysis.[18] In each sequencing run, a [^{35}S]methionine-labeled immunoglobulin κ chain is routinely included as an internal standard. The methionine residues at positions 4, 11, and 13 allow the determination of a repetitive yield. Thiazolinone derivatives are evaporated under nitrogen, scintillation fluid is added, and the radioactivity of the samples is determined. Typical sequencing data for the leucine and histidine assignments for the Qa-2 antigen is shown in Fig. 3. Assignments are based on (1) the appearance of a peak of radioactivity at a given sequencing step and (2) the ability of multiple peaks to fit a standard repetitive yield curve as determined for each run using the [^{35}S]methionine immunoglobulin light-chain standard.[6]

Tryptic peptide map comparisons and NH$_2$-terminal sequence analysis has provided insights into the evolution and possible functions of these antigens.[6,9,10] At present, no techniques are available for the isolation of native, biologically active antigens. However, with the availability of monoclonal antibodies which are reactive with TL and Qa-2 (Table II),

[18] R. G. Cook, M. S. Siegelman, J. D. Capra, J. W. Uhr, and E. S. Vitetta, *J. Immunol.* **122,** 2122 (1979).

the isolation of these antigens by specific affinity chromatography may prove useful. A similar approach has proven invaluable in the structural and functional analysis of the H-2K^k antigens.[19,20]

Acknowledgments

The work described in this chapter was supported by research grants from the National Institutes of Health (AI-13448) and from the Welsh Foundation. In addition, the authors wish to acknowledge Drs. Richard Cook, J. Donald Capra, Katherine McIntyre, and Jonathan Uhr for their contributions, advice, and criticism of this work. The expert technical assistance of Ms. Linda Trahan, Ms. Sandy Graham, Ms. Shirley Shanahan, Ms. Kay Snave, and Ms. Martha Miles is gratefully acknowledged. We also thank Ms. G. A. Cheek for her patient and expert typing skills.

[19] K. C. Stallcup, T. A. Springer, and M. F. Mescher, *J. Immunol.* **127**, 921 (1981).
[20] J. E. Mole, F. Hunter, J. W. Paslay, A. S. Bhown, and J. C. Bennett, *Mol. Immunol.* **19**, 1 (1982).

[51] Purification and Molecular Cloning of Rat Ia Antigens

By W. Robert McMaster

Introduction

The Ia antigens are a set of highly polymorphic cell surface molecules involved in many cell–cell interactions and coded for by the major histocompatibility complex (MHC). The polymorphism of the Ia antigens is directly related to their function, as these molecules are thought to be the products of the immune response genes which genetically determine the outcome of many types of immune responses.[1] Ia antigens have been identified in several different mammalian species and are composed of two noncovalently bonded glycosylated polypeptides referred to as the α and β chains. The Ia glycoproteins show a restricted tissue distribution being present on B lymphocytes, dendritic cells, some macrophages and thymocytes, and on epithelial and reticular cells of thymus and spleen. These molecules are not generally found on the majority of T lymphocytes or other cells. The rat MHC, called the RT1 complex, codes for two sets of Ia antigens referred to as Ia-A and Ia-E which correlate to the mouse H-2 I-A and I-E Ia antigens, respectively.[2] There are at least two

[1] J. Klein, *Science* **203**, 516 (1979).
[2] T. Fukumoto, W. R. McMaster, and A. F. Williams, *Eur. J. Immunol.* **12**, 237 (1982).

sets of Ia antigens encoded by the human HLA complex. The HLA-DR antigens are homologous to rodent Ia-E antigens[3] while the HLA-DC1 (HLA-DS) antigens have recently been shown to be homologous to Ia-A antigens.[4,5]

This chapter describes the purification of Ia glycoproteins from detergent solubilized rat spleen membranes using monoclonal antibody affinity chromatography and the characterization of the purified Ia molecules. Methods are also described for the purification of mRNA from rat spleen and the molecular cloning of cDNA coding for rat Ia-A antigens.

Production of Monoclonal Antibodies to Rat Ia-A Antigens
(This Volume [57])

A series of monoclonal antibodies to rat Ia antigens were derived during a study of the characterization of glycoproteins from rat thymus.[6] Rat thymocyte glycoproteins were prepared by lentil lectin affinity chromatography of membrane proteins solubilized by sodium deoxycholate followed by chromatography on Sephadex G-200. A fraction of Ia-like glycoproteins was then used to immunize mice and monoclonal antibodies prepared essentially as described by Kohler and Milstein.[7] The techniques for producing monoclonal antibodies have recently been reviewed in a previous volume of this series.[8] Monoclonal antibodies which bound to rat lymphocytes were detected using indirect radioimmune cellular binding assays and a fluorescence-activated cell sorter (FACS). The advantage of indirect binding assays for the detection and quantitation of cell surface molecules has been reviewed.[9] Using this approach, a series of monoclonal antibodies of the IgG class, designated MRC OX 3, 4, 5, and 6 were derived. These antibodies detect Ia antigenic determinants present on all rat B lymphocytes and on a subpopulation of rat thymocytes. MRC OX3 antibody detect an Ia antigenic determinant polymorphic in the rat, while MRC OX 4, 5, and 6 detect a determinant common to all rat strains tested. When assayed on spleen cells from recombinant mouse strains, all four antibodies detect polymorphic Ia determinants encoded by the H-2 *I-A* subregion. MRC OX3 antibody reacts with cells of the H-2 haplotype *b* and *s* while MRC OX 4, 5, and 6 antibodies react with cells of the H-2 haplo-

[3] J. Silver, W. A. Russell, B. L. Reis, and J. A. Frelinger, *Proc. Natl. Acad. Sci. U.S.A.* **74,** 5131 (1977).
[4] S. M. Goyert, J. E. Shively, and J. Silver, *J. Exp. Med.* **156,** 550 (1982).
[5] M. R. Bono and J. L. Strominger, *Nature (London)* **299,** 836 (1982).
[6] W. R. McMaster and A. F. Williams, *Immunol. Rev.* **47,** 117 (1979).
[7] G. Kohler and C. Milstein, *Eur. J. Immunol.* **6,** 511 (1976).
[8] G. Galfre and C. Milstein, this series, Vol. 73 [1].
[9] A. F. Williams, *Contemp. Top. Mol. Immunol.* **6,** 83 (1977).

type *k* and *s*. Therefore, MRC OX 3, 4, 5, and 6 antibodies react with the rat homolog of mouse I-A antigens and define Ia-A antigens in the rat.

Purification of Rat Ia Antigens Using Monoclonal Antibody Affinity Chromatography

The methods described are based on techniques first used to purify rat Thy 1 antigen from brain tissue[10] and modified for the use of monoclonal antibodies.

Purification of IgG from Ascities Fluid. CLARIFICATION OF ASCITIES FLUID. Ascities fluid is collected from the peritoneal cavity of mice, injected with Pristane 2 weeks prior to the injection of the appropriate monoclonal antibody cell line, and is stored at −20°. Upon thawing, the ascities fluid is filtered through cotton wool to remove aggregated material and centrifuged at 10,000 *g* for 20 min at 4°. This step should immediately precede ammonium sulfate precipitation as further material aggregates upon storage at 4° or after freeze thawing.

As each monoclonal antibody may have slightly different characteristics, it is important to follow the antibody activity during each step of the purification by an appropriate serological assay. As a safety precaution, all supernatants should be saved until all the antibody activity can be accounted for.

AMMONIUM SULFATE PRECIPITATION. Immunoglobulin preparation from ascities fluid is described starting from an initial volume of 50 ml usually containing between 50 to 250 mg IgG. To 50 ml of clarified ascities fluid 50 ml of phosphate-buffered saline pH 7.3 (PBS: 0.2 g KCl, 0.2 g KH_2PO_4, 1.15 g Na_2HPO_4, 8.0 g NaCl per 1000 ml distilled H_2O) is added. IgG is precipitated by adding dropwise 100 ml of 100% saturated ammonium sulfate pH 7.4 while stirring at 20° (to prepare 100% saturated ammonium sulfate: 800 g of ammonium sulfate is added to 1000 ml of distilled H_2O. The mixture is heated to 100° for 30 min, then cooled, and the pH is adjusted to 7.4). After the saturated ammonium sulfate is added, the mixture is stirred for 30 min at 20°, and then left at 4° for 1 hr without stirring. The precipitate is recovered by centrifugation at 10,000 *g* for 20 min at 4°, dissolved in 50 ml PBS, and reprecipitated by adding 50 ml 100% saturated ammonium sulfate as before. The final precipitate is dissolved in DEAE buffer (0.032 *M* Tris base, 0.025 *M* HCl, 10 m*M* NaN_3, pH 7.3) and dialyzed extensively against the same buffer at 4°. Proteins are determined from the absorbance at 280 nm (1 mg/ml IgG = 1.35 OD units).

DEAE ION EXCHANGE CHROMATOGRAPHY. DEAE-cellulose (DE-52

---

[10] A. N. Barclay, M. Letarte-Muirhead, and A. F. Williams, *Biochem. J.* **151**, 699 (1975).

Whatman) and DEAE-Sephacel (Pharmacia) exhibit similar characteristics. A column of DEAE is prepared using 1 ml of gel for 5 mg protein and equilibrated with DEAE buffer. The column is washed until the pH and/ or conductivity of the effluent is equal to that of the DEAE buffer. The protein sample is applied and the column is washed with DEAE buffer until absorbance at 280 nm is less than 0.05. Four to 5 ml fractions are collected. IgG is eluted with a linear gradient of 0 to 0.1 M sodium chloride in DEAE buffer (300 ml total for a 25 ml column). The column is then eluted with 0.5 M sodium chloride in DEAE buffer. The absorbance at 280 nm is determined for each fraction. Protein-containing fractions are assayed for antibody activity and analyzed for purity by polyacrylamide gel electrophoresis in sodium dodecyl sulfate (SDS–PAGE). IgG-containing fractions are pooled, concentrated by ultrafiltration to 5 mg/ml, and stored at $-20°$. A protein which is occasionally present following this procedure is serum transferrin of molecular weight 85,000. Transferrin can be separated from IgG by ammonium sulfate precipitation at 50% saturation, as described, or by gel chromatography on Sephacryl S-200 or equivalent gel. Methods to prepare $F(ab')_2$ and Fab' fragments by pepsin digestion have been recently reviewed.[11]

Preparation of Monoclonal Antibody Affinity Columns. Sepharose 4B (Pharmacia) activated by reaction with cyanogen bromide has been used for the preparation of monoclonal antibody affinity columns. The following steps must be carried out in a fume hood. To activate Sepharose 4B, 25 ml of wet beads is washed with 250 ml cold distilled H_2O on a sintered glass funnel, dried until just moist, and then transferred to a glass beaker containing 25 ml distilled H_2O. The beaker is placed in an ice bucket and the slurry gently stirred. A pH electrode is placed in the slurry to monitor the reaction. Seven hundred and fifty milligrams of cyanogen bromide is weighed in a fume hood and dissolved in 5 ml dimethylformamide. The dissolved cyanogen bromide is added to the Sepharose 4B dropwise over 1 min. The pH of the reaction is kept between 10 and 11 with 5 M sodium hydroxide for 10 to 15 min. The activated Sepharose 4B is transferred as rapidly as possible to a sintered glass funnel and washed with 250 ml of cold distilled H_2O followed by 250 ml cold sodium borate buffer (0.05 M $Na_2B_4O_7 \cdot 10H_2O$ pH 8.0, 10 mM NaN_3). The activated Sepharose 4B is dried on a sintered glass funnel and then transferred to a tube containing 25 ml of monoclonal antibody (5 mg/ml) in sodium borate buffer. The gel and monoclonal antibody should form a thick slurry which is gently rotated for 18 to 24 hr at $4°$. The gel mixture is returned to a sintered glass

[11] P. Parham, M. J. Androlewicz, F. M. Brodsky, N. J. Holmes, and J. P. Ways, *J. Immunol. Tech.* **53**, R101 (1982).

funnel and filtered. To determine coupling efficiency the absorbance at 280 nm of the filtered supernatant is measured and compared to that of the original solution. The gel is washed with 250 ml of sodium borate buffer and any free reactive groups are blocked by resuspending in 50 ml of 0.15 M ethanolamine in sodium borate buffer pH 8.5. The gel is incubated for 2 hr at 4° and then washed with 10 mM Tris–HCl, 10 mM NaN$_3$ pH 8.0. Prior to use the free IgG that may be present should be washed off the gel with elution buffer.

Preparation of Spleen Membrane. Crude rat spleen membranes are prepared using the Tween 40 method.[12] All steps are carried out at 4°. Whole spleens, usually 150 g or about 150 spleens, are disrupted in 150 ml PBS in a Waring blender at full speed. To minimize proteolytic degradation, 10 mM iodoacetamide and 5 mM diisopropylfluorophosphate (DFP) or 5 mM phenylmethylsulfonyl fluoride (PMSF) are added prior to disruption. An equal volume of 5% (v/v) Tween 40 in PBS is added to the disrupted spleens and the mixture homogenized by four strokes of a Potter-Elvehjem homogenizer with a power driven Teflon pestle. The mixture is stirred for 60 min on ice and nuclei sedimented by centrifugation at 3000 g for 30 min. The supernatant is filtered through two layers of cheesecloth and centrifuged at 60,000 g for 90 min to yield a pelleted crude membrane. The membrane fraction is resuspended in 10 mM Tris–HCl pH 8.0 and centrifuged at 60,000 g for a further 90 min. The resulting membrane fraction is resuspended in a total of 100 ml 10 mM Tris–HCl pH 8.0, 10 mM NaN$_3$, and solubilized as described below or stored at −70°.

Solubilization of Spleen Membrane and Monoclonal Antibody Affinity Chromatography. Spleen membranes (150 ml) are solubilized by the addition of an equal volume (150 ml) of 10% (w/v) sodium deoxycholate in 10 mM Tris–HCl pH 8.0, 10 mM NaN$_3$ and homogenized as described above. Iodoacetamide (10 mM) and 5 mM DFP or 5 mM PMSF are added prior to homogenization. The mixture is stirred on ice for 60 min and centrifuged at 60,000 g for 120 min. The supernatant is passed first through a column of rabbit IgG-Sepharose 4B, then through a monoclonal antibody-Sepharose 4B column (25 ml conjugated Sepharose 4B containing 125 mg IgG) at a flow rate of 10 to 15 ml/hr. It is important to precede the specific affinity column with a nonspecific column as the latter removes any material binding to cyanogen bromide-activated Sepharose 4B or nonspecifically to IgG. Columns are washed with deoxycholate buffer (0.5% w/v sodium deoxycholate in 10 mM Tris–HCl pH 8.0, 10 mM NaN$_3$) until absorbance at 280 nm is less than 0.05. The monoclonal antibody column is then washed with deoxycholate buffer containing 0.2

[12] R. Standring and A. F. Williams, *Biochim. Biophys. Acta* **508,** 85 (1978).

M NaCl and eluted with deoxycholate elution buffer (0.5% w/v sodium deoxycholate in 0.05 M diethylamine–HCl pH 11.5, 10 mM NaN$_3$). Absorbance at 280 nm and the pH are determined. Each fraction (4 ml) is then immediately neutralized by the addition of solid glycine. A high pH elution buffer is used because sodium deoxycholate is insoluble below pH 7. Furthermore, this elution procedure has given at least 50% recovery of antigenic activity in the purification of several different cell surface glycoproteins.

The eluted material is pooled and concentrated by ultrafiltration and chromatographed on a Sephacyl S-200 column (1.6 × 100 cm) eluted with deoxycholate buffer. Each fraction is analyzed by SDS–PAGE and for antigenic activity and those fractions containing activity are pooled and concentrated by ultrafiltration. In order to remove sodium deoxycholate, the protein is precipitated by adding 95% ethanol to a final concentration of 75% and by incubating at −20° for 48 hr. The precipitate is washed in 75% ethanol by centrifugation, dried under vacuum, dissolved in 10 mM Tris–HCl pH 8.0, 10 mM NaN$_3$, and stored at −20°.

Yields of antigenic activity are calculated by carrying out quantitative inhibition assays using cellular radioimmune binding assays. These assays can be used in the presence of detergents by treating target cells with glutaraldehyde prior to use.[9] However, use of some monoclonal antibodies in inhibition assays in the presence of detergents results in an underestimate of actual yields of antigenic activity.[6] This is partly due to differences in the binding properties of some monoclonal antibodies in the presence and absence of detergents. The binding properties of several monoclonal antibodies to cell surfaces have been discussed recently.[13] In the purification of rat Ia antigens, rabbit antibodies prepared by immunization with purified Ia antigen were more reliable in determining the yields and overall purification than monoclonal antibodies.[6] Most of the rat Ia antigenic activity is lost in preparation of membranes where the recovery varies between 30 and 50%. Monoclonal antibody columns retain 90% of the rat Ia antigenic activity present in the sodium deoxycholate solubilized spleen membrane preparation and 70% of the applied activity is eluted with high pH buffer. The purification of rat Ia antigen in the affinity column step is 200- to 400-fold and values of several thousandfold have been obtained in the purification of the rat thymocyte glycoprotein W3/13 antigen.[14]

In the purification of rat Ia antigen, the spleen membranes are solubilized in sodium deoxycholate since this detergent will release DNA from

[13] D. W. Mason and A. F. Williams, *Biochem. J.* **187**, 1 (1980).
[14] W. R. A. Brown, A. N. Barclay, C. A. Sunderland, and A. F. Williams, *Nature (London)* **289**, 456 (1981).

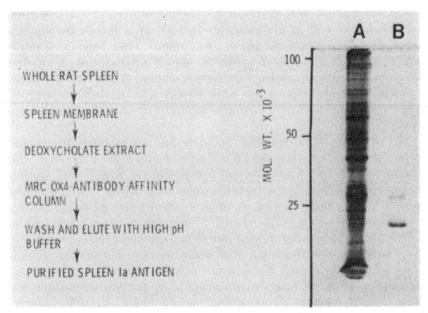

Fig. 1. Purification of Ia-A antigens from spleen membrane of Wistar rats. A summary of the purification steps is shown and analysis by SDS–PAGE of track A, an aliquot (100 µg) of sodium deoxycholate solubilized rat spleen membrane electrophoresed in the presence of 100 mM dithiothreitol; track B, an aliquot (8 µg) of eluted Ia-A antigen electrophoresed without reduction.

the nuclei of whole cells. To minimize the loss of antigen occurring during the membrane preparation, whole rat spleens are solubilized directly with the nonionic detergent Triton X-100. The detergent extract is then applied to a monoclonal antibody column washed with Triton X-100, followed by deoxycholate buffer. The bound Ia antigen is then eluted with deoxycholate buffer pH 11.5. Following this procedure, however, the Ia antigens are contaminated with other proteins which cannot be removed. A procedure has been developed recently for the purification of W3/13 antigen by which whole thymocytes are solubilized first in the nonionic detergent Brij 96, the nuclei removed, and then sodium deoxycholate added to solubilize further any aggregated material.[14] This procedure eliminates the need for preparing membrane and may be suitable for the purification of other membrane proteins.

A summary of the procedures for the purification of Ia-A antigens from rat spleen using MRC 0X4 monoclonal antibody is shown in Fig. 1. SDS–PAGE of an aliquot of the starting sodium deoxycholate extract (track A) and of an aliquot of the eluted material from the antibody column (track B) is shown in Fig. 1. The purified rat Ia-A antigen was composed of two polypeptides, the α and β chains, of molecular weight 30,000 and 25,000 (unreduced), respectively. As shown in Fig. 1, the eluted antigen was free from contaminating proteins and the use of the monoclonal antibody column resulted in a single step purification.

Strategy to Produce Monoclonal Antibodies to a Second Rat Ia Antigen

The first series of monoclonal antibodies (MRC 0X3, 4, 5, and 6) were clearly shown to react with the rat homolog of mouse I-A encoded Ia antigens. Quantitative inhibition studies using rat alloantibodies against Ia antigens and rat Ia antigen purified using MRC 0X4 antibody showed that approximately 50% of these rat alloantibodies reacted with rat Ia-A antigens, while the remainder appeared to react with other rat Ia molecules.[6] To investigate this possibility further, the following approach was used to prepare monoclonal antibodies to these additional rat Ia antigens, which, subsequently, were shown to be homologous to mouse I-E encoded antigens.

Rat spleen membrane glycoproteins are prepared by lentil lectin affinity chromatography and separated according to size by gel filtration in deoxycholate buffer. Fractions containing Ia antigens, as detected by SDS–PAGE analysis, are pooled and Ia-A antigens removed by affinity chromatography on a MRC 0X6 antibody column. The remaining glycoproteins are used to immunize mice and a monoclonal antibody, MRC 0X17, reacting with a second rat Ia antigen is produced.[2] The MRC 0X17 antibody reacts only with B lymphocytes, detects an antigenic determinant common to all rat strains tested, and does not cross-react with mouse spleen cells.

A monoclonal affinity column is prepared using MRC 0X17 antibody following the methods described above and is used to purify the corresponding Ia antigen from sodium deoxycholate solubilized rat spleen membranes. Rabbit antibodies are prepared to the purified Ia antigen which cross-react with mouse spleen cells. The binding pattern of these rabbit antibodies to spleen cells from various mouse strains maps to the mouse H-2 *I-E* subregion thus establishing that the MRC 0X17 Ia antigen is the rat homolog of mouse I-E Ia antigen. The rat Ia-A antigen, purified using MRC 0X4 or 6 antibodies, and the rat Ia-E antigen, purified using MRC 0X17 antibody, can be shown to be encoded by the rat MHC (RT1 complex) by their ability to inhibit the binding of rat alloantibodies. Each Ia antigen used alone inhibits approximately 50% of the binding of the rat alloantibodies; when added together, greater than 90% of the specific binding is inhibited. Quantitative absorption studies using monoclonal and rabbit antibodies show that there is no cross-reactions between the rat Ia-A and Ia-E antigens.[2]

Separation of Rat Ia α and β Chains and Location of Antigenic Determinants

The α and β chain of Ia antigens are strongly associated by noncovalent bonding. The two chains can be dissociated by heating to 100° in the presence of SDS. The α and β chains can then be separated by preparative SDS–PAGE under nonreducing conditions. The following procedure[15] has been used to purify the α and β chains of rat Ia-A and Ia-E antigens previously purified by monoclonal antibody affinity chromatography.

Separation of Ia α and β Chains. One milligram of purified Ia antigen is precipitated with 75% ethanol. The precipitate is dissolved in 0.5 ml 5% SDS prepared in SDS–PAGE stacking gel buffer of Laemmli[16] (0.125 M Tris–HCl pH 8.3, 15% glycerol, 0.1% bromophenol blue) and heated to 100° for 5 min. After cooling, the sample is loaded onto a cylindrical gel composed of a 3% polyacrylamide stacking gel (1 × 1.5 cm) and a 10% polyacrylamide separating gel (1.5 × 10 cm) and electrophoresed for 6 hr. The gel is cut into 1.5 mm slices and each slice individually eluted with 3 × 1 ml of 0.1% SDS in distilled H_2O at 20° for a total of 24 hr. An aliquot of each sample is analyzed by slab SDS–PAGE. The fractions found to contain either α or β chain are pooled, concentrated by ultrafiltration, and dialyzed against 0.1% SDS or distilled H_2O. Yields of between 40 and 70% for each chain are obtained routinely.

In order to determine which chain carries the antigenic determinant, quantitative inhibition assays using monoclonal antibodies are carried out using whole Ia antigen and each separated chain. After absorption at 4° for 16 hr, each sample is centrifuged and the remaining antibodies assayed using indirect binding assays. With some monoclonal antibodies clear cut results are obtained. MRC 0X6 antibody, which detects a common Ia-A antigenic determinant in rats and a polymorphic determinant in mice, reacts specifically with purified rat Ia-A β chain.[15] Assuming that rat and mouse Ia-A antigens are homologous, these results imply that the mouse polymorphism is located in the Ia-A β chain. With MRC 0X3 antibody, which detects an Ia-A polymorphic determinant in both rats and mice, purified rat Ia-A antigen inhibits the binding to rat lymph node cells whereas no inhibition is found with separated α or β chains when added separately or together. The determinant with which MRC 0X3 reacts,

[15] W. R. McMaster, *Immunogenetics* **13,** 347 (1981).
[16] U. K. Laemmli, *Nature (London)* **277,** 680 (1970).

therefore, depends either on the α and β chains being associated together in their native conformation or the determinant being denatured during the separation procedures. Using the same approach, MRC 0X17 antibody, which reacts with a common Ia-E determinant in rats and does not cross-react with mouse spleen cells, reacts specifically with separated rat Ia-E α chain.[2]

Molecular Cloning of Rat Ia Antigens

Cloning of cDNA Coding for Rat Ia Antigens

A possible approach to the determination of the primary structure of rat Ia antigens is to clone and sequence the complementary DNA (cDNA) coding for these molecules. The protein sequence can then be predicted from the nucleotide sequence and these data may be used to predict the domain structure of these molecules. Furthermore, the cloned cDNAs may be used as probes to study the structure and expression of the corresponding genomic DNA. Studies at the DNA level may also lead to a better understanding of the molecular basis of the polymorphisms exhibited by the Ia antigens.

The approach used to clone cDNA coding for rat Ia-A antigens is to purify mRNA from rat spleen and identify specific mRNA coding for Ia-A α and β chains by *in vitro* translations in the presence of radioactive amino acids. Rat Ia polypeptides are immunoprecipitated with rabbit antibodies prepared against separated Ia-A chains and the radioactive proteins are analyzed by SDS–PAGE. Total spleen mRNA is fractionated by sucrose density gradient centrifugation and fractions containing mRNA coding for Ia-A antigens are pooled and used to prepare double-stranded cDNA. cDNA is inserted into the bacterial plasmid, pBR322, and the resulting recombinant plasmids used to transform *E. coli*. Specific cDNA clones are detected using a positive mRNA selection assay. This strategy results in the identification of a cDNA clone coding for a rat Ia-A α chain.

Purification of Spleen mRNA. Since spleen tissue contains high levels of ribonuclease, methods[17] to inhibit this enzyme must be used in order to obtain intact mRNA. Animals are sacrificed by placing them in a desiccator containing CO_2 (dry ice and water) and spleens immediately removed and placed in sterile ice cold PBS. Whole spleens are disrupted using a Polytron homogenizer (Brinkman Instruments) in 10 ml/g of tissue of guanidine–HCl buffer (7.5 M guanidine–HCl, 0.025 M sodium citrate pH 7.0, 10 mM dithiothreitol). After homogenization the volume is measured. Ten percent sodium lauryl sarcosine in distilled H_2O is added to a final

---

[17] J. M. Chirgwin, A. E. Przybyla, R. J. McDonald, and W. J. Rutter, *Biochemistry* **18,** 5294 (1979).

concentration of 0.5%, and the solution centrifuged at 3000 g for 30 min at 0°. The supernatant is transferred to sterile polypropylene 50-ml plastic test tubes and the RNA precipitated by the addition of 95% ethanol to a final concentration of 33%. The ethanol is mixed well and the preparation is left at −20° for 6 to 18 hr. The precipitate is collected by centrifugation (3000 g for 30 min at 0°) and the supernatant is removed. The test tubes are drained thoroughly and any remaining liquid is wiped away with tissue paper. The pellets are dissolved in one-half of the starting volume with guanidine–HCl buffer and centrifuged as above. The RNA is precipitated by adding 95% ethanol to 33% and left at −20° for 4 to 18 hr. The precipitated RNA is then collected as before, dissolved in one-quarter the starting volume with guanidine–HCl buffer, and reprecipitated by adding 95% ethanol to 33%. From this point onward, all buffers must be treated with 0.2% diethyl pyrocarbonate for 20 min at 20° and autoclaved to inactivate ribonuclease. All glassware must be baked overnight or sterile plastic test tubes and pipets must be used. The final pellet of RNA is dissolved in sterile distilled H_2O and centrifuged at 3000 g for 30 min to remove insoluble material. The absorbance at 260 and 280 nm is measured and the RNA concentration is calculated (1 mg/ml RNA = 20 OD units at 260 nm). The ratio of absorbance at 260 nm : 280 nm should be between 1.75 and 2.0. One gram (wet weight) of whole spleen yields approximately 2 mg of total cellular RNA.

Purification of Poly(A) RNA. Polyadenylated RNA is separated from ribosomal RNA by two rounds of affinity chromatography using oligo(dT)-cellulose.[18] All buffers must be treated with 0.2% diethyl pyrocarbonate for 20 min at 20° and autoclaved. RNA is diluted to 1 to 2 mg/ml with sterile distilled H_2O and EDTA pH 7.5 is added from a 0.2 M stock solution to give a final concentration of 10 mM. RNA is heated to 70° for 5 min and cooled rapidly in an ice bath. Sodium acetate pH 7.5 is added from a 2 M stock solution to give a final concentration of 0.4 M and the RNA is applied to an oligo(dT)-cellulose column (Type T-3, Collaborative Research). The RNA solution is passed through the column three times. The column is then washed with 0.4 M sodium acetate pH 7.5, 10 mM EDTA until the absorbance at 260 nm is less than 0.05. Bound poly(A) RNA is eluted with 1 mM EDTA pH 7.5. The eluate is made 0.4 M in sodium acetate pH 7.5 and 10 mM in EDTA and applied again to the oligo(dT)-cellulose column. The column is washed and bound poly(A) RNA is eluted as detailed above. Eluted poly(A) RNA is precipitated by addition of sodium acetate pH 5.0 to 0.2 M and two volumes of 95% ethanol. The precipitate is allowed to form at −20° for 18 hr and is col-

[18] H. Aviv and P. Leder, *Proc. Natl. Acad. Sci. U.S.A.* **69**, 1408 (1972).

lected by centrifugation at 3000 g for 30 min, washed in 70% ethanol, and dried under vacuum. The final pellet of poly(A) RNA is dissolved in sterile distilled H_2O, aliquoted, and stored at $-70°$. Approximately 3% of total spleen RNA is obtained as poly(A) RNA after two rounds of affinity chromatography on oligo(dT) cellulose.

Cell-Free Translation of Rat Spleen mRNA and Immunoprecipitation of Rat Ia-A Polypeptides. The purified mRNA is tested for biological activity by translation in a cell-free rabbit reticulocyte lysate assay supplemented with [^{35}S]methionine or [^{3}H]leucine essentially as described.[19] Commercial kits from New England Nuclear or Amersham have proven to be reliable and convenient to use. Incorporation of radioactive amino acids into proteins is determined by trichloroacetic acid (TCA) precipitation followed by analysis on SDS–PAGE and autoradiography. To increase sensitivity, the polyacrylamide gels are treated with Enhance (New England Nuclear) and exposed to preflashed X-ray film (Kodak X-Omat XAR-2) at $-70°$ as described.[20]

Immunoprecipitations are carried out using monoclonal or rabbit antibodies and formalin-fixed protein A containing *Staphylococcus aureus* (Bethesda Research Laboratories, Bethesda, MD). Prior to use, *S. aureus* is washed two times in 0.5% Triton X-100 in PBS and resuspended in the original volume of the same buffer.

Spleen mRNA, 1 to 2 μg, is translated in a 25 μl lysate reaction in a sterile 1.5 ml microfuge test tube at 37° for 60 min. A 5-μl aliquot is taken for analysis by SDS–PAGE and two 1-μl aliquots are precipitated with TCA to determine total incorporation of radioactivity. The remaining mixture is placed on ice and is precleared by adding 5 μl normal rabbit serum followed by 100 μl of *S. aureus* in 0.5% Triton X-100. The mixture is incubated at 4° for 30 min and then centrifuged in a microfuge for 5 min. The supernatant is transferred to a new microfuge test tube and 5 μl of specific antibody added: rabbit antiserum or 25 μg monoclonal antibody. The antibody is allowed to react for 4 to 6 hr at 4° and 100 μl of *S. aureus* is added. The mixture is incubated for 30 min with occasional mixing followed by the addition of 0.75 ml 0.5% Triton X-100. The mixture is centrifuged in a microfuge for 2 min and the supernatant removed by aspiration. The pellet is vortexed and then 0.75 ml of 0.5% deoxycholate buffer added. The mixture is centrifuged again and washed one additional time with deoxycholate buffer. The final pellet is vortexed, 50 μl of SDS–PAGE sample buffer containing 100 mM dithiothreitol is added, and the sample boiled for 5 min. The material is centrifuged for 5 min and the supernatant analyzed by SDS–PAGE or stored at $-20°$ until used.

[19] H. R. B. Pelham and R. J. Jackson, *Eur. J. Biochem.* **67,** 247 (1976).
[20] R. A. Laskey and A. D. Mills, *FEBS Lett.* **82,** 314 (1977).

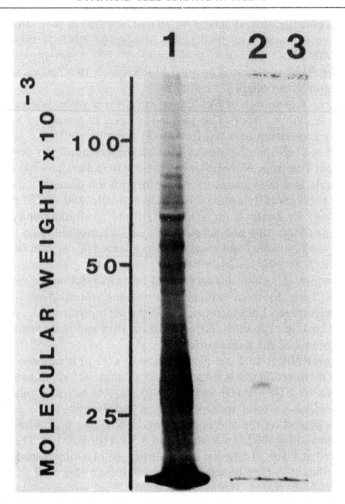

FIG. 2. Cell-free translations of rat spleen mRNA and immunoprecipitation of rat Ia-A α chains. Wistar rat spleen poly(A) RNA was translated in 25 μl rabbit reticulocyte lysate assays containing 20 μCi [^{3}H]leucine (110 Ci/mmol). Translated products were immunoprecipitated using rabbit antibodies against separated rat Ia-A α chains and protein A containing *Staphylococcus aureus*, analyzed by SDS–PAGE, and visualized by fluorography. Track 1, 5 μl total translated products; track 2, immunoprecipitated translated products; track 3, as in track 2 except that immunoprecipitation was carried out in the presence of 1 μg nonradioactive rat Ia-A α chain.

An example of the cell-free translation of total rat spleen poly(A) mRNA and the identification of *in vitro* synthesized rat Ia-A α chains is shown in Fig. 2. In these experiments rat spleen mRNA is translated in a rabbit reticulocyte lysate assay containing [³H]leucine. Rat Ia polypeptides are immunoprecipitated using rabbit or MRC 0X6 antibodies to the isolated α or β chains. Rabbit antibodies are prepared by immunizing rabbits twice at three week intervals with 75 μg of separated Ia-A α or β chains (see above) emulsified in complete Freund's adjuvant. Serum is collected 2 weeks after the last immunization. Samples of the total translation products and immunoprecipitates are analyzed by SDS–PAGE and fluorography. Figure 2 (track 1) shows the total polypeptides synthesized using rat spleen mRNA. A control with no mRNA added shows no bands (data not shown). Track 2 shows the results after immunoprecipitation with rabbit anti Ia-A α chain antibodies. A single band of apparent molecular weight 26,000 is clearly visible. Track 3 shows the results of blocking the immunoprecipitation of radioactive α chain with the addition of 1 μg of purified rat α chain to the reaction mixture prior to the addition of the rabbit antibodies. In this instance there is no radioactive α chain detectable and these results clearly demonstrate that the band in track 2 is the cell-free translation product of mRNA coding for rat Ia-A α chain.

Attempts were also made to identify the cell-free translation product of mRNA coding for rat Ia-A β chain by immunoprecipitation with three different antibody preparations: MRC 0X6 antibody, which reacts with separated Ia-A β chains (see above), rabbit antibodies prepared against separated Ia-A β chain, and rabbit antibodies prepared against reduced and carboxymethylated Ia-A β chain. In no instance is a β chain detectable. Cell-free translations can also be carried out in the presence of dog pancreas microsomes in order for the translation products to be "processed" (leader sequence removed and carbohydrate side chains attached) and membrane proteins inserted into microsomal membranes.[21] After translation the radioactive products are immunoprecipitated with each of the three preparations of antibodies to Ia-A β chains and analyzed by SDS–PAGE. Again no bands are visible using any of these three anti-β chain antibodies. These results could be due to the absence of Ia-A β chain mRNA in the poly(A) preparation of total rat spleen mRNA, the failure of β chain mRNA to be translated in the reticulocyte lysate assay, or to the fact that the *in vitro* synthesized Ia-A β chain products have a different conformation from that of the separated β chain and, therefore, do not react with any of the three different antibody preparations. At this point it is not clear which of these possibilities is occurring. These results

[21] R. J. Jackson and G. Blobel, *Proc. Natl. Acad. Sci. U.S.A.* **74**, 5598 (1977).

point out the difficulties encountered in identifying mRNA to a given protein by cell-free translation and immunoprecipitation using specific antibodies.

Enrichment of Specific mRNA by Sucrose Gradient Centrifugation. Messenger RNA can be fractionated by preparative sucrose gradient centrifugation or by preparative polyacrylamide agarose gel electrophoresis. To carry out sucrose gradient centrifugation either isokinetic (exponential) or linear gradients are used. Isokinetic 5–20% sucrose gradients in 10 mM sodium acetate pH 5.0, 100 mM NaCl, 1 mM EDTA are prepared as described.[22,23] Linear gradients (5–20%) are formed using an appropriate gradient mixer. RNA (250 μg in 50 μl 1 mM EDTA pH 8) is heated to 70° for 5 min and quickly cooled in an ice bath prior to layering over the gradient. Centrifugation is carried out in a Beckman SW41, SW50.1, or equivalent rotor at 75,000 g and 4° for 18 hr. The effluent is collected in 0.4 ml fractions and mRNA in each fraction precipitated by the addition of two volumes of 95% ethanol and incubation at −20° for 18 hr. The precipitate is recovered by centrifugation, washed in 70% ethanol, dried under vacuum, and dissolved in 50 μl sterile distilled H_2O. Aliquots, 2 μl, of each fraction are then translated in a cell-free assay and the radioactive products identified by immunoprecipitation and analysis by SDS–PAGE. Fractions containing specific mRNA are pooled and centrifuged on sucrose gradients as before. Fractions containing specific mRNA are pooled, precipitated as described, and used as templates for cDNA synthesis.

Sucrose gradient fractionation usually results in a 5- to 10-fold enrichment for specific mRNA species. A much higher degree of enrichment can be achieved, in some instances, by immunoprecipitation of polysomes using antibodies reacting with the nascent polypeptide chain.[24–26] These methods have been demonstrated to be very efficient and have resulted in a several hundredfold enrichment of low abundance mRNA species from various tissues or cell lines from which intact polysomes can be prepared. With spleen tissues it has not been possible to prepare intact polysomes (unpublished results) presumably due to high levels of ribonuclease. Therefore, this approach has not been useful in order to enrich mRNA coding for rat Ia antigens.

[22] H. Noll, *Nature (London)* **215**, 360 (1967).
[23] K. S. McCarty, Jr., R. T. Vollmer, and K. S. McCarty, *Anal. Biochem.* **61**, 165 (1974).
[24] R. T. A. MacGillivray, S. J. Friezner Degen, T. Chandra, S. L. C. Woo, and E. W. Davie, *Proc. Natl. Acad. Sci. U.S.A.* **77**, 5153 (1980).
[25] N. M. Gough and J. M. Adams, *Biochemistry* **17**, 5560 (1978).
[26] A. J. Korman, P. J. Knudsen, J. F. Kaufman, and J. L. Strominger, *Proc. Natl. Acad. Sci. U.S.A.* **79**, 1844 (1982).

cDNA Synthesis. Double-stranded cDNA is synthesized essentially as described.[27] Poly(A) RNA (1 mg/ml in distilled H_2O) is heated to 100° for 2 min and quickly cooled on ice. A 200 μl (final volume) reaction mixture is prepared on ice in a siliconized 1.5-ml microfuge tube containing 50 μg mRNA in 50 mM Tris–HCl pH 8.3 at 43° 8 mM $MgCl_2$, 5 mM dithiothreitol, 8 mM sodium pyrophosphate, 500 μM of each deoxyribonucleotide triphosphate, 20 μCi deoxyribocytidine [^{32}P]triphosphate (dCTP), 3000 Ci/mmol (New England Nuclear), 100 μg/ml oligo(dT) 12–18 (Collaborative Research), 100 μg/ml nuclease-free bovine serum albumin (Enzo Biochemicals), 20 units human placental ribonuclease inhibitor (Enzo Biochemicals), 200 units avian myoblastosis virus (AMV) reverse transcriptase (Life Sciences Inc.), and the mixture is incubated for 30 min at 43°. The dCTP specific activity is calculated by determining the total ^{32}P radioactivity in two 1-μl aliquots of the reaction mixture and ^{32}P incorporation into cDNA is calculated on the TCA precipitate of two 1-μl aliquots. The reaction is stopped by the addition of EDTA pH 8.0 to a final concentration of 10 mM, cDNA is heated to 100° for 1 min, quickly cooled on ice, and mRNA hydrolized by adding 4 M NaOH to a final concentration of 0.4 M and incubation at 20° for 16 to 18 hr. The mixture is neutralized by adding 5 M HCl, 50 μg of carrier tRNA (Sigma type X) is added, and the cDNA extracted with one-half volume phenol and one-half volume chloroform: isoamyl alcohol (24:1 v/v). The aqueous layer is chromatographed on a 10 ml column of Sephadex G-50 Fine (Pharmacia) in TEN 8 (10 mM Tris–HCl pH 8.0, 1 mM EDTA, 100 mM NaCl). The cDNA eluting in the excluded volume is precipitated by the addition of NaCl to 0.2 M and two volumes 95% ethanol and incubation at −20° for 18 hr.

Single-stranded cDNA is recovered by centrifugation in a microfuge, dissolved in 50 μl distilled H_2O, and the cDNA concentration calculated by determining the total ^{32}P radioactivity in two 1-μl aliquots. The amount (in terms of micrograms) of cDNA is calculated and the average length estimated by analysis of an aliquot by denaturing gel electrophoresis followed by autoradiography. From these two values the number of picomoles of 3′ ends is determined. A tract of approximately 20 dC nucleotides is then added to each 3′ end. Single-stranded cDNA is heated to 100° for 2 min and quickly cooled on ice. A reaction mixture is prepared on ice (40 μl total volume per 4 pmol 3′ ends) containing the cDNA in 140 mM potassium cacodylate, 30 mM Tris–HCl pH 7.0, 0.1 mM dithiothreitol,

[27] H. Land, M. Grez, H. Hauser, W. Lindenmaier, and G. Schutz, *Nucleic Acid Res.* **9,** 2251 (1981).

1 mM cobalt chloride, 25 μM dCTP, 20 μCi [^{32}P]dCTP (3000 Ci/mmol), and 40 units terminal transferase (Boehringer Mannheim) per 40 μl reaction. The initial cDNA concentration is calculated by determining the ^{32}P radioactivity of two 1-μl aliquots of TCA precipitate. The reaction mixture is incubated at 37° for 20 min. The dCTP specific activity is calculated by determining the total ^{32}P radioactivity in two 1-μl aliquots and the dCTP incorporated is calculated by determining the ^{32}P radioactivity in two 1-μl aliquots of TCA precipitate. The number of dC nucleotides added per 3' end is then calculated. C-tailed cDNA is phenol extracted, chromatographed on Sephadex G-50 Fine, ethanol precipitated, and dissolved in distilled H$_2$O as before.

Second-strand synthesis is primed using oligo(dG) and carried out using AMV reverse transcriptase. C-tailed single-stranded cDNA is heated to 100° for 1 min and quickly cooled on ice. A reaction mixture is prepared on ice containing cDNA (1 μg cDNA/100 μl) in 50 mM Tris–HCl pH 8.3 at 43°, 8 mM MgCl$_2$, 5 mM dithiothreitol, 500 μM of each deoxyribonucleotide triphosphate, 30 μCi [^{3}H]dCTP (20 Ci/mmol), 100 μg/ml bovine serum albumin, 50 μg/ml oligo(dG) 12–18 (Collaborative Research), and AMV reverse transcriptase 100 units/100 μl reaction. The mixture is incubated at 43° for 120 min. Incorporation of [^{3}H]dCTP into second-strand cDNA is calculated by determining the ^{3}H radioactivity in two 1-μl aliquots of TCA precipitate. Double-stranded cDNA is phenol extracted, chromatographed on Sephadex G-50 Fine, and ethanol precipitated as before.

Additional poly(dC) tails of approximately 10 nucleotides in length are added as described for C-tailing first-strand cDNA except that the dCTP is added to a final concentration of 10 μM. C-tailed double-stranded cDNA is then separated by preparative nondenaturing agarose gel electrophoresis and cDNA molecules greater than 600 base pairs in length are eluted as described.[28] The eluted cDNA is ethanol precipitated, dissolved in distilled H$_2$O, and stored at $-20°$.

Annealing and Transformation. The plasmid vector, pBR322,[29] is digested to completion with *Pst*I restriction endonuclease enzyme and dG tails of approximately 5 to 10 nucleotides in length are added using the conditions detailed above for tailing double-stranded cDNA with deoxyriboguanosine triphosphate at a final concentration of 10 μM. Equimolar amounts of *Pst*I digested G-tailed pBR322 (50 ng) and C-tailed double-stranded cDNA (5 to 20 ng) are added to give a final volume of 50 μl in 10 mM Tris–HCl pH 7.5, 1 mM EDTA, 100 mM NaCl. DNA is annealed by heating to 65° for 5 min followed by incubation at 43° for 120 min and

[28] S. C. Girvitz, S. Bacchetti, A. J. Rainbow, and F. L. Graham, *Anal. Biochem.* **106,** 492 (1980).
[29] F. Bolivar and K. Backman, this series, Vol. 68, p. 245.

cooling to room temperature for 120 min. Annealed DNA is placed on ice for 15 min and transformation is carried out as described.[30] Calcium chloride treated *E. coli* strain RR1[29] is prepared as described[30] and 100 μl of *E. coli* added to each 50 μl annealing mixture. Annealed DNA and *E. coli* are kept on ice for 45 min, heated to 37° for 5 min, and kept at room temperature for 5 min. *E. coli* are transferred to a culture tube containing 2 ml LB medium (10 g Difco tryptone, 5 g Difco yeast extract, and 10 g NaCl/liter distilled H_2O)[29] and incubated at 37° for 60 min with occasional shaking. Transformed *E. coli* are then plated onto fresh LB agarplates containing 10 μg/ml tetracycline (0.1 ml cells per plate) and incubated overnight at 37°. Annealed cDNA should give between 50 and 200 × 10³ transformants/μg cDNA. Colonies containing recombinant plasmids are resistant to tetracycline and sensitive to ampicillin.

Identification of Specific cDNA Clones by mRNA Selection

To identify specific cDNA clones an mRNA selection assay essentially as described is used.[31] Recombinant plasmid DNA is immobilized on nitrocellulose filters and hybridized with an excess of mRNA. Filters are washed and bound mRNA eluted and translated in a cell-free system. Specific polypeptides are identified by immunoprecipitation and analysis by SDS–PAGE. Several cDNA clones coding for MHC antigens have been identified by this approach.[32-36] The same strategy can be used to identify cDNA clones which specifically hybridize to mRNA coding for proteins which have a detectable biological activity. In this instance, eluted mRNA is either translated in a cell-free system or injected into frog oocytes and the products assayed using a biological assay. This approach has enabled the identification of cDNA clones coding for interferon[37] and more recently for the T lymphocyte growth factor, interleukin-2.[38]

[30] M. Dagert and S. D. Ehrlick, *Gene* **6**, 23 (1979).

[31] J. Parnes, B. Velan, A. Felsenfeld, L. Ramanathan, U. Ferrini, E. Appella, and J. G. Seidman, *Proc. Natl. Acad. Sci. U.S.A.* **78**, 2253 (1981).

[32] H. L. Ploegh, H. T. Orr, and J. L. Strominger, *Proc. Natl. Acad. Sci. U.S.A.* **77**, 6081 (1981).

[33] S. Kvist, F. Bregegere, L. Rask, B. Cami, H. Garoff, F. Daniel, K. Wiman, D. Larhammar, J. P. Abastado, G. Gachelin, P. A. Peterson, B. Dobberstein, and P. Kourilsky, *Proc. Natl. Acad. Sci. U.S.A.* **78**, 2772 (1981).

[34] J. S. Lee, J. Trowsdale, and W. F. Bodmer, *Proc. Natl. Acad. Sci. U.S.A.* **79**, 545 (1982).

[35] K. Winman, D. Larhammar, L. Claesson, K. Gustafsson, L. Schenning, P. Bill, J. Bohme, M. Denaro, B. Dobberstein, U. Hammerling, S. Kist, B. Servenius, J. Sundelin, P. Peterson, and L. Rask, *Proc. Natl. Acad. Sci. U.S.A.* **79**, 1703 (1982).

[36] C. Wake, E. O. Long, M. Strubin, R. Accolla, S. Carrel, and B. Mach, *Proc. Natl. Acad. Sci. U.S.A.* **79**, 6979 (1982).

[37] S. Nagata, H. Taira, A. Hall, L. Johnsrud, M. Streuli, J. Ecsodi, W. Boll, K. Cantell, and C. Weissman, *Nature (London)* **284**, 316 (1980).

[38] T. Taniguchi, H. Matsui, T. Fujita, C. Takaoka, N. Kashima, R. Yoshimoto, and J. Hamuro, *Nature (London)* **302**, 305 (1983).

Plasmid DNA Preparation. Individual colonies resulting from transformations, as described above, are transferred to 5 ml cultures of LB medium containing tetracycline (10 μg/ml), grown overnight, and stored at 4°. Six individual cultures are used to innoculate (0.1 ml of each culture) 40 ml of LB medium containing tetracycline (10 μg/ml) in 125-ml flasks. The cells are grown with shaking for 3 to 6 hr until the absorbance at 600 nm is between 0.7 and 1.0. Chloramphenicol (160 mg/ml in 95% ethanol) is added to give a final concentration of 200 μg/ml. The cells are grown overnight at 37° with shaking to ensure adequate oxygenation.

Plasmid DNA is prepared essentially as described.[39] Forty milliliters of culture is transferred to sterile 50-ml polycarbonate tubes and the cells collected by centrifugation at 5000 g for 5 min at 4°. The resulting cell pellets are resuspended in 0.9 ml 25 mM Tris–HCl pH 8.0, 10 mM EDTA 50 mM glucose, and 0.1 ml freshly dissolved lysozyme (20 mg/ml) in 50 mM Tris–HCl pH 8.0. Cells are left on ice for 30 min, 1.0 ml of freshly prepared 1% SDS in 0.2 M sodium hydroxide is added to each tube, and the contents mixed by swirling gently. The mixtures are kept on ice for 5 min and 1.5 ml 3 M sodium acetate pH 4.8 added to each test tube, the contents mixed by swirling gently and left on ice for a further 60 min. Lysates are cleared by centrifugation at 22,000 g for 30 min at 4° and the supernatants transferred to 15 ml polypropylene test tubes. DNA is precipitated by adding an equal volume of 100% isopropanol and cooling to −20° for 1 hr. DNA is collected by centrifugation, dissolved in 2 ml TEN 8, and extracted with 1 ml phenol and 1 ml chloroform–isoamyl alcohol (24 : 1 v/v). The aqueous layers are transferred to new test tubes; sodium chloride added to 0.2 M, and plasmid DNA precipitated by addition of two volumes of 95% ethanol and incubation at −20° for 18 hr. Precipitated DNA is collected by centrifugation, washed with 70% ethanol, dried under vacuum, and dissolved in 1 ml 20 mM Tris–HCl, 1 mM EDTA pH 8.0.

Filter Binding and Hybridization. Plasmid DNA dissolved in 1 ml of 20 mM Tris–HCl, 1 mM EDTA pH 8.0 is heated to 100° for 10 min. One milliliter of 1 M sodium hydroxide is added and the mixture is kept at room temperature for 20 min. The DNA is neutralized by adding 6 ml of 1.5 M sodium chloride, 0.15 M sodium citrate, 0.25 M Tris–HCl pH 8.0, 0.25 M HCl, and is then immediately filtered through 13-mm-diameter prewashed nitrocellulose filters (Sartorius 0.2 μm or Schleicher and Schuell BA 85 0.45 μm, numbered with a soft pencil) at a flow rate of 1–2 ml/min by applying a slight vacuum. A sampling manifold which holds 10 individual 13-mm-diameter filters (Amicon) may be used when a large number of preparations are being screened. Each filter with bound DNA

[39] H. C. Birnboim and J. Doly, *Nucleic Acids Res.* **7,** 1513 (1979).

is washed with 30 ml of 6×SSC (0.9 M sodium chloride, 0.09 M sodium citrate), air dried, and baked at 70° for 18 hr. A small disk, 5 mm diameter, is cut from each filter using a paper hole punch. Up to 20 disks are placed in a siliconized glass vial and prehybridized at 42° for 2 hr in 1 ml of 50% deionized formamide (pH 7), 20 mM Pipes–HCl pH 6.4, 0.4 M sodium chloride, 100 μg/ml tRNA (Sigma type X), 0.2% SDS. The prehybridization solution is then removed and replaced with 1 ml of an identical hybridization solution containing poly(A) RNA at 400 μg/ml. Disks are hybridized at 42° for 6 hr with occasional shaking and then transferred to a sterile 50 ml polypropylene test tube. Disks are washed (50 ml per wash) at 42° nine times with 10 mM Tris–HCl pH 8.0, 0.15 M sodium chloride, 1 mM EDTA, 0.5% SDS, and two times with 50 ml of the same buffer without SDS. Two disks are placed in each siliconized and autoclaved 1.5 ml polypropylene microfuge tube. Bound mRNA is eluted with 0.3 ml water containing 30 μg carrier RNA. The tubes are heated at 100° for 2 min, quickly frozen in dry ice ethanol, and thawed at room temperature. The filters are removed, and the mRNA precipitated by adding sodium acetate pH 5.0 to 0.3 M and two volumes of 95% ethanol. Precipitated RNA is kept at −20° for 18 hr, recovered by centrifugation in a microfuge for 10 min, washed with 70% ethanol, dried under vacuum, and dissolved in 4 μl distilled H_2O.

Each disk contains plasmid DNA from 6 individual colonies, therefore, each elution represents 12 individual colonies. Eluted mRNA is translated in a cell-free translation assay and specific polypeptides are identified by immunoprecipitation and analysis by SDS–PAGE, as described above. Once positive pools are identified, the 12 cultures which constituted the original innoculums are individually screened using the same procedure. A large number of individual cDNA transformants can be screened in a relatively short period of time. It is helpful to include a positive control to ensure that all procedures are working and, therefore, all potential specific cDNA clones will be identified.

Molecular Cloning of Rat Ia-A′α Chain cDNA

Cell free translation and immunoprecipitation were used to identify mRNA coding for rat Ia-A α chains from Wistar strain rats (RT1ᵘ) as shown in Fig. 3. Total rat spleen mRNA (prepared as described above) is fractionated on isokinetic 5–20% sucrose density gradients and fractions containing Ia-A α chain mRNA pooled and recentrifuged. The resulting pool of enriched mRNA is used as a template for double-stranded cDNA synthesis as described above, size separated by agarose gel electrophoresis, and cDNA molecules greater than 400 base pairs eluted. An aliquot of this cDNA (20 ng) is annealed to PstI digested, G-tailed pBR322 (65 ng)

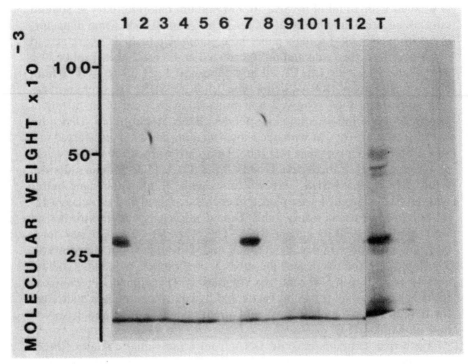

FIG. 3. Identification of cDNA clones coding for rat Ia-A α chain by mRNA selection. Recombinant DNA was prepared from 40 ml cultures, immobilized on nitrocellulose filter, and hybridized with total rat spleen mRNA. After washing, bound mRNA was eluted and translated in 50 μl rabbit reticulocyte lysate assay containing 100 μCi [³⁵S]methionine (500 Ci/mmol). Translated products were immunoprecipitated with rabbit antibodies to separated rat Ia-A α chains, analyzed by SDS–PAGE, and visualized by fluorography. Tracks 1 to 12: immunoprecipitated translation products of selected mRNA representing 12 individual cDNA transformants (12 cultures had previously been used to prepare a pool of plasmid DNA which had positively selected Ia-A α chain mRNA); track T, immunoprecipitated translated products of total spleen mRNA.

and the resulting recombinant plasmids are used to transform *E. coli* strain RR1 as described. From this transformation 1500 independent colonies are usually obtained. Individual colonies are transferred to 5 ml LB medium and used as stock cultures to prepare pools of plasmid DNA. Plasmid DNA is bound to nitrocellulose filters and mRNA selection assays carried out as described. From an initial screen of 192 colonies (16 pools representing 12 colonies) one pool is identified which specifically selects mRNA coding rat Ia-A α chain. The 12 individual cultures are then screened and the results are shown in Fig. 3. From this screen plasmid DNA from two cultures specifically selects mRNA coded for rat Ia-A α chains (Fig. 3, tracks 1 and 7). Analysis of the plasmid DNA from these two positive cultures shows that culture number 7 contains a single recombinant plasmid designated pRIa.1 while culture number 1 contains two plasmids, one of which is pRIa.1 and the other which is negative on rescreening. Restriction endonuclease enzyme analysis shows that the cDNA insert of pRIa.1 is approximately 600 base pairs in length and that one PstI site is not regenerated. A 150 base pair EcoRI-PstI fragment from pRIa.1 is prepared and used to screen another cDNA library prepared from total rat spleen mRNA. From a screen of 2000 independent transformants, two colonies which hybridized to this fragment are obtained. The relative abundance of Ia-A α chain mRNA in this library, prepared from total spleen mRNA, is approximately 0.1%. Of these two positive colonies the recombinant plasmid containing the largest cDNA insert (800 base pairs in length) is designated pRIa.2.

The complete nucleotide sequence of the cDNA insert of pRIa.2, determined by the methods of Maxam and Gilbert,[40] is shown in Fig. 4. The DNA sequence is 779 nucleotides in length and contains a single open reading frame of 387 nucleotides which, when translated into protein sequence, codes for the carboxy-terminal 129 amino acids of a rat Ia-A α chain (Fig. 4).[41] This represents approximately 55% of the mature α chain based on an apparent molecular weight of 30,000 and on the predicted structures of human HLA-DC1[42] and mouse H-2 I-A[43] α chains. Following the open reading frame there is an untranslated region of 290 nucleotides terminating in a tract of poly(A) of 63 nucleotides. Fifteen nucleotides prior to the poly(A) region there is a putative polyadenylation site AATAAA. This is, therefore, the 3' end of the corresponding mRNA.

The protein structure of the rat Ia-A α chain can be divided into different regions or domains by analogy with the domain structure of the HLA DC1 α chain.[42] Based on this sequence the cDNA insert of pRIa-2 codes for amino acids 102 to 232. This represents the majority of the second extracellular domain α 2 (amino acids 104 to 181) which contains a putative disulfide loop of 55 amino acids from Cys-110 to Cys-164 and a connecting peptide of 13 amino acids (182 to 194). Following these two extracellular regions is a hydrophobic transmembrane region of 23 amino acids (195 to 217) and a cytoplasmic region of 15 amino acids (218 to 232).

The rat Ia-A α chain is highly homologous, in terms of both DNA and amino acid sequence identity, to human HLA-DC1 and mouse H-2 I-A α chains. Comparison of the DNA protein coding region shows there is 85%

[40] A. Maxam and W. Gilbert, this series, Vol. 65, p. 499.
[41] A. E. Wallis and W. R. McMaster, Immunogenetics 19, 53 (1984).
[42] C. Auffray, A. J. Korman, M. Roux-Dosseto, R. Bono, and J. L. Strominger, Proc. Natl. Acad. Sci. U.S.A. 79, 6337 (1982).
[43] C. O. Benoist, D. J. Mathis, M. R. Kanter, V. E. Williams, II, and H. O. McDevitt, Proc. Natl. Acad. Sci. U.S.A. 80, 534 (1983).

```
         1                                                              40
(G)    T CAG CCC AAC ACC CTC ATC TGC TTT GTA GAC AAC ATC TTT
     18  GLN PRO ASN THR LEU ILE CYS PHE VAL ASP ASN ILE PHE
        104                                                            116


                                                                        85
       CCT CCT GTG ATC AAT ATC ACA TGG TTG AGA AAC AGC AAG CCA GTC
       PRO PRO VAL ILE ASN ILE THR TRP LEU ARG ASN SER LYS PRO VAL
                                                                       131


                                                                       130
       ACA GAA GGC GTT TAT GAG ACC AGC TTC CTT TCC AAC CCT GAC CAT
       THR GLU GLY VAL TYR GLU THR SER PHE LEU SER ASN PRO ASP HIS
                                                                       146


                                                                       175
       TCC TTC CAC AAG ATG GCT TAC CTC ACC TTC ATC CCT TCC AAC GAC
       SER PHE HIS LYS MET ALA TYR LEU THR PHE ILE PRO SER ASN ASP
                                                                       161


                                                                       220
       GAC ATT TAT GAC TGC AAG GTG GAG CAC TGG GGC CTG GAC GAG CCG
       ASP ILE TYR ASP CYS LYS VAL GLU HIS TRP GLY LEU ASP GLU PRO
                                                                       176


                                                                       265
       GTT CTA AAA CAC TGG GAA CCT GAG GTT CCA GCC CCC ATG TCA GAG
       VAL LEU LYS HIS TRP GLU PRO GLU VAL PRO ALA PRO MET SER GLU
                          181 182                                      191


                                                                       310
       CTG ACA GAG ACT GTG GTC TGT GCC CTG GGG TTG TCT GTG GGC CTC
       LEU THR GLU THR VAL VAL CYS ALA LEU GLY LEU SER VAL GLY LEU
                194 195                                                206


                                                                       355
       GTG GGC ATC GTG GTG GGC ACC ATC TTC ATC ATT CAA GGC CTG CGA
       VAL GLY ILE VAL VAL GLY THR ILE PHE ILE ILE GLN GLY LEU ARG
                                                 217 218               221


                                                                       400
       TCA GAT GGC CCC TCC AGA CAC CCA GGG CCC CTT TGA GTC ACA CCC
       SER ASP GLY PRO SER ARG HIS PRO GLY PRO LEU ***
                                                 232


                                                                       445
       TGG GAA AGA AGG TGC GTG GCC CTC TAC AGG CAA GAT GTA GTG TGA
                                                                       490
       GGG GTG ACC TGG CAC AGT GTG TTT TCT GCC CCA ATT CAT CGT GTT
                                                                       535
       CTT TCT CTT CTC CTG GTG TCT CCC ATC TTG CTC TTC CCT TGG CCC
                                                                       590
       CCA GGC TGT CCA CCT CAT GGC TCT CAC GCC CTT GGA ATT CTC CCC
                                                                       625
       TGA CCT GAG TTT CAT TTT TGG CAT CTT CCA AGT CGA ATC TAC TAT
                                                                       670
       AGA TTC CGA GAC CCT GAT TGA TGC TCC ACC AAA CCA ATA AAC CTC
            681
       TCA TAA GTT GG (A)   (C)
                      63     17
```

FIG. 4. The nucleotide sequence of the cDNA insert of pRIa.2 coding for a rat Ia-A α chain and the predicted amino acid sequence. The DNA sequence is presented 5′ to 3′ and

DNA sequence identity between rat and human α chains and 91% DNA sequence identity between rat and mouse α chains. At the protein level, without the need to introduce any gaps, between rat and human α chains there is 81% protein sequence identity and between rat and mouse α chains there is 91% protein sequence identity. The Ia-A α chains, therefore, have been highly conserved during evolution and are more homologous to each other between species than are the Ia-A and Ia-E α chains within the same species.[41]

The major histocompatibility complex Class I[44] and Ia[45] antigens and rodent Thy-1[46] antigens have been shown to be homologous in terms of sequence identity to immunoglobulin domains. Analysis of the predicted amino acid sequence of the rat Ia-A α chain also shows that this protein shares several conserved amino acids which are characteristic of immunoglobulin constant region domains.[41] These conserved amino acids are thought to be important in maintaining a β-pleated sheet structure in immunoglobulin domains. Therefore, this homology implies that the α 2 domain of Ia-A α chain also resemble an immunoglobulin-like fold. Further analysis of the structure of Ia antigens may determine which domains are involved in cell–cell interactions within the immune system may lead to a better understanding of the molecular basis of the immune response genes.

Acknowledgments

I am grateful to Drs. R. T. A. MacGillivray and A. F. Williams for many discussions and helpful suggestions during the course of these studies. This work was supported in part by grants from the MRC (Canada) and the B.C. Health Care Research Foundation. W. R. McMaster is a Scholar of the MRC.

[44] H. T. Orr, J. A. Lopez de Castro, D. Lancet, and J. L. Strominger, *Biochemistry* **18**, 5711 (1979).

[45] D. Larhammar, K. Gustafsson, L. Claesson, P. Bill, K. Winman, L. Schenning, J. Sundelin, E. Widmark, P. A. Peterson, and L. Rask, *Cell* **30**, 153 (1982).

[46] A. F. Williams and J. Gagnon, *Science* **216**, 696 (1982).

only the coding strand is shown. Numbers above each line correspond to the nucleotide position starting after the poly(G) tail and ending prior to the tract of poly(A). The putative polyadenylation site, AATAAA, is underlined. The predicted amino acid sequence of the carboxy-terminal 129 positions of the rat Ia-A α chain is shown below the nucleotide sequence. Numbers below each line corresponds to the amino acid positions of an HLA-DC1 α chain. (From Auffray et al.[42])

[52] Purification of Human HLA-A and HLA-B Class I Histocompatibility Antigens

By José A. López de Castro

Introduction

Human class I histocompatibility antigens are integral membrane glycoproteins encoded by the HLA-A, -B, and -C loci of the major histocompatibility complex (MHC). They are present at the surface of the great majority of cells and are known to play a role in restricting the recognition and killing of target cells by cytolytic T lymphocytes. Their molecule is composed of a MHC-encoded heavy chain of about 44,000 daltons, noncovalently associated to β_2-microglobulin, an invariant polypeptide of 12,000 daltons. The most conspicuous feature of these antigens is their unparalleled degree of genetic polymorphism. All three HLA -A, -B, and -C loci are multiallelic and the corresponding alleles are codominantly expressed, so that up to six different HLA class I molecules may be present on the surface of a heterozygous cell. Thus, the isolation of homogeneous preparations of these antigens will confront the biochemist with both the difficulties inherent to manipulating membrane proteins, which require the presence of detergents to keep them in solution, and with the problem of separating the various closely related specificities. A number of methods have been developed to circumvent these obstacles by means of a judicious choice of the source of material and solubilization procedures that preserve the native antigenicity. All these methods fall into two main categories. Those based on physicochemical differences and those making use of the antigenic specificity of each molecule.

In general, large-scale purification has been favored because milligram amounts of protein are desirable for most biochemical studies.

This chapter will not attempt a comprehensive review of isolation procedures of HLA class I antigens. Instead, it will be centered on the description of those of well established value in providing homogeneous protein suitable for structural and functional studies. Their success is emphasized by the fact that histocompatibility antigens remain today one of the few well-characterized eukaryotic membrane protein structures.

Source of Material

Lymphoblastoid cell lines, cadaveric spleens, and platelets are the main sources for the purification of HLA class I antigens. B-cell lines

METHODS IN ENZYMOLOGY, VOL. 108

obtained by Epstein–Barr virus transformation of peripheral blood lymphocytes (PBLs) present distinct advantages over other materials. They grow actively in culture providing a stable and reproducible system. The amount of HLA-A and -B antigens is selectively increased 20- to 50-fold over that in normal PBLs on a per cell basis,[1] constituting about 0.5% of total membrane protein.[2] HLA-C antigens are expressed to a much lower extent. Although no precise quantitation has been carried out it has been roughly estimated to represent less than 10% the amount of HLA-A plus HLA-B antigens.[3] Suitable B cell lines for the isolation of a given antigen may be obtained after transformation of PBLs from selected individuals. A great variety of HLA typed cells are usually available from cell banks and research laboratories. Many of these lines are derived from offsprings of consanguineous marriages or other individuals that have been selected for being homozygous at one or more HLA loci. Hence, by appropriate selection of the cell line, it is possible to reduce the number of different HLA molecular species, considerably simplifying subsequent purification of individual specificities. Major drawbacks of using lymphoblastoid cell lines for biochemical work are the elevated costs involved in growing large amounts of cells and the requirement of large-scale tissue culture facilities. These factors may impose limitations to many laboratories. Of course, they are a problem only when milligram amounts of homogeneous protein are desired, but not for radiochemical work, where only small amounts of cells are required. Either spleens or platelets may be used as alternative sources when large amounts of cultured cells are not available, because both materials are inexpensive and easy to obtain. Although chemical amounts of HLA class I antigens have been purified from these sources, they express much less of the HLA antigens than do lymphoblastoid cells. In addition, they do not constitute reproducible sources of material and final preparations are composed of a mixture of HLA-A, -B, and -C specificities. This chapter will concentrate mainly on the description of large-scale isolation procedures which use lymphoblastoid cell lines as starting materials. Current methods for purification of HLA antigens from spleens and platelets have been developed by Peterson and colleagues[4,5] and will be detailed below.

[1] J. M. McCune, R. E. Humphreys, R. R. Yocum, and J. L. Strominger, *Proc. Natl. Acad. Sci. U.S.A.* **72**, 3206 (1975).

[2] J. L. Strominger, D. L. Mann, P. Parham. R. Robb, T. Springer, and C. Terhorst, *Cold Spring Harbor Symp. Quant. Biol.* **45**, 323 (1976).

[3] D. Snary, C. J. Barnstable, W. F. Bodmer, and M. J. Crumpton, *Eur. J. Immunol.* **8**, 580 (1977).

[4] L. Trägårdh, B. Curman, K. Wiman, L. Rask, and P. A. Peterson, *Biochemistry* **18**, 2218 (1979).

[5] L. Trägårdh, L. Klareskog, B. Curman, L. Rask, and P. A. Peterson, *Scand. J. Immunol.* **9**, 303 (1979).

Preparation of Membranes

Crude membranes are obtained after hypotonic lysis of cells by the following procedure which is specified for batches of 100 g of frozen cells.

1. Thaw out cells in a 37° bath and place them on ice.
2. Add 150 ml of ice-cold 10 mM Tris–HCl buffer, pH 8.0, containing 0.2 mM dithiothreitol (DTT) and 0.1 mM phenylmethylsulfonyl fluoride (PMSF) (lysis buffer). Adjust pH to 8.0 at 4° with 2 M Tris base. PMSF must be added from a 12 mg/ml fresh solution in ethanol. Homogenize with a Teflon pestle homogenizer.
3. Centrifuge for 10 min at 4000 g and 4°. Save the supernatant and proceed with the pellet to step 4.
4. Take the low-speed pellets and repeat steps 2 and 3 six times. Pool the supernatants and discard the pellet, composed mainly of nuclei.
5. Centrifuge the combined supernatants for 1 hr at 105,000 g and 4°. Discard the high-speed supernatant, which contains the soluble material.
6. Wash the pellet once by resuspending in 100 ml of lysis buffer.
7. Centrifuge for 1.5 hr at 105,000 g and 4°. Resuspend the pellet in 100 ml of lysis buffer and save it frozen at −80°.

Further purification of crude membranes may be achieved by centrifugation on sucrose as follows (in which case steps 6 and 7 are not necessary).

8. Homogenize crude membrane pellets from step 5 in 100 ml of ice-cold lysis buffer with a Potter-Elvehjem homogenizer.
9. Layer the suspension in two tubes over 20 ml of 10% sucrose in lysis buffer and centrifuge for 2 hr at 100,000 g and 4°.
10. Wash out sucrose by homogenizing the pellet in 100 ml of ice-cold lysis buffer and centrifuge at 105,000 g for 1 hr at 4°. This last centrifugation may be carried out at 40,000 g for 20 min (4°) if further resuspension of the membrane pellet by vortexing is desired (see below). The pellet constitutes the purified membrane preparation. Store at −80° in lysis buffer.

Pober and colleagues[6] have shown that detergent-solubilized HLA antigens from lymphoblastoid cell membranes are usually associated with variable amounts of actin. They have developed a method for obtaining actin-free HLA antigens by removing it from cell membranes prior to detergent solubilization. The following procedure is detailed for batches of 100 g of cells. All solutions must contain 0.1 mM PMSF.

[6] J. S. Pober, B. C. Guild, J. L. Strominger, and W. R. Veatch, *Biochemistry* **20**, 5625 (1981).

11. Resuspend (by vortexing) the pellet of purified membranes as obtained in step 10 in 200 ml of ice-cold 1 mM (Na)$_2$EDTA, pH 7.4.

12. Collect membranes by centrifugation at 40,000 g for 20 min and 4°.

13. Wash the pellet once by repeating steps 11 and 12.

14. Homogenize the pellet from the previous step in 200 ml of an ice-cold buffer containing 2 mM Tris–HCl, pH 8.0, 0.2 mM NaATP, 0.2 mM Mg$_2$Cl, 0.5 mM DTT, and 0.002% sodium azide.

15. Dialyze against 4 liters of the same solution for 14–20 hr at 4°.

16. Centrifuge at 40,000 g for 20 min and 4°. Discard the supernatant containing the crude cytoskeletal extract.

17. Wash the membrane pellet once as in step 13. Resuspend it in 10 mM Tris–HCl, pH 8.0 and save it frozen at −80°.

Solubilization Methods

Alloantigenically active HLA antigens may be released from cell membranes by means of neutral detergents or by treatment of membranes with proteolytic enzymes, specially papain. The choice of either procedure will determine the physicochemical behavior of the solubilized product, therefore dictating subsequent purification strategies.

Detergent Solubilization

Choice of Detergent. A number of detergents have been examined as agents for the solubilization of HLA antigens.[7] A major consideration is their capacity to preserve the alloantigenic activity of the solubilized antigen, which is crucial for biological assays and immunochemical purification steps. Therefore, detergents which do not preserve alloantigenic activity such as sodium dodecyl sulfate will not be discussed here. The use of this detergent is limited mostly to the analysis of HLA fractions by polyacrylamide gel electrophoresis (SDS–PAGE). The physicochemical properties of the detergent are important in conditioning sample manipulations during purification. Chief among these are stability in a variety of buffer solutions, charge, absorbance at 280 nm, micelle size, and critical micelle concentration (CMC). Charged detergents interfere with separation by ion-exchange chromatography or isoelectric focusing; detergents with extinction coefficient at 280 nm may preclude spectrophotometric detection of protein; micelle size is a limiting factor in gel filtration procedures and low CMC prevents detergent exchange or elimination by dialysis.

[7] T. A. Springer, D. L. Mann, A. L. DeFranco, and J. L. Strominger, *J. Biol. Chem.* **252,** 4682 (1977).

Those detergents which preserve HLA antigenicity are known not to bind to the extracellular portion of the molecule but probably only to the hydrophobic intramembrane segment.[8,9] These detergents may be classified into two groups according to their properties and to the use they have found in HLA biochemistry. The first group includes bile salts (mainly sodium deoxycholate) and the nonionic detergent octyl glucoside. These compounds have in common their optical transparency at 280 nm and their high CMC. Because of this last property which allows rapid removal of detergent by dialysis they have been used in reconstitution of HLA antigens into liposomes.[10,11] The solubilizing capacity of octyl glucoside for histocompatibility antigens has not been sufficiently explored. Bile salts are excellent solubilizing agents,[7] but their use in biochemical purification of HLA antigens has been limited because they tend to gel or to precipitate below pH 7.8, at high salt concentrations, or in the presence of divalent ions. Furthermore, their ionic nature precludes the use of separation procedures based on charge.

The second group of nondenaturing agents is formed by a number of polyoxyethylene glycol derivatives, such as Nonidet P-40 (NP-40), Triton X-100, and Brij detergents. Although these compounds share the inconvenience of having large micelle size and low CMC, because of their nonionic nature, stability in aquous buffers and excellent solubilizing properties they are widely used for solubilization and purification of HLA antigens. Some members of the Brij series, namely Brij 99 and Brij 97, are most suitable for large scale purification. Being alkyl (oleyl alcohol) derivatives they have low absorbance at 280 nm. This is an advantage over aryl derivatives such as NP-40 and Triton X-100, the hydrophobic moiety of which (p-tert-octylphenol) has a high extinction coefficient at that wavelength. In addition, Brij detergents have a certain selectivity in the solubilization of HLA antigens over the bulk of membrane proteins,[7] a property which allows differential extraction from lymphoblastoid cell membranes. For radiochemical work, where these considerations are of secondary importance, NP-40 is the most commonly used detergent.

Procedure. The following procedure, first introduced by Springer *et al.*,[7] is generally used in isolation of HLA antigens from lymphoblastoid cells.

[8] T. A. Springer, J. L. Strominger, and D. L. Mann, *Proc. Natl. Acad. Sci. U.S.A.* **71**, 1539 (1974).
[9] T. A. Springer and J. L. Strominger, *Proc. Natl. Acad. Sci. U.S.A.* **73**, 2481 (1976).
[10] V. H. Engelhard, B. G. Guild, A. Helenius, C. Terhorst, and J. L. Strominger, *Proc. Natl. Acad. Sci. U.S.A.* **75**, 3230 (1978).
[11] L. Klareskog, G. Banck, A. Forsgren, and P. A. Peterson, *Proc. Natl. Acad. Sci. U.S.A.* **75**, 6197 (1978).

1. Prepare a mixture of Brij 99 : Brij 97 in a proportion 2 : 1. (These detergents are available from Atlas Chemical Division, ICI America Inc., Wilmington, DE or from Emulsion Engineering, Elk Grove Village, IL.) Since the degree of purity may be variable, it is advisable to select among different lots for those with lowest absorbance at 280 nm. Also, those lots which precipitate at 4° must be discarded.

2. Homogenize membrane pellets in ice-cold 10 mM Tris–HCl, pH 8.0 to give a protein concentration of 11 mg/ml. (A rough estimate of membrane protein is obtained by boiling an aliquot in 1% SDS and measuring its absorbance at 280 and 310 nm. One unit of the difference $A_{280} - A_{310}$ is approximately 0.6 mg/ml.)

3. Add a 20% solution of Brij 99 : Brij 97 to the membranes while stirring, to give a ratio of detergent to protein 2 : 1. Add 1 mM DTT to avoid formation of disulfide-bonded oligomeric forms of HLA antigens.[12] Stir gently for 30 min at 4°.

4. Centrifuge at 105,000 g for 1 hr and 4°. Carefully remove the supernatant solution from the detergent-insoluble residue, avoiding a viscous layer on top of the pellet.

Papain Solubilization

Treatment of cell membranes with papain results in the solubilization of a fragment of the histocompatibility antigens which contains the antigenic determinants of the native molecule and consists of their essentially intact extracellular portion.[9,13] Carboxypeptidase analyses combined with sequence studies[14,15] suggest that papain cleaves the HLA molecule at several peptide bonds closely before the hydrophobic transmembrane segment. HLA antigens solubilized in this way behave as stable globular proteins, readily soluble in aqueous buffers which may be purified in the absence of detergent by conventional chromatographic techniques. For this reason, papain solubilization has been profusely employed in structural studies of HLA antigens. It allows the isolation of milligram amounts of homogeneous protein suitable for sequence determination,[14,16,17] circu-

---

[12] T. Springer, R. J. Robb, C. Terhorst, and J. L. Strominger, *J. Biol. Chem.* **252**, 4694 (1977).

[13] P. Cresswell, M. J. Turner, and J. L. Strominger, *Proc. Natl. Acad. Sci. U.S.A.* **70**, 1603 (1973).

[14] H. T. Orr, J. A. López de Castro, D. Lancet, and J. L. Strominger, *Biochemistry* **18**, 5712 (1979).

[15] M. Malissen, B. Malissen, and B. Jordan, *Proc. Natl. Acad. Sci. U.S.A.* **79**, 893 (1982).

[16] J. A. López de Castro, H. T. Orr, R. J. Robb, T. G. Kostyk, D. L. Mann, and J. L. Strominger, *Biochemistry* **18**, 5704 (1979).

[17] J. A. López de Castro, J. L. Strominger, D. M. Strong, and H. T. Orr, *Proc. Natl. Acad. Sci. U.S.A.* **79**, 3813 (1982).

lar dichroism,[4,18] and X-ray diffraction studies which are usually difficult to carry out with most membrane proteins.

Sanderson and Batchelor[19] first used papain for solubilization of HLA molecules from human spleens, following the pioneering studies of Nathenson and Shimada on murine H-2 antigens.[20] The method was later adapted to HLA purification from lymphoblastoid cells.[21] Parham *et al.*[22] introduced further modifications which allowed a nearly quantitative solubilization of HLA antigens from cell membranes. Their procedure has been adopted, with minor modifications, for subsequent biochemical studies of papain-solubilized HLA antigens from lymphoblastoid cells.

Procedure

1. Homogenize the membrane pellet from 100 g of cells with ice-cold 10 mM Tris–HCl, pH 8.0, 2.5 mM EDTA to a final volume of 100 ml. This gives approximately 20 mg/ml of membrane protein.

2. Warm the suspension at 37° for 30 min in a water bath.

3. Activate 400 mg (6000 units) of papain (Worhington, 2× crystallized) by adding 66 mg of cysteine free base from an aquous solution of 20 mg/ml and incubating for 5 min at 37°.

4. Add activated papain to the warmed membrane suspension. Incubate at 37° for 1 hr with occasional shaking. Stop the reaction by adding 100 ml of ice-cold 10 mM Tris–HCl, pH 8.0 and placing the reaction flask on ice.

5. Centrifuge at 105,000 g for 1 hr at 4° and discard the insoluble material at the pellet. Perform all subsequent steps at 4°.

6. Immediately apply the supernatant to a DEAE-cellulose column (Whatman DE-52) to remove papain. The column must contain a minimum bed of 1 ml/100 mg of membrane protein or 15 ml/100 g of cells. The column is previously equilibrated with 10 mM Tris–HCl pH 8.0 and run with at least 10 column volumes of the same buffer. HLA antigens remain bound whereas papain flows through the column and is removed at this stage, thus obviating the use of alkylating reagents[21] that could lead to partial modification of methionine residues in HLA and have an adverse effect on CNBr cleavage.

[18] D. Lancet, P. Parham, and J. L. Strominger, *Proc. Natl. Acad. Sci. U.S.A.* **76,** 3844 (1979).
[19] A. R. Sanderson and J. R. Batchelor, *Nature (London)* **219,** 184 (1968).
[20] S. G. Nathenson and A. Shimada, *Transplantation* **6,** 662 (1968).
[21] M. J. Turner, P. Cresswell, P. Parham. J. L. Strominger, D. L. Mann, and A. R. Sanderson, *J. Biol. Chem.* **250,** 4512 (1975).
[22] P. Parham, B. N. Alpert, H. T. Orr, and J. L. Strominger, *J. Biol. Chem.* **252,** 7555 (1977).

7. Elute retained protein with 10 mM Tris–HCl buffer, pH 8.0 containing 0.25 M NaCl. Collect 5 ml fractions in plastic tubes and read absorbance at 280 nm of 20-μl aliquots of each fraction in 1 ml of water.

8. Pool protein fractions without concentrating for further purification steps.

Purification Methods

Isolation of Detergent-Solubilized HLA Antigens

In general, the presence of detergent has precluded the use of conventional separation procedures for purification of HLA antigens. Although successful separation of HLA-A and -B specificities from a cell line by repeated agarose gel filtration and isoelectric focusing has been reported,[7] this procedure has found little application, partially because biochemical purification of papain-solubilized antigens is more suitable as a routine procedure for large-scale work, but mainly because immunoaffinity methods provide a more rapid and specific alternative.

The use of specific antibodies for isolation of HLA antigens has traditionally been limited because human alloantisera are weak and therefore inappropriate for immunochemical purification. Affinity chromatography using anti-β_2-microglobulin immunoabsorbents[23] provides a rapid way of obtaining milligram amounts of detergent-solubilized HLA antigens. Unfortunately this method does not allow separation of specificities, a drawback which limits its applicability in biochemical work. Monoclonal antibodies constitute excellent immunochemical reagents because of their high affinity and monospecificity. Homogeneous preparations of individual HLA specificities may be obtained by a straightforward, high-yield procedure which uses a combination of allospecific monoclonal antibody immunoaffinity columns. The main limitation of this method is the difficulty in obtaining monoclonal antibodies of restricted specificity. Most anti-HLA monoclonal antibodies show a great deal of cross-reactivity and allospecific or nearly allospecific antibodies are not yet available for many HLA specificities. The following procedure was developed by Parham[24] for isolation of milligram amounts of alloantigenically active HLA-A2 and HLA-B7 proteins from a doubly homozygous lymphoblastoid cell line and was slightly modified later by Pober *et al.*[6] It may be readily adapted to the isolation of any HLA protein for which specific monoclonal antibodies are available.

[23] R. J. Robb, J. L. Strominger, and D. L. Mann, *J. Biol. Chem.* **251,** 5427 (1976).
[24] P. Parham, *J. Biol. Chem.* **254,** 8709 (1979).

Procedure. The following specifications are for batches of 150 g of cells.

1. Precipitate PA2.1 (anti-HLA-A2) or BB7.1 (anti-HLA-B7) antibodies from ascitic fluid by adding an equal volume of saturated $(NH_4)_2SO_4$.[24a] Normal mouse immunoglobulin is obtained in the same way from mouse serum.

2. After centrifugation (4,000 g, 20 min), redissolve the precipitate in 0.14 M NaCl, 10 mM sodium phosphate, pH 7.4 and exhaustively dialyze against the same buffer.

3. Couple crude antibody preparations to CL-4B Sepharose (Pharmacia) by means of CNBr activation[25] at a coupling density of about 2 mg of protein/ml of packed Sepharose. The binding capacity of this immunoabsorbent is about 0.5 mg of HLA/ml of derivatized Sepharose.

4. Prepare a series of four 25-ml columns connected in the following sequence: (I) Sepharose CL-4B. (II) Normal mouse immunoglobulin-Sepharose (human immunoglobulin may also be used). (III) PA2.1-Sepharose. (IV) BB7.1-Sepharose. Column I is to remove particulate material. Column II is intended to retain Fc receptors and all materials which bind nonspecifically to immunoglobulins. Columns III and IV retain HLA-A2 and -B7, respectively. All subsequent steps must be performed at 4°.

5. Preelute columns with 0.05 M diethylamine containing 0.2% NP-40.

6. Equilibrate columns with 0.1 M Tris–HCl pH 8.0, 1% NP-40.

7. Apply the detergent solubilized protein and wash the sample through the series of columns with 100 ml of 0.1 M Tris–HCl pH 8.0, 1% NP-40.

8. Separate the columns and wash separately each monoclonal antibody column with 10 column volumes of 1 M Tris–HCl, pH 8.0, 1% NP-40, and with 10 column volumes of 20 mM Tris–HCl, pH 8.0, 0.2% NP-40. High salt wash produces some HLA bleeding from the column but it drastically removes nonspecifically bound proteins.

9. Elute each monoclonal antibody column with 100 ml of 50 mM diethylamine (pH 11.5), 0.2% NP-40. Immediately neutralize each fraction by addition of 0.1 volume of 2 M Tris–HCl, pH 8.0. This short exposure to high pH produces little permanent denaturation of HLA antigens.

10. Reequilibrate columns with 0.1 M Tris–HCl, pH 8.0, 1% NP-40, and store them in this buffer at 4°.

11. Concentrate antigen containing fractions by ultrafiltration and dia-

---

[24a] For preparation of monoclonal antibodies see P. Parham and W. F. Bodmer, *Nature (London)* **276,** 397 (1978).

[25] S. C. March, I. Parikh, and P. Cuatrecasas, *Anal. Biochem.* **60,** 149 (1974).

lyze against 10 mM Tris–HCl, pH 8.0. Store HLA antigens frozen at −70°.

HLA antigens prepared from crude membranes by this procedure are consistently contaminated with variable amounts of actin, probably due to the existence of HLA–actin complexes. Preparation of pure, actin-free HLA antigens may be accomplished by disrupting membrane–cytoskeleton interactions prior to detergent solubilization as detailed above. Under these circumstances yields are of about 4 mg of each HLA specificity per 100 g of homozygous lymphoblastoid cells.

Isolation of Papain-Solubilized HLA Antigens

The stability of papain-solubilized HLA antigens in aqueous solutions has made it possible to apply conventional chromatographic techniques to the purification of these molecules. In the absence of high affinity allospecific antibodies it was not possible to separate individual specificities by immunochemical methods. For this reason, a great deal of effort was committed, mainly by Strominger and colleagues, to the development of large-scale biochemical procedures for obtaining homogeneous HLA antigens. Using doubly homozygous cell lines at the HLA-A and -B loci reduces the problem to the separation of single HLA-A and -B products which may be carried out on the basis of charge differences. Turner et al.[21] originally developed a method which was considerably improved by Parham et al.[22] to reduce the number of manipulations while increasing the yield of pure antigen. This method, with minor modifications, has been established as a routine purification procedure. Details are given below for preparations derived for 100 g of cells.

Procedure. All steps must be carried out at 4°.

1. Prepare a 5 × 90 cm column of Sephacryl S-200 (Pharmacia), equilibrated with 0.14 M NH$_4$HCO$_3$ and 0.02% sodium azide.

2. Apply unconcentrated papain-solubilized membrane proteins as obtained after chromatographic removal of papain (see step 8 of papain solubilization precedure) to the gel filtration column. Collect 10 ml fractions in plastic tubes and read absorbance at 280 and 230 nm.

3. Take 200-μl aliquots of appropriate fractions, lyophilize, and subject them to slab SDS–PAGE[26] to detect HLA-containing fractions.

4. Pool fractions containing HLA antigens and dialyze against 5 mM Tris-phosphate buffer, pH 8.0 (0.343 ml/liter of 85% phosphoric acid; pH is adjusted with saturated Tris base). pH should be checked at 4°. Conductivity of this buffer at 4° must be 350 × 10^{-6} Ω^{-1}.

5. Apply dialyzed sample to a 9.5 ml DEAE-cellulose (Whatman DE-52) column made in a 10 ml plastic pipet, equilibrated with 5 mM Tris-

[26] U. K. Laemmli, *Nature (London)* **227** 680 (1970).

phosphate, pH 8.0. Elute with a linear gradient formed with 125 ml of this buffer and 125 ml of 0.1 M Tris-phosphate, pH 5.6 (6.86 ml/liter of 85% phosphoric acid. Adjust pH at 4° with saturated Tris base). Conductivity of this buffer at 4° should be 2350 × 10^{-6} Ω^{-1}. Collect 90 fractions of 2 ml in plastic tubes, then 20 fractions with the final buffer at pH 5.6 and 20 additional fractions with starting buffer at pH 8.0, containing 0.25 M NaCl. Read absorbance at 280 and 230 nm.

6. HLA fractions are identified by SDS–PAGE (12%) of aliquots desalted by precipitation with trichloroacetic acid. Separation of HLA-A and -B antigens is achieved in step 5 with HLA-A antigens eluting first in all combinations of specificities tested so far. Pools containing no cross-contaminated individual specificities are stored frozen at −70°. Usually some fractions are obtained between those containing HLA-A and -B which contain a mixture of both. These fractions are discarded.

The yield of this procedure is about 4 mg of each specificity per 100 g of homozygous cells as estimated spectrophotometrically (assuming that A_{280} for HLA antigens at 1 mg/ml is 1.5) or by Lowry assay.[27] The separation of HLA-A and -B antigens by ion-exchange chromatography as described in this method has been successful for all combinations of antigens that have been tested such as HLA-A2 and -B7 or -B12,[21] and HLA-A28 and -B40,[17] but there is no evidence that any given pair of HLA-A and HLA-B molecules will be separated in this way.

The availability of allospecific monoclonal antibodies provides the possibility of performing large-scale purification of individual HLA specificities by immunoaffinity methods. The procedure described above for separation of detergent-solubilized antigens by monoclonal antibody columns is readily applicable to purification of papain-solubilized molecules[24] with the obvious omission of detergent from all buffer solutions. Reported yields of pure antigens with this method are at least equal to those obtained with nonimmunochemical procedures and it is much shorter to perform. In addition, immunoaffinity methods may be used for separation of individual specificities from heterozygous cell lines, provided that the appropriate monoclonal antibody is available. This is an important advantage because allospecificities from the same HLA locus are not separated by ion-exchange chromatography in the procedure described above. Thus, affinity chromatography with monoclonal antibodies may provide a more general way for large-scale purification of papain-solubilized HLA antigens. Unfortunately, its applicability is limited by the difficulty in obtaining HLA monoclonal antibodies of highly restricted specificity.

[27] O. H. Lowry, R. J. Rosebrough, A. L. Farr, and R. J. Randall, *J. Biol. Chem* **193**, 265 (1951).

Isolation of HLA Antigens from Other Sources

If the isolation of pure specificities is not required, milligram amounts of HLA class I antigens may be obtained from sources other than cultured cells, mainly platelets and spleens. These materials are available to laboratories which lack large tissue culture facilities and may therefore constitute a suitable alternative for many purposes. However, it should be stressed that because of heterogeneity of the starting material separation of individual specificities may not be achieved and final preparations will be composed of a mixture of HLA-A, -B and -C antigens.

Detergent-Solubilized HLA Antigens from Platelets

Large amounts of outdated platelets may be obtained from blood centers. The following method was developed by Trägårdh et al.[5] for the preparation of deoxycholate-solubilized HLA antigens. Highly purified material can be isolated from crude membranes in reasonably good yield by a four-step chromatographic procedure.

1. Lyse platelets by repeatedly freeze-thawing platelet concentrates in the presence of 1 mM PMSF. This protease inhibitor must be added to all buffers in further steps.

2. Centrifuge at 20,000 g for 1 hr at 4°. The pellet constitutes the crude membrane preparation.

3. Resuspend the membrane pellet in 0.02 M Tris–HCl buffer, pH 8.0, containing 10 mM sodium deoxycholate (Merck), at a concentration of 10 mg/ml of protein, as estimated by a Lowry assay.[28] Incubate at 4° for 30 min. About 20 to 25% of total protein is solubilized in this way.

4. Centrifuge at 105,000 g for 1 hr at 4° to remove particulate material. Collect the supernatant.

All subsequent chromatographic steps must be carried out at 4°.

5. Pass the protein supernatant through a lentil-lectin affinity chromatography column equilibrated with 0.02 M Tris–HCl, pH 8.0, 10 mM sodium deoxycholate. Wash with at least 10 column volumes of the same buffer. Elute the glycoprotein fraction, containing HLA antigens, with equilibration buffer supplemented with 10% α-methyl mannoside (Sigma).

6. Subject the glycoprotein fraction to gel filtration on Sephadex G-200 in 0.02 M Tris–HCl, pH 8.0, 10 mM sodium deoxycholate. HLA antigenic activity is recovered in two peaks at this stage, which correspond to oligomeric and monomeric forms of the HLA molecule, respectively. Pool the major peak, containing the monomeric form.

---

[28] J. R. Dulley and P. A. Greve, *Anal. Biochem.* **64**, 136 (1975).

7. Apply the HLA fraction to a rabbit anti-human β_2-microglobulin[4] column equilibrated with 0.02 M Tris–HCl, pH 8.0, containing 0.05 M NaCl and 0.1% Tween-80. Wash with several column volumes of the same buffer. Detergent is exchanged at this step. Elute bound protein with equilibration buffer containing 2 M potassium thiocyanate.

8. Dialyze the eluted HLA fraction at 4° against 0.02 M Tris–HCl, pH 8.0, 0.05 M NaCl, 0.01% Tween-80. Pass dialyzed material through a DEAE-Sephadex column, equilibrated with the same buffer. Contaminating rabbit IgG, which may leak off the immunosorbent column in the previous step, flows through and is removed at this stage. HLA material remains bound and is eluted with equilibration buffer containing 0.5 M NaCl.

HLA antigens obtained by this procedure are pure as estimated by SDS–PAGE. Recovery is about 6% of the HLA class I antigens present in the crude platelet membranes or 3 mg/15 g of total membrane protein.

Papain-Solubilized HLA Antigens from Spleen

The following procedure, as originally described by Trägårdh et al.,[4] allows the isolation of pure papain-solubilized HLA class I antigens from crude spleen cell membranes. All steps are carried out at 4°, unless otherwise specified.

1. Mince and homogenize spleen tissue (750 g) in 1000 ml of 0.02 M Tris–HCl buffer pH 7.4, 0.15 M NaCl.

2. Centrifuge at 10,000 g for 10 min. Discard the pellet.

3. Centrifuge the supernatant at 105,000 g for 1 hr.

4. Resuspend the pellet in 500 ml of the above Tris–HCl buffer and centrifuge again as in step 3. The washed pellet constitutes the crude membrane fraction.

5. Resuspend the membrane pellet in 0.02 M Tris–HCl buffer, pH 7.4, 0.15 M NaCl, 0.05 M cysteine, and 0.05 M EDTA. Adjust protein concentration (as estimated by Lowry assay) to 10 mg/ml.

6. Add papain (60 units/mg) to a final concentration of 3 mg of enzyme per ml. Incubate at 37° for 1 hr.

7. Terminate digestion by adding 0.055 M final concentration of iodoacetic acid. Adjust pH to 8.0 with NaOH.

8. Centrifuge at 105,000 g for 1 hr. Collect the supernatant containing papain solubilized membrane protein.

9. Dialyze the supernatant against 0.02 M Tris–HCl, pH 8.0, 0.05 M NaCl.

10. Load the dialyzed sample onto a DEAE-Sephadex column (40 × 6 cm for about 3 g of protein) previously equilibrated with dialysis buffer.

Elute at 60 ml/hr with 1200 ml of the same buffer, followed by a 2000-ml linear gradient from 0.05 to 0.6 M NaCl and by 500 ml of 1 M NaCl in 0.02 M Tris–HCl, pH 8.0. Pool HLA containing fractions and concentrate by ultrafiltration.

11. Dialyze the HLA pool against 0.02 M Tris–HCl, pH 7.2, 0.05 M NaCl. Load the sample on a second DEAE-Sephadex column (26 × 6 cm) equilibrated with the dialysis buffer. Elute with 1500 ml of a linear gradient from 0.05 to 0.5 M NaCl. Pool HLA material.

12. Pass the HLA fraction through a Sephadex G-200 column (146 × 2 cm for 60 mg of protein) equilibrated with 0.02 M Tris–HCl, pH 8.0, 0.15 M NaCl. Pool HLA containing fractions.

13. Load HLA material on a rabbit anti-human β_2-microglobulin column equilibrated with the same buffer as in step 12. Wash out unbound proteins with several column volumes of the equilibration buffer. Elute the retained HLA fraction with equilibration buffer, containing 3 M MgCl$_2$.

14. Dialyze HLA sample against 0.02 M Tris–HCl, pH 8.0, 0.15 M NaCl. Concentrate by ultrafiltration and pass through a Sephadex G-100 column (100 × 1 cm) equilibrated with the same buffer. Pool the HLA fraction.

The purified material obtained after step 14 is homogeneous by SDS–PAGE. The yield of the procedure is over 40% of the HLA-A,B,C antigens present in the papain-solubilized membrane protein fraction, or about 3 mg/13 g of total solubilized protein.

Detection Assays

Monitoring purification of HLA antigens requires methods for detecting the presence of these proteins through the separation procedures. Radioimmunoassay[29,30] and, more recently, solid phase enzyme immunoassay[31] may be conveniently used for this purpose. Nevertheless, the most widely used methods are the inhibition of antibody-mediated immune cytolysis and SDS–PAGE. The former is a sensitive and specific assay for measuring HLA allospecific activity, which does not require a knowledge of the structure of the molecule. It is generally employed when biologically active HLA antigens are to be obtained. However, this assay

[29] P. E. Evrin, P. A. Peterson, L. Wide, and I. Berggård, *Scand. J. Clin. Lab. Invest* **28**, 439 (1971).

[30] F. A. Lemonnier, N. Rebai, P. P. Le Bouteiller, B. Malissen, D. M. Caillol, and F. M. Kourilsky, *J. Immunol. Methods* **54**, 9 (1982).

[31] A. Ferreira, Y. Revilla, A. Bootello, and P. Gonzalez-Porqué, *J. Immunol. Methods* **65**, 373 (1983).

provides no information about the nature and amount of contaminants in HLA fractions and is tedious to perform as a routine procedure. On the other hand, SDS–PAGE is a straightforward and highly sensitive technique which readily informs about the presence of HLA proteins as well as any protein contaminants in the samples. Because of its relative simplicity and sensitivity it is systematically used to estimate the purity of HLA preparations. Obviously, it requires a previous knowledge of the band patterns typical of detergent- and papain-solubilized HLA class I antigens and bears no information regarding alloantigenic activity in HLA fractions. Hence, it is not a substitute for biological assays.

Inhibition of Antibody-Mediated Cytotoxicity by HLA Antigens

In this assay, solubilized HLA antigens bind to HLA antibodies, preventing their binding to lymphocyte target cells of appropriate HLA specificity. In the presence of complement, the fraction of antibodies not bound to soluble antigens will lyse ^{51}Cr-labeled target cells, releasing intracellular isotope. The amount of radioactivity released is proportional to the antigenic activity of HLA fractions. The procedure described below is based on the standard cytotoxicity test[32] as modified for microtiter plates.[33] It is directly applicable to papain-solubilized antigens, but it must be modified to avoid detergent mediated lysis of target cells when detergent-solubilized antigens are assayed. This modification includes a preincubation of antigen samples with serum albumin,[7] which is very effective in preventing detergent lysis of cells without affecting the lytic capacity of antisera and complement.

Procedure for Papain-Solubilized Antigens

1. Typed cells are obtained directly from lymphoblastoid cell cultures, or from peripheral blood by Ficoll–Hypaque gradient centrifugation.[34] In this later case, wash cells once with dextran/gelatin/veronal/saline (DGV buffer), whose composition is the following: $CaCl_2 \cdot 2H_2O$, 26 mg/liter; $MgSO_4$ (anhydrous), 59 mg/liter; NaCl, 8500 mg/liter; dextrose, 10,000 mg/liter; gelatin, 600 mg/liter; sodium veronal, 380 mg/liter; veronal, 580 mg/liter.

2. Resuspend cells in 100 μl of supernatant culture medium (for cultured cells) or autologous serum (for PBLs).

[32] A. R. Sanderson, *Nature (London)* **215**, 23 (1967).
[33] N. Tanigaki, Y. Miyakawa, Y. Yagi, and D. Pressman, *J. Immunol.* **107**, 402 (1971).
[34] A. Boyum, *Scand. J. Clin. Lab. Invest, Suppl.* **21**, 31 (1968).

3. Add 50 μl of sodium [^{51}Cr]chromate (10 mCi/ml) in 0.9% saline (New England Nuclear) and incubate for 2 to 3 hr at 37°, gently resuspending the cells every 30 min.

4. Prepare a dilution of antiserum which kills 75% of the cells. Add 5 μl of this diluted serum to each well of a microtiter plate.

5. Prepare serial dilutions of antigen in DGV buffer and add 5 μl of each dilution to the corresponding wells. Each dilution may be assayed in duplicate. Mix (by scratching the plate) and incubate for 1 hr at 37°.

6. After ^{51}Cr incubation (step 3), centrifuge cells at 400 g for 5 min. Remove supernatant with a Pasteur pipet.

7. Wash cells 6 times with DGV buffer and adjust final cell concentration to 1×10^6 cells/ml in the same buffer.

8. Add 5 μl of cells to each well, mix, and incubate for 1 hr at 37°.

9. Reconstitute lyophilized rabbit complement and add 5 μl per well. Mix and incubate at 37° for 30 min.

10. Dilute a 0.1 M EDTA solution 15-fold into DGV buffer. Add 150 μl of this diluted solution per well.

11. Centrifuge plates at 400 g for 5 min. Transfer 100 μl of supernatant from each well to glass tubes and count radioactivity in a gamma-counter.

The inhibitory titer is the inverse of the dilution of antigen in the assay volume after target cell addition, which gives 50% inhibition of ^{51}C release. Controls are set up as indicated in the table.

Procedure for Detergent-Solubilized Antigens. Steps 1 to 4 are performed as in the previous procedure.

5. Mix antigen samples with 30% bovine serum albumin in 0.9% NaCl, 0.5% sodium azide at a proportion of 10 volumes of albumin solution per volume of 1% detergent. Then, prepare serial dilutions in DGV buffer, add 5 μl of each dilution to the corresponding wells and incubate for 2 hr at 37°.

CONTROLS

Control	Antibody	Antigen	Cells (μl)	Complement	EDTA (μl)
(a) Maximum ^{51}Cr release[a]	DGV	DGV	5	DGV	150
(b) Spontaneous release	DGV	DGV	5	DGV	150
(c) Lysis by complement	DGV	DGV	5	5 μl	150
(d) Lysis by antibody	5 μl	DGV	5	DGV	150
(e) Lysis by Ab + C	5 μl	DGV	5	5 μl	150
(f) Lysis by Ag + C	DGV	5 μl	5	5 μl	150

[a] For control a, freeze-thaw rapidly three times using a dry ice/ethanol bath.

6–7. Perform steps 6 and 7 as in the procedure above.

8. Add 5 μl of cells to each well. Mix and incubate at 37° for 30 min.

9. Reconstitute lyophilized rabbit complement and dilute 1:4 with 0.14 M NaCl, 0.3 mM CaCl$_2$, 1 mM MgCl$_2$, 0.01 M Tris–HCl, pH 8.0. Add 100 μl of this diluted complement per well, mix, and incubate at 37° for 30 min. This large volume prevents nonspecific inhibition of complement by albumin or by other materials in the antigen solution, without lowering the titer of HLA sera.

10. Prepare a solution containing DGV buffer and 20 mM EDTA in 0.14 M NaCl, 10 mM sodium phosphate, pH 7.4, in a proportion 1:1. A suspension of erythrocytes from the Ficoll–Hypaque gradient may be added to this solution (1 drop/20 ml) as a marker for the cell pellet. Add 50 μl of this solution per well and mix.

11. Centrifuge plates at 400 g for 5 min. Remove 100 μl of supernatant from each well, transfer to glass tubes, and count radioactivity.

Set controls as in the previous procedure.

Polyacrylamide Gel Electrophoresis of HLA Antigens

Slab SDS–PAGE as described by Laemmli[26] is commonly used for monitoring purification of HLA class I molecules. Since it is a standard procedure it will not be described here and only some comments about the band patterns to be expected will be made. Routine gels are made of 8-cm-long slabs, 12% polyacrylamide. Under these conditions detergent-solubilized HLA-A and -B antigens migrate as a set of two bands of apparent molecular weight 44,000 and 12,000, corresponding to the heavy chain and to β_2-microglobulin, respectively. HLA specificities are not separated, so that a single 44K band will be seen even if a mixture of alloantigens is being analyzed. More important is the fact that actin is not separated in this gel system from the HLA heavy chain band, so that its presence may go undetected. For this reason, analysis of detergent solubilized HLA samples is better performed using longer (30 cm) 7–15% polyacrylamide gradient gels. In this system HLA heavy chain is clearly separated from actin and splitting of the 44K band is often seen in the form of a more or less defined doublet. However, heterogeneity of this band does not necessarily reflect the presence of mixed alloantigens, since pure specificities may show microheterogeneity[6] possibly arising from differences in glycosylation.[35]

Unless care is taken to inactivate intracellular proteases during detergent solubilization and subsequent purification steps, a 39K band may

[35] M. S. Krangel, H. T. Orr, and J. L. Strominger, *Cell* **18**, 979 (1979).

appear as a contaminant of HLA samples. This band arises from proteolytic cleavage of the heavy chain at a highly susceptible site between the hydrophobic transmembrane segment and the C-terminal intracytoplasmic region.[9] A second protease susceptible site exists between the extracytoplasmic portion and the transmembrane segment, which cleavage generates a 34K band, corresponding to the extracellular portion of the heavy chain. But cleavage at this point is slower and this band is seldom seen in detergent-solubilized samples.

Papain-solubilized antigens show a pattern of two bands of 34K and 12K corresponding to the heavy chain and β^2-microglobulin, respectively. At least some HLA-A and -B specificities have slightly different mobility and the corresponding heavy chain bands are separated even in short 12% gels. A very weak band of around 22K is often seen when papain-solubilized antigens are being analyzed in this way. The nature of this polypeptide has not been explored. There is no evidence of its being a non-HLA component because purified HLA samples show homogeneous N-terminal sequences. It has been shown that stringent protease treatment may produce a low-yield cleavage of the heavy chain, generating fragments of an apparent molecular weight of 20,000 and 12,000.[36] Thus, it is conceivable that the 22K band may arise from secondary cleavage of HLA molecules by the large amounts of papain employed in the solubilization procedure.

A Note on Radiochemical Purification of HLA Antigens

Purification of biosynthetically labeled histocompatibility antigens by immunochemical methods has some significant advantages over large-scale procedures. First, only small numbers of cells are needed for biosynthetic labeling, obviating the requirement of large tissue culture facilities. Second, given the appropriate antibody, single specificities may be isolated even from heterozygous cell lines. Third, by using double labeling with isotopes of different energy (i.e., ^3H and ^{14}C) isolated antigens are amenable to direct structural comparison by peptide mapping. Radiochemical methods have been extensively used in the biochemical analysis of H-2 antigens and the procedures involved are described in this volume [48, 50]. Their application to HLA biochemistry is more recent because the weakness of HLA alloantisera has precluded the use of immunoprecipitation or immunoaffinity chromatography for isolation of radiolabeled HLA antigens prior to the development of the much stronger mouse anti-HLA monoclonal antibodies. The potential of these methods

---

[36] L. Trägårdh, K. Wiman, L. Rask, and P. A. Peterson, *Biochemistry* **18**, 1322 (1979).

for the biochemical characterization of HLA polymorphism is indicated by recent studies on the structure of HLA variants and mutants.[37,38]

Acknowledgments

This work was supported in part by grants from the Fondo de Investigaciones Sanitarias de la Seguridad Social of Spain (N. 10/81 and 12/81). The author thanks Drs. T. A. Springer and J. Pober for critical comments on the manuscript and Dr. R. Bragado for helpful discussions.

[37] M. S. Krangel, S. Taketani, W. E. Biddison, D. M. Strong, and J. L. Strominger, *Biochemistry* **21,** 6313 (1982).

[38] M. S. Krangel, S. Taketani, D. Pious, and J. L. Strominger, *J. Exp. Med.* **157,** 324 (1983).

[53] Use of Monoclonal Antibody Immunoaffinity Columns to Purify Subsets of Human HLA-DR Antigens

By DAVID W. ANDREWS, M. ROSA BONO, JAMES F. KAUFMAN,
PETER KNUDSEN, and JACK L. STROMINGER

The control of immune responsiveness lies in the molecules encoded within the major histocompatibility complex (MHC).[1,2,2a] This collection of loci, called HLA in humans, encodes at least 6 cell surface glycoproteins and 3 serum complement components. The loci have been subdivided into three classes, based on similarities in structure, function, or both.[3] The Class I genes in humans encode the HLA-A, -B, and -C polypeptide chains of approximately 44,000 daltons, which form a heterodimer with β_2-microglobulin (12,000 daltons). This functional unit is distributed on virtually all nucleated cells. The Class II genes in humans are called HLA-D. This region contains multiple loci which encode sets of 2

[1] J. L. Strominger, V. H. Engelhard, A. Fuks, B. C. Guild, F. Hyafil, J. F. Kaufman, A. J. Korman, T. G. Kostyk, M. S. Krangel, D. Lancet, J. A. López de Castro, D. G. Mann, H. T. Orr, P. R. Parham, K. C. Parker, H. L. Ploegh, J. S. Pober, R. J. Robb, and D. A. Shackelford, *in* "The Role of the Major Histocompatibility Complex in Immunobiology" (M. E. Dorf, ed.), p. 115. Garland Press, New York, 1981.

[2a] Abbreviations: MHC, major histocompatibility complex; SDS, sodium dodecyl sulfate; DTT, dithiothreitol; Tris, tris(hydroxymethyl)aminomethane; PMSF, phenylethylsulfonyl fluoride; DOC, deoxycholate; WGA, wheat germ agglutinin; Con A, concanavalin A.

[2] B. Benacerraf, *Science* **212,** 1229 (1981).

[3] J. Klein, A. Juxetic, C. Baxevanis, and Z. Nagy, *Nature (London)* **291,** 455 (1981).

polypeptide chains (α and β) of different sizes, which are expressed primarily on lymphocytes and monocytes, but have been observed on a variety of other cells.[4] The apparent molecular weight of these chains varies among species and sometimes between loci within species. In humans, the HLA-DR glycoproteins are 34,000 daltons (α chain) and 29,000 daltons (β chain), nevertheless, their primary structure is remarkably similar to the HLA-A and -B molecules.[5,6] The Class III genes encode complement components, which function in the immune system, but whose structural relationship to the Class I and II proteins is unknown. The HLA-A, -B, -C, and -D molecules are serologically and structurally highly polymorphic. Ten alleles have been defined by cellular reactivity in the HLA-D region and 12 in the HLA-DR region, defined by alloantisera.[7] This review addresses the isolation and purification of these molecules.

The polymorphic variability of the HLA-D region mentioned above has been reflected in the apparent multiplicity of the α and β chain species observed in two-dimensional gels, detected following immunoprecipitation with a variety of antisera and monoclonal antibodies.[4,8–15] While some of this complexity may be artifactual[10] or attributable to posttranslational modification, it has become clear that different gene products can be isolated by immunoprecipitation procedures. We, as well as others,[16–19] have used immunoaffinity chromatography to isolate D-region molecules, and then separated the α and β chains by chromatography on hydroxylap-

[4] D. A. Shackelford, J. F. Kaufman, A. J. Korman, and J. L. Strominger, *Immunol. Rev.* **66**, 133 (1982).

[5] J. F. Kaufman and J. L. Strominger, *Nature (London)* **297**, 694 (1982).

[6] A. J. Korman, C. A. Auffray, A. Schamboeck, and J. L. Strominger, *Proc. Natl. Acad. Sci. U.S.A.* **79**, 6013 (1982).

[7] P. I. Terasaki, *in* "Histocompatibility Testing-1980," p. 18. UCLA Press, Los Angeles, California, 1980.

[8] M. L. Markert and P. Cresswell, *J. Immunol.* **128**, 1999 (1982).

[9] M. L. Markert and P. Cresswell, *J. Immunol.* **128**, 2004 (1982).

[10] D. A. Shackelford and J. L. Strominger, *J. Exp. Med.* **151**, 141 (1980).

[11] D. A. Shackelford and J. L. Strominger, *J. Immunol.* **130**, 274 (1983).

[12] D. A. Shackelford, L. A. Lampson, and J. L. Strominger, *J. Immunol.* **127**, 1403 (1981).

[13] D. A. Shackelford, D. L. Mann, J. J. van Rood, G. B. Ferrara, and J. L. Strominger, *Proc. Natl. Acad. Sci. U.S.A.* **78**, 4566 (1981).

[14] T. A. de Kretser, M. J. Crumpton, J. G. Bodmer, and W. F. Bodmer, *Eur. J. Immunol.* **12**, 214 (1982).

[15] R. W. Karr, C. C. Kannapell, J. A. Stein, H. M. Gebel, D. L. Mann, R. J. Duquesnoy, T. C. Fuller, G. E. Rodey, and B. D. Schwartz, *J. Immunol.* **128**, 1809 (1982).

[16] S. M. Goyert, J. E. Shively, and J. Silver, *J. Exp. Med.* **156**, 550 (1982).

[17] R. Bono and J. L. Strominger, *Nature (London)* **299**, 836 (1982).

[18] D. W. Andrews, M. R. Bono, and J. L. Strominger, *Biochemistry* **21**, 6625 (1982).

[19] L. E. Walker, R. Hewick, M. W. Hunkapillar, L. E. Hood, W. J. Dreyer, and R. A. Reisfeld, *Biochemistry* **22**, 185 (1983).

atite for amino acid sequence analysis. Our procedure, as well as other separation methods, are described herein.

Isolation of DR and DR-Related Antigens

Detergent solubilized membrane proteins are prepared from a homozygous B lymphoblastoid cell, LB (DRw6,6). Membrane proteins are solubilized by incubating cells for 30 min at 4° in a buffer consisting of 2% NP-40, 0.01 M Tris, pH 7.8, 0.14 M NaCl, 1 mM MgCl$_2$, and 0.1 mM PMSF. The ratio of cells to buffer is 1 g of cells/4 ml of buffer. After incubation, the mixture is centrifuged at 150,000 g for 1 hr, and the supernatant is recovered and centrifuged for an additional 20 min at 12,000 g. The subsequent steps involving immunoaffinity chromatography are carried out at 4°. In order to remove actin, the supernatant is passed over a column of Sepharose 4B-CL (25 ml) coupled to normal rabbit serum. The lysate (for 20 g of cells, about 80 ml) is then passed over a column (10 ml) of the monoclonal antibody LKT111[20] immobilized on Affi-Gel 10 (Bio-Rad). The effluent from this column is recycled over the LKT111 column for 24 hr at a flow rate of no greater than 5 ml/hr. The effluent is then applied on a column of the monoclonal antibody L243[21] under the same conditions. The columns are then washed with no more than 5 column volumes of 0.01 M Tris buffer, pH 8.0 containing 0.14 M NaCl and 0.1% deoxycholate. Bound proteins are then eluted with 0.1 M Tris buffer, pH 11.5 containing 0.5% deoxycholate and 5% glycerol. The fractions are collected, neutralized with 1 N HCl, and aliquots are taken and treated with a 5-fold excess (v/v) of acetone to precipitate protein. The resulting precipitates are analyzed by SDS–polyacrylamide gel electrophoresis to locate the DR molecules.[22] Fractions containing the DR proteins are dialyzed against 0.1% SDS and lyophilized, dialyzed against the hydroxylapatite buffer (vide infra), or precipitated with acetone. The separation of two immunochemically distinct subsets of DR antigens is illustrated in Fig. 1. The pool of protein that is not bound to the LKT111 or L243 affinity columns is used to isolate the closely related, but immunochemically distinct molecule, DC-1, in a similar way.[17] This pool is cycled over a column of the monoclonal antibody Genox 3.53,[23] which is then washed and eluted in the same manner used for the LKT111 and L243 columns. During elution of the Genox 3.53 column, fractions are collected, neutralized, and examined by SDS–polyacrylamide gel electrophoresis to locate

[20] M. R. Bono, F. Hyafil, J. Kalil, V. Koblar, J. Weils, E. Wollman, C. Mawas, and M. Fellous, *Transplant. Clin. Immunol.* **11,** 109 (1979).

[21] L. Lampson and R. Levy, *J. Immunol.* **125,** 293 (1980).

[22] U. K. Laemmli, *Nature (London)* **227,** 680 (1970).

[23] F. M. Brodsky, P. Parham, and W. F. Bodmer, *Tissue Antigens* **16,** 30 (1980).

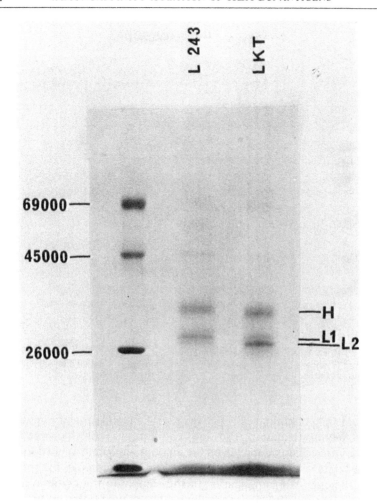

FIG. 1. SDS–polyacrylamide gel electrophoretic analysis of isolated DR molecules eluted from LKT111 and L243 monoclonal antibody columns (see text for details of procedure). The band labeled "H" corresponds to the DR α chain. The bands labeled "L1" and "L2" represent DR β chains of differing mobility which constitute the native $\alpha\beta$ dimers recognized by the different monoclonal antibodies. Gels were 10% polyacrylamide.

DC-1 molecules. Those fractions containing DC-1 are prepared for the separation of α and β chains, as described for the DR molecules.

Separation of α and β Chains

The various steps of this procedure are carried out at 37°. The lyophilized DR or DC-1 antigens are dissolved in 0.1 M sodium phosphate

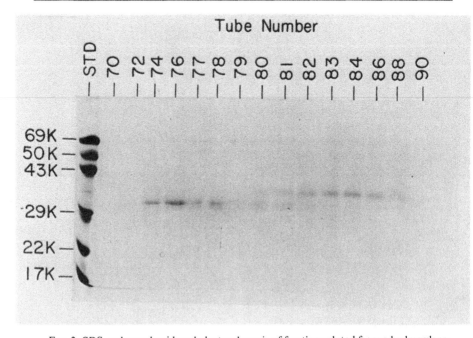

FIG. 2. SDS–polyacrylamide gel electrophoresis of fractions eluted from a hydroxylapatite column. Fractions 74–78 contain β chain. Fractions 83–88 contain α chain. The intermediate fractions contain a mixture of α and β chains. For further details, see text.

buffer, pH 6.4, containing 0.1% SDS and 1 mM DTT and applied to a column of hydroxlyapatite (Bio-Rad) equilibrated with the same buffer. The size of the column depends on the amount of protein applied to it. For the material derived from 20 g of cells, a 10 ml of column of hydroxylapatite is sufficient. After application of the sample to the column, a linear gradient is applied (0.1–0.5 M sodium phosphate), followed by elution with 0.5 M sodium phosphate, pH 6.4, 0.1% SDS, and 1 mM DTT. This procedure results in the separation of α and β chains, the β chain eluting at lower ionic strength.[24,25] Fractions are collected and aliquots taken, diluted with Laemmli sample buffer, and subjected to SDS–polyacrylamide gel electrophoresis (Fig. 2). Fractions containing the respective chains are dialyzed vs 0.1% SDS and lyophilized. Chains isolated in this manner are homogeneous and can be easily sequenced.[17,18]

Another method of separation of α and β chains involves the preparation of a series of monoclonal antibodies to distinguish denatured α and β

[24] J. H. Freed, *Mol. Immunol.* **17**, 453 (1980).
[25] J. F. Kaufman and J. L. Strominger, *Biochemistry* (submitted for publication).

chains.[26] In this procedure, mice are immunized with DR antigens isolated from membranes on Con A-agarose. The DR isolate is denatured by treatment at 100° in 2% SDS followed by reduction with dithiothreitol and alkylation with iodoacetic acid. This mixture is used to immunize 5 week old BALB/c female mice. Screening of antibody-producing cells by a plate binding assay reveals clones that recognize isolated DR α and β chains.

After reduction and alkylation, the DR mixture is dialyzed vs 0.1% SDS, 10 mM morpholinopropane sulfonate, pH 8.0 in order to remove the dithiothreitol. The sample is then diluted with one volume of 2% Brij 97:99, 0.1 M NaCl, 50 mM morpholinopropane sulfonate, pH 8.0 and the solution is applied to a column of HC2.1 (an antibody that recognizes denatured α chains and *in vitro* translated nascent α chains: described by Knudsen *et al.*[26]) coupled to Sepharose. The column is washed with 15 column volumes of 0.2% Brij 97:99, 50 mM morpholinopropane sulfonate, 0.1 M NaCl before eluting the DR α chain with 10 mM HCl, 0.2% Brij 97:99. In this way, the α chain can be separated from the β chain which can be recovered from the column flow through.

Other Methods for the Isolation of D-Region Antigens

Physicochemical methods have been successfully used to isolate D-region molecules from several different human cell lines.[19,25,27] In our laboratory,[25] the following method is used. The first step utilizes lectin affinity chromatography of lysates derived from papain-treated membranes or untreated cells. Among a number of lectins examined (including Con A, WGA, ricin A and B, gorse agglutinin), Con A-Sepharose (Pharmacia) gives the highest recovery of class II antigens. Since class I antigens are not bound by this lectin, this is a useful step for separating human class I from class II antigens.

Attempts to purify the separated chains of detergent-solubilized class II antigens by the methods used for detergent-solubilized class I antigens were unsuccessful. The prime difficulty was aggregation in urea or guanidine–HCL. SDS gel electrophoresis is a procedure which resolves the chains, but preparative gel electrophoreses requires steps to prevent oxidative degradation and is cumbersome. Therefore a column separation procedure in SDS was devised. Samples of DR antigens are precipitated with acetone, resuspended in 2% SDS and 10% 2-mercaptoethanol, and

[26] P. J. Knudsen, J. McLean, and J. L. Strominger, *J. Immunol.* (submitted for publication).
[27] H. Kratzin, C. Y. Yang, H. Gotz, E. Pauley, S. Kolbel, G. Egert, F. P. Thinnes, P. Wernet, P. Altevogt, and N. Hilschmann, *Hoppe-Seyler's Z. Physiol. Biochem.* **362,** 1665 (1981).

briefly boiled. The samples are then applied to a column of Sephadex G-200 (or Sephacryl S-300) equilibrated with 0.1 M sodium phosphate buffer, pH 6.4, 1 mM DTT, 0.1% SDS. The resulting fractions are analyzed by absorbance at 280 nM and by SDS gel electrophoresis. The fractions containing class II light and heavy chains are pooled and separated using hydroxylapatite column chromatography, as previously described.

Hilschmann's group purified the D-region molecules in a 3-step procedure (although no details have been published): (1) ion-exchange chromatography in the presence of detergent (CM-cellulose/NP-40); (2) gel filtration on Sephacryl S-300 in 0.6% SDS; (3) chromatography on hydroxylapatite in 0.1% SDS to separate the α and β chains.[27,28] This procedure yields sufficient material for the determination of the complete sequence of the extracellular domains of the α and β chains.

Another method for the isolation of D-region antigens[19] involves the following steps: (1) chromatography on lentil lectin-gel and elution of the bound glycoproteins with 0.1 M Tris buffer, pH 8.0, containing 0.15 M NaCl, 0.001 M MgCl$_2$, 0.02% NaN$_3$, 0.5 mM phenylmethylsulfonyl fluoride, and 1% Renex-30 (Accurate Chem. Co., Hicksville, NY). (2) The proteins eluted from the lentil lectin column are concentrated and heated in SDS at 100° for 2 min. (3) Fractionation of the denatured material by gel filtration on HPLC. The pooled fractions containing the antigens may be used for the separation of the α and β chains by hydroxylapatite chromatography in the presence of SDS. The isolated α and β chains may be further purified by gel filtration on an HPLC column.

Acknowledgments

It is a pleasure to acknowledge the patience and skill of Alice Furumoto-Dawson in the preparation and typing of this manuscript. We are grateful to Dr. Joan M. Gorga for her careful editing. Supported by research grants from NIH (AI-10736).

[28] C. Y. Yang, H. Kratzin, H. Gotz, F. P. Thinnes, T. Kruse, G. Egert, E. Pauley, S. Kolbel, P. Wernet, and N. Hilschmann, *Hoppe-Seyler's Z. Physiol. Chem.* **363,** 671 (1982).

[54] Use of the HLA-DR Antigens Incorporated into Liposomes to Generate HLA-DR Specific Cytotoxic T Lymphocytes

By Joan C. Gorga, Judith Foran, Steven J. Burakoff, and Jack L. Strominger

In order to elucidate the molecular mechanisms involved in the induction of cytolytic T lymphocytes (CTLs)[1] in the immune response, it is helpful to be able to separate the antigens expressed on the cell surface, and to test their individual abilities to act as stimulators of and targets for the action of CTLs. Human histocompatibility antigens at various stages of purification have been used to generate a CTL response. Isolated lymphoblastoid cell membranes and the dialyzed, detergent-soluble fraction from isolated membranes are both effective in eliciting CTLs.[1a,2] The class I major histocompatibility antigens (HLA-A, -B, and -C), purified by immunoaffinity chromatography[3,4] and reconstituted into liposomes[5] were found to induce secondary CTLs.[1a,4] The class II antigens (primarily HLA-DR), purified by lentil lectin chromatography as the remaining glycoprotein fraction after removal of the class I antigens, and reconstituted into liposomes, also induce secondary CTLs,[2] and these CTLs were shown to be distinct from those induced by the reconstituted class I antigens.[4] Recently, HLA-DR antigens (DR) have been purified by immunoaffinity chromatography and used to generate CTLs.[6] The use of purified DR to generate DR-specific CTLs reduces the chances that minor contaminants of the preparation, such as class I antigens, lentil lectin

---

[1] Abbreviations: CTL, cytolytic T lymphocyte; DR, HLA-DR antigens; Tris–Cl, tris(hydroxymethyl)aminomethane hydrochloride; DTT, dithiothreitol; PMSF, phenylmethylsulfonyl fluoride; (Na)$_2$EDTA, disodium ethylenediaminetetraacetic acid; NP40, Nonidet P-40; DOC, sodium deoxycholate; MOPS, morpholinopropanesulfonic acid; SDS, sodium dodecyl sulfate.

[1a] V. H. Engelhard, J. L. Strominger, M. Mescher, and S. J. Burakoff, *Proc. Natl. Acad. Sci. U.S.A.* **75**, 5688 (1978).

[2] S. J. Burakoff, V. H. Engelhard, J. Kaufman, and J. L. Strominger, *Nature (London)* **283**, 495 (1980).

[3] R. J. Robb, J. L. Strominger, and D. A. Mann, *J. Biol. Chem.* **251**, 5427 (1976).

[4] V. H. Engelhard, J. F. Kaufman, J. L. Strominger, and S. J. Burakoff, *J. Exp. Med.* **152**, 54s (1980).

[5] V. H. Engelhard, B. C. Guild, A. Helenius, C. Terhorst, and J. L. Strominger, *Proc. Natl. Acad. Sci. U.S.A.* **75**, 3230 (1978).

[6] J. C. Gorga, P. J. Knudsen, J. Foran, J. L. Strominger, and S. J. Burakoff, in preparation.

(which acts as a mitogen), or cytoskeletal proteins, will affect the results. Purified antigen also facilitates analysis of both the requirements for generating CTLs and their specificities. One can begin to add back components such as cytoskeletal proteins, which have been shown to interact with the class I antigen HLA-A2[7] and have been found to enhance the stimulation by reconstituted H-2,[8] and one can investigate the effectiveness of other class II antigens and DR subsets in generating CTLs.

Purification of HLA-DR

The DR used for incorporation into liposomes is purified by a procedure which involves the selective removal of cytoskeletal proteins from the membranes prior to detergent solubilization.[7] DR is purified from the lymphoblastoid cell lines JY (DR4,w6)[4] or LB (DRw6,w6).[4] All steps are carried out at 4°.

Isolation and Solubilization of Membranes

Frozen cells (50 g), stored at $-70°$, are thawed and lysed by incubation in 125 ml of 10 mM Tris–Cl/1 mM dithiothreitol (DTT)/0.1 mM phenylmethylsulfonyl fluoride (PMSF), pH 8.0, for 30 min. Cellular debris is removed by centrifugation at 4000 g_{max} for 5 min, and the pellet is repeatedly washed with lysis buffer until the supernatant is no longer turbid. The combined supernatants are centrifuged at 100,000 g_{max} for 1 hr. The pellet is homogenized in 100 ml of lysis buffer and centrifuged for 20 min at 48,000 g_{max}. The crude membrane pellet is resuspended by vortexing in 100 ml 1 mM (Na)$_2$EDTA/0.1 mM PMSF, pH 7.5, centrifuged for 20 min at 48,000 g_{max}, and washed again with 1 mM (Na)$_2$EDTA/0.1 mM PMSF, pH 7.5, in the same manner. The washed pellet is homogenized in 100 ml of 2 mM Tris–Cl/200 μM NaATP/200 μM MgCl$_2$/500 μM DTT/0.002% NaN$_3$/0.1 mM PMSF, pH 8.0, and dialyzed against 2 liters of the same buffer for 14–20 hr in Spectrapor 2 dialysis membranes (MW cutoff 12,000). The material is harvested by centrifugation for 20 min at 48,000 g_{max}, washed twice with 1 mM (Na)$_2$EDTA/0.1 mM PMSF, pH 7.5, and homogenized in 80 ml distilled H$_2$O. A 20-ml aliquot of 20% Brij (2 : 1 Brij 99 : 97) is added. The suspension is gently stirred for 30 min and centrifuged for 2 hr at 100,000 g_{max}. The clear supernatant is retained for purification of histocompatibility antigens. The supernatant is applied sequentially to two 12-ml columns preequilibrated with 10 mM Tris–Cl/1%

---

[7] J. S. Pober, B. C. Guild, J. L. Strominger, and W. R. Veatch, *Biochemistry* **20**, 5625 (1981).

[8] S. H. Hermann and M. F. Mescher, *Proc. Natl. Acad. Sci. U.S.A.* **78**, 2488 (1981).

Nonidet P-40 (NP40), pH 8.0. These columns (Sepharose CL-4B and normal human immunoglobulin coupled to Sepharose CL-4B, respectively) remove material which binds nonspecifically to the gel and to immunoglobulins (actin, in particular). The columns are washed with 40 ml of 100 mM Tris–Cl/1% NP40, pH 8.0, and the entire effluent is collected and used for isolation of DR antigens on immunoaffinity columns.

Preparation of DR Immunoaffinity Columns

DR immunoaffinity columns are formed by coupling DR-specific monoclonal antibodies (primarily LB3.1,[6] but L243[9] and L227[10] have also been used successfully) to Affi-Gel 10 (Bio-Rad). Cyanogen bromide-activated Sepharose CL-4B can also be used. Antibody is twice precipitated from mouse ascites fluid by addition of ammonium sulfate to 45% saturated. Fifty milligrams of the partially purified antibody is coupled to 5 g of Affi-Gel 10 in a volume of 12.5 ml with gentle shaking for 12–16 hr at 4°. The coupling efficiency is about 70%. The gel is pelleted by centrifugation for 5 min at 500 g_{max}, the supernatant is removed, and any remaining reactive groups are blocked by incubation for 1 hr with 10 ml of 1 M glycine ethyl ester, pH 8.0. The column is purged with 50 mM glycine/ 0.1% sodium deoxycholate (DOC), pH 11.5, and equilibrated with 20 mM morpholinopropanesulfonic acid (MOPS)/150 mM NaCl/0.1% DOC, pH 8.0.

Immunoaffinity Purification of DR

The Brij/NP40 effluent from the Sepharose columns is divided into aliquots, and an amount equivalent to 5 g of cells (i.e., one-tenth of the total volume of the effluent) is passed through the DR immunoaffinity column at a flow rate of approximately 10 ml/hr. The column is washed with 50 ml of the equilibration buffer and then eluted with 10 ml of 50 mM glycine/0.1% DOC, pH 11.5, at a flow rate of approximately 40 ml/hr. Each fraction is immediately neutralized with 2 M glycine, pH 2. The absorbance at 280 nm is read, and the peak fractions are combined and dialyzed in Spectrapor 1 dialysis membranes (MW cutoff 6000) against at least 50 volumes of 20 mM MOPS/150 mM NaCl/0.1% DOC, pH 8.0, for 16 hr at 4°. The dialyzed eluate is concentrated 10- to 20-fold by Amicon filtration using a PM-30 membrane, and the concentrated DR is divided into aliquots and stored at −20° at a concentration of about 1 mg/ml protein, as measured by Pederson's modification[11] of the Lowry

[9] D. A. Shackelford, L. A. Lampson, and J. L. Strominger, *J. Immunol.* **130**, 289 (1983).
[10] L. A. Lampson and R. Levy, *J. Immunol.* **125**, 293 (1980).
[11] G. L. Pederson, *Anal. Biochem.* **83**, 346 (1977).

method.[12] The capacity of the column is about 1 mg of DR. Approximately 1.5 mg of DR is obtained from 5 g of cells. With this rapid immunoaffinity procedure, optimal yields of DR may not be obtained, but the more important consideration is whether the purified antigen retains its native conformation. After this procedure, the purified DR is found to be in its native conformation as indicated by the fact that the light and heavy chains remain complexed upon sodium dodecyl sulfate (SDS)–polyacrylamide gel electrophoresis[13] unless the protein is boiled in SDS sample buffer.[14]

Reconstitution of HLA-DR into Liposomes

The purified, detergent-solubilized DR is combined with detergent-solubilized JY or LB membrane lipids, and liposomes are formed upon removal of the detergent by dialysis, as follows. Lipids are extracted from JY or LB cell membranes with chloroform/methanol according to the method of Bligh and Dyer,[15] and are stored at approximately 40 mg/ml in 2 : 1 chloroform : methanol at $-20°$ under N_2. Lipids (8 mg) are combined with 20 mg of DOC in methanol (5% w/v). The solvent is evaporated in an N_2 stream, and residual solvent is removed by placing the tube in a desiccator under vacuum and over P_2O_5 for a minimum of 2 hr. Buffer (20 mM MOPS/150 mM NaCl/0.1% DOC, pH 8.0 at 4°) and purified DR in buffer are added to give 400 μg DR and a total volume of 2 ml. The solution is vortexed to dissolve the lipid/DOC residue, sterilized by passage through a 0.22-μm filter, and dialyzed against 5 × 4 liters of 140 mM NaCl/10 mM Tris–Cl, pH 8.0 at 4°, in autoclaved Spectrapor 2 dialysis membranes with buffer changes every 12 hr. The resulting liposomes are pelleted by centrifugation at 100,000 g_{max} for 1 hr at 4° in sterile centrifuge tubes, and resuspended in 2 ml of sterile 140 mM NaCl/10 mM Tris–Cl, pH 8.0 at 4°. The resuspended liposomes are divided into aliquots, frozen at $-70°$, and thawed just before use.

The recoveries of protein, measured as described above,[11,12] and lipid, measured by phospholipid phosphorus[16] are approximately 50 and 75%, respectively. The optimal protein : lipid ratio is in the range of 20–100 μg protein : mg lipid. The freeze-thaw step appears to be important for effec-

[12] O. H. Lowry, N. J. Rosebrough, A. L. Farr, and R. J. Randall, J. Biol. Chem. 193, 265 (1951).

[13] U. K. Laemmli, Nature (London) 227, 680 (1970).

[14] T. A. Springer, J. F. Kaufman, L. A. Siddoway, D. L. Mann, and J. L. Strominger, J. Biol. Chem. 252, 6201 (1977).

[15] E. G. Bligh and W. J. Dyer, Can. J. Biochem. Physiol. 37, 911 (1959).

[16] M. A. Zoccoli, S. A. Baldwin, and G. E. Lienhard, J. Biol. Chem. 253, 6923 (1978).

CTL ASSAYS OF DR LIPOSOMES WITH AND
WITHOUT FREEZE-THAW[a]

Stimulator	Specific release (%)
—	10.1
JY cells (1×10^5)	84.4
JY membranes (10 μg)	82.1
DR liposomes (3.6 μg)	15.1
DR liposomes (3.6 μg) + freeze-thaw	72.8

[a] DR liposomes were prepared in one batch as described and divided in half. One aliquot was frozen at $-70°$ and thawed before use in the CTL assay, while the other was assayed directly. The data are for cytotoxicity of ^{51}Cr-labeled JY target cells at an effector:target ratio of 70:1.

tive stimulation of CTLs. We have compared the stimulation of CTLs by DR in liposomes that were not frozen with that of DR in liposomes that were frozen and thawed (see the table). The frozen-and-thawed liposomes were more effective, presumably because they presented a larger surface area to the cells. By electron microscopy, the nonfrozen liposomes were found to be small and homogeneous (40–80 nm in diameter), while the frozen-and-thawed liposomes were heterogeneous in size (40–500 nm in diameter) with many very large structures.

CTL Assay

Induction of Secondary CTLs

For the induction of secondary cytolytic T lymphocytes, C57BL/6 mice are primed by intraperitoneal injection of 2×10^7 viable JY cells and are boosted at least 1 month later in the same way.[1] One week following the boost, the spleens are removed and teased to give a single cell suspension, and CTLs are generated *in vitro*. Spleen cells (7×10^6) are cultured with 1×10^5 irradiated (5000 rads) JY cells, JY cell membranes (10 μg of protein), HLA-DR liposomes (1–50 μg of protein), or no additions, in a final volume of 2 ml in 24-well plates in a humid atmosphere of 95% air/5% CO_2 at 37°. The culture medium consists of RPMI 1640 with 100 units/ml penicillin, 100 μg/ml streptomycin, 2 mM glutamine, 50 μM 2-mercapto-ethanol, and 10% heat-inactivated fetal bovine serum. After 5 days, the cells (effectors) are reisolated and tested for cytotoxicity on ^{51}Cr-labeled targets.

Assay of Cytotoxicity (This Volume [21])

Target cells are prepared by incubating 1×10^7 cells with 0.1 ml of $Na^{51}CrO_4$ (1 mCi/ml, Amersham) for 60 min at 37° and then washing three times with medium. Usually, JY or Daudi (typed as DRw6, with no class I antigens expressed) are used as targets, but others have been used with varying degrees of success.[4] In antibody blocking experiments, the target cells are preincubated at this stage for 30 min with the desired dilutions of antiserum. One-hundred-microliter aliquots of the target cell suspension, containing 1×10^4 labeled cells, are placed in microtiter wells, and effector cells (normal spleen cells or the immune spleen cells cultured as described above) are added in a volume of 100 μl to give effector:target ratios in the range of 10–100:1. The cultures are incubated for 4 hr at 37° with intermittent shaking in an atmosphere of 83% N_2/10% CO_2/7% O_2. After incubation, the plates are centrifuged at 800 g_{max} for 15 min. Radioactivity is determined on a 100-μl aliquot of each supernatant to determine isotope release. The percentage of specific release is calculated using the equation $[(E - C)/(Ft - C)]100$ where E is the isotope release from wells containing immune effector cells and target cells, C is the isotope release from wells containing normal spleen cells and target cells (spontaneous release), and Ft is the maximum isotope release determined after four cycles of freezing and thawing target cells.

Antibody Blocking

The specificities of the CTLs generated are assessed by antibody blocking of the target cells and with a variety of target cells expressing different DR allospecificities. Because the immunoglobulin concentrations of various antisera are different, it is usually helpful to test the antiserum blocking over a range of final dilutions, for example, from 1:1 to 1:64, to determine the effective blocking concentration for a particular antiserum. DR-specific CTLs induced by DR liposomes were inhibited by the rabbit heteroserum anti-p29,34, which recognizes class II antigens,[14] at a dilution of 1:10, but not by the monoclonal antibody W6/32, which recognizes class I antigens,[17] at a dilution of 1:4, a concentration sufficient to block CTLs induced by HLA-A,B liposomes.[4] Although we have not done so, it might be preferable to use purified antibody, rather than antiserum or ascites fluid for this purpose, in order to be able to quantitate the amounts of immunoglobulin added.

---

[17] C. J. Barnstable, W. F. Bodmer, G. Brown, G. Galfre, C. Milstein, A. F. Williams, and A. Ziegler, *Cell* **14**, 9 (1978).

Target Cell Panels

The allospecificities of the DR-specific CTLs can be assessed by using a panel of target cells expressing different DR allospecificities.[4] This analysis is complicated when using DR purified from JY cells because JY expresses DR4 and w6. It is essential to measure the spontaneous release of ^{51}Cr and the lysis by immune spleen cells incubated with no additions *in vitro*, as these values vary considerably between different target cell lines, and must be taken into account when comparing the results. In several experiments testing the lysis of target cells typed as homogeneous for DRs 1–8 by CTLs stimulated with DR from JY cells, preferential lysis of cells expressing DR4 and/or 6 was not observed (unpublished results). These results may indicate that in this system we are working with a mixture of CTLS with different specificities. While some of the CTLs may be allospecific, others may recognize what appear to be supertypic determinants in the DR region, such as MT2.[18] Engelhard and Benjamin[19] found by clonal analysis of CTLs generated by incubation of spleen cells with irradiated JY cells that the majority of DR-reactive clones exhibited broad cross-reactivity on targets of different DR types, while only a few were highly specific. Thus, in our assay system, lysis by CTLs with broad cross-reactivity will mask lysis by allospecific CTLs.

It has become apparent that other loci such as DC[20] and SB[21] exist in the DR region. We are investigating the abilities of these antigens to stimulate specific CTLs.

Acknowledgments

Supported by research Grants AM-13230 and CA-34129 from the National Institutes of Health. SJB is a recipient of an American Cancer Society Faculty Research Award. JCG is a postdoctoral fellow of the Juvenile Diabetes Foundation International.

[18] D. A. Shackelford, J. F. Kaufman, A. J. Korman, and J. L. Strominger, *Immunol. Rev.* **66**, 133 (1982).

[19] V. H. Engelhard and C. Benjamin, *J. Immunol.* **129**, 2621 (1982).

[20] R. Tosi, N. Tanigaki, D. Centis, G. B. Ferrara, and D. Pressman, *J. Exp. Med.* **148**, 1592 (1978).

[21] S. Shaw, A. H. Johnson, and G. M. Shearer, *J. Exp. Med.* **152**, 565 (1980).

[55] HLA Antigens in Serum

By M. A. PELLEGRINO, C. RUSSO, and J. P. ALLISON

Serologic and Chemical Properties of Serum HLA Antigens

Human histocompatibility antigens are sets of highly polymorphic gly-coproteins encoded within the MHC region that play a critical role in cooperation between immunocompetent cells. HLA-A,B,C antigens are expressed on most, if not all, nucleated cells while Ia antigens have a restricted tissue distribution, as they are mainly expressed on cells associated with immune functions. However, in man, as in other species, histocompatibility antigens are also found in body fluids, such as serum, urine, amniotic fluid, and seminal plasma.

Only a limited number of studies have compared the expression of HLA-A,B,C allospecificities in serum with the HLA typing of peripheral lymphocytes from the same donor.[1,2] No absolute correspondence has been found and the occurrence of both false positives and false negatives has been encountered. False negatives may occur because the level of soluble HLA antigens in some individuals is below that detectable by the assay methods used. False positives could occur due to the well-known phenomenon of cross-reactivity among certain groups of HLA-A,B,C allospecificities.

Different allospecificities have been detected in different individuals in varying levels: it is not clear whether levels vary in different individuals irrespective of the HLA allospecificity involved, or whether variation is a characteristic of alleles rather than individuals. It is noteworthy that the allospecificity HLA-A9 is found in serum in significantly greater amounts than any other allospecificity.[1]

No studies have been published on detection of Ia allospecificities in human sera. In fact, Ia molecules have been shown to be present in human sera by their reactivity with polyclonal and monoclonal xenoanti-bodies directed to common determinants of Ia antigens.[3,4]

---

[1] M. A. Pellegrino, S. Ferrone, A. Pellegrino, S. K. Oh, and R. A. Reisfeld, *Eur. J. Immunol.* **4**, 250 (1974).

[2] B. D. Tait, R. I. Finlay, and M. J. Simons, *Tissue Antigens* **17**, 129 (1981).

[3] B. S. Wilson, F. Indiveri, M. A. Pellegrino, and S. Ferrone, *J. Immunol.* **122**, 1967 (1979).

[4] M. S. Sandrin, M. M. Henning, H. A. Vaughan, I. F. C. McKenzie, and C. R. Parish, *JNCI, J. Natl. Cancer Inst.* **66**, 279 (1981).

HLA-A,B antigens in serum exist as lipid–glycoprotein complexes, probably as a result of shedding of these integral membrane proteins from the cell surface with their associated boundary lipid intact.[5] The behavior of these serum antigens upon ultracentrifugation at different densities indicates that the MHC antigens coisolate with the high density lipoproteins. The molecular structure of HLA-A,B antigens derived from the serum is similar to that of antigens isolated from lymphoid cells: the HLA-A,B antigens are glycoproteins and have a two subunit structure with a polymorphic MW 45,000 heavy chain covalently associated with a 12,000 MW β_2-microglobulin. The size of the heavy chain of serum HLA-A,B antigens is similar to that of HLA-A,B antigens isolated from cells with detergents and larger than that of HLA-A,B antigens solubilized with papain.

In the human as in the murine system, the majority (>95%) of the Ia antigenic activity exists as a low-molecular-weight (2000–5000) form.[6] These antigens are oligosaccharide in nature and seem to be expressed on glycolipid molecules found both on the cell surface and in serum. However, a small portion (<5%) of the Ia antigenic activity in serum is present as a high-molecular-weight form, being associated with a 500,000 molecular weight glycoprotein(s) that has an α_2-macroglobulin-like electrophoretic mobility.

Serum HLA-A,B antigens are immunogenic in xenogeneic combinations, since they have been used successfully to elicit anti-HLA-A9 antibodies in rabbits.[5] More importantly, serum HLA-A,B antigens are immunogenic in allogeneic combinations.[7] In fact, infusion of cell free plasma elicits anti-HLA-A,B antibodies in patients without prior exposure to allogeneic HLA antigens, boosts anti-HLA-A,B cytotoxic antibodies in patients who had formed cytotoxic antibodies following blood transfusions,[7] and affects the survival of skin allograft in humans as well as in rats.[8,9] Therefore, it is possible that histocompatibility antigens in serum may play a role in clinical medicine. The improved outcome of renal graft recipients who have received multiple blood transfusions prior to transplantation has attracted much discussion. The blood components responsible for this effect are still to be identified, and plasma, which

[5] J. P. Allison, M. A. Pellegrino, S. Ferrone, G. N. Callahan, and R. A. Reisfeld, J. Immunol. **118**, 1004 (1977).

[6] M. S. Sandrin, I. F. C. McKenzie, T. J. Higgins, and C. R. Parish, Mol. Immunol. **18**, 513 (1981).

[7] M. A. Pellegrino, F. Indiveri, U. Fagiolo, A. Antonello, and S. Ferrone, Transplantation **33**, 530 (1982).

[8] J. J. van Rood, A. van Leeuwen, C. T. Koch, and E. Frederiks, in "Histocompatibility Testing-1970" (P. I. Terasaki, ed.), p. 483. Copenhagen, Munksgaard, 1970.

[9] M. Hasek, J. Chutna, V. Holan, and M. Sladacek, Nature (London) **262**, 295 (1976).

contains about 15% of the HLA activity in human blood,[10] remains a possible candidate.

Changes in HLA-A,B allospecificities were found when sera from patients with melanoma at various stages of the disease were analyzed. Sera from patients with stage III melanoma displayed a reproducible abnormal reactivity with anti-HLA-B5, B7, and B12 alloantisera.[11] The relationship between the extent of the disease and the appearance of HLA-A,B allospecificities nondetectable on autologous lymphocytes (i.e., "alien" HLA-A,B allospecificities) may indicate that they are shed by the tumor cells. These findings raise the possibility that fluctuations of serum MHC product levels (both "alien" and/or "self") in diseases may be used for diagnosis, monitoring of therapy, determining recurrences, and assessing prognosis.

This is supported by alterations observed in the levels of human Ia antigens in tumor-bearing patients: the serum levels in patients with cancer of breast, colon, lung, ovary, pancreas, and prostate were reduced by 60–100% when compared to normal controls.[4] On the other hand, it has been reported that the level of serum Ia antigens increased in leukemia patients.[3]

Thus, there are several reasons of clinical relevance why the presence of HLA antigens in serum demands close attention. Furthermore, detection of specific HLA antigenic activity by serum typing offers potential advantages over cellular typing in several circumstances, such as when patient lymphocytes are not readily available in retrospective studies and in studies of diseases with a fatal outcome.

Detection of HLA Antigens in Serum

The methods used to detect soluble HLA antigens *in vitro* are adaptations of the methods used for the serological characterization of HLA antigens on lymphocytes and are based on the ability of soluble HLA antigens to combine specifically with corresponding antibodies and to decrease the antibody activity as measured by an indicator system. The complement-dependent lymphomicrocytotoxicity test is the most widely used among the several methods developed for the serological characterization of HLA antigens[12] and the lymphomicrocytotoxicity-inhibition test is the assay generally used to detect soluble HLA antigens. The indirect rosette inhibition assay has been used successfully to detect Ia antigens in serum and, at that, only to detect the level of all the Ia anti-

[10] R. H. Aster, B. H. Miskovich, and G. E. Rodey, *Transplantation* **16**, 205 (1973).
[11] S. D. Nathanson, M. A. Pellegrino, and S. Ferrone, *Transplant. Proc.* **13**, 1935 (1981).
[12] P. I. Terasaki and J. D. McClelland, *Nature (London)* **204**, 998 (1965).

gens, regardless of their allospecificity, through the use of polyclonal and monoclonal xenoantibodies to common determinants of the Ia molecules.[3,4]

The techniques used thus far have allowed for qualitative determinations: quantitative measurements are possible only in comparing one sample relative to another.

Lately, with the advent of hybridoma technology[13] and the production of monoclonal antibodies, a double determinant immune assay (DDIA) has been developed. In the histocompatibility system, owing to the availability of monoclonal antibodies directed to allospecificities and to framework determinants of the HLA antigens, it is possible to detect certain single HLA-A,B allospecificities, and/or total level of HLA-A,B and Ia antigens in serum.[14,15] Widescale application of this technique for tissue typing, however, is still far in the future, as monoclonal antibodies directed against all the different allospecificities are not available yet.

A classical radioimmunoassay has been developed for the detection of a limited number of HLA-DR antigens but has not been used thus far for determination of HLA-DR antigens in serum.[16] However generalized application of a radioimmunoassay to the histocompatability system is not a simple feat since the high degree of polymorphism of the HLA system would require the purification of a tremendous number of antigens.

The Cytotoxicity-Inhibition Test (This Volume [21])

Principle. The complement-dependent lymphocytotoxicity test relies upon alterations in permeability of the cell membrane induced by the action of complement on target lymphocytes which have been exposed to cytotoxic antibodies. The immunological lesion of the lymphocyte membrane is shown by the uptake of supravital dyes. In the inhibition test, the alloantisera or polyclonal and monoclonal xenoantibodies are incubated first with the inhibitor source and then their residual cytolytic activity is tested in a regular complement-dependent lymphomicrocytotoxicity assay.

Procedure. Using a 50-μl Hamilton syringe fitted with a repeating dispenser (Hamilton Co., Reno, NE) 1 μl of selected dilutions of alloantiserum (see below) is incubated with 1 μl of 2-fold progressive dilutions [in Hanks' balanced salt solution (HBSS)] of HLA antigens in disposable polystyrene trays (Microtest Plate, No. 3034, Falcon Plastics, Oxnard,

[13] G. Kohler and C. Milstein, *Nature (London)* **256,** 495 (1975).
[14] C. Russo, M. A. Pellegrino, and S. Ferrone, *Transplant. Proc.* **15,** 57 (1983).
[15] C. Russo, M. A. Pellegrino, and S. Ferrone, *Transplant. Proc.* **15,** 66 (1983).
[16] N. Tanigaki and R. Tosi, *Tissue Antigens* **20,** 1 (1982).

CA) filled with mineral oil. Mineral oil is used to avoid evaporation of the reagents used in so small volumes. Incubations are carried out at room temperature for 60 min since previous experiments have shown that shorter incubation periods do not accomplish a complete blocking of the antibody. Sixty minutes are sufficient to reach a maximal effect. After the addition of alloantiserum and soluble antigen, the target cells, suspended in HBSS (1 μl containing 1000 cells) are added to the mixture; after a 30 min incubation at room temperature, 3 μl of rabbit serum as source of complement is added and the mixture is incubated for an additional 180 min, also at room temperature. Eosin (5% in saline; 2 μl) is added to each microdroplet, and after 2 min the reaction is stopped by the addition of 2 μl of 36% formalin. The microdroplets are examined for cell survival by inverted phase contrast microscopy at a magnification of 10–40×. Decreased potency of alloantibody caused by the specific reaction with soluble HLA antigens is detected as an increased percentage of viable cells. The percentage of inhibition is calculated using the following formula:

$$\% \text{ inhibition} = 100 - \frac{\begin{array}{c}\% \text{ cells killed in} \\ \text{presence of inhibitor}\end{array}}{\begin{array}{c}\% \text{ cells killed in} \\ \text{absence of inhibitor}\end{array}} \times 100$$

The percentage of inhibition of a cytotoxic alloantiserum is plotted on an arithmetic scale against the amount of antigen added to the antiserum. The relationship between the two parameters is expressed in sigmoidal fashion. From this curve, it is possible to calculate the amount of antigen required for a 50% reduction of the cytotoxic activity of the alloantiserum. This parameter is called inhibition dosage and designated ID_{50}. The ID_{50} value determined in this way is highly reproducible and indicates (1) presence or absence of any HLA allospecificity under study, (2) potency of the soluble antigen preparation, and (3) degree of purification of an antigenic preparation, as less protein will be necessary to achieve a 50% reduction of the cytotoxic activity of the alloantiserum when purity is increased.

Antisera. For the assay of the soluble HLA antigens, operationally monospecific HLA alloantisera are required to avoid nonspecific reactions in the inhibition test; the presence in the antisera of a mixture of alloantibodies directed against other known or unknown HLA and/or non-HLA specificities may lead to erroneous results. To prove the specificity of the assay, one must show that a given alloantiserum which recognizes a given HLA allospecificity on the cell surface is not inhibited by sera obtained from donors whose lymphocytes lack the specificity against which the alloantiserum is directed. Some sera routinely and efficiently used in several laboratories for typing purposes may not be suitable for the inhibition test since at any concentration of the antigenic preparation employed there is a proportion of dead target cells.

Each antiserum must be used at the highest dilution at which 95% cell death occurs, since this has been found most sensitive for the detection of soluble HLA antigens. In fact, when more concentrated alloantiserum is used, larger quantities of antigens are required to reach the same degree of inhibition, and determination of 50% of inhibition is less precise. On the other hand, higher dilutions of antiserum produce a lower percentage of dead cells (ranging between 40 and 70%) and results are less reproducible.

Target Cells. As target cells, peripheral blood lymphocytes from subjects typed for all recognized HLA-A,B,C specificities are used. The target cells should be an almost pure suspension of lymphocytes: viability of each batch of cells should be determined and should be greater than 95%. There are many acceptable methods for the isolation of viable lymphocytes.[1,2,12] Contamination of the lymphocyte preparation by erythrocytes and polymorphonuclear leukocytes makes reading more difficult, but does not materially affect sensitivity of the assay. Contamination of the lymphocyte suspension by platelets could mimic the blocking effect, since platelets express HLA-A,B,C antigens and therefore compete for antibody.

Complement. Rabbit serum is the most efficient source of complement in the lymphocytotoxicity test for HLA typing: the rabbit complement used is a pool of sera from a large number of rabbits, screened for lack of spontaneous cytotoxicity to peripheral lymphocytes. Good rabbit complement for HLA typing is commercially available (i.e., Pel-Freez, Rogers, AR).

Controls. Extensive controls are necessary to assess technical performance and reagent stability and to assure specificity of a given blocking reaction. Control tests with (1) cells and HBSS, (2) cells and antigen, (3) cells and complement, should give less than 5% dead cells, ensuring against cytotoxicity of the rabbit complement, of the antigen preparation, and death of the target cells through procedural handling. A positive control of antiserum (at the selected titer), cells, and complement—with no antigen—should show 95–100% cell death; a similar control with more dilute antiserum should show proportionately less cell death. As a routine specificity control, an antiserum and target cell combination detecting a specificity putatively not present in the antigenic preparation should also be included.

The Indirect Rosette-Inhibition Test (This Volume [7])

Principle. In the indirect rosette test, the binding of xenoantibodies to HLA antigens (i.e., murine and rabbit antibodies) to a target cell through the corresponding HLA antigens is evidenced by formation of "rosettes" upon addition of sheep red blood cells (SRBC) chemically coupled with goat anti-murine or anti-rabbit Ig antibodies. In the inhibition test, the polyclonal or monoclonal antibodies are incubated with the inhibitor source, then tested for residual unbound antibodies in the indirect rosette test.

Preparation of the Antibody-Coupled SRBC. Sheep red blood cells collected in Alsever's solution and not more than 2 weeks old are washed three times in saline [0.85% NaCl (w/v) with a 400 g, 10 min centrifugation after each wash]. One volume of packed SRBC is mixed with one volume of goat anti-murine, or goat anti-rabbit Ig (1 mg/ml) and one volume of chromic chloride [0.1% (w/v) $CrCl_3 \cdot 6H_2O$ in saline]. After 5 min at room temperature, the cells are washed three times in saline, resuspended to 2% (v/v) in saline, and stored at 4° up to 1 week. In conjugating antibodies to SRBC, the following should be observed:

1. Goat anti-murine and goat anti-rabbit purified by positive affinity chromatography are required. They are commercially available through sources such as TAGO, Inc., Burlingame, CA.

2. Phosphate-containing buffer must be avoided in the chromic chloride coupling of the red blood cells.

3. The chromic chloride solution has to be aged. The aging protocol is as follows: a 1% (w/v) $CrCl_3 \cdot 6H_2O$ in saline is immediately adjusted to pH 5 with 1 M NaOH. The solution is stored at room temperature for 3 weeks, its pH being readjusted to 5.0 with 1 M NaOH three times weekly. It is then ready for use as coupling reagent: it is used after 10-fold dilution with saline, so that it does not aggregate SRBC. The stock solution is stable at least 6 months without further pH readjustment. The chemical basis of the aging process is poorly understood as is the mechanism of chromic chloride coupling of proteins.

Procedure. In preliminary experiments, the polyclonal and monoclonal antibodies directed to HLA antigens are carefully titered against cells expressing the antigen(s) under study with the use of the indirect rosette test, to select a dilution for the inhibition assays.

In the inhibition test, equal volumes (20 μl final volume) of dilutions of the sera to be tested as source of HLA antigens and of the selected appropriate dilutions of antibodies to HLA antigens are incubated in a microtiter plate at 4° overnight. Then 20 μl of the target cells (5 × 10⁶ cells/ml in minimum essential medium) is added, the mixture is incubated at 4° for 60 min, and the cells are washed six times with minimum essential medium. The washed cells are resuspended in 20 μl of minimum essential medium and mixed with 40 μl of a 2% suspension of SRBC coated with goat anti-rabbit (or murine) Ig antibodies. This mixture is centrifuged at

200 g for 5 min, resuspended by vigorous pipetting, and stained with toluidine blue (20 μl of a 1% solution in saline). The percentage of rosettes is determined by microscopic examination of 200 cells. The degree of inhibition is calculated in terms of inhibitory activity per milliliter of serum sample and represents the reciprocal titer required for 20% inhibition of the maximum rosette-forming cells.

The Double Determinant Immune Assay (DDIA)

Principle. In this assay, an insolubilized monoclonal antibody directed to an HLA allodeterminant is used to "fish out" the corresponding HLA molecules. The level of HLA molecules bound to the first antibody is then measured by the binding of an ^{125}I-labeled monoclonal antibody reacting with a distinct, sterically independent epitope carried by all HLA molecules.

This assay has been so far used for detection of HLA-A2, A28 molecules in serum by using murine monoclonal antibody CR11-351, recognizing an epitope common to HLA-A2 and A28 molecules, as the insolubilized antibody, and murine monoclonal antibody NAMB-1, recognizing human β_2-microglobulin, as the tracer antibody.[15]

Product, Purification and Radioiodination of Murine Monoclonal Antibodies. To produce HLA monoclonal antibodies mice are immunized with cells[14,15] or with purified antigens,[16a] and splenocytes are fused to myeloma cells with polyethylene glycol as described.[13] Hybridomas secreting anti-HLA antibodies are expanded by forming ascites in syngeneic mice. The monoclonal antibodies are purified from ascitic fluid by a one step procedure using the caprilic acid method. Ascitic fluid is centrifuged at 10,000 rpm for 20 min to remove cells, cellular debris, and/or fibrin clots. Ten milliliters of ascitic fluid is diluted with 20 ml of 0.06 M acetate buffer, pH 4. After adjusting the pH of the solution to 4.8 with 1 N HCl, 330 μl of caprilic acid is added dropwise with vigorous stirring. The solution is stirred for 30 min at room temperature and centrifuged at 10,000 rpm for 30 min at 10°. The supernatant is collected, adjusted to pH 5.7 with 1 N NaOH, and dialyzed against 0.015 M acetate buffer, pH 5.7. The supernatant is stored at -70°.

The purified monoclonal antibodies are radiolabeled with ^{125}I using the chloramine-T method.[17]

Procedure. One hundred microliters of a purified monoclonal antibody (100 μg/ml of NaHCO$_3$ buffer 0.1 M, pH 9.5) is added to each well of a flexible polyvinyl 96-well plate and dried by overnight incubation at room

[16a] F. M. Brodsky, P. Parham, C. J. Barnstable, M. J. Crumpton, and W. F. Bodmer, *Immunol. Rev.* **47**, 3 (1979).

[17] W. M. Hunter and F. C. Greenwood, *Nature (London)* **194**, 495 (1962).

temperature. The plate is then washed three times with saline containing 0.05% Tween-20 (TSB). Then 50 μl of neat human serum or serum diluted in TSB containing 10% calf serum is added and incubated for 2 hr at 37°. If necessary, the plates can be reincubated with more human serum any number of times, to allow detection of antigens present in serum in very low concentrations. Following three washings with TSB, ^{125}I-labeled monoclonal antibodies (5 × 10^4 cpm) are added to each well. At the end of a 2 hr incubation at 37°, wells are washed five times with TSB. Wells are then cut with a surgical scalpel and the radioactivity determined in a gamma counter. Nonspecific binding is assessed by incubating the serum under investigation and the ^{125}I-labeled antibody in wells coated with an irrelevant monoclonal antibody. Results are expressed as cpm bound by the insolubilized antibody reacted with the serum minus the nonspecifically bound cpm.

A panel of 200 donors has been tested by serum typing with DDIA for HLA-A2,A28 positivity, and the correlation between serum typing by DDIA and the conventional cellular typing by lymphomicrocytotoxicity test was 98% (C. Russo, unpublished results).

Isolation and Purification of Serum HLA Antigens

The initial steps in the purification of HLA-A,B antigens from serum take advantage of the fact that serum HLA-A,B antigens are glycoprotein complexes associated with high-density lipoproteins (HDL). Several of the different techniques used to prepare HDL fractions [i.e., ultracentrifugal flotation (d 1.08–1.21), precipitation with the polyanion dextran sulfate or with phosphotungstic acid] were compared to obtain HLA-A,B antigens from serum. The best results were obtained by subjecting the fraction obtained by phosphotungstate precipitation to a single ultracentrifugal flotation of d 1.23.[5] This two-step procedure allows rapid isolation of serum HLA-A,B antigens with minimum ultracentrifugation time. Further purification is accomplished by cleavage of HLA-A,B molecules from the lipid component by digestion with papain and then sequential gel filtration on Sephadex G-150, ion exchange chromatography on DE-50, and affinity chromatography on Con A-Sepharose.[5,18] The amount of HLA-A9 antigenic activity and the degree of purification at each step of the isolation procedure are reported in the table.

Purification and Isolation of HDL. Very low density lipoproteins (VLDL) and LDL are precipitated by adding dropwise 10 ml of 4% sodium phosphotungstate (pH 7.3) and 2.5 ml of 2 M MgCl$_2$ to 100 ml of serum or plasma. After removal of the precipitate by centrifugation at

[18] G. N. Callahan, S. Ferrone, J. P. Allison, and R. A. Reisfeld, *in* "The Handbook of Cancer Immunology" (H. Waters, ed.), Vol. 8, p. 79. Garland Press, New York, 1981.

PURIFICATION OF HLA-A9 FROM SERUM

Step	Volume (ml)	Protein (mg/ml)	HLA activity $ID_{50}/ml \times 10^{-3}$		Yield (%)	Purification factor
			A1	A9		
Serum[a]	1260	64	0	32	100	—
HDL	116	12.9	0	256	74	40
Papain digest	140	4.9	0	128	44	52
Sephadex G-150	199	0.018	0	64	32	7,196
DE-52	38	0.043	0	256	24	11,765
Con A-Sepharose	9.3	0.012	0	512	11.8	82,075

[a] Serum was a pool from donors carrying the specificity A9.

6000 g for 10 min in a Sorvall RC2B centrifuge, the supernatant is mixed with 90 ml of 4% sodium phosphotungstate (pH 7.3) and the resulting precipitate is removed by centrifugation as above. Then 17.5 ml of 2 M $MgCl_2$ is added to the second supernatant and after 2 hr the HDL fraction is collected by centrifugation at 20,000 g for 20 min. The second precipitate which contains HDL, is resuspended in 20 ml phosphate-buffered saline (PBS: 0.15 M NaCl, 0.01 M sodium phosphate, pH 7.3), and 10% sodium carbonate is added dropwise until dissolution is achieved. The HDL fraction is finally dialyzed against 10% NaCl and against PBS containing 0.01% sodium azide, brought to d 1.21 with solid KBr (density is checked with a pycnometer at 25°) and centrifuged at 170,600 g for 24 hr at 4°. The tubes are then sectioned in the middle clear zone with a tube slicer (Nuclear Supply and Service, Co., Washington, D.C.) and the top fraction is dialyzed against Tris-buffered saline (TBS: 0.15 M NaCl, 0.01 M Tris–HCl, pH 8.0) containing 0.01% sodium azide.

Papain Digestion. Papain digestion is performed by adding 60 mg of papain (Worthington, Diagnostic Systems, Freehold, NJ) to 30 ml of HDL (10 mg/ml in TBS) prewarmed at 37°; 0.5 ml of 0.1 M L-cysteine is then added. The solution is mixed continuously for 30 min and then the digestion is terminated by addition of 0.6 ml of 0.13 M iodoacetamide, with rapid cooling in an ice bath. The precipitate which forms upon chilling is removed by centrifugation at 10,000 g for 15 min; the supernatant is dialyzed against TBS.

Gel Filtration. Sephadex G-150 (Pharmacia, Piscataway, NJ) is prepared according to the manufacturer's instructions. Molecular weight markers are blue dextran 2000 (MW $> 1 \times 10^6$), γ-globulin (MW 160,000), ovalbumin (MW 43,000), and equine heart cytochrome c (MW 12,000). After dialysis against TBS the papain digested material is applied to a 1.5 × 92.5-cm column of Sephadex G-150 equilibrated with TBS and eluted at 10 ml/hr. The HDL-associated peak emerges at the void volume of the column, while the HLA antigens are detected as inhibitory activity of the corresponding alloantisera at an elution volume corresponding to a molecular weight of 46,000.

DEAE Chromatography. Fractions containing HLA activity from the Sephadex G-150 column are pooled and dialyzed against 10 mM potassium phosphate, pH 8.0, containing 0.1% NaN$_3$ and 10% glycerol. The dialyzed sample is then applied to a 2.5 × 25 cm DE-52 column equilibrated with dialysis buffer and eluted in a stepwise manner with the following buffers, all containing 0.1% NaN$_3$ and 10% glycerol: dialysis buffer, 10 mM potassium phosphate, pH 8.0, 20 mM potassium phosphate, pH 7.5, 40 mM potassium phosphate, pH 7.0, 60 mM potassium phosphate, pH 7.0, and 100 mM potassium phosphate, pH 7.0. The bulk of HLA-A9 activity is eluted at 40–60 mM. Other specificities may vary. Active fractions are pooled and concentrated.

Affinity Chromatography on Concanavalin A Sepharose. Pooled, concentrated samples from DEAE chromatography are dialyzed against sodium phosphate buffer (0.01 M, pH 7.3) containing 1 M NaCl and applied to 1.5 × 10 cm column of Con A-Sepharose (2 mg/ml, coupled by the method of Cuatrecasas[19]) equilibrated in the same buffer until all unbound protein (as monitored by absorbance at 280 nm) is eluted. Bound glycoproteins, containing HLA antigens, are eluted with the same buffer containing 0.1 M α-methyl mannoside.

[19] S. C. March, I. Parikh, and P. Cuatrescasas, *Anal. Biochem.* **60**, 149 (1974).

[56] Identification and Characterization of Human T Lymphocyte Antigens

By PAUL L. ROMAIN, ORESTE ACUTO, and STUART F. SCHLOSSMAN

Introduction

T lymphocytes play a central role in the immune response by virtue of their ability to recognize antigens with a high degree of specificity, to act as effector cells, and to regulate not only the intensity, but also the nature of the immune response.[1] In recent years advances in diverse areas of basic research and medical technology have enormously enhanced the capacity to study the human lymphocyte. These studies have led to the detection and characterization of functionally distinct regulatory subpopulations of T lymphocytes and to important insights regarding the structures on T cells involved in cellular interactions and in the recognition of specific antigens.

The ability to begin to define regulatory and effector functions of human T cells at a molecular level has been based on the detection and detailed characterization of a number of surface molecules found on the T cell. A sequence of steps, each involving multiple laboratory methodologies, has made definition of these surface antigens possible. These steps include the development of reagents which identify such surface antigens, the functional characterization of the cells on which they are found, analysis of the distribution of these surface markers during T cell differentiation, the biochemical characterization of the surface structures themselves, and finally, an analysis of the precise role of the given surface antigen in T lymphocyte function. We will attempt in this chapter to briefly review our present understanding of the T cell surface antigens which have been identified and then to discuss the overall approach and general principles our laboratory has followed to characterize these antigens. In addition, we will describe in greater detail the specific methodologies we have employed for generating monoclonal antibodies which recognize human T lymphocyte surface antigens and for the isolation of these surface structures for biochemical studies.

Background

The foundation for the more recent work which has heavily utilized monoclonal antibodies was laid in the "premonoclonal" era by studies which made extensive use of heteroantisera[2,3] raised against cell surface antigens on normal, transformed, or malignant cells, of naturally occurring autoantibodies[4] and alloantisera produced against T cell subsets by programmed immunization.[5] The approach used to develop two of these

[1] E. L. Reinherz and S. F. Schlossman, *Cell* **19**, 821 (1980).

[2] R. L. Evans, J. M. Breard, H. Lazarus, S. F. Schlossman, and L. Chess, *J. Exp. Med.* **145**, 221 (1977).

[3] R. L. Evans, H. Lazarus, A. C. Penta, and S. F. Schlossman, *J. Immunol.* **120**, 1423 (1978).

[4] A. J. Strelkauskas, V. Schauf, B. S. Wilson, L. Chess, and S. F. Schlossman, *J. Immunol.* **120**, 1278 (1978).

[5] G. B. Ferrara, A. J. Strelkauskas, A Longo, J. McDowell, E. J. Yunis, and S. F. Schlossman, *J. Immunol.* **123**, 1272 (1979).

reagents is illustrative of these early studies. An example of one such heteroreagent is the anti-TH$_2$ heteroantiserum, initially produced in the rabbit and subsequently in the horse, by immunizing with purified human T cells or thymocytes.[3] Once antisera were obtained, they were absorbed first with AB$^+$ human erythrocytes and B cell lymphoblastoid lines; the resultant antisera being reactive with all T cells. Then, a human T-CLL tumor population, presumed to represent a clonal expansion of a restricted subset of mature T lymphocytes, was used to absorb the heteroantisera in an attempt to derive a subset specific reagent. This approach led to the production of a heteroantiserum which after extensive evaluation was shown to be reactive with 70–80% of thymocytes and 20–30% of human peripheral blood T cells but not with B or null cells, and to identify the subpopulation of peripheral blood T cells responsible for cytotoxic effector function and suppression of B cell immunoglobulin (Ig) secretion. The reciprocal TH$_2^-$ population, on the other hand, contained the helper population necessary for both Ig synthesis and required for the generation of the cytotoxic cell. Another useful reagent was obtained by screening of patients who exhibited autoantibodies directed against T cells.[4] A number of patients with active juvenile chronic arthridites were observed whose sera contained complement fixing autoantibodies directed against approximately 30% of normal T cells, all of which were nonreactive with anti-TH$_2$. These "JRA$^+$" cells represented an immunoregulatory subset which has subsequently been shown to contain inducers of suppressor effector cells, while the reciprocal JRA$^-$ population contains helper cells for B cell Ig secretion.[6] This earlier work was aided by advances in the development of methods for isolation and characterization of human lymphocytes by virtue of other cell surface markers and physical properties of cells such as the presence of sIg on B cells, the ability of sheep erythrocytes to form rosettes with human T cells, and the presence of Fc receptors on the surfaces of certain cells. These various approaches and reagents and, in particular, the use of flow cytometry combined with immunofluorescence techniques, had provided evidence that it was possible to separate functionally distinct lymphocyte populations by virtue of their cell surface antigenic characteristics.[7] Moreover, they provided useful tools as a basis for the subsequent work using monoclonal antibodies.

Monoclonal antibodies have largely replaced heteroantisera, programmed immunization, and autoantibodies for the characterization and isolation of subsets of T cells. They do not require absorption with irrele-

[6] C. Morimoto, E. L. Reinherz, Y. Borel, and S. F. Schlossman, *J. Immunol.* **130,** 157 (1983).

[7] E. L. Reinherz, A. J. Strelkauskas, C. O'Brien, and S. F. Schlossman, *J. Immunol.* **123,** 83 (1979).

vant tissues and, thus, theoretically unlimited quantities of high titer reagents can be produced.[8] Their homogeneity and availability make them ideal for clinical use.[9,10]

Distribution and Immunologic Function of T Cell Antigens

Surface structures identified in our laboratory by monoclonal antibodies which have proven most useful in the analysis of human T lymphocyte function are listed in Table I, with some basic information about their known characteristics. These surface antigens can be broken down into several broad categories. The antigens in the first group are found on mature T lymphocytes, including virtually all peripheral blood T cells. T1,T3,and T12 are found in addition on medullary thymocytes and T1 and T3 have also been detected on a minor population of cortical thymocytes.[11] The acquisition of these mature T cell antigens is associated with the development of immunocompetence.[12] T1 appears to be a homolog of murine Lyt-1; its precise function is unknown.[13] T3 is linked to antigen specific T cell responsiveness.[14] Biochemical analyses of the major 20 kilodalton T3 molecule and noncovalently linked 25 kilodalton protein suggest that they are nonpolymorphic.[15,16] Anti-T3 will inhibit antigen-specific T cell proliferative responses and cytotoxic effector function and enhance responses to interleukin 2.[14,17] The designation Ti, in contrast, refers to a structure which is detected in association with T3,[17] but which does exhibit polymorphism.[18] Antibodies against this molecule have thus

[8] G. Galfre and C. Milstein, this series, Vol. 73, p. 3.

[9] L. M. Nadler, J. Ritz, J. D. Griffin, R. F. Todd, III, E. L. Reinherz, and S. F. Schlossman, *Prog. Hematol.* **12,** 187 (1981).

[10] E. L. Reinherz, R. Geha, J. M. Rappeport, M. Wilson, A. C. Penta, R. E. Hussey, K. A. Fitzgerald, J. F. Daley, H. Levine, F. S. Rosen, and S. F. Schlossman, *Proc. Natl. Acad. Sci. U.S.A.* **79,** 6047 (1982).

[11] E. L. Reinherz, P. C. Kung, G. Goldstein, and S. F. Schlossman, *Proc. Natl. Acad. Sci. U.S.A.* **77,** 1588 (1980).

[12] T. Umiel, J. F. Daley, A. K. Bhan, R. H. Levey, S. F. Schlossman, and E. L. Reinherz, *J. Immunol.* **129,** 1054 (1982).

[13] E. L. Reinherz, P. C. Kung, G. Goldstein, and S. F. Schlossman, *J. Immunol.* **123,** 1312 (1979).

[14] E. L. Reinherz, S. Meuer, K. A. Fitzgerald, R. E. Hussey, H. Levine, and S. F. Schlossman, *Cell* **30,** 735 (1982).

[15] A. van Agthoven, C. Terhorst, E. L. Reinherz, and S. F. Schlossman, *Eur. J. Imunol.* **11,** 18 (1981).

[16] J. Borst, M. A. Prendiville, and C. Terhorst, *J. Immunol.* **128,** 1560 (1982).

[17] S. C. Meuer, K. A. Fitzgerald, R. E. Hussey, J. C. Hodgdon, S. F. Schlossman, and E. L. Reinherz, *J. Exp. Med.* **157,** 705 (1983).

[18] O. Acuto, S. C. Meuer, J. C. Hodgdon, S. F. Schlossman, and E. L. Reinherz, *J. Exp. Med.* **158,** 1368 (1983).

TABLE I

HUMAN T LYMPHOCYTE SURFACE ANTIGENS RECOGNIZED BY MONOCLONAL ANTIBODIES

| T cell surface antigen | Molecular weight of molecules ($\times 10^3$) | | T cell population defined | Comments | Commercial names of MoAb which recognize antigen[a] |
	Nonreduced	Reduced			
T1	67	67	All mature T cells and medullary thymocytes express T1, T3, and T12. T1 and T3 are also expressed in low density on cortical thymocytes	Homologous to murine Lyt-1, modulates	Anti-T1$_A$, Leu1, OKT1
T3[b]	20,25	20,25		Antibody is mitogenic for resting T cells and induces modulation of T3 and of Ti, with which it is membrane associated	Anti-T3$_A$, Leu4, OKT3
T12		120		Antibody does not inhibit or mimic antigen-specific T cell responses	
Ti	90	49–51 + 41–43	Specific for an individual T cell clone (clonotype)	T cell antigen receptor. Membrane associated with T3. Antibodies have same effect as anti-T3 but only on individual clone with which it reacts	
T4	62	62	Majority of thymocytes and 60–65% of peripheral T cells	T4+ peripheral T cells contain all inducer functions and class II MHC specific CTL	Anti-T4$_A$, Leu3a,b, OKT4

T8 (T5)	76	33 + 31	Majority of thymocytes and 30–35% of peripheral T cells	T8+ peripheral T cells contain all suppressor function and class I MHC specific CTL	Anti-T8$_A$, Leu2a,b, OKT8
T6	49	49	70–80% of thymocytes (cortical only)	β_2-M associated, homologous to murine TL	Anti-T6$_A$, Leu6, OKT6, Nal/34
T9	190	94	10% of thymocytes, activated T cells	Transferrin receptor; not T cell specific	OKT9,5E9
T10	37	45	All thymocytes, activated T cells	Also on plasma cells, not T cell specific	OKT10
T11	50	50	All thymocytes and T cells	E rosette associated protein; greatest density on thymocytes and suppressor T cells; also on many null cells	Anti-T11,9,6, Leu5, OKT11
TQ1	80	80	50% of peripheral T cells, majority of which are T4+	Majority of T4+TQ1+ are JRA+ containing the inducer of suppression; T4+TQ1− contain majority of inducers of help for Ig secretion	

[a] Anti-T designations available through Coulter Electronics, Hialeah, FL; Leu designations available through Becton-Dickinson, Mountain View, CA; OK designations available through Ortho Pharmaceutical, Raritan, NJ; Nal/34 available through Accurate Chemicals, NJ; 5E9 available from NIAID monoclonal antibody serum bank, Bethesda, MD.

[b] Major 20 kilodalton protein recognized by anti-T3 is associated noncovalently with a 25 kilodalton protein.

far been specific for unique T cell clones. Anti-Ti antibodies have similar effects to anti-T3 but only on the individual clones with which they react. The T3 and Ti surface structures will comodulate upon binding with their respective ligands under the proper conditions. Substantial data now suggest that the T3–Ti complex represents the T cell receptor for antigen. Finally, anti-T12 detects an additional distinct structure found on mature T cells.[11,19] This latter antibody has proven extremely useful in the prevention of graft-versus-host disease and renal allograft rejection.[10]

The next set of surface antigens distinguish the two reciprocal functionally distinct subsets of peripheral blood T cells. T4 is found on approximately two-thirds of peripheral T cells which comprise the inducer cells for T–T, T–B, and T–Mϕ interactions.[20,21] In addition, T4$^+$ cells contain cytotoxic T lymphocytes (CTL), specific for class II major histocompatibility (MHC) determinants (HLA-Dr, SB, etc.) and anti-T4 inhibits such cytotoxic effector functions at the recognition stage on T4$^+$ cells only.[19,22,23] In contrast, T8$^+$ peripheral T cells[21,24,25] contain the suppressor effector population as well as class I MHC (HLA-A, B, C) specific CTL.[19] Anti-T8 inhibits cytotoxic effector function of T8$^+$ cells only and appears homologous to the murine Lyt-2 antigen.[26] In addition to their reciprocal presence on mature cells (including medullary thymocytes), T4 and T8 are coexpressed on the vast majority of cortical thymocytes.[11]

Cortical thymocytes which coexpress T4 and T8 also bear the T6 antigen.[11] T6 is the homolog of the murine TL antigen and is associated with β_2-microglobulin.[27] This antigen was previously identified by a heteroantiserum termed anti-HTL.[11,28]

[19] S. C. Meuer, S. F. Schlossman, and E. L. Reinherz, *Proc. Natl. Acad. Sci. U.S.A.* **79**, 4395 (1982).

[20] E. L. Reinherz, P. C. Kung, G. Goldstein, and S. F. Schlossman, *Proc. Natl. Acad. Sci. U.S.A.* **76**, 4061 (1979).

[21] C. Terhorst, A. van Agthoven, E. L. Reinherz, and S. F. Schlossman, *Science* **209**, 520 (1980).

[22] W. E. Biddison, R. E. Rao, M. A. Talle, G. Goldstein, and S. Shaw, *J. Exp. Med.* **156**, 1065 (1982).

[23] A. M. Krensky, C. Clayberger, C. S. Reiss, J. L. Strominger, and S. J. Burakoff, *J. Immunol.* **129**, 2001 (1982).

[24] E. L. Reinherz, P. C. Kung, G. Goldstein, and S. F. Schlossman, *J. Immunol.* **124**, 1301 (1980).

[25] R. L. Evans, D. W. Wall, C. D. Platsoucas, F. P. Siegal, S. M. Fikrig, C. M. Testa, and R. A. Good, *Proc. Natl. Acad. Sci. U.S.A.* **78**, 544 (1981).

[26] J. A. Ledbetter, R. L. Evans, M. Lipinski, C. Cunningham-Rundles, R. A. Good, and L. A. Herzenberg, *J. Exp. Med.* **153**, 310 (1981).

[27] C. Terhorst, A. van Agthoven, K. LeClair, P. Snow, E. L. Reinherz, and S. F. Schlossman, *Cell* **23**, 771 (1981).

[28] S. F. Schlossman, L. Chess, R. E. Humphreys, and J. L. Strominger, *Proc. Natl. Acad. Sci. U.S.A.* **73**, 1288 (1976).

Other antigens detected on thymocytes by monoclonal antibodies have proven useful but are widely distributed. T9 is detectable on stage I thymocytes (approximately 10%) which lack T6 and T3, normal and malignant populations of non-T cells including bone marrow erythroid precursors, and on some activated mature T cells.[11,29] It has been shown to be the receptor for transferrin.[30] T10, in contrast, is found on all thymocytes, on activated T and null cells and on plasma cells.[11,27,29] T11 is an extremely important antigen present on 100% of T cells including all thymocytes and peripheral blood T cells. T11 defines the receptor for sheep erythrocytes (SRBC), the E rosette receptor, and can block SRBC-T cell binding at very high titers.[31,32]

The study of surface antigens uniquely characteristic of activated T cells is an important emerging area. Antibodies directed against Ia[33,34] and other surface antigens[35] which appear on activated T cells (but are not T lineage restricted) have been of interest; antibodies against surface structures found uniquely on activated T cells will be of obvious importance. One example of such an antibody is the monoclonal described by Uchiyama and co-workers[36] termed anti-Tac which reacts with activated functionally mature T cells and which appears to recognize the receptor[37] for human T cell growth factor.

Finally, the last category includes an antibody termed anti-TQ1 which is reactive with only 50% of peripheral T cells, the majority of which are contained within the T4$^+$ subset and are reactive with the JRA autoantibody discussed above.[38] The T4$^+$TQ1$^+$JRA$^+$ population contains the inducer of suppression and the majority of responders in the autologous mixed lymphocyte reaction, while the T4$^+$TQ1$^-$JRA$^-$ cells are inducers of help and contain the majority of cells which proliferate directly in re-

[29] T. Hercend, J. Ritz, S. F. Schlossman, and E. L. Reinherz, *Hum. Immunol.* **3,** 247 (1981).
[30] D. R. Sutherland, D. Delia, C. Schneider, R. A. Newman, J. Kemshead, and M. F. Greaves, *Proc. Natl. Acad. Sci. U.S.A.* **78,** 4515 (1981).
[31] M. Kamoun, P. J. Martin, J. A. Hansen, M. A. Brown, A. W. Siadak, and R. C. Nowinski, *J. Exp. Med.* **153,** 207 (1981).
[32] W. Verbi, M. F. Greaves, C. Schneider, K. Koubek, G. Janossy, H. Stein, P. Kung, and G. Goldstein, *Eur. J. Immunol.* **12,** 81 (1982).
[33] E. L. Reinherz, P. C. Kung, J. M. Pesando, J. Ritz, G. Goldstein, and S. F. Schlossman, *J. Exp. Med.* **150,** 1472 (1979).
[34] H. S. Ko, S. M. Fu, R. J. Winchester, D. T. Y. Yu, and H. G. Kunkel, *J. Exp. Med.* **150,** 246 (1979).
[35] T. Hercend, L. M. Nadler, J. M. Pesando, E. L. Reinherz, S. F. Schlossman, and J. Ritz, *Cell Immunol.* **64,** 192 (1981).
[36] T. Uchiyama, S. Broder, and T. A. Waldmann, *J. Immunol.* **126,** 1393 (1981).
[37] W. J. Leonard, J. M. Depper, T. Uchiyama, K. A. Smith, T. A. Waldmann, and W. C. Greene, *Nature* (London) **300,** 267 (1982).
[38] E. L. Reinherz, C. Morimoto, K. A. Fitzgerald, R. E. Hussey, J. F. Daley, and S. F. Schlossman, *J. Immunol.* **128,** 463 (1982).

sponse to soluble antigen.[6,38,39] Leu8, which has recently been described, appears similar to anti-TQ1.[40] It is reactive with only a subset of peripheral T cells but its function has not yet been described in great detail. A number of experiments emphasize the functional heterogeneity within the inducer (T4[+]) and suppressor/cytotoxic (T8[+]) subsets, and the identification and characterization of surface antigens limited to subpopulations within these subsets, such as those mentioned above, are a matter of great interest.

In order to delineate the approach which can be taken to identify and characterize important human T lymphocyte antigens, we will outline several strategies for the characterization of human T cell surface structures. In particular, we will focus on our immunization and screening strategies for the development of monoclonal antibodies against T cell surface components and on the use of such antibodies to specifically precipitate these antigens in highly purified form for further biochemical characterization. We will also briefly mention approaches to the characterization of cell structures defined by these antibodies and the functional program of such cells, but a detailed discussion of these aspects is beyond the scope of this chapter.

Identification of Human T Lymphocyte Antigens

Production of Monoclonal Antibodies

Monoclonal antibodies are produced by an adaptation of the method of Kohler and Milstein.[41] These methods have been fully described in great detail in recent volumes in this series[8,42] and the reader can refer to these articles for a fuller discussion of both the principles and a detailed description of the methods. We will briefly outline below the approach which has been taken in our laboratory with emphasis on those techniques of particular use for studies of human T lymphocytes.

Immunization and Initial Growth of Hybrids

The following approach is currently used in this laboratory:

1. Four 8-week-old female BALB/cJ mice (Jackson Laboratories, Bar Harbor, ME) are immunized intraperitoneally with 1–2 × 10[7] lympho-

[39] P. L. Romain, C. Morimoto, J. M. Daley, L. S. Palley, E. L. Reinherz, and S. F. Schlossman, Clin. Immunol. Immunopathol. **30,** 117 (1984).

[40] P. A. Gatenby, G. S. Kansas, C. Y. Xian, R. L. Evans, and E. G. Engleman, J. Immunol. **129,** 1997 (1982).

[41] G. Kohler and C. Milstein, Nature (London) **256,** 495 (1975).

[42] This series, Vol. 92.

cytes in PBS from a selected cell population. Use of a cell population enriched for the particular antigen of interest is obviously preferable where feasible, though not absolutely necessary.

2. One to two weeks after the initial immunization, two mice are again injected ip with the cells as before, while two mice are given half the cells ip and the other half iv by tail vein injection. Three days following the second immunization, the spleens of the ip/iv primed mice are removed and teased under aseptic conditions to prepare single cell suspensions. This timing enriches for the recently activated splenic B lymphoblasts, which fuse relatively efficiently.

3. Cells from two spleens (approximately $3-4 \times 10^8$ cells) are fused in 30% polyethylene glycol (PEG 1000) in serum-free RPMI 1640 (Gibco, Grand Island, NY) with P3/NS1/1-Ag1 murine plasmacytoma cells in a ratio of 5 spleen cells per myeloma cell. The murine plasmacytoma cells are most effective if they are in their exponential growth phase at the time of fusion.

4. Cells are resuspended with irradiated (2500 rad) feeder cells from female BALB/cJ retired breeders at a ratio of one feeder per myeloma cell in hypoxanthine/aminopterin/ and thymidine (HAT) medium. Thirty $\times 10^6$ cells/plate are then placed into 96-well microtiter plates (about 3×10^5 spleen cells/well) for culture at 37° with 6% CO_2 in a humid atmosphere.

5. Within approximately 10–14 days, the supernatants of growing clones are selected for testing. Growing clones are usually detectable at this point by a yellowing of the color of their growth medium (due to pH changes). They can also be detected by inspection of microwells under low power on an inverted microscope.

Once growth is clearly detected, supernatants should be promptly screened to allow detection and subcloning before overgrowth of a nonsecreting hybrid occurs. In addition, intitial screening should be done at least at two points in time since some hybrids of interest may grow more slowly than others and be missed at the first screening. Screening will be discussed in greater detail below.

It is usually sufficient to employ either thymocytes or E rosette-positive peripheral blood lymphocytes for the generation of either pan-T reagents or anti-T4 antibodies, while peripheral blood T lymphocytes may not be as effective as thymocytes if an anti-T8 antibody is desired. If a homogeneous cell line (such as the T cell chronic lymphocytic leukemia lines which have been used to generate anti-T8 antibodies) or T cell clones are available, these have great utility for generating specific reagents. Cloned T cells have proven extremely effective in the generation of antibodies which have highly limited reactivity.[17] These clonotypic reagents have allowed for the identification of the Ti surface structure which appears to have a role in antigen specific cellular recognition. It is possible to use T cell growth factor dependent T cell clones as immunogens in somewhat lower numbers than heterogeneous T cell populations since they are highly enriched for antigen-positive cells.

Variations in the protocols depend, among other things, on the particular immunogen, the availability of given tissues and cell lines, the number of cloned cells available at a given time, the strain of mouse employed, and the investigator's needs and schedule. The two mice which have been primed but not used in the initial fusion are ready if needed after only a single ip and iv boost if the first fusion does not produce the desired antibody, if too many or too few hybrids are produced by the protocol and ratios used, if something unanticipated occurs, or simply if more hybrid antibodies are desired.

Screening and Maintenance of Clones

There are a number of methods which can be used for the screening of hybrids.[42] Some of these are described in Table II. Because of its rapidity, the ability to detect murine Ig of any subclass (unlike complement dependent methods, for example) its potential for discerning both quantitative and qualitative characteristics of antigen distribution, and the lack of need for radioactive reagents, we screen by indirect immunofluorescent staining using flow cytometry[13,43] (this volume [19]).

Screening of antibodies is done in several phases. First, supernatants with reactivity against the immunogen are selected for further study, then screening is performed with the initial positives on an EBV transformed B lymphoblastoid cell line to eliminate non-T cell-specific antibodies. When T cell clones or other activated T cells are used as the immunogens, this step is particularly useful to identify antibodies directed at common determinants such as Ia antigen. In this situation, 50–60% of the growing hybrids reactive with the immunogen may be eliminated at this step. Careful screening must be done to detect the distribution of the surface antigen which a given monoclonal identifies. Among the cells and tissues most useful at this step are peripheral T cells and T cell subpopulations, thymus, macrophages, and activated T cells (or clones where indicated). Antigens on activated cells should be screened on activated B cells, plasma cells, and macrophages as well. Leukemia and lymphoma cell lines may also be useful for detailed characterization of given antibodies. For most antibodies, screening is usually not required at this stage on erythrocytes, neutrophils, platelets, or whole bone marrow. After a given

---

[43] L. A. Herzenberg and L. A. Herzenberg, in "Handbook of Experimental Immunology" (D. M. Weir, ed.), 3rd ed., Chapter 22. Blackwell, Oxford, 1978.

TABLE II
SCREENING ASSAYS FOR MURINE HYBRIDOMA IMMUNOGLOBULIN PRODUCTION

Technique	Advantages	Disadvantages
Indirect immunofluorescence	Rapidity. Not class-specific with proper developing reagents, gives information on antigen density, proportion of positive cells, and size of antibody-reactive cells	Must control for low level of nonspecific binding. Need expensive specialized equipment
Radioimmunoassay	Rapid. Not class-specific with proper developing reagents. Can assay large numbers at one time	Requires use of radiolabeled reagents, difficult with viable cell preparations, not informative regarding proportion of population, size of cells reactive with antibody or relative antigen density
ELISA	Rapid. Not class-specific with proper developing reagents. Can assay large numbers at one time	Requires enzyme-linked developing reagent. Requires specialized equipment, not informative regarding proportion of population, size of cells reactive with antibody, or relative antigen density
Complement mediated lysis	Relatively inexpensive, no special equipment, gives information regarding proportion of cells reactive with complement-fixing antibody	Only identifies murine hybrid antibodies of subclasses which fix complement well. No information regarding antigen density

hybrid is determined to be secreting an antibody of potential interest (positive on the immunogen and absent from the B cell lines), it should be rescreened on the immunogen before more detailed evaluation. The hybrids of interest are cloned as soon as possible by limiting dilution in the presence of feeder cells, then recloned to assure monoclonality.[44] After such clones have been selected and isolated, they are transferred to produce malignant ascites by injection of $1-2 \times 10^6$ cells intraperitoneally into BALB/cJ mice which have been primed with 0.5 ml pristane (Aldrich Chemical Co., Milwaukee, WI) 3–5 days previously. Ascites produced by

[44] D. E. Yelton, B. A. Diamond, S. P. Kwan, and M. D. Scharff, *Curr. Top. Microbiol. Immunol.* **81**, 1 (1978).

growing cloned hybrids will contain approximately 2–20 mg/ml of mono-specific high titer antibody. The concentration of antibody in the original hybridoma supernatant will in general be two logs less.

The following procedure is used for screening of cells by indirect immunofluorescence:

1. Supernatants (60–100 μl) from cultures containing growing hybrids to be tested are incubated at 4° for 30 min in 12 × 75-mm round bottom polystyrene centrifuge tubes with 10^6 cells which have been washed in medium containing 2% AB serum (all washes contain 2% AB serum to diminish nonspecific Fc binding).

2. Incubated cells are then washed twice with medium.

3. To the washed pellet, 100 μl of a combination of fluoresceinated goat anti-mouse IgG and anti-mouse IgM (Meloy Laboratories, Springfield, VA) is added and the cells are resuspended and incubated an additional 30 min at 4°. Although the dilution used needs to be titred for each lot, a 1 : 40–1 : 50 dilution of antibody is usually adequate.

4. After two additional washes, cells are analyzed by flow cytometry. We currently use an Epics V cell sorter (Coulter Electronics, Hialeah, FL) or a FACS I (Becton Dickinson, Mountain View, CA).

5. Intensity of fluorescence is determined for 10,000 cells in each population and evaluated for displacement of the histogram of the test antibody in comparison to an unreactive control antibody. For each positive test sample the proportion of cells which is positive is noted and the displayed histogram may be recorded for comparison with antibodies of known distribution. From 75 to 150 samples per hour may be tested depending on the number of histogram tracings recorded.

For accurate characterization of the distribution of a given cell surface antigen recognized by a monoclonal antibody, a carefully planned sequence of screening is of paramount importance. Furthermore, until an antibody has been carefully evaluated for its reactivity against a wide variety of non-T cells, it cannot be assumed that such a reagent is T cell specific. This is of particular importance with antibodies raised against activated T cells and T cell clones. In addition, cells from a number of different individuals may be used as targets in an effort to detect antibodies which might identify polymorphic cell surface determinants.

Further Characterization and Purification of the Monoclonal Antibodies

These procedures have been widely discussed in detail elsewhere and are not specific to antibodies detecting T cell antigens. Briefly, murine Ig isotype and subclass can be determined by either standard Ouchterlony analysis using commercially available antisera or by use of subclass specific fluorescein-labeled goat anti-mouse Ig reagents by indirect immunofluorescence. Monoclonal antibodies of the IgG isotypes can then be purified from ascites by DEAE cellulose anion exchange chromatography or where possible, by protein A Sepharose affinity chromatography. Monoclonal antibodies of the IgM isotypes can be purified by sizing exclusion chromatography on Sephadex G-200. Purified antibody can be stored at $-70°$ in phosphate-buffered saline. If Fab or $F(ab')_2$ fragments are needed, antibodies can be treated by papain or pepsin digestion,[45,46] respectively, as discussed elsewhere.

Characterization of Human T Lymphocyte Antigens

Once a surface antigen has been identified as being specific for T lymphocytes, there are two major aspects to the characterization of such antigens. One area is the functional analysis of the cells bearing such antigens (particularly in the case of antigens detected on only a subset of T cells), and determination of the functional role of the surface structure itself in cellular interactions and effector functions; and the other is the purification and biochemical analysis of the structure itself. In the remaining portion of this chapter, we will first mention some general approaches to the former area and then, in greater detail, will describe the methodology for specific immunoprecipitation of purified antigens for T cells using monoclonal antibodies.

Functional Analysis of T Cell Populations and Surface Antigens Defined by Monoclonal Antibodies

To evaluate the function of cells bearing a given surface antigen identified by a monoclonal antibody cells bearing such an antigen must first be isolated, then tested *in vitro*. The specific *in vitro* assays used will depend on which broad functional capacities the cell can be expected to have by virtue of the known surface markers which it bears.[1] For a detailed discussion of methods for cell culture in general and for the isolation of subpopulations of human peripheral blood lymphocytes, the reader should refer to Vol. 58 in this series and Section II of this volume [9], respectively. Using such standard methods to isolate T cells it would then be possible to either positively or negatively select subpopulations en-

[45] R. R. Porter, *Biochem. J.* **73**, 119 (1959).
[46] A. Nisonoff, F. C. Wissler, L. N. Lipman, and D. L. Woernley, *Arch. Biochem. Biophys.* **89**, 230 (1960).

riched for the cells to be studied. If the monoclonal antibody is comple- ment fixing, this property can be taken advantage of to eliminate the cell in order to examine the effects on the functional capacities of the remain- ing population. Regardless of its ability to fix complement, cells reactive with a given monoclonal can also be depleted by use of a fluorescence- activated cell sorter,[20,24] by a rosette depletion method,[47] or by an immune adherence technique.[48] These three latter methods can be used con- versely to positively select for a specific subpopulation of cells which bind antibody. Once a particular subpopulation is isolated its ability to respond to various stimuli including mitogens, soluble antigens, alloantigen, and autologous non-T cells can then be tested. In addition, the ability of cells activated by these various stimuli to function as inducers or effectors of either help or suppression of either proliferative responses or cytotoxic- ity, and as inducers or suppressors of immunoglobulin production can be carefully evaluated.[1,6,7,20,24,33,38,39]

The role of a particular surface structure itself in cellular interactions and effector functions can be studied by use of the monoclonal antibody *in vitro* during given assays to determine the effects on the cell of binding to the surface antigen.[14,17–19,49] For antigens which serve as specific cell sur- face receptors such binding may have a profound functional effect, caus- ing cell activation or, alternatively, blocking cell–cell interaction.[30–32,37,49] Since the binding to different epitopes of a surface molecule may clearly have different effects, the use of such reagents allows for the fine dissec- tion of the various regions and related components of a structure which may be involved in the binding of regulatory molecules and in effector functions. Blocking experiments are also important in detecting antibod- ies reactive with identical surface structures. Moreover, analysis of sur- face molecules which undergo modulation upon binding to antibody or during cell activation can lead to the detection of structurally or function- ally associated antigens, such as the clonotypic antigen binding struc- ture.[17,18] Of particular use in blocking studies are interleukin-2-dependent T cell clones.[19,49] Such clones also provide a homogeneous source of cells for biochemical studies and can be used to generate clonotypic reagents in an attempt to identify antigen binding structures on T cells. Finally, iso- type identical antibodies, which preferably also bind to the same cells, are also important controls in any blocking experiments.

[47] J. D. Griffin, R. Beveridge, and S. F. Schlossman, *Blood* **60,** 30 (1982).
[48] E. L. Reinherz, A. C. Penta, R. E. Hussey, and S. F. Schlossman, *Clin. Immunol. Immunopathol.* **21,** 257 (1981).
[49] E. L. Reinherz, S. C. Meuer, and S. F. Schlossman, *Immunol. Rev.* **74,** 83 (1983).

Immunoprecipitation of T Cell Antigens for Purification and
Biochemical Characterization

For biochemical studies of T cell antigens, the specificity of monoclonal antibodies is of obvious utility because of the purity with which antigens can be obtained for analytical and preparative use. Detailed characterization of surface antigens using monoclonal antibodies, therefore, allows for analysis of the fine structure of these molecules, and by performing immunoprecipitations after preclearing with selected antibodies, clues can be obtained to the relationships between and among different families of surface molecules. Specific immunoprecipitation of T cell antigens as described will provide the materials needed for more extensive biochemical analysis.

Techniques used for the radiolabeling of T cell antigens are fairly standard (see the detailed discussions in previous volumes of this series and in other chapters in this volume). We generally prefer to utilize the glucose oxidase/lactoperoxidase technique for external labeling (as discussed by Morrison[50]) because it is reproducible and potentially less disruptive of the membrane antigens under study. The method described below for the specific immunoprecipitation of labeled cell structures is useful for either externally or internally labeled proteins.

Immunoprecipitation of T Cell Antigens Using Monoclonal Antibodies

Materials and Reagents

Lysis buffer: 10 mM NaH$_2$PO$_4$, pH 7.4; 150 mM NaCl; 1 mM EDTA, 1 mM EGTA; 1 mM NaF; 1% Triton X-100 (or NP-40); 1 mM PMSF and 2 μg/ml ovomucoid trypsin inhibitor (Sigma) (these last two reagents should be added to make up fresh buffer on the day of the procedure)

NaDOC (sodium deoxycholate) 10% solution

BSA (10 mg/ml stock solution)

Staphylococcus aureus, 10% in suspension (BRL, Gaithersburgh, MD)

Sodium dodecyl sulfate, 10% (SDS)

Monoclonal antibodies coupled to Sepharose 4B by BrCN activation[18,51,52] (BrCN-activated Sepharose obtained from Pharmacia, Piscataway, NJ)

[50] M. Morrison, this series, Vol. 70, p. 214.
[51] S. C. Meuer, J. C. Hodgdon, R. E. Hussey, J. P. Protentis, S. F. Schlossman, and E. L. Reinherz, *J. Exp. Med.* **158,** 988 (1983).
[52] R. Axén, J. Poráth, and S. Ernback, *Nature (London)* **214,** 1302 (1967).

Sample buffer for Laemmli SDS–PAGE[53]
15-ml conical centrifuge tubes (Falcon #2095)
Eppendorf 1.5 ml polypropylene microcentrifuge tubes

Equipment

Microcentrifuge (Eppendorf model 5414)
Air-driven table top ultracentrifuge (Beckman "Airfuge")
Rotator

Procedure

1. Lyse pellet of radiolabeled cells with lysis buffer, using 100 μl buffer per each 10^7 labeled cells. (We have usually labeled $1–5 \times 10^7$ cells using $1–2$ mCi $Na^{125}I$.) The pellet should be gently loosened before the addition of lysis buffer.

2. Vortex the cells, then leave them on ice for 30–45 min.

3. Centrifuge the lysates for 10 min in the Eppendorf microcentrifuge ($10,000\ g$) to remove nuclei. (At this point, supernatants can be frozen and stored at $-70°$ if necessary before proceeding.)

4. Ultracentrifuge in airfuge at 30 psi (equivalent to $120,000\ g$) for 20–30 min to remove insoluble cytoskeletal proteins.

5. Divide the ultracentrifuged supernatant into aliquots each equivalent to the lysates from $3–5 \times 10^6$ cells (the larger amount of cells is preferred).

6. Add BSA to the aliquots to a final concentration of 1 mg/ml. (This serves to reduce the nonspecific absorption of irrelevant molecules by Sepharose.)

The subsequent steps for the specific immunoprecipitation can be performed in one of several ways. We prefer to use a purified monoclonal antibody coupled to Sepharose because in our hands the precipitates obtained have the least background. Alternatively, Staph A (formalin-fixed bacteria) or protein A Sepharose (either requiring use of an antibody which is protein A binding) or preformed complexes of rabbit anti-mouse Ig and the mouse monoclonal antibody can be used.

7. Wash the Staph A with lysis buffer (approximately 1 ml buffer per 50 μl) at least once before use.

8. Bring volume of each aliquot up to 200 μl with lysis buffer and add 50 μl 10% Staph A.

9. Incubate at $4°$ on a rotator for 30 min.

10. Centrifuge in the microcentrifuge for 5 min.

---

[53] U. K. Laemmli, *Nature (London)* **227**, 680 (1970).

11. Transfer supernatants to fresh tubes.

12. Add 40 μl of 50% suspension of an irrelevant monoclonal antibody (isotype identical to the antigen specific antibody to be used) coupled to Sepharose 4B beads (by standard methods) to each supernatant aliquot.

13. Centrifuge for 3 min in the microcentrifuge.

14. Transfer the supernatants to fresh tubes.

15. Repeat steps 12–14 twice.

16. To the precleared supernatants add 5–20 μl packed specific antibody coupled Sepharose beads (or 40 μl of 50% Sepharose, etc.).

17. Incubate for 2 hr at 4° on rotator.

18. Centrifuge in Eppendorf for 5 min.

19. The supernatants may be pooled and stored at −70° for up to 2 weeks if needed.

20. Add 200 ml lysis buffer (without NaCl) to each pellet.

21. Transfer pellets to 15 ml tubes (can use a 20–200 μl pipettor with standard plastic tips which have had about 5–10 mm cut off distally to make a larger hole). Rewash if beads still stuck to tubes.

22. Add 10 ml lysis buffer (without NaCl) + 0.5% NaDOC to each 15-ml tube.

23. Centrifuge at 1750 g for 5 min.

24. Aspirate the supernatant and discard.

25. Add to each tube 10 ml of lysis buffer without NaCl + 0.5% NaDOC + 0.05% SDS.

26. Repeat steps 23 and 24.

27. Add 10 ml of lysis buffer + 0.5 M NaCl + 0.5% NaDOC to each tube.

28. Repeat steps 23 and 24.

29. Add 1 ml lysis buffer to each tube to transfer back to 1.5-ml tubes.

30. Centrifuge for 5 min in microfuge.

31. Resuspend pellets in 50 μl of Laemmli sample buffer with or without 5% 2-mercaptoethanol for reducing or nonreducing conditions, respectively.

32. Incubate for 5 min in a heating block at 100° or in boiling water.

33. The resultant solution can be applied to a gel for analysis or stored at 70°.

Characterization of Purified Surface Antigens

Detailed discussion of various electrophoretic procedures used to characterize isolated surface components are beyond the scope of this chapter. Briefly, however, in addition to one-dimensional SDS–PAGE on a vertical slab gel,[53] 2-D gels using isoelectric focusing in the first dimension and SDS–PAGE in the second[54] may be useful for more detailed molecular characterization. Proteins containing interchain disulfide bonds can be resolved from nondisulfide bonded proteins by "diagonal" 2-D gels in which the first dimension is nonreducing SDS–PAGE and the second is under reducing conditions.[55] This can be useful as a one step procedure for distinguishing homodimers from heterodimers.

A more detailed analysis of protein heterogeneity, particularly at the level of primary structural differences in variable regions of polymorphic determinants, is most powerfully approached through the use of peptide mapping,[56] in particular, 2-D peptide mapping on thin layers (microfingerprinting). These and sequence data are beyond the scope of this chapter.

Acknowledgments

The authors wish to thank Dr. Ellis L. Reinherz for his thoughtful comments and suggestions.

This work was supported in part by National Institutes of Health Grant AI 12069.

Paul Romain is the recipient of National Institutes of Health Clinical Investigator Award K08-AMO1181. Oreste Acuto is a recipient of a Cancer Research Institute J. Morton (Davidowitz) Davis and Rosalind Davis Fellowship and is on leave of absence from the Istituto di Fisiologia Generale, Cattedra di Bialogio Molecolare Facolta di Scienze, Università di Roma, Italy.

[54] P. H. O'Farrell, *J. Biol. Chem.* **250,** 4007 (1975).
[55] L. J. Takemoto, T. Miyakawa, and C. F. Fox, *in* "Cell Shape and Surface Architecture" (J. P. Ravel, U. Henning, and C. F. Fox, eds.) p. 606. Alan R. Liss, Inc., New York, 1977.
[56] K. Weber and M. Osborn, *in* "The Proteins" (H. Neurath and R. Hill, eds.), 3rd ed., Vol. 1, p. 179. Academic Press, New York, 1975.

[57] Thy-1.1 and Thy-1.2 Alloantigens: An Overview

By MICHELLE LETARTE

Discovery and Gene Mapping

Immunization of C3H mice with thymocytes of AKR mice led to the discovery by Reif and Allen of a strong antigen associated with AKR thymocytes.[1] The antigen was shown to be present in large amounts not only on thymus, but on brain cells and on some leukemias of AKR mice.[2]

[1] A. E. Reif and J. M. V. Allen, *Nature (London)* **200,** 1332 (1963).
[2] A. E. Reif and J. M. V. Allen, *Nature (London)* **203,** 886 (1964).

Thymocytes of RF mice were lysed by C3H anti-AKR alloserum and complement whereas thymocytes from several other strains including C3H, BALB/c, C57BL/6, C57BL/10, DBA/1, DBA/2, and A were not.[2] A reciprocal alloantiserum, AKR anti-C3H, lysed thymocytes from C3H strain but not from AKR and RF strains. This strain distribution indicated that the antigen recognized was neither H-2 (since AKR and C3H both express H-2^k) nor any of the mouse histocompatibility antigens H-1 to H-6 known at the time. The determinants defined by the alloantisera C3H anti-AKR and AKR anti-C3H were designated θ-AKR and θ-C3H, respectively.[2] Inbred mice are homozygous and express one of the two alleles whereas heterozygous mice express both alleles.[3,4] Backcrosses between (RF × DBA/2)F$_1$ and DBA/2 mice established that θ was in linkage group II on chromosome 9 at a distance of 16.8 ± 3.6 units from the d (dilute coat color) marker.[5] Further mapping using backcrosses between (RF × CBA)F$_1$ and CBA mice established that the order of genes on chromosome 9 is *centromere–Thy-1–d–Mod-1–Trf*.[6,7]

θ antigen is now designated Thy-1 antigen. The two alleles of the *Thy-1* gene are referred to as *Thy-1^a* and *Thy-1^b*. The product encoded by the *Thy-1^a* allele is the Thy-1.1 alloantigen (θ-AKR) and the product of the *Thy-1^b* allele is the Thy-1.2 alloantigen (θ-C3H). Table I illustrates the distribution of Thy-1.1 and Thy-1.2 alloantigens in different murine strains. Congenic lines differing at the *Thy-1* locus have been developed: A-Thy-1^a and A.AL-old carrying the *Thy-1^a* allele on A background and B10.PL carrying the *Thy-1^a* allele on the B10 background (taken from Zaleski and Klein[8]). Such congenic lines have established the Thy-1.1 and Thy-1.2 allospecificities of allosera and of monoclonal antibodies.

Tissue Distribution of Thy-1 Alloantigens

Thy-1.1 and Thy-1.2 antigens are present in large amounts in brain and thymus of mice[2–4] but in smaller amounts on thymus-derived peripheral T cells. Lymph node, spleen, and bone marrow cell suspension have anti-Thy-1 absorptive capacities of 14, 7, and 1%, respectively relative to

[3] A. E. Reif and J. M. V. Allen, *Nature (London)* **209**, 521 (1966).
[4] A. E. Reif and J. M. V. Allen, *J. Exp. Med.* **120**, 413 (1964).
[5] K. Itakura, J. J. Hutton, E. A. Boyse, and L. J. Old, *Nature (London) New Biol.* **230**, 126 (1971).
[6] E. P. Blankenhorn and T. C. Douglas, *J. Hered.* **63**, 259 (1972).
[7] P. L. Altman and D. D. Katz, "Biological Handbook III." Fed. Am. Soc. Exp. Biol. Bethesda, Maryland, 1979.
[8] M. Zaleski and J. Klein, *Immunol. Rev.* **38**, 120 (1978).

TABLE I
STRAIN DISTRIBUTION OF Thy-1.1 AND
Thy-1.2 ALLOANTIGENS[a]

Allele	Specificity	Strain distribution
Thy-1[a]	Thy-1.1	AKR, BDP, MA/MyJ, PL, RF, RIII/2J, SEA/GnJ
Thy-1[b]	Thy-1.2	A, BALB/c, CBA/H-T6, CBA/J, CE, C3H/An, C3H/He, C3H$_f$/Bi, C57BL/6, C57BR, C57BR/cd, C57L, C58, DA, DBA/1, DBA/2, FL/2Re, GR/A, HRS, HSFS/N (Swiss), I, LP, NZB, SEC/1ReJ, SJL, SM/J, ST/bJ, STS/A, SWR, WB, and 129

[a] From Altman and Katz.[7]

thymocytes.[9] Sixty to eighty-five percent of lymph node and thoracic duct lymphocytes, 30–50% of spleen lymphocytes, and more than 90% of thymocytes express Thy-1 on their surface. Although the levels of Thy-1 are lower on cells of peripheral lymphoid organs than on thymocytes, the presence of Thy-1 on murine lymphocytes is associated with the T cell phenotype.[10] Thy-1 is a general T cell marker only in the murine species. The appearance of Thy-1 on mouse lymphocytes is generally associated with maturation in the thymus. Thymectomized and nude mice show a marked reduction in the number of Thy-1 bearing cells in lymph nodes and spleen.[10–12] In thymic nude mice, however, immunocompetent cells expressing low levels of Thy-1 are present.[13–15]

Thy-1 bearing cells have also been obtained from bone marrow progenitors in long-term cultures.[16] Thus, the acquisition of Thy-1 within the

[9] A. F. Williams, A. N. Barclay, M. Letarte-Muirhead, and R. J. Morris, *Cold Spring Harbor Symp. Quant. Biol.* **41,** 51 (1977).
[10] M. C. Raff and H. H. Wortis, *Immunology* **18,** 931 (1970).
[11] M. C. Raff, *Immunology* **19,** 637 (1970).
[12] I. L. Weissman, M. Small, C. G. Fathman, and L. A. Herzenberg, *Fed. Proc. Fed. Am. Soc. Exp. Biol.* **34,** 141 (1975).
[13] F. Loor and G. E. Roelants, *Nature (London)* **251,** 229 (1974).
[14] H. Ishikawa and K. Saito, *J. Exp. Med.* **151,** 965 (1980).
[15] M. M. Feneck, M. Letarte, and D. Osoba, *Cell Immunol.* **57,** 339 (1981).
[16] E. V. Jones-Villeneuve, J. J. Rusthoven, R. G. Miller, and R. A. Phillips, *J. Immunol.* **124,** 597 (1980).

murine T cell lineage is associated with a maturational event which normally, but not exclusively, occurs within the thymus. More recently, with the aid of monoclonal antibodies to Thy-1 and with a biotin-avidin amplification immunofluorescence detection system, Thy-1.1 and Thy-1.2 alloantigens have been observed on 25–30% of mouse bone marrow cells.[17] Sorting of Thy-1⁺ mouse bone marrow cells with the flow cytometer yields populations enriched in multipotential stem cells, prothymocytes, and some B cell precursors.[17] Thus, immature murine hemopoietic cells can also express Thy-1 alloantigens, but at a lower density than mature T lymphocytes and thymocytes.

Monoclonal Antibodies to Thy-1.1 and Thy-1.2 Alloantigens

Immunization Conditions

Mouse strains differ in the level of antibody they produce against Thy-1 alloantigen. The antibody response to the Thy-1.1 alloantigen is measured by assessing the titer of the sera of immunized mice in the cytotoxicity assay and by enumerating the number of plaque-forming cells (PFC) in the spleens of immunized mice.[8] The genetic control of the response to Thy-1.1 is complex. Segregation analysis of crosses between high and low responders suggests one major codominant gene, *Ir-Thy-1*, associated with the H-2 complex and at least one minor gene (*Ir-5*) outside of H-2.[8] Monoclonal antibodies to Thy-1.1 are best made in high responding strains, such as CBA/J. For example, when immunized with AKR thymocytes, BALB/c mice, which are low responders, produce 982 ± 108 PFC per spleen, while CBA/J mice, which are high responders, produce $14,150 \pm 170$ PFC per spleen.[8] The response to Thy-1.2 alloantigen is also genetically controlled but the differences in the level of antibody produced between high and low responders are less pronounced.[8]

The immunogenicity of Thy-1 alloantigens, in addition to being controlled by the *Ir* genes of the recipient, is also dependent on the presence of helper determinants on the surface of the thymocytes used for immunization.[18] To examine these effects, AKR/J (Thy-1.1) mice are immunized with thymocytes from the closely related strain AKR/Cum or from the unrelated CBA strain (both Thy-1.2). CBA thymocytes (4×10^7, intravenously) elicit 2241 ± 252 PFC, whereas AKR/Cum thymocytes induce 15 ± 8 PFC per spleen in the immunized AKR/J mice, 7 days after immunization. Thus, an incompatibility at the *Thy-1* locus alone (AKR/Cum vs

---

[17] R. S. Basch and J. W. Berman, *Eur. J. Immunol.* **12,** 359 (1982).
[18] P. Lake and T. C. Douglas, *Nature (London)* **275,** 220 (1978).

AKR/J) does not elicit an antibody response. Other alloantigens present on CBA thymocytes must act, therefore, as helper determinants in the induction of the anti-Thy-1.2 response.[18]

These observations have been considered in the production of monoclonal antibodies. Although one might expect to derive hybridomas producing anti-Thy-1.1 antibodies even in a low responder strain, the frequency of clones producing such antibodies is related to the input of PFC in the fusion mixture.[19] Therefore, PFC responses to Thy-1.1 and the frequency of hybridomas can be optimized by (1) immunizing an intermediate or high responder strain, e.g. (BALB/c × BALB/K)F_1 or 129 mice,[8,20,21] (2) using spleen cells at the peak of the PFC response, 4 days after the secondary challenge,[19] (3) injecting the mice with cells carrying helper determinants such as leukemic cells or rat thymocytes,[19,21] (4) using secondary rather than primary responses,[19,20,22] and (5) using high ratios (5 : 1) of myeloma cells to spleen cells during the fusion.[19]

A monoclonal antibody to Thy-1.1, MRC 0X7, is derived from an immunization of BALB/c mice with purified rat thymus Thy-1. Helper determinants on the rat Thy-1.1 molecule, which differ from mouse Thy-1.1 molecule by a few amino acid residues, may have increased the immune response observed in a strain which is a low responder to murine Thy-1.1.[22]

Monoclonal antibodies to Thy-1.2 alloantigen can also be produced.[19,20,23] The number of cells required to prime AKR mice is critical. Maximum PFC response is obtained with 1×10^7 thymocytes.[19] The procedures for fusion, growth, and selection of hybridomas are discussed in references 19–23. They are essentially as described originally by Köhler and Milstein.[24,25]

Antigenic Specificity of the Monoclonal Antibodies

Monoclonal antibodies to Thy-1.1 and Thy-1.2 should show the same strain and tissue reactivity as conventional alloantisera when measured by cytotoxicity assays, cellular radioimmunoassays, or immunofluorescence. For example, from fusion between spleen cells derived from CBA/ mice immunized with rat thymocytes and P3/NSI/1-Ag4-1 myeloma cells, anti-Thy-1.1 antibodies are obtained. IgM from six hybridoma lines, in the

[19] P. Lake, E. A. Clark, M. Khorshidi, and G. H. Sunshine, *Eur. J. Immunol.* **9**, 875 (1979).
[20] A. Marshak-Rothstein, P. Fink, T. Gridley, D. H. Ravlet, M. J. Bevan, and M. L. Gefter, *J. Immunol.* **122**, 2491 (1979).
[21] L. L. Houston, R. C. Nowinski, and I. D. Bernstein, *J. Immunol.* **125**, 837 (1980).
[22] D. W. Mason and A. F. Williams, *Biochem. J.* **187**, 1 (1980).
[23] P. Draber, J. Zikan, and M. Vojtiskova, *J. Immunogenet.* **7**, 455 (1980).
[24] G. Köhler and C. Milstein, *Nature (London)* **256**, 495 (1975).
[25] G. Köhler and C. Milstein, *Eur. J. Immunol.* **6**, 511 (1976).

presence of complement, specifically kill 95% of AKR/J (Thy-1.1) thymocytes but no AKR/Cum (Thy-1.2) thymocytes. IgG_2 from 3 hybridoma producing lines specifically kill Thy-1.1 thymocytes but with a 100-fold lower titer than the IgM antibodies.[19] All these hybridoma antibodies lyse rat thymocytes (95%) and 60 and 40% of AKR lymph node cells and spleen cells, respectively, as expected for anti-Thy-1.1 antibodies.[19] Figure 1 illustrates the reactivity of F7D5, an IgM hybridoma-producing line, which shows specificity similar to that of conventional anti-Thy-1.2 allosera, although at a higher titer. This antibody is produced by a hybridoma resulting from a fusion between P3/NSI/1-Ag4-1 and spleen cells from AKR mice immunized with CBA thymocytes. Line F7D5 is stable and produces high levels of cytotoxic antibody. Lysis of thymus, lymph node, and spleen lymphocytes is 95, 60–65, and 35–40%, respectively. The ascites fluid produced in (AKR × BALB/c)F_1 mice is lyophilized, reconstituted with H_2O, diluted 20-fold in 1% BSA/PBS, and stored at $-70°$ in small aliquots. In a ^{51}Cr cytotoxicity assay, a 250,000-fold dilution of this antibody is sufficient to cause 50% lysis of B10.BR thymocytes (Thy-1.2). No significant release of ^{51}Cr is observed with rat thymocytes (Fig. 1a).[26]

The specificity of F7D5 antibody is demonstrated in a cellular radioimmunoassay. The antibody at a 10,000-fold dilution specifically binds to B10.A thymocytes (Thy-1.2). This is 18-fold above the binding found with AKR thymocytes (Thy-1.1) (Fig. 1b).[26] The binding of the IgM anti-Thy-1.2 antibody is detected with ^{125}I-labeled F(ab')$_2$ rabbit-IgG anti-mouse IgG-Fab fragment (RAM-Fab), which reacts with mouse IgM because of its anti-light chain activity. The sensitivity of detection of F7D5 binding is increased by using 2 ng of RAM-Fab (16 hr incubation) or 10 ng (2 hr incubation) instead of the usual 25 ng (2 hr incubation).[26]

Competitive binding between monoclonal antibodies can also be studied by cellular radioimmunoassay. Competitive binding between ^{125}I-labeled purified IgG of an anti-Thy-1.2 monoclonal antibody (e.g., HO-13-4) and varying dilutions of unlabeled monoclonal antibodies (e.g., HO-13-4, HO-15-2, or HO-15-7) using thymocytes as target cells show that these three antibodies recognize Thy-1.2 determinant.[20] HO-13-4 and HO-15-2 bind with approximately equal avidity to C3H thymocytes, while HO-15-7 binds with less avidity than HO-13-4.[20]

Quantitation of Antigenic Sites per Cell

The number of antigenic sites on the surface of a cell is best quantitated in a one-step assay using monovalent and divalent fragments of purified antibodies. The number of Thy-1.1 antigenic sites on the surface of mouse and rat thymocytes has been estimated using F(ab')$_2$, the diva-

[26] M. Letarte, J. Addis, M. E. MacDonald, A. Bernstein, and P. Lake, *Can. J. Biochem.* **58,** 1026 (1980).

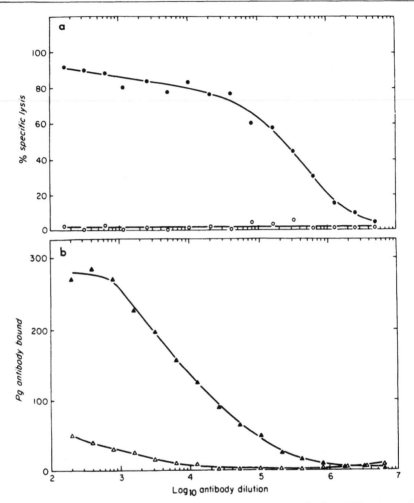

FIG. 1. Specific reactivity of F7D5 anti-Thy-1.2 monoclonal antibody. (a) The cytotoxic activity of F7D5 hybridoma antibody is measured on B10.BR thymocytes (Thy-1.2) (●) and on rat thymocytes (Thy-1.1) (○) with the microcytotoxicity assay; 5×10^4 ^{51}Cr-labeled thymocytes (50 μl) are incubated with 50 μl of reconstituted ascites fluid at the dilutions shown on the abscissa; "Low Tox" complement (Cedar Lane Laboratories, Hornby, Ontario) is used at a dilution of 1 : 15. Percentage specific lysis = (cpm released with antibody − spontaneous cpm released)/(maximum cpm released − spontaneous cpm released). (b) The specific binding of F7D5 antibody is measured in the radioimmunoassay with 2×10^6 glutaraldehyde-fixed B10.A (Thy-1.2) (▲) or AKR (Thy-1.1) (△) thymocytes. Dilutions of ascites (25 μl) are incubated for 16 hr with the target cells (25 μl). Following washings in 0.1% BSA/PBS, cells are incubated for 16 hr with ^{125}I-labeled RAM-Fab (2 ng, 75,000 cpm). Results are expressed in picograms of RAM-Fab bound to the target cells. Adapted from Letarte *et al.*[26]

lent pepsin fragment, or Fab', the monovalent pepsin fragment of purified MRC OX7 anti-Thy-1.1 antibody.[22] To prepare these fragments, MRC OX7 IgG is incubated with 2% pepsin (w/w) at pH 4 in sodium acetate buffer for 20 hr at 37°. F(ab')₂ is separated from undigested IgG on Sephacryl S-200 and shown to be homogeneous by SDS–PAGE.[22] F(ab')₂ (280 μg; 100 μl) is labeled with ^{125}I using chloramine-T in the presence of DMSO and separated from unreacted ^{125}I on Sephadex G-50 with a yield of 80–95%.[22] ^{125}I-labeled F(ab')₂ is reduced with 0.02 M mercaptoethanol at 37° for 1 hr, and alkylated with iodoacetamide. The resulting monovalent pepsin fragment, ^{125}I-labeled Fab', is fractionated on Sephacryl S-200.[22]

The number of molecules of anti-Thy-1.1 antibody bound per cell at saturation is estimated at concentrations of labeled F(ab')₂ and Fab' fragments of the antibody at which a 2-fold increase in concentration gives < 5% increase in binding. The number of molecules bound per cell is calculated from the cpm bound, the specific radioactivity of the labeled antibody (cpm/ng protein), and the cell number. The number of molecules of Fab' antibody bound per cell at saturation is equal to the number of molecules of antigen accessible on the cell surface. Table II shows that rat thymocytes bind 1040×10^3 molecules of Fab' antibody. However, when the binding is measured with the divalent pepsin fragment, only 704×10^3 molecules are bound indicating that 50% of the F(ab')₂ antibodies is involved in divalent binding (1 molecule of antibody binding to 2 molecules of Thy-1.1). The number of Fab' anti-Thy-1.1 bound to AKR thymocytes is 637×10^3 molecules per cell while the number of F(ab')₂ molecules is 318×10^3 molecules per cell indicating that 100% of the binding of F(ab')₂ anti-Thy-1.1 to AKR thymocytes is bivalent. The density of Thy-1 on mouse thymocytes is 64% of that calculated for rat thymocytes (Table II). Bivalently bound antibody dissociates more slowly than univalently bound antibody. The fact that all F(ab')₂ binds to AKR thymocytes in a bivalent mode whereas 50% of it binds to rat thymocytes in a univalent mode is due to a dissociation rate 10 times greater for the univalent antibody from mouse thymocytes than from rat thymocytes.[22]

Kinetics of anti-Thy-1.1 binding to target cells show that differences in antibody affinities arise mainly through variations in dissociation rates. These considerations are particularly important in the analysis of competitive inhibition studies between different monoclonal antibodies directed at similar or related antigenic determinants.

The Biochemical Nature of Thy-1 Alloantigens

Anti-Thy-1.1 reactivity, defined by the binding of C3H anti-AKR serum to AKR or rat thymocytes, is absorbed completely with a glycopro-

TABLE II

SATURATION BINDING OF ^{125}I-LABELED F(ab')$_2$ AND
^{125}I-LABELED Fab' ANTI-Thy-1.1 TO
LYMPHOID CELLS[a]

Molecules of antibody bound per cell $\times 10^{-3}$	Rat thymocytes	AKR thymocytes
Fab' anti-Thy-1.1	1040 ± 75	637 ± 79
F(ab')$_2$ anti-Thy-1.1	704 ± 43	318 ± 28
Fab'/F(ab')$_2$ ratio	1.48	1.98

[a] Fresh thymocytes (1 × 10^6 per assay) are incubated in duplicate with 40 μl of ^{125}I-labeled F(ab')$_2$ at 16 and 8 μg/ml or with 75 μl of ^{125}I-labeled Fab' at 8 and 4 μg/ml for 1 hr on ice; 1.5 ml 0.5% BSA/PBS 0.01 M NaN$_3$ is added and the cells are pelleted by centrifugation at 1500 g for 2.5 min. Controls include cells preincubated with 10 μl (30 μg/ml) of unlabeled F(ab')$_2$ anti-Thy-1.1 for 45 min prior to addition of labeled antibodies. The binding is the same at both antibody concentrations tested indicating saturation of sites. The values shown are the means of 7 experiments for rat thymocytes and 4 for mouse thymocytes with control values subtracted. Data taken from Mason and Williams.[22]

tein[27,28] purified from thymus or brain of rats which migrates as a single band of apparent MW 24,000 on SDS–PAGE. When eluted from the gel, this material reacts with anti-Thy-1.1 and rabbit anti-Thy-1 antibodies.[27,28] This glycoprotein is designated Thy-1 glycoprotein. Chemical analysis of rat brain and thymus Thy-1 glycoproteins (which appear antigenically identical) reveals similar amino acid compositions but different carbohydrate compositions[29] suggesting that the antigenic determinants, including Thy-1.1, are carried by the polypeptide chain. Furthermore, antigenic activities of the Thy-1 glycoprotein are destroyed by heating at 70–80° for 10 min or by pronase digestion for 24 hr at 37°.[9,29] Thy-1.2 antigenicity expressed on a murine lymphoblastoid cell line is also destroyed by proteolytic digestion with papain and/or protease.[30]

[27] M. Letarte-Muirhead, A. N. Barclay, and A. F. Williams, Biochem. J. 151, 685 (1975).
[28] A. N. Barclay, M. Letarte-Muirhead, and A. F. Williams, Biochem. J. 151, 699 (1975).
[29] A. N. Barclay, M. Letarte-Muirhead, A. F. Williams, and R. A. Faulkes, Nature (London) 263, 563 (1976).
[30] U. N. Kucich, J. C. Bennett, and B. J. Johnson, J. Immunol. 115, 626 (1975).

Thy-1 glycoproteins, purified from brain or thymus of Thy-1.1 and Thy-1.2 mice, block the reactivity of monoclonal anti-Thy-1.1 and Thy-1.2 antibodies, respectively. The binding of the F(ab')$_2$ fragments of MRC OX7 IgG (anti-Thy-1.1 antibody) to rat thymocytes is completely inhibited by purified rat brain Thy-1 glycoprotein. The amounts needed for complete inhibition correspond to the Thy-1.1 antigenic density expected on the target cells.[22]

Thy-1 glycoprotein purified from C57BL/10 mouse brain blocks both binding and cytotoxic activities of F7D5 IgM anti-Thy-1.2 antibody toward Thy-1.2 bearing thymocytes.[26] Purified Thy-1.2 glycoprotein (5 μg/ml) fully absorbs the same cytotoxic activity absorbed by 2×10^8 thymocytes. Knowing that the molecular weight of brain Thy-1 is 17,500 and that 6.4×10^5 molecules of Thy-1 are expressed per mouse thymocyte (assuming that similar amounts of Thy-1 are present in Thy-1.2 and Thy-1.1 thymocytes), one can calculate that 2×10^8 thymocytes correspond to 3.8 μg of Thy-1.

Eighty percent of the binding of F7D5 antibody, at a dilution of 10,000-fold, is prevented by purified Thy-1 glycoprotein (20 μg/ml). From this point of view cells are more efficient than purified glycoprotein, since 5×10^7 cells (corresponding to 1 μg of Thy-1) are sufficient to block binding. These results suggest that F7D5 dissociates faster from soluble Thy-1 than it does from cells.

The amino acid analyses of Thy-1.1 and Thy-1.2 glycoproteins from AKR and C3H (or CBA) mice are similar.[31,32] The Thy-1.1 glycoprotein has one more arginine residue and one less glutamic acid/glutamine residue.[31,32]

Figure 2 illustrates the fractionation of the tryptic peptides of Thy-1.1 and Thy-1.2 glycoproteins.[32] The profiles obtained are identical except for the T_{7b} peptide of Thy-1.2 which has no corresponding peak in Thy-1.1. The latter contains instead 2 other peaks, T_8 and T_9. The T_{7b} peptide and the T_8 peptides are labeled indicating that they contained half-cystine residues. When the peptides are further fractionated by HPLC and sequenced, there is no indication of more than one residue at any position. The complete sequence of Thy-1.1 and Thy-1.2 is shown in Fig. 3. The T_{7b} peptide (Ile-68 to Arg-106) of Thy-1.2 corresponds to two peptides in Thy-1.1: the T_8 peptide (Ile-68 to Arg-89) carrying Cys-86 and the T_9 peptide (Val-90 to Arg-106).[32] Thus, the only difference in the amino acid sequence of Thy-1.1 and Thy-1.2 is the substitution of arginine-89 (Thy-1.1) for glutamine-89 (Thy-1.2).[32] Rat Thy-1 which carries the Thy-1.1 allele

[31] S. F. Cotmore, S. A. Crowhurst, and M. D. Waterfield, *Eur. J. Immunol.* **11**, 597 (1981).
[32] A. F. Williams and J. Gagnon, *Science* **216**, 696 (1982).

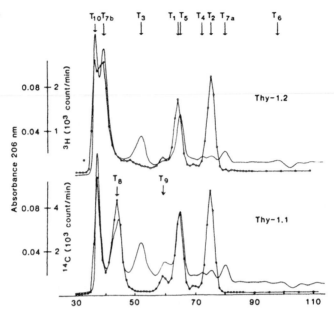

FIG. 2. Separation of tryptic peptides from mouse Thy-1.2 and Thy-1.1 glycoproteins. The glycoproteins are reduced and alkylated (with [³H]- or [¹⁴C]iodoacetic acid) and succinylated. The digestion is carried out in Brij 96 detergent which allows the recovery of the COOH-terminal tryptic peptide which binds to the detergent micelle. The fractionation of tryptic digests on BioGel P-10 in 0.1 *M* ammonium bicarbonate is shown for Thy-1.1 (10 nmol) and Thy-1.2 (8 nmol). The absorbance of 206 nm (—) and the radioactivity (○) incorporated into *S*-carboxymethylcysteine are shown. The arrows indicate the elution positions of the tryptic peptides shown in Fig. 3. (From Williams and Gagnon.[32])

has an arginine residue at position 88. Mouse Thy-1 has one additional residue due to insertion of serine-29.

Thus, in terms of the Thy-1 polypeptide chain, a single substitution accounts for alloantigenic differences. In a model drawn for mouse Thy-1, based on the folding of the polypeptide in an Ig domain-like structure, residue 89 is exposed on the surface of the molecule and would be accessible for antibody binding.[32] The fact that a single amino acid is responsible for the alloantigenicity of Thy-1 can explain the requirement for helper determinants in the production of antibodies to Thy-1 alloantigens.[18,19]

Functional Studies Using Monoclonal Antibodies to Thy-1

Since Thy-1 is present on functional T cells in the mouse, monoclonal anti-Thy-1 antibodies in the presence of complement abrogate some T cell functions. *In vitro* C57BL/10 spleen cells treated with HO-13-4 anti-Thy-1.2 monoclonal antibody and complement lose their ability to re-

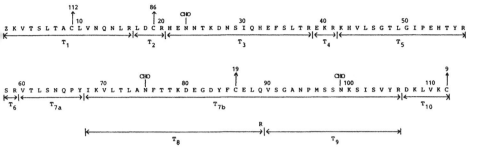

FIG. 3. Sequence of mouse Thy-1.1 and Thy-1.2 glycoproteins. Peptides are prepared for sequencing by digesting 100 nmol of Thy-1.1 and Thy-1.2 with trypsin and fractionating on BioGel P-10. They are further purified by high-performance liquid chromatography (HPLC) on a Waters μBondapak C_{18} column in 0.1% ammonium bicarbonate or 0.2% formic acid buffers with a gradient of acetonitrile. The peptides are identical for Thy-1.1 and Thy-1.2, except for T_{7b}, T_8, and T_9, which differ because of the presence of Arg at position 89 in Thy-1.1 compared with glutamine for Thy-1.2. The tryptic peptides from both Thy-1.1 and Thy-1.2 are sequenced in a Beckman automatic sequencer (with Polybrene), and the phenyl-thiohydantoin derivatives are identified by HPLC. The peptides linked to Cys-9 and Cys-112 bind to the detergent micelles and are eluted at the front of a BioGel P-10 column in 0.1 M acetic acid while the peptides linked by Cys-19 and Cys-86 are retarded. The peptides are oxidized and purified by gel filtration and HPLC. (From Williams and Gagnon.[32])

spond to the T cell mitogen concanavalin A, but not to the B cell mitogen lipopolysaccharide.[20] Cytotoxic activity of Thy-1.2 effector T cells *in vitro* is also abolished by treatment with HO-13-4 antibody and complement.[20] Cells expressing Thy-1.2 are absorbed specifically to HO-13-4 purified IgG coupled to Sepharose 6MB and viable cells are recovered in good yield from the gel.[20]

The *in vivo* localization of 19-12, a monoclonal anti-Thy-1.1 antibody, was studied in normal AKR/J and AKR/Cum mice.[21] The F(ab′)$_2$ fragment of 19-12 purified IgG$_{2a}$ is labeled with ^{125}I and injected iv (10^9 cpm) into age-matched AKR/Cum and AKR/J mice. Radiolabeled 19-12 IgG$_{2a}$-F(ab′)$_2$ is cleared from the blood of AKR/Cum mice (which do not react with this antibody) with a half-life of 6 days. In contrast, 90% is cleared from AKR/J mice with an estimated half-life of 0.58 days. This more rapid clearance from the circulation is associated with uptake by Thy-1.1 bearing tissues.[21] Specific localization of the antibody to lymph nodes and spleen of AKR/J mice is 475 and 250 times, respectively, that observed in AKR/Cum mice. The failure to observe the uptake of antibody into the brain is probably due to the blood–brain transport barrier for macromolecules. The lack of uptake by thymus is probably due to a slower rate of transport of the antibody across the capillary linings of the organ. With injection of saturating amounts (10 mg) of the monoclonal antibody, virtually all thymocytes are coated with anti-Thy-1 antibodies.

[58] Isolation and Characterization of the Thy-1 Glycoproteins of Rat Brain and Thymus

By MICHELLE LETARTE

Thy-1 antigens have been identified in mouse, rat, human, and dog. Their tissue distribution varies between species.[1-11] Thy-1 antigens are abundant in adult brain and thymus of rat and mouse. Thy-1 is also present on mouse peripheral T cells and constitutes the T cell marker of that species.[1] In the rat, Thy-1 is not a general T cell marker. However, it is expressed on a subpopulation of bone marrow cells.[2,3] In man and dog, brain is the only tissue which contains large amounts of Thy-1.[4-6] Canine Thy-1 is present in small amounts on all thymocytes and peripheral T cells whereas human Thy-1 is virtually absent from bone marrow, thymus, and peripheral T cells.[5,6] Thy-1 is also found in small amounts on dog and human kidney cells,[5] mouse and human fibroblasts,[7,8] mouse epidermal cells,[9] mammary cells,[10] and immature skeletal muscle.[11]

Murine Thy-1 antigen exists in two allotypic forms, Thy-1.1 and Thy-1.2[1] Thy-1.1, but not Thy-1.2 is found in the rat.[2,3] Other antigenic determinants on the Thy-1 molecule have been recognized with heteroantisera produced in rabbits. Certain determinants recognized by the heterosera are species-specific while others are cross-reactive between species. The existence of antibodies to species cross-reactive determinants led to the identification of the canine and human homologs of rat Thy-1.[3-5] Purification and characterization of the Thy-1 antigens of thymus and brain have been achieved primarily in the rat.

---

[1] A. E. Reif and J. M. V. Allen, *J. Exp. Med.* **120**, 413 (1964).
[2] T. C. Douglas, *J. Exp. Med.* **136**, 1054 (1972).
[3] A. F. Williams, A. N. Barclay, M. Letarte-Muirhead, and R. J. Morris, *Cold Spring Harbor Symp. Quant. Biol.* **41**, 51 (1977).
[4] R. Arndt, R. Stark, and H.-G. Thiele, *Immunology* **33**, 101 (1977).
[5] R. Dalchau and J. W. Fabre, *J. Exp. Med.* **149**, 576 (1979).
[6] J. L. McKenzie and J. W. Fabre, *J. Immunol.* **126**, 843 (1981).
[7] P. L. Stern, *Nature (London)* **246**, 76 (1973).
[8] S. F. Cotmore, S. A. Crowhurst, and M. D. Waterfield, *Eur. J. Immunol.* **11**, 597 (1981).
[9] M. Scheid, E. A. Boyse, E. A. Carswell, and L. J. Old, *J. Exp. Med.* **135**, 938 (1972).
[10] V. A. Lennon, M. Unger, and R. Dulbecco, *Proc. Natl. Acad. Sci. U.S.A.* **75**, 6093 (1978).
[11] J. F. Lesley and V. A. Lennon, *Nature (London)* **268**, 163 (1977).

Assay for the Detection of Thy-1 Antigenic Activities

The purification of Thy-1 molecules is monitored by a cellular radioimmunoassay using glutaraldehyde-fixed target cells.[12,13]

The preparation of glutaraldehyde-fixed cells is as follows: cells washed in phosphate-buffered saline (PBS) are resuspended at 1×10^8/ml and an equal volume of 0.25% glutaraldehyde in PBS is added; after incubation for 5 min at 23°, the reaction is stopped by the addition of 5% bovine serum albumin (BSA)-PBS and the cells are washed with 0.5% BSA-PBS. Cells can be stored at −70° for several years without loss of antigenicity. We have used this procedure in our laboratory with lymphocytes, erythrocytes, leukemic cells, and lymphoblastoid cell lines of different species. All antigens on glutaraldehyde-fixed cells have retained their reactivity with antibody. Cross-linking of the membrane proteins with glutaraldehyde increases, in some cases, the number of accessible antigenic sites.

The amount of Thy-1 antigenic activity present at different stages of the isolation procedure is measured by inhibition of the binding of anti-Thy-1 antibody to glutaraldehyde-fixed mouse or rat thymocytes.[12-14] The incubation of antibodies with target cells is performed in microtiter plates. Absorption of anti-Thy-1 antibodies with soluble antigenic fractions prior to the binding steps can also be performed in the plates. All assays and dilutions are performed in 0.5% BSA-PBS and at 4°. The rabbit or mouse anti-Thy-1 antibodies (25 μl), at a predetermined concentration corresponding to the linear portion of their titration curves, are incubated (3 to 16 hr) with 25 μl of serially diluted antigenic fraction to be tested. Glutaraldehyde-fixed thymocytes (1×10^6) (25 μl) are added for 1 hr and washed 3 times with 0.1% BSA-PBS. The anti-Thy-1 antibody not bound by the soluble fraction during the absorption step will be bound to the target cells and measured by a 2-hr incubation with ^{125}I-labeled F(ab')$_2$ rabbit IgG-anti-mouse IgG in the case of mouse anti-Thy-1 antibodies or with ^{125}I-labeled F(ab')$_2$ horse IgG anti-rabbit IgG (25 ng/well; 100,000 cpm) in the case of rabbit anti-Thy-1 antibodies.[13,14] The preparation and radiolabeling of these pepsin fragments of affinity purified reagents are described elsewhere.[15,16]

The percentage of antibody bound to the target cells is then plotted as a function of the concentration of antigenic fraction added. A relative unit

[12] A. F. Williams, *Eur. J. Immunol.* **3**, 628 (1973).
[13] M. Letarte-Muirhead, R. T. Acton, and A. F. Williams, *Biochem. J.* **143**, 51 (1974).
[14] M. Letarte-Muirhead, A. N. Barclay, and A. F. Williams, *Biochem. J.* **151**, 685 (1975).
[15] J. C. Jensenius and A. F. Williams, *Eur. J. Immunol.* **4**, 91 (1974).
[16] M. Letarte, H.-S. Teh, and G. Meghji, *J. Immunol.* **125**, 370 (1980).

of activity is defined as the concentration of antigen giving 50% inhibition of the antibody binding under the experimental conditions chosen. The total antigenic activity recovered in each fraction during the isolation procedure is assessed and the yield estimated relative to the total antigenic activity present in the starting material. Specific antigenic activity is expressed in units per milligram of protein. A relative unit of activity is defined for each antibody–target cell combination. Yield and relative specific antigenic activity can be compared. Inhibition of cellular radioimmunoassays have been used to monitor the purification of several antigens.[3,13,14,16–18]

Inhibition assays with detergent-solubilized antigens are only accurate if the complex of antibody and soluble antigen does not dissociate significantly in the period of time required by target cells to bind the uncomplexed free antibody. Antibody binds univalently to soluble antigen but can bind bivalently to membrane-bound antigen. If target cells bind antibody as it dissociates from soluble antigen, the amount of antibody not bound to the soluble antigen would be overestimated. If the dissociation rate of the antibody from the soluble antigen is small, the competitive assay described above is valid. In several cases, it has been shown that the capacity of membrane antigens to bind antibody is not affected by their solubilization in deoxycholate.[3,13,18,19] For a discussion of the kinetics of antibody binding to membrane antigens in solution and at the cell surface, refer to Mason and Williams.[20]

Purification of Thy-1 from Rat Thymus

Figure 1 summarizes the steps of the initial purification of Thy-1 antigen.[14] Thymus glands are obtained from 70 to 100 rats (10^{11} thymocytes) and crude membranes are prepared at 4° by resuspending the thymocytes at 2.5×10^8 cells/ml in 2% Tween-40 in 0.01 M Tris–HCl buffer, pH 8.0 0.15 M NaCl plus 0.02% NaN_3.[21] Thy-1 activity is released in a 50% yield, on a membrane fragment sedimenting at 50,000 g for 90 min. The crude membranes are solubilized in 50 ml of 2% deoxycholate in 0.01 M Tris–HCl buffer, pH 8.0 and centrifuged at 50,000 g for 120 min; 28% of the original activity is recovered at this stage. The soluble deoxycholate extract (50 ml) is applied to a Sephadex G-200 column (5 × 90 cm), equilibrated, and run in 0.5% deoxycholate in the same buffer (deoxycholate buffer). Thy-1.1 and Thy-1 xenoantigenic activities elute as a single peak

[17] A. N. Barclay, M. Letarte-Muirhead, and A. F. Williams, *Biochem. J.* **151**, 699 (1975).

[18] M. Letarte and G. Meghji, *J. Immunol.* **121**, 1718 (1978).

[19] R. J. Morris and A. F. Williams, *Eur. J. Immunol.* **7**, 360 (1977).

[20] D. W. Mason and A. F. Williams, *Biochem. J.* **187**, 1 (1980).

[21] R. J. Morris, M. Letarte-Muirhead, and A. F. Williams, *Eur. J. Immunol.* **5**, 282 (1975).

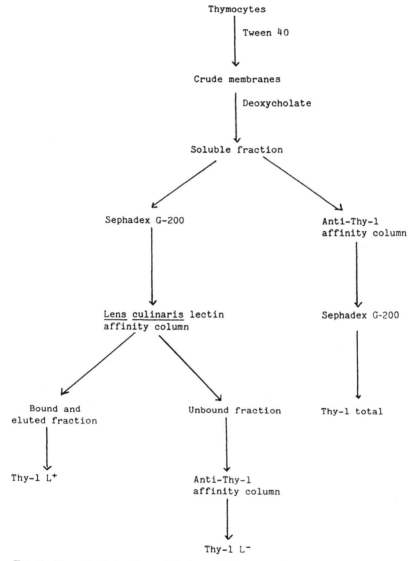

Fig. 1. Flow sheet for the purification of Thy-1 from rat thymus. (From Letarte-Muirhead *et al.*[14])

at the same position as ovalbumin. The total yield of Thy-1 activity is 15% and the specific activity is increased 30-fold.[14] This material is then fractionated on a *Lens culinaris* lectin Sepharose 4B column, equilibrated in deoxycholate buffer, under conditions which do not saturate the column (35 mg lectin, 14 mg protein). Bound glycoproteins are eluted with 0.5 *M* methyl α-D-glucopyranoside in deoxycholate buffer; 0.9 mg protein is recovered, enriched 260-fold in Thy-1 activity. This fraction is referred to as Thy-1 L$^+$ fraction as it contains the Thy-1 molecules binding to *Lens culinaris* lectin.

The sample is incubated for 1 hr in 0.1% sodium dodecyl sulfate (SDS) 0.01 M Tris, pH 8.0 and then run on SDS–polyacrylamide gel electrophoresis (PAGE). One major band is found with apparent molecular weights of 32,000 (5.6% polyacrylamide) and 25,300 (12.5% polyacrylamide).[14] To verify that Thy-1 antigenic activities coincide with the major band, material may be eluted from the gel slices and the activity determined with the radioimmunoassay described above. To this end, the gels are cut into 3-mm slices and crushed in 0.05 ml of 0.05% SDS in 0.01 M Tris pH 8.0. The final concentration of SDS in the inhibition assays is 0.025%. This concentration of SDS inhibits 90% of Thy-1 antigenic activity. Sufficient activity, however, remains to establish that the major band on the gel contains both activities.[14]

To purify the fraction of Thy-1 which does not bind to *Lens culinaris* lectin (Thy-1 L$^-$), an immunoaffinity step with anti-rat brain Thy-1 column is introduced. In order to minimize nonspecific binding of material to the anti-Thy-1 column, the sample is passed first through a rabbit IgG Sepharose 4B column. The Na$_2$SO$_4$ (16%) precipitate from a rabbit anti-rat brain Thy-1 is coupled to Sepharose 4B (20–35 mg of protein/ml). The anti-Thy-1 column (9–15 ml) is prewashed with 3 volumes of 0.05 M diethylamine–HCl pH 11.5, and washed with deoxycholate buffer. Both the *Lens culinaris* lectin unbound fraction and the deoxycholate soluble extract are bound to the anti-Thy-1 column. Material bound to the anti-Thy-1 column in deoxycholate buffer is eluted with 0.05 M diethylamine–HCl, pH 11.5, plus 0.5% deoxycholate. The eluted fractions are immediately brought to pH 8.3 with solid glycine and dialyzed overnight against deoxycholate buffer. These fractions are further purified by gel filtration on Sephadex G-200. The Thy-1 L$^-$ fraction contains a major band with an apparent molecular weight of 27,200 (12.5% acrylamide) slightly higher than the Thy-1 L$^+$ fraction (25,300). Both fractions are sufficiently homogeneous to permit chemical analysis.[22]

The availability of monoclonal antibodies has greatly facilitated the purification of antigens from soluble extracts. The elution conditions described above (i.e., 0.05 M diethylamine pH 11.5 plus 0.5% deoxycholate or cholate, at 4° followed by immediate neutralization with glycine) give a good recovery of antigenically active fractions.

[22] A. N. Barclay, M. Letarte-Muirhead, A. F. Williams, and R. A. Faulkes, *Nature (London)* **263,** 563 (1976).

The purification of Thy-1.1 from the murine BW5147 lymphoblastoid line has been reported using similar procedures.[23] One milligram of protein is obtained, enriched 2139-fold in Thy-1 activity with a yield of 10.6%.

Purification of Thy-1 from Rat Brain

Crude membrane fractions of brain homogenate are extracted with 2% Lubrol-PX before solubilization of the membrane pellet with deoxycholate. Thy-1 is purified by affinity on *Lens culinaris* lectin. Unlike rat thymus Thy-1, most of the rat brain Thy-1 binds to the lectin. After Sephadex G-200 chromatography, Thy-1.1 and Thy-1 xenoantigenic activities are enriched 440-fold (12% yield) and 720-fold (20% yield), respectively. Two milligrams of purified protein is obtained from 7300 mg of brain homogenate proteins.[17] Thy-1 is also purified by a similar procedure from brain of Thy-1.2-bearing mice with a 923-fold purification in xenoantigenic activity and a yield of 26%.[18] Both rat and mouse brain Thy-1 migrate as a single band of apparent molecular weight 24,000 on SDS–PAGE.

The procedure for the purification of large amounts of Thy-1 from rat brain has been modified to include a chloroform–methanol extraction step and affinity chromatography with a monoclonal antibody column.[24] The chloroform–methanol (2 : 1) powder derived from 200 g of brain (wet weight) is homogenized in a Waring Blender with 600 ml of cold 5% deoxycholate in 0.05 M Tris–HCl, pH 8.0 followed by stirring overnight at 4°. The recovery of Thy-1 xenoantigenic activity in the soluble deoxycholate extract is 48% with a 4.4-fold purification.[24] The deoxycholate extract derived from two such preparations (approximately 1.3 liter from 400 g of brain wet weight) is fractionated on an anti-Thy-1 (MRC OX7) monoclonal antibody affinity column[24] prepared as follows: IgG from 150 ml of ascitic fluid is precipitated with 16% Na_2SO_4 and fractionated on a DEAE-cellulose column equilibrated with 0.025 M Tris–HCl pH 7.5, 0.05 M NaCl. The anti-Thy-1.1 IgG (930 mg) is eluted with 0.07 M NaCl and shown to be homogeneous by SDS–PAGE electrophoresis.[20] The purified anti-Thy-1.1 IgG is coupled to Sepharose 4B at 10 mg/ml of beads.[24]

The deoxycholate extract of rat brain is fractionated first on a rabbit-IgG Sepharose 4B column and then an anti-Thy-1.1 IgG column at a rate of 12 ml/hr. After elution from the affinity column the Thy-1-containing sample is neutralized as described above for thymus Thy-1. Sixty percent of Thy-1 activity is recovered, with a purification factor of

[23] R. K. Zwerner, P. A. Barstad, and R. T. Acton, *J. Exp. Med.* **146**, 986 (1977).
[24] D. G. Campbell, J. Gagnon, K. B. M. Reid, and A. F. Williams, *Biochem. J.* **195**, 15 (1981).

TABLE I
MOLECULAR WEIGHT OF RAT Thy-1 GLYCOPROTEIN[a]

Method used for estimation of molecular weight	Source of Thy-1 glycoprotein		
	Thymus		
	Thy-1L[+]	Thy-1L[−]	Brain Thy-1
SDS electrophoresis			
12.5% acrylamide	25,300	27,200	24,100
5.6% acrylamide	31,600	34,800	31,300
Sedimentation analysis and gel filtration	28,000		27,000
Sedimentation equilibrium			
Purified glycoprotein	18,700		17,500
Polypeptide portion only	12,500		12,500

[a] From Letarte-Muirhead et al.[13,14,17] and Kuchel et al.[32]

770-fold. After gel filtration on Sephacryl S-200, the purification factor is increased to 1000-fold. From 400 g of rat brain, 10.7 mg of pure Thy-1 is obtained.[24]

The absorbance in deoxycholate buffer at 278 nm is 2.5 ± 0.3 for a solution containing 10 mg of Thy-1/ml (11 determinations).[24] The purification of Thy-1-like glycoproteins from human brain and fibroblasts using monoclonal antibodies to human Thy-1 as immunoadsorbents[6,8,25] has been described. The proteins purified are similar in apparent molecular weight and in amino acid composition to mouse and rat Thy-1.[8]

The Molecular Weight of Thy-1 Glycoproteins

Apparent molecular weights are determined by SDS–PAGE of reduced and alkylated Thy-1 molecules.[26] Glycoproteins may behave anomalously on SDS–PAGE and generally larger than predicted particularly at low acrylamide concentrations.[27,28] As shown in Table I, thymus and brain Thy-1 migrate less in 5.6% than in 12.5% polyacrylamide. The molecular weight of brain Thy-1 is consistently lower than that of thymus Thy-1. The Thy-1 L[−] component is more heterogeneous than the Thy-1 L[+] component resulting in a broader band on the gels. It also has a higher

[25] R. C. Seeger, Y. L. Danon, S. A. Rayner, and F. Hoover, J. Immunol. 128, 983 (1982).

[26] K. Weber and M. Osborn, J. Biol. Chem. 244, 4406 (1969).

[27] H. Glossmann and D. M. Neville, J. Biol. Chem. 246, 6339 (1971).

[28] J. P. Segrest, R. L. Jackson, E. P. Andrews, and V. T. Marchesi, Biochem. Biophys. Res. Commun. 44, 390 (1971).

molecular weight than the Thy-1 L^+ component. These differences in apparent molecular weight between brain and thymus Thy-1 and between Thy-1 L^+ and Thy-L^- components of thymus are due to differences in glycosylation and not in polypeptide structure.[22].

The partial specific volume of Thy-1 plus bound deoxycholate is 0.70 ml/g, as established by sedimentation in 5 to 20% sucrose gradients in H_2O and in D_2O.[13,29] The apparent $s_{20,w}$ value is 2.4 ± 0.04 and the Stokes radius determined by gel filtration on Sephadex G-200 in deoxycholate buffer[13] is 3.1 nm.[30] From these values, the molecular weight of Thy-1 plus bound deoxycholate is estimated to be 28,000 and the frictional coefficient 1.57.[13] These results suggest that Thy-1 is released from the membrane by deoxycholate in a monomeric form.

The molecular weights of membrane molecules in detergents can be determined by using sedimentation–equilibrium methods[31] if the amount of detergent bound can be assessed. The amount of deoxycholate bound to thymus and brain Thy-1 may be measured by equilibrium dialysis at 21° in microcells using highly purified sodium deoxycholate and [^{14}C]deoxycholate.[32] Thy-1 glycoproteins, purified by fractionation procedures that used twice recrystallized deoxycholate, are precipitated with ethanol, resuspended, and dialyzed against 0.01 M Tris–HCl, pH 8.0, 0.005 M NaN_3 at 0.8 mg/ml prior to equilibrium dialysis or ultrancentrifugation. Thy-1 glycoproteins bind little deoxycholate at low concentrations of detergent but binding increases sharply at the critical micellar concentration of deoxycholate. At concentrations above 2.5 mg/ml, no increase in deoxycholate binding occurs.[32] The amount of deoxycholate bound at equilibrium at 21° is about 24 ± 1 μg/100 μg of Thy-1 glycoprotein. It appears that Thy-1 binds to deoxycholate micelles and not to detergent monomers as suggested by the very rapid increase in binding at the critical micellar concentration, i.e., between 1 and 2 mg/ml. The binding curve is very similar to that observed with cytochrome b_5.[33,34]

After correction for the amount of deoxycholate bound, molecular weights of 18,700 and 17,500 can be calculated for brain and thymus Thy-1, respectively.[32] Since the carbohydrate content is 32 and 29%, respectively, the polypeptide chain of both thymus and brain Thy-1 has a molecular weight of 12,500,[32] in good agreement with the value obtained from

[29] J. C. Meunier, R. W. Olsen, and J. P. Changeux, *FEBS Lett.* **24**, 63 (1972).

[30] L. M. Siegel and K. J. Monty, *Biochim. Biophys. Acta* **112**, 346 (1966).

[31] C. Tanford, Y. Nozaki, J. A. Reynolds, and S. Makino, *Biochemistry* **13**, 2369 (1974).

[32] P. W. Kuchel, D. G. Campbell, A. N. Barclay, and A. F. Williams, *Biochem. J.* **169**, 411 (1978).

[33] N. C. Robinson and C. Tanford, *Biochemistry* **14**, 369 (1975).

[34] C. Tanford and J. Reynolds, *Biochim. Biophys. Acta* **457**, 133 (1976).

the amino acid sequence. Without detergent, thymus and brain Thy-1 have molecular weights of 300,000 and 270,000, respectively, corresponding to 16 monomers in a roughly spherical aggregate with a frictional ratio of 1.35.[32] Values derived by other methods are overestimated, as shown in Table I.

Determination of the Amino Acid Sequence of Rat Brain Thy-1

The amino acid sequence of rat brain Thy-1 may be obtained from studies of the peptides generated by digestion of the entire glycoprotein with either trypsin or V8 proteinase. Thy-1 glycoprotein (5–10 mg/ml) is dissolved in 7 M guanidinium chloride, 0.5 M Tris–HCl, pH 8.0. The protein is reduced with a 40-fold molar excess of dithiothreitol, alkylated with an 8-fold molar excess of iodo[2-14C]acetic acid, an excess of unlabeled iodoacetic acid is added, and the sample is dialyzed against 0.1 M ammonium carbonate.[24] Succinylation is carried out by addition of five portions of 20 mg of succinic anhydride at 30 min intervals to a solution containing 10 mg/ml of reduced and alkylated Thy-1 dissolved in 1.0 M Tris–HCl pH 10.5.[24] The succinylated, reduced, and alkylated Thy-1 is subjected to trypsin digestion while the reduced and alkylated Thy-1 is treated with V8 proteinase. The peptides are separated on BioGel P-10, followed by paper electrophoresis when necessary and sequenced by automated Edman degradation. These procedures are described in detail by Campbell et al.[24,35]

Figure 2 presents the sequence of rat brain Thy-1 polypeptide. The characteristic features of the sequence are the following: (1) The amino terminus is blocked and is a pyroglutamic acid residue, as shown by results of digestion with pyroglutamate aminopeptidase.[24] (2) Two disulfide bonds are present. No free cysteine residue is found as shown by the lack of incorporation of [14C]iodoacetic acid. The half-cystine at position 19 is linked to the half-cystine at position 85 and Cys-111 is linked to Cys-9.[24] (3) Three N-linked glycosylation sites exist: Asn (23)-Asn-Thr, Asn (74)-Phe-Thr, and Asn (98)-Lys-Thr.[35] (4) The C-terminal peptides are unusual: they are obtained in an aggregated form and can be purified only after binding to Brij 96 micelles. They appear to have hydrophobic properties and yet do not contain any extended sequence of hydrophobic residues. The C terminal residue has not been identified directly. Cys-111 is the last residue found during sequencing. The C-terminal peptides contain ethanolamine plus some galactosamine and glucosamine. It has been postulated that a lipid is attached covalently to the carboxyl groups of Cys-111 since the sulfhydryl group is linked to Cys-9 and the ε-amino

[35] D. G. Campbell, A. F. Williams, P. M. Bayley, and K. B. M. Reid, Nature (London) 282, 341 (1979).

                              10   20
‹Glu-Arg-Val-Ile-Ser-Leu-Thr-Ala-Cys-Leu-Val-Asn-Gln-Asn-Leu-Arg-Leu-Asp-Cys-Arg-His-Glu-

CHO   30   40
|
Asn-Asn-Thr-Asn-Leu-Pro-Ile-Gln-His-Glu-Phe-Ser-Leu-Thr-Arg-Glu-Lys-Lys-Lys-His-Val-Leu-

              50   60
Ser-Gly-Thr-Leu-Gly-Val-Pro-Glu-His-Thr-Tyr-Arg-Ser-Arg-Val-Asn-Leu-Phe-Ser-Asp-Arg-Phe-

         70   CHO   80
                                  |
Ile-Lys-Val-Leu-Thr-Leu-Ala-Asn-Phe-Thr-Thr-Lys-Asp-Glu-Gly-Asp-Tyr-Met-Cys-Glu-Leu-Arg-

    90   CHO   100   110
                                           |
Val-Ser-Gly-Gln-Asn-Pro-Thr-Ser-Ser-Asn-Lys-Thr-Ile-Asn-Val-Ile-Arg-Asp-Lys-Leu-Val-Lys-Cys

FIG. 2. Amino acid sequence of rat brain Thy-1 glycoprotein. (From Campbell *et al*[24]).

group of Lys-110 can be succinylated. This lipid could be responsible for the anchorage of Thy-1 to the membrane and would explain the hydrophobic properties of the C terminal peptides.[24,35]

The most striking observation revealed by structural analysis of Thy-1 polypeptide is its homology with immunoglobulin domains.[35–37] The size of Thy-1 polypeptides (111 amino acids) is that of an Ig domain and the two disulfide bonds are consistent with the folding observed within the Ig domain. Thy-1 contains anti-parallel β strands which form two β-pleated sheets held together by hydrophobic interactions and by the Cys-19–Cys-85 bond. The circular dichroism spectrum of Thy-1 shows a negative peak at 217 nm, similar to that of Ig, and characteristic of β-pleated sheet structure.[35] The greatest homology is between mouse brain Thy-1 and the variable region immunoglobulin domains. There is also good homology with Ig constant domains and with β_2-microglobulin.[37] It has been suggested that Thy-1 may be the primordial Ig domain. However, since the expression of Thy-1 is conserved on neuronal cells and not on lymphoid cells, it might be associated more directly with a set of molecules involved in neuronal recognition rather than with the V-region molecules implicated in lymphoid recognition.[37]

The sequence of mouse Thy-1 is very similar to that of rat Thy-1 except for two blocks of residues (26 to 29 and 63 to 67). Mouse Thy-1 has 112 residues due to insertion of a serine at position 29. Thus, the disulfide bonds of mouse Thy-1 are between Cys-9–Cys-112 and Cys-19–Cys-86 and the asparagine N-linked carbohydrate side chains are at positions 23, 76, and 99.[37] When the amino acid analyses of Thy-1 from human brain,

[36] F. E. Cohen, J. Novotny, M. J. E. Sternberg, D. G. Campbell, and A. F. Williams, *Biochem. J.* **195**, 31 (1981).
[37] A. F. Williams and J. Gagnon, *Science* **216**, 696 (1982).

human fibroblasts, rat and mouse brains are compared, they reveal few differences suggesting that the Thy-1 like proteins are highly conserved throughout evolution.[8]

A cDNA clone encoding the rat thymus Thy-1 antigen has been isolated and sequenced; the transcript corresponds to the first 103 amino acids of Thy-1, followed directly by a poly(A) tract.[38] The DNA sequence of the clone thus ends 8 amino acids earlier than predicted from the amino acid sequence. However, this might be due to the presence in the coding sequence of a presumptive polyadenylation signal, found 12 nucleotides upstream from the poly(A) tract.[38]

The Carbohydrate Structure of Thy-1 Glycoproteins

Neutral sugars and sialic acid of rat brain Thy-1 and of the Thy-1 L^+ and Thy-1 L^- forms of rat thymus are measured by gas–liquid chromatography and amino sugars are determined by ion-exchange chromatography.[22] Marked differences are observed between brain and thymus Thy-1, as shown in Table II. Galactosamine is found only in brain Thy-1 and, as mentioned earlier, might be associated with the C-terminal peptide. Sialic acid content is very low in brain Thy-1, but two molecules per molecule of Thy-1 are present in thymus Thy-1. The amounts of fucose, glucose, and galactose differ by a factor of 2 or more between brain and thymus Thy-1.

The differences between Thy-1 L^+ and Thy-1 L^- are smaller than those found between brain and thymus Thy-1. They are likely to represent microheterogeneity within the glycopeptide structure of Thy-1 molecules from thymus Thy-1. The specificity of *Lens culinaris* lectin has been elucidated.[39] The presence of a fucose attached to the asparagine N-linked acetylglucosamine residue is essential for binding to the lectin. Furthermore, two α-mannosyl residues are required for binding and substitution of one α-mannosyl residue at both C2 and C4 prevents binding.[39] Thus, the Thy-1 L^- molecules, which do not bind to *Lens culinaris* lectin, might have extra substitution (tertiary versus biantennary complex structure) since no difference in fucose content is apparent between Thy-1 L^+ and Thy-1 L^-.

Lens culinaris binds 50% of the Thy-1 molecules. Wheat germ agglutinin binds 25% and concanavalin A binds all Thy-1 molecules.[40] The Thy-1 glycoprotein, purified by affinity to concanavalin A, shows on isoelectric focusing at least six charged variants (p*I* from 5 to 9). Treatment of the isolated variants with neuraminidase indicates that the charge heterogene-

[38] T. Moriuchi, H.-C. Chang, R. Denome, and J. Silver, *Nature (London)* **301**, 80 (1983).

[39] K. Kornfeld, M. L. Reitman, and R. Kornfeld, *J. Biol. Chem.* **256**, 6633 (1981).

[40] S. Carlsson and T. Stigbrand, *Eur. J. Biochem.* **123**, 1 (1982).

TABLE II
CARBOHYDRATE COMPOSITION OF RAT Thy-1
GLYCOPROTEINS[a]

| | Molecules of carbohydrate per molecule of Thy-1 | | |
| | | Thymus | |
Carbohydrate	Brain Thy-1	Thy-1 L[+]	Thy-1 L[−]
Fucose	2.0	1.1	1.0
Mannose	13.3	11.8	10.4
Galactose	2.0	6.1	7.7
Glucose	0.7	1.4	1.2
Glucosamine	9.2	10.4	13.0
Galactosamine	1.1	0.0	0.0
Sialic Acid	0.2	2.0	2.4
Percentage by weight of carbohydrate	29	32	35

[a] Adapted from Barclay et al.[22] Data expressed in molecules of carbohydrate per 100 amino acid residues were converted to molecules per molecule of Thy-1 using a factor of 1.11.

ity is due to different amounts of sialic acid. However, desialylation of Thy-1 does not alter its binding to either concanavalin A, *Lens culinaris,* or wheat germ agglutinin.[40] These results imply that the microheterogeneity is due to variations in the core sugars of Thy-1 as well as to different degrees of sialylation.

Three N-linked glycosylation sites have been identified for rat brain Thy-1 at Asn-23, Asn-74, Asn-98.[24] Mouse brain Thy-1.1 and Thy-1.2 also have 3 glycosylation sites at Asn-23, Asn-75, and Asn-99.[35]

A series of mutants which do not express Thy-1 on their surface have been derived from the BW5147 mouse lymphoma cell line. These mutants synthesize an abnormal Thy-1 glycoprotein of short half-life.[41] The studies of these mutants and of the wild type parental cell have shown that Thy-1 carries two types of oligosaccharides: a higher molecular weight species containing galactose, mannose, and glucosamine representative of N-linked complex type structure and a smaller molecular weight species containing only mannose and glucosamine and representative of the N-linked high-mannose type structure.

[41] I. S. Trowbridge, R. Hyman, and C. Mazauskas, *Cell* **14,** 21 (1978).

[59] Lyt-1, Lyt-2, and Lyt-3 Antigens

By PAUL D. GOTTLIEB, EDWARD B. REILLY, PAUL J. DURDA, and
MALGORZATA NIEZGODKA

The Lyt-1, Lyt-2, and Lyt-3 alloantigens defined by Boyse, Old, and co-workers[1,2] reside on the surface of mouse thymocytes and on subpopulations of peripheral lymphocytes. Interest in the Lyt-2 and Lyt-3 cell surface molecules has been kindled by the finding that they are T cell specific, and they mark certain functional subpopulations of T cells (see Cantor and Boyse[3] for review). The Lyt-2 and Lyt-3 alloantigens are present on cytotoxic killer and suppressor T cells but not on T cells which mediate helper function or delayed type hypersensitivity (DTH). The Lyt-1 alloantigen was originally thought to be specific for T cells which mediate helper and DTH functions.[4,5] However, it has since been shown to be present in much smaller quantity on cytotoxic T cells,[6,7] on certain B cell lymphomas,[8] and on what appears to be a unique subpopulation of B lymphocytes.[9–11]

The presence of Lyt-2 and Lyt-3 on only certain functional subsets of T cells has raised the possibility that they may participate in the specialized function of the cells that bear them.[12] Close linkage of Lyt-2 and Lyt-3 to the immunoglobulin light chain locus raised the possibility that these

[1] E. A. Boyse, M. Miyazawa, T. Aoki, and L. J. Old, *Proc. R. Soc. London, Ser. B* **170**, 175 (1968).

[2] E. A. Boyse, K. Itakura, E. Stockert, C. A. Iritani, and M. Miura, *Transplantation* **11**, 351 (1971).

[3] H. Cantor and E. A. Boyse, *Cold Spring Harbor Symp. Quant. Biol.* **41**, 23 (1977).

[4] H. Cantor and E. A. Boyse, *J. Exp. Med.* **141**, 1376 (1975).

[5] H. Cantor and E. A. Boyse, *J. Exp. Med.* **141**, 1390 (1975).

[6] H. Shiku, P. Kisielow, M. A. Bean, T. Takahashi, E. A. Boyse, H. F. Oettgen, and L. J. Old, *J. Exp. Med.* **141**, 227 (1975).

[7] E. Nakayama, W. Dippold, H. Shiku, E. Stockert, H. F. Oettgen, and L. J. Old, *Proc. Natl. Acad. Sci. U.S.A.* **77**, 2890 (1980).

[8] L. L. Lanier, N. L. Warner, J. A. Ledbetter, and L. A. Herzenberg, *J. Exp. Med.* **153**, 998 (1981).

[9] V. Manohar, E. Brown, W. M. Leiserson, and T. Chused, *J. Immunol.* **129**, 532 (1982).

[10] R. R. Hardy, K. R. Hayakawa, J. Haaijman, and L. A. Herzenberg, *Ann. N.Y. Acad. Sci.* (in press).

[11] K. Hayakawa, R. R. Hardy, D. R. Parks, and L. A. Herzenberg, *J. Exp. Med.* **157**, 202 (1983).

[12] E. Nakayama, H. Shiku, E. Stockert, H. F. Oettgen, and L. J. Old, *Proc. Natl. Acad. Sci. U.S.A.* **76**, 1977 (1979).

molecules might be structurally related to immunoglobulins and might contribute in some way to the specific antigen receptor on these T cells.[13,14]

Lyt-1, Lyt-2, and Lyt-3 Genetic Loci

Two alleles have been described at each locus (e.g., *Lyt-2*[a] and *Lyt-2*[b]) and each allele determines an antigenic specificity (Lyt-2.1 and Lyt-2.2, respectively). The *Lyt-1* locus (originally called Ly A) has been mapped to chromosome 19, linkage group XII.[15] The *Lyt-2* (previously Ly-B and Ly-2) and *Lyt-3* (previously Ly C) loci are closely linked to each other on chromosome 6, linkage group IX.[16] No genetic recombination has been observed between the *Lyt-2* and *Lyt-3* genetic loci. However, biochemical analysis of molecules bearing Lyt-2 and Lyt-3 antigenic specificities suggest that they reside on different polypeptides which are associated with each other.[17–19]

The *Lyt-2* and *Lyt-3* loci have been found to be closely linked to loci which determine the expression of certain immunoglobulin kappa light chain polymorphisms.[13,14] These light chain polymorphisms lie in the variable regions, and can be detected by several methods including peptide mapping,[20] isoelectric focussing,[21–24] and their influence upon the expression of idiotypes.[25,26]

Lyt Alloantisera and Monoclonal Antibodies

A variety of alloantisera and monoclonal antibodies have been described with specificity for Lyt-1, Lyt-2, and Lyt-3. Shen and co-work-

[13] P. D. Gottlieb, *J. Exp. Med.* **140,** 1432 (1974).
[14] P. D. Gottlieb and P. J. Durda, *Cold Spring Harbor Symp. Quant. Biol.* **41,** 805 (1977).
[15] K. Itakura, J. J. Hutton, E. A. Boyse, and L. J. Old, *Nature (London), New Biol.* **230,** 126 (1971).
[16] K. Itakura, J. J. Hutton, E. A. Boyse, and L. J. Old, *Transplantation* **13,** 239 (1972).
[17] P. J. Durda and P. D. Gottlieb, *J. Immunol.* **121,** 983 (1978).
[18] E. B. Reilly, K. Auditore-Hargreaves, U. Hammerling, and P. D. Gottlieb, *J. Immunol.* **125,** 2245 (1980).
[19] J. A. Ledbetter, W. E. Seaman, T. T. Tsu, and L. A. Herzenberg, *J. Exp. Med.* **153,** 1503 (1981).
[20] G. M. Edelman and P. D. Gottlieb, *Proc. Natl. Acad. Sci. U.S.A.* **67,** 1192 (1970).
[21] D. Gibson, *J. Exp. Med.* **144,** 298 (1976).
[22] D. M. Gibson, B. A. Taylor, and M. Cherry, *J. Immunol.* **121,** 1585 (1978).
[23] J. L. Claflin, *Eur. J. Immunol.* **6,** 666 (1976).
[24] J. L. Claflin, B. A. Taylor, M. Cherry, and M. Cubberley, *Immunogenetics* **6,** 379 (1978).
[25] J. A. Laskin, A. Gray, A. Nisonoff, N. R. Klinman, and P. D. Gottlieb, *Proc. Natl. Acad. Sci. U.S.A.* **74,** 4600 (1977).
[26] E. A. Dzierzak, C. A. Janeway, Jr., R. W. Rosenstein, and P. D. Gottlieb, *J. Exp. Med.* **152,** 720 (1980).

TABLE I
CONVENTIONAL LYT ALLOANTISERA

Specificity	Recipient	Donor	Reference
Anti-Lyt-1.1	(BALB/c × B6)F$_1$	B6-Lyt-1^a	a
Anti-Lyt-1.2	C3H/An	C3H.CE-Lyt-1.2 : DS	b
Anti-Lyt-2.1	B6-H-2^k	B6-H-2^k.CE-Lyt-2.1 : DS	b
Anti-Lyt-2.2	(C3H/An × B6-Lyt-2^a)F$_1$	B6	a
Anti-Lyt-3.1	(CBA/H-T6J × SJL/J)F$_1$	C58	c
Anti-Lyt-3.1	B6-Lyt-2^a	B6-Lyt-2^a, Lyt-3^a	d
Anti-Lyt-3.2	C58	C58.CE-Lyt-3.2 : DS	b

^{a} F. W. Shen, E. A. Boyse, and H. Cantor, *Immunogenetics* **2**, 581 (1976).
^{b} F. W. Shen, *Immunogenetics* **5**, 291 (1977).
^{c} E. A. Boyse, K. Itakura, E. Stockert, C. A. Iritani, and M. Miura, *Transplantation* **11**, 351 (1971).
^{d} S. C. Boos, J. J. Monaco, and P. D. Gottlieb, *Immunogenetics* **7**, 165 (1978).

TABLE II
LYT-CONGENIC STRAINS OF MICE

Strain	Lyt-1	Lyt-2	Lyt-3	H-2	Donor	Backcross generation	Reference
B6	b	b	b	b	—	—	—
B6-Lyt-1^a	a	b	b	b	C3H/An	17	a
B6-Lyt-2^a	b	a	b	b	C3H/An	16	a
B6-Lyt-2^a, Lyt-3^a	b	a	a	b	RF	16	a
B6.PL-Lyt-2^a, Lyt-3^a/Cy	b	a	a	b	PL	10	b
B6-Lyt-2^a, Lyt-3^a, H-2^k	b	a	a	k	B6-H-2^k	—	c
BALB/cAn	b	b	b	d	—	—	—
C.C58^f	b	a	a	d	C58/J	18	d
C.AKR^f	b	a	a	d	AKR/J	19	d
AKR/J	b	a	a	k	—	—	—
AKR.C	b	b	b	k	BALB/cJ	18	d
C58/J	b	a	a	k	—	—	—
C58.C	b	b	b	k	BALB/cJ	15	e

^{a} E. A. Boyse, K. Itakura, E. Stockert, C. A. Iritani, and M. Miura, *Transplantation* **11**, 351 (1971).
^{b} D. M. Gibson, B. A. Taylor, and M. Cherry, *J. Immunol.* **121**, 1585 (1978).
^{c} S. C. Boos, J. J. Monaco, and P. D. Gottlieb, *Immunogenetics* **7**, 165 (1978).
^{d} P. D. Gottlieb, A. Marshak-Rothstein, K. Auditore-Hargreaves, D. B. Berkoben, D. A. August, R. M. Roche, and J. D. Benedetto, *Immunogenetics* **10**, 545 (1980).
^{e} P. D. Gottlieb, unpublished.
^{f} BALB/cAn genetic background.

TABLE III
MONOCLONAL Lyt ANTIBODIES

Specificity	Name	Species	Class	Cytotoxic	Protein A-binding	Reference
Lyt-1.1	7-20.6/3	Mouse	IgG_{2a}	+	+	a
Lyt-1	53-7.3	Rat	IgG_{2a}	+/−	−	b
Lyt-2	53-6.7	Rat	IgG_{2a}	+/−	+/−	b
Lyt-2.2	HO-2.2	Mouse	IgM	+	(−)	c
Lyt-2.2	19/178	Mouse	IgG_{2a}	+	+	d
Lyt-3.1	HO-3.1	Mouse	IgM	+	(−)	c
Lyt-3	53-5	Rat	IgG_1	−	−	b, e
Lyt-3.2	M-Lyt-3.2	Mouse	IgM	+	(−)	f

[a] P. M. Hogarth, T. A. Potter, F. N. Cornell, R. McLachlan, and I. F. McKenzie, *J. Immunol.* **125**, 1618 (1980).

[b] J. A. Ledbetter and L. A. Herzenberg, *Immunol. Rev.* **47**, 63 (1979).

[c] P. D. Gottlieb, A. Marshak-Rothstein, K. Auditore-Hargreaves, D. B. Berkoben, D. A. August, R. M. Roche, and J. D. Benedetto, *Immunogenetics* **10**, 545 (1980).

[d] E. B. Reilly, K. Auditore-Hargreaves, U. Hammerling, and P. D. Gottlieb, *J. Immunol.* **125**, 2245 (1980).

[e] J. A. Ledbetter, E. E. Seaman, T. T. Tsu, and L. A. Herzenberg, *J. Exp. Med.* **153**, 1503 (1981).

[f] E. Nakayama, W. Dippold, H. Shiku, E. Stockert, H. F. Oettgen, and L. J. Old, *Proc. Natl. Acad. Sci. U.S.A.* **77**, 2890 (1980).

ers[27,28] have described alloantisera specific for each antigenic determinant (Table I). A list of existing Lyt congenic strains is presented in Table II.

A variety of monoclonal antibodies specific for Lyt-1, Lyt-2, and Lyt-3 antigens have also been produced (Table III). Some of these are produced by alloimmunization and thus distinguish alleles at the respective locus. The HO-3.1 IgM monoclonal antibody is highly specific for Lyt-3.1 when tested by complement-dependent cytotoxicity, but shows some reactivity with the Lyt-3.2 molecular species when tested by immunoprecipitation.[29] Some anti-Lyt monoclonal antibodies are produced by rat anti-mouse immunizations, and in general, they recognize species differences and do not distinguish alleles at a locus.[30] However, the 53-5 rat monoclonal antibody, reported to be reactive with Lyt-3,[19] appears to have a 200-fold greater affinity for Lyt-3.2-positive cells than for Lyt-3.1-positive cells.

[27] F. W. Shen, E. A. Boyse, and H. Cantor, *Immunogenetics* **2**, 591 (1976).

[28] F. W. Shen, *Immunogenetics* **5**, 291 (1977).

[29] P. D. Gottlieb, A. Marshak-Rothstein, K. Auditore-Hargreaves, D. B. Berkoben, D. A. August, R. M. Roche, and J. D. Benedetto, *Immunogenetics* **10**, 545 (1980).

[30] J. A. Ledbetter and L. A. Herzenberg, *Immunol. Rev.* **47**, 63 (1979).

Structural Analysis of Lyt Antigens

In order to identify the molecular species which bear the Lyt-1, Lyt-2, and Lyt-3 antigenic determinants, lymphocytes from various sources are labeled at the cell surface with either ^{125}I and lactoperoxidase[31] or NaB^3H_4 and galactose oxidase.[32] Biosynthetic labeling with [^{35}S]methionine is also used. Cells are then solubilized with a nonionic detergent, generally Nonidet P-40 (NP-40),[33] nuclei are removed by centrifugation, and extracts are precipitated with conventional or monoclonal antibody.

Precipitates are obtained by either a sandwich method using antimouse immunoglobulin reagents, or with protein A-positive *Staphylococcus aureus,* Cowan I strain (SaCI).[34] Analyses have involved two-dimensional polyacrylamide gel electrophoresis, with nonequilibrium pH gradient electrophoresis (NEPHGE)[35,36] in the first dimension and SDS–PAGE in the second.

Surface Labeling, Solubilization, and Preparation for
  Gel Electrophoresis

Iodination of surface components with ^{125}I is performed essentially as described by Haustein *et al.*[31] Aliquots containing 10^7 thymocytes are centrifuged in capped conical polystyrene tubes for 8 min at 140 *g*. One such aliquot is iodinated for each specific or control precipitation to be performed. Supernates are removed and 20 μl of lactoperoxidase (A_{280} = 2.80) at a concentration of 2.0 mg/ml in PBS is added. Thirty microliters of neutralized ^{125}I (300 μCi) in PBS is then added to each tube. The iodination reaction, initiated by addition of 10 μl of 0.03% hydrogen peroxide in PBS (diluted from 30% stock solution just before use) is carried out at 30°. The reaction is generally terminated after 5 min by addition of 2 ml of cold PBS containing 5 m*M* KI.

After centrifugation at 140 *g* for 8 min, the cells are washed three times in 2 ml of PBS-KI by centrifugation and resuspension. After the second and third centrifugation steps the cells are transferred to new polystyrene tubes. After the final wash, cell pellets are resuspended in 75 μl of a solution containing 0.15 *M* NaCl, 0.005 *M* EDTA, 0.05 *M* Tris, 0.02%

[31] D. Haustein, J. J. Marchalonis, and A. W. Harris, *Biochemistry* **14,** 1826 (1975).
[32] C. Gahmborg, and S.-I. Hakamori, *Proc. Natl. Acad. Sci. U.S.A.* **70,** 3329 (1973).
[33] Definitions used in this review: NP-40, Nonidet P-40 detergent; SaCI, *Staphylococcus aureus,* Cowan I strain: SDS–PAGE, polyacrylamide gel electrophoresis in sodium dodecyl sulfate; NEPHGE, non-equilibrium pH gradient electrophoresis; M$_r$, molecular weight (relative); NMS, normal mouse serum.
[34] S. W. Kessler, *J. Immunol.* **115,** 1617 (1975).
[35] P. Z. O'Farrell, H. M. Goodman, and P. H. O'Farrell, *Cell* **12,** 1133 (1977).
[36] P. P. Jones, *J. Exp. Med.* **146,** 1261 (1977).

sodium azide, pH 7.4, and 0.5% NP-40 detergent (0.5% NP-NET buffer). Cell suspensions are then allowed to stand at room temperature for 15 min. NP-40 extracts from the several parallel iodinations are then pooled into a 1-ml polyethylene tube to be realiquoted for subsequent precipitations to ensure that each sample initially has equal amounts of ^{125}I label. Nuclei and cellular debris are removed by centrifugation in an Eppendorf 3200 centrifuge at 3200 g for 15 min. The supernate is then divided into the appropriate number of aliquots for precipitation.

In preparation for immunoprecipitation, frozen aliquots of fixed *Staphylococcus aureus*, Cowan I strain (SaCI), prepared essentially as described by Kessler,[34] are thawed and centrifuged at 2000 g for 20 min. Pellets are resuspended by vigorous vortexing in 0.5% NP-NET buffer, recentrifuged, washed once in 0.05% NP-NET buffer, and then resuspended in one-half the volume of the original suspension (i.e., to 20% v/v) in 0.05% NP-NET buffer containing ovalbumin (1 mg/ml) and KI (0.005 M).

To aliquots of NP-40 extracts of 10^7 labeled cells, appropriate quantities of normal mouse serum, antiserum, or monoclonal antibody preparation are added. After incubation for 30 min at 4°, a volume of washed SaCI suspension is added which is twice that of the antiserum used. Samples are incubated for an additional 5 min at 4° and then centrifuged at 2000 g for 6 min, after which pellets are washed three times with ice cold 0.5% NP-NET buffer. In general, samples are preprecipitated with normal mouse serum (NMS) and SaCI as described above at least once before precipitation with either NMS or specific antiserum in preparation for polyacrylamide gel electrophoresis.

To solubilize immune complexes bound to SaCI for electrophoresis on Weber-Osborn gels,[37] each washed pellet is suspended in a solution containing SDS (4% w/w), urea (6.6 M), and Tris (0.05 M), pH 8.4 by vigorous vortexing and placed in a boiling water bath for 3 min. After removal of SaCI by centrifugation (5000 g for 6 min), samples to be reduced before electrophoresis receive 5 μl of 2-mercaptoethanol and a 0.05% solution of bromophenol blue and the boiling step is repeated. If electrophoresis is to be performed on Laemmli[38] slab gels, samples are solubilized in 0.0625 M Tris–HCl, pH 6.8, 10% (w/v) glycerol, 5% (v/v) 2-mercaptoethanol, 2.3% (w/v) SDS.

For two-dimensional analysis on nonequilibrium pH gradient electrophoresis gels (NEPHGE gels) as described by O'Farrell *et al.*,[35] immune precipitates are solubilized in 50 μl of isoelectric focusing (IEF) buffer [9.5 M urea, 2% (w/v) NP-40, 1.6% pH 5 to 7, and 0.4% pH 3 to 10 ampho-

[37] K. Weber and M. Osborn, *J. Biol. Chem.* **244**, 4406 (1969).
[38] U. K. Laemmli, *Nature (London)* **227**, 680 (1970).

lytes, 5% 2-mercaptoethanol). Samples are then electrophoresed in the first dimension in cylindrical gels at 3000 V/hr toward the cathode (−). Gels are then equilibrated in Laemmli SDS sample buffer (see above) and embedded in Laemmli 10% acrylamide slab gels with a 4.75% stacking gel for electrophoresis in the second dimension.[36] After electrophoresis for 4 hr at 20 mA, gels are fixed, stained, and destained, dried and exposed to Kodak XR film with Du-Pont Lightening-Plus intensifying screens at −70° for autoradiography for 5 to 7 days.

The labeling of surface galactosyl residues with NaB^3H_4 is carried out essentially as described by Gahmborg and Hakamori.[32] Aliquots of 10^7 thymocytes in 0.5 ml PBS are treated with 20 μl of galactose oxidase (10 units) in PBS at 22° for 2 hr. Cells are pelleted, washed once with 1 ml of PBS, and resuspended in 0.5 ml of PBS. Five microliters of a solution containing NaB^3H_4 (1 mCi) in 0.1 M NaOH is added, and the samples are incubated at room temperature for 10 min. A second 5-μl aliquot of NaB^3H_4 is added and the incubation continued for another 20 min. Cells are then pelleted by centrifugation at 140 g, washed once with 1 ml of PBS, and NP-40 extraction is performed as described above in preparation for electrophoresis on Weber-Osborn polyacrylamide gels.

For study of the Lyt-2 and Lyt-3 antigens by either ^{125}I-labeling or metabolic labeling, extracts are often subjected to anion exchange chromatography on a column of DEAE-Sephadex A-50 equilibrated with 0.005 M phosphate buffer, pH 7.0, containing 0.5% NP-40. Labeled extracts are dialyzed against the same buffer, applied to the column, and nonadherent material (approximately 25% of the applied radioactivity in the case of ^{125}I-labeled material; less for metabolically labeled material) is used for immunoprecipitation. This step allows enrichment for Lyt-2 and Lyt-3 antigens.[18]

Metabolic Labeling

Metabolic labeling with [^{35}S]methionine is performed essentially as described by Rothenberg and Triglia.[39] Aseptically prepared thymocytes are suspended in Dulbecco's modified Eagle's (DME) medium at a density of 5–8 × 10^7 cells/ml. Cells are starved for 20 min in methionine-free DME medium and then labeled in methionine-free DME medium containing 50 μM 2-mercaptoethanol and 2% dialyzed heat-inactivated fetal calf serum. [^{35}S]Methionine in sterile aqueous solution is added at 500 μCi/ml and labeling is performed for 10 min at 37°. This short labeling time is the most useful for studying the Lyt-2 and Lyt-3 antigens. Cells are then washed three times with 15 ml volumes of ice-cold PBS and extracted

---

[39] E. Rothenberg and D. Triglia, *J. Exp. Med.* **157**, 365 (1983).

with lysis buffer (PBS, pH 7.2, containing 1 mM EDTA, 1% NP-40, 0.5% sodium deoxycholate, and 0.1% SDS) at a concentration of 1–2 × 10⁸ cells/ml. Lysates are centrifuged at 100,000 g at 4° for 1 hr prior to immunoprecipitation.

Preprecipitation of metabolically labeled material is performed with 200 μl of a 30% suspension of Sepharose 4B-conjuated normal rat IgG washed in lysis buffer and then incubated for 15 min at 4° in lysis buffer containing 0.2% bovine serum albumin to block nonspecific adsorption. Two preprecipitations are performed, each for 1–2 hr. Precipitation with Sepharose 4B-coupled specific antibody is performed overnight at 4°. Precipitates are washed four times with PBS containing 1% NP-40, 0.5% sodium deoxycholate, and 0.1% SDS, and once with 10 mM Tris–HCl, pH 7.5, 0.1% NP-40. Bound antigen is released from the immunoabsorbant by heating for 10 min at 80° in sample buffer containing 2.4% SDS, 10% glycerol, 5% 2-mercaptoethanol, 0.065 M Tris–HCl, pH 6.8 prior to electrophoresis on polyacrylamide slab gels as described by Laemmli.[38]

Structural Analysis of the Lyt-1 Antigen

Conventional anti Lyt-1 serum (Table I) is used to precipitate extracts of ¹²⁵I-labeled B6-Lyt-1ᵃ (Lyt-1.1-positive) and B6 (Lyt-1.1-negative) thymocytes,[40] and the precipitated products are analyzed by SDS–PAGE. A labeled component of 67,000 (67K) M_r is observed only with Lyt-1.1-positive extracts under both reducing and nonreducing conditions, suggesting that the antigen exists as a monomer of this molecular weight. When cells are labeled with NaB³H₄ and galactose oxidase, the 67K component is again the major labeled species, but a significant amount of labeled component of M_r 87K is also specifically precipitated. It is not known whether these components are structurally related to each other.

When analyzed on two-dimensional NEPHGE polyacrylamide gels, the rat anti-Lyt-1 monoclonal antibody (53-7.3) (Table III) precipitates a protein of M_r 70K under both reducing and nonreducing conditions which exhibits extensive charge heterogeneity.[30] This antibody does not distinguish between products of the *Lyt-1ᵃ* and *Lyt-1ᵇ* alleles. An identical pattern is observed with conventional anti-Lyt-1.1 serum and ¹²⁵I-labeled B6-Lyt-1ᵃ thymocytes but not with B6 thymocytes (Lyt-1.2-positive).

The sensitivity of the 67K component to proteolytic digestion on the cell surface is monitored by treating ¹²⁵I-labeled thymocytes with trypsin, making an NP-40 extract, and precipitating with conventional anti Lyt-1.1.serum.[40] Results indicate little or no effect of trypsin treatment upon subsequent Lyt-1.1 precipitation.

[40] P. J. Durda, C. Shapiro, and P. D. Gottlieb, *J. Immunol.* **120**, 53 (1978).

Finally, Lanier *et al.*[8] have precipitated Lyt-1 from extracts of [³⁵S]methionine-labeled WEHI-55 cells, a surface IgM and Lyt-1 positive, Thy-1 and Lyt-2/3-negative murine B lymphoma. As assessed by one-dimensional SDS–PAGE, the Lyt-1 molecule from this B lymphoma appears as a single diffuse band of approximately the same average molecular weight as that seen on thymocytes. The authors suggest the possibility that glycosylation differences may exist between the B and T cell forms of Lyt-1.

Characterization of Lyt-2 and Lyt-3 Antigens

Structural studies of the Lyt-2 and Lyt-3 antigens involve radiolabeling the surface of mouse thymocytes (B6-Lyt-2ᵃ, Lyt-3ᵃ) with ¹²⁵I and lactoperoxidase (see above) (or NaB³H₄ and galactose oxidase), solubilization of the plasma membrane with NP-40, and precipitation with conventional or monoclonal antibodies.[17-19] Analysis of this material on two-dimensional gels (NEPHGE and SDS–PAGE) under reducing and nonreducing conditions led to the conclusion that three molecular species of these antigens with molecular weight of 35,000 (35K), 30,000 (30K), and <30,000 (<30K) exist. These three forms can combine to give 65,000 dalton dimers, as shown in Fig. 1.[18] ¹²⁵I-labeled material, isolated from one-dimensional SDS–PAGE gels was digested with trypsin and subjected to analysis on high-pressure liquid chromatography (HPLC). While very similar peptide maps were obtained with the 35K and 30K material, the <30K material showed a clearly different map. The results of immunoprecipitation carried out with monoclonal antibodies directed to the Lyt-2.2 (19/178) and to the Lyt-3 antigen (53-5) on mildly reduced and alkylated extracts of ¹²⁵I-labeled thymocytes, followed by one-dimensional reducing and nonreducing SDS–PAGE suggest that the Lyt-2 antigenic determinants are present on the 35K and 30K species, while the Lyt-3 determinants are present on the <30K species[19] (Fig. 1).

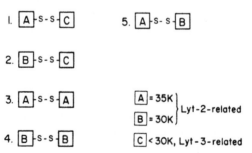

FIG. 1. Proposed species of apparent MW 65,000 bearing Lyt-2 and/or Lyt-3 antigenic determinants. All are dimers with subunits joined by disulfide bond(s). The predominant species are 1 and 2 which bear both Lyt-2 and Lyt-3 antigenic determinants.

Sequential Precipitation with Anti-Lyt-2.1 and Anti-Lyt-3.1 Sera

To explore the relationship between molecules bearing Lyt-2 and Lyt-3 antigenic determinants, sequential precipitation experiments were performed in which the effect of precipitation of extracts of ^{125}I-labeled thymocytes with one anti-Lyt serum upon subsequent precipitation with a second antiserum was explored.[17] Exhaustive precipitation of B6-Lyt-2[a], Lyt-3[a] extracts with anti-Lyt-2.1 serum slightly reduces the amount of labeled species subsequently precipitated with anti-Lyt-3.1 serum. In contrast, no labeled components are precipitated with anti-Lyt-2.1 serum after precipitation with anti-Lyt-3.1 serum. This suggested two kinds of molecular entities present in extracts—one with both Lyt-2.1 and Lyt-3.1 determinants accessible to antibody, and the other with only the Lyt-3.1 determinants accessible.

In order to further examine the physical relationship of molecules bearing Lyt-2 and Lyt-3 antigenic determinants, ^{125}I-labeled B6-Lyt-2[a], Lyt-3[a] thymocytes were treated with low concentrations of trypsin before detergent extraction and immunoprecipitation.[17] Precipitation with anti-Lyt-2.1 serum is unaffected by this treatment, whereas the quantity of ^{125}I-labeled components precipitated by anti-Lyt-3.1 serum is dramatically reduced. Results of sequential precipitation experiments performed with extracts of similarly trypsinized cells indicate *independent* precipitation of Lyt-2.1-bearing molecules and the small amount of Lyt-3.1-bearing species remaining after trypsin treatment.[17] It was therefore concluded that Lyt-3.1-bearing polypeptides are sensitive to trypsin on the intact cell, and that Lyt-2.1 antigenic determinants reside on a different polypeptide which is resistant to mild trypsin digestion on the cell surface. However, as shown in the sequential precipitations of extracts of untreated cells (see above), polypeptides bearing Lyt-2.1 and Lyt-3.1 antigenic determinants are clearly associated with each other in NP-40 extracts and probably on the surface of intact thymocytes.

Preliminary Model for Association of Lyt-2 and Lyt-3 Antigens

A preliminary model for the association of Lyt-2 and Lyt-3 molecular species is presented in Fig. 2. Of the five types of dimers shown in Fig. 1, the weight of evidence indicates that dimers 1 and 2, composed of one Lyt-2-related subunit (labeled A or B) and one Lyt-3-related subunit (labeled C) joined by (a) disulfide bond(s), are by far the predominant molecular species. While only these dimers are shown in Fig. 2, the others, if they exist, are likely to behave similarly. It is suggested that these dimers can associate in the NP-40 extract and on the cell surface to form the higher aggregates seen on polyacrylamide gels of immune precipitates

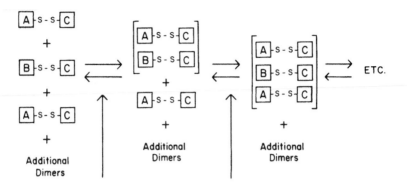

Dissociation can be enhanced by anti-Lyt-3.2 antibody

FIG. 2. Model for association-dissociation of 65,000 MW dimers bearing both Lyt-2 and Lyt-3 polypeptides into tetramers, hexamers, and higher multimers. Only species 1 and 2 of Fig. 1 are shown, but the other species, if they exist, are likely to behave similarly. It is suggested that dissociation of multimeric forms can be promoted by anti-Lyt-3.2 serum (see text).

performed without reduction.[18,19] Certain anti-Lyt-3.2 antibodies appear to promote dissociation of noncovalently aggregated dimers and thereby expose buried Lyt-2 antigenic determinants for interaction with anti-Lyt-2 antibody.

Studies on the distribution and mobility of Lyt-2 and Lyt-3 antigens on the cell surface indicate that Lyt-2 and Lyt-3 molecules detected by monoclonal antibodies (53-6.7) and (53-5) are fully mobile on the thymocyte surface, and the extent and/or strength of their oligomerization (e.g., Fig. 2) on the cell surface is not sufficient to allow them to be patched or capped by one of the antibodies tested. However, both antibodies together are sufficient to cause extensive immobilization, patching and capping.[41,42] The ability to induce patching with anti-Lyt-2 and anti-Lyt-3 antibodies simultaneously is eliminated by mild reduction and alkylation of cell surface sulfhydryl groups, or by sodium azide. However, mobility of antigen is found to be restricted in the presence of azide. This suggests that short range crosslinking is still occurring, and it is only the long range rearrangements needed for patching and capping that are inhibited by azide. The effects of reduction and alkylation are likely to reflect, at least in part, disruption of Lyt-2–Lyt-3 molecular complexes, either multimers of 65K units (if indeed these are disulfide bonded) or the individual 65K dimers themselves. This might be expected to result in ineffi-

[41] F. D. Howard, J. A. Ledbetter, D. P. Carter, L. M. Smith, and H. M. McConnell, *Mol. Immunol.* **11**, 1481 (1982).
[42] H. R. Petty, L. M. Smith, D. T. Fearon, and H. M. McConnell, *Proc. Natl. Acad. Sci. U.S.A.* **77**, 6587 (1980).

cient crosslinking and thereby inhibit patch formation. It is also possible, however, that the inhibition of patch formation reflects modification of some other cell surface component(s) by the reduction and alkylation.

APPENDIX: Listing of Ly Antigens

Ly Antigens

A large number of antigens have been defined on the surface of murine hematopoietic cells by antibodies. Many of these, like the Lyt-1, Lyt-2, and Lyt-3 alloantigens[1,2] described by Boyse, Old, and co-workers on murine T lymphocytes, are defined by alloantisera. More recently, xenoantisera and hybridomas in which the immune partner is either mouse or rat have been used to define cell surface antigens on a wide variety of hematopoietic cells. Many of these alloantisera and monoclonal antibodies have been invaluable in distinguishing and in some cases separating different functional subsets of lymphocytes, and have therefore contributed significantly to our knowledge of immune function.

The number of antigens defined on murine hematopoietic cells, excluding platelets and red blood cells, is now greater than 50, and includes antigens defined on B and T lymphocytes, macrophages, and natural killer cells. This does not include antigens encoded by the *T* and *Tla* regions of chromosome 17, plasma cell antigens, surface immunoglobulins, and antigens determined by endogenous viruses. There has been some discussion of establishing a uniform nomenclature for the approximately 50 antigens described to date on lymphocytes, natural killer cells, and macrophages and for any others defined in the future. All such antigens would be called Ly antigens, regardless of the precise cell type on which they reside, and the designation Ly would be followed by a number (e.g., Ly-1, Ly-38). Such a scheme would facilitate computer-assisted analyses of apparent new antigenic specificities as they are discovered and would introduce some order into the identification of new antigens, but would have the disadvantage that the designations would not be as informative as they are at present (e.g., Lyt-2, Mac-1).

Table IV contains a listing of murine lymphocyte antigens which investigators have generally agreed to designate as Ly antigens. Many more antigens exist on lymphocytes and other hematopoietic cells which have not been given the designation "Ly antigen," and these are not included in the present list. The information on Ly antigens which appears in Table IV is taken largely from a more complete listing of murine cell surface antigens by Flaherty and Forman.[43]

---

[43] L. Flaherty and J. Forman, *in* "Handbook of Experimental Immunology" (D. M. Weir, ed.), 3rd ed. Blackwell, Oxford (in press).

TABLE IV
MURINE Ly ANTIGENS

Ly antigen	Syn.	Similar or identical molecules[a]	Chromosome	Numbers of alleles	Monoclonal	Tissue distribution[b]	Kilodaltons	Functional marker[c]	References
Ly-1	LyA, Mu, Lyt-1		19	2	Yes	T, LN, S	67	T helper, T cytotoxic, B cell subset	1, 4, 8, 15, 40
Ly-2	LyB, Lyt-2		6	2	Yes	T, LN, S	30, 35	T cytotoxic/suppressor	1, 4, 17–19
Ly-3	LyC, Lyt-3		6	2	Yes	S, T, LN	<30	T cytotoxic/suppressor	1, 2, 4, 17–19
Ly-4		H-3, β_2M, Lym-11	2	2		E?, T?, LN S, O?			47–49[g]
Ly-5	Lyt-4	T200, B220	1	2	Yes	T, LN, S, O	200, 205, 220	Hematopoietic cells	50–56
Ly-6		Ala-1, Ly-8	2 or 9[d]	2	Yes	T, LN, S, O	33.5	Some T, some B	57–64
Ly-7		Ly-7, Ala-1 DAG	2 or 9	3	Yes	T, LN, S			65–67
Ly-8				2		T, LN, S, O			60, 61
Ly-9	T100, Lgp100		1	2	Yes	T, LN, S	100		44–46, 68
Lym-10			19	2	Yes	T, LN, S, O		Some T, some B	69
Ly-11(a)	Ly-10		2	2		T, LN, S		Prothymocytes, NK	70–72

Antigen	Similar molecules[a]	Chromosome	No.	Alloantigen	Distribution[b]	Mol. wt.	Cell type	References
Lym-11(b)	M, H-3, Ly-4	2	2	Yes	E?, T, LN, S, O	12		49,73
Ly-11(c)			2		T, LN, S, O	18–22?		74
Ly-12			2		T, LN, S			74
Ly-13			2		E, T, LN, S, O			74
Ly-14		2	2		T, LN, S, O	73?		74
Ly-15	Ly-18(a)	12	2	Yes	T, LN, S, O			75
Ly-16		1	2		S?		T cytotoxic	76
Ly-17	Lym-20				T, LN, S, O			77
Ly-18	Lym-18(b)	12	2	Yes	T, LN, S, O			78
Ly-19(a)		4	2		T, LN, S			79
Lym-19(b)			2	Yes	T, LN, S, O			80
Lym-20	Ly-17, Lym-1, M1s	1	2	Yes	T, LN, S, O		B cells?	81
Lym-21		7	2	Yes	T, LN, S, O		T cells	82
Ly-22(a)		4	2		T, LN, S			83
Lym-22(b)		1	2	Yes	T, LN, S			84
Lyb-2	Lyb-4, Lyb-6	4	3		T?, LN, S	40–50	B cells?	85–87
Lyb-3	Lyb-5	X?	2		LN, S	68	B cells	88, 89
Lyb-4	Lyb-2, Lyb-6	4	2		LN, S	44?	B cells?	90, 91
Lyb-5	Lyb-3	X?	2		LN, S		B cells	92
Lyb-6	Lyb-2, Lyb-4	4	2		LN, S	44?	B cells	91
Lyb-7		12	2	Yes	LN, S		B cells	93
Lyb-8		7	2		LN, S	95	B cells	94
LyM-1	M1s	1	3?		LN, S		B cells?	95, 96

[a] Molecules or alloantigens which have similar genetic and/or biochemical properties.
[b] E, Erythrocytes; T, thymus; LN, lymph node; S, spleen and/or bone marrow; O, other.

(continued)

TABLE IV (*continued*)

c Only alloantigens which have been used to separate functionally distinct subpopulations of hematopoietic cells are indicated. Separation could be either by selective elimination by complement or by cell sorting.

d There is currently some debate in the literature concerning the chromosomal position of the Ly-6 family of alloantigens including Ly-6, Ly-8, DAG, Ala-1, H9/25, and ThB.[97,98] They are tightly linked on either chromosome 2 or 9.

e Only on activated spleen cells.

f Lyb-3 and Lyb-5 are missing in the CBA/N strain of mice with a defective X-linked gene. The structural gene controlling these antigens has not been mapped.

g References 47–98:

47 G. Snell, M. Cherry, I. McKenzie, and D. Bailey, *Proc. Natl. Acad. Sci. U.S.A.* **70**, 1108 (1973).

48 J. Michaelson, E. Rothenberg, and E. A. Boyse, *Immunogenetics* **11**, 93 (1980).

49 N. Tada, S. Kimura, A. Hatzfeld, and U. Hammerling, *Immunogenetics* **11**, 441 (1980).

50 K. Komuro, E. Itakura, E. A. Boyse, and M. John, *Immunogenetics* **1**, 452 (1975).

51 M. P. Scheid and D. Triglia, *Immunogenetics* **9**, 423 (1979).

52 J. Michaelson, M. P. Scheid, and E. A. Boyse, *Immunogenetics* **9**, 193 (1979).

53 A. W. Siadak and R. C. Nowinski, *J. Immunol.* **125**, 1400 (1980).

54 M. B. Omary, I. S. Trowbridge, and M. P. Scheid, *J. Exp. Med.* **151**, 1311 (1980).

55 R. L. Coffman and I. L. Weissman, *Nature (London)* **289**, 681 (1981).

56 F.-W. Shen, *in* "Monoclonal Antibodies and T-Cell Hybridomas" (G. J. Hammerling, U. Hammerling, and J. F. Kearney, eds.), p. 25. Elsevier/North-Holland Biomedical Press, Amsterdam, 1981.

57 I. F. C. McKenzie, M. Cherry, and G. D. Snell, *Immunogenetics* **5**, 25 (1977).

58 A. J. Feeney and U. Hammerling, *Immunogenetics* **3**, 369 (1976).

59 A. J. Feeney, *Immunogenetics* **7**, 537 (1978).

60 J. A. Frelinger and D. B. Murphy, *Immunogenetics* **3**, 481 (1976).

61 M. A. Horton and J. A. Sachs, *Immunogenetics* **9**, 273 (1979).

62 F. Takei, G. Galfre, T. Alderson, E. S. Lennox, and C. Milstein, *Eur. J. Immunol.* **10**, 241 (1980).

63 H. Auchinclos, K. Ozato, and D. H. Sachs, *J. Immunol.* **127**, 1839 (1981).

64 A. Matossian-Rogers, P. Rogers, J. A. Ledbetter, and L. A. Herzenberg, *Immunogenetics* **15**, 591 (1982).

65 I. F. C. McKenzie, J. Gardiner, M. Cherry, and G. D. Snell, *Transplant. Proc.* **9**, 667 (1977).

66 L. L. Lanier and N. L. Warner, *Hybridoma* **1**, 227 (1982).

[67] L. L. Lanier and N. L. Warner, *Hybridoma* **2**, 177 (1983).

[68] P. M. Hogarth, J. Craig, and I. F. C. McKenzie, *Immunogenetics* **11**, 65 (1980).

[69] S. Kimura, N. Tada, and U. Hammerling, *Immunogenetics* **10**, 363 (1980).

[70] D. Meruelo, A. Paolino, N. Flieger, and M. Offer, *J. Immunol.* **125**, 2713 (1980).

[71] D. Meruelo, A. Paolino, N. Flieger, J. Dworkin, M. Offer, N. Hirayoma, and Z. Ovary, *J. Immunol.* **125**, 2719 (1980).

[72] D. Meruelo, A. Paolino, N. Flieger, and J. Dworkin, *Fed. Proc., Fed. Am. Soc. Exp. Biol.* **39**, 798 (1980).

[73] K. Tomonari, N. Tada, S. Kimura, U. Hammerling, and E. Weksler, *Immunogenetics* **15**, 605 (1982).

[74] T. A. Potter and I. F. C. McKenzie, *Immunogenetics* **12**, 351 (1981).

[75] T. A. Potter, P. M. Hogarth, and I. F. C. McKenzie, *Transplantation* **31**, 339 (1981).

[76] A. Finnegan and F. L. Owen, *J. Immunol.* **127**, 1947 (1981).

[77] F.-W. Shen and E. A. Boyse, *Immunogenetics* **11**, 315 (1980).

[78] S. Kimura, N. Tada, Y. Liu, and U. Hammerling, *Immunogenetics* **12**, 547 (1981).

[79] A. Finnegan and F. L. Owen, *ICN-UCLA Symp. Immunoglobulin Idiotypes and Their Expression Abstract*, p. 51 (1981).

[80] N. Tada, S. Kimura, Y. Liu, B. Taylor, and U. Hammerling, *Immunogenetics* **13**, 539 (1981).

[81] S. Kimura, N. Tada, E. Nakayama, Y. Liu, and U. Hammerling, *Immunogenetics* **14**, 3 (1981).

[82] J. Kinnard and D. Meruelo, *Immunogenetics* **15**, 239 (1982).

[83] D. Meruelo, M. Offer, and N. Flieger, *J. Immunol.* **130**, 946 (1983).

[84] N. Tada, S. Kimura, L. Liu-Lam, and U. Hammerling, *Hybridoma* **2**, 29 (1983).

[85] H. Sato and E. A. Boyse, *Immunogenetics* **3**, 565 (1976).

[86] J. S. Tung, J. Michaelson, H. Sato, E. Vitetta, and E. A. Boyse, *Immunogenetics* **5**, 485 (1977).

[87] F.-W. Shen, M. Spanondis, and E. A. Boyse, *Immunogenetics* **5**, 481 (1977).

[88] B. Huber, R. K. Gershon, and H. Cantor, *J. Exp. Med.* **145**, 10 (1977).

[89] R. E. Cone, B. Huber, H. Cantor, and R. K. Gershon, *J. Immunol.* **120**, 1733 (1978).

[90] J. G. Freund, A. Ahmed, R. E. Budd, M. E. Dorf, K. W. Sell, W. E. Vannier, and R. E. Humphreys, *J. Immunol.* **117**, 1903 (1976).

[91] S. Kessler, A. Ahmed, and I. Scher, *in* "B Lymphocytes in the Immune Response" (M. Cooper, D. E. Mosier, I. Scher, E. Vitetta, eds.), p. 47. Elsevier/North-Holland Biomedical Press, Amsterdam, 1979.

[92] A. Ahmed, I. Scher, S. O. Sharrow, A. H. Smith, W. E. Paul, D. H. Sachs, and K. W. Sell, *J. Exp. Med.* **145**, 101 (1977).

[93] B. Subbarao, A. Ahmed, W. E. Paul, I. Scher, R. Lieberman, and D. E. Mosier, *J. Immunol.* **122**, 2279 (1979).

[94] F. W. Symington, B. Subbarao, D. E. Mosier, and J. Sprent, *Immunogenetics* **16**, 381 (1982).

[95] S. L. Tonkonogy and H. J. Winn, *Immunogenetics* **5**, 57 (1977).

[96] H. Sato, S. Kimura, and K. Itakura, *J. Immunol.* **8**, 27 (1981).

[97] M. A. Horton and C. M. Hetherington, *Immunogenetics* **11**, 521 (1980).

[98] D. Meruelo, M. Offer, and A. Rossomando, *Proc. Natl. Acad. Sci. U.S.A.* **79**, 7460 (1982).

Antibody Reagents

Alloantisera and xenoantisera thought to be specific for a particular antigen often contain autoantibodies and other contaminating antibodies to unsuspected specificities such as viral components and alloantigens. Absorption tests as well use of negative controls, preferably from a congenic strain, are necessary to confirm the specificity of such antisera. Different pools of antisera may have different amounts and specificities of contaminating antibodies. Thus care must be exercised in the use of such antisera.

Monoclonal antibodies can be produced in large quantities, and they have the advantage that laboratories around the world can produce and employ identical reagents, presumably for an indefinite period of time. This is in contrast to alloantisera which generally are produced in limited quantity, require absorptions, differ from batch to batch, and must be continually produced and screened for specificity. However, monoclonal antibodies can also exhibit major cross-reactions, which are not abolished by absorption. Care must therefore be exercised in the choice of monoclonal antibody to be employed and in its application.

Molecular Characterization of Ly Antigens

Antibodies have been used as specific reagents to isolate the antigenic molecules to which they bind. In most cases, these molecules have proved to be glycoproteins. Isolation of these molecules in this manner has provided information on their subunit structure, chemical and physical properties, and, in some instances, allowed determination of their amino acid sequence.

Different Antibody Reagents Often Recognize the Same Antigen

In a number of instances, different investigators have produced or otherwise identified antibodies which recognized the same antigenic species in different ways—for example, in the case of Ly-9, by two different alloantisera[44,45] and a rat-mouse hybridoma.[46] Initially, such situations have led to some confusion since the investigators assign different names to their antigens when, in fact, they are dealing with the same antigen. However, tissue distribution, representation on T and B lymphocytes,

[44] P. J. Durda, S. C. Boos, and P. D. Gottlieb, *J. Immunol.* **122,** 1407 (1979).

[45] B. J. Mathieson, S. O. Sharrow, K. Bottomly, and B. J. Fowlkes, *J. Immunol.* **125,** 2127 (1980).

[46] J. A. Ledbetter, J. W. Goding, T. T. Tsu, and L. A. Herzenberg, *Immunogenetics* **8,** 347 (1979).

biochemical characterization, strain distribution (for alloantigens), and genetic crosses and backcrosses will reveal the identity of the two independently identified antigens, and the investigators then agree upon a single designation.

Acknowledgments

Work described in this review and preparation of the manuscript were sponsored by Public Health Service Awards CA15808 and CA30147 from the National Cancer Institute to Paul D. Gottlieb, by Grant CA14051 from the National Cancer Institute to Massachusetts Institute of Technology Center for Cancer Research, by American Cancer Society Grant IM-113 to Paul D. Gottlieb, and by Robert A. Welch Foundation Grant F-884 to Paul D. Gottlieb. Edward B. Reilly was supported by a National Needs Postdoctoral Fellowship SP-791888 from the National Science Foundation. Malgorzata Niezgodka was supported by grant F-884 from the Robert A. Welch Foundation.

We are grateful to Dr. Lorraine Flaherty and Dr. James Forman for providing us with their listing of murine cell surface antigens before publication.

[60] Detection by Immunochemical Techniques of Cell Surface Markers on Epidermal Langerhans Cells

By PAUL R. BERGSTRESSER and DOLORES V. JUAREZ

Introduction

Mammalian epidermis consists of three major cell populations: Keratinocytes, melanocytes, and Langerhans cells. Within the last decade considerable attention has been paid to the functional properties of epidermal Langerhans cells, the least well characterized of these three, and in so doing the hypothesis that these dendritic bone marrow derived cells perform critical antigen presenting function in skin has been tested.[1,2] The evidence supporting this hypothesis is beyond the scope of this review, but the experimental strategies used for its development have required reliable methods for Langerhans cell identification. These strategies have included enzymatic disaggregation of epidermis, followed by *in vitro* separation procedures designed to produce cell suspensions which are relatively enriched and relatively depleted of Langerhans cells.[3] In

---

[1] G. B. Toews, P. R. Bergstresser, and J. W. Streilein, *J. Immunol.* **124**, 445 (1980).

[2] G. Stingl, S. I. Katz, L. Clement, I. Green, and E. M. Shevach, *J. Immunol.* **121**, 2005 (1978).

[3] S. Sullivan, P. R. Bergstresser, and J. W. Streilein, *J. Invest. Dermatol.* **80**, 343 (abstr.) (1983).

addition, the emigration of Langerhans cells into allografts of skin have required, on a genetic basis, the identification of the source (host or donor) of such cells.[4] And finally, we have employed several detection techniques to identify cutaneous sites which are naturally depleted of Langerhans cells and experimental procedures which deplete epidermis of normal Langerhans cells.[5]

Specimen Preparation

Whole Epidermis. The enumeration of Langerhans cells as they occur *in vivo* has been facilitated greatly by techniques in which whole mounts of epidermis are examined "en face" by light or fluorescence microscopy. For this work we have used succesfully[5] the method of Baumberger *et al.*,[6] in which the chelation of divalent cations by EDTA leads to a selective separation of dermis from epidermis. This procedure employs a buffered EDTA medium consisting of 116 mM NaCl, 2.6 mM KCl, 8 mM Na_2HPO_4, 1.4 mM KH_2PO_4, with 8.2 g/liter added $Na_4EDTA \cdot 2H_2O$ (20 mM). Just prior to use, the pH of this medium is adjusted to 7.3 with CO_2 gassing. For *human skin,* razor shaved, split thickness skin specimens are prepared with a commercial dermatome or keratotome such as the Castroviejo (Storz Instrument Co., St. Louis, MO). Alternatively, punch or excisional skin biopsies may be obtained with standard surgical techniques and then a surgical separation made through the dermis in a free hand fashion.[7] Finally, we have utilized epidermal blisters directly for Langerhans cell identification.[8] For each of the first two techniques, dermal thickness must be kept to a minimum to facilitate penetration of EDTA through the dermis to the dermal–epidermal junction. This will ensure more rapid and more even separation. For *rodent skin,* razor-shaved abdominal wall, tail, or ear skin is used after excision from underlying tissues. We have occasionally employed the keratotome to prepare rodent skin for separation, although the dermis in these species is frequently thin enough to permit adequate penetration of EDTA after surgical excision alone.

To separate dermis from epidermis, small specimens of skin, up to 2.5 × 2.5 cm, are incubated for 2–4 hr in a Petri dish containing the EDTA medium, placed in a 5% CO_2 incubator (37°). Gentle agitation within the

[4] J. W. Streilein, L. W. Lonsberry, and P. R. Bergstresser, *J. Exp. Med.* **155,** 863 (1983).

[5] P. R. Bergstresser, C. R. Fletcher, and J. W. Streilein, *J. Invest. Dermatol.* **74,** 77 (1980).

[6] J. P. Baumberger, V. Suntzeff, and E. V. Cowdry, *J. Natl. Cancer Inst. (U.S.)* **2,** 413 (1942).

[7] H. Beerman, *in* "Skin Surgery" 3rd ed. (E. Epstein, ed.), p. 82. Thomas, Springfield, Illinois, 1970.

[8] R. D. Sontheimer and P. R. Bergstresser, *J. Invest. Dermatol.* **79,** 237 (1982).

incubator or periodic swirling of the Petri dish will facilitate the process of separation. At the appropriate time, each specimen is grasped gently by the dermal surface with fine forceps such as those used in handling specimens for electron microscopy. Overlying epidermis is grasped with a second pair of forceps at the most lateral edge and carefully folded back as a single sheet. When the incubation is insufficient, epidermis will fail to separate or will tend to shred off in small pieces. When this occurs, specimens should be incubated for additional time. We have observed that species and anatomic sites having numerous hairs, such as murine abdominal wall skin, are quite difficult to separate. More uniform results have been obtained by applying a thin layer of cyanoacrylic cement (Loctite Group; Cleveland, OH) to the dry epidermal surface prior to incubation.[5] This prevents hydration of the stratum corneum during incubation and provides rigid structural support for the epidermis during separation. After separation, specimens are ready for Langerhans cell identification procedures.

Epidermal Cell Suspensions. Numerous strategies for the identification of Langerhans cell function use single-cell preparations.[2,3] To obtain epidermal cells from murine truncal skin, the following procedure has been employed with success. Reagents include (1) 0.3% trypsin in GNK buffer: 880 mg NaCl, 40 mg KCl, 100 mg glucose, 95 ml distilled water; 0.3 g of trypsin is then added and the pH adjusted to 7.6 with 7.5% (w/v) $NaHCO_3$ in distilled water. Final volume is brought to 100 ml; (2) bovine pancreatic DNase, Type 1, salt free (ICN, Nutritional Biochemicals, Cleveland, OH) 10 mg in 10 ml sterile 6.5 mM phosphate-buffered saline (PBS); (3) minimal essential medium (MEM) (Eagle) (320-1575; Gibco Laboratories, Grand Island NY) with 10% heat-inactivated fetal calf serum (FCS) (230-6140; Gibco).

Preparation of disaggregated epidermal cells from *rodent skin* is accomplished with the following procedure. (1) The rodent (which in our studies is most frequently a mouse) is sacrificed with ether anesthesia and the abdominal hair is carefully removed with a dry razor. (2) The mouse skin is soaked in PBS for 5 min to hydrate the stratum corneum and then is pinned, with abdomen exposed, to a wooden board. (3) Cellophane tape (Scotch #810; 3M, St. Paul, MN) is applied and removed from the shaved abdominal wall site until the surface glistens (5 to 10 applications), and then the last tape application is left in place to prevent desiccation of the exposed epidermal surface. (4) The skin is excised and dissected away from underlying fat, removing excess fat by blunt dissection. (5) The tape is removed and 4 × 4-cm specimens of skin are incubated in 10 ml of trypsin solution at 37° for 30 to 60 min with periodic gentle agitation. (6) DNase solution (1.5 ml) is added and the incubation is continued for 5–10 additional minutes. (7) The entire suspension is shaken to release epidermal cells from the residual dermis, the dermis is removed and 10 ml of MEM with 10% FCS is added. (8) The cell suspension is pipetted gently to break up aggregates and then filtered through a Dispo-culture tube (Sera Separa #208-3084-020, Evergreen Co., Los Angeles, CA). (9) Cells are washed in MEM with 10% FCS twice at 400 g for 10 min (4°) and resuspended in 10 ml of media. (10) Cells are counted in a hemocytometer and viability is assessed by trypan blue exclusion (630-5250; Gibco).

To obtain single-cell preparations of epidermal cells from *human skin,* one begins with a keratotome-prepared split thickness specimen which is then cut into 2.5 × 2.5-cm squares. Once again, dermal thickness will limit the rate of penetration of trypsin through the dermis so that the manner in which the specimen is obtained will have a significant influence on epidermal cell recoveries. To disaggregate epidermal cells from human skin one begins at Step 5 above and continue through Step 10. However, at Step 7 the remaining dermis as well as the remaining sheet of stratum corneum must be removed from the cell suspension. in order to identify and classify epidermal cells recovered during disaggregation, they are resuspended at 2.5×10^5/ml. After this, 1×10^5 cells in 0.4 ml are centrifuged onto acetone-cleaned glass slides (600 rpm; 5 min) (Cytospin; Shandon, Sewickley, PA). Cells are allowed to air dry thoroughly and then fixed in acetone for 30 min at room temperature prior to the staining procedure.

Cryostat Sections. Langerhans cells were first identified by histochemical techniques which employed cryostat preparations of frozen skin.[9] In recent years, however, the use of frozen sections has been replaced to a large extent by studies in which whole mounts of epidermis are used. In whole mounts, greater numbers of cells may be seen, surface densities may be counted without statistical correction,[10] and the full outline of dendritic Langerhans cells may be seen. To prepare sections of skin, 4 × 4-mm specimens of whole skin are placed into embedding medium (Tissue-Tek II; Lab-Tek Products, Naperville, IL) and then frozen at −20° in a cryostat. Sections measuring between 4 and 20 μm in thickness are prepared with the microtome set at −20°. Prepared sections are placed on acetone-cleaned glass microscope slides and allowed to air dry thoroughly prior to the staining procedure.

Adenosine Triphosphatase

Epidermal Langerhans cells possess relatively high amounts of cell surface nucleotide phosphatases and the presence of adenosine triphos-

[9] K. Wolff, *Arch. Klin. Exp. Dermatol.* **218,** 446 (1964).

[10] M. Abercrombie, *Anat. Rec.* **94,** 239 (1946).

phatase (ATPase) on Langerhans cells has been used to distinguish them from other epidermal cells.[9,11] The majority of investigators who have worked with cell surface ATPase have used techniques derived from the original procedure of Wachstein and Meisel,[12] and they have used cryostat-prepared sections of frozen skin. In this technique, adenosine triphosphate (ATP) at physiologic pH is hydrolyzed by cell surface ATPase in the presence of $Pb(NO_3)_2$. Released phosphate precipitates with Pb^{2+} as white $Pb_3(PO_4)_2$. For histochemical purposes, specimens are then incubated in $(NH_4)_2S$ to substitute PbS for $Pb_3(PO_4)_2$, producing a black to brown precipitate which identifies the site of enzyme activity.

Fixation and identification procedures for ATPase, except for the time of incubation, are independent of the type of preparation. We have followed the method of Juhlin and Shelley[13] as adapted from Wachstein and Meisel.[12] Buffer A (wash solution) consists of 0.2 M mono[tris(hydroxymethyl)aminomethane] maleate (Trismal maleate; Sigma), pH 7.3 with 6.87% (w/v) added sucrose. Buffer (B) used during ATP hydrolysis is identical except that it contains 8.55% (w/v) added sucrose. Cacodylate-formaldehyde fixation solution consists of 6.85% (w/v) sucrose and 4% formaldehyde in 0.08 M cacodylic acid. The ATP–$Pb(NO_3)_2$ solution consists of 42 ml of buffer B with 10 mg added ATP (Sigma), 5 ml of 5% (w/v) $MgSO_4$, and 3 ml of 2% (w/v) $Pb(NO_3)_2$ all added just prior to use. The ammonium sulfide solution is a 1 : 50 dilution of 22% $(NH_4)_2S$ in distilled water.

For staining tissue, either as whole specimens of epidermis or as cytocentrifuged cells on a microscope slide, the following procedure is employed.

1. The specimen is washed in three changes of buffer A for a total of 20 min (4°).
2. Fixed in cacodylic acid-formaldehyde for 20 min (4°).
3. Washed in three changes of buffer A for a total of 30 min (4°).
4. Incubated in freshly prepared ATP–$Pb(NO_3)_2$ solution for about 30 min (37°). The time of incubation may range from 15 to 60 min as determined by experience.
5. Washed in three changes of buffer A for a total of 20 min (23°).
6. The color is developed in $(NH_4)_2S$ solution for 5 min (23°).
7. Washed in three changes of distilled water for 10 min (23°).

Specimens of epidermis may be mounted, dermal aspect up, on a glass microscope slide and covered with a 90%/10% solution of glycerol/PBS,

[11] K. Wolff and R. K. Winkelmann, *J. Invest. Dermatol.* **48**, 50 (1967).
[12] M. Wachstein and E. Meisel, *Am. J. Clin. Pathol.* **27**, 13 (1957).
[13] L. Juhlin and W. B. Shelley, *Acta Derm. Venereol.* **57**, 289 (1977).

for examination by light microscopy. Alternatively, specimens may be dehydrated through graded alcohols,[14] followed by xylene and mounted with Permount (Fisher Scientific Co., Fair Lane, NJ). Cytocentrifuged cells are covered with a drop of glycerol/PBS followed by a cover slip to be followed by examination by light microscopy. Alternatively, the specimen is allowed to dry after the first wash, covered with about 0.02 ml of xylene followed by a single drop of Permount and then a cover slip. Specimens mounted in glycerol will fade over several days while those mounted in Permount will last without change for up to 5 years. Xylene-based mounting media such as Permount produce minimal light scattering and will consequently make unstained cells in cytocentrifuge preparations difficult to see. This can be corrected by the judicious use of 0.01% w/v toluidine blue in distilled water as a counterstain, followed by a brief wash in distilled water prior to drying and Permount application.

Cyanoacrylic cement, when too thick, will make stained epidermal specimens difficult to manipulate. The cement may be removed after staining and before dehydration by washing in two changes of acetone. From studies on the substrate specificity of Langerhans cell surface adenosine phosphatases,[15] it has been concluded that adenosine diphosphate (ADP) is a more specific substrate for the identification of Langerhans cells and we frequently use it rather than ATP, particularly when background staining is excessive.

When whole mounts of normal epidermis are examined by light microscopy, ATPase-positive Langerhans cells are seen as having a dendritic configuration with two to five cytoplasmic processes radiating out from central cell bodies. In some preparations these processes may extend for distances many times greater than one keratinocyte diameter. In suspensions of cells however, all cells, including Langerhans cells, become rounded. Those Langerhans cells which remain associated with small clusters of epidermal cells tend to envelop adjacent cells, demonstrating the strong affinity between these disparate cell types.

Ia Antigens

Langerhans cells, exclusively among epidermal cells, bear cell surface proteins and surface receptors usually found on cells of macrophage/monocyte lineage. Most important for their identification is the presence of Class II alloantigens. In mice Langerhans cells bear Ia antigens and in

---

[14] P. R. Bergstresser, R. J. Pariser, and J. R. Taylor, *J. Invest. Dermatol.* **70**, 280 (1978).
[15] B. Chaker, M. D. Tharp, and R. P. Bergstresser, *J. Invest. Dermatol.* **82**, 496 (1984).

SELECTED IMMUNOREAGENTS EMPLOYED FOR
LANGERHANS CELL IDENTIFICATION

Reagent	Conjugate	Dilution[a]
Monoclonal anti-mouse		
Ia^k(2) (BD)[b]	None or biotin	1 : 20
Ia^d (BD)	None	1 : 20
Thy-1.2^c (NEN)	None	1 : 100
Thy-1.2 (BD)	Biotin	1 : 20
Monoclonal anti-human		
HLA-DR (BD)	None	1 : 30
T6 (O)	None	1 : 30
Avidin (BD)	Fluorescein	1 : 100
	Rhodamine	1 : 100
Goat anti-mouse		
IgG (7S) (M)	Fluorescein	1 : 10

[a] Reagents diluted in 6.5 mM phosphate-buffered saline.

[b] Manufacturers: BD, Becton-Dickinson Monoclonal Center, Inc.; NEN, New England Nuclear; O, Ortho Diagnostics; M, Meloy Laboratories, Inc.

[c] Thy-1 antigen does not occur on Langerhans cells but may be found on recently observed dendritic cells population in murine epidermis.[25]

guinea pigs they bear B cell alloantigens.[16–18] Investigators first made use of alloantisera for the detection of Class II determinants, but more recently monoclonal reagents have been employed for this purpose. When conducting such studies, care should be exercised in identifying Ia-bearing cells in skin which has been manipulated, since reports now indicate that keratinocytes will express their own Ia antigens in certain immunologically mediated inflammatory conditions.[19–21] The procedure described below which identifies Langerhans cells on the basis of cell surface Ia antigens uses selected commercial monoclonal antibodies as immunoreagents since commercial reagents are frequently well characterized and are available to all investigators. The table lists some of the reagents used

[16] G. Rowden, T. M. Philips, and T. L. Delovitch, *Immunogenetics* **7**, 465 (1978).

[17] G. Stingl, S. I. Katz, L. D. Abelson, and D. L. Mann, *J. Immunol.* **120**, 661 (1978).

[18] G. Stingl, S. I. Katz, E M. Shevach, E. Wolff-Schreiner, and I. Green, *J. Immunol.* **120**, 570 (1978).

[19] S. M. Breathnach and S. I. Katz, *J. Invest. Dermatol.* **80**, 314 (abstr.) (1983).

[20] G. G. Krueger, M. Eman, L. K. Roberts, and R. A. Daynes, *J. Invest. Dermatol.* **80**, 315 (abstr.) (1983).

[21] B. Volc-Platzer, O. Majdic, K. Wolff, W. Knapp, and G. Stingl, *J. Invest. Dermatol.* **80**, 315 (abstr.) (1983).

successfully in our laboratory, although there are obviously other sources.

Whole mounts of epidermis are prepared as described previously. For *murine skin* from mice of H-2^k haplotype such as C3H, the following two reagents are used. (1) Anti-Ia^k (specificity 2) (Becton-Dickinson, Sunnyvale, CA) and (2) goat anti-mouse IgG (7 S) (Meloy Laboratories; Springfield, VA) The procedure follows. (1) Epidermal specimens that are approximately 4×4 mm in size are fixed in 2 ml of acetone for 30 min (23°). Those specimens which were separated with the aid of cyanoacrylic cement will require several changes of acetone to remove all residual cement. (2) Using a small glass vial, specimens are washed three times in 2 ml PBS for a total of 30 min (4°). (3) They are then incubated overnight in 200 μl of a 1 : 20 dilution of monoclonal anti-Ia in PBS (4°). (4) Each is then washed three times in 2 ml PBS for a total of 90 min on a rocker platform (23°). (5) Specimen are incubated for 2 hr on a rocker platform in 200 μl of a 1 : 10 dilution of goat anti-mouse IgG (23°). (6) They are washed three times on a rocker platform in PBS for a total of 90 min (23°). (7) Specimens are finally mounted, dermal side up, on microscope slides in a 90%/10% solution of glycerol/PBS. For *human skin* the staining procedure is identical aside from the requirement for different reagents.[22,23]

Specimens of epidermis or epidermal cells which have been treated with fluorescent dye-conjugate immunoreagents should be examined with a fluorescence microscope equipped for epiillumination. Langerhans cells are identified in epidermis by the dendritic configurations of their cell processes, although they may also exhibit an intense spot of fluorescence at a perinuclear location within the cytoplasm. Prepared specimens of whole epidermis will remain readable for several days and may last as long as several weeks when stored at 4° in the dark. We have occasionally used successfully the recommendation to include p-phenylenediamine (1 mg/ml) in mounting media to prevent ultraviolet light-induced fading of fluorescein under the microscope.[24] In our experience, biotin-conjugated primary reagents followed by fluorescein, rhodamine, or horseradish peroxidase-labeled avidin (not described) provide greater specificity with less background contamination than do conjugated antiimmunoglobulins.

Disaggregated epidermal cells and sections of skin may be stained directly on microscope slides in the same manner. The procedure is identical except that the time of incubation in the monoclonal reagent is shortened to 30 min. Slides are incubated flat in a plastic humid shaker to prevent drying of reagents. Movement of reagents across the slide can be

[22] R. D. Sontheimer and P. R. Bergstresser, *J. Invest. Dermatol.* **79**, 237 (1983).
[23] C. A. Elmets, P. R. Bergstresser, and J. W. Streilein, *J. Invest. Dermatol.* **79**, 340 (1982).
[24] J. C. Huff, W. L. Weston, and K. D. Wanda, *J. Invest. Dermatol.* **78**, 449 (1982).

prevented by encircling the specimen with a wax pencil or a diamond pen. Among disaggregated cells the intensity of fluorescence is frequently diminished significantly from that seen within the epidermis. Fluorescence fades within several days on disaggregated cells.

Comment

It is important to recognize that the detection of Ia antigens or ATPase activity on subpopulations of epidermal cells from unperturbed skin identifies rather reliably those cells which are Langerhans cells or other cell types.[25] On the other hand, the absence of these immunochemical markers does not establish the absence of such cells.

Acknowledgments

This work was supported in part by USPHS Grant AI17363. We gratefully acknowledge the expert assistance of Mrs. Betty Janes in preparing this manuscript.

[25] P. R. Bergstresser, R. E. Tigelaar, J. H. Dees, and J. W. Streilein, *J. Invest. Dermatol.* **81,** 286 (1983).

Author Index

Numbers in parentheses are footnote reference numbers and indicate that an author's work is referred to although the name is not cited in the text.

Subject Index

A

preparation, for antiglobulin rosetting
reactions, 390, 391
protein A coated, 408, 409
sheep, antibody coupling, 620
SpA-coated
applications in detection of antigens,
413
detection of surface IgG, 411–413
Esterase stain
of Kupffer cells, 291, 292
preparation, 291, 292

F

F9 teratocarcinoma cell, 503
Fab', 649
F(ab')$_2$, 647–649
in immunofluorescence, 384
of membrane components, 425
region, protein A binding site, 470–472
Fc receptor, fluorescent antibody binding,
425
Ferritin, linked to protein A, 408
Fibroblast, monolayers, 126
Ficoll, density separation of lymphocytes
on, 105
Ficoll-Hypaque, 143
separation, of human blood, 244, 245
Fingerprint, of cell suspensions, 86, 87
Flow cytometry, 197, 198, 303, 626. *See
also* Fluorescence-activated cell
sorting
excitation sources, 213
of fluorescent dextran uptake, 345–347
fluorescent sources, 208
instrumentation, 199
photomultiplier tube, 212, 214
Flow microfluorimetry, 170
Fluid pinocytosis, 336–347
fluorescent dextran marker, 337–339
horseradish peroxidase marker, 337–
339
substrate choice, 336, 337
Fluorescein, 209, 210
plus phycoerythrin, plus allophycocyanin, immunofluorescence
measurement system, 215, 216
plus propidium iodide
immunofluorescence measurement
system, 214–216
plus Texas Red, immunofluores-

cence measurement system,
215, 216
Fluorescein isothiocyanate, 414
conjugates
characterization, 420, 421
fluorochrome concentration, 420
fluorochrome/protein ratio, 420, 421
protein concentration, 420
purification, 419
storage, 421
coupling to protein, 417–419
coupling to protein A, 407
Fluorescence, patterns, 423
caps, 372–375
classification, 372–375
ring, 372, 373
spots/patches, 327, 373
Fluorescence-activated cell sorter, 154, 170
separation of thymocytes with PNA in,
178, 179
Fluorescence-activated cell sorting, 197–
241, 248. *See also* Flow cytometry
amplification of gene for dihydrofolate
reductase, 237, 238
analysis of HLA mRNA in trophoblast
cells, 240, 241
analytical data, 219–225
applications, in lymphoid cell analysis,
232–237
auxiliary information, 223
cell sorter, 200–202
aligning, 219
basic components, 200–203
elaboration and variation, 203, 204
tuning, 219
with more excitation sources and
detectors, 203, 204
cell-staining procedures, 208–212
cell-staining reagents, 208–212
characterization of lymphoid cell subpopulations, 232–235
data processing, 219–225
data recovery, 223
data storage format, 222
detection of Ia-A antigens, 559
detection of small numbers of monoclonal B lymphocytes in human
blood, 237
detectors, 212–214
dual immunofluorescence measurements, 234

Isopaque-Ficoll technique, of cell separation, 91–95

J

Jerne plaque assay, 83
JRA cell, 626

K

Keratinocyte, 683
Kupffer cell, 313. *See also* Macrophage
  characterization, 291–293
  isolation, 285
    by adhesion, 285
  phagocytosis, 291–293
  studies, 293, 294
  yield, from rats, 293

L

Lactate dehydrogenase, in assay of macrophage cytotoxicity, 325, 332
Lactoperoxidase, in iodination of protein A, 407
Langerhans cell, 313, 445
  epidermal
    adenosine triphosphatase activity, 686–688, 691
    functional properties, 683
    immunochemical detection, 683–691
      specimen preparation, 684–686
    isolation, 357–362
  Ia antigen source, 529
  identification, 683, 684
  immunologic function, 358
  properties, 357
  separation, 358
  source, identification, 684
L cell, mouse
  fingerprint, 87
  sedimentation profile in asynchronous exponential growth, 85–87
Lectin. *See also specific lectin*
  in cell characterization and separation, 153–155
  cell separation by, 168–179
  for characterization of lymphocytes, 167, 168
  conjugation with fluorochrome, 419
  for flow cytometry, 208
  sources, 170

Lectin affinity chromatography, 153, 154, 170
  applications, 167
  of H-2 antigen preparations, 510
  of HLA antigens, 624
  of HLA-DR antigen, 605, 606
  of Ia antigen, 538, 539
  of Thy-1 antigen, 657, 658
Lectin-Sepharose gel, 156–158
Leprosy, in nude mouse, 355
Leukemia
  chronic lymphocytic, 154
    T cell lines, 633
  chronic lymphoid, immunoglobulin localization study, 400–405
  Ia antigen in, 616
  phenotyping, 235–237
Leukemogenesis, after thymectomy, 14
Leukocyte, fluid pinocytosis, 336–347
LFA-1 antigen, isolation, 469
Light extinction, measurement, in flow cell geometries, 205
Light scatter
  analysis
    in flow cytometry, 205–208
    uses, 205
  forward, 206, 212
  from nonspherical cells, 206
  standards, 207, 208
  theory, 205, 206
  at two or more wavelengths, 207
Limulus polyphemus agglutinin, 167
Listeria monocytogenes, immunity, adoptive transfer, 38
Low molecular weight protein, isolation, 469
Lubrol WX, 466, 539
Ly antigen, 677–683
  antibody reagents, 682
    that recognize the same antigen, 682, 683
  distribution, 677
  Ly-1, 678
    isolation, 469
  Ly-2, 678
    isolation, 469
  Ly-3, 678
    isolation, 469
  Ly-4, 678
  Ly-5, 678
  Ly-6, 678

Mixed leukocyte reaction, 262, 267
Moloney sarcoma virus, immunity, adoptive transfer, 38
Monoclonal antibody, 198, 250
  19/178, 669
  53-10.1, 234
  53-5, 669
  53-6.7, 669
  53-7.3, 669
  7-20.6/3, 669
  to additional rat antigens, 565, 566
  affinity chromatography, of rat Ia
    antigens, 560–565
  affinity columns, preparation, 561, 562
  against human T-cell antigens, 627–
    632
  against T cell surface components,
    development, 632–636
  against *Tla* region gene products, 551,
    553, 557, 558
  vs. alloantisera, 491
  antiimmunoglobulin, 314
  anti-Lyt-2, 121
  antimacrophage, 312, 313
    10-75-3, 319, 320
    1-21J, 315, 316
    2.4G2, 317
    54-2, 317
    63D3, 319, 324
    AcM.1, 317
    anti-Mac-1, 324
    anti-Mo-1, 318, 320, 324
    anti-Mo-2, 318
    anti-Mo-3, 318
    C10H5, 319, 324
    D5D6, 319, 324
    F4/80, 317
    M43, 317, 324
    M57, 317, 322
    M102, 317, 322
    M143, 317
    M1/70, 314, 316, 318, 324
    M206, 319, 324
    M3/31, 316
    M3/37, 317
    M3/38, 316
    M3/84, 316
    Mac-120, 318, 324
    MMA, 319, 324
    mouse antigens identified by, 314–
      317

MφP-9, 319
MφP-15, 319
MφS-1, 319
MφS-39, 319
OKM1, 318
preparation, 313, 314
UC45, 319
use, 313–324
anti-T11,9,6, 629
anti-T1A, 628
anti-T3A, 628
anti-T4A, 628
anti-T6A, 629
anti-T8, 633
anti-T8A, 629
anti-Tac, 631
anti-TQ1, 631, 632
available through ATCC, 492–494
for characterization of T-cell subsets,
  626, 627
cocktail, 384
CR11-351, in HLA detection assay,
  621
in FACS, 232
for flow cytometry, 208
HcLF7D5, 647, 648
HO-13-4, 647, 652, 653
HO-15-2, 647
HO-15-7, 647
HO-2.2, 669
HO-3.1, 669
in hybridoma culture fluid, 480
to Ia antigens, 689, 690
in immune precipitation of Ia antigen,
  536, 537
immunoaffinity columns
  preparation for HLA-DR purification, 609
  for purification of HLA-DR antigens, 600–606
for immunofluorescence, 384
in immunoprecipitation of human T cell
  antigens, 639–641
internal labeling, 485
L243, 602, 603
for Langerhans cell identification, 689,
  690
in large-scale purification of HLA
  antigen, 592
Leu1, 628
Leu2a,b, 629

Printed and bound by CPI Group (UK) Ltd, Croydon, CR0 4YY

03/10/2024

01040410-0017